Lecture Notes in Computer Science 12118

Advanced Research in Computing and Software Science

Subline of Lecture Notes in Computer Science

More information about this subseries at http://www.springer.com/series/7407

Yoshiharu Kohayakawa ·
Flávio Keidi Miyazawa (Eds.)

LATIN 2020: Theoretical Informatics

14th Latin American Symposium
São Paulo, Brazil, January 5–8, 2021
Proceedings

Editors
Yoshiharu Kohayakawa
University of São Paulo
São Paulo, Brazil

Flávio Keidi Miyazawa
University of Campinas
Campinas, Brazil

ISSN 0302-9743 ISSN 1611-3349 (electronic)
Lecture Notes in Computer Science
ISBN 978-3-030-61791-2 ISBN 978-3-030-61792-9 (eBook)
https://doi.org/10.1007/978-3-030-61792-9

LNCS Sublibrary: SL1 – Theoretical Computer Science and General Issues

This Springer imprint is published by the registered company Springer Nature Switzerland AG
The registered company address is: Gewerbestrasse 11, 6330 Cham, Switzerland

Preface

We are very pleased to present this volume with the papers accepted to the 14th Latin American Theoretical INformatics Symposium (LATIN 2020). The conference was scheduled to run in May 2020, in São Paulo, Brazil, the birthplace of this series of meetings, but owing to the global COVID-19 pandemic, it was postponed to 2021. Previous editions of LATIN took place in São Paulo, Brazil (1992), Valparaíso, Chile (1995), Campinas, Brazil (1998), Punta del Este, Uruguay (2000), Cancún, Mexico (2002), Buenos Aires, Argentina (2004), Valdivia, Chile (2006), Búzios, Brazil (2008), Oaxaca, Mexico (2010), Arequipa, Peru (2012), Montevideo, Uruguay (2014), Ensenada, Mexico (2016), and Buenos Aires, Argentina (2018).

The conference received 136 submissions from around the world. Each submission was reviewed by three Program Committee members, often with the help of additional external referees. After an intense reviewing process, the Program Committee selected 50 submissions for presentation.

We are delighted that the following distinguished speakers accepted our invitation to give a plenary lecture at LATIN 2020: Maria-Florina Balcan (Carnegie Mellon University, USA), Nikhil Bansal (Centrum Wiskunde & Informatica and Eindhoven University of Technology, The Netherlands), Maria Chudnovsky (Princeton University, USA), Nicole Immorlica (Microsoft Research, USA) Eduardo Sany Laber (Pontifical Catholic University of Rio de Janeiro, Brazil), Alexander Razborov (The University of Chicago, USA), Luca Trevisan (Bocconi University, Italy), and Bianca Zadrozny (IBM Research, Brazil). Furthermore, Nikhil Bansal kindly accepted to give the course "Algorithmic Discrepancy and Applications" in LATIN 2020. We are very grateful to all the speakers for accepting to support and enrich LATIN 2020. The title and abstract of the keynote talks as well as the course are available on the conference website.

LATIN 2020 featured two awards: the Imre Simon Test-of-Time Award and the Alejandro López-Ortiz Best Paper Award. In this edition, the Imre Simon Test-of-Time Award winner is Anne Brüggemann-Klein, for her paper "Regular expressions into finite automata," which appeared in LATIN 1992. For the Alejandro López-Ortiz Best Paper Award, the Program Committee selected the paper "Monotone Circuit Lower Bounds from Robust Sunflowers," by Bruno Pasqualotto Cavalar, Mrinal Kumar, and Benjamin Rossman. We thank Springer for supporting both awards.

In order to facilitate and promote attendance, poster sessions were planned for LATIN 2020 to encourage theory students and young researchers to report their ongoing research and preliminary findings. We thank the Posters Committee members for their work.

Our heartfelt thanks go to the authors for their excellent papers and cooperation, to the Program Committee members for the insightful discussions, to the subreferees for their careful reports, and to the Steering Committee for their valuable advice and feedback.

We are very grateful to the organizing team for all their support and commitment in these challenging times, which demanded inventiveness and all sorts of additional work. Last but not least, we warmly thank all our academic and corporate sponsors: B2W, CAPES, CNPq, FAPESP, Google, IME-USP, SBC, Springer, UFABC, UNICAMP, and USP.

August 2020

Yoshiharu Kohayakawa
Flávio Keidi Miyazawa

The Imre Simon Test-of-Time Award

The winner of the 2020 Imre Simon Test-of-Time Award, considering papers up to the 2010 edition of the Latin American Theoretical INformatics Symposium (LATIN), is:

"Regular expressions into finite automata" by Anne Brüggemann-Klein, LATIN 1992, LNCS 583, 87–98, 1992,

which has also been published in *Theoretical Computer Science*, vol. 120 (1993), 197–213, in its more detailed journal version.

Not many results in (theoretical) computer science are considered as being as basic and fundamental as Kleene's theorem. This result states the equality of the family of regular languages, that is, the family of languages denoted by regular expressions, with the family of languages accepted by finite automata.

If the fact that languages accepted by finite automata are regular is mostly of theoretical, or aesthetical, interest, the one that regular languages are accepted by finite automata, that is, more precisely, that a regular expression can be turned into an equivalent finite automaton, is of paramount importance and ubiquitously implemented in countless pieces of software. Every textbook in computer science, especially those dealing with parsing and compiler construction, has a chapter where such algorithms are described. They are due to Thompson, Glushkov, Brzozowski, or Antimirov, to name a few.

The contribution of the awarded paper in the abundant literature on the subject is original and unique in the sense that it does not offer a new algorithm or an improvement of an algorithm that computes an automaton from an expression. It describes a syntactic transformation of a regular expression E into an equivalent expression, called the 'star normal form of E'. An expression is in star normal form (SNF) if the star operator applies only to subexpressions that denote languages which do not contain the empty word.

The computation of the SNF of an expression is achieved by a recursive traversal of the syntactic tree of the expression, hence with a linear complexity in the size of the expression. The paper then establishes that the 'Glushkov construction' applied to an expression in SNF is of quadratic complexity – in contrast with the general case, which is of cubic complexity. And the two properties together thus yield the first quadratic algorithm for computing an automaton from an expression.

The interest of the concept of SNF goes beyond this complexity breakthrough which already brings the paper numerous quotations and references. It first allows one to clarify the notions of weak and strong unambiguity (an expression is weakly unambiguous iff its SNF is strongly unambiguous). More important, the definition of SNF underlies those of 1-unambiguous expressions and languages, developed in a subsequent paper by the same author (together with D. Wood) and that were motivated by the study of the grammars appearing in markup languages. This is another strong background for the long-lasting interest in the paper selected for the 2020 Imre Simon Test-of-Time Award.

January 2020

Marcos Kiwi
Conrado Martínez
Jacques Sakarovitch

Organization

Program Committee Chairs

Yoshiharu Kohayakawa	University of São Paulo, Brazil
Flávio Keidi Miyazawa	University of Campinas, Brazil

Steering Committee

Kirk Pruhs	University of Pittsburgh, USA
Michael Bender	Stony Brook University, USA
Cristina Fernandes	University of São Paulo, Brazil
Joachim von zur Gathen	Bonn-Aachen International Center for Information Technology, Germany
Evangelos Kranakis	Carleton University, Canada
Alfredo Viola	Universidad de la República, Uruguay

Program Committee

Andris Ambainis	University of Latvia, Latvia
Frédérique Bassino	CNRS and Université Sorbonne Paris Nord, France
Flavia Bonomo	Universidad de Buenos Aires, Argentina
Prosenjit Bose	Carleton University, Canada
Olivier Carton	Université Paris Diderot, France
Ferdinando Cicalese	University of Verona, Italy
Jose Correa	Universidad de Chile, Chile
Pierluigi Crescenzi	Université Paris Diderot, France
Luc Devroye	McGill University, Canada
Martin Dietzfelbinger	Technische Universität Ilmenau, Germany
David Fernández-Baca	Iowa State University, USA
Esteban Feuerstein	Universidad de Buenos Aires, Argentina
Eldar Fischer	Technion - Israel Institute of Technology, Israel
Pierre Fraigniaud	CNRS and Université Paris Diderot, France
Martin Fürer	Penn State University, USA
Anna Gál	The University of Texas at Austin, USA
Ron Holzman	Technion - Israel Institute of Technology, Israel
Marcos Kiwi	Universidad de Chile, Chile
Yoshiharu Kohayakawa	University of São Paulo, Brazil
Teresa Krick	Universidad de Buenos Aires, Argentina
Cláudia Linhares Sales	Federal University of Ceará, Brazil
Kazuhisa Makino	Kyoto University, Japan
Conrado Martínez	Universitat Politècnica de Catalunya, Spain
Flávio Keidi Miyazawa	University of Campinas, Brazil

Marco Molinaro	PUC-Rio, Brazil
Veli Mäkinen	University of Helsinki, Finland
Gonzalo Navarro	Universidad de Chile, Chile
Rolf Niedermeier	Technische Universität Berlin, Germany
Rafael Oliveira	University of Toronto, Canada
Roberto Oliveira	IMPA, Brazil
Daniel Panario	Carleton University, Canada
Alessandro Panconesi	Sapienza University of Rome, Italy
Pan Peng	The University of Sheffield, UK
Ely Porat	Bar-Ilan University, Israel
Paweł Prałat	Ryerson University, Canada
Pavel Pudlák	Czech Academy of Sciences, Czech Republic
Svetlana Puzynina	Saint Petersburg State University, Russia
Sergio Rajsbaum	Universidad Nacional Autónoma de Mexico, Mexico
Andrea Richa	Arizona State University, USA
Rahul Santhanam	University of Oxford, UK
Asaf Shapira	Tel Aviv University, Israel
Alistair Sinclair	University of California, Berkeley, USA
Mohit Singh	Georgia Institute of Technology, USA
Maya Stein	Universidad de Chile, Chile
Jayme Szwarcfiter	Federal University of Rio de Janeiro, Brazil
Eli Upfal	Brown University, USA
Jorge Urrutia	Universidad Nacional Autónoma de México, Mexico
Brigitte Vallée	Université de Caen, France
Mikhail Volkov	Ural Federal University, Russia
Raphael Yuster	University of Haifa, Israel

Organizing Committee

José C. de Pina (Co-chair)	University of São Paulo, Brazil
Carla N. Lintzmayer (Co-chair)	Federal University of ABC, Brazil
Guilherme O. Mota (Co-chair)	Federal University of ABC, Brazil
Yoshiko Wakabayashi (Co-chair)	University of São Paulo, Brazil
Marcel K. de Carli Silva	University of São Paulo, Brazil
Carlos E. Ferreira	University of São Paulo, Brazil
Cristina G. Fernandes	University of São Paulo, Brazil
Arnaldo Mandel	University of São Paulo, Brazil
Daniel M. Martin	Federal University of ABC, Brazil
Lehilton L. C. Pedrosa	University of Campinas, Brazil
Sinai Robins	University of São Paulo, Brazil
Cristiane M. Sato	Federal University of ABC, Brazil
Rafael C. S. Schouery	University of Campinas, Brazil
Eduardo C. Xavier	University of Campinas, Brazil

Additional Reviewers

Yehuda Afek
Matteo Almanza
Simon Apers
Julio Aracena
Julio Araujo
Diego Arroyuelo
Srinivasan Arunachalam
Yeganeh Bahoo
Niranjan Balachandran
Gill Barequet
Zuzana Bednarova
Alexander Belov
Omri Ben-Eliezer
Fabrício Benevides
Matthias Bentert
Benjamin Bergougnoux
René Van Bevern
Ahmad Biniaz
Hans Bodlaender
Ilario Bonacina
Anthony Bonato
Nicolas Bousquet
Marco Bressan
Karl Bringmann
Shaowei Cai
Xing Cai
Victor Campos
Clément Canonne
Bastien Cazaux
Dibyayan Chakraborty
Erika Coelho
Cyrus Cousins
Gabriel Coutinho
Anthony D'Angelo
Konrad Dabrowski
Joshua Daymude
Marcelo de Carvalho
Mateus de Oliveira Oliveira
Pedro de Rezende
Lorenzo De Stefani
Antoine Deza
Giuseppe Di Luna
Josep Diaz
Benjamin Doerr
Vinicius dos Santos
Jean-Philippe Dubernard
Enrica Duchi
Philippe Duchon
Guillaume Ducoffe
Tinaz Ekim
Sergi Elizalde
Matthias Englert
Massimo Equi
Thomas Erlebach
Luerbio Faria
Andreas Feldmann
Cristina Fernandes
Gabriele Fici
Till Fluschnik
Lukas Folwarczny
Travis Gagie
Pu Gao
Konstantinos Georgiou
Daniel Gibney
Shay Golan
Claude Gravel
Jarosław Grytczuk
Sylvain Guillemot
Shahrzad Haddadan
Pooya Hatami
Benjamin Hellouin de Menibus
Anne-Sophie Himmel
Cecilia Holmgren
Janis Iraids
Vesna Iršič
Svante Janson
Gwenaël Joret
Bogumił Kamiński
Leon Kellerhals
Michael Khachay
Sang-Sub Kim
Konstantin Kobylkin
Tomasz Kociumaka
Martins Kokainis
Christian Komusiewicz
Dmitry Kosolobov

O-Joung Kwon
Elmar Langetepe
Carlos Lima
Paloma Lima
Min Chih Lin
Sylvain Lombardy
Raul Lopes
Hosam Mahmoud
Konstantin Makarychev
Arnaldo Mandel
Andrea Marino
Leonardo Martinez
Paolo Massazza
Pedro Matias
Alessio Mazzetto
Alessia Milani
Marni Mishna
Hendrik Molter
Guilherme Mota
Matthias Müller-Hannemann
Marcelo Mydlarz
André Nichterlein
Prajakta Nimbhorkar
Hugo Nobrega
Thomas Nowak
Pascal Ochem
Carlos Ochoa
Stephan Olariu
Juan Manuel Ortiz de Zarate
Ayoub Otmani
Linda Pagli
Prakash Panangaden
Olga Parshina
Matthew Patitz
Lehilton Pedrosa
Pablo Pérez-Lantero
Martin Pergel
Miguel Pizaña
Daniel Posner
Dominique Poulalhon
Ali Pourmiri
Lionel Pournin
Thomas Prellberg
Krisjanis Prusis
Artem Pyatkin
Arash Rafiey
Giuseppe Re
Vinod Reddy
Malte Renken
Susanna Rezende
Assaf Rinot
Matteo Riondato
Trent Rogers
Javiel Rojas-Ledesma
Massimiliano Rossi
Pablo Rotondo
Ville Salo
Rudini Sampaio
Mathieu Sassolas
Ignasi Sau
Saket Saurabh
Joe Sawada
Christian Scheffer
Kevin Schewior
Ulrich Schmid
Ben Seamone
Ohad Shamir
Saswata Shannigrahi
Arseny Shur
Ana Silva
Ronan Soares
José Soto
Uéverton Souza
Joachim Spoerhase
Jakob Spooner
Alejandro Strejilevich de Loma
Xiaoming Sun
Jay Tenenbaum
Hans Tiwari
Gabriel Tolosa
Nadi Tomeh
Iddo Tzameret
Eleni Tzanaki
Pavel Valtr
Victor Verdugo
Sergey Verlan
José Verschae
Laurent Viennot

Jevgenijs Vihrovs
Yoshiko Wakabayashi
Haitao Wang
Jamison Weber
Fan Wei
Stu Whittington
Sebastian Wiederrecht
Andreas Wiese
Nengkun Yu
Philipp Zschoche

Posters Committee

Lehilton L. C. Pedrosa (Chair)	University of Campinas, Brazil
Rafael C. S. Schouery	University of Campinas, Brazil
Yoshiko Wakabayashi	University of São Paulo, Brazil
Eduardo C. Xavier	University of Campinas, Brazil

Sponsoring Institutions

Brazilian Research Agencies and Academic Institutions

CAPES	Coordination for the Improvement of Higher Education Personnel
CNPq	National Council for Scientific and Technological Development
FAPESP	São Paulo Research Foundation
IME-USP	Institute of Mathematics and Statistics, USP
SBC	Brazilian Computer Society
UFABC	Federal University of ABC
UNICAMP	University of Campinas
USP	University of São Paulo

Corporate Sponsors
B2W
Google
Springer

Contents

Algorithms and Data Structures

Computational Geometry

Complexity Theory

Quantum Computing

Neural Networks and Biologically Inspired Computing

Randomization

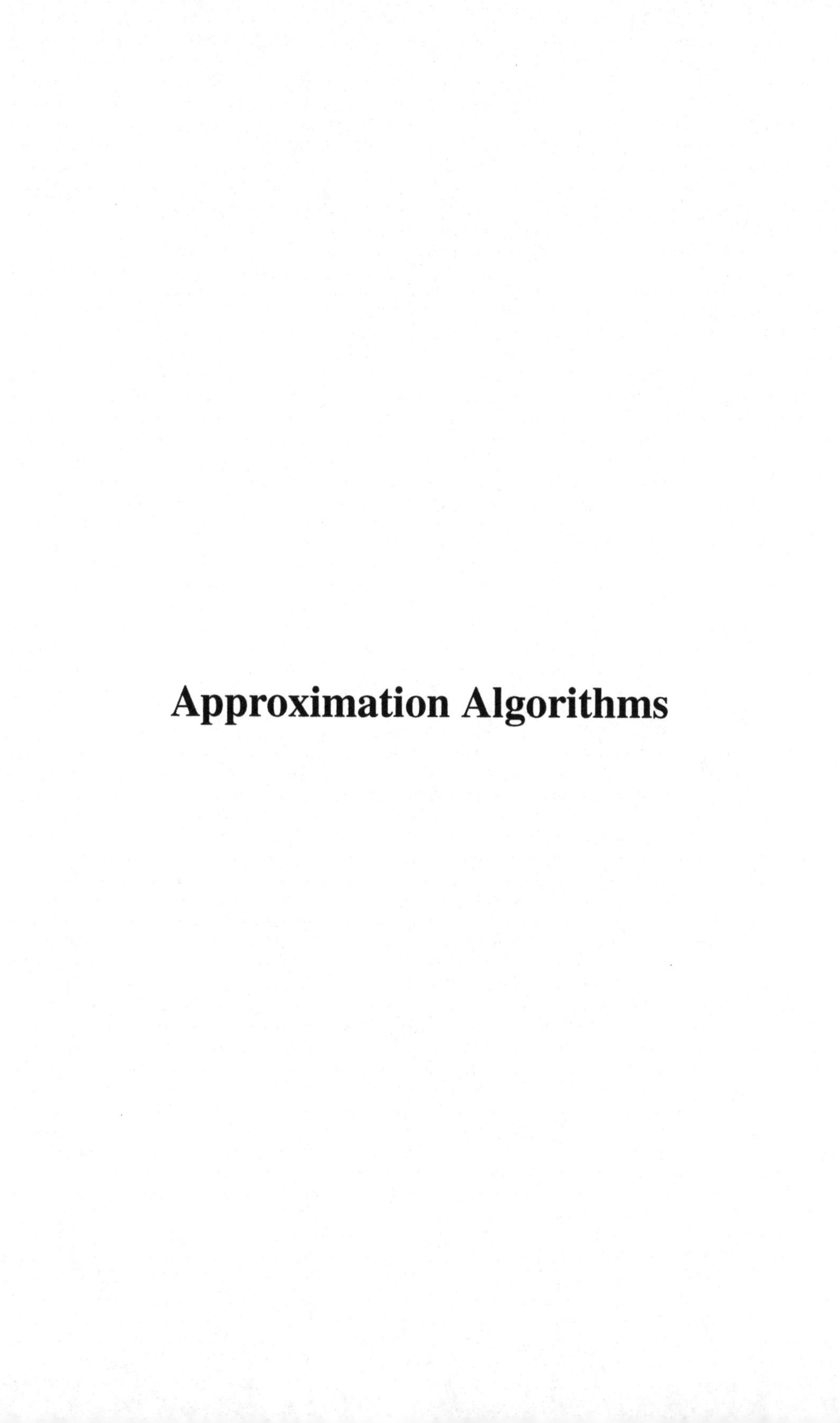

Approximation Algorithms

PTAS for Steiner Tree on Map Graphs

Jarosław Byrka[1], Mateusz Lewandowski[1(✉)], Syed Mohammad Meesum[1], Joachim Spoerhase[2], and Sumedha Uniyal[2]

[1] Institute of Computer Science, University of Wrocław, Wrocław, Poland
mlewandowski@cs.uni.wroc.pl
[2] Aalto University, Espoo, Finland

Abstract. We study the Steiner tree problem on map graphs, which substantially generalize planar graphs as they allow arbitrarily large cliques. We obtain a PTAS for Steiner tree on map graphs, which builds on the result for planar edge weighted instances of Borradaile et al.

The Steiner tree problem on map graphs can be casted as a special case of the planar node-weighted Steiner tree problem, for which only a 2.4-approximation is known. We prove and use a contraction decomposition theorem for planar node weighted instances. This readily reduces the problem of finding a PTAS for planar node-weighted Steiner tree to finding a spanner, *i.e.*, a constant-factor approximation containing a nearly optimum solution. Finally, we pin-point places where known techniques for constructing such spanner fail on node weighted instances and further progress requires new ideas.

1 Introduction

The Steiner tree problem has been recognized by both theorists and practitioners as one of the most fundamental problems in combinatorial optimization and network design. In this classical NP-hard problem we are given a graph $G = (V, E)$ and a set of terminals R. The goal is to find a tree connecting all the terminals of minimum cost. A long sequence of papers established the current best approximation ratio of 1.386 [10].

The node-weighted Steiner tree problem (NWST) is a generalization of the above problem. This can be easily seen by placing additional vertices in the middle of edges. Moreover, an easy reduction shows that this variant is as difficult to approximate as the Set Cover problem. Indeed, there are greedy $O(\log n)$ approximation algorithms [20, 24] matching this lower bound.

Much research has been devoted to studying combinatorial optimization problems on planar graphs, *i.e.* graphs that can be drawn on a plane without crossings. This natural restriction allows for better results, especially in terms of approximation algorithms. To this end, multiple techniques have been developed using the structural properties of planar graphs, including balanced separators

The first three authors were supported by the NCN grant number 2015/18/E/ST6/00456.

Y. Kohayakawa and F. K. Miyazawa (Eds.): LATIN 2020, LNCS 12118, pp. 3–14, 2020.
https://doi.org/10.1007/978-3-030-61792-9_1

[1,5,25], bidimensionality [17], local search [12,15], shifting technique [2]. Such techniques are immediately applicable to a wide range of problems.

Steiner tree problems, however, require more involved construction. The already established framework for approximation schemes for Steiner problems on planar graphs can be summarized as follows:

1. **Construct a spanner**
   A spanner is a subgraph of the input graph satisfying two properties: (1) total cost of the spanner is at most $f(\epsilon)$ times the cost of the optimum solution and (2) the spanner preserves a nearly-optimum solution, *i.e.* there exist a solution of cost $(1+\epsilon) \cdot \mathrm{OPT}$. Planarity of the input graph is heavily used to find such a spanner.
2. **Apply the contraction decomposition theorem**
   The edges of the spanner are partitioned into k sets, such that contracting each set results in a graph of constant treewidth. Because we started with a cheap spanner, there is a choice of k for which the cheapest such set of edges has cost $\varepsilon \cdot \mathrm{OPT}$. This partitioning is given by a contraction decomposition theorem [23] (also known as *thinning*) which can be obtained by applying Baker's shifting technique [2] to the dual graph.
3. **Solve bounded-treewidth instances**
   The remaining instance is solved exactly (or in some cases approximately) in polynomial time via dynamic programming.

Indeed, the PTAS construction for the Steiner tree problem due to Borradaile, Klein and Mathieu [8] uses exactly this framework. Follow-up results for related problems such as Steiner forest [6,19], prize-collecting Steiner tree [3], group Steiner tree [4] successfully follow the same approach (although adding new important ingredients like spanner bootstrapping).

On the other hand—despite many efforts—the status of the node-weighted Steiner tree problem is not yet decided on planar graphs. The state-of-the-art algorithms achieve only constant factor approximations. A GW-like primal-dual method gives a ratio of 6 [16], which was further simplified and improved to 3 by Moldenhauer [26]. The current best result is a more involved 2.4-approximation by Berman and Yaroslavtsev [7]. However, as the integrality gap of the LP used by the above primal-dual algorithms is lower-bounded by 2, this approach does apparently not lead to an approximation scheme.

1.1 Motivation for Map Graphs

The problems tractable on planar graphs are often considered also in more general classes of graphs. Most common such classes include bounded genus graphs and even more general H-minor-free graphs. In this work however, we focus on a different generalization, *i.e.* map graphs introduced by Chen et al. [13]. They are defined as intersection graphs of internally disjoint connected regions in the plane. Unlike for planar graphs, two regions are adjacent if they share at least one point (see Fig. 1). Notably, map graphs are not H-minor-free as they may contain arbitrarily large cliques as minors.

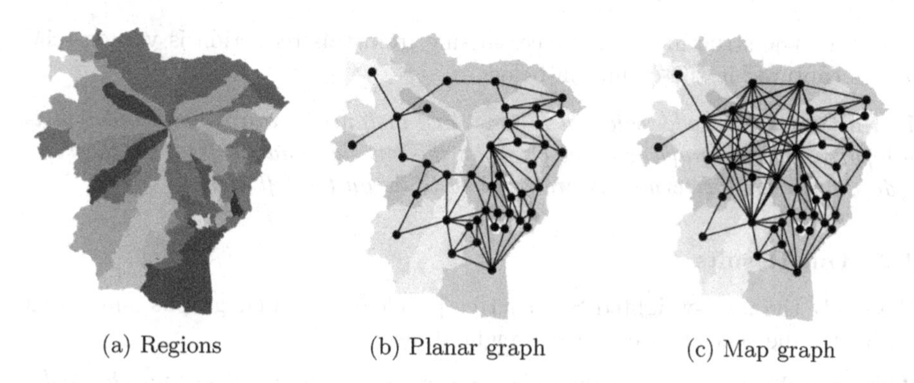

(a) Regions (b) Planar graph (c) Map graph

Fig. 1. (a) Some municipalities of the province of Catania (Sicily, Italy). The vertices are representing connected regions. (b) The planar graph has edges between two regions if they share a border. (c) The map graph has edges whenever regions share at least a single point.

It is useful to characterize map graphs as *half-squares* of bipartite planar graphs. A half-square of a bipartite graph $W = (V \cup U, E_W)$ is a graph $G = (V, E)$ where we have an edge between a pair of vertices, whenever the distance between these vertices in W is equal to two. If W is planar, then it is called a witness graph of map graph G. See Fig. 2 for a witness graph (solid edges) and the corresponding map graph (dashed edges).

We are the first to study the Steiner tree problem on map graphs. We study the case when all edges have uniform cost; otherwise the map graphs would capture the general case. To see this, observe that a clique K_n is a map graph and putting large costs on some edges mimics any arbitrary graph.

On the other hand, the case of map graphs with uniform edge costs is still more general than arbitrary edge-weighted planar graphs for the Steiner tree problem. This follows from the fact that subdividing edges preserves planarity, and we can reduce planar graphs to the uniform case (loosing a factor $1 + \epsilon$ by guessing the heaviest edge in an optimum solution, pruning heavier and contracting cheap edges).

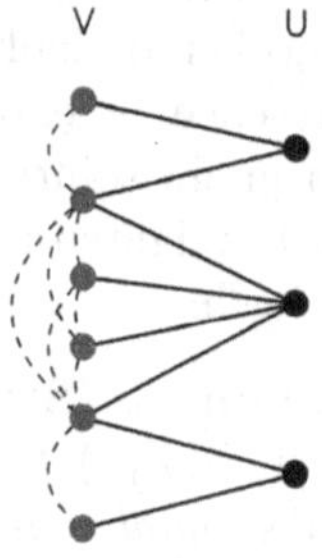

Fig. 2. Map graph and its witness

Therefore it is natural to ask if there is a PTAS for the Steiner tree in our setting. This question gets even more compelling upon realizing that this is a special case of the node-weighted problem on planar graphs. To see this, consider the following reduction: take the witness graph $W = (V \cup U, E_W)$ of the uniformly edge-weighted map graph and put weight 1 on the vertices in V and weight 0 on the vertices in U. The terminals are kept at the corresponding vertices in V. The solutions for the resulting node-weighted problem can be easily translated back to the initial instance. The validity of the reduction is assured by the simple fact that the number of vertices in a tree is equal to the number of edges in this tree plus one. The structure of instances arising from this reduction is very special and is captured in the definition below.

Definition 1. *A node-weighted Steiner tree instance is* ***map-weighted*** *if it is a bipartite planar graph with weight* 1 *on the left side and weight* 0 *on the right side. Moreover, the terminals are required to lie on the left side.*

1.2 Our Results

We study the node-weighted Steiner tree problem on planar graphs and give a PTAS for the special case of map-weighted instances.

Theorem 1. *There is a polynomial-time approximation scheme for the node-weighted Steiner tree problem on map-weighted instances.*

By the reduction described above, we immediately obtain a PTAS for the edge-weighted Steiner tree problem on uniform map graphs.

Corollary 1. *There is a polynomial-time approximation scheme for the Steiner tree problem on uniform map graphs.*

In the proof of Theorem 1 we adopt the framework for constructing PTASs and the brick-decomposition of Borradaile et al. [8]. However, we need to tackle additional obstacles related to high-degree vertices in the node-weighted setting.

The first difficulties emerge in the Spanner construction. In the *cutting-open* step, the duplication of high-degree vertices may make the cost unbounded. Another difficulty is bounding the number of portals needed. Essentially, the presence of expensive high-degree vertices excludes the existence of nearly-optimum solution with bounded number of joining vertices. The properties of map-weighted instances allow us to overcome multiple difficulties and prove the following.

Lemma 1 (Steiner-Tree Spanner). *Given a map-weighted instance* $W = (V \cup U, E_W)$ *for a map graph* G*, where* $R \subseteq V$ *are terminal nodes, there is a polynomial time algorithm that outputs a spanner subgraph* $H \subseteq W$ *containing all the terminals* R *which has the following properties:*

(i) (shortness property) $w(H) \leq f(\varepsilon) \cdot OPT(W, R)$
(ii) (spanning property) $OPT(H, R) \leq (1 + \varepsilon) \cdot OPT(W, R)$

where $f(\varepsilon)$ *is a function that depends only on* ε *and* $OPT(G, R)$ *is the cost of an optimal Steiner tree for graph* G *and the set* $R \subseteq V(G)$ *of terminals.*

A different issue arises in the use of the Contraction Decomposition Theorem with node weights. A naive approach could be to move the costs of vertices to edges by setting the cost of each edge to be the sum of costs of its endpoints and then using the contraction decomposition theorem as it is. However—again due to high-degree nodes—the total cost of edges would no longer be a constant approximation of OPT. Therefore we cannot directly use the existing contraction decomposition theorem.

To handle the last issue, we develop a new decomposition theorem with the additional property that each vertex participates in a limited number of sets.

Lemma 2 (Node-weighted Contraction Decomposition). *There is a polynomial time algorithm that given a planar embedding of a graph H and an integer k, finds k sets $E_0, E_1, \ldots, E_{k-1} \subseteq E(H)$ such that:*

(i) contracting each E_i results in a graph with treewidth $O(k)$, and
(ii) for each vertex v, all the incident edges of v are in at most two sets E_i, E_j.

We note that our decomposition can be applied to any node-weighted *contraction-closed* problem, *i.e.* the problem for which contracting edges and setting the weight of resulting vertex to 0 does not increase the value of optimum solution. Therefore the lemma above adds a novel technique to the existing framework for planar approximation schemes.

Finally, using standard techniques, we give a dynamic programming algorithm for the node-weighted Steiner tree problem on bounded treewidth instances (see Appendix A of the extended version of the paper [11]).

Lemma 3 (Bounded Treewidth NWST). *An optimal node-weighted Steiner tree can be found in time $2^{\mathcal{O}(t \log t)} \cdot n^{\mathcal{O}(1)}$, where t is the treewidth of the input graph with n vertices.*

We note that Lemma 2 and Lemma 3 work for arbitrary node-weights. Only the spanner construction of Lemma 1 uses properties of the map-weighted instances.

In the next section we give the details of the Spanner construction. In Sect. 3 we prove Lemma 2 and show how the combination of the three above lemmas yields the main result. In the last section we conclude with a puzzling open problem.

2 Spanner Construction for Map-Weighted Graph

In this section we describe how we construct the spanner for a map-weighted planar witness graph W and prove Lemma 1. For convenience, instead of $(1+\varepsilon)$, we will prove property (ii) for $(1 + c\varepsilon)$ where $c \geq 0$ is some fixed constant. For any given $\varepsilon > 0$, running the construction for $\tilde{\varepsilon} = \varepsilon/c$ gives the precise result. From now on, we will work with a fixed embedding of the witness graph W.

Notations: For any map-weighted graph W, we define $d_W\colon V^2 \to \mathbb{R}$ to be the function giving the node-weighted length of the shortest-path between any two vertices using only the edges from W (including the end vertices weights). Let $P_W(u, v) \subseteq W$ be an arbitrary path of cost $d_W(u, v)$. Similarly, let $\ell_W\colon V^2 \to \mathbb{R}$ be the length of the unweighted shortest-path ignoring the node-weights between any two vertices using only the edges from W. Similarly we define for any path $P \subseteq W$, $c(P)$ to be the cost of the path corresponding to the map-weights (including the end vertices) and $\ell(P)$ to be the length of the unweighted-path ignoring the node-weights. Analogously, for any graph $H \subseteq W$, we define $c(H)$ to be the total weight of nodes of H and $\ell(H)$ to be the number of edges of H.

For any path P and $u, v \in V(P)$, we define $P[u, v]$ to be the sub-path starting at u and ending at v (including u and v). If $P[u, v]$ has at least one internal node then $P(u, v)$ denotes the sub-path starting at u and ending at v but *excluding* u and v. We refer to any path/cycle with no edges and one vertex as *singleton* path/cycle and the ones containing at least one edge as *non-singleton* path/cycle respectively.

The spanner construction is summarized in Algorithm 1.

Algorithm 1. Spanner construction

1: Start with a 2.4-approximate node-weighted Steiner tree solution for graph W and terminal set R using [7].
2: Cut open the corresponding solution tree ST in W to create another graph W' which has an outer face with boundary $\mathcal{D}$ of cost at most $10 \cdot OPT$.
3: Build the mortar graph MG on the cut-open graph W' using the procedure in Section 6 of [8], ignoring the weights on the nodes and using $\ell(e) = 1$ for each $e \in E(W')$ and $\varepsilon' := \varepsilon/4$
4: Construct the set $P(B) \subseteq \partial B$ of portals for each brick $B \in \mathcal{B}$.
5: For each brick $B \in \mathcal{B}$ and for each subset $X \subseteq P(B)$, run the generalized Dreyfus-Wagner algorithm [18,9] to compute the optimal Steiner tree on terminal set X in map-weighted graph B in time $3^{|X|}n^{\mathcal{O}(1)}$.
6: Return the union of MG along with all the trees found in the previous step.

Before proving Lemma 1, we elaborate on the steps of Algorithm 1 that require more detailed explanation and state the key properties of the construction.

Cutting-Open operation. We start with a 2.4-approximate node-weighted Steiner tree solution ST for our node-weighted plane graph W and terminal set R using [7]. Using tree ST, we perform an *cut-open* operation as in [8] to create a new map-weighted planar graph W' whose outer face is a simple cycle $\mathcal{D}$ arising from ST.

Since we are dealing with node weights and the node degrees are unbounded, we need an additional argument to bound the cost of $\mathcal{D}$ as compared to the edge-weighted case. A crucial property used to prove the observation is that all the leaves of ST have weight one.

Lemma 4. *(Cut-Open) The cost $c(\mathcal{D})$ of the boundary $\mathcal{D}$ is at most* $10 \cdot \mathrm{OPT}$. *Moreover $R \subseteq V(\mathcal{D})$.*

Mortar Graph Construction. We apply the construction of a mortar graph along with a brick decomposition as described in [8] as a black box. Here, we state the properties of the mortar graph that we need in our work without referring to the details of the algorithm that constructs it.

(i) The mortar graph MG is a subgraph of the cut-open graph W' and vertices of MG contain all the terminals.

(ii) Let f be a face of the mortar graph. A brick B (corresponding to f) is the subgraph of W' enclosed by the boundary ∂f of f. Specifically, the boundary ∂B of B is precisely ∂f. Let $\mathcal{B}$ denote the set of bricks in W'.
(iii) The collection of all bricks covers the cut-open graph W'.
(iv) The mortar graph is "grid-like" in the following sense. The boundary ∂B of each brick B can be decomposed into a western part W_B, a southern part S_B, an eastern part E_B and a northern part N_B. Each of these parts is close to be a shortest path, *i.e.* each subpath is at most $1 + \epsilon$ times more expensive than the shortest path between the endpoints.

The construction of the mortar graph and the corresponding brick decomposition as described in [8] has two parameters. An error parameter ε' and an edge-weight function ℓ. We invoke their construction procedure of the mortar graph as a black box using error parameter $\varepsilon' = \varepsilon/4$ and unit edge-weights $\ell(e) = 1$ for all edges $e \in E(W')$. Note that the node weights are ignored in this construction.

In what follows, we will prove certain properties of the mortar graph MG about its node weights and error parameter ε based on the fact that similar properties hold with respect to the unit edge weights and error parameter $\varepsilon' = \varepsilon/4$.

The following technical lemma tells us that the node weight of a path is roughly half its edge length apart from a small additive offset. It turns out convenient for the shortness properties of the spanner that this offset is the same for any two paths sharing their end nodes.

Lemma 5. *Let P, P' be two paths sharing both of their end points u and v. Then the following properties hold.*

(i) There is $b \in \{0, -1, -2\}$ such that $\ell(P) = 2c(P) + b$ and $\ell(P') = 2c(P') + b$.
(ii) $\ell(P)/2 \leq c(P) \leq \ell(P)/2 + 1$
(iii) $\ell_W(u, v)/2 \leq d_W(u, v) \leq \ell_W(u, v)/2 + 1$
(iv) P is a shortest path under ℓ_W if and only if it is a shortest path under d_W.

The following lemma gives cost bounds on the mortar graph. In contrast to [8], we have to exclude singleton boundaries in property (i) in order to avoid a cost explosion. To account for the singleton boundaries in the shortness property of Lemma 1 we bound their total number separately. (See proof of Lemma 1.)

Lemma 6. *The mortar graph MG has the following two properties.*

(i) The total cost $\sum_{B \in \mathcal{B}: E(W_B) \neq \emptyset} c(W_B) + \sum_{B \in \mathcal{B}: E(E_B) \neq \emptyset} c(E_B)$ of all the non-singleton western and eastern boundaries of all bricks is bounded by $O(\varepsilon) \cdot$ OPT.
(ii) The total cost $c(MG)$ of the mortar graph is $O(1/\varepsilon) \cdot \mathrm{OPT}$.

Designating Portals. For finding the portals, we cannot directly use the same greedy procedure as in Step 3(a) [8]. It does not work for bricks having a boundary with small cost, because of the additive one in Lemma 5 bound. To circumvent this issue, we pick *all* the vertices to be the set of portals when the boundary has small cost. And then for any remaining brick, the boundary cost is bounded from below. For these bricks the greedy procedure works, as the additive plus can be absorbed in the big-Oh by creating a factor 3 gap in the number of portals and the cost bounds, which is sufficient.

By balancing all the parameters, we get that for any brick $B \in \mathcal{B}$ there exists at most 3τ portals $P(B)$ such that each vertex on the boundary of B lies within a distance of at most $c(\partial B)/\tau$ from some portal. Here $\tau = \tau(\varepsilon) = \Theta(g(\varepsilon)\varepsilon^{-2})$, where $g(\varepsilon)$ is defined in Lemma 16 of the extended version of the paper [11].

Lemma 7. *Given a brick $B \in \mathcal{B}$, there exists a set of vertices $P(B) \subseteq \partial B$, such that:*

1. *(**Cardinality Property**) $|P(B)| \leq 3\tau$*
2. *(**Coverage Property**) For any $u \in V(\partial B)$, there exists $v \in P(B)$, such that $d_{\partial B}(u, v) \leq c(\partial B)/\tau$ and $\ell_{\partial B}(u, v) \leq \ell(\partial B)/(3\tau)$*

Now we can sketc.h the proof of Lemma 1.

Proof (Proof of Lemma 1).
(i) **Shortness property.** We have to bound the total cost of H, which consists of the mortar graph and optimal Steiner trees added in step 5 of Algorithm 1. By Lemma 6 the cost of the mortar graph is $O(1/\varepsilon) \cdot \mathrm{OPT}$. We bound the cost of Steiner trees analogously as in the Lemma 4.1 [8], *i.e.*, we charge it to the cost of the mortar graph (losing a large constant). However, we have to take extra care to not overcharge vertices adjacent to multiple bricks.

Consider any brick B and any tree connecting portals of B added in step 5. The cost of this tree can be upper-bounded by the cost of the boundary of the brick $c(\partial B)$. Since there is a constant number of such trees (this follows from Lemma 7), the total cost of the trees added is a constant times the cost of the boundary of the brick. Now, if every vertex belonged to the boundary of a constant number of bricks, this would imply that the total cost of all Steiner trees is bounded by constant times OPT. Below we show that if this is not the case for some vertices, then we have a different way to pay for the cost incurred by these vertices.

We say that a vertex v is a *corner* of a brick B if it belongs to the intersection of N_B (or S_B) with E_B (or W_B). In the special case in which E_B or W_B is empty, we call any v that belongs to the intersection $N_B \cap S_B$ a *trivial corner.* We also say, that v is a *regular* boundary vertex of a brick if it is not a corner of this brick.

Observe that v can be a regular boundary vertex of at most two bricks. It remains to show how to charge corner vertices. For trivial corners, observe that there are as many unique pairs (corner vertex, corresponding brick) as there were strips during creation of the mortar graph. Note that there are $O(f(\varepsilon)OPT)$ strips (see Lemma 10 in Appendix B of extended paper [11]). This, together with the fact that weight of each vertex is at most 1 implies that we charge at most constant times OPT for trivial corners.

The charging for non-trivial corners is different. By the first property of Lemma 6, we know that the sum of the costs of non-singleton west and east boundaries for all bricks is bounded by $O(\varepsilon)$OPT. As non-trivial corners belong to W_B or E_B, the total cost incurred by charging to non-trivial corners is also bounded by constant times OPT. This finishes the proof of the shortness property.

(ii) **Spanning property** The proof of the spanning property is similar in the spirit to the proof of Structural Theorem 3.2, and Lemma 4.2 in [8]. However, we cannot use their approach of portal-connected graph via the *brick insertion* operation, as in the node weighted setting this would destroy the structure of the optimum solution. Therefore we give a slightly more direct proof, where we avoid the portal-connected graph at all.

Moreover, we have to take extra care when showing structural lemmas. These proofs do not transfer immediately to the node-weighted instances. For example, we have to heavily use special structure of map-weighted instances to bound the number of *joining vertices.* Due to the high technicality of the arguments and lack of space, the details are explained in the extended version of the paper [11]. □

3 Node-Weighted Contraction Decomposition

In this section we give a reduction of a spanner to graphs with bounded treewidth. The input to our reduction is a spanner (*e.g.* the one constructed in the previous section, see Lemma 1), *i.e.* a graph H of cost $f(\epsilon) \cdot \mathrm{OPT}$ that approximately preserves an optimum solution.

We apply Lemma 2 (proven later in this section) with $k = \frac{2 \cdot f(\epsilon)}{\epsilon}$ to graph H and obtain sets $E_0, E_1, \ldots, E_{k-1}$. Now, define the cost of the set of edges to be the total weight of vertices incident to edges in this set. Because every vertex belongs to at most two sets, the total cost of all the edge sets is at most $2f(\epsilon) \cdot \mathrm{OPT}$ and therefore, the cheapest set, say E_c has cost at most $\epsilon \cdot \mathrm{OPT}$.

We now contract E_c to obtain graph H'. We assign weight 0 to the vertices resulting from contraction, while the weight of the other stays untouched. It is clear that after this operation the value of the optimum solution will not increase. Now we solve the node-weighted Steiner tree problem for H' using Lemma 3 (see Appendix A of the extended version of a paper [11]). We can do this in polynomial time, since the treewidth of H' is at most k. We include the set of edges E_c in the obtained solution for H' to get the final tree of cost $(1 + \epsilon) \cdot \mathrm{OPT}$.

Therefore we are left with proving Lemma 2. Here, we build on Klein's [23] contraction decomposition and modify it to our needs.

Proof (Proof of Lemma 2).

At first, we triangulate the dual graph H^* by adding an artificial vertex in the middle of each face of the dual graph and introducing artificial edges (see Fig. 3). This is the crucial step which—as explained later—enables us to control the level of edges in the breadth-first search tree.

Let J^* be the graph after above modification of H^*. Now, from some arbitrary node r we run breadth-first search on J^*, this gives us a partition of $V(J^*) =$

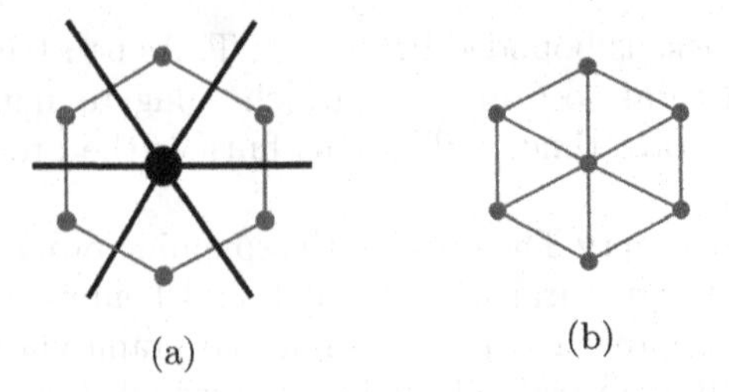

Fig. 3. (a) A vertex of a graph (in black) and the dual. The dual is shown in blue. (b) Triangulation of the dual graph. (Color figure online)

$L_0 \cup L_1 \ldots \cup L_d$, for some $d \in \mathbb{Z}^+$. Fix an $i \in \{0, \ldots, k-1\}$, and let E_i^* be the set of edges with one endpoint in L_p and the other endpoint in L_{p+1}, for all p congruent to i modulo k. Also, let E_i be the set of primal edges of H corresponding to the dual edges in E_i^*. Note that we do not include in E_i the artificial edges. We claim that sets E_i satisfy both the requirements.

First we show that each vertex participates in at most two sets.

Lemma 8. *For each vertex $v \in H$, all the incident edges of v are contained in at most two sets E_i, E_j.*

Proof. Consider the faces $f_1, f_2, \ldots f_l$ incident to a vertex v (in the clockwise order of appearing in the planar embedding). Each f_i has a corresponding vertex in the dual graph H^* and therefore also in J^*. Moreover, there is a cycle on these vertices. Call the edges of this cycle $e_1^*, e_2^*, \ldots e_l^*$. These edges correspond to all primal edges $e_1, e_2, \ldots e_l$ incident to vertex v in H. However, J^* has also additional vertex g adjacent to all vertices f_i. Therefore, the distance between any two f_i and f_j in J^* is at most 2. Hence, the vertices f_i will be in at most three consecutive layers of BFS ordering. Therefore, all the incident edges of v will be contained in at most two of the E_i^*'s, and hence be contained in at most two corresponding E_i's. □

We are left with showing that contracting each set reduces the treewidth. In essence, we use the argument of Klein. We only need to take care of artificial edges.

Lemma 9. *The graph H after contracting E_i has treewidth $O(k)$.*

Proof. Let J be a dual graph of J^*. We will call J the primal of J^*.

By directly applying the result of Klein, contracting all the primal edges of E_i^* from J results in a graph X of treewidth $O(k)$. It is easy to see, that contracting all other artificial edges in X results exactly in a graph $H_{/E_i}$. Since contraction of edges does not increase treewidth, the lemma follows. □

The algorithm described above together with Lemma 8 and Lemma 9 completes the proof of Lemma 2. □

4 Conclusion

We reduced the node-weighted Steiner tree problem on planar graphs to the problem of finding a spanner. The main obstacles in constructing such general spanner are caused by the high-degree vertices. The first difficulty arises already in the *cutting-open* step. The second issue is related to bounding the number of joining vertices.

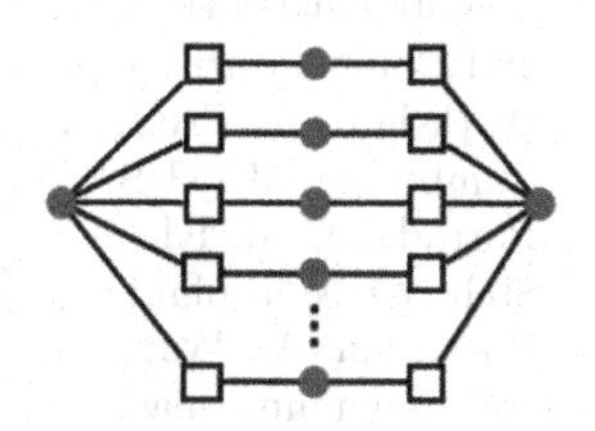

Fig. 4. Node-weighted subset spanner does not exists. Red vertices are of cost 1, squares are terminals of cost 0.

As we have shown, both of the difficulties are solvable in map-weighted graphs. However, we pose an interesting open problem: decide the existence of a PTAS for node-weighted Steiner tree problem on map-weighted graphs where terminals are allowed to lie also on vertices of weight 0.

There are two reasons why the above open problem is compelling. First, it nicely isolates the first difficulty—the latter issue is not present on such graphs. Second, a node-weighted *subset spanner* does not exist for these instances. A subset spanner is a cheap subgraph (in terms of optimum Steiner tree) that approximately preserves distances between pairs of terminals. In contrast, a subset spanner construction exists in the edge-weighted case [22].

Figure 4 gives an example of such a node-weighted instance. The cheapest Steiner tree has cost 3. However, any subset spanner would have to use all the red central vertices.

References

1. Arora, S., Grigni, M., Karger, D.R., Klein, P.N., Woloszyn, A.: A polynomial-time approximation scheme for weighted planar graph TSP. In: Proceedings of the Ninth Annual ACM-SIAM Symposium on Discrete Algorithms, 25–27 January 1998, San Francisco, California, USA, pp. 33–41 (1998)
2. Baker, B.S.: Approximation algorithms for np-complete problems on planar graphs. J. ACM **41**(1), 153–180 (1994)
3. Bateni, M., Chekuri, C., Ene, A., Hajiaghayi, M.T., Korula, N., Marx, D.: Prize-collecting Steiner problems on planar graphs. In: Proceedings of the Twenty-Second Annual ACM-SIAM Symposium on Discrete Algorithms, SODA 2011, San Francisco, California, USA, January 23–25, 2011, pp. 1028–1049 (2011)
4. Bateni, M., Demaine, E.D., Hajiaghayi, M., Marx, D.: A PTAS for planar group Steiner tree via spanner bootstrapping and prize collecting. In: Proceedings of the 48th Annual ACM SIGACT Symposium on Theory of Computing, STOC 2016, Cambridge, MA, USA, 18–21 June 2016, pp. 570–583 (2016)
5. Bateni, M., Farhadi, A., Hajiaghayi, M.: Polynomial-time approximation scheme for minimum k-cut in planar and minor-free graphs. In: Proceedings of the Thirtieth Annual ACM-SIAM Symposium on Discrete Algorithms, SODA 2019, San Diego, California, USA, 6–9 January 2019, pp. 1055–1068 (2019)
6. Bateni, M., Hajiaghayi, M.T., Marx, D.: Approximation schemes for Steiner forest on planar graphs and graphs of bounded treewidth. J. ACM **58**(5), 21:1–21:37 (2011)

7. Berman, P., Yaroslavtsev, G.: Primal-dual approximation algorithms for node-weighted network design in planar graphs. In: Approximation, Randomization, and Combinatorial Optimization. Algorithms and Techniques - 15th International Workshop, APPROX 2012, and 16th International Workshop, RANDOM 2012, Cambridge, MA, USA, 15–17 August 2012. Proceedings, pp. 50–60 (2012)
8. Borradaile, G., Klein, P., Mathieu, C.: An o(n log n) approximation scheme for Steiner tree in planar graphs. ACM Trans. Algorithms **5**, 31:1–31:31 (2009)
9. Buchanan, A., Wang, Y., Butenko, S.: Algorithms for node-weighted Steiner tree and maximum-weight connected subgraph. Networks **72**(2), 238–248 (2018)
10. Byrka, J., Grandoni, F., Rothvoß, T., Sanità, L.: Steiner tree approximation via iterative randomized rounding. J. ACM **60**(1), 6:1–6:33 (2013)
11. Byrka, J., Lewandowski, M., Meesum, S.M., Spoerhase, J., Uniyal, S.: PTAS for steiner tree on map graphs. CoRR abs/1912.00717 (2019), http://arxiv.org/abs/1912.00717
12. Cabello, S., Gajser, D.: Simple ptas's for families of graphs excluding a minor. Discrete Appl. Math. **189**, 41–48 (2015)
13. Chen, Z., Grigni, M., Papadimitriou, C.H.: Map graphs. J. ACM **49**(2), 127–138 (2002)
14. Chimani, M., Mutzel, P., Zey, B.: Improved Steiner tree algorithms for bounded treewidth. J. Discrete Algorithms **16**, 67–78 (2012)
15. Cohen-Addad, V., Klein, P.N., Mathieu, C.: Local search yields approximation schemes for k-means and k-median in euclidean and minor-free metrics. SIAM J. Comput. **48**(2), 644–667 (2019)
16. Demaine, E.D., Hajiaghayi, M.T., Klein, P.N.: Node-weighted Steiner tree and group Steiner tree in planar graphs. ACM Trans. Algorithms **10**(3), 13:1–13:20 (2014)
17. Demaine, E.D., Hajiaghayi, M.: The bidimensionality theory and its algorithmic applications. Comput. J. **51**(3), 292–302 (2008)
18. Dreyfus, S.E., Wagner, R.A.: The Steiner problem in graphs. Networks **1**(3), 195–207 (1971)
19. Eisenstat, D., Klein, P.N., Mathieu, C.: An efficient polynomial-time approximation scheme for Steiner forest in planar graphs. In: Proceedings of the Twenty-Third Annual ACM-SIAM Symposium on Discrete Algorithms, SODA 2012, Kyoto, Japan, 17–19 January 2012, pp. 626–638 (2012)
20. Guha, S., Moss, A., Naor, J., Schieber, B.: Efficient recovery from power outage (extended abstract). In: Proceedings of the Thirty-First Annual ACM Symposium on Theory of Computing, 1–4 May 1999, Atlanta, Georgia, USA, pp. 574–582 (1999)
21. Kammer, F., Tholey, T.: Approximate tree decompositions of planar graphs in linear time. Theoret. Comput. Sci. **645**, 60–90 (2016)
22. Klein, P.N.: A subset spanner for planar graphs: with application to subset TSP. In: Proceedings of the 38th Annual ACM Symposium on Theory of Computing, Seattle, WA, USA, 21–23 May 2006, pp. 749–756 (2006)
23. Klein, P.N.: A linear-time approximation scheme for TSP in undirected planar graphs with edge-weights. SIAM J. Comput. **37**(6), 1926–1952 (2008)
24. Klein, P.N., Ravi, R.: A nearly best-possible approximation algorithm for node-weighted Steiner trees. J. Algorithms **19**(1), 104–115 (1995)
25. Lipton, R.J., Tarjan, R.E.: Applications of a planar separator theorem. SIAM J. Comput. **9**(3), 615–627 (1980)
26. Moldenhauer, C.: Primal-dual approximation algorithms for node-weighted Steiner forest on planar graphs. Inf. Comput. **222**, 293–306 (2013)

Near-Linear Time Algorithm for Approximate Minimum Degree Spanning Trees

Ran Duan, Haoqing He(✉), and Tianyi Zhang

Institute for Interdisciplinary Information Sciences,
Tsinghua University, Beijing, China
{hehq13,tianyi-z16}@mails.tsinghua.edu.cn, duanran@mail.tsinghua.edu.cn

Abstract. Given a graph $G = (V, E)$, we wish to compute a spanning tree whose maximum vertex degree, i.e. tree degree, is as small as possible. Computing the exact optimal solution is known to be NP-hard, since it generalizes the Hamiltonian path problem. For the approximation version of this problem, a $\tilde{O}(mn)$ time algorithm that computes a spanning tree of degree at most $\Delta^* + 1$ is previously known [Fürer & Raghavachari 1994]; here Δ^* denotes the minimum tree degree of all the spanning trees. In this paper we give the first near-linear time approximation algorithm for this problem. Specifically speaking, we propose an $\tilde{O}(\frac{1}{\epsilon^7}m)$ time algorithm that computes a spanning tree with tree degree $(1+\epsilon)\Delta^* + O(\frac{1}{\epsilon^2}\log n)$ for any constant $\epsilon \in (0, \frac{1}{6})$. Thus, when $\Delta^* = \omega(\log n)$, we can achieve approximate solutions with constant approximate ratio arbitrarily close to 1 in near-linear time.

Keywords: Approximate algorithm · Graph algorithm · Minimum degree spanning tree

1 Introduction

Computing minimum degree spanning trees is a fundamental problem that has inspired a long time of research. Let $G = (V, E)$ be an undirected graph, and we wish to compute a spanning tree of G whose tree degree, or maximum vertex degree in the tree, is the smallest. Clearly this problem is NP-hard as the Hamiltonian path problem can be reduced to it, and so we could only hope for a good approximation in polynomial time. The optimal approximation of this problem was achieved in [7] where the authors proposed a [1]$\tilde{O}(mn)$ time algorithm that computes a spanning tree of tree degree $\leq \Delta^* + 1$; conventionally $n = |V|, m = |E|$ and Δ^* denotes the minimum tree degree of all the spanning trees. For convenience, in this paper the degree of a vertex usually means its tree degree in the current spanning tree.

This work has been supported in part by the Zhongguancun Haihua Institute for Frontier Information Technology.

[1] $\tilde{O}(\cdot)$ hides poly-logarithmic factors.

Y. Kohayakawa and F. K. Miyazawa (Eds.): LATIN 2020, LNCS 12118, pp. 15–26, 2020.
https://doi.org/10.1007/978-3-030-61792-9_2

Our Result. The major result of this paper is a near-linear time algorithm for computing minimum degree spanning trees in undirected graphs. Formally we propose the following statement.

Theorem 1. *For any constant* $\epsilon \in (0, \frac{1}{6})$, *there is an algorithm that runs in* $O(\frac{1}{\epsilon^7} m \log^7 n)$ *time which computes a spanning tree with tree degree at most* $(1+\epsilon)\Delta^* + \frac{576}{\epsilon^2}\log n$.

The core argument of Theorem 1 is that, starting from an arbitrary spanning tree, we repeatedly search for a sequence of distinct non-tree edges, named as *augmenting sequence*, to modify the current spanning tree which immediately reduces the degree of some high-degree vertex. The idea of augmenting sequence is similar to [7], that is, given a fixed degree bound k, an augmenting sequence w.r.t. the current spanning tree and k is a sequence of vertex-disjoint non-tree edges $(w_1, z_1), (w_2, z_2), \cdots, (w_h, z_h)$ such that $w_1, w_2, \cdots, w_{h-1}$ have tree degree $k-1$ and $w_h, z_1, z_2, \cdots, z_h$ have tree degree $< k-1$. Also there is a vertex w_0 with tree degree $\geq k$ on the tree path between w_1 and z_1, and w_i for $1 \leq i < h$ is on the tree path between w_{i+1} and z_{i+1} but not on the tree path between w_j and z_j for $j > i+1$. Then we can add theses edges $(w_1, z_1) \cdots, (w_h, z_h)$ to the spanning tree and delete the edges associated with $w_0, \cdots, w_{h-1}$ on the cycles formed, so the total degree of vertices with degree $\geq k$ will decrease by 1 but more degree-$(k-1)$ vertices may emerge.

In our process of searching, similar to the blocking flow approach [3] for max-flow, we first construct a layering of the graph by the shortest length of augmenting sequences, then each time find a shortest augmenting sequence in the layering and do such tree modification by this augmenting sequence, thus after near-linear time the shortest length of augmenting sequences would increase. We repeat this until the length of the shortest augmenting sequence is longer than $\frac{1}{\epsilon}\log n$. When this happens, the number of layers also exceeds $\frac{1}{\epsilon}\log n$, so there are two adjacent layer whose ratio is at most $1+\epsilon$, then if the number of augmenting sequences we found are not too large (not too many new degree-$(k-1)$ vertices emerge), we can argue a $1+O(\epsilon)$ approximation for the optimal solution Δ^*. In the whole procedure of our algorithm, we can let $k = (1-O(\epsilon))\Delta$ for the degree Δ of the current spanning tree, and make k increase by one after each iteration until in some iteration the sum of degree of all the vertices with degree $\geq k$ is not significantly decreased. See Sect. 3.2.

Related Work. There is a line of works that are concerned with low-degree trees in weighted undirected graphs. In this scenario, the target low-degree that we wish to compute is constrained by two parameters: an upper bound B on tree degree, an upper bound C on the total weight summed over all tree edges. The problem was originally formulated in [4]. Two subsequent papers [10,11] proposed polynomial time algorithms that compute a tree with cost $\leq wC$ and degree $\leq \frac{w}{w-1}bB + \log_b n$, $\forall b, w > 1$. This result was substantially improved by [2]; using certain augmenting path technique, their algorithm is capable of finding a tree with cost $\leq C$ and degree $B + O(\log n / \log\log n)$. Results and techniques from [2] might sound similar to ours, but in undirected graphs we are actually faced with different technical difficulties. [2]'s result was improved by [8] where for all k, a spanning tree of degree $\leq k+2$ and of cost at most the cost of the optimum spanning tree of maximum degree at most k can be computed in polynomial time. The degree bound was later further improved from $k+2$ to the optimal $k+1$ in [14]. There are also many works studying the minimum degree Steiner tree problem in undirected graphs [5,7,13], and the minimum degree spanning tree or Steiner tree problems in directed graphs [1,6,9,12,15].

2 Preliminary

Let $G = (V, E)$ be the graph we consider, and we assume G is a connected graph. Logarithms are taken at base 2. During the execution of our algorithm, a spanning tree T will be maintained. For every $u \in V$, let $\deg(u)$ be the tree degree of u in T, and the degree of the spanning tree T is defined as $\Delta = \max_{u \in V} \deg(u)$. Our algorithm will repeatedly modify T to reduce its degree Δ. Let Δ^* denote the minimum degree of all the spanning trees. For each pair $u, v \in V$, let $\rho_{u,v}$ be the unique tree path that connects u and v in T. For each $1 \leq k \leq n$, define $S_k = \{u \mid \deg(u) \geq k\}$ to be the set of vertices of degree at least k, $N_k = \{u \mid \deg(u) = k\}$ to be the set of vertices of degree exactly k, and $d_k = \sum_{u \in S_k} \deg(u)$ to be the sum of degrees of vertices in S_k.

Boundary Edge and Boundary Set. Boundary edge and boundary set are important concepts to get the lower bound of Δ^*.

Definition 1. *For a graph $G = (V, E)$ and a sequence of disjoint vertex subsets $V_1, V_2, \cdots, V_l \subseteq V$, an edge $(u, v) \in E$ is called a boundary edge if $u \in V_i, v \in V_j$ for $1 \leq i \neq j \leq l$, or $u \in V_i$ for some i but $v \notin V_1 \cup \cdots \cup V_l$. A vertex set W is called a boundary set (with respect to $V_1, V_2, \cdots, V_l$), if for every boundary edge (u, v), at least one of u, v belongs to W.*

Lemma 1. *Let $V_1, V_2, \cdots, V_l \subseteq V$ be a sequence of disjoint vertex subsets, W be a boundary set and Δ^* be the minimum degree of all the spanning tree in G. Then, $\Delta^* \geq \frac{l-1}{|W|}$.*

Proof. By Definition 1, every set V_i can only be connected to other vertices by boundary edges, so for any spanning tree T of G, there are at least $l-1$ boundary edges connecting $V_1, V_2, \cdots, V_l$ in T. Then for any boundary edge (u, v), at least one of u, v belongs to W. Thus by the pigeon-hole principle, there exists a $u \in W$ whose tree degree is $\geq \frac{l-1}{|W|}$.

3 A $(1+\epsilon)\Delta^* + O(\frac{1}{\epsilon^2}\log n)$ Approximation

Let $\epsilon \in (0, \frac{1}{48})$ be a fixed parameter. This algorithm starts from an arbitrary spanning tree T and keeps modifying T to decrease its tree degree Δ. It consists of two phases: the *large-step phase* and the *small-step phase.*

- In the large-step phase, as long as $\Delta \geq \frac{10\log^2 n}{\epsilon^3}$, we repeatedly apply a near-linear time subroutine that, either Δ is reduced to $\leq (1-\epsilon)\cdot\Delta$ or a spanning tree T is returned with the guarantee that $\Delta = (1+O(\epsilon))\Delta^*$.
- In the small-step phase, we deal with the situation where $\frac{9\log n}{\epsilon^2} \leq \Delta < \frac{10\log^2 n}{\epsilon^3}$, and a spanning tree T is returned with the guarantee that $\Delta \leq (1+O(\epsilon))\Delta^* + O(\frac{\log n}{\epsilon^2})$.

For the rest of this section, we first propose and analyze the degree reduction algorithm $\mathsf{AugSeqDegRed}$ which efficiently reduces the total degree of vertices with degree $\geq k$ by 1 using an augmenting sequence technique. After that we specify how the large-step phase work. Due to page limit, the small-step phase is omitted and can be found in the full version of this paper.

3.1 Degree Reduction via Augmenting Sequences

In this algorithm, we continue to explore possibilities of improving the tree structure using the idea of augmenting sequence as in [7]. For a non-tree edge (u, v) that connects two different components C_u, C_v of $\mathsf{T} \setminus S_k$ where $deg(u) = k-1$, we try to add (u, v) to T and delete some edge incident on S_k to eliminate cycles. At the same time, as $deg(u)$ increases to k, we keep looking for a sequence of distinct non-tree edges inside C_u to add to T and delete a sequence of tree edges to eliminate cycles.

A difficulty is that when the degrees of some original $(k-1)$-degree vertices decrease, it is hard to make the layering of the graph stable. Therefore, we define *marked vertices* instead of the concept of the vertices with degree $\geq k-1$. Given a degree threshold $k \leq \Delta$, a vertex gets marked whenever its tree degree becomes $k-1$, and it stays marked even if its tree degree becomes below $k-2$ afterwards. We only re-initialize the set of marked vertices when we change k in Sect. 3.2. Then we can define *augmenting sequence* formally.

Definition 2 (augmenting sequence). *An h-length augmenting sequence consists of a sequence of vertex-disjoint non-tree edges $(w_1, z_1), \cdots, (w_h, z_h) \in E$ with the following properties.*

(i) $\exists w_0 \in \rho_{w_1,z_1} \cap S_k$, and for all $0 \leq i < h, w_i \in \rho_{w_{i+1},z_{i+1}} \setminus (\bigcup_{j=i+2}^{h} \rho_{w_j,z_j})$.
(ii) All z_i's are unmarked ($\forall 1 \leq i \leq h$); w_i's are marked for all $1 \leq i < h$ and w_h is unmarked.

The tree can be modified by the augmenting sequence $(w_1, z_1), \cdots, (w_h, z_h)$:

Lemma 2 (tree modification). *Given an augmenting sequence $(w_1, z_1), \cdots, (w_h, z_h) \in E$, one can modify T such that d_k decreases and no vertices are added to S_k. Also Δ cannot increase.*

Proof. We modify T in an inductive way. For $i = h-1, h-2, \cdots, 0$, as $w_i \in \rho_{w_{i+1},z_{i+1}}$, we can take an arbitrary tree edge $(w_i, x) \in \rho_{w_{i+1},z_{i+1}}$, and then perform an update $\mathsf{T} \leftarrow \mathsf{T} \cup \{(w_{i+1}, z_{i+1})\} \setminus \{(w_i, x)\}$ which guarantees that T

Algorithm 1: Layering

```
1 B_0 ← S_k, h ← 0;
2 while h < 1 + log_{1+ε} n do
3     compute the forest {C_u^h} spanned by T \ (∪_{i=0}^h B_i);
4     if exists unmarked u, v such that (u, v) ∈ E, C_u^h ≠ C_v^h then
5         break;
6     else
7         compute B_{h+1} to be the set of all marked vertices u ∈ V \ (∪_{i=0}^h B_i)
          such that there exists an unmarked adjacent vertex v with C_u^h ≠ C_v^h;
8         h ← h + 1;
9 return h and B_0, B_1, ··· , B_h;
```

is still a spanning tree. Because $w_j \notin \rho_{w_{i+1},z_{i+1}}$ for $0 \leq j \leq i-1$, tree update $\mathsf{T} \leftarrow \mathsf{T} \cup \{(w_{i+1}, z_{i+1})\} \setminus \{(w_i, x)\}$ does not change the connected components of $\mathsf{T} \setminus \{w_j\}$, so the property that $w_j \in \rho_{w_{j+1},z_{j+1}} \setminus (\bigcup_{l=j+2}^{h} \rho_{w_l,z_l}), \forall 0 \leq j < i$ is preserved.

During the process, if for any z_i, $\deg(z_i)$ $(1 \leq i \leq h)$ becomes $k-1$ during the process, mark z_i. By definition, d_k decreases as w_0 loses a tree neighbour; plus, no vertices are newly added to S_k because all $\deg(w_i), 1 \leq i < h$ are unchanged and $\deg(w_h) \leq k-2$, $\deg(z_i) \leq k-2, \forall 1 \leq i \leq h$. Also vertices in S_k can only lose tree neighbors so Δ cannot increase.

Now, back to the AugSeqDegRed algorithm. The core of this algorithm is that, if the currently shortest augmenting sequences have length h $(h < 1+\log_{1+\epsilon} n)$, it searches for augmenting sequences of length h and applies Lemma 2 to decrease d_k. When there is no augmenting sequence of length h, it repeats this process for some larger h. Finally this algorithm terminates when $h \geq 1 + \log_{1+\epsilon} n$ and we prove a lower bound on Δ^* based on the structure of T.

First, we introduce the Layering algorithm which computes an auxiliary layering of the graph that will also help tree modification later. Initially set $B_0 \leftarrow S_k$. Inductively, suppose $B_0, B_1, \cdots, B_h, h \geq 0$ is already computed, then we compute the forest spanned by $\mathsf{T} \setminus (\bigcup_{i=0}^{h} B_i)$; for each $u \in V \setminus (\bigcup_{i=0}^{h} B_i)$, let C_u^h be the connected component of $\mathsf{T} \setminus (\bigcup_{i=0}^{h} B_i)$ that contains u. If there exists an edge $(u, v) \in E$ such that both u, v are unmarked vertices, and that $C_u^h \neq C_v^h$, then the algorithm terminates and reports that the shortest length of augmenting sequences is equal to $h+1$; otherwise, we compute B_{h+1} to be the set of all marked vertices $u \in V \setminus (\bigcup_{i=0}^{h} B_i)$ such that there exists an unmarked adjacent vertex v with $C_u^h \neq C_v^h$, and then continue until $h > 1 + \log_{1+\epsilon} n$. Note that whenever $B_h = \emptyset$, $B_{h+1}, \cdots, B_{\lceil 1+\log_{1+\epsilon} n \rceil}$ are all empty. The pseudo code is shown in the Layering algorithm 1.

After we have invoked Layering and computed a sequence of vertex subsets $B_0, B_1, \cdots, B_h$ which naturally divides the graph into $h+2$ layers (including a layer of other vertices), every time we will find a length-$(h+1)$ augmenting

Algorithm 2: AugDFS(i,(u,v))

```
1  if i = 1 then
2      return (u, v);
3  for untagged w ∈ ρ_{u,v} ∩ B_{i-1} do
4      for unmarked z such that (w, z) is untagged and C_z^{i-2} ≠ C_w^{i-2} do
5          p_{i-1} ← AugDFS(i-1,(w,z));
6          tag (w, z);
7          if p_{i-1} ≠ null then
8              let p_i be p_{i-1} plus (u, v);
9              return p_i;
10     tag w;
11 return null;
```

sequence $(w_1, z_1), (w_2, z_2), \cdots, (w_{h+1}, z_{h+1})$ such that $w_i \in B_i$ for $1 \le i \le h$, then apply tree modifications of Lemma 2 by this augmenting sequence. Repeat this until there is no more length-$(h+1)$ augmenting sequences any more. The difficulty in searching for the shortest augmenting sequences is that, for a search that starts from a pair of adjacent and unmarked vertices u, v satisfying $C_u^h \neq C_v^h$ and goes up the layers $B_h, B_{h-1}, \cdots, B_1, B_0$, not every route can reach the top layer B_0 because some previous $(h+1)$-length augmenting sequences have already blocked the road. Therefore, a depth-first search needs to be performed. To save running time, some tricks are needed: if a certain vertex has been searched before by some previous $(h+1)$-length augmenting sequences and has failed to lead a way upwards to B_0, then we *tag* this vertex so that future depth-first searches may avoid this tagged vertex; if a certain edge has been searched before, then we *tag* this edge whatsoever. The AugDFS algorithm may be a better illustration of this algorithm. The recursive algorithm AugDFS takes the layer number i and an edge (u, v) as input and keeps searching for edges between a vertex $w \in (u, v) \cap B_{i-1}$ and an unmarked vertex z. If such an edge is found, invoke AugDFS with the parameter $(i-1, (w, z))$ and return the result plus (u, v). The pseudo code is shown in the AugDFS algorithm 2. Later we will prove that AugDFS(h+1,(u,v)) always returns an augmenting sequence if exists.

The upper-level AugSeqDegRed algorithm repeatedly applies Layering followed by several rounds of AugDFS. Each time AugDFS returns an augmenting sequence p, modify T by Lemma 2 via p. The repeat-loop ends when $h \ge 1 + \log_{1+\epsilon} n$. The pseudo code is shown in the AugSeqDegRed algorithm 3.

Before proving termination of AugSeqDegRed, we first need to argue some properties of Layering. The following lemma and corollary will serve as the basis for our future proof.

Lemma 3 (the blocking property). *Throughout each iteration of the repeat-loop in AugSeqDegRed, for any* $1 \le i < h$ *and any two adjacent vertices* $u, v \in$

Algorithm 3: AugSeqDegRed(k)

```
1  mark all degree k − 1 vertices, unmark other vertices;
2  repeat
3      run Layering which computes h and B_0, B_1, ··· , B_h;
4      untag all vertices and edges;
5      for (u, v) ∈ E such that u, v are unmarked and adjacent, and that C_u^h ≠ C_v^h
       do
6          p ←AugDFS(h+1,(u,v));
7          if p ≠ null then
8              modify T by augmenting sequence p via Lemma 2;
9  until h ≥ 1 + log_{1+ε} n;
10 return T;
```

$V \setminus (\bigcup_{j=0}^{i} B_j)$ *such that* u *is unmarked and* $C_u^i \neq C_v^i$, *then* $v \in B_{i+1}$. *(Recall that* C_u^h *is the connected component of* $\mathsf{T} \setminus (\bigcup_{i=0}^{h} B_i)$ *that contains* u*.)*

Proof. By rules of Layering, this blocking property holds right after Layering outputs them. This claim continuous to hold afterwards because tree modifications only merge components C_u^i's and never split any C_u^i's.

Here is an important corollary of this Lemma 3, whose proof is in the full version of this paper.

Corollary 1. *Throughout each iteration of the repeat-loop, for any* $w \in B_i, 1 \leq i \leq h$, *suppose* w *is adjacent to an unmarked* z *such that* $C_w^{i-1} \neq C_z^{i-1}$. *Then* $\rho_{w,z}$ *only contains vertices from* $V \setminus (\bigcup_{j=0}^{i-2} B_j)$.

Now we have the following lemmas:

Lemma 4. *If AugDFS(h+1, (u,v)) returns a sequence of edges* $(w_1, z_1), \cdots, (w_{h+1}, z_{h+1})$, *then* $w_i \in B_i$ *for* $1 \leq i \leq h$, *and* $w_{h+1}, z_1, \cdots z_{h+1}$ *are unmarked, also the edges are vertex-disjoint.*

Proof. The initial u, v are unmarked. From the algorithm, when calling AugDFS(i, (u,v)), we find a $w \in \rho_{u,v} \cap B_{i-1}$ and z is unmarked, so the corresponding $w_i \in B_i$ for $1 \leq i \leq h$, and $w_{h+1}, z_1, \cdots z_{h+1}$ are unmarked, also the vertices $\{w_i | 1 \leq i \leq h\}$ are distinct. To see that $w_{h+1}, z_1, \cdots z_{h+1}$ are distinct, we argue that in one execution of AugDFS(i, (u,v)), w and z have $C_z^{i-2} \neq C_w^{i-2}$ but $C_z^{i-3} = C_w^{i-3}$, since if $C_z^{i-3} \neq C_w^{i-3}$, w would be in B_{i-2} by the algorithm Layering. Thus $w_{h+1}, z_1, \cdots z_{h+1}$ are in distinct components in $\mathsf{T} \setminus (\bigcup_{i=0}^{h} B_i)$.

Lemma 5. *In the AugSeqDegRed algorithm, AugDFS(h+1, (u,v)) returns either null or an augmenting sequence.*

Proof. Assume a sequence of edges $(w_1, z_1), \cdots, (w_{h+1}, z_{h+1})$ is returned by AugDFS(h+1, (u,v)). Property (ii) in Definition 2 is proved by Lemma 4.

Now let us focus on property (i). We can take an arbitrary $w_0 \in \rho_{w_1,z_1} \cap B_0$ since $C^0_{w_1} \neq C^0_{z_1}$ by the algorithm. Also since $w_i \in B_i, \forall 0 \leq i \leq h$, by Corollary 1 we know ρ_{w_i,z_i} does not contain any $w_j, 0 \leq j \leq i-2$, so property (ii) holds.

The following statement concludes the AugSeqDegRed algorithm will terminate quickly.

Lemma 6. *In the AugSeqDegRed algorithm, h is increased by at least one during each repeat-loop, except the last one.*

Proof. By the rules of Layering, it is easy to see that at the beginning when Layering outputs $B_0, B_1, \cdots, B_h$, the shortest length of augmenting sequence is equal to $h+1$. So it suffices to prove that by the end of this iteration the shortest augmenting sequence has length $> h+1$.

First we need to characterize all augmenting sequences using $B_0, B_1, \cdots, B_h$. Let the sequence (w_1, z_1), $(w_2, z_2), \cdots, (w_l, z_l)$ be an arbitrary augmenting sequence and let w_0 be the B_0-vertex on ρ_{w_1,z_1}. We argue that $l \geq h+1$, and more importantly, if $l = h+1$, it must be that $w_i \in B_i, \forall 0 \leq i \leq h$. We inductively prove that $w_i \in \bigcup_{j=0}^{i} B_j$ for $i = 0, 1, \cdots, l-1$. The basis is obvious as is required by property (i) in Definition 2. Now assume $w_i \in B_r$ for some $r \leq i$. Then, from algorithm Layering, it would not be hard to see $w_{i+1} \in \bigcup_{j=0}^{r+1} B_j \subseteq \bigcup_{j=0}^{i+1} B_j$. Now, since components $\{C^r_u\}$ for $r \leq h-1$ are not connected by edges whose both endpoints are unmarked by Lemma 3, so $\rho_{w_l,z_l} \cap \bigcup_{j=0}^{h-1} B_j = \emptyset$, and on the other hand $w_{l-1} \in \rho_{w_l,z_l} \cap \bigcup_{j=0}^{l-1} B_j$, so $l \geq h+1$. Plus, we can see from the induction that, when $l = h+1$ it must be that $w_i \in B_i, \forall 0 \leq i \leq h$.

For any unmarked and adjacent vertices u, v such that $C^h_u \neq C^h_v$, consider the instance of AugDFS with input $(h+1, (u, v))$. We make two claims.

(1) If there is an $(h+1)$-length augmenting sequence ending with (u, v), AugDFS would succeed in finding one.
(2) If it has returned null, then there would be no $(h+1)$-augmenting sequence ending with (u, v) throughout the entire repeat-loop iteration.

If (1), (2) can be proved, then by the end of this repeat-loop iteration, there would be no $(h+1)$-length augmenting sequences because such augmenting sequence should end with a pair of adjacent unmarked vertices. Next we come to prove (1), (2).

(1) The depth-first search of AugDFS exactly coincides with the conditions that $w_i \in B_i$, except that it skips all tagged vertices and edges. Now we prove that omitting tagged vertices and edges does not miss any $(h+1)$-length augmenting sequences. For an edge (w, z) to be tagged, either a further recursion AugDFS has succeeded or failed in finding an augmenting sequences; in the former case, C^{i-2}_w and C^{i-2}_z has been merged, and so the condition $C^{i-2}_w \neq C^{i-2}_z$ would be violated afterwards; in the latter case, we would not need to recur on (w, z) since the components w.r.t. $B_0, \cdots, B_{i-2}$ also can only merge. For a vertex w to be tagged, we must have enumerated all of its untagged edges (w, z) but failed to find any augmenting sequences, and therefore any future depth-first searches on w would still end up in vain.

(2) If AugDFS has once failed to find any augmenting sequences starting with (u, v), then all vertices $w \in \rho_{u,v} \cap B_h$ visited by this instance of AugDFS should be tagged and they would be omitted by all succeeding instances of AugDFS. Therefore $\rho_{u,v} \cap B_h$ would stay unchanged since then. Hence, if we re-run AugDFS with $h+1, (u, v)$, it will return null without any recursion because all vertices in $\rho_{u,v} \cap B_h$ are tagged.

Suppose AugSeqDegRed has terminated with $B_0, B_1, \cdots, B_{\lceil \log_{1+\epsilon} n+1 \rceil}$. We introduce the notion of a *clean component*, a sequence of disjoint vertex subsets, and apply Lemma 1 to get the lower bound on Δ^*.

Definition 3. *After an instance of AugSeqDegRed has been executed, for any vertex $u \in V \setminus (\bigcup_{i=0}^{h} B_i)$, an arbitrary component $C_u^h, 0 \leq h \leq \lceil \log_{1+\epsilon} n + 1 \rceil$ is called* clean *if all vertices in C_u^h are unmarked.*

Lemma 7. *For any $0 \leq h < \lceil \log_{1+\epsilon} n + 1 \rceil$, suppose $\mathsf{T} \setminus (\bigcup_{i=0}^{h} B_i)$ has l clean components, then a lower bound holds that $\Delta^* \geq \frac{l-1}{\sum_{i=0}^{h+1} |B_i|}$.*

Proof. Since $h < \lceil \log_{1+\epsilon} n + 1 \rceil$, B_h is not the last one, so there is no edge connecting two unmarked vertices in different components of $\mathsf{T} \setminus (\bigcup_{i=0}^{h} B_i)$. By Lemma 3, any edge that connects a clean components of $\mathsf{T} \setminus (\bigcup_{i=0}^{h} B_i)$ outwards must be incident on a vertex in $\bigcup_{i=0}^{h+1} B_i$, so $\bigcup_{i=0}^{h+1} B_i$ is a boundary set w.r.t. clean components. Therefore by Lemma 1 we have $\Delta^* \geq \frac{l-1}{|\bigcup_{i=0}^{h+1} B_i|} = \frac{l-1}{\sum_{i=0}^{h+1} |B_i|}$

Lemma 8. *There is an implementation of procedure AugSeqDegRed that runs in $O(\frac{1}{\epsilon^2} m \log^2 n)$ time.*

(Proof omitted.)

3.2 Large-Step Phase

The large-step phase are described in the ImprovedMDST algorithm 4, in which we deal with the case $\Delta \geq \frac{10 \log^2 n}{\epsilon^3}$ (the small-step phase for $\frac{9 \log n}{\epsilon^2} \leq \Delta < \frac{10 \log^2 n}{\epsilon^3}$ is in the full version of this paper). It works by invoking AugSeqDegRed with an incremental parameters k from $(1-2\epsilon)\Delta + 1$ if $d_{k-1} \leq 2d_k$. Within each iteration, if AugSeqDegRed fails to reduce d_k by a factor of $(1 - \frac{\epsilon^2}{2 \log n})$, then the algorithm reports a lower bound on Δ^* and returns T immediately. Otherwise, increase k by 1 and continue until d_k becomes 0. Since $d_{k+1} \leq d_k$, d_k will become 0 in at most $O(\log^2 n/\epsilon^2)$ iterations. Once $d_k = 0$, Δ must have decreased and repeat the while-loop. (Note that by Lemma 2, Δ cannot increase during the whole algorithm.)

Running Time. In the large-step phase, every iteration d_k shrinks by a factor of $\leq (1 - \frac{\epsilon^2}{2 \log n})$, so d_k will become zero in $O(\log^2 n/\epsilon^2)$ iterations. We have:

Lemma 9. *The running time of the large-step phase is $O(\frac{1}{\epsilon^5} m \log^5 n)$.*

Algorithm 4: ImprovedMDST

```
1  Let T be a spanning tree of G with tree degree Δ;
   /* Large-step phase                                          */
2  while Δ ≥ 10 log^2 n / ε^3 do
3  |   k = (1 − 2ε)Δ + 1;
4  |   while d_k > 0 do
5  |   |   if d_{k−1} ≤ 2d_k then
6  |   |   |   d ← d_k;
7  |   |   |   run AugSeqDegRed(k);
8  |   |   |   if d_k > (1 − ε^2 / (2 log n)) · d then
9  |   |   |   |   return T;
10 |   |   k = k + 1;
11 |   update the tree degree Δ;
12 return T;
```

Proof. From the previous subsection we already know that AugSeqDegRed runs in $O(\frac{1}{\epsilon^2} m \log^2 n)$ time, so here we only need to upper bound the total number of times AugSeqDegRed gets invoked before $\Delta < \frac{10\log^2 n}{\epsilon^3}$ or a spanning tree T is returned within a while-loop. Next we only focus on the previous cases because it takes a longer running time. In this case, at the end of each iteration, $d_k \le (1 - \frac{\epsilon^2}{2\log n}) \cdot d$. The inside while-loop would break when $k > (1-2\epsilon)\Delta + \frac{2\log^2 n}{\epsilon^2}$ because by the time $d_k \le \left(1 - \frac{\epsilon^2}{2\log n}\right)^{\frac{2\log^2 n}{\epsilon^2}} \cdot d_{(1-2\epsilon)\Delta} \le \frac{d_{(1-2\epsilon)\Delta}}{n} < 1$. As $(1-2\epsilon)\Delta + \frac{2\log^2 n}{\epsilon^2} \le (1-\epsilon)\Delta$ when $\Delta \ge \frac{10\log^2 n}{\epsilon^3}$, which means Δ has been reduced by a factor of at most $1-\epsilon$ in the end of each while-loop and there are at most $O(\frac{1}{\epsilon}\log n)$ while-loops. In summary, the total running time of the large-step phase is $O\left(\frac{\log^2 n}{\epsilon^2} m \times \frac{\log^2 n}{\epsilon^2} \times \frac{\log n}{\epsilon}\right) = O\left(m \cdot \frac{\log^5 n}{\epsilon^5}\right)$.

Approximation Guarantee. When a spanning tree T is returned within the large-step phase, the vertex subsets $B_0, B_1, \cdots, B_{\lceil 1+\log_{1+\epsilon} n\rceil}$ created by AugSeqDegRed satisfies the blocking property (see Lemma 3). By Lemma 7, there is a lower bound on Δ^* for each vertex set $B_h, 0 \le h < \lceil 1 + \log_{1+\epsilon} n \rceil$ as long as we get the lower bound on the number of clean components in $\mathsf{T} \setminus (\bigcup_{i=0}^{h} B_i)$. The following two statements show the lower bound on Δ^*.

Lemma 10. *For any vertex subset B and any spanning tree T, the number of connected components in $T \setminus B$ is at least $\sum_{u\in B} \deg(u) - 2|B| + 2$.*

(Proof omitted and can be found in the full version of this paper.)

Lemma 11. *If a spanning tree T is returned within the large-step phase and k is the parameter of the last invoked AugSeqDegRed, for any $0 \le h < \lceil 1+\log_{1+\epsilon} n\rceil$,* *the number of clean components in* $\mathsf{T}\setminus(\bigcup_{i=0}^{h} B_i)$ *is at least* $k\cdot(1-4\epsilon)\sum_{i=0}^{h}|B_i|+1$ *for* $\epsilon \in (0, \frac{1}{48})$. *Furthermore,*

$$\Delta^* \geq k(1-4\epsilon)\cdot\frac{\sum_{i=0}^{h}|B_i|}{\sum_{i=0}^{h+1}|B_i|}$$

Proof. By Lemma 10, the number of tree components in $\mathsf{T}\setminus(\bigcup_{i=0}^{h} B_i)$ is at least $\sum_{u\in\bigcup_{i=0}^{h} B_i}\deg(u) - 2\left|\bigcup_{i=0}^{h} B_i\right| + 2$. Let d'_k, d'_{k-1}, S'_{k-1} and S'_k be snapshots of d_k, d_{k-1}, S_{k-1} and S_k right before the last instance of AugSeqDegRed started and let M be the set of all marked vertices $\notin S'_{k-1}$ (i.e., vertices that are initially unmarked) by the end of AugSeqDegRed. Then, the number of clean components in $\mathsf{T}\setminus(\bigcup_{i=0}^{h} B_i)$ is at least

$$\sum_{u\in\bigcup_{i=0}^{h} B_i}\deg(u) - 2\sum_{i=0}^{h}|B_i| + 2 - |M\cup S'_{k-1}|$$

We have the following lower bound on $\sum_{u\in\bigcup_{i=0}^{h} B_i}\deg(u)$ and upper bound on $|M\cup S'_{k-1}|$, whose detailed proof is in the full version.

$$\sum_{u\in\bigcup_{i=0}^{h} B_i}\deg(u) \geq (k-1)\sum_{i=1}^{h}|B_i| + \left(1-\frac{3\epsilon^2}{2\log n}-\frac{\epsilon}{2}\right)d'_k$$

$$|M\cup S'_{k-1}| \leq 2\epsilon\cdot d'_k$$

Then for $n > 2$ and $\epsilon \in (0, \frac{1}{48})$, $k \geq \frac{9\log n}{\epsilon^2} - \log n$ and $d'_k \geq k|B_0|$, it is not hard to see the number of clean components in $\mathsf{T}\setminus(\bigcup_{i=0}^{h} B_i)$ is at least $k(1-4\epsilon)\cdot\sum_{i=0}^{h}|B_i| + 2$. Apply Lemma 7 and conclude the proof

$$\Delta^* \geq \frac{k(1-4\epsilon)\cdot\sum_{i=0}^{h}|B_i|+1}{\sum_{i=0}^{h+1}|B_i|} \geq k(1-4\epsilon)\cdot\frac{\sum_{i=0}^{h}|B_i|}{\sum_{i=0}^{h+1}|B_i|}$$

In the following statement, we combine all the inequalities for each B_h and get the upper bound on Δ with Δ^*.

Lemma 12. *When a spanning tree* T *is returned within the large-step phase, it must be that* $\Delta \leq (1+8\epsilon)\cdot\Delta^*$ *for* $\epsilon \in (0, \frac{1}{48})$.

Proof. Consider the most recent execution of AugSeqDegRed before returning. By the previous subsection, this instance of AugSeqDegRed has created a sequence of disjoint vertex subsets $B_0, B_1, \cdots, B_{1+\log_{1+\epsilon} n}$ that satisfy the blocking property. By the pigeon-hole principle, there exists an h such that $\frac{\sum_{i=0}^{h}|B_i|}{\sum_{i=0}^{h+1}|B_i|} \geq \frac{1}{1+\epsilon}$. Then by Lemma 11, (recall that in the large-step phase $k > (1-2\epsilon)\Delta$)

$$\Delta^* \geq k(1-4\epsilon)\cdot\frac{1}{1+\epsilon} > \frac{1-6\epsilon+8\epsilon^2}{1+\epsilon}\Delta$$

or equivalently, $\Delta \leq \frac{1+\epsilon}{1-6\epsilon+8\epsilon^2}\Delta^* < (1+8\epsilon)\Delta^*$ when $\epsilon \in (0, \frac{1}{48})$.

We claim that, for any constant $\epsilon \in (0, \frac{1}{6})$, the ImprovedMDST algorithm computes a spanning tree with tree degree at most $(1+\epsilon)\Delta^* + O(\frac{\log^2 n}{\epsilon^3})$ in $O(\frac{1}{\epsilon^5} m \log^5 n)$ time (by resetting $\epsilon \to 8\epsilon'$ where ϵ' is the ϵ in previous analysis). In the full paper, we will see that with the small-step phase, we can compute a spanning tree with tree degree $(1+\epsilon)\Delta^* + \frac{576}{\epsilon^2} \log n$ in $O(\frac{1}{\epsilon^7} m \log^7 n)$ time.

References

1. Bansal, N., Khandekar, R., Nagarajan, V.: Additive guarantees for degree-bounded directed network design. SIAM J. Comput. **39**(4), 1413–1431 (2009)
2. Chaudhuri, K., Rao, S., Riesenfeld, S., Talwar, K.: What would edmonds do? augmenting paths and witnesses for degree-bounded MSTs. In: Chekuri, C., Jansen, K., Rolim, J.D.P., Trevisan, L. (eds.) APPROX/RANDOM -2005. LNCS, vol. 3624, pp. 26–39. Springer, Heidelberg (2005). https://doi.org/10.1007/11538462_3
3. Dinitz, Y.: Algorithm for solution of a problem of maximum flow in networks with power estimation. Soviet Math. Dokl. **11**, 1277–1280 (1970)
4. Fischer, T.: Optimizing the degree of minimum weight spanning trees. Cornell University, Technical report (1993)
5. Fraigniaud, P.: Approximation algorithms for minimum-time broadcast under the vertex-disjoint paths mode. In: Proceedings of the 9th Annual European Symposium on Algorithms, pp. 440–451 (2001)
6. Fürer, M., Raghavachari, B.: An NC approximation algorithm for the minimum degree spanning tree problem. In: Proceedings of the 28th Annual Allerton Conference on Communication, Control and Computing, pp. 274–281 (1990)
7. Fürer, M., Raghavachari, B.: Approximating the minimum-degree Steiner tree to within one of optimal. J. Algorithms **17**(3), 409–423 (1994)
8. Goemans, M.X.: Minimum bounded degree spanning trees. In: Proceedings of the 47th Annual IEEE Symposium on Foundations of Computer Science, FOCS 2006, pp. 273–282. IEEE (2006)
9. Klein, P.N., Krishnan, R., Raghavachari, B., Ravi, R.: Approximation algorithms for finding low-degree subgraphs. Networks **44**(3), 203–215 (2004)
10. Könemann, J., Ravi, R.: A matter of degree: improved approximation algorithms for degree-bounded minimum spanning trees. In: Proceedings of the 32nd Annual ACM Symposium on Theory of Computing, pp. 537–546. ACM (2000)
11. Könemann, J., Ravi, R.: Primal-dual meets local search: approximating mst's with nonuniform degree bounds. In: Proceedings of the 35th Annual ACM Symposium on Theory of Computing, pp. 389–395. ACM (2003)
12. Krishnan, R., Raghavachari, B.: The directed minimum-degree spanning tree problem. In: Hariharan, R., Vinay, V., Mukund, M. (eds.) FSTTCS 2001. LNCS, vol. 2245, pp. 232–243. Springer, Heidelberg (2001). https://doi.org/10.1007/3-540-45294-X_20
13. Ravi, R.: Rapid rumor ramification: approximating the minimum broadcast time. In: Proceedings of the 35th Annual Symposium on Foundations of Computer Science, pp. 202–213. IEEE (1994)
14. Singh, M., Lau, L.C.: Approximating minimum bounded degree spanning trees to within one of optimal. In: Proceedings of the 39th Annual ACM Symposium on Theory of Computing, pp. 661–670. ACM (2007)
15. Yao, G., Zhu, D., Li, H., Ma, S.: A polynomial algorithm to compute the minimum degree spanning trees of directed acyclic graphs with applications to the broadcast problem. Discrete Math. **308**(17), 3951–3959 (2008)

Approximation Algorithms for Cost-Robust Discrete Minimization Problems Based on Their LP-Relaxations

Khaled Elbassioni(✉)

Khalifa University of Science and Technology, Abu Dhabi, UAE
khaled.elbassioni@ku.ac.ae

Abstract. We consider robust discrete minimization problems where uncertainty is defined by a convex set in the objective. Assuming the existence of an integrality gap verifier with a bounded approximation guarantee for the LP relaxation of the non-robust version of the problem, we derive approximation algorithms for the robust version under different types of uncertainty, including polyhedral and ellipsoidal uncertainty.

Keywords: Approximation algorithms · Discrete optimization · Linear programming · Randomized rounding · Robust optimization

1 Introduction

Standard optimization algorithms assume precise knowledge of their inputs, and find optimal or near-optimal solutions under this assumption. However, in real-life applications, the input data may be known up to a limited precision with errors introduced possibly due to inaccuracy in measurements or lack of exact information about the precise value of the input parameters. Clearly, an optimization algorithm designed based on such distorted data to optimize a certain objective function would not yield reliable results, if no special consideration of such uncertainty is taken. Several approaches to deal with uncertainty in data have been introduced, including *stochastic optimization* (see e.g., [11]), where certain probabilistic assumptions are made on the uncertainty and the objective is to optimize the average-case or the probability of a certain desirable event, and *robust optimization* (see, e.g., [4]), where some deterministic assumptions are made on the uncertain parameters, and the objective is to optimize over the worst-case these parameters can assume[1].

In this paper, we consider a class of *robust* discrete optimization (DO) problems, where uncertainty is assumed to be only *in the objective* (called sometimes *cost-robust* optimization problems). Given a discrete set of feasible solutions, one

[1] Yet, there is a third (intermediate) approach, namely, *distributionally robust optimization* (see, e.g., [14]), in which one optimizes the expectation over the worst-case choice from a set of distributions on the uncertain parameters.

Y. Kohayakawa and F. K. Miyazawa (Eds.): LATIN 2020, LNCS 12118, pp. 27–37, 2020.
https://doi.org/10.1007/978-3-030-61792-9_3

is interested in maximizing/minimizing a linear objective function over this set; it is assumed that the objective function is not explicitly given, but is known to belong to a *convex* uncertainty set. The requirement is to solve the optimization problem in the *worst-case* scenario that the objective assumes in the uncertainty set. Our goal is to show how an approximation algorithm, based on the linear programming (LP) relaxation for the nominal version of a discrete optimization problem, can be used to derive an approximation algorithm for the robust version. We will focus on minimization problems, even though some of the results can be extended to maximization problems.

1.1 Integrality Gap Verifiers

More formally, we consider a minimization problem over a discrete set $\mathcal{S} \subseteq \mathbb{Z}_+^n$ and a corresponding LP-relaxation over $\mathcal{Q} \subseteq \mathbb{R}_+^n$:

$$\text{OPT} = \min \quad c^T x \qquad \text{s.t.} \quad x \in \mathcal{S} \tag{1}$$

$$z^* = \min \quad c^T x \qquad \text{s.t.} \quad x \in \mathcal{Q}, \tag{2}$$

where $c \in \mathbb{R}_+^n$. We will be mainly working with discrete optimization problems for which there is an approximation algorithm that rounds any feasible LP solution to a discrete one with a bounded approximation ratio. This is formulated in the following definition.

Definition 1. *For $\alpha \geq 1$, a (deterministic) α-integrality gap verifier $\mathcal{A} = \mathcal{A}(c,x)$ for (1)–(2), w.r.t. a class $\mathscr{C} \subseteq \mathbb{R}_+^n$ of objectives is a polytime algorithm that, given any $c \in \mathscr{C}$ and any $x \in \mathcal{Q}$ returns an $\widehat{x} \in \mathcal{S}$ such that $c^T\widehat{x} \leq \alpha \cdot c^T x$. An integrality gap verifier $\mathcal{A}$ is said to be* oblivious *(see, e.g., [18]) if $\mathcal{A}(c,x) = \mathcal{A}(x)$ does not depend on the objective c. When the class of objectives is $\mathscr{C} = \mathbb{R}_+^n$, we simply call $\mathcal{A}$ an (oblivious) integrality gap verifier.*

A *randomized α-integrality gap verifier* is the same as in Definition 1 except that it returns a random $\widehat{x} \in \mathcal{S}$ such that $\mathbb{E}[c^T\widehat{x}] \leq \alpha c^T x$. We will consider a special class of randomized integrality gap verifiers that are given by the following definition.

Definition 2. *For $\alpha \geq 1$ and $x \in \mathcal{Q}$, an α-approximate semi-negatively correlated randomized rounding, denoted α-*ANCRR*, of x is an $\widehat{x}$ such that: (i) $\Pr[\widehat{x} \in \mathcal{S}] = 1 - o(1)$; (ii) $\mathbb{E}[c^T\widehat{x}] \leq \alpha c^T x$; and (iii) for any $S \subseteq [n]$:*

$$\Pr\Big[\bigwedge_{i\in S}(\widehat{x}_i = 1)\Big] \leq \prod_{i\in S}\Pr[\widehat{x}_i = 1]. \tag{3}$$

*An α-*ANCRR *integrality gap verifier is a polytime algorithm that, given any $x \in \mathcal{Q}$, returns an α-*ANCRR *of x.*

Remark 1. Consider a minimization problem (1) and its LP relaxation (2). By Markov's inequality and probabilistic amplification, given an α-randomized intergality gap verifier $\mathcal{A}$, $x \in \mathcal{Q}$, $c \in \mathbb{R}_+^n$ and $\epsilon > 0$, we can get in $O(\frac{\log n}{\epsilon})$ calls to $\mathcal{A}$ an $\widehat{x} \in \mathcal{S}$ such that $c^T\widehat{x} \leq (1+\epsilon)\alpha \cdot c^T x$ holds with probability $1 - o(1)$.

1.2 Robust Discrete Optimization Problems

In the framework of robust optimization (see, e.g. [4,6]), we assume that the objective vector c is *not known exactly*. Instead, it is given by a *convex uncertainty set* $\mathcal{C} \subseteq \mathbb{R}^n_+$. It is required to find a (near)-optimal solution for the DO problem under the *worst-case* choice of objective $c \in \mathcal{C}$. Typical examples of uncertainty sets $\mathcal{C}$ include:

- Polyhedral uncertainty: $\mathcal{C} := \mathcal{P}(A, b, c^0) := \{c \in \mathbb{R}^n_+ : A(c - c^0) \leq b\}$, for given matrix $A \in \mathbb{R}^{m \times n}_+$, vector $b \in \mathbb{R}^m_+$ and (nominal) vector $c^0 \in \mathbb{R}^n_+$.
- Ellipsoidal uncertainty: $\mathcal{C} := \mathbf{E}(c^0, D) := \{c \in \mathbb{R}^n_+ : (c - c^0)^T D^{-2}(c - c^0) \leq 1\}$, for given positive definite matrix $D \in \mathbb{R}^{m \times n}$ and vector $c^0 \in \mathbb{R}^n_+$.

More generally, we will consider a class of uncertainty sets defined by *affine perturbations* around a nominal vector $c^0 \in \mathbb{R}^n_+$ (see, e.g., [4]):

$$\mathcal{C} = \mathcal{C}(c^0, c^1, \ldots, c^r; \mathcal{D}) := \left\{ c := c^0 + \sum_{r=1}^{k} \delta_r c^r \, : \, \delta = (\delta_1, \ldots, \delta_k) \in \mathcal{D} \right\}, \tag{4}$$

where $c^0, c^1, \ldots, c^k \in \mathbb{R}^n_+$ and $\mathcal{D} \subseteq \mathbb{R}^k$ is a *convex* perturbation set:

- Polyhedral perturbation $\mathcal{D} = \mathcal{P}(A, b, 0) := \{\delta \in \mathbb{R}^k_+ : A\delta \leq b\}$, for given matrix $A \in \mathbb{R}^{m \times k}_+$ and vector $b \in \mathbb{R}^m_+$.
- Ellipsoidal perturbation: $\mathcal{D} = \mathbf{E}(0, D) := \{\delta \in \mathbb{R}^k_+ : \delta^T D^{-2} \delta \leq 1\}$, for a given positive definite matrix $D \in \mathbb{R}^{k \times k}$.

The vectors $c^1, \ldots, c^k \in \mathbb{R}^n_+$ will be called the *generators* of the perturbation set $\mathcal{D}$. Note that a polyhedral uncertainty set $\mathcal{P}(A, b, c^0)$ can be described in the form (4) by setting $\mathcal{C} := \mathcal{C}(c^0, \mathbf{1}^1, \ldots, \mathbf{1}^n; \mathcal{D})$ for the polyhedral perturbation set $\mathcal{D} := \mathcal{P}(A, b, 0)$, where $\mathbf{1}^j$ denotes the jth unit vector in $\mathbb{R}^n$. Similarly, an ellipsoidal uncertainty set $E(c^0, D)$ can be described in the form (4) by setting $\mathcal{C} := \mathcal{C}(c^0, \mathbf{1}^1, \ldots, \mathbf{1}^n; \mathcal{D})$ for the ellipsoidal perturbation set $\mathcal{D} := \mathbf{E}(0, D)$.

1.3 Convex Relaxation for the Robust DO Problem

We can model the robust DO problem and its convex relaxation as follows:

$$\text{OPT}_R = \min_{x \in \mathcal{S}} \max_{c \in \mathcal{C}} \; c^T x, \tag{5}$$

$$z^*_R = \min_{x \in \mathcal{Q}} \max_{c \in \mathcal{C}} \; c^T x. \tag{6}$$

Equivalenlty, we can write (5)–(6) as

$$\text{OPT}_R = \min \; z \tag{7}$$
$$\text{s.t.} \;\; c^T x \leq z \quad \forall c \in \mathcal{C} \tag{8}$$
$$x \in \mathcal{S}.$$

$$z^*_R = \min \; z \tag{9}$$
$$\text{s.t.} \;\; c^T x \leq z \quad \forall c \in \mathcal{C} \tag{10}$$
$$x \in \mathcal{Q}.$$

Note that (6) amounts to a convex programming problem that can be solved (almost to optimality) in polynomial time (see, e.g., [21]). *Near-optimal* solutions can also be found more *efficiently*, based on the semi-infinite LP formulation (9), using the *multiplicative weight updates method* [17].

1.4 Approximation Guarantees for a Robust DO Problem

We consider both deterministic and randomized algorithms for the robust optimization problem (5) (see, e.g., [8,23]):

Definition 3. *For $\alpha \geq 1$, a randomized approximation algorithm $\mathcal{B}$ for the robust DO problem (5) is said to be:*

- α-robust-in-expectation *(w.r.t. the uncertainty set $\mathcal{C}$), if the expected objective in the uncertainty set $\mathcal{C}$, w.r.t. the output solution, over the random choices of the algorithm, is within a factor of α from the optimum solution:*

$$\mathbb{E}_{\widehat{x}\sim\mathcal{B}}[c^T\widehat{x}] \leq \alpha \cdot \mathrm{OPT}_R \quad \forall c \in \mathcal{C};$$

- α-robust-with-high-probability, *if with probability approaching* 1, *all objectives in the uncertainty set $\mathcal{C}$, w.r.t. the output solution, are within a factor of α from the optimum solution:*

$$\Pr_{\widehat{x}\sim\mathcal{B}}[c^T\widehat{x} \leq \alpha \cdot \mathrm{OPT}_R \quad \forall c \in \mathcal{C}] = 1 - o(1);$$

- α-deterministically robust *if it is α-robust with probability* 1, *i.e., it outputs a vector $\widehat{x} \in \mathcal{S}$ such that:*

$$c^T\widehat{x} \leq \alpha \cdot \mathrm{OPT}_R \quad \forall c \in \mathcal{C}.$$

Clearly, the notion of α-deterministically robust is stronger than that of α-robust-with-high-probability, which is, in turn, (more or less[2]) stronger than that of α-robust-in-expectation.

1.5 Summary of Main Results

To describe the results we obtain in this paper, let us consider the polyhedral/ellipsoidal uncertainty sets:

$$\mathcal{C}_1 := \{c := c^0 + u \mid u \in \mathbb{R}^n_+,\ u \leq d, \quad Au \leq b\} \tag{11}$$

$$\mathcal{C}_2 := \{c := c^0 + \mathbf{C}\delta \mid \delta \in \mathbb{R}^k_+,\ A\delta \leq b\} \tag{12}$$

$$\mathcal{C}_3 := \{c := c^0 + u \mid u \in \mathbb{R}^n_+,\ \|D^{-1}u\|_2 \leq 1\} \tag{13}$$

$$\mathcal{C}_4 := \{c := c^0 + \mathbf{C}\delta \mid \delta \in \mathbb{R}^k_+,\ \|D^{-1}\delta\|_2 \leq 1\}. \tag{14}$$

Assume the matrices $A, b, d, \mathbf{C}$ are non-negative and D is positive definite, where $\mathbf{C} \in \mathbb{R}^{n\times k}_+$ is the matrix whose columns are $c^1, \ldots, c^k$. Let m be the number of rows of A, $\beta := \min_j \max_i a_{ij}$ and $\gamma := \max_{i,j} a_{ij}$, $c_{\min} := \min_{r\neq 0, j:\ c_j^r>0} c_j^r$ and $c_{\max} := \max_{r\neq 0, j} c_j^r$. Our results are summarized in Table 1. The first column describes the restrictions on the discrete set $\mathcal{S}$ (if any): $\mathcal{S}$ is *binary* if $\mathcal{S} \subseteq \{0,1\}^n$ and *covering* if $x \in \mathcal{S}$ and $y \geq x$ implies $y \in \mathcal{S}$. In the second column, we

[2] Indeed, if $\mathcal{B}$ is α-robust-with-high-probability, then for any $c \in \mathcal{C}$, $\mathbb{E}_{\widehat{x}\sim\mathcal{B}}[c^T\widehat{x}] \leq \alpha \cdot \mathrm{OPT}_R + o(1) \max_{c\in\mathcal{C},\ x\in\mathcal{S}} c^T x = \alpha \cdot \mathrm{OPT}_R + o(1)$.

describe the type of uncertainty set considered, and the conditions on it (if any). The third column gives the type of approximation algorithm which we assume available for the nominal problem, while the fourth column gives the guarantee for the corresponding robust version which we obtain in this paper. As can be seen from the table, except for the first two results, the approximation factors we obtain depend on the "width" of the uncertainty set as described by the ratios $\frac{\gamma}{\beta}$ and $\frac{c_{\max}}{c_{\min}}$ for polyhedral uncertainty, and $\frac{\lambda_{\max}(D)}{\lambda_{\min}(D)}$ for ellipsoidal uncertainty. The approximation ratio is also proportional to the square root of the number of generators in the perturbation set. Whether these bounds can be significantly improved remains an interesting open question.

1.6 Some Related Work

While there is an extensive body of work on robust *continuous* optimization problems (see, e.g., [4–7,10,16,24]), much less is known in the *discrete* case, where most work has considered special uncertainty sets or specific discrete problems. In [8], Bertsimas and Sim consider the minimization problem (5) with *budget uncertainly*, where at most k components of the objective are allowed to increase; for *binary* optimization problems they gave an α-deterministically robust approximation algorithm for the robust version which is obtained by making $n+1$ calls to any α-approximation algorithm for the non-robust version. Some generalizations of this result to the *non-binary* case were obtained in [20], and other improvements and generalizations were obtained in [3]. In Sect. 2.1 below, we show that the number of calls to the approximation algorithm can be made significantly smaller and also extend the result to any *constant* number of budget constraints. For *uncorrelated ellipsoidal uncertainty* (where the uncertainty set is an axis-aligned ellipsoid), Bertsimas and Sim [9] also gave a *pseudo polynomial-time* reduction from solving a robust version of a DO problem over a binary set $\mathcal{S}$ to a linear optimization problem over the same set. As observed in [22, Chapter 2], when specialized to *ball uncertainty*, this yields a polynomial time algorithm for solving the robust problem, whenever the nominal version can be solved in polynomial time. This should be contrasted with our result above, where an $O(\alpha\sqrt{n})$-approximation for the robust problem with ellipsoidal uncertainty, satisfying $D \succ 0$, over an arbitrary discrete set, can be obtained from any α-integrailty gap verifier for the nominal problem.

More recently, Kawase and Sumita (2018) gave robust-in-expectation algorithms for special problems such as the knapsack problem and the maximum independent set problem in the *intersection of r matroids*, among others. We note, however, that their results are not of the "reduction" type, that is, they provide algorithms that are specific to each problem. We note also that some of these results can be derived from our reduction in Sect. 2. Finally, it is worth noting that there is a number of results on special problems, such as ShortestPath [1], MinCostFlow [8], MachineScheduling [12], VehicleRouting [2], two-stage robust optimization [15,19], mostly under a class of budget uncertainty. In general, this seems to be a growing area of research, see, e.g., the theses by Poss [24] and Ilyina [22].

Table 1. Summary of the reductions.

$\mathcal{S}$	Uncertainty set	Available Approx. Alg.	Approximation guarantee
General	General convex set	General α-integrality gap verifier	α-robust-in-expectation
Binary	$\mathcal{C}_1$; $m = O(1)$	General α-approx. Alg.	$O(\alpha)$-deterministically robust
Binary & Covering	$\mathcal{C}_1$	General α-integrality gap verifier	$O(\alpha + \sqrt{\frac{\alpha\gamma n}{\beta}})$-deterministically robust
Binary	$\mathcal{C}_2$	α-ANCRR integrality gap verifier	$O(\alpha\sqrt{k\log(k)\frac{\gamma}{\beta}\frac{c_{\max}}{c_{\min}}})$-robust-with-high-probability
General	$\mathcal{C}_4$; $DC \geq 0$	General α-integrality gap verifier	$O(\alpha\sqrt{k})$-deterministically robust-
Binary & Covering	$\mathcal{C}_3$; $D^{-1} > 0$	General α-integrality gap verifier	$O(\alpha + \sqrt{\frac{\alpha\lambda_{\max}(D)n}{\lambda_{\min}(D)}})$-deterministically robust
Binary	$\mathcal{C}_4$; $D^{-1} > 0$	α-ANCRR integrality gap verifier	$O(\alpha\sqrt{k\log(k)\frac{\lambda_{\max}(D)}{\lambda_{\min}(D)}\frac{c_{\max}}{c_{\min}}})$-robust-with-high-probability

Outline of the Techniques. Almost all the results in Table 1 are based on solving the convex relaxation for the robust optimization problem (in some form), then rounding the obtained fractional solution. A useful tool that we rely on, first proved by Carr and Vempala [13], allows one to turn a given non-oblivious integrality gap verifier for the LP-relaxation into an oblivious one. Another ingredient of our proofs is the use of *strong LP-duality* to go from a maxmin-optimization problem to a purely minimization problem; this was the approach used by Bertsimas and Sim in [8], which we push further by combining it with randomized rounding techniques (see, e.g., [25]), and using a *dual-fitting* argument to bound the approximation guarantee on the rounded solution. First, we describe this approach for polyhedral uncertainty, then it would not be hard to extend the results to ellipsoidal uncertainty (which are omitted die to lack of space), by envisioning an ellipsoid as a polytope with *infinitely many* linear inequalities.

2 A Robust-in-Expectation Approximation Algorithm

We first observe simply that an *oblivious* intergality gap verifier for the nominal problem implies an α-robust-in-expectation algorithm for the robust version.

Lemma 1. *Consider a combinatorial minimization problem (1) and its LP relaxation (2), admitting an oblivious α-integrality gap verifier $\mathcal{A}$ w.r.t. a class $\mathscr{C}$ of objectives. Then there is a polytime α-robust-in-expectation algorithm for the robust version (7) w.r.t. to any convex uncertainty set $C \subseteq \mathscr{C}$.*

Carr and Vempala [13] gave a decomposition theorem that allows one to use an α-integrality gap verifier for a given LP-relaxation of a combinatorial minimization problem, to decompose a given fractional solution to the LP into a convex combination of integer solutions that is dominated by α times the fractional solution. We can restate their result as follows.

Theorem 1 [13]. *Consider a discrete minimization problem (1) and its LP relaxation (2), admitting an α-integrality gap verifier $\mathcal{A}$. Then there is a polytime algorithm that, for any given $x^* \in \mathcal{Q}$, finds a set $\mathcal{X} \subseteq \mathcal{S}$, of polynomial size, and a set of convex multipliers $\{\mu_x \in \mathbb{R}_+ : x \in \mathcal{X}\}$, $\sum_{x \in \mathcal{X}} \mu_x = 1$, such that*

$$\alpha x^* \geq \sum_{x \in \mathcal{X}} \mu_x x. \tag{15}$$

We obtain the following (known) corollary of Theorem 1.

Corollary 1. *Consider a discrete minimization problem (1) and its LP relaxation (2), admitting an α-integrality gap verifier $\mathcal{A}$. Then (2) admits an oblivious α-integrality gap verifier $\mathcal{A}'$.*

From Lemma 1 and Corollary 1, we obtain an α-robust-in-expectation algorithm for (5) from an α-integrality gap verifier for (1)–(2).

Theorem 2. *Consider a discrete minimization problem (1) and its LP relaxation (2), admitting an α-integrality gap verifier $\mathcal{A}$. Then there is a polytime α-robust-in-expectation algorithm for the robust version (7) w.r.t. to the any convex uncertainty set $\mathcal{C} \subseteq \mathbb{R}^n_+$.*

We emphasize that, in Theorem 2, the integrality gap verifier must be defined with w.r.t. the *whole* class $\mathscr{C} = \mathbb{R}^n_+$ of objectives. Finally, we note that the results in this section can be extended, in a straightforward way, to maximization problems.

2.1 A Deterministically Robust Algorithm for a Class of Polyhedral Uncertainty

In [8], Bertsimas and Sim considered the minimization version of the DO problem (1), when the set $\mathcal{S} \subseteq \{0,1\}^n$ and the (budget) uncertainty set $\mathcal{C}$ is given by

$$\mathcal{C} = \left\{ c := c^0 + d \circ u \;\middle|\; u \in \mathbb{R}^n_+,\ u_i \leq 1, \forall i \in [n],\ \sum_{i=1}^n u_i \leq k \right\}, \tag{16}$$

where $c^0, d \in \mathbb{R}^n_+$ are given non-negative vectors, $k \in \mathbb{Z}_+$ is a given positive integer, and $d \circ y$ is the n-dimensional vector with components $(d \circ u)_i := d_i u_i$, for $i = 1, \dots, n$. The constraints in (16) describe the situation when the uncertainty in each component of the objective vector c is described by an interval $[c^0_j, c^0_j + d_j]$ and at most k components are allowed to change. It was shown in [8] that an α-deterministically robust approximation algorithm for the minimization version of (5) with the uncertainty set given in (16), can be obtained from $n+1$ calls to an α-approximation algorithm for the nominal problem (1).

In this section, we extend this result as follows. Consider a polyhedral uncertainty set given by

$$\mathcal{C} = \{c := c^0 + u \mid u \in \mathbb{R}^n_+,\ u \leq d,\quad Au \leq b\}, \tag{17}$$

where $d \in \mathbb{R}^n_+$, $b \in \mathbb{R}^m_+$ are given non-negative vectors and $A \in \mathbb{R}^{m \times n}_+$ is a given non-negative matrix. Note that the uncertainty set $\mathcal{C}$ in (16) can be written in the form (17) by replacing $d \circ u$ by u and setting $A := \left[\frac{1}{d_1} \cdots \frac{1}{d_n}\right] \in \mathbb{R}^{1 \times n}_+$, $b := \left[k\right] \in \mathbb{R}^1_+$ (assuming w.l.o.g. that $d_i > 0$ for all i).

Fix an $\epsilon > 0$. As we shall see below, we may assume, w.l.o.g., that $b_i > 0$ for all $i \in [m]$. Define

$$L(A, c^0, d) := n \cdot \max\left\{\frac{\max_j c^0_j}{\min_j c^0_j}, \frac{(m+n)}{\epsilon} \cdot \max\left\{\frac{\max_{i,j} a_{ij}/b_i}{\min_j \max_i a_{ij}/b_i}, \frac{\max_j d_j}{\min_j d_j}\right\}\right\}. \tag{18}$$

Theorem 3. *Consider the DO problem (1), when the set $\mathcal{S} \subseteq \{0,1\}^n$ and the uncertainty set $\mathcal{C}$ is given by (17). Then, for any given $\epsilon > 0$, there is a $(1+\epsilon)\alpha$-deterministically robust approximation algorithm for the cost-robust version (5), which can be obtained from $O(\frac{\log L(A,c^0,d)}{\epsilon}(\log \frac{(1+\epsilon)m}{\epsilon})^m)$ calls to an α-approximation algorithm for the nominal problem (1).*

Note that, if both $\frac{\max_j c^0_j}{\min_j c^0_j}$ and $\frac{\max_j d_j}{\min_j d_j}$ are bounded by poly(n), then Theorem 3 requires only polylog(n) number of calls to the integraliy gap verifier, which is an *exponential* improvement over the result in [8] in such a case.

A set $\mathcal{S} \subseteq \{0,1\}^n$ is said to be *covering* if $x \in \mathcal{S}$ implies that $y \in \mathcal{S}$ for any $y \geq x$. For instance, if the set $\mathcal{S}$ represents subgraphs (say, as edge sets) of a given graph satisfying a certain *monotone* property (such as connectivity or containment), then $\mathcal{S}$ is covering. Theorem 3 gives a reduction from any α-approximation algorithm to a $(1+\epsilon)\alpha$-deterministically robust approximation algorithm, assuming $m = O(1)$. When m is not a constant, and the set $\mathcal{S}$ is of the covering type, we have the following result.

Theorem 4. *Consider the DO problem (1), when the set $\mathcal{S} \subseteq \{0,1\}^n$ is a covering set and the uncertainty set $\mathcal{C}$ is given by (17). Then, there is an $(\alpha + 2\sqrt{\frac{\alpha\gamma n}{\beta}})$-deterministically robust approximation algorithm for the robust version (5), which can be obtained by a polynomial number of calls to an α-integrality gap verifier for the nominal problem (1).*

3 A Robust-with-high-probability Approximation Algorithm for Polyhedral Uncertainty

Next, we consider the case when the uncertainty set $\mathcal{C}$ is given by (4) and $\mathcal{D} = \{\delta \in \mathbb{R}^k_+ : A\delta \leq b\}$. Let $\beta := \min_j \max_i a_{ij}$ and $\gamma := \max_{i,j} a_{ij}$, $c_{\min} := \min_{r \neq 0, j:\ c^r_j > 0} c^r_j$, $c^r_{\max} := \max_j c^r_j$ and $c_{\max} := \max_{r \neq 0} c^r_{\max}$.

Theorem 5. *Consider the DO problem (1), when $\mathcal{S} \subseteq \{0.1\}^n$ and the uncertainty set $\mathcal{C}$ is given by (4) and $\mathcal{D} = \{\delta \in \mathbb{R}^k_+ : A\delta \leq b\}$. Then, there is an $O(\alpha\sqrt{k\log(k)\frac{\gamma}{\beta}\frac{c_{\max}}{c_{\min}}})$-robust-with-high-probability approximation algorithm for the robust version (5), which can be obtained by a polynomial number of calls to an α-*ANCRR *integrality gap verifier for the nominal problem (1).*

Proof. Assume the availability of an α-ANCRR integrality gap verifier for the nominal problem (1). We assume w.l.o.g. that $b = \mathbf{1}_m$. Note that the robust DO problem (5) in this case takes the form:

$$\text{OPT}_R = \min_{x\in\mathcal{S}}\left\{(c^0)^T x + \max_{\delta\in\mathbb{R}^k_+:\ A\delta\leq\mathbf{1}_m} x^T\mathbf{C}\delta\right\}, \tag{19}$$

where $\mathbf{C} \in \mathbb{R}^{n\times k}_+$ is the matrix whose columns are $c^1, \ldots, c^k$. Let us consider the inner maximization problem in (19) and its dual (for a given $x \in \{0,1\}^n$):

$$\begin{aligned} z^*(x) = \max \quad & x^T\mathbf{C}\delta && (20)\\ \text{s.t.} \quad & A\delta \leq \mathbf{1}_m, && (21)\\ & \delta \in \mathbb{R}^k_+ \end{aligned}$$

$$\begin{aligned} z^*(x) = \min \quad & \mathbf{1}_m^T\theta && (22)\\ \text{s.t.} \quad & A^T\theta \geq \mathbf{C}^T x, && (23)\\ & \theta \in \mathbb{R}^m_+. && (24) \end{aligned}$$

Note that if $\mathbf{C}^T x = 0$ for $x \in \{0,1\}^n$, then $z^*(x) = 0$ and $x_j = 0$ for all $j \in J := \{j \in [n] \mid \exists r \in [k] :\ c^r_j > 0\}$. Thus, by considering the relaxation (2) with $c = c^0$ and $\mathcal{Q}$ replaced by $\mathcal{Q}' := \{x \in \mathcal{Q} :\ x_j = 0\ \ \forall j \in J\}$, and calling the integrality gap verifier on the obtained optimal fractional solution x^*, we can find an $\widehat{x}$ that belongs to $\mathcal{S}$ with prob. $1 - o(1)$ such that $\mathbb{E}[(c^0)^T\widehat{x}] \leq \alpha(c^0)^T x^*$ (or discover that none exist if the relaxation is infeasible). In view of Remark 1, this expectation guarantee can be turned into a high-probability guarantee without sacrificing much the approximation ratio, that is, we can get a solution $\widehat{x}^0$ such that, with probability $1 - o(1)$, we have $\widehat{x}^0 \in \mathcal{S}$ and $(c^0)^T\widehat{x}^0 \leq (1+\epsilon)(c^0)^T x^*$, for any given $\epsilon > 0$. We will assume therefore in the following that $\mathbf{C}^T x \neq 0$ for all $x \in \mathcal{S}$, as we will return the minimum of the solution obtained under this assumption and $(c^0)^T\widehat{x}^0$.

Claim 1. *For any $x \in \{0,1\}^n$ such that $\mathbf{C}^T x \neq 0$, we have $z^*(x) \geq \frac{c_{\min}}{\gamma}$.*

Let z^*_R be the value of the relaxation for (19), that is,

$$z^*_R = \min_{x\in\mathcal{Q}}\left\{(c^0)^T x + \max_{\delta\in\mathbb{R}^k_+:\ A\delta\leq\mathbf{1}_m} x^T\mathbf{C}\delta\right\}. \tag{25}$$

Using strong LP duality (20)–(22), we may rewrite (25) as

$$\begin{aligned} z^*_R = \min \quad & (c^0)^T x + \mathbf{1}_m^T\theta && (26)\\ \text{s.t.} \quad & A^T\theta \geq \mathbf{C}^T x, && (27) \end{aligned}$$

$$\theta \in \mathbb{R}_+^m, x \in \mathcal{Q}.$$

Let (x^*, θ^*) be an optimal solution for the LP (26). We call the α-ANCRR integrality gap verifier on x^* to get an α-ANCRR $\widehat{x}$. Let $\tau \in (0, 1)$ be a number to be chosen later, and define $R := \{r \in [k] : (c^r)^T x^* \geq \tau c^r_{\max}\}$.

Claim 2. *For* $\rho \geq 1$, $\Pr\left[\forall r \in R: (c^r)^T \widehat{x} \leq (1+\rho)\alpha(a^r)^T\theta^*\right] \geq 1 - ke^{-\rho\alpha\tau/3}$.

For $r \notin R$, define $i(r)$ to be the smallest $i \in [m]$ such that $i \in \text{argmax}_{i'} a_{i'r}$. Let us next choose $\rho := \frac{6\ln(2k)}{\tau} > 1$ and define the dual solution $\widehat{\theta} \in \mathbb{R}^m_+$ as follows:

$$\widehat{\theta}_i := (1+\rho)\alpha\theta_i^* + \frac{1}{\beta} \sum_{r \notin R:\ i=i(r)} (c^r)^T \widehat{x}, \qquad \text{for } i = 1, \ldots, m.$$

Let us fix an arbitrary constant $\epsilon \in (0, 1)$.

Claim 3. *With probability* $1 - o(1)$, $(\widehat{x}, \widehat{\theta})$ *is feasible for* (26) *and* $(c^0)^T\widehat{x} + \mathbf{1}_m^T\widehat{\theta} \leq \left((1+\rho) + \frac{(1+\epsilon)\gamma\tau c_{max} k}{\beta c_{\min}}\right)\alpha\text{OPT}_R$.

It follows from Claim 3 that, with probability $1 - o(1)$, for any $c = c^0 + \mathbf{C}\delta \in \mathcal{C}$, given by (4) with $\mathcal{D} = \{\delta \in \mathbb{R}^k_+ : A\delta \leq \mathbf{1}_m\}$, we have

$$c^T\widehat{x} = (c^0)^T\widehat{x} + (\mathbf{C}\delta)^T\widehat{x} \leq (c^0)^T\widehat{x} + z^*(\widehat{x}) \leq (c^0)^T\widehat{x} + \mathbf{1}_m^T\widehat{\theta}$$
$$\leq \left((1+\rho) + \frac{(1+\epsilon)\gamma\tau c_{max} k}{\beta c_{\min}}\right)\alpha\text{OPT}_R.$$

The theorem follows by choosing $\tau := \sqrt{\frac{6\beta\ln(2k)c_{\min}}{(1+\epsilon)\gamma c_{max} k}}$.

References

1. Pessoa, A.A., Pugliese, L.D.P., Guerriero, F., Poss, M.: Robust constrained shortest path problems under budgeted uncertainty. Networks **66**(2), 98–111 (2015)
2. Agra, A., Santos, M., Nace, D., Poss, M.: A dynamic programming approach for a class of robust optimization problems. SIAM J. Optim. **26**(3), 1799–1823 (2016)
3. Álvarez-Miranda, E., Ljubić, I., Toth, P.: A note on the Bertsimas & sim algorithm for robust combinatorial optimization problems. 4OR **11**(4), 349–360 (2013)
4. Ben-Tal, A., El Ghaoui, L., Nemirovski, A.S.: Robust Optimization. Princeton Series in Applied Mathematics, Princeton University Press, October 2009
5. Ben-Tal, A., Nemirovski, A.: Robust convex optimization. Math. Oper. Res. **23**(4), 769–805 (1998)
6. Ben-Tal, A., Nemirovski, A.: Robust optimization - methodology and applications. Math. Program. **92**(3), 453–480 (2002)
7. Bertsimas, D., Brown, D., Caramanis, C.: Theory and applications of robust optimization. SIAM Rev. **53**(3), 464–501 (2011)
8. Bertsimas, D., Sim, M.: Robust discrete optimization and network flows. Math. Program. **98**(1), 49–71 (2003)

9. Bertsimas, D., Sim, M.: Robust discrete optimization under ellipsoidal uncertainty sets. Technical report, Technical report, MIT (2004)
10. Bertsimas, D., Sim, M.: Tractable approximations to robust conic optimization problems. Math. Program. **107**(1–2), 5–36 (2006)
11. Birge, J.R., Louveaux, F.: Introduction to Stochastic Programming, 2nd edn. Springer, New York (2011). https://doi.org/10.1007/978-1-4614-0237-4
12. Bougeret, M., Pessoa, A.A., Poss, M.: Robust scheduling with budgeted uncertainty. Discrete Appl. Math. **261**, 93–107 (2019)
13. Carr, R.D., Vempala, S.: Randomized metarounding. Random Struct. Algorithms **20**(3), 343–352 (2002)
14. Delage, E., Ye, Y.: Distributionally robust optimization under moment uncertainty with application to data-driven problems. Oper. Res. **58**(3), 595–612 (2010)
15. Dhamdhere, K., Goyal, V., Ravi, R., Singh, M.: How to pay, come what may: approximation algorithms for demand-robust covering problems. In: Proceedings of the 46th Annual IEEE Symposium on Foundations of Computer Science, FOCS 2005, pp. 367–378 (2005)
16. El Ghaoui, L., Oustry, F., Lebret, H.: Robust solutions to uncertain semidefinite programs. SIAM J. Optim. **9**(1), 33–52 (1998)
17. Elbassioni, K., Makino, K., Najy, W.: A multiplicative weight updates algorithm for packing and covering semi-infinite linear programs. Algorithmica (2019)
18. Feige, U., Feldman, M., Talgam-Cohen, I.: Oblivious Rounding and the Integrality Gap. In: Approximation, Randomization, and Combinatorial Optimization. Algorithms and Techniques, APPROX/RANDOM 2016. Leibniz International Proceedings in Informatics (LIPIcs), vol. 60, pp. 8:1–8:23 (2016)
19. Feige, U., Jain, K., Mahdian, M., Mirrokni, V.: Robust combinatorial optimization with exponential scenarios. In: Fischetti, M., Williamson, D.P. (eds.) IPCO 2007. LNCS, vol. 4513, pp. 439–453. Springer, Heidelberg (2007). https://doi.org/10.1007/978-3-540-72792-7_33
20. Goetzmann, K.-S., Stiller, S., Telha, C.: Optimization over integers with robustness in cost and few constraints. In: Solis-Oba, R., Persiano, G. (eds.) WAOA 2011. LNCS, vol. 7164, pp. 89–101. Springer, Heidelberg (2012). https://doi.org/10.1007/978-3-642-29116-6_8
21. Grötschel, M., Lovász, L., Schrijver, A.: Geometric Algorithms and Combinatorial Optimization, Algorithms and Combinatorics, second corrected edn., vol. 2. Springer, Heidelberg (1993). https://doi.org/10.1007/978-3-642-97881-4
22. Ilyina, A.: Combinatorial optimization under ellipsoidal uncertainty. Technischen Universität Dortmund. Ph.D. Thesis (2017)
23. Kawase, Y., Sumita, H.: Randomized strategies for robust combinatorial optimization. In: Proceedings of the 33rd AAAI Conference on Artificial Intelligence, AAAI, pp. 7876–7883 (2019)
24. Poss, M.: Contributions to robust combinatorial optimization with budgeted uncertainty. Université de Montpellier. Operations Research [cs.RO] (2016 tel-01421260)
25. Srinivasan, A.: Improved approximation guarantees for packing and covering integer programs. SIAM J. Comput. **29**(2), 648–670 (1999)

Scheduling on Hybrid Platforms: Improved Approximability Window

Vincent Fagnon[1](✉), Imed Kacem[2], Giorgio Lucarelli[2], and Bertrand Simon[3]

[1] University Grenoble Alpes, CNRS, Inria, Grenoble INP, LIG, 38000 Grenoble, France
vincent.fagnon@univ-grenoble-alpes.fr
[2] LCOMS, University of Lorraine, Metz, France
{imed.kacem,giorgio.lucarelli}@univ-lorraine.fr
[3] Universität Bremen, Bremen, Germany
bsimon@uni-bremen.de

Abstract. Modern platforms are using accelerators in conjunction with standard processing units in order to reduce the running time of specific operations, such as matrix operations, and improve their performance. Scheduling on such hybrid platforms is a challenging problem since the algorithms used for the case of homogeneous resources do not adapt well. In this paper we consider the problem of scheduling a set of tasks subject to precedence constraints on hybrid platforms, composed of two types of processing units. We propose a $(3+2\sqrt{2})$-approximation algorithm and a conditional lower bound of 3 on the approximation ratio. These results improve upon the 6-approximation algorithm proposed by Kedad-Sidhoum *et al.* as well as the lower bound of 2 due to Svensson for identical machines. Our algorithm is inspired by the former one and distinguishes the allocation and the scheduling phases. However, we propose a different allocation procedure which, although is less efficient for the allocation sub-problem, leads to an improved approximation ratio for the whole scheduling problem. This approximation ratio actually decreases when the number of processing units of each type is close and matches the conditional lower bound when they are equal.

Keywords: Approximation algorithms · Scheduling · Precedence constrains · CPU/GPU

1 Introduction

Nowadays, more and more High Performance Computing platforms use special purpose processors in conjunction with classical Central Processing Units (CPUs) in order to accelerate specific operations and improve their performance. A typical example is the use of modern Graphics Processing Units (GPUs) which can accelerate vector and matrix operations.

Y. Kohayakawa and F. K. Miyazawa (Eds.): LATIN 2020, LNCS 12118, pp. 38–49, 2020.
https://doi.org/10.1007/978-3-030-61792-9_4

Due to the heterogeneity that introduce this kind of accelerators, the scheduling problem on such hybrid platforms becomes more challenging. Several experimental results as well as theoretical lower bounds [1] show that the decision of the allocation of a task to the type of processors is crucial for the performance of the system. Specifically, classical greedy policies, such as Graham's List Scheduling [10], which perform well in the case of identical computing resources, fail to generalize on hybrid platforms. For this reason, all known algorithms for hybrid platforms [1,5,7,11] choose the type of the resource for each task before deciding its scheduling in the time horizon.

In this paper, we focus on the problem of scheduling an application on such an hybrid platform consisting of m identical CPUs and k identical GPUs. An application is described as a set of n mono-processor tasks V which are linked through precedence dependencies described by a directed acyclic graph $G = (V, E)$. This means that a task can start being executed only after all of its predecessors are completed. The processing time of task j on a CPU (resp. on a GPU) is denoted by $\overline{p_j}$ (resp. by $\underline{p_j}$), and we do not assume any relation between $\overline{p_j}$ and $\underline{p_j}$. This is justified in real systems where tasks performing for instance matrix operations can be executed much more efficiently on a GPU, while the execution of tasks which need to communicate often with the file system is faster on a CPU. Therefore, we can assume without loss of generality than $m \geq k$.

We are interested in designing polynomial-time algorithms with good performance guarantees in the worst case. As performance measure we use the well-known approximation ratio which compares the solution of an algorithm and the optimal solution with respect to an objective function. In this paper, we study the *makespan* objective, that is we aim at minimizing the completion time of the last task. Extending the Graham notation, we will denote this problem as $(Pm, Pk) \mid prec \mid C_{\max}$.

For this problem, a 6-approximation algorithm named HLP (Heterogeneous Linear Program) has been proposed by Kedad-Sidhoum *et al.* [11]. This algorithm has two phases. In the first phase a "good" allocation of each task either on the CPU or on the GPU side is decided. This decision is based on an integer linear program which uses a 0–1 decision variable x_j for each task j: x_j will be equal to one if j is assigned to the CPU side, and to zero otherwise. This integer linear program does not model the whole scheduling problem but only the allocation decision, trying to balance the average load on the CPUs and GPUs as well as the critical path length. The fractional relaxation of this program is solved and the allocation of each task j is determined by a simple rounding rule: it is assigned to GPUs if $x_j < 1/2$, and to CPUs otherwise. In the second phase, the greedy List Scheduling algorithm is used to schedule the tasks respecting the precedence constraints and the allocation defined in the first phase.

The authors in [11] prove that the value of 1/2 chosen is best possible with respect to the linear program used in the first phase. In a sense, they prove that the integrality gap of the linear program relaxation is 2. Furthermore, given this simple rounding rule based on 1/2, Amaris *et al.* [1] present a tight example of HLP which asymptotically attains an approximation ratio of 6, even if another scheduling algorithm is used in the second phase. Despite both previous negative results, we show that HLP can achieve a better approximation ratio by using a different rounding procedure. Indeed, even though we use a rounding which is not the best possible with respect to the allocation problem solved in the first phase, this rounding allows us to obtain stronger guarantees on the scheduling phase and therefore improve the approximation ratio.

The main difference with HLP is that we allocate task j to the fastest processor type if x_j is close to 1/2 in the fractional relaxation solution. We then achieve an approximation ratio smaller than $3 + 2\sqrt{2}$ and that tends towards 3 when m/k is close to 1.

The best known lower bound on the approximation ratio is the same as for identical machines, *i.e.*, 4/3 [13], but can be improved to 2 by assuming a variant of the unique games conjecture [15]. Our second contribution is to improve this conditional lower bound to 3 for any value of m/k assuming a stronger conjecture introduced by Bazzi and Norouzi-Fard [3]. This conditional lower bound is therefore tight when $m = k$.

Organization of the Paper

In Sect. 2 we give a literature review by positioning our problem with respect to closely related ones and by presenting several known approximability results. In Sect. 3 we present our adapted algorithm for the problem of scheduling on hybrid platforms as well as its analysis which leads to an approximation ratio of 5.83. In Sect. 4, we prove a conditional lower bound of 3 on the approximation ratio. Finally, we conclude in Sect. 5.

2 Related Work

The problem of scheduling on hybrid platforms consisting of two sets of identical processors is a generalization of the classical problem of scheduling on parallel identical processors, denoted by $P \mid prec \mid C_{\max}$. On the other hand, our problem is a special case of the problem of scheduling on unrelated processors (denoted by $R \mid prec \mid C_{\max}$), where each task has a different processing time on each processor. Moreover, in the case of scheduling on related processors (denoted by $Q \mid prec \mid C_{\max}$), each processor has its specific speed and the processing time of each task depends on the speed of the assigned processor. This problem is more general than $P \mid prec \mid C_{\max}$ in the sense that the processing time of a task is different on each processor. However, in the former problem all tasks are accelerated or decelerated by the same factor when using a specific processor, while in our case two tasks do not necessarily have the same behavior (acceleration or deceleration) if they are scheduled on a CPU or a GPU.

For $P \mid prec \mid C_{\max}$, the greedy List Scheduling algorithm proposed by Graham [10] achieves an approximation ratio of $(2 - \frac{1}{m})$, where m is the number of the processors. Svensson [15] proved that this is the best possible approximation result that we can expect, assuming $P \neq NP$ and a variant of the unique games conjecture introduced by Bansal and Khot [2]. Note that this negative result holds also for our more general problem. For $Q \mid prec \mid C_{\max}$, a series of algorithms with logarithmic approximation ratios are known (see for example [6,8]), while Li [14] has recently proposed a $O(\log(m)/\log(\log(m)))$-approximation algorithm which is the current best known ratio. On the negative side, Bazzi and Norouzi-Fard [3] show that it is not possible to have a constant approximation ratio assuming the NP-hardness of some problems on k-partite graphs. No result is actually known for $R \mid prec \mid C_{\max}$. However, there are few approximation algorithms for special classes of precedence graphs (see for example [12]).

For the problem $(Pm, Pk) \mid prec \mid C_{\max}$, targeting hybrid platforms, Kedad-Sidhoum *et al.* [11] presented a 6-approximation algorithm as we reported before by separating the allocation and the scheduling phases. Amaris *et al.* [1] proposed small improvements on both phases, without improving upon the approximation ratio. However, they show that using the rounding proposed in [11], any scheduling policy cannot lead to an approximation ratio strictly smaller than 6. In the absence of precedence constraints, a polynomial time approximation scheme has been proposed by Bleuse *et al.* [4].

The problem of scheduling on hybrid platforms has been also studied in the online case. If the tasks are not subject to precedence relations, then a 3.85-competitive algorithm has been proposed in [7], while the authors show also that no online algorithm can have a competitive ratio strictly less than 2. In the presence of precedence constraints, Amaris *et al.* [1] consider that tasks arrive in an online order respecting the precedence relations and they give a $(4\sqrt{m/k})$-competitive algorithm. This result has been improved by Canon *et al.* [5] who provide a $(2\sqrt{m/k}+1)$-competitive algorithm, while they show that no online algorithm can have a competitive ratio smaller than $\sqrt{m/k}$.

3 A 5.83-Approximation Algorithm

In this section we present the improved approximation algorithm and its analysis for the problem $(Pm, Pk) \mid prec \mid C_{\max}$. Although several ingredients of our algorithm have been already presented in [11], we present here all the steps of the algorithm for the sake of completeness.

3.1 The Algorithm HLP-*b*

As explained in introduction, the algorithm HLP-b has two phases: the allocation phase and the scheduling one. The allocation phase is based on an integer linear program. For each task $j \in V$, let x_j be a decision variable which is equal to 1 if task j is assigned to the CPU side, and to 0 otherwise. Moreover, let C_j be a variable corresponding to the completion time of task j. Finally, let C_{max} be a variable that indicates the maximum completion time over all tasks. For the sake of simplicity, we add in G a fictive task 0 with $\overline{p_0} = \underline{p_0} = 0$ which precedes all other tasks. Consider the following integer linear program similarly to Kedad-Sidhoum *et al.* [11].

$$\text{Minimize } C_{\max}$$

$$\frac{1}{m}\sum_{j\in V}\overline{p_j}x_j \leq C_{max} \tag{1}$$

$$\frac{1}{k}\sum_{j\in V}\underline{p_j}(1-x_j) \leq C_{max} \tag{2}$$

$$C_i + \overline{p_j}x_j + \underline{p_j}(1-x_j) \leq C_j \qquad \forall (i,j)\in E \tag{3}$$

$$0 \leq C_j \leq C_{max} \qquad \forall j \in V \tag{4}$$

$$x_j \in \{0,1\} \qquad \forall j \in V \tag{5}$$

Constraints (1) and (2) imply that the makespan of any schedule cannot be smaller than the average *load* on the CPU and GPU sides, respectively. Constraints (3) and (4) build up the *critical path* of the precedence graph, i.e., the path of G with the longest total completion time. In any schedule, the critical path length is a lower bound of the makespan. Note that the critical path of the input instance cannot be defined before the allocation decision for all tasks since the exact processing time of a task depends on this allocation. Constraint 5 is the integrality constraint for the decision variable x_j. In what follows, we relax the integrality constraint and we replace it by $x_j \in [0,1]$ for each task j in V, in order to get a linear program which we can solve in polynomial time. The above integer linear program is not completely equivalent to our scheduling problem, but the objective value of its optimal solution is a lower bound of any optimal schedule.

The rounding procedure of HLP-b is based on a parameter $b \geq 2$. We will show in Sect. 3.2 that the best choice is $b = 1 + \sqrt{\frac{2+k/m}{1-k/m}}$. Let x_j^R be the value of the decision variable for task j in an optimal solution of the above linear program relaxation. We define x_j^A to be the value of the decision variable for task j in our algorithm's schedule, that is the value of the decision variable obtained by the rounding procedure. The allocation phase of our algorithm rounds the optimal relaxed solution $\{x_j^R\}$ to the feasible solution $\{x_j^A\}$ as follows:

- if $x_j^R \geq 1 - \frac{1}{b}$, then $x_j^A = 1$;
- if $x_j^R \leq \frac{1}{b}$, then $x_j^A = 0$;
- if $\frac{1}{b} < x_j^R < 1 - \frac{1}{b}$ and $\overline{p_j} \geq \underline{p_j}$, then $x_j^A = 0$;
- if $\frac{1}{b} < x_j^R < 1 - \frac{1}{b}$ and $\overline{p_j} < \underline{p_j}$, then $x_j^A = 1$.

Intuitively, if the linear program solution is close to an integer ($x_j \leq \frac{1}{b}$ or $x_j \geq 1 - \frac{1}{b}$) then we follow its proposal, else we choose the processor type with the smallest processing time: the task is allocated to a CPU (*i.e.*, $x_j^A = 1$), if $\overline{p_j} < \underline{p_j}$ and to a GPU otherwise.

Given the allocation obtained by the previous procedure, HLP-b proceeds to the scheduling phase. The classical List Scheduling algorithm is applied respecting the allocation $\{x_j^A\}$ and the precedence constraints: tasks are allocated to the earliest available processor of the correct type in a topological order.

3.2 Analysis of the Algorithm HLP-b

We begin the analysis of HLP-b with two lemmas stating properties of the rounding procedure.

Lemma 1. *For each task* $j \in V$ *we have* $(1 - x_j^A)\underline{p_j} \leq b \cdot (1 - x_j^R)\underline{p_j}$.

Proof. Consider any task $j \in V$. Note first that if j is assigned to the CPU side by the algorithm then $x_j^A = 1$ and the lemma directly holds since $x_j^R \leq 1$. Then, we assume that j is assigned to the GPU side, that is $x_j^A = 0$. Hence, $x_j^R \leq (1 - \frac{1}{b})$. Therefore, we conclude as $b \cdot (1 - x_j^R)\underline{p_j} \geq \underline{p_j} = (1 - x_j^A)\underline{p_j}$. □

Lemma 2. *For each task* $j \in V$ *we have:*

$$x_j^A\overline{p_j} + (1 - x_j^A)\underline{p_j} \leq \frac{b}{b-1}(x_j^R\overline{p_j} + (1 - x_j^R)\underline{p_j}).$$

Proof. Consider any task $j \in V$. We have the following three cases.

– If $x_j^R \leq \frac{1}{b}$, then $x_j^A = 0$ and we have:

$$(1 - x_j^R)\underline{p_j} \geq (1 - \frac{1}{b})(1 - x_j^A)\underline{p_j}$$
$$(1 - x_j^R)\underline{p_j} + x_j^R\overline{p_j} \geq (1 - \frac{1}{b})\left((1 - x_j^A)\underline{p_j} + x_j^A\overline{p_j}\right).$$

– If $x_j^R \geq \left(1 - \frac{1}{b}\right)$, then $x_j^A = 1$ and we have:

$$x_j^R\overline{p_j} \geq (1 - \frac{1}{b})x_j^A\overline{p_j}$$
$$(1 - x_j^R)\underline{p_j} + x_j^R\overline{p_j} \geq (1 - \frac{1}{b})\left((1 - x_j^A)\underline{p_j} + x_j^A\overline{p_j}\right).$$

– If $\frac{1}{b} < x_j^R < (1 - \frac{1}{b})$, then we have:

$$x_j^R\overline{p_j} + (1 - x_j^R)\underline{p_j} \geq \min(\overline{p_j}, \underline{p_j}) = x_j^A\overline{p_j} + (1 - x_j^A)\underline{p_j}.$$

Therefore, combining the three cases, we obtain the lemma as $b/(b-1) \geq 1$. □

Based on the two previous lemmas, the following theorem gives the approximation ratio of our algorithm HLP-b.

Theorem 1. HLP-b *achieves an approximation ratio of* $3 + 2\sqrt{2 - \frac{k}{m} - \frac{k}{m}^2}$, *which is upper bounded by* $3 + 2\sqrt{2} \leq 5.83$.

Proof. We first define CP^A the value of the critical path length of G after the allocation phase of HLP-b. Denoting by $\mathcal{P}$ the set of paths in G, this value equals:

$$CP^A = \max_{p \in \mathcal{P}} \Big\{ \sum_{j \in p} (\overline{p_j} x_j^A + \underline{p_j}(1 - x_j^A)) \Big\}.$$

Furthermore, let $C_{\max}^A$, $C_{\max}^R$ and $C_{\max}^*$ be respectively the makespan of the schedule created by HLP-b, the objective value in an optimal solution of the linear program relaxation and the makespan of an optimal solution for our problem.

Following the same arguments as in [1,11] and since HLP-b is a List Scheduling algorithm, the total time during which there is at least one idle CPU and one idle GPU is upper-bounded by CP^A. Moreover, the total time during which no CPU (resp. no GPU) is idle in our schedule is upper-bounded by the average workload assigned to CPUs (resp. GPUs). So we have:

$$\begin{aligned} C_{max}^A &\leq \frac{1}{m} \sum_{j \in V} (\overline{p_j} x_j^A) + \frac{1}{k} \sum_{j \in V} (\underline{p_j}(1 - x_j^A)) + CP^A \\ &= \frac{k}{mk} \cdot \sum_{j \in V} (\overline{p_j} x_j^A + (\underline{p_j}(1 - x_j^A)) + \frac{m-k}{mk} \sum_{j \in V} (\underline{p_j}(1 - x_j^A)) + CP^A \end{aligned}$$

Using Lemmas 1 and 2, we obtain:

$$\begin{aligned} C_{max}^A \leq \frac{b}{b-1} \frac{1}{m} \sum_{j \in V} \left(x_j^R \overline{p_j} + (1 - x_j^R) \underline{p_j} \right) + b \frac{m-k}{mk} \sum_{j \in V} (1 - x_j^R) \underline{p_j} \\ + \frac{b}{b-1} \max_{p \in \mathcal{P}} \left\{ \sum_{j \in p} \left(x_j^R \overline{p_j} + (1 - x_j^R) \underline{p_j} \right) \right\} \end{aligned}$$

Now, the constraints (1) to (4) of the linear program relaxation give us:

$$C_{max}^A \leq \frac{b}{b-1} \frac{mC_{max}^R + kC_{max}^R}{m} + b \frac{m-k}{mk} kC_{max}^R + \frac{b}{b-1} C_{max}^R$$

Since $C_{\max}^R \leq C_{\max}^*$ we get:

$$\frac{C_{max}^A}{C_{max}^*} \leq \frac{b}{b-1} \cdot \frac{m+k}{m} + b \cdot \frac{m-k}{m} + \frac{b}{b-1} = b + 2 \cdot \frac{b}{b-1} - \frac{k}{m}(b - \frac{b}{b-1}).$$

The minimum of this ratio is reached by choosing $b = 1 + \sqrt{\frac{2+k/m}{1-k/m}} > 1 + \sqrt{2}$, which gives:

$$\frac{C_{max}^A}{C_{max}^*} \leq 3 + 2\sqrt{2 - \frac{k}{m} - \frac{k}{m}^2} \leq 3 + 2\sqrt{2} \approx 5.83.$$

□

4 Conditional Lower Bound on the Approximation Factor

In this section, we extend the results of Bazzi and Norouzi-Fard [3] in our setting. Assuming Hypothesis 1 (see below), they show that it is NP-hard to approximate $Q \mid prec \mid C_{\max}$ within a constant factor. If we focus on only two types of processors, their result implies a lower bound of 2 on the approximation ratio, therefore not improving on Svensson's result [15]. We improve their result to obtain a conditional lower bound of 3 stated in Theorem 2, which therefore also holds in our more restricted setting $(Pm, Pk) \mid prec \mid C_{\max}$ in which the processing times on both processor types can be arbitrary. Due to lack of space, we do not discuss further the relevance of Hypothesis 1 or its link to the weaker Unique Games Conjecture and refer the reader to [3] for more details.

Theorem 2. *Assuming Hypothesis 1 and $P \neq NP$, there exist no polynomial-time $(3-\alpha)$-approximation, for any $\alpha > 0$, for the problem $(Pm, Pk) \mid prec \mid C_{\max}$, even if the processors are related.*

Hypothesis 1 (q-partite problem). *For every small $\varepsilon, \delta > 0$, and every integral constant $q, Q > 1$, the following problem is NP-hard: given a q-partite graph $G_q = (V_1, \ldots, V_q, E_1, \ldots, E_{q-1})$ with $|V_i| = n$ for all $1 \leq i \leq q$ and E_i being the set of edges between V_i and V_{i+1} for all $1 \leq i < q$, distinguish between the two following cases:*

- *YES Case: every V_i can be partitioned into $V_{i,0}, \ldots V_{i,Q-1}$, such that:*
  - *there is no edge between V_{i,j_1} and V_{i+1,j_2} for all $1 \leq i < q$, $j_1 > j_2$.*
  - *$|V_{i,j}| \geq \frac{1-\varepsilon}{Q} n$, for all $1 \leq i \leq q$, $0 \leq j \leq Q-1$.*
- *NO Case: for every $1 \leq i < q$ and every two sets $S \subseteq V_i$, $T \subseteq V_{i+1}$ such that $|S| = |T| = \lfloor \delta n \rfloor$, there is an edge between S and T.*

We start by fixing several values: an integer q multiple of 3, an integer Q, $\delta \leq 1/(2Q)$ and $\varepsilon \leq 1/Q^2$. We consider the q-partite problem parameterized by Q, ε, δ, which is assumed to be NP-hard under Hypothesis 1.

Reduction. We define a reduction from $G_q = (V_1, \ldots, V_q, E_1, \ldots, E_{q-1})$, a q-partite graph where for each i, $|V_i| = n > Q$, to a scheduling instance $\mathcal{I}$. The instance consists of $m = \lceil (1+Q\varepsilon)n^4 \rceil$ CPUs and $k = \lceil (1+Q\varepsilon)n^2 \rceil$ GPUs and uses two types of tasks: CPU tasks verifying $\overline{p_j} = n\underline{p_j} = 1$, and GPU tasks verifying $\overline{p_j} = n\underline{p_j} = n$. The tasks and edges (*i.e.*, precedence constraints) are defined as follows. For each $0 \leq z < q/3$, and for each:

- vertex $v \in V_{3z+1}$, create a set $\mathcal{J}_{3z+1,v}$ of $Qn - Q$ GPU tasks (type a).
- vertex $v \in V_{3z+2}$, create a set $\mathcal{J}_{3z+2,v}$ of Qn^3 CPU tasks (type b).
- vertex $v \in V_{3z+3}$, create a set $\mathcal{J}_{3z+3,v}$ of $Q-2$ GPU tasks (type c) indexed $J^1_{3z+3,v}, \ldots J^{Q-2}_{3z+3,v}$, and an edge from $J^{\ell}_{3z+3,v}$ to $J^{\ell+1}_{3z+3,v}$ for ℓ from 1 to $Q-3$.
- edge $(v, w) \in E_i$, create all edges from the set $\mathcal{J}_{i,v}$ to the set $\mathcal{J}_{i+1,w}$.

Intuitively, the tasks corresponding to each set V_i of G_q can be computed in Q time slots. To achieve this, each set of type b requires almost all the CPUs, each set of type a requires almost all but n GPUs, and each set of type c requires n GPUs. On a YES instance, it is possible to progress simultaneously on the tasks corresponding to three consecutive sets V_i, by pipe-lining the execution, thus obtaining a makespan close to $qQ/3$. For example, it is possible to execute $V_{i,1}$ at some time step, and then to execute $V_{i+1,1}$ and $V_{i,2}$ in parallel. On a NO instance, the tasks corresponding to each V_i have to be scheduled almost independently, thus not efficiently using the processing power: there are too few GPUs to process a significant amount of CPU tasks, and CPUs are too slow to process GPU tasks. The minimum possible makespan is then close to qQ. The two following lemmas state these results formally.

Lemma 3 (Completeness). *If G_q corresponds to the YES case of the q-partite problem, then instance $\mathcal{I}$ admits a schedule of makespan $(q+3)Q/3$.*

Proof. Suppose that G_q corresponds to a YES instance of the q-partite problem, and let $V_{i,j}$ for $1 \leq i \leq q$ and $j < Q$ be the associated partition of the sets V_i. Note that the size of any set $V_{i,j}$ of the partition is at most $(1+Q\varepsilon)n/Q$, since $\sum_{j=0}^{Q-1}|V_{i,j}| = |V_i| = n$ and, by definition, in a YES instance it holds that $|V_{i,j}| \geq \frac{1-\varepsilon}{Q}n$. We next partition the tasks of $\mathcal{I}$ into sets $S_{i,j}$. For each z, $0 \leq z < q/3$, and j, $0 \leq j \leq Q-1$, we define:

- type A: $S_{zQ+1,j} = \bigcup_{v \in V_{3z+1,j}} \mathcal{J}_{3z+1,v}$, and thus $|S_{zQ+1,j}| \leq (Qn-Q)(1+Q\varepsilon)n/Q \leq k(1-1/n)$.
- type B: $S_{zQ+2,j} = \bigcup_{v \in V_{3z+2,j}} \mathcal{J}_{3z+2,v}$, and thus $|S_{zQ+2,j}| \leq Qn^3(1+Q\varepsilon)n/Q \leq (1+Q\varepsilon)n^4 \leq m$.
- type C: for $1 \leq \ell \leq Q-2$, $S_{zQ+2+\ell,j} = \bigcup_{v \in V_{3z+3,j}} \{J^{\ell}_{3z+3,v}\}$, and thus $|S_{zQ+2+\ell,j}| = (1+Q\varepsilon)n/Q \leq k/(nQ)$.

Let $\mathcal{T}_t$ be the union of all $S_{i,j}$ with $t = i+j$, $1 \leq i \leq Qq/3$ and $0 \leq j \leq Q-1$. We create a schedule for instance $\mathcal{I}$ as follows: at the time slot $[t-1, t)$, we schedule the tasks of set $\mathcal{T}_t$. A sketc.h of the beginning of this schedule is given in Table 1. The type and the number of machines (CPUs or GPUs) for executing each set of tasks $S_{i,j}$ is also given in this table. Note that the tasks of the second triplet $\langle V_4, V_5, V_6\rangle$ start executing from time slot $[Q, Q+1)$: specifically, $S_{Q+1,0}$ contains tasks in V_4. Moreover, the execution of some tasks of the first triplet $\langle V_1, V_2, V_3\rangle$ takes place after time $Q+1$: specifically, the last tasks in this triplet belong to the set $S_{Q,Q-1}$ and they are executed in the time slot $[2Q-2, 2Q-1)$. However, there is no a time slot in which 3 triplets are involved.

In the last time slot of the created schedule we execute the tasks in $\mathcal{T}_t$ with $t = i+j$, $i = Qq/3$ and $j = Q-1$. Hence, the makespan is $Qq/3+Q-1 < Qq/3+Q$. It remains to prove the feasibility of the created schedule: the precedence constraints are satisfied and there are enough machines to perform the assigned tasks at each time slot.

Consider first the precedence constraints inside each set $\mathcal{J}_{3z+3,v}$, $0 \leq z < q/3$ and $v \in V_{3z+3}$, that is the arc from the task $J^{\ell}_{3z+3,v}$ to the task $J^{\ell+1}_{3z+3,v}$, for all ℓ, $1 \leq \ell \leq Q-3$. By construction, $J^{\ell}_{3z+3,v} \in S_{zQ+2+\ell,j}$ and $J^{\ell+1}_{3z+3,v} \in S_{zQ+2+\ell+1,j}$.

Table 1. A sketch of the beginning of the schedule for the tasks in $\mathcal{I}$.

	CPU	GPU							
	m	$k(1-1/n)$	$k/(nQ)$	$k/(nQ)$	$k/(nQ)$	...	$k/(nQ)$	$k/(nQ)$	$k/(nQ)$
$[0,1)$		$s_{1,0}$							
$[1,2)$	$s_{2,0}$	$s_{1,1}$							
$[2,3)$	$s_{2,1}$	$s_{1,2}$	$s_{3,0}$						
$[3,4)$	$s_{2,2}$	$s_{1,3}$	$s_{4,0}$	$s_{3,1}$					
$[4,5)$	$s_{2,3}$	$s_{1,4}$	$s_{5,0}$	$s_{4,1}$	$s_{3,2}$				
...	...	...	...	...	...				
$[Q-1,Q)$	$s_{2,Q-2}$	$s_{1,Q-1}$	$s_{Q,0}$	$s_{Q-1,1}$	$s_{Q-2,2}$	...	$s_{3,Q-3}$		
$[Q,Q+1)$	$s_{2,Q-1}$	$s_{Q+1,0}$		$s_{Q,1}$	$s_{Q-1,2}$	...	$s_{4,Q-3}$	$s_{3,Q-2}$	
$[Q+1,Q+2)$	$s_{Q+2,0}$	$s_{Q+1,1}$			$s_{Q,2}$	...	$s_{5,Q-3}$	$s_{4,Q-2}$	$s_{3,Q-1}$
$[Q+2,Q+3)$	$s_{Q+2,1}$	$s_{Q+1,2}$				...	$s_{6,Q-3}$	$s_{5,Q-2}$	$s_{4,Q-1}$
$[Q+3,Q+4)$	$s_{Q+2,2}$	$s_{Q+1,3}$	$s_{Q+3,0}$			...	$s_{7,Q-3}$	$s_{6,Q-2}$	$s_{5,Q-1}$
$[Q+4,Q+5)$	$s_{Q+2,3}$	$s_{Q+1,4}$	$s_{Q+4,0}$	$s_{Q+3,1}$		...	$s_{8,Q-3}$	$s_{7,Q-2}$	$s_{6,Q-1}$
...									

Thus, $J^{\ell}_{3z+3,v}$ is executed in the time slot $zQ + 2 + \ell + j$, while $J^{\ell+1}_{3z+3,v}$ in the time slot $zQ+2+\ell+1+j > zQ+2+\ell+j$, and hence this kind of precedence constraints are satisfied.

Consider now the precedence constraint from a task $J \in \mathcal{J}_{i,v}$ corresponding to $v \in V_{i,j_1} \subset V_i$ to a task $J' \in \mathcal{J}_{i+1,w}$ corresponding to $w \in V_{i+1,j_2} \subset V_{i+1}$. By construction and due to the fact that G_q is a YES instance, an arc from J to J' exists only if $j_1 \leq j_2$. Assume that J belongs to the set S_{i_1,j_1}, while J' belongs to the set S_{i_2,j_2}. By the definition of the sets $S_{i,j}$, we have that $i_1 < i_2$. Thus, $i_1 + j_1 < i_2 + j_2$ which means that J is executed in a time slot before J', and hence this kind of precedence constraints are also satisfied.

It remains to show that each set $\mathcal{T}_t$ is composed of at most m CPU tasks and k GPU tasks, so can be computed in a single time slot. In a given set $\mathcal{T}_t$, there can be at most one set of type A, one set of type B and $Q-2$ sets of type C. As explained in the definition of the sets $S_{i,j}$, each set of type B is composed of at most m CPU tasks. Moreover, each set of type A is composed of at most $k(1-1/n)$ GPU tasks, while each of the $Q-2$ sets of type C is composed of at most k/nQ GPU tasks. In total, there are $k(1-1/n) + (Q-2)k/nQ < k$ GPU tasks, and the lemma follows. □

Lemma 4 (Soundness). *If G_q corresponds to the NO case of the q-partite problem, then all schedules of instance $\mathcal{I}$ have a makespan at least $f(Q)qQ$, where f tends towards 1 when Q grows.*

Proof. (Proof sketch, full version in [9]*).* Suppose that G_q corresponds to a NO instance of the q-partite problem, and consider a schedule of $\mathcal{I}$ that minimizes the makespan. Consider the tasks associated to V_i and V_{i+1}, for some i. The execution of these two sets cannot significantly overlap: because of the precedence constraints in a NO instance, we need to complete a fraction $(1-\delta)$ of the tasks associated to V_i before starting more than a fraction δ of the tasks associated to V_{i+1}.

Moreover, because CPUs are slower than GPUs by a factor $n > Q$, and because there are far fewer GPUs than CPUs (by a factor n^2), executing the tasks associated to a single set V_i on all processors takes a time close to Q.

Overall, we have to execute the q sets V_i nearly sequentially, and each one needs almost Q time steps to be processed, so the total makespan tends towards qQ when Q is large. □

We are now ready to complete the proof.

Proof. (Proof of Theorem 2). Let $\alpha > 0$ and choose q and Q such that $f(Q)qQ > (3-\alpha)(q+3)Q/3$. Consider an instance G_q of the corresponding q-partite problem, with $n > Q$. Because of Lemmas 3 and 4, if G_q is a YES instance, then its optimal makespan is at most $(q+3)Q/3$, and otherwise, its makespan is at least $f(Q)qQ > (3-\alpha)(q+3)Q/3$.

Therefore, an algorithm approximating the scheduling problem within a factor $3-\alpha$ also solves the q-partite problem in polynomial time, which contradicts Hypothesis 1 and $P \neq NP$. □

We can furthermore adapt this proof to show the following result:

Corollary 1. *Assuming Hypothesis 1 and $P \neq NP$, the problem $(Pm, Pk) \mid prec \mid C_{\max}$ has no $3-\alpha$-approximation, for any $\alpha > 0$ and any value of m/k.*

Proof (Proof sketch). Define CPU tasks as $\overline{p_j} = 1$ and $\underline{p_j} = \infty$, and GPU tasks as $\overline{p_j} = \infty$ and $\underline{p_j} = 1$. The value of k is the same as before, but we now consider any value of $m \geq k$, and we define the sets of type b as containing $n_b = \lfloor Qmn/k \rfloor$ tasks instead of Qn^3. The completeness lemma is still valid as $(1+Q\varepsilon)n \cdot n_b \leq m$ and the soundness lemma holds as tasks cannot be processed on the other resource type. □

This result is interesting as the competitive ratio of the algorithms known for $(Pm, Pk) \mid prec \mid C_{\max}$ both in the offline $(3+2\sqrt{2-\frac{k}{m}-\frac{k}{m}^2})$ and in the online $(1+2\sqrt{m/k}$ [5]) setting tend towards 3 when m/k is close to 1, so there is no gap between the conditional lower bound and the upper bound for this case. Note that this hardness result also holds if an oracle provides the allocation (CPU or GPU for each task), in which case List Scheduling is 3-competitive [5, Theorem 7]. Therefore, the gap between the conditional lower bound and the algorithm HLP-b is mainly due to the difficulty of the allocation.

5 Conclusion

We propose a $(3+2\sqrt{2})$-approximation algorithm HLP-b for the $(Pm, Pk) \mid prec \mid C_{\max}$ problem. Our algorithm improves the approximation ratio upon the previous 6-approximation algorithm known in the literature, by using a different rounding procedure, which although is not optimal for the allocation phase, leads to a better worst-case ratio for the whole problem. We also show a conditional lower bound of 3 on the approximation ratio for this problem, assuming a generalized variant of the unique games conjecture, improving over the previous result of 2. The approximation ratio of HLP-b actually decreases towards 3 when m and k are close, thus closing the gap with the lower bound for $m = k$. The natural objective would be to close this gap for all values of m and k.

References

1. Amaris, M., Lucarelli, G., Mommessin, C., Trystram, D.: Generic algorithms for scheduling applications on hybrid multi-core machines. In: Rivera, F.F., Pena, T.F., Cabaleiro, J.C. (eds.) Euro-Par 2017. LNCS, vol. 10417, pp. 220–231. Springer, Cham (2017). https://doi.org/10.1007/978-3-319-64203-1_16
2. Bansal, N., Khot, S.: Optimal long code test with one free bit. In: Proceedings of the 50th Annual IEEE Symposium on Foundations of Computer Science, pp. 453–462. IEEE (2009)
3. Bazzi, A., Norouzi-Fard, A.: Towards tight lower bounds for scheduling problems. In: Bansal, N., Finocchi, I. (eds.) ESA 2015. LNCS, vol. 9294, pp. 118–129. Springer, Heidelberg (2015). https://doi.org/10.1007/978-3-662-48350-3_11
4. Bleuse, R., Kedad-Sidhoum, S., Monna, F., Mounié, G., Trystram, D.: Scheduling independent tasks on multi-cores with GPU accelerators. Concurr. Comput. Pract. Exp. **27**(6), 1625–1638 (2015)
5. Canon, L.C., Marchal, L., Simon, B., Vivien, F.: Online scheduling of task graphs on heterogeneous platforms. IEEE Trans. Parallel Distrib. Syst. **31**, 721–732 (2020)
6. Chekuri, C., Bender, M.: An efficient approximation algorithm for minimizing makespan on uniformly related machines. J. Algorithms **41**(2), 212–224 (2001)
7. Chen, L., Ye, D., Zhang, G.: Online scheduling of mixed CPU-GPU jobs. Int. J. Found. Comput. Sci. **25**(6), 745–762 (2014)
8. Chudak, F.A., Shmoys, D.B.: Approximation algorithms for precedence-constrained scheduling problems on parallel machines that run at different speeds. J. Algorithms **30**(2), 323–343 (1999)
9. Fagnon, V., Kacem, I., Lucarelli, G., Simon, B.: Scheduling on hybrid platforms: improved approximability window. arXiv preprint: 1912.03088 (2019)
10. Graham, R.L.: Bounds on multiprocessing timing anomalies. SIAM J. Appl. Math. **17**(2), 416–429 (1969)
11. Kedad-Sidhoum, S., Monna, F., Trystram, D.: Scheduling tasks with precedence constraints on hybrid multi-core machines. In: Proceedings of the 2015 IEEE International Parallel and Distributed Processing Symposium Workshop, IPDPSW 2015, pp. 27–33 (2015)
12. Kumar, V.A., Marathe, M.V., Parthasarathy, S., Srinivasan, A.: Scheduling on unrelated machines under tree-like precedence constraints. Algorithmica **55**(1), 205–226 (2009)
13. Lenstra, J.K., Rinnooy Kan, A.: Complexity of scheduling under precedence constraints. Oper. Res. **26**(1), 22–35 (1978)
14. Li, S.: Scheduling to minimize total weighted completion time via time-indexed linear programming relaxations. In: Proceedings of the 58th IEEE Annual Symposium on Foundations of Computer Science, FOCS 2017, pp. 283–294. IEEE (2017)
15. Svensson, O.: Hardness of precedence constrained scheduling on identical machines. SIAM J. Comput. **40**(5), 1258–1274 (2011)

Leafy Spanning Arborescences in DAGs

Cristina G. Fernandes[1] and Carla N. Lintzmayer[2](✉)

[1] Instituto de Matemática e Estatística, Universidade de São Paulo, São Paulo, Brazil
cris@ime.usp.br

[2] Centro de Matemática, Computação e Cognição, Universidade Federal do ABC, Santo André, Brazil
carla.negri@ufabc.edu.br

Abstract. Broadcasting in a computer network is a method of transferring a message to all recipients simultaneously. It is common in this situation to use a tree with many leaves to perform the broadcast, as internal nodes have to forward the messages received, while leaves are only receptors. We consider the subjacent problem of, given a directed graph D, finding a spanning arborescence of D, if one exists, with the maximum number of leaves. In this paper, we concentrate on the class of rooted directed acyclic graphs, for which the problem is known to be MaxSNP-hard. A 2-approximation was previously known for this problem on this class of directed graphs. We improve on this result, presenting a 3/2-approximation. We also adapt a result for the undirected case and derive an inapproximability result for the vertex-weighted version of Maximum Leaf Spanning Arborescence on rooted directed acyclic graphs.

Keywords: Maximum leaf spanning arborescence · Directed acyclic graphs · Maximum leaf weighted spanning arborescence · Approximation algorithms

1 Introduction

The problem of, given a connected undirected graph, finding a spanning tree with the maximum number of leaves is well known in the literature, appearing as one of the NP-hard problems in [12]. With many applications in network design problems, the best known result for it is a long standing 2-approximation proposed by Solis-Oba [20,21]. In the literature, a directed version of this problem has also been considered.

For network broadcast, one looks for a directed spanning tree rooted at a source node, in which all arcs are directed away from the source. Broadcast trees with many leaves are preferable in this situation [16,18]. Internal nodes

This research was conducted while the authors were attending the 3rd WoPOCA: "Workshop Paulista em Otimização, Combinatória e Algoritmos". C. G. Fernandes was partially supported by CNPq (Proc. 308116/2016-0 and 423833/2018-9).

Y. Kohayakawa and F. K. Miyazawa (Eds.): LATIN 2020, LNCS 12118, pp. 50–62, 2020.
https://doi.org/10.1007/978-3-030-61792-9_5

have to forward the messages received, while leaves are only receptors. Also, in some applications, it is interesting to build a more robust backbone tree, and possibly less expensive links to reach the endpoint clients. The cost of such a backbone tree is usually related to its number of arcs. By maximizing the number of leaves in a rooted directed spanning tree, we are minimizing the number of arcs in the tree obtained from removing the arcs incident to the leaves, which can be seen as a backbone tree for the network. To define the directed version of the problem precisely, we introduce some notation.

Let D be a directed graph. A vertex r in D is a *root* if there is a directed path in D from r to every vertex in D. If r is a root in D, then we say D is *r-rooted*, or simply *rooted*. We say D is *acyclic* if there is no directed cycle in D. A directed acyclic graph is called a *dag*, for short. Note that any rooted dag has only one root. An *arborescence* is an r-rooted dag T for which there is a unique directed path from r to every vertex in T. The *out-degree* of a vertex in a directed graph is the number of arcs that start in that vertex, while the *in-degree* of a vertex is the number of arcs that end in that vertex. A vertex of out-degree 0 in an arborescence is called a *leaf*.

The MAXIMUM LEAF SPANNING ARBORESCENCE is the problem of, given a rooted directed graph D, finding a spanning arborescence of D with the maximum number of leaves. Let $\text{opt}(D)$ denote the number of leaves in such an arborescence.

Given an undirected graph G, one can consider the digraph D obtained by substituting each edge by two arcs, one in each direction. With this construction, it is easy to deduce that the MAXIMUM LEAF SPANNING ARBORESCENCE is NP-hard, as its undirected version. Alon et al. [1] showed that the MAXIMUM LEAF SPANNING ARBORESCENCE remains NP-hard on dags. They were in fact investigating whether the MAXIMUM LEAF SPANNING ARBORESCENCE is fixed parameter tractable [7], and they gave a positive answer for strongly connected digraphs, as well as for dags. Fernau et al. [10] provided a cubic size kernel for the MAXIMUM LEAF SPANNING ARBORESCENCE, and Daligault and Thomassé [8] improved on this result, providing a quadratic size kernel. It is worth mentioning that a linear size kernel is known for the undirected version of the problem.

As a byproduct, Daligault and Thomassé [8] derived a 92-approximation for the MAXIMUM LEAF SPANNING ARBORESCENCE in general rooted directed graphs. This turns into a 24-approximation when the digraph has no digon (directed cycle of length two). More recently, Schwartges, Spoerhase, and Wolff [19] described a 2-approximation for the case in which the digraph is acyclic, and proved that this restricted version of the MAXIMUM LEAF SPANNING ARBORESCENCE is MaxSNP-hard. Their algorithm is inspired on a greedy 3-approximation by Lu and Ravi [17] for the undirected version of the problem.

Sections 2 and 3 present a $\frac{3}{2}$-approximation algorithm for the MAXIMUM LEAF SPANNING ARBORESCENCE on rooted dags. Our algorithm is somehow inspired on Solis-Oba's algorithm, in the sense that it prioritizes certain expansion rules. However, there is a key difference: in one of the rules, the number of expansions can be optimized. Section 4 explores the relation of our algorithm with matchings. Section 5 shows an inapproximability result for the vertex-weighted version of Maximum Leaf Spanning Arborescence on rooted dags.

2 The Algorithm

A *branching* is a forest of arborescences. A vertex that is not a leaf in a branching is called *internal*. For a positive integer t, a t-*branching* is a branching all of whose internal vertices have out-degree at least t. See Fig. 1.

For a directed graph D, we denote by $V(D)$ and $A(D)$ the set of vertices and arcs of D respectively. For a vertex v in $V(D)$, we denote by $d^+_D(v)$ its out-degree in D and by $d^-_D(v)$ its in-degree in D. The *out-neighbors* of v are the extreme vertices of arcs that start at v. We say a spanning t-branching is maximal if, for any vertex of out-degree 0, its set of out-neighbors with in-degree 0 contains less than t vertices. The spanning branchings in Fig. 1 are maximal.

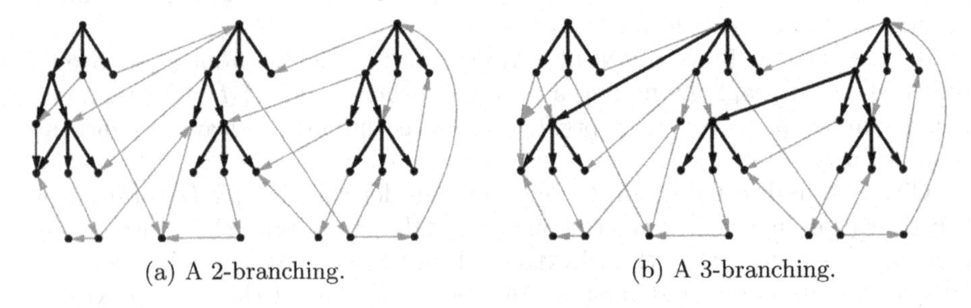

(a) A 2-branching. (b) A 3-branching.

Fig. 1. The bold arcs show two different branchings in a rooted dag.

Algorithm 1 presents GreedyExpand(D, t, F), the heart of our approximation. Given a rooted dag D, a positive integer t, and a spanning $(t+1)$-branching F of D, it returns a maximal spanning t-branching of D containing F.

Algorithm 1. GreedyExpand(D, t, F)

Input: rooted dag D, a positive integer t, and a spanning $(t+1)$-branching F of D
Output: a maximal spanning t-branching of D containing F

```
F' ← F
for each v ∈ V(D) such that d⁺_F'(v) = 0 do
    U_v ← {vu ∈ A(D) : d⁻_F'(u) = 0}
    if |U_v| ≥ t then F' ← F' + U_v
return F'
```

Let us argue that the call GreedyExpand(D, t, F) produces a maximal t-branching. Indeed, the returned F' is spanning because F' contains the spanning branching F. Besides, all internal vertices of F' have in-degree at most one and out-degree at least t. So F' is a t-branching and is clearly maximal. For instance, the branchings in Fig. 1 would be possible outputs of the calls GREEDYEXPAND(D, 2, F) and GREEDYEXPAND(D, 3, F), respectively, when D is the depicted dag and F is the spanning branching of D with no arcs.

We observe that the GREEDYEXPAND is an extension of the EXPANSION algorithm by Schwartges, Spoerhase, and Wolff [19]. Particularly, if F is the spanning branching of D with no arcs, then GREEDYEXPAND(D, 2, F) behaves as EXPANSION(D) on any rooted dag D.

Next we present our approximation for the MAXIMUM LEAF SPANNING ARBORESCENCE on rooted dags, on Algorithm 2, named MAXLEAVES. It uses twice the GREEDYEXPAND previously presented. Algorithm MAXLEAVES also uses an algorithm MAXEXPAND(D, F) that receives a rooted dag D and a maximal spanning 3-branching F of D, and returns a maximum spanning 2-branching of D containing F. Algorithm MAXEXPAND(D, F) will be described after MAXLEAVES.

Algorithm 2. MAXLEAVES(D)

Input: rooted acyclic directed graph D
Output: spanning arborescence with at least $\frac{3}{2}$ opt(D) leaves
let F_0 be the spanning branching with no arcs
$F_1 \leftarrow$ GREEDYEXPAND(D, 3, F_0)
$F_2 \leftarrow$ MAXEXPAND(D, F_1)
$T \leftarrow$ GREEDYEXPAND(D, 1, F_2)
return T

The call GREEDYEXPAND(D, 1, F) returns a maximal 1-branching of the rooted dag D containing F, that is, a spanning arborescence of D containing F. So MAXLEAVES(D) indeed produces a spanning arborescence of D. See Fig. 2. In the next section, we will prove that algorithm MAXLEAVES is a $\frac{3}{2}$-approximation for the MAXIMUM LEAF SPANNING ARBORESCENCE on rooted dags.

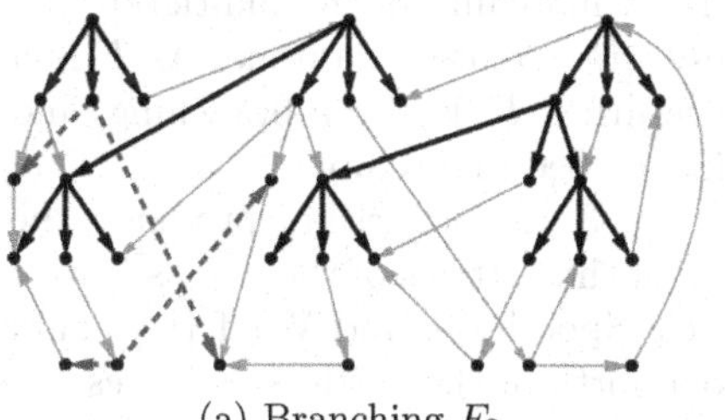

(a) Branching F_2.

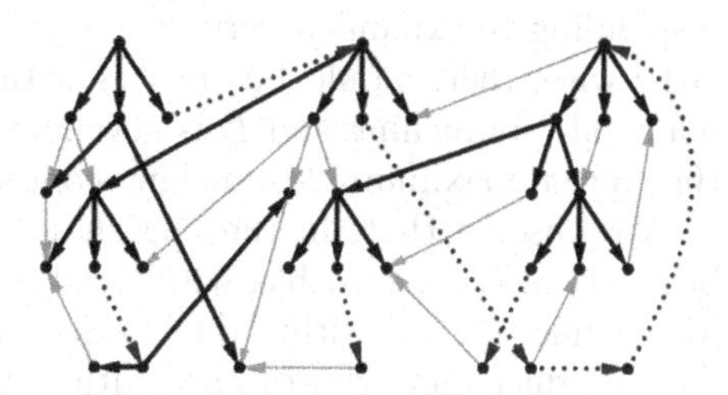

(b) Spanning arborescence T.

Fig. 2. In the left, the bold arcs represent a possible maximal 3-branching F_1 and the dashed arcs were added to obtain F_2. In the right, the bold arcs represent F_2 and the dotted arcs were added to obtain T.

The MaxExpand(D, F) procedure, presented in Algorithm 3, is an optimized version of GreedyExpand(D, 2, F). It uses an algorithm MaximumMatching that receives an undirected multigraph G and returns a maximum matching in G. Polynomial-time algorithms for this are known in the literature [9].

Algorithm 3. MaxExpand(D, F)

Input: rooted dag D and a maximal spanning 3-branching F of D
Output: a maximum spanning 2-branching of D containing F

```
for each v ∈ V(D) such that d⁺_F(v) = 0 do
    U_v ← {vu ∈ A(D) : d⁻_F(u) = 0}
Candidates ← {v ∈ V(D) : d⁺_F(v) = 0 and |U_v| = 2}
V' ← {u ∈ V(D) : d⁻_F(u) = 0}
E' ← {e_v = uw : v ∈ Candidates and U_v = {vu, vw}}
let G be the undirected multigraph (V', E')
M ← MaximumMatching(G)
F' ← F
for each e_v ∈ M do
    F' ← F' + U_v
return F'
```

The call MaxExpand(D, F) produces a maximum spanning 2-branching of D containing F. It does this by constructing an undirected multigraph G whose vertices are vertices of in-degree 0 in F and an edge uw exists in G if u and w are the only out-neighbors of in-degree 0 in F of some vertex v of out-degree 0 in F. Thus, edge uw of G represents an expansion that can be performed on vertex v of D. The fact that more than one vertex of out-degree 0 in F may have vertices u and w of in-degree 0 in F as their out-neighbors shows the need for a multigraph. Independent edges in this undirected multigraph correspond to compatible expansions, so a maximum matching gives the maximum number of expansions that can be performed in D. See Fig. 3a.

Indeed, note that, for the returned F' to be a branching, the edges e_v corresponding to expanded vertices v must form a matching in the multigraph G. Otherwise, there would be vertices with in-degree greater than one. As F is a maximal 3-branching and D is acyclic, the returned F' is also a branching, and therefore a maximum 2-branching containing F. See Figure 3b.

We observe that, in the dag shown in Fig. 3, our algorithm produces the best arborescence possible, with roughly half of the vertices of the dag as leaves. Meanwhile, the algorithm due to Schwartges, Spoerhase, and Wolff [19] could have produced an arborescence with only one forth of the vertices as leaves.

3 Approximation Ratio

Let F_1, F_2, and T be the branchings produced during the call MaxLeaves(D). For $i = 1, 2$, let k_i be the number of non-trivial components of F_i and N_i be the

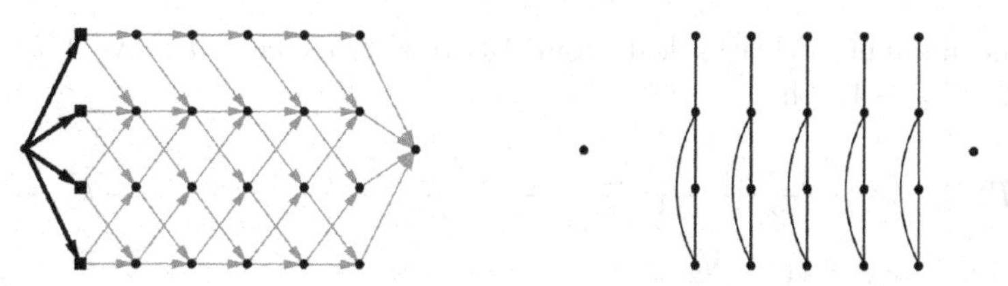

(a) The four bold arcs incident to the root of the dag are a maximal 3-branching. The round vertices form the set V, and the corresponding multigraph G is in the right.

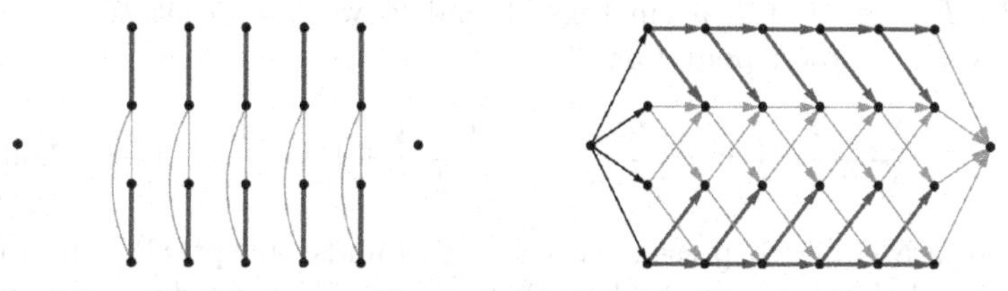

(b) On the left, a maximum matching in bold. On the right, the corresponding expansions in bold.

Fig. 3. Example of an execution of MAXEXPAND.

number of vertices in such components. We denote by $\ell(F)$ the number of leaves in any branching F.

For example, if D is the dag depicted in Fig. 1, then F_1 could be the spanning 3-branching depicted in Fig. 1b, F_2 could be the spanning 2-branching depicted in Fig. 2a, and T could be the arborescence in Fig. 2b. In this example, we have $k_1 = 3$, $N_1 = 25$, $k_2 = 4$, and $N_2 = 30$.

Lemma 1. *Let T be the arborescence produced by* MAXLEAVES*(D). Then*

$$\ell(T) \geq \frac{N_1 - k_1}{6} + \frac{N_2 - k_2}{2} + 1 .$$

Proof. Let n be the number of vertices of D. Let $T_1, \ldots, T_{k_1}$ be the non-trivial arborescences in F_1. Note that $\ell(T_j) \geq \frac{1+2|V(T_j)|}{3}$ because all internal vertices of T_j have out-degree at least 3. Therefore,

$$\begin{aligned}\ell(F_1) &= n - N_1 + \sum_{j=1}^{k_1} \ell(T_j) \geq n - N_1 + \sum_{j=1}^{k_1} \frac{1 + 2|V(T_j)|}{3} \\ &= n - N_1 + \frac{2N_1}{3} + \frac{k_1}{3} = n - \frac{N_1 - k_1}{3} .\end{aligned}$$

The number of components in F_i is $n - N_i + k_i$ for $i = 1, 2$. Hence, the number of leaves lost from F_1 to F_2 is exactly

$$\frac{(n - N_1 + k_1) - (n - N_2 + k_2)}{2} = \frac{N_2 - k_2}{2} - \frac{N_1 - k_1}{2} .$$

Also, the number of leaves lost from F_2 to T is exactly $n - N_2 + k_2 - 1 = n - (N_2 - k_2) - 1$. Thus

$$\begin{aligned}\ell(T) &\geq n - \frac{N_1 - k_1}{3} - \left(\frac{N_2 - k_2}{2} - \frac{N_1 - k_1}{2}\right) - (n - (N_2 - k_2) - 1) \\ &= \frac{N_1 - k_1}{6} + \frac{N_2 - k_2}{2} + 1\,. \end{aligned}$$

□

For D, F_1, F_2, and T as in Figs. 1 and 2, we have that $\ell(T) = 18$ and Lemma 1 gives as lower bound on $\ell(T)$

$$\frac{N_1 - k_1}{6} + \frac{N_2 - k_2}{2} + 1 = \frac{25 - 3}{6} + \frac{30 - 4}{2} + 1 = \frac{11}{3} + 14 = 17.666\ldots$$

Now we are going to present two upper bounds on $\mathrm{opt}(D)$. The following upper bound holds because the branching F_2 could be produced as output of the Expansion algorithm from Schwartges, Spoerhase, and Wolff [19].

Lemma 2 (Lemma 5 [19]). *It holds that* $\mathrm{opt}(D) \leq N_2 - k_2 + 1$.

The next lemma is the key for the approximation ratio analysis.

Lemma 3. *It holds that* $\mathrm{opt}(D) \leq \dfrac{N_1 - k_1}{2} + \dfrac{N_2 - k_2}{2} + 1$.

Proof. We apply on F_1 the same definition of witness that Schwartges, Spoerhase, and Wolff [19] used in their proof of Lemma 2. Let T^* be a spanning arborescence of D with the maximum number of leaves. Call R the set of all roots of non-trivial components of F_1. Call L the set of leaves of T^* that are isolated vertices of F_1. Let $Z := L \cup R \setminus \{r\}$, where r is the root of D. See Fig. 4a. The witness of a vertex $z \in Z$ is the closest proper predecessor $q(z)$ of z in T^* which is in a non-trivial component of F_1. Note that each witness is an internal vertex of T^*. These witnesses will not necessarily be pairwise distinct, as in [19]. See Fig. 4b.

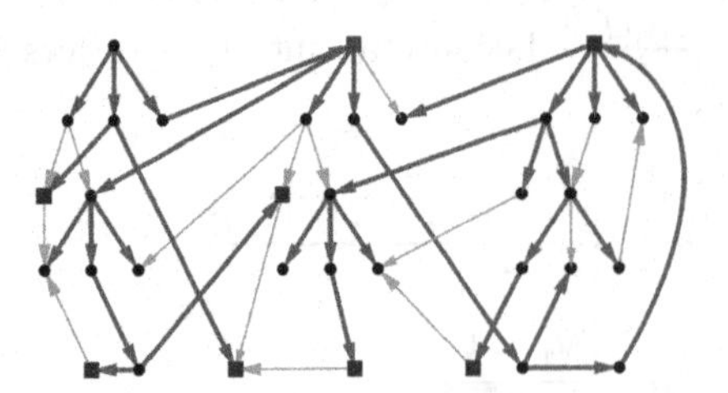

(a) The red and bold arcs show T^*, and blue and square vertices show the set Z.

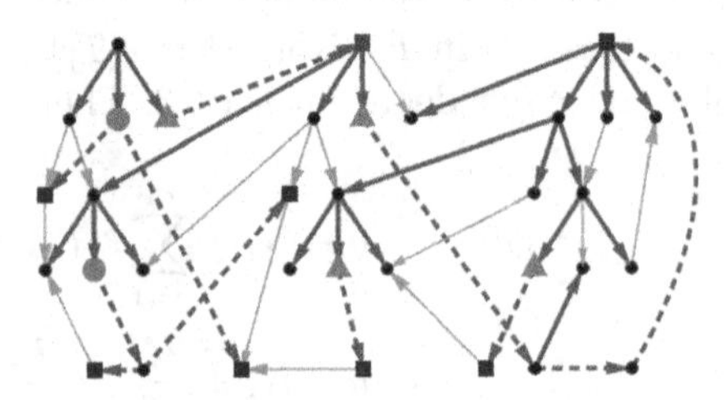

(b) Path between each vertex in Z and its witness in dashed arcs.

Fig. 4. The green triangular vertices are the witnesses for a vertex in Z and the green big circles mark two vertices that are the witnesses for two vertices in Z. (Color figure online)

We will prove that the number w of distinct witnesses is

$$w \geq |Z| - \frac{N_2 - k_2}{2} + \frac{N_1 - k_1}{2} = k_1 - 1 + |L| - \frac{N_2 - k_2}{2} + \frac{N_1 - k_1}{2}. \quad (1)$$

From this, because each witness lies in a non-trivial component of F_1 and is internal in T^*, we deduce that

$$\begin{aligned} \mathrm{opt}(D) &\leq N_1 - \left(k_1 - 1 + |L| - \frac{N_2 - k_2}{2} + \frac{N_1 - k_1}{2}\right) + |L| \\ &= N_1 - k_1 + \frac{N_2 - k_2}{2} - \frac{N_1 - k_1}{2} + 1 = \frac{N_1 - k_1}{2} + \frac{N_2 - k_2}{2} + 1. \end{aligned}$$

It remains to prove (1).

For a witness s, let $Z_s := \{z \in Z : q(z) = s\}$. Let T_s^* be the subarborescence of T^* induced by the union of all paths in T^* from s to each vertex in Z_s. The number of such arborescences T_s^* is exactly w. Note that the only internal vertex of T_s^* that is in a non-trivial component of F_1 is its root, which is necessarily a leaf of F_1 (because there is no arc from an internal vertex of F_1 to vertices in Z). Thus the maximum out-degree in T_s^* is at most two.

First let us argue that no z in Z_s is a predecessor in T_s^* of another z' in Z_s. Suppose by contradiction that z is in the path from s to z'. Then z is not a leaf of T^* and therefore z is in R, and thus in a non-trivial component of F_1. This leads to a contradiction because z, and not s, would be the witness for z'. Therefore T_s^* has exactly $|Z_s|$ leaves.

Let B be the set of vertices v such that $e_v \in M$, where M is the maximum matching in the multigraph G computed during the execution of MAX-EXPAND(D, F_1). Observe that $|M|$ is exactly the number of leaves lost from branching F_1 to F_2, so

$$|M| = |B| = \frac{N_2 - k_2}{2} - \frac{N_1 - k_1}{2}. \quad (2)$$

Now let us argue that the vertices with out-degree two in T_s^* are all in the set *Candidates*. Let v be one such vertex. Either v is an isolated vertex or v is a leaf of a non-trivial component of F_1. Therefore $d_{F_1}^+(v) = 0$. As the two children of v in T_s^* have in-degree 0 in F_1, the arcs from v to both are in U_v. Hence $v \in$ *Candidates*.

Let C_s be the set of vertices of *Candidates* with out-degree two in T_s^* and $C = \cup_s C_s$. Then the number of leaves in T_s^* is $|Z_s| = |C_s| + 1$. The set of internal vertices of T_s^* and of $T_{s'}^*$ are disjoint for distinct witnesses s and s'. Thus the sets C_s and $C_{s'}$ are disjoint. Let M_C be the set of edges of G corresponding to the vertices in C. Note that M_C is a matching, so $|C| = |M_C| \leq |M| = |B|$. Hence

$$|Z| = \sum_s |Z_s| = \sum_s (|C_s| + 1) = |C| + w \leq |B| + w.$$

Therefore $w \geq |Z| - |B| = k_1 - 1 + |L| - (\frac{N_2 - k_2}{2} - \frac{N_1 - k_1}{2})$, as in (1). □

Continuing with our example, if D is the dag depicted in Fig. 1, then Lemma 2 implies that $\mathrm{opt}(D) \leq 27$, while Lemma 3 implies that $\mathrm{opt}(D) \leq 25$.

Theorem 1. *Algorithm* MaxLeaves *is a* $\frac{3}{2}$*-approximation for the* Maximum Leaf Spanning Arborescence *on rooted directed acyclic graphs.*

Proof. For a rooted dag D, let T be the output of MaxLeaves(D). Then

$$\begin{aligned} \ell(T) &\geq \frac{N_1-k_1}{6}+\frac{N_2-k_2}{2}+1 & (3)\\ &= \frac{N_1-k_1}{6}+\frac{N_2-k_2}{6}+\frac{N_2-k_2}{3}+1 \\ &\geq \frac{\mathrm{opt}(D)-1}{3}+\frac{\mathrm{opt}(D)-1}{3}+1 & (4)\\ &> 2\,\frac{\mathrm{opt}(D)}{3}\,, \end{aligned}$$

where (3) holds by Lemma 1 and (4) holds by Lemmas 2 and 3. □

The bound given in Theorem 1 is tight. Indeed, an example similar to the one by Schwartges, Spoerhase, and Wolff [19] for their algorithm proves that algorithm MaxLeaves can achieve ratios arbitrarily close to 3/2. See Fig. 5.

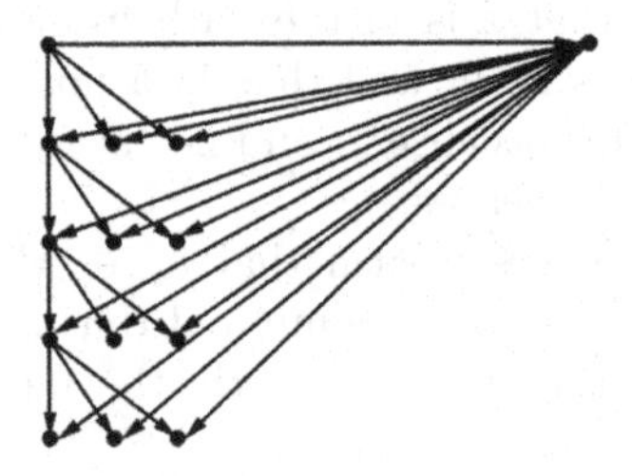

Fig. 5. A rooted dag with $n = 3k + 2$ vertices, for $k = 4$, where MaxLeaves can produce an arborescence with $2k + 2$ leaves and opt $= 3k$ leaves.

4 Using Approximations for 3-Dimensional Matching

The problem known as 3D-Matching, from 3-dimensional matching, consists in the following. Given a finite set U and a collection $\mathcal{S}$ of subsets of U with three elements each, find a collection $\mathcal{S}' \subseteq \mathcal{S}$ of pairwise disjoint sets with as many sets as possible. The name of the problem comes from the fact that one can define a 3-regular hypergraph on the vertex set U whose edges are the sets in $\mathcal{S}$, and the collection $\mathcal{S}'$ is a maximum matching in such hypergraph.

This problem is NP-hard [12], and there is a $\frac{4}{3}$-approximation for it [6,11] as well as a $(2+\epsilon)$-approximation, for any $\epsilon > 0$, for its weighted variant [2,3].

The strategy in MaxLeaves can be generalized by using an approximation algorithm for 3D-Matching. One possibility is, for a rooted dag D, to call GreedyExpand(D, 4, F_0) with the empty spanning branching F_0, obtaining F_1, then to use the $\frac{4}{3}$-approximation for 3D-Matching to expand F_1 with a good set of 3-expansions, resulting in a branching F_2. Then we can proceed as in MaxLeaves, that is, calling MaxExpand(D, F_2) to obtain F_3, and GreedyExpand(D, 1, F_3) to obtain the final arborescence T. So far we have not been able to analyze this algorithm. But, while trying to analyze it, we thought of a weighted variant for it. This weighted variant does not give an improvement, but it can have implications for Maximum Leaf Spanning Arborescence on rooted dags if a better approximation for the weighted variant of 3D-Matching is designed. So we briefly describe it ahead.

The second possibility we investigated makes use of weights. We start by calling GreedyExpand(D, 4, F_0) with the empty spanning branching F_0, obtaining F_1. After that, we create an instance of the weighted 3D-Matching where feasible 3-expansions turn into sets of weight two and feasible 2-expansions turn into sets of weight one. We then use an approximation for the weighted 3D-Matching to obtain a branching F_2 from F_1. We finish by calling GreedyExpand(D, 1, F_2) to obtain an arborescence. Name the resulting algorithm MaxLeaves-W3DM. Due to the space restrictions, we omit the proof of the next result.

Theorem 2. *If the approximation used for the weighted* 3D-Matching *has ratio* $\alpha > 1$, *then* MaxLeaves-W3DM *is a* $\max\{\frac{4}{3}, \alpha\}$*-approximation for the* Maximum Leaf Spanning Arborescence *on rooted directed acyclic graphs.*

At the moment, as the best approximation for the weighted 3D-Matching has ratio greater than 2, this does not provide any improvement on the previously best known ratio for Maximum Leaf Spanning Arborescence. Now, only a ratio better than 3/2 for the weighted 3D-Matching would provide an improvement.

5 Inapproximability of the Vertex-Weighted Version

A vertex-weighted generalization of the maximum leaf spanning tree (the undirected version of our problem) was considered in the literature. In such generalization, one is given a connected vertex-weighted graph and the goal is to find a spanning tree whose sum of leaf weights is maximum.

Jansen [14] proved that, unless P = NP, this version of the problem does not admit a polynomial-time factor $O(n^{\frac{1}{2}-\epsilon})$ or a $O(\text{opt}^{\frac{1}{3}-\epsilon})$-approximation for any $\epsilon > 0$, where n is the number of vertices of the given graph. His reduction is from the Independent Set problem. A straightforward modification of his reduction shows the same inapproximability results for the vertex-weighted version of Maximum Leaf Spanning Arborescence on rooted dags. Next we describe his reduction adapted to produce rooted dags with binary weights.

The Independent Set problem consists of the following: given a graph G, find an independent set in G with as many vertices as possible.

Let G be an instance of the INDEPENDENT SET problem. Let D be the rooted dag that has as vertices the vertices of G, a new vertex r as its root, and a vertex e for each edge e of G. There is an arc from r to each vertex of G in D. For each edge $e = uv$ of G, there is an arc from u to e and an arc from v to e in D. So, if G has n vertices and m edges, D has $n+m+1$ vertices and $n+2m$ arcs. See Fig. 6. Note that D is r-rooted and acyclic and that, in any spanning arborescence in D, the vertices corresponding to edges of G are leaves, because they have out-degree 0 in D. Because the complement of an independent set is an edge cover, it is not hard to see that the following holds.

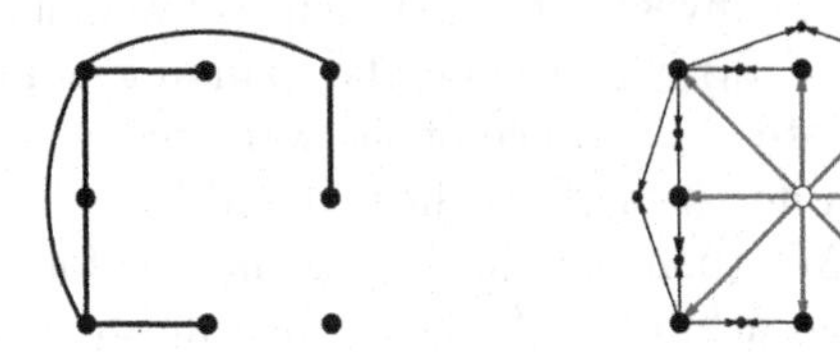

Fig. 6. An instance of the INDEPENDENT SET problem and the corresponding rooted dag. The white vertex is the root of the dag.

Lemma 4. *Set S is an independent set in G if and only if there is a spanning arborescence in D that has $S \cup E(G)$ as leaves.*

So, we assign weights to the vertices of D as follows: vertices of G have weight 1 while vertices corresponding to edges of G have weight 0. The root r may have an arbitrary weight, because it will never be a leaf in a spanning arborescence of D. For any $\epsilon > 0$, there is no polinomial-time $O(n^{1-\epsilon})$-approximation for INDEPENDENT SET unless P = NP [13], where n is the number of vertices of the given graph G. Hence, using this and Lemma 4, we derive the following.

Theorem 3. *The vertex-weighted* MAXIMUM LEAF SPANNING ARBORESCENCE *on directed acyclic graphs with binary weights and maximum in-degree 2 does not have a polinomial-time $O(n^{1-\epsilon})$-approximation for any $\epsilon > 0$, unless $P = NP$, where n is the number of weight-one vertices of the given directed graph.*

To avoid using weight zero, a similar result can be obtained by assigning weights m and 1 instead of 1 and 0, respectively, where m is the number of edges in G. For this case, Lemma 4 implies that an independent set of size t in G corresponds to a spanning arborescence of leaf weight $(t+1)m$, and a similar inapproximability result holds, as Jansen [14] proved for the undirected version.

6 Future Directions

Improving on the 92-approximation for the general directed case would be very interesting. A major difficulty is that greedy strategies do not apply so easily, because not every branching can be extended to a spanning branching in an arbitrary rooted digraph. The strategy used by Daligault and Thomassé [8] consists of a series of reductions, and some of them end up with a dag. It is tempting to try to use an approximation for dags within their algorithm to achieve an improved ratio, however we did not succeed in doing that so far.

Directed acyclic graphs have directed tree width zero [15]. Is it possible to extend our approximation or any greedy algorithm for MAXIMUM LEAF SPANNING ARBORESCENCE to address directed graphs with bounded directed tree width?

It is natural to wonder if there is a way to optimize one of the expansions used in Solis-Oba's algorithm to achieve a better approximation ratio for the undirected case. Also, for the undirected case, there are better approximations for cubic graphs [4,5]. Maybe one can obtain better bounds on the approximation ratio for dags with out-degree bounded by three or two.

References

1. Alon, N., Fomin, F., Gutin, G., Krivelevich, M., Saurabh, S.: Spanning directed trees with many leaves. SIAM J. Disc. Math. **23**(1), 466–476 (2009)
2. Arkin, E.M., Hassin, R.: On local search for weighted k-set packing. Math. Oper. Res. **23**(3), 640–648 (1998)
3. Berman, P.: A $d/2$ approximation for maximum weight independent set in d-claw free graphs. Nord. J. Comput. **7**(3), 178–184 (2000)
4. Bonsma, P., Zickfeld, F.: A 3/2-approximation algorithm for finding spanning trees with many leaves in cubic graphs. SIAM J. Disc. Math. **25**(4), 1652–1666 (2011)
5. Correa, J., Fernandes, C.G., Matamala, M., Wakabayashi, Y.: A 5/3-approximation for finding spanning trees with many leaves in cubic graphs. In: Proceedings of the 5th International Workshop on Approximation and Online Algorithms, WAOA 2007. LNCS, vol. 4927, pp. 184–192 (2008)
6. Cygan, M.: Improved approximation for 3-dimensional matching via bounded pathwidth local search. In: Proceedings of the IEEE 54th Annual Symposium on Foundations of Computer Science, FOCS 2013, pp. 509–518 (2013)
7. Cygan, M., et al.: Parameterized Algorithms. Springer, Cham (2015). https://doi.org/10.1007/978-3-319-21275-3
8. Daligault, J., Thomassé, S.: On finding directed trees with many leaves. In: Chen, J., Fomin, F.V. (eds.) IWPEC 2009. LNCS, vol. 5917, pp. 86–97. Springer, Heidelberg (2009). https://doi.org/10.1007/978-3-642-11269-0_7
9. Edmonds, J.: Paths, trees, and flowers. Can. J. Math. **17**, 449–467 (1965)
10. Fernau, H., Fomin, F.V., Lokshtanov, D., Raible, D., Saurabh, S., Villanger, Y.: Kernel(s) for problems with no kernel: On out-trees with many leaves. In: Proc. of the 26th International Symposium on Theoretical Aspects of Computer Science, STACS 2009. LIPIcs, vol. 3, pp. 421–432 (2009)
11. Fürer, M., Yu, H.: Approximating the k-set packing problem by local improvements. In: Fouilhoux, P., Gouveia, L.E.N., Mahjoub, A.R., Paschos, V.T. (eds.) ISCO 2014. LNCS, vol. 8596, pp. 408–420. Springer, Cham (2014). https://doi.org/10.1007/978-3-319-09174-7_35
12. Garey, M.R., Johnson, D.S.: Computers and Intractability. W.H. Freeman and Co., New York (1979)

13. Håstad, J.: Clique is hard to approximate within $n^{1-\epsilon}$. Acta Math. **182**(1), 105–142 (1999)
14. Jansen, B.M.P.: Kernelization for maximum leaf spanning tree with positive vertex weights. J. Graph Algorithms Appl. **16**(4), 811–846 (2012)
15. Johnson, T., Robertson, N., Seymour, P.D., Thomas, R.: Directed tree-width. J. Combinat. Theory, Ser. B **82**, 138–154 (2001)
16. Jüttner, A., Magi, A.: Tree based broadcast in ad hoc networks. Mobile Netw. Appl. **10**(5), 753–762 (2005)
17. Lu, H., Ravi, R.: Approximating maximum leaf spanning trees in almost linear time. J. Algorithms **29**(1), 132–141 (1998)
18. Pope, J., Simon, R.: Efficient one-to-many broadcasting for resource-constrained wireless networks. In: Proceedings of the 40th IEEE Conference on Local Computer Networks, LCN 2015 (2015), 8 p.
19. Schwartges, N., Spoerhase, J., Wolff, A.: Approximation algorithms for the maximum leaf spanning tree problem on acyclic digraphs. In: Solis-Oba, R., Persiano, G. (eds.) WAOA 2011. LNCS, vol. 7164, pp. 77–88. Springer, Heidelberg (2012). https://doi.org/10.1007/978-3-642-29116-6_7
20. Solis-Oba, R.: 2-approximation algorithm for finding a spanning tree with maximum number of leaves. In: Bilardi, G., Italiano, G.F., Pietracaprina, A., Pucci, G. (eds.) ESA 1998. LNCS, vol. 1461, pp. 441–452. Springer, Heidelberg (1998). https://doi.org/10.1007/3-540-68530-8_37
21. Solis-Oba, R., Bonsma, P., Lowski, S.: A 2-approximation algorithm for finding a spanning tree with maximum number of leaves. Algorithmica **77**, 374–388 (2017)

Approximating Routing and Connectivity Problems with Multiple Distances

Lehilton L. C. Pedrosa and Greis Y. O. Quesquén(✉)

Institute of Computing, University of Campinas, Campinas, SP, Brazil
lehilton@ic.unicamp.br, greis.quesquen@students.ic.unicamp.br

Abstract. We consider routing and connectivity problems for which the input includes a complete graph G and multiple edge-weight functions $d_1, d_2, \ldots, d_r$. In each case, a solution is a minimum-cost subgraph H satisfying the constraints of the particular problem. The cost of each edge of H is determined by any chosen function d_i, but there is a service fee $g \geq 0$ for each maximal connected component formed by edges associated with the same function. This is motivated by applications for which a solution can be split between multiple providers, each corresponding to a distance d_i. One example is the Traveling Car Renter Problem (CaRS), which is a generalization of the Traveling Salesman Problem (TSP) whose goal is to visit a set of cities by renting cars from multiple companies. In this paper, we give $\mathcal{O}(\log n)$-approximations for the generalizations with multiple distances of several problems (Steiner TSP, Profitable Tour Problem, and Constrained Forest Problem). This factor is the best-possible unless P = NP.

Keywords: Approximation algorithm · Multiple distance functions

1 Introduction

Travel and tourism are two of the largest world economic sectors, as they comprise a large part the global economy, by generating jobs, driving imports and exports, and stimulating the tourism growth [16]. Transportation is an important and closely related sector, and, when considered from the perspectives of traveling and tourism, it refers to carrying people from their place of living to points of interest [15]. Among the most common means of transport, those based on car rental play an important role. Indeed, the global car rental market is expected to have annual growth rate of 13.55% over the next 5 years [14].

While the majority of the works in the literature focus on maximizing the income of a rental company, car rental decisions also appear in the form of challenging problems for the customers, be they tourists or tourism agencies.

Supported by grant #2015/11937-9, São Paulo Research Foundation (FAPESP), grants #425340/2016-3, #313026/2017-3, #422829/2018-8, National Council for Scientific and Technological Development (CNPq), and Coordenação de Aperfeiçoamento de Pessoal de Nível Superior - Brasil (CAPES) - Finance Code 001.

Y. Kohayakawa and F. K. Miyazawa (Eds.): LATIN 2020, LNCS 12118, pp. 63–75, 2020.
https://doi.org/10.1007/978-3-030-61792-9_6

One such a problem is the Traveling Car Renter Problem (CaRS) [7], which can be illustrated as follows. A traveler wants to visit a set of cities by renting one or more cars, and forming a closed route. The cost incurred by each rental corresponds to the distance traveled by the car plus a return fee, both of which depend on the selected car. The objective is to minimize the total cost.

Note that CaRS is a generalization of the Traveling Salesman Problem (TSP). In fact, TSP is the case of CaRS in which there is only one car, and the return fee is zero. The distinguishing feature of CaRS is that an instance is composed of multiple distance functions, one for each car, thus the traveler not only has to find a route, but has to decide which car to use in each part of the route. Pedrosa et al. [12] observed that, if a solution is required to be a Hamiltonian cycle, or if the distance functions are arbitrary, then CaRS cannot be approximated by any computable function, unless $P = NP$. Then, they considered the Uniform CaRS (UCaRS), that is the case in which the return fee is fixed and the distance functions are metric, and presented an $\mathcal{O}(\log n)$-approximation algorithm. Even for this version, no $o(\log n)$-approximation exists, unless $P = NP$.

Unlike TSP, for which every city is visited by the same car, the main difficulty of UCaRS appears to be selecting which car is used to visit each city. What stops one from renting a distinct car in each city is that exchanging cars incurs an additional cost corresponding to the return fee. Observe that, if a subset of cities is visited sequentially in the route, then finding the best order to visit these cities corresponds to the path version of TSP. This means that a solution of the generalization of TSP with multiple distances corresponds to multiple solutions of the classical path TSP. Many classical routing and connectivity problems fit into this framework, such as Steiner TSP (STSP) [11], Profitable Tour Problem (PTP) [2,5,6], Constrained Forest Problem (CFP) [4,6,8], and many others.

Contributions. In this paper, we discuss several generalizations of classical routing and connectivity problems whose instances are composed of multiple distance functions. More specifically, we consider three related problems: Steiner UCaRS (SUCaRS), Profitable UCaRS (PUCaRS), and Multiple Constrained Forest Problem (MCFP). The first two are routing problems and require a closed walk as a solution; they generalize STSP and PTP, respectively. The last one is a connectivity problem and asks for a spanning forest; it generalizes CFP.

The routing problems can be informally described as follows. In SUCaRS, the goal is to visit only a given subset of cities of the graph, called terminals, by renting one or more of the available cars. To this, the traveler can use intermediate connection cities, called Steiner cities, to minimize the route's total cost. The additional source of hardness in this problem is choosing one of the many combinations of Steiner cities. In PUCaRS, any subset of cities may be visited, but each city that is not visited incurs a penalty. The goal is to minimize the route's total cost plus the penalties associated with non-visited cities. Note that SUCaRS is a special case of PUCaRS in which each terminal has infinite penalty cost, and all other cities have zero penalty cost.

The describe the connectivity problem, we first need to define its classical variant. In CFP, given a set of cities and an integer k, a solution is a minimum-weight spanning forest such that each tree of the forest contains at least k cities. Each edge of the tree represents a cable connecting two cities, and the edge cost is determined by a distance function given in the input. In MCFP, the cables can have different types, each one corresponding to a distinct distance function, but each vertex connecting cables of distinct types incurs additional switching costs. The goal is to minimize the sum of cable and switching costs.

The ideas of the algorithms can be outlined as follows. For SUCaRS, we consider a reduction to UCaRS which preserves the approximation factor, and implies an $\mathcal{O}(\log n)$-approximation for SUCaRS. For PUCaRS, we create instances of two distinct problems. The first is an instance of UCaRS as before, but the second is an instance of PTP, which is the classical version of the problem with a single distance function. Since UCaRS has an $\mathcal{O}(\log n)$-approximation, and PTP has a constant-factor approximation, combining solutions of these instances leads to an $\mathcal{O}(\log n)$-approximation for PUCaRS. Similarly, for MCFP, we create instances of UCaRS and CFP, which lead to an $\mathcal{O}(\log n)$-approximation also for MCFP. We observe that a $o(\log n)$-approximation for any of the three problems implies an $o(\log n)$-approximation for UCaRS, thus our algorithms are asymptotically optimal unless P = NP.

These algorithms illustrate two main tools to solve problems with multiple distance functions. The first is used for SUCaRS and consists of reducing the problem to a particular case whose instance is still composed of multiple distances. The second tool is used for PUCaRS and MCFP and consists of separating the problem of selecting which distance function is assigned to each city from the problem of satisfying the connectivity constraints. The main assumption of our techniques is having a constant-factor approximation to the classical version of the problem, thus they can be used to design approximations for generalizations of other similar routing and connectivity problems.

Related Works. TSP is among the most well-studied combinatorial optimization problems. Karp showed that TSP is NP-hard [9]. For the case in which the distance satisfies the triangle inequality, called Metric TSP, the so called Double-MST algorithm is already a 2-approximation, and the best known factor is 3/2 due to Christofides [3].

In the Steiner TSP (STSP) [11], given a subset of cities, called terminals, the goal is to find a minimum-cost route that visits each terminal at least once. Note that STSP can be reduced to TSP, preserving the approximation factor [10].

A variant is the Profitable Tour Problem (PTP) [5], also known as the Prize-Collecting TSP [2,6]. In this problem, each city is given a penalty, and a solution is a route that visits any subset of cities. The goal is to minimize the route cost plus the penalties of non-visited cities. Bienstock et al. [2] gave a 5/2-approximation, and there is a $(2 - \frac{1}{n-1})$-approximation via primal-dual [6].

A related problem is the Prize-Collecting Steiner Tree Problem introduced by Balas [1]. As in the case of PTP, each city has an associated penalty. The goal is to find a tree which minimizes the cost of the edges plus the penalties of non-visited cities. Bienstock et al. [2] gave an approximation with factor 3, and Goemans and Williamson [6] obtained a $(2 - \frac{1}{n-1})$-approximation. Another connectivity problem is the Constrained Forest Problem (CFP) [8]. Given a graph with positive edge weights and an integer k, the objective is to find a minimum-weight spanning forest such that each of its trees contains at least k vertices. This problem is NP-hard for $k \geq 4$, and a greedy algorithm achieves an approximation factor of 2 [8]. Couëtoux [4] improved this factor to 3/2 with another greedy algorithm which run in $\mathcal{O}(nm)$.

Common Definitions. In each of the considered problems, we are given a set of cities $V = \{1, 2, \ldots, n\}$, a set of distance functions indexed by $C = \{1, 2, \ldots, r\}$, and a service fee $g \geq 0$. For each index $i \in C$, the distance between two cities u and v according to i is denoted by $d_i(u, v) \geq 0$. We assume that each distance function is symmetric and satisfies the triangle inequality, i.e., $d_i(u, v) = d_i(v, u)$ and $d_i(u, v) \leq d_i(u, w) + d_i(w, v)$ for every $i \in C$ and $u, v, w \in V$. Also, given an undirected graph G, we denote by $E(G)$ the set of edges of G and by $V(G)$ the set of vertices of G.

2 Steiner UCaRS (SUCaRS)

An instance of SUCaRS consists of a set of cities $V = \{1, 2, \ldots, n\}$; a subset of cities $T \subseteq V$, called *terminals*; a set of cars $C = \{1, 2, \ldots, r\}$; and a return fee $g \geq 0$. Each car $i \in C$ is associated with a distance function d_i. A solution is a closed walk P on V containing all cities in T and an assignment φ such that each edge (u, v) of P is associated with car $\varphi(u, v) \in C$. Denote by $M(P)$ the number of vertices of P whose incident edges are associated with distinct cars. Note that $M(P)$ corresponds to the number of times which the used car is exchanged, and that a city may be counted more than once. Also, denote by $L(P)$ the length of P, i.e., let $L(P) = \sum_{(u,v) \in E(P)} d_{\varphi(u,v)}(u, v)$. The objective is to find a solution which minimizes the value of P, which is defined as $\mathsf{val}(P) = M(P) \cdot g + L(P)$. UCaRS is the particular case in which $T = V$, and STSP is the particular case in which $|C| = 1$ and $g = 0$.

As mentioned above, this problem asks for a closed walk over the terminals as a solution. A natural question that arises is whether we may delete the non-terminal cities and execute the UCaRS algorithm. This may lead to arbitrarily bad solutions, because there are multiple distance functions. For example, given two terminals u and v, it could be more costly going straight from u to v using the same car than renting different cars at a non-terminal city on the way from u to v. However, the cheapest way to go from a city u to a city v can be found in polynomial time with the aid of an auxiliary graph, defined next.

Given a set of cities V and a set of cars C with corresponding distance functions, the *support graph* of (V, C) is the graph H such that, for each $v \in V$ and $i \in C$, there is a vertex v_i in $V(H)$. The are two kinds of edges, horizontal and vertical. For each pair $u, v \in V$ and each $i \in C$, there is a horizontal edge (u_i, v_i) with cost $w(u_i, v_i) = d_i(u, v)$, and for each $u \in V$ and each pair $i, j \in C$, there is a vertical edge (u_i, u_j) with cost $w(u_i, u_j) = g$. In Fig. 1, we can see an example of H for 3 vertices and 3 cars. For each pair $u, v \in V$ and each pair

$i, j \in C$, we denote by $\mathsf{dist}(u_i, v_j)$ the length of a shortest path from u_i to v_j. Note that a walk in H induces a walk on V in which horizontal edges correspond to moving between cities, and vertical edges corresponds to exchanging the car.

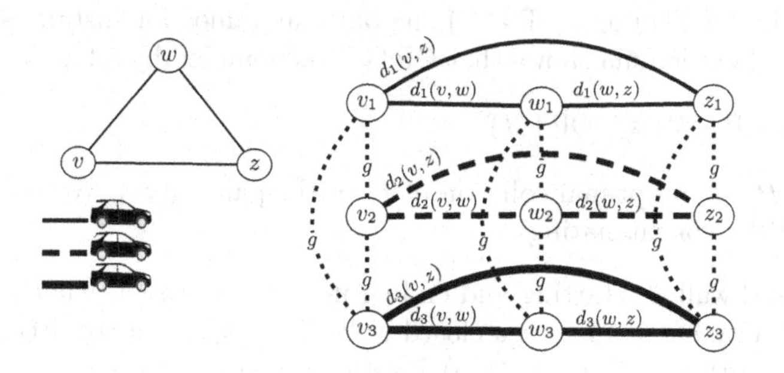

Fig. 1. Support graph.

Using the support graph H, we can reduce an instance of SUCaRS to an instance of UCaRS which contains only terminals and cars corresponds to pairs (i, j) of $C \times C$. A solution for the reduced instance corresponds to a walk on H, which induces a solution for the instance of SUCaRS. This leads to the following.

Theorem 1. *If there exists an α-approximation for UCaRS, then there exists an α-approximation for SUCaRS.*

3 Profitable UCaRS (PUCaRS)

An instance of PUCaRS consists of a set of cities $V = \{1, 2, \ldots, n\}$; a set of cars $C = \{1, 2, \ldots, r\}$; a return fee $g \geq 0$; and, for each city $u \in V$, a penalty $\pi(u) \geq 0$. Each car $i \in C$ is associated with a distance function d_i. A solution is a closed walk P on V containing any subset of cities and an assignment φ such that each edge (u, v) of P is associated with car $\varphi(u, v) \in C$. Define $M(P)$ and $L(P)$ as in the case of SUCaRS and, for some set U, let $\pi(U) = \sum_{u \in U} \pi(u)$. The objective is to find a solution which minimizes the value of P, which is defined as $\mathsf{val}(P) = M(P) \cdot g + L(P) + \pi(V \setminus V(P))$. Note that PTP is the particular case in which $|C| = 1$ and $g = 0$.

Given an instance I of PUCaRS, we create an instance I' of UCaRS as follows. The instance I' is composed of cities V, return fee g, and cars $C' = C \cup \{0\}$, where index 0 corresponds to a new car. For each car $i \in C$, we use the same distance function as in the original instance, d_i. For the car 0, we let d_0 be a new distance function, which will account for penalties of non-visited cities. To define d_0, we create an auxiliary weighted graph Q on V. Assume, without loss of generality, that there is an optimal solution of I for which a car exchange occurs at a city $s \in V$. If a car exchange does not occur in s, then we could try each city of V in polynomial time, and, if no car exchange occurs, then we can simply use an approximation algorithm for PTP for each distance function. Let Q be a star centered at s and connected to each city $u \in V\backslash\{s\}$ by an edge of weight $\pi(u)/2$. Then, for each pair $u, v \in V$, $d_0(u, v)$ is the length of a shortest path between u and v in Q. This completes the definition of I'.

Denote by $\mathsf{OPT}(I)$ and $\mathsf{OPT}(I')$ the optimal values for instances I and I', respectively. Next lemma shows that $\mathsf{OPT}(I')$ is bounded by a factor of $\mathsf{OPT}(I)$.

Lemma 1. $\mathsf{OPT}(I') \leq 2\,\mathsf{OPT}(I)$.

Proof. Let P be an optimal solution of I which spans city s. We create a solution P' of I' by concatenating

(a) the closed walk P starting and ending in s, whose cost is $\mathsf{val}(P)$;
(b) for each city $v \in V \setminus V(P)$, a closed walk $T_v = (s, v, s)$ associated with car 0 whose length is $L(T_v) = d_0(s, v) + d_0(v, s) = \pi(v)$.

Note that P' visits all the cities, then P' is a feasible solution for I'. Since each walk T_v uses the same car, after visiting all cities of P, only one car exchange is necessary. Thus, the cost of P' can be bounded as

$$\begin{aligned}\mathsf{val}(P') &= M(P) \cdot g + L(P) + \textstyle\sum_{v \in V \setminus V(P)} L(T_v) + g \\ &= M(P) \cdot g + L(P) + \textstyle\sum_{v \in V \setminus V(P)} \pi(v) + g \\ &= \mathsf{val}(P) + g \leq \mathsf{OPT}(I) + \mathsf{OPT}(I),\end{aligned}$$

where in the last inequality holds because $\mathsf{val}(P) = \mathsf{OPT}(I)$ and $g \leq \mathsf{OPT}(I)$, as solution P exchanges car at s. Therefore, $\mathsf{OPT}(I') \leq 2\,\mathsf{OPT}(I)$. □

Conversely, a solution of instance I' of UCaRS can contain many sub-walks that are traveled using car 0. This entails paying return fees each time a car is exchanged for car 0. So, we will construct a new solution for which we pay the return fee for car 0 only once. To bound this new-solution cost, we will create an instance of PTP, which is the corresponding problem with only one distance function, and no exchange cost. We do this in the following lemma.

Lemma 2. *Given a solution P' for I', one can construct, in polynomial time, a solution P for I such that* $\mathsf{val}(P) \leq 2\,\mathsf{val}(P') + 4\,\mathsf{OPT}(I)$.

Proof. Since the walk P' visits all cities, we have $s \in V(P')$. Suppose that we are given an edge (u, v) of $E(P')$ such that $\varphi(u, v) = 0$. If such an edge is not incident with s, then we can add (u, s) and (s, v) and remove (u, v) without changing the solution cost, since $d_0(u, v) = d_0(u, s) + d_0(v, s)$. By doing this for every edge associated with car 0, we can assume that each such an edge is incident with s. Therefore, the edges associated with car 0 induce a connected component which contains a star centered at s. By doubling the edges of this star, we build a closed walk B whose edges are associated with car 0.

If we remove the vertices of B from P', then we obtain a set of walks, $W_1, W_2, \ldots, W_q$. By doubling the edges of each such a walk W_ℓ, we construct a

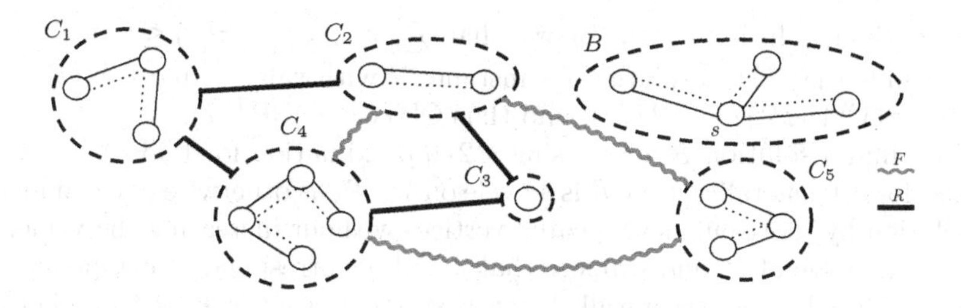

Fig. 2. Connecting components.

closed walk W'_ℓ. The union of all constructed closed walks induces a set of connected components $C_1, C_2, \ldots, C_k$, such that $k \le q$ and each of which is a closed walk itself. Note that the collection of components $V(B), V(C_1), \ldots, V(C_k)$ partition the set of cities. In Fig. 2, we can see an example of this partition.

We bound the overall cost of the closed walks. Each edge of P' appears at most twice as an edge of a closed walk B or C_ℓ. Therefore, $\sum_{\ell=1}^{k} L(C_\ell) + L(B) \le 2L(P')$. Note that the cities of distinct components C_ℓ and $C_{\ell'}$ are connected in P' only through edges associated with car 0, then P' has at least two car exchanges for each of the k components, one for an edge entering the component, and other for an edge leaving the component. Since no car exchange in some walk W_ℓ involves an edge associated with car 0, we have $2k + \sum_{\ell=1}^{k} M(W_\ell) \le M(P')$. As we have doubled the number of car exchanges in each walk W_ℓ, it follows that $4k + \sum_{\ell=1}^{k} M(C_\ell) \le 2M(P')$. Note that each city in $V(B)$ is connected to s, thus $L(B) = 2\sum_{u \in V(B)} d_0(s,u) = 2\sum_{u \in V(B)} \pi(u)/2 = \pi(V(B))$. Combining with the inequalities,

$$(4k + \textstyle\sum_{\ell=1}^{k} M(C_\ell)) \cdot g + \sum_{\ell=1}^{k} L(C_\ell) + \pi(V(B)) \le 2\,\mathsf{val}(P'). \tag{1}$$

Let H be the support graph of (V, C) as defined in Sect. 2, and let dist be the distance function associated with graph H. We construct an instance J of PTP. Start by creating a complete graph G on $V \setminus V(B)$, such that the weight of each edge $(u,v) \in E(G)$ is the cost of the shortest path to go from u to v, i.e., define $\mathrm{d_{sp}}(u,v) = \min\{\mathsf{dist}(u_i, v_j) : i,j \in C\}$. Then, for each $1 \le \ell \le k$, contract the vertices in $V(C_\ell)$, removing loops, but preserving parallel edges. The penalties of each vertex w of G corresponding to a component C_ℓ for some $1 \le \ell \le k$ is defined as $\pi(w) = \sum_{u \in C_\ell} \pi(u)$. Next, we would like to bound the optimal value of J, which is denoted by $\mathsf{OPT}(J)$.

Let P^* be an optimal solution of I, and note that it induces a closed walk W on H with weight $M(P^*) \cdot g + L(P^*)$. This walk may contain vertices corresponding to B or to components $C_1, C_2, \ldots, C_k$. Contract the vertices of each component C_ℓ in W and remove loops. Now, each two vertices u and v of W corresponding to consecutive components are connected by a walk on H with weight at least $\mathrm{d_{sp}}(u,v)$. Therefore, W induces a closed walk F on a subset of $V(G)$ with length $\sum_{(u,v) \in E(F)} \mathrm{d_{sp}}(u,v) \le M(P^*) \cdot g + L(P^*)$. If a vertex $u \in V(G)$ corresponding to a component C_ℓ is not in $V(F)$, then no vertex of C_ℓ is visited by the optimal solution P^*. It follows that $\sum_{u \in V(G) \setminus V(F)} \pi(u) \leq \pi(V \setminus V(P^*))$. We conclude that F is a solution for instance J with value at most $M(P^*) \cdot g + L(P^*) + \pi(V \setminus V(P^*)) = \mathsf{OPT}(I)$, and thus $\mathsf{OPT}(J) \leq \mathsf{OPT}(I)$.

Now, find a solution R for J using a 2-approximation for PTP [6]. Assume, without loss of generality, that R is a cycle on $V(G)$, as otherwise one can modify the solution by short-cutting repeated vertices without increasing the value. Let (v, v') be an edge of R, and suppose that v and v' correspond to components C_ℓ and $C_{\ell'}$. This edge induces a walk D' on V starting at a vertex of C_ℓ and ending at a vertex of $C_{\ell'}$ with $M(D') \cdot g + L(D') \leq \mathrm{d_{sp}}(v, v')$. By doubling the edges of D', we construct a closed walk D such that $M(D) \cdot g + L(D) \leq 2\,\mathrm{d_{sp}}(v, v')$. Let $D_1, D_2, \ldots, D_m$ be the set of constructed closed walks for every edge of R, and note that $m \leq k$, since R is a cycle of G and $V(G) = k$. As the value of R is at most $2\,\mathsf{OPT}(J) \leq 2\,\mathsf{OPT}(I)$, and we have doubled the induced walks,

$$\textstyle\sum_{\ell=1}^{m} M(D_\ell) \cdot g + \sum_{\ell=1}^{m} L(D_\ell) + \pi(V(G) \setminus V(R)) \leq 4\,\mathsf{OPT}(I). \qquad (2)$$

Finally, we create a solution P for I by joining the all connected components C_ℓ which correspond to vertices of R, then inserting the closed walks $D_1, D_2, \ldots, D_m$ which correspond to edges of R. Observe that P is a feasible solution. Indeed, because R is cycle, P is connected, and because each connected component C_ℓ is a closed walk, P is also a closed walk. Each closed walk D_t connects a vertex $v \in C_\ell$ to some vertex $v' \in C_{\ell'}$, with $\ell \neq \ell'$. Thus, introducing closed walk D_t adds four car exchanges, and therefore at most $4m \leq 4k$ exchanges are added in total. Besides theses, there are the car exchanges in internal vertices of each walk C_ℓ and D_ℓ. The length of P is simply the sum of the lengths of walks C_ℓ and D_ℓ. Also, each vertex which is not visited by P is either in B or is a component corresponding to a vertex not visited by R, thus the incurred penalty of P corresponds to penalties of vertices in B and the penalty for solution R. Adding up all costs and combining with inequalities (1) and (2), we conclude that $\mathsf{val}(P) \leq 2\,\mathsf{val}(P') + 4\,\mathsf{OPT}(I)$.

Combining the previous lemma, we can derive an approximation for PUCaRS.

Theorem 2. *If there exists an α-approximation for UCaRS, then there exists an $\mathcal{O}(\alpha)$-approximation for PUCaRS.*

Proof. Recall that $\mathsf{OPT}(I') \leq 2\,\mathsf{OPT}(I)$ by Lemma 1. If P' is an α-approximation for I', then, using Lemma 2, we compute a solution P whose value is

$$\mathsf{val}(P) \leq 2\alpha\,\mathsf{OPT}(I') + 4\,\mathsf{OPT}(I) \leq 4\alpha\,\mathsf{OPT}(I) + 4\,\mathsf{OPT}(I). \qquad \square$$

4 Multiple Constrained Forest Problem (MCFP)

An instance of MCFP consists of a set of cities $V = \{1, 2, \ldots, n\}$; a set of cable types $C = \{1, 2, \ldots, r\}$; a switch cost $g \geq 0$; and an integer number $k \geq 1$. Each cable type $i \in C$ is associated with a distance function d_i. A *spanning forest* is a collection of sub-trees $\mathcal{T} = \{T_1, T_2, \ldots, T_s\}$ such that each city is in at least one sub-tree. The join of these sub-trees can form disjoint-connected components. We say that a spanning forest is *constrained* if each connected component has at least k vertices. A solution of MCFP is a constrained spanning forest $\mathcal{T}$ and an assignment φ such that each tree T in $\mathcal{T}$ is associated with cable type $\varphi(T) \in C$. For each tree $T \in \mathcal{T}$, denote by $L(T)$ the length of T, i.e., let $L(T) = \sum_{(u,v)\in E(T)} d_{\varphi(T)}(u, v)$. The objective is to find a solution which minimizes the cost of $\mathcal{T}$ defined as $\mathsf{val}(\mathcal{T}) = |\mathcal{T}|\cdot g + \sum_{T\in\mathcal{T}} L(T)$. Note that CFP is the particular case in which $|C| = 1$ and $g = 0$.

Consider an instance I for MCFP, and let $\mathrm{d}_{\min}(u, v)$ denote the cheapest weight between two cities u and v among all cable types, i.e., for each $u, v \in V$, we define $\mathrm{d}_{\min}(u, v) = \min\{d_i(u, v) : i \in C\}$. The *bottleneck graph* of I is the graph on V, such that, for each pair $u, v \in V$, there is an edge between them if $\mathrm{d}_{\min}(u, v) \le g$. Denote by $\mathsf{OPT}(I)$ the value of an optimal solution for I.

Lemma 3. *Let G be the bottleneck graph of I, and $S_1, S_2, \ldots, S_h$ be the sets of vertices corresponding to connected components of G. There exists a solution $\mathcal{T}$ for I with $\mathsf{val}(\mathcal{T}) \le 7\,\mathsf{OPT}(I)$ and such that,*

(i) vertices in S_ℓ are contained in a connected component which is induced by joining trees in $\mathcal{T}$, for each $1 \le \ell \le h$; and
(ii) each tree in $\mathcal{T}$ is contained in a set S_ℓ for some ℓ or has exactly one edge.

Proof. Consider an optimal solution $\mathcal{T}^*$ and denote by R the set of all edges of trees in $\mathcal{T}^*$ which have extremes in two distinct sets S_ℓ and S'_ℓ. Also, let $\mathcal{C}^*$ be the collection of sets corresponding to the connected components which are induced by joining the trees of $\mathcal{T}^*$. We say that a *sub-component* of a component C in $\mathcal{C}^*$ is a set of vertices which corresponds to a connected component of the graph obtained from $G[C]$ by removing edges R. Notice that a sub-component is contained in some set S_ℓ.

In Fig. 3, we illustrate the sub-components corresponding to a bottleneck graph G. A set S_ℓ is represented by a dashed circle and a sub-component is represented by a solid circle. The circles corresponding to sub-components of a component in $\mathcal{C}^*$ are filled using a common pattern. Observe that the sub-components of a single component are connected by edges of R. Also, each edge $(u, v) \in R$ is such that $\mathrm{d}_{\min}(u, v) > g$.

We will modify the solution $\mathcal{T}^*$ and construct the required solution $\mathcal{T}$. To satisfy property (i), we connect sub-components of the set S_ℓ for each $1 \le \ell \le h$. Let $D_1, D_2, \ldots, D_k$ be the sub-components contained in S_ℓ, and create a graph H_ℓ as follows: start with the induced graph $G[S_\ell]$, then contract each set of vertices D_t, for $1 \le t \le k$. Since $G[S_\ell]$ is connected, so is H_ℓ, and we can find an arbitrary spanning tree H'_ℓ whose edges connect all the sub-components in S_ℓ.

For each $1 \le \ell \le h$, an edge $(u, v) \in E(H'_\ell)$ corresponds to an edge of G, thus $\mathrm{d}_{\min}(u, v) \le g$. For each such an edge, let $i \in C$ be a cable type with $d_i(u, v) = \mathrm{d}_{\min}(u, v)$, and create a new tree T consisting only of (u, v) and associated with cable type i. Note that the cost of each created tree T is at most

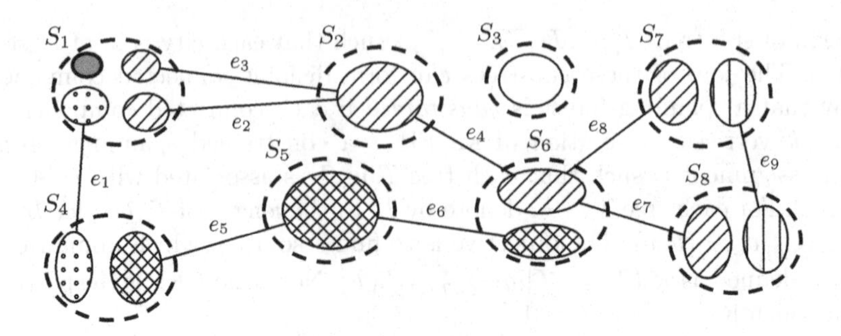

Fig. 3. Sub-components.

$g + L(T) \leq 2g$. The total number of trees added is at most the number of sub-components, and thus is bounded by the number of edges in R plus the number of components of $\mathcal{C}^*$. Then, the solution cost increases by at most $(|R|+|\mathcal{C}^*|)\cdot 2\,g$.

Now, to satisfy property (ii), we break trees containing edges in R. For each edge (u, v) in R, find the tree T in $\mathcal{T}$ which contains this edge, and let $i \in C$ be the cable type associated with T. Note that the graph $T - (u, v)$ is a forest composed of two trees, T_1 and T_2. We remove T from $\mathcal{T}$ and replace it by other trees, all associated with the same cable type i: T_1, T_2 and a new tree composed only of (u, v). Note that the value of the solution increases by $2g$, since the number of trees increase by 2, but the total length of the solution has not changed. Then, the solution cost increases by $|R| \cdot 2\,g$.

Finally, the solution $\mathcal{T}$ satisfy (i) and (ii). Since for each edge $(u, v) \in R$, we have $\mathrm{d_{min}}(u, v) > g$, the total increase is bounded by

$$\begin{aligned}(|R| + |\mathcal{C}^*|) \cdot 2\,g + |R| \cdot 2\,g &= 4|R| \cdot g + 2|\mathcal{C}^*| \cdot g \\ &\leq 4 \textstyle\sum_{(u,v)\in R} \mathrm{d_{min}}(u, v) + 2|\mathcal{C}^*| \cdot g \\ &\leq 4\,\mathsf{OPT}(I) + 2\,\mathsf{OPT}(I).\end{aligned}$$ □

The solution given in Lemma 3 induces a solution of UCaRS for an instance whose set of cities corresponds to S_ℓ. Thus, by solving UCaRS for each S_ℓ, we are left with a spanning forest whose components may have less than k vertices. To fix this, we create an instance of the Constrained Forest Problem (CFP).

Theorem 3. *If there exists an α-approximation for UCaRS, then there exists an $\mathcal{O}(\alpha)$-approximation for MCFP.*

Proof. Consider an instance I of MCFP and let $S_1, S_2, \ldots, S_h$ and $\mathcal{T}$ be as defined by Lemma 3. For each $1 \leq \ell \leq h$, let $\mathcal{T}_\ell$ be the set of trees in $\mathcal{T}$ whose vertices are contained in S_ℓ. For each ℓ, construct an instance I_ℓ of UCaRS whose set of cities is S_ℓ, the set of cars is C, and the return fee is g.

We claim that $\mathsf{OPT}(I_\ell) \leq 2\,\mathsf{val}(\mathcal{T}_\ell)$. To show this, we note that trees in $\mathcal{T}_\ell$ induce a solution P for I_ℓ. For each $T \in \mathcal{T}_\ell$, construct a closed walk D on the vertices of T by doubling the edges of T and finding an Eulerian tour.

Each edge of D is associated with car $\varphi(T)$. By joining all the closed walks, we obtain a closed walk P which visits all cities in S_ℓ and whose length is $L(P) \le 2\sum_{T\in\mathcal{T}_\ell} L(T)$. Note that, for each joined closed walk, we add at most two car exchanges in P, and thus $M(P) \le 2|\mathcal{T}_\ell|$. Therefore, $\mathsf{OPT}(I_\ell) \le \mathsf{val}(P) = M(P)\cdot g + L(P) \le 2|\mathcal{T}_\ell|\cdot g + 2\sum_{T\in\mathcal{T}_\ell} L(T) = 2\,\mathsf{val}(\mathcal{T}_\ell)$.

Now, create an instance J of CFP with the same parameters V and k as the instance I, i.e., the objective is to find a spanning forest on V such that each connected component has at least k vertices. Instance J has switch cost zero and only one cable type, whose corresponding distance function is d', defined next. For each pair $u,v \in V$, $d'(u,v) = 0$ if $u,v \in S_\ell$ for some $1 \le \ell \le h$; otherwise, $d'(u,v) = \mathrm{d_{min}}(u,v)$. This completes the description of J.

We claim that $\mathsf{OPT}(J) \le \mathsf{val}(\mathcal{T})$. Let $\mathcal{R}$ be the set of trees T in $\mathcal{T}$ whose vertices are not contained in S_ℓ for any ℓ. Property (ii) of Lemma 3 implies that each tree of $\mathcal{R}$ is an edge between two sets S_ℓ and $S_{\ell'}$. Let R be the set of all such edges. Observe that, for each $(u,v)\in R$, we have $\mathrm{d_{min}}(u,v) > g$, as (u,v) connects two connected components of G. Thus, $|R|\cdot g < \sum_{(u,v)\in R} \mathrm{d_{min}}(u,v) \le \mathsf{val}(\mathcal{T})$. Moreover, a feasible solution for J can be built joining R and edges with weight zero, thus $\mathsf{OPT}(J) \le \sum_{(u,v)\in R} \mathrm{d_{min}}(u,v) \le \mathsf{val}(\mathcal{T})$.

We execute the following algorithm to find a solution for I. First, build a set of trees $\mathcal{P}$ which connect each set S_ℓ. To this, find a closed walk P_ℓ which is an α-approximation for each instance I_ℓ of UCaRS. Recall that P_ℓ is a sequence of $M(P_\ell)$ paths. Each such a path D is also a tree and is associated with a given car i, thus define $\varphi(D) = i$ and add D to $\mathcal{P}$. Note that $\mathsf{val}(\mathcal{P})$ is bounded by

$$\sum_{\ell=1}^{h}(M(P_\ell)\cdot g + L(P_\ell)) \le \sum_{\ell=1}^{h} \alpha\,\mathsf{OPT}(I_\ell) \le 2\alpha\sum_{\ell=1}^{h} \mathsf{val}(\mathcal{T}_\ell) \le 2\alpha\,\mathsf{val}(\mathcal{T}).$$

The trees in $\mathcal{P}$ induce a spanning forest, but there may be components with less than k vertices. Then, we find a complementary set of trees $\mathcal{F}$ to connect small components. To this, obtain a solution F which is a 3/2-approximation algorithm for the instance J of CFP [4]. Now, consider an each edge (u,v) of F. If $u,v\in S_\ell$ for some ℓ, then vertices u and v are already connected by the trees in $\mathcal{P}$, thus the edge is not necessary. Otherwise, $d'(u,v) = \mathrm{d_{min}}(u,v)$, thus there is a cable type i such that $\mathrm{d_{min}}(u,v) = d_i(u,v)$. We create a tree D consisting of edge (u,v), define $\varphi(D) = i$ and add D to $\mathcal{F}$. Let E' be the set edges of F which were not discarded. Note that for each $(u,v)\in E'$, $d'(u,v) = \mathrm{d_{min}}(u,v) > g$. Thus, $\mathsf{val}(\mathcal{F})$ is bounded by

$$\sum_{(u,v)\in E'}(g + d'(u,v)) \le 2(\sum_{(u,v)\in E'} d'(u,v)) \le 2(3/2\,\mathsf{OPT}(J)) \le 3\,\mathsf{val}(\mathcal{T}).$$

Since F is a feasible solution for J, we conclude that $\mathcal{P}\cup\mathcal{F}$ is a feasible solution for I with value $\mathsf{val}(\mathcal{P}) + \mathsf{val}(\mathcal{F}) \le 2\alpha\,\mathsf{val}(\mathcal{T}) + 3\,\mathsf{val}(\mathcal{T}) \le (2\alpha+3)7\,\mathsf{OPT}(I)$. □

5 Final Remarks

Since there is an $\mathcal{O}(\log n)$-approximation for UCaRS [12], the results above imply that there is an $\mathcal{O}(\log n)$-approximation for SUCaRS, PUCaRS and MCFP.

These factors are asymptotically tight unless P = NP, because a $o(\log n)$-approximation for any of them would imply a $o(\log n)$-approximation for UCaRS. This is clear for SUCaRS and PUCaRS, as they generalize UCaRS. For MCFP, note that a solution $\mathcal{T}$ for MCFP with $k = |V|$ can be converted into a solution for UCaRS whose value is within a constant factor of $\mathsf{val}(\mathcal{T})$.

The developed techniques can be extended to other routing and connectivity problems. As an example, an $\mathcal{O}(\log n)$-approximation can be obtained for generalized versions of the Steiner Tree Problem or the Prize-Collecting Steiner Tree Problem. However, the techniques seem insufficient to tackle the generalization of certain cardinality problems, such as k-MST [13], which asks for one minimum-cost tree spanning k vertices. We left open whether the generalization of k-MST with multiple distances admits an $\mathcal{O}(\log n)$-approximation.

References

1. Balas, E.: The prize collecting traveling salesman problem. Networks **19**(6), 621–636 (1989)
2. Bienstock, D., Goemans, M.X., Simchi-Levi, D., Williamson, D.: A note on the prize collecting traveling salesman problem. Math. Progr. **59**(1–3), 413–420 (1993)
3. Christofides, N.: Worst-case analysis of a new heuristic for the travelling salesman problem. Technical report, Carnegie-Mellon Univ Pittsburgh (1976)
4. Couëtoux, B.: A 3/2-approximation for a constrained forest problem. In: Proceedings of 19th Annual European Symposium on Algorithms, vol. 6942, pp. 652–663 (2011)
5. Dell'Amico, M., Maffioli, F., Värbrand, P.: On prize-collecting tours and the asymmetric travelling salesman problem. Int. Trans. Oper. Res. **2**(3), 297–308 (1995)
6. Goemans, M.X., Williamson, D.P.: A general approximation technique for constrained forest problems. SIAM J. Comput. **24**(2), 296–317 (1995)
7. Goldbarg, M.C., Asconavieta, P.H., Goldbarg, E.F.G.: Memetic algorithm for the traveling car renter problem: an experimental investigation. Memetic Comput. **4**(2), 89–108 (2011)
8. Imielińska, C., Kalantari, B., Khachiyan, L.: A greedy heuristic for a minimum-weight forest problem. Oper. Res. Lett. **14**(2), 65–71 (1993)
9. Karp, R.M.: Reducibility among combinatorial problems. In: Miller, R.E., Thatcher, J.W., Bohlinger, J.D. (eds.) Complexity of Computer Computations, pp. 85–103. Springer, Boston (1972). https://doi.org/10.1007/978-1-4684-2001-2_9
10. Letchford, A.N., Nasiri, S.D., Theis, D.O.: Compact formulations of the Steiner traveling salesman problem and related problems. Eur. J. Oper. Res. **228**(1), 83–92 (2013)
11. Orloff, C.S.: A fundamental problem in vehicle routing. Networks **4**(1), 35–64 (1974)
12. Pedrosa, L.L.C., Quesquén, G.Y.O., Schouery, R.C.S.: An asymptotically optimal approximation algorithm for the travelling car renter problem. In: Proceedings of the 19th Symposium on Algorithmic Approaches for Transportation Modelling, Optimization, and Systems, vol. 75, pp. 14:1–14:15. Schloss Dagstuhl (2019)
13. Ravi, R., Sundaram, R., Marathe, M.V., Rosenkrantz, D.J., Ravi, S.S.: Spanning trees—short or small. SIAM J. Discrete Math. **9**(2), 178–200 (1996)

14. Research, Markets: Car rental market – growth, trends, and forecast (2019–2024) (2019). https://www.researchandmarkets.com/
15. Sorupia, E.: Rethinking the role of transportation in tourism. In: Proceedings of the Eastern Asia Society for Transportation Studies, vol. 5, pp. 1767–1777 (2005)
16. Travel, W., (WTTC), T.C.: Travel & tourism economic impact 2019 world (2018). https://www.wttc.org/

A 2-Approximation for the k-Prize-Collecting Steiner Tree Problem

Lehilton L. C. Pedrosa and Hugo K. K. Rosado(✉)

Institute of Computing, University of Campinas, Campinas, Brazil
{lehilton,hugo.rosado}@ic.unicamp.br

Abstract. We consider the k-prize-collecting Steiner tree problem. An instance is composed of an integer k and a graph G with costs on edges and penalties on vertices. The objective is to find a tree spanning at least k vertices which minimizes the cost of the edges in the tree plus the penalties of vertices not in the tree. This is one of the most fundamental network design problems and is a common generalization of the prize-collecting Steiner tree and the k-minimum spanning tree problems. Our main result is a 2-approximation algorithm, which improves on the currently best known approximation factor of 3.96 and has a faster running time. The algorithm builds on a modification of the primal-dual framework of Goemans and Williamson, and reveals interesting properties that can be applied to other similar problems.

Keywords: Approximation algorithm · Primal-dual · k-prize-collecting Steiner tree · k-MST · Prize-collecting steiner tree

1 Introduction

In many network design problems, the input consists of an edge-weighted graph, and the output is a minimum-cost tree connecting a certain subset of vertices. Two of the most fundamental NP-hard variants are the *prize-collecting Steiner tree* (PCST) and the *k-minimum spanning tree* (k-MST). For PCST, a solution may contain any subset of vertices, but any not spanned vertex incurs a penalty which is added to the objective function. For k-MST, the output tree is required to contain at least k vertices.

We consider the *k-prize-collecting Steiner tree problem* (k-PCST), which is a common generalization of PCST and k-MST. An instance consists of a connected undirected graph $G = (V, E)$, a root vertex r, and a non-negative integer $k \leq |V|$. Each edge $e \in E$ has a non-negative cost c_e, and each vertex $v \in V$ has a non-negative penalty π_v. A solution is a tree spanning at least k vertices, including r, and minimizing the cost of edges of the tree plus the penalties of vertices not spanned by the tree. Without loss of generality, we assume that $\pi_r = \infty$.

Supported by São Paulo Research Foundation (FAPESP) grant #2015/11937-9 and National Council for Scientific and Technological Development (CNPq) grants #425340/2016-3, #313026/2017-3, #422829/2018-8, and #140552/2019-7.

Y. Kohayakawa and F. K. Miyazawa (Eds.): LATIN 2020, LNCS 12118, pp. 76–88, 2020.
https://doi.org/10.1007/978-3-030-61792-9_7

The particular case in which $k = 0$ is PCST. This is the prize-collecting problem considered by Goemans and Williamson, who presented a 2-approximation based on a seminal primal-dual scheme for constrained forest problems [6]. The currently best-known factor is $1.96 + \epsilon$ by Archer et al. [1]. The particular case in which $\pi_v = 0$ for every vertex v is k-MST, and the first constant-factor approximation is a primal-dual algorithm by Blum et al. [2]. This algorithm has been improved by a series of works, leading to a 2-approximation by Garg [5], which is based on a sophisticated use of the primal-dual scheme.

To our knowledge, the first constant-factor approximation for k-PCST is due to Han et al. [7] and has factor 5. They gave a primal-dual algorithm based on the Lagrangean relaxation of a linear program. Later, Matsuda and Takahashi [9] derived a 4-approximation by combining solutions for the underlying instances of PCST and k-MST. The algorithm's running time is $\mathcal{O}(|V|^4|E|\log|V|)$ and is bottlenecked by Garg's 2-approximation, which is used to solve k-MST. By using the $1.96+\epsilon$-approximation for PCST, the approximation factor for k-PCST can be improved to $3.96 + \epsilon$, with a significant increase in the running time.

Our main contribution is a 2-approximation for k-PCST. More precisely, we present an algorithm with running time $\mathcal{O}(|V|^2|E|^2 + |V|^4\log^2|V|)$ that finds a tree T such that $\sum_{e\in E(T)} c_e + 2 \cdot \sum_{v \in V\setminus V(T)} \pi_v \le 2 \cdot \mathrm{opt}$, where opt is the optimal value. This improves on both the approximation factor and the time complexity of the previously best-known algorithms. Our 2-approximation is based on a modified version of the Goemans and Williamson's algorithm, and our analysis reveals many interesting properties of the primal-dual scheme, which might give insights to other problems with similar constraints.

Johnson et al. [8] also considered the quota version of k-MST, which asks for a minimum-cost tree with any number of vertices, but whose vertex weight is at least some given quota. A small modification of our algorithm also leads to a 2-approximation for the quota variant of k-PCST.

Algorithm's Overview. Our algorithm successively executes a modified version of the primal-dual scheme for PCST due to Goemans and Williamson [6]. Their algorithm is divided into a *growth-phase* and a *pruning-phase*. In the growth-phase, it computes a feasible dual solution y such that, for each subset of vertices S, y_S is a non-negative value. It also outputs a tree T and a collection $\mathcal{B}$ of subsets of V, whose edges and subsets correspond to tight dual inequalities of an LP formulation. In the pruning-phase, the algorithm deletes from T the subsets of vertices in $\mathcal{B}$ which do not disconnect the graph, resulting in a pruned tree $\widehat{T}$. To derive a 2-approximation, they bound the value of $\widehat{T}$ by a factor of the dual objective function; in our algorithm, we compare it with an optimal solution directly.

In our modification, the growth-phase receives two new arguments, a potential λ and a tie-breaking list τ. The potential is a uniform increase on the penalties of each vertex. For larger values of λ, the output tree $\widehat{T}$ spans more vertices. During the growth-phase, there might be concurrent events, thus there are multiple choices for the execution path. Usually, these choices are determined by some fixed lexicographic order. Our algorithm, on the contrary, relies on a tie-breaking list τ to control the priority among concurrent events. The i-th element in this list dictates which event gets the highest priority in the i-th iteration of the algorithm. This allows us to control the execution path of the algorithm.

The use of potential λ is built on Garg's arguments for the 2-approximation for k-MST [5], which can be described as follows. If, for some λ, the pruned tree $\widehat{T}$ spans exactly k vertices, this leads to a 2-approximation by using the Lagrangean relaxation strategy (see, e.g., [3]). However, it might be the case that no such λ exists; thus, the idea is to find a particular value of λ such that, for sufficiently small ϵ, using potential $\lambda - \epsilon$ leads to a pruned tree $\widehat{T}_-$ spanning less than k vertices, and using potential $\lambda + \epsilon$ leads to a pruned tree $\widehat{T}_+$ spanning at least than k vertices. The tree $\widehat{T}_-$ is constructed by pruning a tree T_- using a collection $\mathcal{B}_-$. Similarly, $\widehat{T}_+$ is constructed by pruning a tree T_+ using a collection $\mathcal{B}_+$. On the one hand, $\widehat{T}_+$ is a feasible solution, but its cost cannot be bounded in terms of vector y. On the other hand, the value of $\widehat{T}_-$ can be bounded, but it is not a feasible solution.

In Garg's algorithm, the trees T_- and T_+ and the collections $\mathcal{B}_-$ and $\mathcal{B}_+$ might be very different. Thus, to obtain a tree with k vertices and whose cost can be bounded, his algorithm iteratively transforms T_- into T_+ and $\mathcal{B}_-$ into $\mathcal{B}_+$ by replacing one edge of T_- or one subset in $\mathcal{B}_-$ at a time. At some iteration, pruning the current tree using the current collection must result in a tree spanning at least k vertices. Before this step, instead of performing the operation, one augments the current pruned tree by adding a sequence of edges whose corresponding dual restrictions are tight, picking up to k vertices.

Our algorithm also considers similar trees T_- and T_+ and corresponding collections $\mathcal{B}_-$ and $\mathcal{B}_+$. However, both trees are constructed by executing the growth-phase using a single potential λ. To differentiate between the cases, we take into account a tie-braking list τ and its maximal proper prefix $\widetilde{\tau}$. We show how to compute a special tuple (λ, τ), called the *threshold-tuple*, such that executing the growth-phase using τ results in a pruned tree with less than k vertices, while using $\widetilde{\tau}$ results in a pruned tree with at least k vertices (or vice-versa).

We show that the trees and collections output by the two executions of the growth-phase are only slightly different. Indeed, a key ingredient of our analysis (given in Lemma 9) shows that one of the two following scenarios hold:

(i) $\mathcal{B}_-$ and $\mathcal{B}_+$ are equal, and trees T_- and T_+ differ in exactly one edge; or
(ii) T_- and T_+ are equal, and collections $\mathcal{B}_-$ and $\mathcal{B}_+$ differ in exactly one subset.

Moreover, we show that the vector y output in both executions of the growth-phase are the same. This leads to a straightforward way of augmenting the pruned tree $\widehat{T}_-$, by picking a sequence of edges of T_- or T_+ whose corresponding inequalities are tight, without the need for a step-by-step transformation.

Although the computed y satisfy a set of inequalities, these inequalities do not correspond to an LP dual formulation for k-PCST, hence we cannot use weak duality to bound the value of an optimal solution. Instead, we show (in Lemma 15) that either our algorithm returns a 2-approximate solution, or it identifies a non-empty subset of the vertices which are not spanned by any optimal solution. Thus, we either find the desired solution, or can safely reduce the size of the instance. Therefore, by running the algorithm at most $|V| - k$ times, we find a 2-approximation.

Text Organization. In Sect. 2, we introduce the terminology used in the text. In Sect. 3, we detail the modified primal-dual scheme. In Sect. 4, we formally define a threshold-tuple, and show how it can be computed. In Sect. 5, we show that, given a threshold-tuple, one can build a tree spanning exactly k vertices. In Sect. 6, we bound the cost of the computed tree and give our 2-approximation.

2 Definitions and Preliminaries

We say that a nonempty collection $\mathcal{L} \subseteq 2^V$ is *laminar* if, for any $L_1, L_2 \in \mathcal{L}$, either $L_1 \cap L_2 = \emptyset$, or $L_1 \subseteq L_2$, or $L_2 \subseteq L_1$. A laminar collection $\mathcal{L}$ is *binary* if for every $L \in \mathcal{L}$ with $|L| \geq 2$, there are non-empty disjoint $L_1, L_2 \in \mathcal{L}$ such that $L = L_1 \cup L_2$. We denote the collection of inclusion-wise maximal subsets of a collection $\mathcal{L}$ by $\mathcal{L}^*$. Observe that subsets in $\mathcal{L}^*$ are disjoint if $\mathcal{L}$ is laminar.

Let $\mathcal{P}$ be a partition of V, and consider an edge e with extremes on V. If a set of $\mathcal{P}$ contains an extreme of e, then we call this set an *endpoint* of e. We say that an edge e is *internal* in $\mathcal{P}$ if e has only one endpoint, and we say that e is *external* in $\mathcal{P}$ if e has two distinct endpoints. Also, two external edges are said to be *parallel* in $\mathcal{P}$ if they have the same pair of endpoints.

Given a graph H and a subset $L \subseteq V$, we say that H is *L-connected* if $V(H) \cap L = \emptyset$ or if the induced subgraph $H[V(H) \cap L]$ is connected. For some collection $\mathcal{L}$ of subsets of the vertices, we say that H is *$\mathcal{L}$-connected* if H is L-connected for every $L \in \mathcal{L}$. Moreover, we denote by $\delta_H(L)$ the set of edges of H with exactly one extreme in L, and say that L has *degree* $|\delta_H(L)|$ on H. For the case in which $H = G$, we drop the subscript and write just $\delta(L)$.

For a subset $L \subseteq V$, its new penalty is defined as $\pi_L^\lambda = \sum_{v \in L} \pi_v + \lambda |L|$, where λ is a non-negative value which we call *potential.* Note that, for any subset L containing r, we have $\pi_L^\lambda = \infty$. Let y be a vector such that, for each $L \subseteq 2^V$, the entry y_L is a non-negative variable. We say that y *respects* c and π^λ if

$$\sum_{L : e \in \delta(L)} y_L \leq c_e \qquad \text{for every edge } e \in E, \text{ and} \qquad (1)$$

$$\sum_{S : S \subseteq L} y_S \leq \pi_L^\lambda \qquad \text{for every subset } L \subseteq V. \qquad (2)$$

We say that an edge e is *tight* for (y, λ) if the inequality corresponding to e in (1) holds with equality. Analogously, a subset $L \subseteq V$ is tight for (y, λ) if the inequality corresponding to L in (2) is satisfied with equality. If the pair (y, λ) is clear from context, we simply say that e and L are tight. The inequalities are similar to the ones in the dual formulation for PCST [6], with the difference that these include terms for subsets containing r. However, these inequalities do not correspond to the dual of an LP formulation for k-PCST.

Denote by $\mathcal{L}(S)$ the collection of subsets in $\mathcal{L}$ which contain some, but not all vertices of S. Moreover, let $c_{E'} = \sum_{e \in E'} c_e$ for $E' \subseteq E$, and $\pi_L = \sum_{v \in L} \pi_e$ for $L \subseteq V$. To bound the value of an optimal solution T^*, we use next lemma.

Lemma 1. *Let L^* be the minimal subset containing $V(T^*)$ in a laminar collection $\mathcal{L}$. If y respects c and π^λ, then*

$$\textstyle\sum_{L \in \mathcal{L}(L^*)} y_L - \lambda|L^* \setminus V(T^*)| \leq c_{E(T^*)} + \pi_{L^* \setminus V(T^*)}.$$

3 Modified Growth and Pruning Phases

In the following we detail our modification of the primal-dual scheme due to Goemans and Williamson for the prize-collecting Steiner tree problem [6]. The algorithm is composed of two main routines: a clustering algorithm, also known as the growth-phase, and a cleanup algorithm, known as the pruning-phase.

3.1 Modified Clustering Algorithm

The modified *growth-phase* is denoted by $\mathtt{GP}(\lambda, \tau)$. The algorithm maintains a binary laminar collection $\mathcal{L} \subseteq 2^V$, such that $\mathcal{L}^*$ partitions V, and a vector y which respects c and π^λ. It iteratively constructs a forest $F \subseteq G$ and a subcollection of *processed* subsets $\mathcal{B} \subseteq \mathcal{L}$, such that edges of F and subsets in $\mathcal{B}$ are tight for (y, λ). In each iteration, either an edge is added to F, or a subset is included in $\mathcal{B}$.

The algorithm begins by defining $\mathcal{L} = \{\{v\} : v \in V\}$ and $y_S = 0$, for each $S \subseteq V$ (implicitly), and by letting $F = (V, \emptyset)$, and $\mathcal{B} = \emptyset$. Once initialized, it starts the iteration process. At a given moment, a maximal subset $L \in \mathcal{L}^*$ is said to be *active* if it has not been processed yet, i.e., if $L \in \mathcal{L}^* \setminus \mathcal{B}$. In each iteration, we increase the variable y_L of every active subset L uniformly until one of the following events occur:

- ▷ an external edge e with endpoints $L_1, L_2 \in \mathcal{L}^*$ becomes tight, in which case e is added to F, and the union $L_1 \cup L_2$ is included in $\mathcal{L}$; or
- ▷ an active subset L becomes tight, in which case L is included in $\mathcal{B}$.

We note that multiple edges and subsets might become tight simultaneously. In our modified algorithm, we use the *tie-breaking list* τ to decide the order in which the events are processed. A tie-breaking list τ with size $|\tau|$ is a (possibly empty) sequence of edges and subsets. For each $i = 1, 2, \ldots, |\tau|$, the i-th element of the list is denoted by τ_i. In iteration i, the event to be processed is determined according to the following order:

(i) if $i \leq |\tau|$, then the event corresponding to τ_i has the highest priority;
(ii) followed by events corresponding to edges;
(iii) and finally by events corresponding to subsets.

Priority between events of the same type is determined by a lexicographic order.

The algorithm stops when V is the only active subset in $\mathcal{L}^*$, at which point F is a tree. Then, it defines $T = F$ and outputs the pair $(T, \mathcal{B})$. Next lemma collects basic invariants of the growth-phase.

Lemma 2. *At the beginning of any iteration of* $\mathtt{GP}(\lambda, \tau)$*, the following holds:*

(gp1) $\mathcal{L}$ *is a binary laminar collection,* $\bigcup_{L \in \mathcal{L}^*} L = V$*, and* $\emptyset \notin \mathcal{L}$*;*
(gp2) y *respects* c *and* π^λ*;*
(gp3) F *is an* $\mathcal{L}$*-connected forest and every edge* $e \in E(F)$ *is tight for* (y, λ)*;*
(gp4) $\mathcal{B} \subseteq \mathcal{L}$*, and every* $B \in \mathcal{B}$ *is tight for* (y, λ)*, and no* $B \in \mathcal{B}$ *contains* r*.*

3.2 Modified Pruning Algorithm

The modified *pruning-phase* is denoted by $\mathtt{PP}(H, \mathcal{B})$. The algorithm receives a graph H and exhaustively deletes from it processed subsets $B \in \mathcal{B}$ with degree one on H. We say that we *prune* H using $\mathcal{B}$ to mean that we execute algorithm $\mathtt{PP}(H, \mathcal{B})$, and say that a graph H is *pruned* with $\mathcal{B}$ if $|\delta_H(B)| \neq 1$ for every $B \in \mathcal{B}$. We assume that H is connected and contains r, and that $\mathcal{B}$ is a laminar collection of subsets of $V \setminus \{r\}$. Thus, it outputs a connected graph containing r.

Lemma 3. *At the beginning of any iteration of* $\mathtt{PP}(H, \mathcal{B})$*, the following holds:*

(pp1) H *is connected and* $r \in V(H)$*;* *(pp2)* $H[V(H) \cap B]$ *is connected.*

Next lemma implies that the algorithm behaves as a monotonic operation over the input. As a consequence, pruning using a fixed collection $\mathcal{B}$ is monotonic with respect to the subgraph relation, and pruning a fixed graph H is monotonically reversing with respect to the subcolletion relation.

Lemma 4. *Consider connected graphs* D *and* H*. Assume that* D *is pruned with* $\mathcal{B}$ *and let* H' *be the graph output by* $\mathtt{PP}(H, \mathcal{B})$*. If* $D \subseteq H$*, then* $D \subseteq H'$*.*

Corollary 5. *Let* D' *and* H' *be the graphs output by* $\mathtt{PP}(D, \mathcal{B})$ *and* $\mathtt{PP}(H, \mathcal{B})$*, respectively. If* $D \subseteq H$*, then* $D' \subseteq H'$*; and if* $H' \subseteq D \subseteq H$*, then* $H' = D'$*.*

Corollary 6. *Let* H_1 *and* H_2 *be the graphs output by* $\mathtt{PP}(H, \mathcal{B}_1)$ *and* $\mathtt{PP}(H, \mathcal{B}_2)$*, respectively. If* $\mathcal{B}_1 \subseteq \mathcal{B}_2$*, then* $H_2 \subseteq H_1$*.*

3.3 Modified Goemans-Williamson Algorithm

The modified *Goemans-Williamson* algorithm wraps up the growth and the pruning-phases, and it is denoted by $\mathtt{GW}(\lambda, \tau)$. First, the algorithm executes $\mathtt{GP}(\lambda, \tau)$ to obtain a pair $(T, \mathcal{B})$. Then, it executes $\mathtt{PP}(T, \mathcal{B})$ and returns the pruned tree $\widehat{T}$. One important property of $\mathtt{GW}(\lambda, \tau)$ is that, if $\lambda > c_E$, then no subset is ever processed, and the output tree $\widehat{T}$ spans the whole set of vertices V.

4 The Threshold-Tuple

We run the modified algorithm using potential zero and passing an empty tie-breaking list, i.e., we run $\mathtt{GW}(0, \emptyset)$. Observe that the returned tree $\widehat{T}$, is a 2-approximation for the corresponding PCST instance [4,8]. Thus, if $\widehat{T}$ spans at least k vertices, it is also a 2-approximation for k-PCST.

Lemma 7. *If executing* $\texttt{GW}(0, \emptyset)$ *outputs a tree* $\widehat{T}$ *such that* $V(\widehat{T}) \geq k$, *then* $c_{E(\widehat{T})} + 2\pi_{V \setminus V(\widehat{T})} \leq 2\left(c_{E(T^*)} + \pi_{V \setminus V(T^*)}\right)$.

In the remaining of this section, we assume that executing $\texttt{GW}(0, \emptyset)$ returns a tree spanning less than k vertices. We would like to find λ and associated τ such that the returned tree spans exactly k vertices, but it might be the case that no such pair exists. Instead, our goal will be finding a special tuple (λ, τ), called the *threshold-tuple*, which will be defined below.

First, we need to introduce some notation. Note that, in the i-th iteration of $\texttt{GP}(\lambda, \tau)$, the edge or subset corresponding to τ_i is not necessarily tight. We say that a tie-breaking list τ is *respected* by potential λ if the sequence of edges and subsets of V processed in the first $|\tau|$ iterations of $\texttt{GP}(\lambda, \tau)$ corresponds to τ. Also, we denote by $\widetilde{\tau}$ the prefix of τ with size $|\tau| - 1$.

Definition 8. *Let* $\widehat{T}$ *be the tree returned by* $\texttt{GW}(\lambda, \tau)$, *and* $\widehat{T}'$ *be the tree returned by* $\texttt{GW}(\lambda, \widetilde{\tau})$. *We say that* (λ, τ) *is a* threshold-tuple *if*

(i) τ *is respected by* λ, *and* *(ii)* $|V(\widehat{T})| \geq k > |V(\widehat{T}')|$ *or* $|V(\widehat{T})| < k \leq |V(\widehat{T}')|$.

Next lemma summarizes the main properties of a threshold-tuple (λ, τ). It states that, although the pruned trees span different numbers of vertices, the difference in the outputs of growth-phases $\texttt{GP}(\lambda, \tau)$ and $\texttt{GP}(\lambda, \widetilde{\tau})$ is very small.

Lemma 9. *Let* (λ, τ) *be a threshold-tuple, and let* $\sigma = \tau_{|\tau|}$. *Also, let* T, $\mathcal{B}$ *and* y *be the output computed by* $\texttt{GP}(\lambda, \tau)$, *and let* T', $\mathcal{B}'$ *and* y' *be the output computed by* $\texttt{GP}(\lambda, \widetilde{\tau})$. *Then,* $y = y'$, *and*

(i) if σ *is an edge, then* $\mathcal{B} = \mathcal{B}'$, $\sigma \notin E(T')$, *and* $T \subseteq T' + \sigma$;
(ii) if σ *is a subset, then* $T = T'$, $\sigma \notin \mathcal{B}'$, *and* $\mathcal{B} = \mathcal{B}' \cup \{\sigma\}$.

Proof sketch. Assume σ is an edge and that a different edge σ' is processed in iteration $|\tau|$ of $\texttt{GP}(\lambda, \widetilde{\tau})$ (the other cases follow from similar arguments).

Since τ is respected by λ, so is $\widetilde{\tau}$. This implies that at the start of the $|\tau|$-th iteration, F, $\mathcal{L}$, $\mathcal{B}$, and y are the same in both executions of the growth-phase. Thus, the sets of tight external edges in these executions are also the same, say $\mathcal{Y}$. Because the priority of edges is higher, each edge in $\mathcal{Y}$ is either processed or becomes internal before any subset is processed. Let $(\sigma' = e'_0, e'_1, \ldots, e'_m)$ and $(\sigma = e_0, e_1, \ldots, e_m)$ be the sequences of edges in $\mathcal{Y}$ which are processed by $\texttt{GP}(\lambda, \widetilde{\tau})$ and by $\texttt{GP}(\lambda, \tau)$. Since edges are processed in a fixed lexicographic order, there is some index ℓ such that $e'_i = e_{i+1}$ for each $0 \leq i < \ell$, and such that e'_ℓ is parallel to e_0 at iteration $|\tau| + \ell$ of $\texttt{GP}(\lambda, \widetilde{\tau})$. Thus, at the end of iteration $|\tau| + \ell$ of both executions, vector y has not changed, and the collections of maximal subsets are the same. It follows that the succeeding iterations process the same sequence of edges and subsets, and the lemma follows when σ and σ' are distinct edges.

Figure 1 illustrates the execution of $\texttt{GP}(\lambda, \widetilde{\tau})$ and $\texttt{GP}(\lambda, \tau)$. Solid contours represent subsets in $\mathcal{L}$, and filled contours represent processed subsets in $\mathcal{B}$. In (a), we have the state at the beginning of iteration $|\tau|$, where only tight external edges are drawn and indices correspond to the lexicographic order. In (b), we show the resulting state after $\mathtt{GP}(\lambda, \widetilde{\tau})$ processed the sequence $(\sigma' = e_0, e_1, e_2, e_3, e_4)$, and, in (c), the state after $\mathtt{GP}(\lambda, \tau)$ processed the sequence $(\sigma = e_5, e_0, e_1, e_2, e_3)$. □

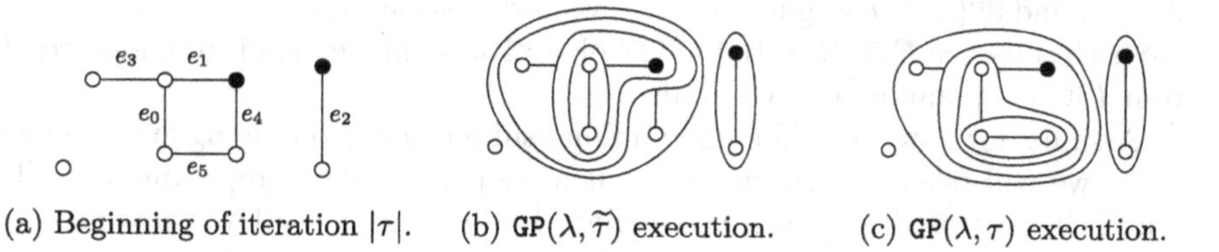

(a) Beginning of iteration $|\tau|$. (b) $\mathtt{GP}(\lambda, \widetilde{\tau})$ execution. (c) $\mathtt{GP}(\lambda, \tau)$ execution.

Fig. 1. Difference between the execution of $\mathtt{GP}(\lambda, \widetilde{\tau})$ and of $\mathtt{GP}(\lambda, \tau)$.

4.1 The Threshold-Tuple Search

To find a threshold-tuple, we run a search algorithm, denoted by TS. Start with an empty tie-breaking list τ, and initialize variables $a = 0$ and $b = c_E + 1$. We will extend τ by adding one entry per iteration and maintaining two invariants: (i) τ is respected by λ for every $\lambda \in [a, b]$; (ii) $\mathtt{GW}(a, \tau)$ returns a tree spanning less than k vertices, and $\mathtt{GW}(b, \tau)$ returns a tree spanning at least k vertices.

At the beginning of iteration i of TS, τ has size $i - 1$. Since τ is respected by λ for every $\lambda \in [a, b]$, we know that, at the beginning of iteration i of $\mathtt{GP}(\lambda, \tau)$, the collection of maximal subsets and the set of external edges are the same for every $\lambda \in [a, b]$. As a consequence, one can show that the increase on y variables necessary for each such edge or subset σ to become tight is a function $\epsilon_\sigma(\lambda)$ linear in $\lambda \in [a, b]$. We say that $p \in (a, b)$ is a *diverging potential* if $\epsilon_\sigma(p) = \epsilon_{\sigma'}(p)$ for distinct ϵ_σ and $\epsilon_{\sigma'}$. Using line-intersection algorithms, we can identify an edge or subset σ and consecutive diverging potentials a', b' such that $\mathtt{GW}(a', \tau)$ returns less than k vertices, $\mathtt{GW}(b', \tau)$ returns at least k vertices, and σ becomes tight at iteration i of $\mathtt{GP}(\lambda, \tau)$ for every $\lambda \in [a', b']$. We extend τ by defining $\tau_i = \sigma$ and, if either (a', τ) or (b', τ) is a threshold-tuple, then we are done. Otherwise, updating $a = a'$ and $b = b'$ maintains both invariants, and we repeat the process.

Observe that this procedure ends. Otherwise, the same sequence of edges and subsets is processed by $\mathtt{GP}(a, \tau)$ and $\mathtt{GP}(b, \tau)$, as the sequence is finite, and $\mathtt{GW}(a, \tau)$ and $\mathtt{GW}(b, \tau)$ would return identical trees, contradicting the invariants.

Lemma 10. *If executing* $\mathtt{GW}(0, \emptyset)$ *returns a tree spanning less than k vertices, then* TS *finds a threshold-tuple.*

5 Finding a Solution with a Threshold-Tuple

Assume that we are given a threshold-tuple (λ, τ). Then, by executing $\mathtt{GW}(\lambda, \tau)$ and $\mathtt{GW}(\lambda, \widetilde{\tau})$, we obtain two trees, one which spans less than k vertices, and the other, at least k vertices. Let $\widehat{T}_-$ and $\widehat{T}_+$ be the trees with less than k vertices, and at least k vertices, respectively, and let $(T_-, \mathcal{B}_-)$ and $(T_+, \mathcal{B}_+)$ denote the corresponding pairs computed in the growth-phase.

Denote by $\mathcal{L}_-$ and $\mathcal{L}_+$ the laminar collections computed in the growth-phase. While these laminar collections might be (slightly) different, Lemma 9 states that $\texttt{GP}(\lambda, \tau)$ and $\texttt{GP}(\lambda, \widetilde{\tau})$ compute identical vectors y. Moreover, each edge of $T_- \cup T_+$ and each subset in $\mathcal{B}_- \cup \mathcal{B}_+$ is tight. The objective of this section is to find a tree T from $T_- \cup T_+$ spanning at least k vertices.

Assume, for now, that V is the minimal subset in $\mathcal{L}_-$ containing every vertex of T^*; we will relax this assumption when we present the 2-approximation. To bound the cost of T^*, one can use Lemma 1, which gives the lower bound

$$\left(\sum_{L \in \mathcal{L}_-(V)} y_L - \lambda |V \setminus V(T^*)|\right) \leq c_{E(T^*)} + \pi_{V \setminus V(T^*)}.$$

To bound the cost of T, we need to prove an inequality analogous to the one given by Goemans and Williamson's analysis [6], i.e., we want to show that

$$c_{E(T)} + 2\pi_{V \setminus V(T)} \leq 2\left(\sum_{L \in \mathcal{L}_-(V)} y_L - \lambda |V \setminus V(T)|\right).$$

5.1 Selecting k Vertices

The previous inequalities suggest that it is sufficient to find a solution T such that $|V(T)| \leq |V(T^*)|$. But, since a feasible solution spans at least k vertices, our goal is restricted to computing a tree which spans exactly k vertices. Given a threshold-tuple, such a tree is constructed by the *picking-vertices* algorithm, denoted by $\texttt{PV}(\lambda, \tau)$. This algorithm follows some ideas due to Garg [5].

Let $H = T_- \cup T_+$ and $\widehat{H}$ be the resulting tree obtained by pruning H using $\mathcal{B}_+$. Let σ be the edge or subset processed at iteration $|\tau|$ of $\texttt{GP}(\lambda, \tau)$. If σ is a subset, then Lemma 9 implies that $H = T_- = T_+$, and then $\widehat{T}_+ = \widehat{H}$. Also, collections $\mathcal{B}_-$ and $\mathcal{B}_+$ differ in exactly one subset, which is σ. Since $\widehat{T}_-$ spans fewer vertices than $\widehat{T}_+$, collection $\mathcal{B}_-$ must contain more subsets than collection $\mathcal{B}_+$, by the monotonicity of the pruning operation. Thus, in this case, $\mathcal{B}_- = \mathcal{B}_+ \cup \{\sigma\}$.

We would like to obtain a tree from H spanning exactly k vertices. To use Goemans and Williamson's analysis, this tree needs to be pruned with a collection of tight subsets. Pruning H using $\mathcal{B}_+$ results in $\widehat{H}$, which spans too many vertices, but, if we prune $\widehat{H}$ using $\mathcal{B}_-$, then the pruning algorithm would delete a sequence of disjoint subsets $D_1, \ldots, D_\ell$, until finding the tree $\widehat{T}_-$. This sequence induces a path on $\widehat{H}$, as illustrated in Fig. 2(a). Observe that each subset D_i, for $1 \leq i \leq \ell$, is contained in a corresponding subset $B_i \in \mathcal{B}_-$, and recall that invariant (*pp2*) states that D_i induces a connected subgraph of $\widehat{H}$.

When σ is an edge, Lemma 9 states that $\mathcal{B}_- = \mathcal{B}_+$ and trees T_- and T_+ differ in exactly one edge. Thus, T_+ has an edge e_+ such that $H = T_- + e_+$. Notice that H contains a unique cycle C, which includes e_+. We note that $\widehat{H}$ also contains C.

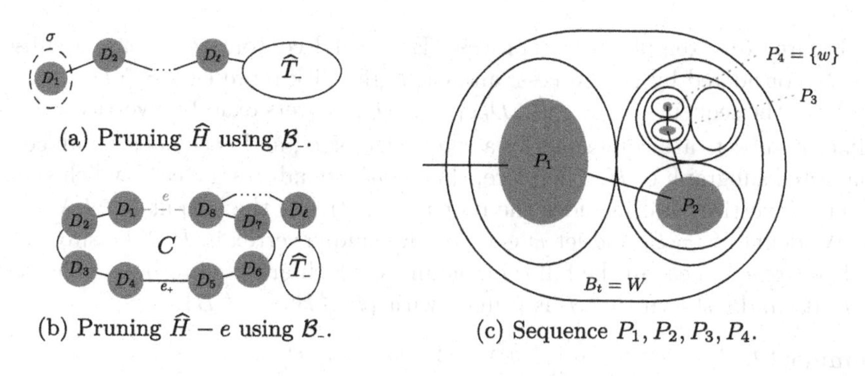

Fig. 2. Sequence of subsets computed in $\mathtt{PV}(\lambda, \tau)$.

Lemma 11. *If σ is an edge, then $\widehat{H}$ contains the cycle C.*

Since $\widehat{H} - e_+$ is a subgraph of T_-, pruning $\widehat{H} - e_+$ using $\mathcal{B}_-$ results in $\widehat{T}_-$, by monotonicity. In this case, however, the deleted subsets might not induce a path on $\widehat{H}$. But Lemma 11 implies that no subset in $\mathcal{B}_-$ contains all vertices of C, thus these subsets induce a cycle on $\widehat{H}$. Therefore, we can find an edge e of C such that pruning on $\widehat{H} - e$ also results in the tree $\widehat{T}_-$, and the sequence of deleted subsets $D_1, \dots, D_\ell$ induces a path on $\widehat{H}$. This is shown in Fig. 2(b).

Define $D_{\ell+1} = V(\widehat{T}_-)$ and notice that, since $D_{\ell+1}$ covers less than k vertices, but sequence $D_1, \dots, D_{\ell+1}$ covers at least k vertices, there is an index t such that $D_{t+1}, \dots, D_{\ell+1}$ covers less than k vertices and $D_t, \dots, D_{\ell+1}$ covers at least k vertices. If $D_{t+1}, \dots, D_{\ell+1}$ covers exactly $k - m$ vertices, then we would like augment this sequence by iteratively picking subsets from D_t which add up to m vertices. The goal is to find a sequence of subsets $P_1, P_2, \dots, P_s$ in $\mathcal{L}_-$ such that: each subset P_i induces a connected subgraph in $\widehat{H}$; P_1 is connected to D_{t+1}; adjacent subsets are connected by an edge of $\widehat{H}$; and $|P_s| = 1$.

This can be done as follows. Suppose that we already have computed a sequence $P_1, P_2, \dots, P_{i-1}$, and want to pick m vertices in $S \cap D_t$ for some subset $S \in \mathcal{L}_-$. Also, suppose there is an edge connecting P_{i-1} to some vertex $v \in S \cap D_t$. To initialize the process, let $P_0 = D_{t+1}$ and $S = B_t$, where B_t is the subset in $\mathcal{B}_-$ corresponding to D_t. Note that an edge connects D_{t+1} to some $v \in D_t$.

If $m = 1$, then define $P_i = \{v\}$, and we are done. Otherwise, we have $m \geq 2$, thus S contains at least two vertices. Since $\mathcal{L}_-$ is binary laminar, this implies that there are disjoint subsets S_1 and S_2 with $S = S_1 \cup S_2$, and such that $v \in S_1$.

If $|S_1 \cap D_t| \geq m$, then just make $S = S_1$, and repeat the process. This does not change the assumptions, except that it makes S smaller. Otherwise, we have $|S_1 \cap D_t| < m$, but this implies $|S_2 \cap D_t| \geq m - |S_1 \cap D_t|$ because $|S \cap D_t| \geq m$. It follows that $\widehat{H}$ spans vertices in both S_1 and S_2, and, since $\widehat{H}$ is S-connected, there must be an edge connecting a vertex $v_1 \in S_1$ to a vertex $v_2 \in S_2$. In this case, we define $P_i = S_1 \cap D_t$, update the variables by making $m = m - |P_i|$, $S = S_2$ and $v = v_2$, and repeat the process for $i + 1$. Note that P_i induces a connected subgraph in $\widehat{H}$ because $\widehat{H}$ is S_1-connected.

Figure 2(c) exemplifies this process. Each solid contour represents a subset $S \in \mathcal{B}_{-}$ considered by the process, and each gray disk represents $S \cap D_t$.

Now, the sequence $P_s, \ldots, P_1, D_{t+1}, \ldots, D_{\ell+1}$ covers exactly k vertices. Also, adjacent subsets are connected by a tight edge of $\widehat{H}$, and each subset induces a connected subgraph of $\widehat{H}$. Therefore, this sequence induces a tree T which spans exactly k vertices and includes the root r. This tree is the output of $\texttt{PV}(\lambda, \tau)$.

We define $W = B_t$ and let $w \in W$ be the unique vertex in P_s. The subset W and vertex w appear in the following lemmas, which are used to bound the cost of T. It can be shown that T is pruned with $\{B \in \mathcal{B}_{-} : w \notin B\}$.

Lemma 12. *Let $B \in \mathcal{B}_{-}$. If $|\delta_T(B)| = 1$, then $w \in B$.*

Denote by $\mathcal{L}_w[W]$ the collection of $S \in \mathcal{L}$ such that $w \in S$ and $S \subseteq W$.

Lemma 13. *Consider the execution of* GP *which returned $(T_{-}, \mathcal{B}_{-})$, and let $\mathcal{A}$ be the collection of active subsets that contain some vertex of T at the beginning of an iteration. Then, $\sum_{A \in \mathcal{A}} |\delta_T(A)| \leq 2(|\mathcal{A}| - |\mathcal{A}_w[W]|)$.*

Proof sketch. Let $\mathcal{I} \subseteq \mathcal{B}_{-}$ be the collection of maximal subsets in $\mathcal{L}^*_{-}$ which are not active and contain some vertex of T, then $\mathcal{A} \cup \mathcal{I}$ partitions $V(T)$. Therefore, at most one subset in $\mathcal{A} \cup \mathcal{I}$ contains w, which implies $|\mathcal{A}_w[W]| \leq 1$.

Let $\mathcal{I}_1$ be the collection of subsets in $\mathcal{I}$ with degree one on T, and consider the graph T' obtained from T by contracting each subset in $\mathcal{A} \cup \mathcal{I}$. If σ is a subset, then T is $\mathcal{L}_{-}$-connected, thus T' is a tree. If σ is an edge, then each connected component of $T - e_{+}$ is $\mathcal{L}_{-}$-connected, and one can show that T' is either a tree or it has at most one cycle whose edges are also in C (the unique cycle of H).

Assume T' is a tree. Note that $|\mathcal{I}_1| + |\mathcal{A}_w[W]| \leq 1$, as $w \in I$ for each $I \in \mathcal{I}_1$, by Lemma 12. As a vertex of T' corresponding to a subset S has degree $|\delta_T(S)|$,

$$\begin{aligned}\textstyle\sum_{A \in \mathcal{A}} |\delta_T(A)| + 2|\mathcal{I}| - |\mathcal{I}_1| \leq \sum_{A \in \mathcal{A}} |\delta_T(A)| + \sum_{I \in \mathcal{I}} |\delta_T(I)| &= 2(|\mathcal{A}| + |\mathcal{I}| - 1) \\ &\leq 2(|\mathcal{A}| + |\mathcal{I}| - |\mathcal{I}_1| - |\mathcal{A}_w[W]|).\end{aligned}$$

Assume T' has one cycle. Then, there is a maximal subset L in $\mathcal{A} \cup \mathcal{I}$ which induces a disconnected subgraph of T. Since no maximal subset in $\mathcal{A} \cup \mathcal{I}$ contains all the vertices of C, it can be shown that W is a proper subset of L, and thus $w \in L$. Because L is not a subset of W, we have $|\mathcal{A}_w[W]| = 0$, and, because L has degree at least two on T, Lemma 12 implies $|\mathcal{I}_1| = 0$. It follows that

$$\textstyle\sum_{A \in \mathcal{A}} |\delta_T(A)| + 2|\mathcal{I}| \leq \sum_{A \in \mathcal{A}} |\delta_T(A)| + \sum_{I \in \mathcal{I}} |\delta_T(I)| = 2(|\mathcal{A}| + |\mathcal{I}| - |\mathcal{A}_w[W]|).$$

In each of the cases, we have $\sum_{A \in \mathcal{A}} |\delta_T(A)| \leq 2(|\mathcal{A}| - |\mathcal{A}_w[W]|)$. □

6 The 2-Approximation

We bound the tree output of PV. In this section, we take $\mathcal{L}$ to be the collection $\mathcal{L}_{-}$, and denote by $\mathcal{L}[S]$ the collection of subsets in $\mathcal{L}$ which are subsets of S.

Lemma 14. *If* $\mathtt{PV}(\lambda,\tau)$ *outputs* T*, then* $c_{E(T)} + 2\pi^{\lambda}_{V\setminus V(T)} \leq 2\sum_{L\in\mathcal{L}(V)} y_L$.

Proof sketch. The lemma follows by combining the bounds

$$c_{E(T)} \leq 2\sum_{L\in\mathcal{L}(V(T))} y_L - 2\sum_{L\in\mathcal{L}_w[W]} y_L, \quad \text{and}$$
$$2\pi^{\lambda}_{V\setminus V(T)} \leq 2\sum_{L\in\mathcal{L}[V\setminus V(T)]} y_L + 2\sum_{L\in\mathcal{L}_w[W]} y_L.$$

Since the edges of T are tight for (y,λ), the first bound is equivalent to $c_{E(T)} = \sum_{L\in\mathcal{L}(V(T))} |\delta_T(L)| y_L \leq 2\sum_{L\in\mathcal{L}(V(T))} y_L - 2\sum_{L\in\mathcal{L}_w[W]} y_L$. We prove by induction. The inequality holds when the algorithm starts, thus, suppose it is valid at the beginning of an iteration, and let $\mathcal{A}$ denote the active subsets which contain a vertex of T. For each $A \in \mathcal{A}$, the variable y_A is increased by Δ. If some subset $A \in \mathcal{A}$ contains $V(T)$, then $W \subsetneq A$, because $\mathcal{L}$ is laminar and $w \in V(T)$, thus neither side of the inequality changes. Otherwise, the left-hand side increases by $\sum_{A\in\mathcal{A}(V(T))} |\delta_T(A)|\Delta$, and the right-hand side increases by $2(|\mathcal{A}| - |\mathcal{A}_w[W]|)\Delta$. By Lemma 13, the inequality holds at the end of the iteration.

Proving the second bound is more involved. One can show that, for each $v \in V \setminus V(T)$, either v is contained in a tight processed subset of $V \setminus V(T)$; or every subset in $\mathcal{L}[W]$ which contains v and a vertex of T also contains w. Vertices of the former kind can be paid by y variables of subsets in $\mathcal{L}[V \setminus V(T)]$; and vertices of the latter kind can be paid by y variables of subsets in $\mathcal{L}[V \setminus V(T)] \cup \mathcal{L}_w[W]$ which were not used to pay the vertices of the first kind. □

As the tree output by $\mathtt{PV}(\lambda,\tau)$ spans exactly k vertices and y respects c and π^{λ}, using Lemmas 1 and 14, we derive the final ingredient of our 2-approximation.

Lemma 15. *Let* T *be the output of* $\mathtt{PV}(\lambda,\tau)$ *and* L^* *be the minimal subset in* $\mathcal{L}$ *containing* $V(T^*)$*. If* $L^* = V$*, then* $c_{E(T)} + 2\pi_{V\setminus V(T)} \leq 2\left(c_{E(T^*)} + \pi_{V\setminus V(T^*)}\right)$.

We now present our 2-approximation, which is denoted by $\mathtt{2\text{-}APPROX}$. The algorithm computes a series of trees, the best of which becomes the output. In each iteration, start by computing a tree executing $\mathtt{GW}(0,\emptyset)$, and, if it spans at least k vertices, then store this tree and stop the iteration process. Otherwise, a threshold-tuple (λ,τ) can be computed by $\mathtt{TS}$. Now, store the tree returned by $\mathtt{PV}(\lambda,\tau)$ and let $\mathcal{L}$ be the laminar collection used by $\mathtt{PV}(\lambda,\tau)$ when computing this tree. Let $L_r \in \mathcal{L}$ be the inclusion-wise maximal proper subset of V containing r. If $|L_r| < k$, then stop; if not, repeat the iteration process with $G[L_r]$.

Observe that at least one vertex is deleted in each iteration, and the reduced graph is connected because it is $\mathcal{L}$-connected. Thus, the algorithm stops, and the output is the computed tree T which minimizes the cost with respect to the original graph G. We argue why T is a 2-approximation. Consider the last iteration which processed some subgraph G' containing T^*. If the algorithm stopped in this iteration after computing a tree using $\mathtt{GW}(0,\emptyset)$, then this tree spans at least k vertices, and Lemma 7 implies that this is a 2-approximate solution with respect to G'. Otherwise, the inclusion-wise minimal subset containing T^* is $V(G')$, and Lemma 15 implies that the tree computed by $\mathtt{PV}(\lambda,\tau)$ is a 2-approximation with respect to G'. Now, note that a 2-approximation with respect to G' is also a 2-approximation with respect to G.

Theorem 16. *Let T be the tree returned by* 2-APPROX, *then*

$$c_{E(T)} + 2\pi_{V \setminus V(T)} \leq 2\left(c_{E(T^*)} + \pi_{V \setminus V(T^*)}\right).$$

References

1. Archer, A., Bateni, M., Hajiaghayi, M., Karloff, H.: Improved approximation algorithms for prize-collecting Steiner tree and TSP. SIAM J. Comput. **40**(2), 309–332 (2011)
2. Blum, A., Ravi, R., Vempala, S.: A constant-factor approximation algorithm for the k-MST problem. J. Comput. Syst. Sci. **58**(1), 101–108 (1999)
3. Chudak, F.A., Roughgarden, T., Williamson, D.P.: Approximate k-MSTs and k-Steiner trees via the primal-dual method and Lagrangean relaxation. Math. Program. **100**(2), 411–421 (2004)
4. Feofiloff, P., Fernandes, C.G., Ferreira, C.E., de Pina, J.C.: A note on Johnson, Minkoff and Phillips' algorithm for the prize-collecting Steiner tree problem. arXiv e-prints p. 1004.1437 (2010)
5. Garg, N.: Saving an epsilon: A 2-approximation for the k-MST problem in graphs. In: Proceedings of the 37th Annual ACM Symposium on Theory of Computing, pp. 396–402 (2005)
6. Goemans, M.X., Williamson, D.P.: A general approximation technique for constrained forest problems. SIAM J. Comput. **24**(2), 296–317 (1995)
7. Han, L., Xu, D., Du, D., Wu, C.: A 5-approximation algorithm for the k-prize-collecting Steiner tree problem. Optim. Lett. **13**(3), 573–585 (2017)
8. Johnson, D.S., Minkoff, M., Phillips, S.: The prize collecting Steiner tree problem: theory and practice. In: Proceedings of the 11th Annual ACM-SIAM Symposium on Discrete Algorithms, pp. 760–769 (2000)
9. Matsuda, Y., Takahashi, S.: A 4-approximation algorithm for k-prize collecting Steiner tree problems. Optim. Lett. **13**(2), 341–348 (2018)

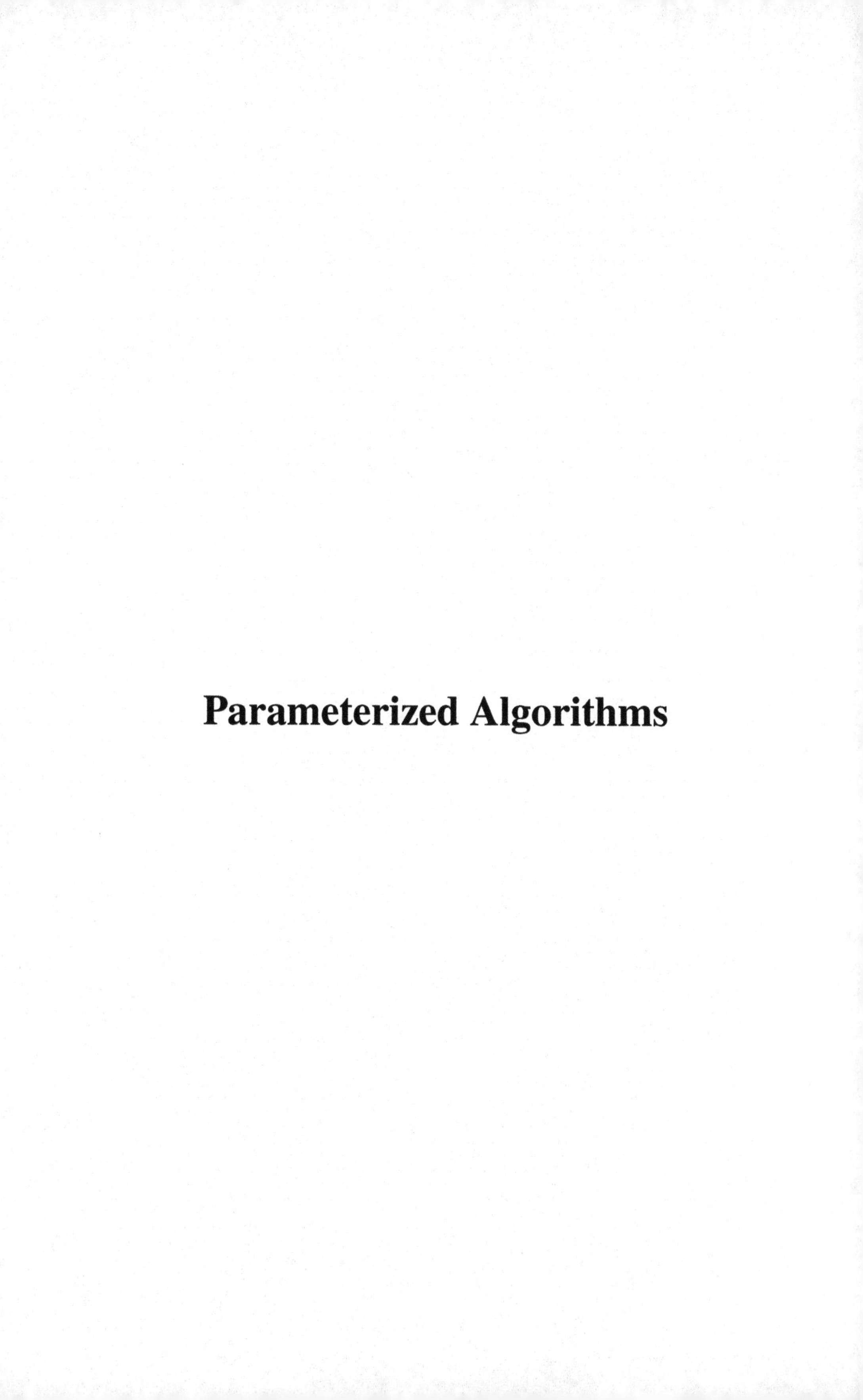
Parameterized Algorithms

Maximizing Happiness in Graphs of Bounded Clique-Width

Ivan Bliznets and Danil Sagunov(✉)

St. Petersburg Department of Steklov Mathematical Institute of Russian Academy of Sciences, 27 Fontanka, St. Petersburg, Russia
iabliznets@gmail.com, danilka.pro@gmail.com

Abstract. Clique-width is one of the most important parameters that describes structural complexity of a graph. Probably, only treewidth is more studied graph width parameter. In this paper we study how clique-width influences the complexity of the Maximum Happy Vertices (MHV) and Maximum Happy Edges (MHE) problems. We answer a question of Choudhari and Reddy '18 about parameterization by the distance to threshold graphs by showing that MHE is NP-complete on threshold graphs. Hence, it is not even in XP when parameterized by clique-width, since threshold graphs have clique-width at most two. As a complement for this result we provide a $n^{\mathcal{O}(\ell \cdot \mathrm{cw})}$ algorithm for MHE, where ℓ is the number of colors and cw is the clique-width of the input graph. We also construct an FPT algorithm for MHV with running time $\mathcal{O}^*((\ell+1)^{\mathcal{O}(\mathrm{cw})})$, where ℓ is the number of colors in the input. Additionally, we show $\mathcal{O}(\ell n^2)$ algorithm for MHV on interval graphs.

1 Introduction

Clique-width is one of the most important parameters that describe structural complexity of a graph. Probably, only treewidth is more studied graph width parameter. We note that one can treat clique-width as some generalization of treewidth as graphs of bounded treewidth have bounded clique-width. Hence, the existence of an FPT algorithm parameterized by clique-width is a stronger result than the existence of an FPT algorithm parameterized by treewidth. Complexity of many problems were studied parameterized by the clique-width parameter, including Max-Cut [11], Edge Dominating Set [11], Hamiltonian Path [10], Graph k-Colorability [12,16], computation of the Tutte polynomial [13], Dominating Set [16,17], computation of chromatic polynomial [23], and Target Set Selection [14]. In this paper, we continue the line of the research and investigate computational and parameterized complexity of the Maximum Happy Vertices and Maximum Happy Edges problems parameterized by clique-width of the input graph.

Before defining Maximum Happy Vertices and Maximum Happy Edges, we need to define what a happy vertex or a happy edge is.

This research was supported by the Russian Science Foundation (project 16-11-10123-П).

Y. Kohayakawa and F. K. Miyazawa (Eds.): LATIN 2020, LNCS 12118, pp. 91–103, 2020.
https://doi.org/10.1007/978-3-030-61792-9_8

Definition 1. Let G be a graph and let $c : V(G) \to [\ell]$ be a coloring of its vertices. We say that an edge $uv \in E(G)$ *is happy with respect to* c (or simply *happy*, if c is clear from the context) if its endpoints share the same color, i.e. $c(u) = c(v)$. We say that a vertex $v \in V(G)$ *is happy with respect to* c if all its neighbours have the same color as v, i.e. $c(v) = c(u)$ for each neighbour u of v in G.

We now give the formal definition of both problems.

Maximum Happy Vertices (MHV)	
Input:	A graph G, a partial coloring of vertices $p : S \to [\ell]$ for some $S \subseteq V(G)$ and an integer k.
Question:	Is there a coloring $c : V(G) \to [\ell]$ extending partial coloring p such that the number of happy vertices with respect to c is at least k?

Maximum Happy Edges (MHE)	
Input:	A graph G, a partial coloring of vertices $p : S \to [\ell]$ for some $S \subseteq V(G)$ and an integer k.
Question:	Is there a coloring $c : V(G) \to [\ell]$ extending partial coloring p such that the number of happy edges with respect to c is at least k?

Maximum Happy Vertices and Maximum Happy Edges were introduced by Zhang and Li in 2015 [27], motivated by their study of algorithmic aspects of homophyly law in large networks. These problems recently attracted a lot of attention from different lines of reseach. From the parameterized point of view, the problems were studied in [1–6,24]. Works [25–28] are devoted to approximation algorithms for MHV and MHE. Finally, Lewis et al. [19] study the problems from experimental perspective.

Before we state our results we mention some previously known results under different parameterizations. Aravind et al. [3] constructed $\mathcal{O}^*(\ell^{\text{tw}})$ and $\mathcal{O}^*(2^{\text{nd}})$ algorithms for both MHV and MHE, where tw is the treewidth of the input graph and nd is the neighbourhood diversity of the input graph. Misra and Reddy [24] constructed $\mathcal{O}^*(\text{vc}^{\mathcal{O}(\text{vc})})$ algorithms for both problems, where vc is the vertex cover number of the input graph.

Our Results: Below cw is the clique-width of the input graph, ℓ is the number of colors in the input precoloring and n is the number of vertices in the input graph. In the paper we prove the following results for the Maximum Happy Edges problem:

- MHE admits an XP-algorithm with running time $n^{\mathcal{O}(\ell \cdot \text{cw})}$, if a cw-expression of the input graph is given;
- MHE does not admit an XP-algorithm parameterized by clique-width alone, unless P = NP (by showing that MHE is NP-complete on threshold graphs).

Note that the question of the complexity of MHE on the class of threshold graphs was asked explicitly by Choudhari and Reddy in [6].

For the Maximum Happy Vertices problem we establish the following results:

- MHV admits an FPT algorithm with $\mathcal{O}^*((\ell+1)^{\mathcal{O}(\mathrm{cw})})$ running time, if a cw-expression of the input graph is given (note that MHV parameterized by clique-width alone is W[2]-hard [5]);
- Additionally, MHV is solvable on the class of interval graphs in time $\mathcal{O}(\ell n^2)$.

Our work shows that clique-width is a parameter under which computational complexity of problems MHV and MHE differ most significantly. On graphs of bounded clique-width, MHV admits an FPT algorithm with running time $\mathcal{O}^*((\ell+1)^{\mathcal{O}(\mathrm{cw})})$, while MHE is NP-complete on graphs of clique-width two and does not admit even an XP-algorithm when parameterized by cw, however, we show that there is an XP-algorithm for the extended parameter $\mathrm{cw}+\ell$. Note that when parameterized by treewidth, neighbourhood diversity or vertex cover, the problems are known to have similar complexity. We believe that the FPT algorithm for MHV parameterized by $\mathrm{cw}+\ell$ is the most interesting result of this paper. Unfortunately, due to the tight page limit, the proof of this result is omitted to the full version for the sake of describing an answer to the open question of Choudhari and Reddy from [6] almost completely.

After establishing existence of polynomial algorithms for problems on graphs of bounded clique-width, it is natural to investigate complexity of problems on minimal hereditary classes of unbounded clique-width. Unit interval graphs is one of such graph classes [20]. We show that MHV is polynomially solvable on the class of interval graphs, which is a wider graph class. So we think that our result for interval graphs nicely complements our understanding of computational complexity of MHV parameterized by clique-width. We note that interval graphs also separate MHV and MHE, as MHE is NP-complete on threshold graphs, that are a subclass of interval graphs.

2 Preliminaries

Basic Notation. We denote the set of positive integer numbers by $\mathbb{N}$. For each positive integer k, by $[k]$ we denote the set of all positive integers not exceeding k, $\{1,2,\dots,k\}$. We use $\sqcup$ for the disjoint union operator, i.e. $A \sqcup B$ equals $A \cup B$, with an additional constraint that A and B are disjoint.

We use the traditional $\mathcal{O}$-notation for asymptotical upper bounds. We additionally use the $\mathcal{O}^*$-notation that hides polynomial factors. We investigate MHV and MHE mostly from the parameterized point of view. For a detailed survey in parameterized algorithms we refer to the book of Cygan et al. [8]. Throughout the paper, we use standard graph notation and terminology, following the book of Diestel [9]. All graphs in our work are undirected simple graphs.

Graph Colorings. When dealing with instances of MHV or MHE, we use a notion of colorings. A *coloring* of a graph G is a function that maps vertices of the graph to the set of colors. If this function is partial, we call such coloring *partial*. If not stated otherwise, we use ℓ for the number of distinct colors, and assume that colors are integers in $[\ell]$. A partial coloring p is always given as a part of the input for both problems, along with graph G. We also call p a *precoloring* of the graph G, and use (G, p) to denote the graph along with the precoloring. The goal of both problems is to extend this partial coloring to a specific coloring c that maps each vertex to a color. We call c a *full coloring* (or simply, a coloring) of G that extends p. We may also say that c is a coloring of (G, p). For a full coloring c of a graph G by $\mathcal{H}(G, c)$ we denote the set of all vertices in G that are happy with respect to c.

Clique-Width. In order to define *cliquewidth* we follow definitions presented by Lackner et al. in their work on MULTICUT parameterized by clique-width [18].

To define clique-width, we need to define k-expressions first. For any $k \in \mathbb{N}$, a *k-expression* Φ describes a graph G_Φ, whose vertices are labeled with integers in $[k]$. k-expressions and its corresponding graphs are defined recursively. Depending on its topmost operator, a k-expression Φ can be of four following types.

1. *Introducing a vertex.* $\Phi = i(v)$, where $i \in [k]$ is a label and v is a vertex. G_Φ is a graph consisting of a single vertex v with label i, i.e. $V(G_\Phi) = \{v\}$.
2. *Disjoint union.* $\Phi = \Phi' \oplus \Phi''$, where Φ' and Φ'' are smaller subexpressions. G_Φ is a disjoint union of the graphs $G_{\Phi'}$ and $G_{\Phi''}$, i.e. $V(G_\Phi) = V(G_{\Phi'}) \sqcup V(G_{\Phi''})$ and $E(G_\Phi) = E(G_{\Phi'}) \sqcup E(G_{\Phi''})$. The labels of the vertices remain the same.
3. *Renaming labels.* $\Phi = \rho_{i \to j}(\Phi')$. The structure of G_Φ remains the same as the structure of $G_{\Phi'}$, but each vertex with label i receives label j.
4. *Introducing edges.* $\Phi = \eta_{i,j}(\Phi')$. G_Φ is obtained from $G_{\Phi'}$ by connecting each vertex with label i with each vertex with label j.

Clique-width of a graph G is defined as the smallest value of k needed to describe G with a k-expression and is denoted as $\mathrm{cw}(G)$, or simply cw.

There is still no known FPT-algorithm for finding a k-expression of a given graph G. However, there is an FPT-algorithm that decides whether $\mathrm{cw}(G) > k$ or outputs $(2^{3k+2} - 1)$-expression of G. For more details on clique-width we refer to [15].

Due to the space restrictions, we omit proofs of some theorems and lemmata. We mark such theorems and lemmata with the '$\star$' sign. Missing proofs can be found in the full version of the paper.

3 Maximum Happy Edges

This section is dedicated to the MAXIMUM HAPPY EDGES problem parameterized by clique-width. We start with showing that MAXIMUM HAPPY EDGES is NP-complete on graphs of clique-width at most two.

In [6], Choudhari and Reddy proved that MHV is polynomially solvable on the class of threshold graphs (that have clique-width at most two [21]) and questioned the complexity of MHE on the same graph class. We answer their question by showing that Maximum Happy Edges is NP-complete on threshold graphs. To prove this, we require the following useful characterization of threshold graphs.

Lemma 1 [22]. *Threshold graphs are graphs that can be partitioned in a clique $K = \{u_1, u_2, \ldots, u_k\}$ and an independent set I, such that $N[u_i] \subseteq N[u_{i+1}]$ holds for every $i \in [k-1]$.*

We now prove the abovementioned hardness of MHE.

Theorem 1. Maximum Happy Edges *is* NP*-complete on the class of threshold graphs.*

Proof. We reduce from SAT, that is a classical NP-complete problem. Let F be a boolean formula on n variables in conjunctive normal form $F = C_1 \wedge C_2 \wedge \ldots \wedge C_m$. C_i is a clause being a disjunction of distinct literals, so it can be represented as $C_i = l_{i,1} \vee l_{i,2} \vee \ldots \vee l_{i,k_i}$, where each literal $l_{i,t}$ is either a variable x_j or its negation $\overline{x_j}$ for some $j \in [n]$.

We show how to, given F, construct an instance (G, p, k) of Maximum Happy Edges, such that F is satisfiable if and only if (G, p, k) is a yes-instance of MHE. Moreover, G is a threshold graph and the construction can be done in polynomial-time.

Let F be a boolean formula on n variables in CNF, consisting of m clauses. We construct (G, p, k) as follows.

G will be a threshold graph. So it will consist of two parts: a clique K and an independent set I. Firstly, we introduce the clique vertices in G. For each clause C_i of F we introduce a new vertex c_i in G. For each variable x_j of F we introduce m^2 new vertices $v_{j,1}, v_{j,2}, \ldots, v_{j,m^2}$ in G. We introduce all possible edges between these $m + nm^2$ vertices in G so these vertices form the clique K in the partition of G.

Before we proceed, let us give an intuition of the further construction. Each color we use in p corresponds to a literal in F, i.e. to an element in $L = \{x_1, x_2, \ldots, x_n, \overline{x_1}, \overline{x_2}, \ldots, \overline{x_n}\}$. Thus, we use $2n$ colors in p. For convenience, we use corresponding literals to denote colors instead of the numbers in $[2n]$. We want each clause vertex c_i to be colored with a color corresponding to one of its literals, i.e. one of the colors $l_{i,1}, l_{i,2}, \ldots, l_{i,k_i}$ in any optimal coloring. Similarly, we want each variable vertex $v_{j,t}$ corresponding to the variable x_j to be colored with one of the colors corresponding to the literals of x_j, i.e. either x_j or $\overline{x_j}$. For each vertex $u \in K$, we denote the set of required colors as $\mathcal{L}(u)$, i.e. $\mathcal{L}(c_i) = C_i = \{l_{i,1}, l_{i,2}, \ldots, l_{i,k_i}\}$ for clause vertices, and $\mathcal{L}(v_{j,t}) = \{x_j, \overline{x_j}\}$ for variable vertices. The purpose of the remaining independent set of G is exactly to ensure that the vertices of the clique are colored with the required colors.

Our graph is a threshold graph. It means that it is possible to find an order $u_1, u_2, \ldots, u_{|K|}$ of the vertices in K such that $N[u_i] \subseteq N[u_{i+1}]$ for each $i \in [|K| - 1]$, i.e. satisfy the condition of Lemma 1. The order we obtain is the following: $u_i = c_i$ for every $i \in [m]$, and $u_{m+jm^2+t} = v_{j+1,t}$ for each $j \in \{0, 1, \ldots, n-1\}$

and each $t \in [m^2]$. The condition of Lemma 1 is satisfied as we step by step add vertices to I. The ith step will correspond to the vertex $u_i \in K$. At this step we introduce all neighbours of u_i in I and their colors in the precoloring p. For convenience we denote $N(u_i) \cap I$ by P_i.

At first, we construct P_1 in the following way. For each $l \in \mathcal{L}(u_1)$, add exactly $m + nm^2$ vertices to P_1 and color them with the color l. No more vertices are added to P_1, so $|P_1| = |\mathcal{L}(u_1)| \cdot (m + nm^2)$. Then, for each $i \in [2, m + nm^2]$, we construct P_i by adding new vertices to P_{i-1} and precoloring them. By doing so we satisfy condition $N[u_i] \subseteq N[u_{i+1}]$ for each $i \in [m + nm^2 - 1]$. The process of this construction is described below and illustrated in Fig. 1.

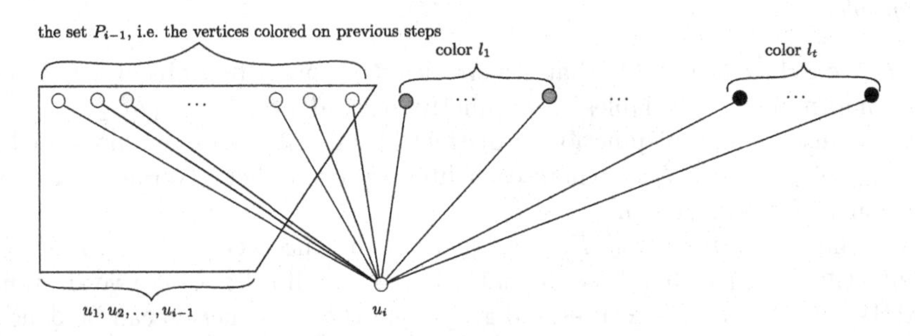

Fig. 1. Step i. That is addition of the vertex u_i and construction of the set P_i. Let $\mathcal{L}(u_i) = \{l_1, l_2, \dots l_t\}$. Vertex u_i is connected to all vertices in P_{i-1}. Moreover, for each $j \in \{1, 2, \dots t\}$ we introduce $i(m + nm^2) - |\mathcal{N}_p(P_{i-1}, l_j)|$ vertices precolored in color l_j and connect them to u_i. Recall that all bottom vertices, i.e. u_q for $q \in [i]$, are also pairwise connected (this is not shown in the figure as well as edges from P_{i-1} to $u_1, u_2, \dots, u_{i-1}$).

Let $\mathcal{N}_p(P_i, l)$ be the number of vertices in P_i that are precolored with the color l, i.e. $\mathcal{N}_p(P_i, l) = |\{u \in P_i \mid p(u) = l\}|$. For each i we require that the vertices in P_i are precolored mostly with required colors for u_i, that is, colors in the set $\mathcal{L}(u_i)$. Formally, for every $l \in L$, we require

$$\begin{aligned} &\mathcal{N}_p(P_i, l) = i(m + nm^2), && \text{if } l \in \mathcal{L}(u_i), \\ &\mathcal{N}_p(P_i, l) \leq (i-1) \cdot (m + nm^2), && \text{if } l \notin \mathcal{L}(u_i). \end{aligned} \tag{*}$$

Note that P_1 satisfies (*).

Now let P_{i-1} be constructed and satisfy (*). We construct P_i that also satisfies the constraint. We start with $P_i = P_{i-1}$. Then for each $l \in \mathcal{L}(u_i)$ we introduce $i(m + nm^2) - \mathcal{N}_p(P_{i-1}, l)$ new vertices precolored with color l to P_i. For every $l \in \mathcal{L}(u_i)$, $\mathcal{N}_p(P_i, l) = \mathcal{N}_p(P_{i-1}, l) + i(m + nm^2) - \mathcal{N}_p(P_{i-1}, l) = i(m + nm^2)$. On the other hand, for every $l \notin \mathcal{L}(u_i)$, $\mathcal{N}_p(P_i, l) = \mathcal{N}_p(P_{i-1}, l) \leq (i-1) \cdot (m + nm^2)$. Hence, P_i also satisfies (*).

The construction of G is finished. Let us remark again that K forms a clique in G and $I = P_{m+nm^2}$ forms an independent set in G. We constructed graph in a way that $N[u_i] \subseteq N[u_{i+1}]$ (as $P_i \subseteq P_{i+1}$) for each $i \in [m + nm^2 - 1]$

where $K = \{u_1, u_2, \ldots, u_{m+nm^2}\}$. Thus, by Lemma 1, G is a threshold graph. Moreover, construction of G is done in polynomial time.

We finally set the number of required happy edges

$$k = (m + nm^2) \cdot \binom{m + nm^2 + 1}{2} + n \cdot \binom{m^2}{2} + m^3$$

and argue that F is satisfiable if and only if (G, p, k) is a yes-instance of MHE.

Let F be satisfiable, that is, F has a satisfying assignment $\sigma : x_j \mapsto \{0, 1\}$. We construct coloring c of G extending p that yields at least k happy edges as follows.

For each $j \in [n]$ and $t \in [m^2]$, set the color of the vertex $v_{j,t}$ corresponding to the variable x_j with the color corresponding to the literal of x_j that evaluates to 1 with respect to σ, i.e. $c(v_{j,t}) = x_j$, if $\sigma(x_j) = 1$, or $c(v_{j,t}) = \overline{x_j}$, if $\sigma(x_j) = 0$.

For each $i \in [m]$, there is at least one variable satisfying clause C_i. In other words, there exists $j \in [n]$, such that either $x_j \in C_i$ and $\sigma(x_j) = 1$, or $\overline{x_j} \in C_i$ and $\sigma(x_j) = 0$. Choose any such j and color the corresponding clause vertex c_i with the color corresponding to the literal of x_j that evaluates to true. That is, $c(c_i) = x_j$ if $\sigma(x_j) = 1$, or $c(c_i) = \overline{x_j}$ if $\sigma(x_j) = 0$. There is no any uncolored vertex left, so the construction of c is finished.

Claim 1 ($\star$). *There are at least k happy edges in G with respect to c.*

Hence, we showed that if F is satisfiable, then (G, p, k) is a yes-instance of MHE. We now give a proof in the other direction.

Let c be a coloring of G extending p such that at least k edges of G are happy with respect to c. We assume that c is optimal, i.e. it yields the maximum number of happy edges in G. We make the following claims and then show how to construct a satisfying assignment σ of F.

Claim 2 ($\star$). *In any optimal coloring c of G extending p, $c(u_i) \in \mathcal{L}(u_i)$ for every $u_i \in K$.*

Claim 3. *In any optimal coloring c of G extending p, all variable vertices corresponding to the same variable are colored with the same color. Formally, $c(v_{j,t_1}) = c(v_{j,t_2})$ for every $j \in [n]$ and $t_1, t_2 \in [m^2]$.*

Proof of the Claim. Suppose that c is an optimal coloring extending p, but $c(v_{j,t_1}) \neq c(v_{j,t_2})$ for some $j \in [n]$, $t_1, t_2 \in [m^2]$. By Claim 2, $c(v_{j,t_1})$ and $c(v_{j,t_2})$ are distinct literals of variable x_j, and $v_{j,t_1} = u_{m+(j-1)m^2+t_1}$ and $v_{j,t_2} = u_{m+(j-1)m^2+t_2}$ are incident to exactly $(m + (j-1) \cdot m^2 + t_1) \cdot (m + nm^2)$ and $(m + (j-1) \cdot m^2 + t_2) \cdot (m + nm^2)$ happy edges going in I respectively according to (*).

Let h_1 and h_2 be the number of vertices in K that are colored with colors $c(v_{j,t_1})$ and $c(v_{j,t_2})$, respectively. Thus, v_{j,t_1} and v_{j,t_2} are incident to exactly $h_1 - 1$ and $h_2 - 1$ happy edges in $G[K]$, respectively. Note that the edge between v_{j,t_1} and v_{j,t_2} is not happy.

Without loss of generality, $h_1 \geq h_2$. Change the color of v_{j,t_2} in c to $c(v_{j,t_1})$. Since $c(v_{j,t_2})$ is still a literal of x_j, hence $c(v_{j,t_2}) \in \mathcal{L}(v_{j,t_2})$, the number of happy edges connecting v_{j,t_2} and I does not change, even though the set of such happy edges becomes different. Consider edges in $G[K]$. v_{j,t_2} is now adjacent to h_1 neighbours of the same color, as the edge between v_{j,t_1} and v_{j,t_2} also becomes happy. Since $h_1 > h_2 - 1$, we have increased the total number of happy edges in G with respect to c. This contradicts the optimality of c.

We now use the above claims to construct σ from an optimal coloring c yielding at least k happy edges. By Claim 2, there are exactly $(m + nm^2) \cdot (1 + 2 + \ldots + (m + nm^2)) = (m + nm^2) \cdot \binom{m+nm^2+1}{2}$ happy edges between K and I with respect to c. By Claim 3, there are exactly $n \cdot \binom{m^2}{2}$ happy edges between all variable vertices. There are exactly m clause vertices in G, hence there are at most $\binom{m}{2}$ happy edges between all clause vertices. The only edges left are the edges between clause and variable vertices, hence there are at least

$$k - (m + nm^2) \cdot \binom{m + nm^2 + 1}{2} - n \cdot \binom{m^2}{2} - \binom{m}{2} = m^3 - \binom{m}{2}$$

happy edges between clause and variable vertices.

Construct σ according to the colors of variable vertices, so that the literal corresponding to $c(v_{j,t})$ evaluates to 1 with respect to σ. Formally, for each $j \in [n]$, $\sigma(x_j) = 1$ if $c(v_{j,1}) = x_j$, and $\sigma(x_j) = 0$ if $c(v_{j,1}) = \overline{x_j}$. We now argue that each clause $C_i \in F$ contains a literal that evaluates to 1 with respect to σ, and that this literal is $c(c_i)$.

Suppose that it is not true, and there is a clause C_i so that $c(c_i)$ is a literal that evaluates to 0 with respect to σ. By construction of σ, there are no happy edges between c_i and the variable vertices. c_i corresponds to a literal evaluating to 0, but all colors of variable vertices are literals that evaluates to 1 with respect to σ. Moreover, any other clause vertex $c_{i'}$ is adjacent to either 0 or m^2 variable vertices of color $c(c_{i'})$. For each literal, there are either 0 or m^2 variable vertices colored correspondingly to this literal.

There are exactly $m-1$ clause vertices apart from c_i, hence at most $(m-1) \cdot m^2$ edges between clause vertices and variable vertices are happy with respect to c. But $(m-1) \cdot m^2 = m^3 - m^2 < m^3 - \binom{m}{2}$ for any $m > 0$, a contradiction. Thus, each clause C_i contains a literal that evaluates to 1 with respect to σ, i.e. σ is a satisfying assignment of F. We proved that if (G, p, k) is a yes-instance of MHE, then F is satisfiable. The proof is complete. □

Corollary 1. There is no XP-algorithm for Maximum Happy Edges parameterized by clique-width, unless P = NP.

Proof. Suppose there is an XP-algorithm for Maximum Happy Edges parameterized by clique-width, i.e. there is an algorithm with running time $n^{f(\mathrm{cw})}$ for some function f for MHE. Threshold graphs are a subclass of cographs [21], that is, graphs of clique-width at most two [7]. Hence, MHE on threshold graphs can be solved in $n^{f(2)} = n^{\mathcal{O}(1)}$. Then, by Theorem 1, problem that is solvable in polynomial time is NP-hard, hence P = NP. □

We have shown that MHE parameterized by clique-width alone is hard. Following known results on the existence of $\mathcal{O}^*(\ell^{\mathcal{O}(\mathrm{pw})})$ and $\mathcal{O}^*(\ell^{\mathcal{O}(\mathrm{tw})})$ running time algorithms for both MHV and MHE parameterized by pathwidth or treewidth combined with the number of colors ℓ [1,2,24], it is reasonable to ask the complexity of MHE parameterized by cw+ℓ. We now show that MHE parameterized by cw + ℓ admits an XP-algorithm.

Theorem 2. *There is an algorithm for* Maximum Happy Edges *with* $n^{\mathcal{O}(\ell \cdot \mathrm{cw})}$ *running time, if a* cw*-expression of* G *is given.*

Proof. The algorithm is by standard dynamic programming on a given w-expression Ψ of G. We assume that Ψ is a nice w-expression of G, i.e. no edge is introduced twice in Ψ. For each subexpression Φ of Ψ,

$$OPT(\Phi, n_{1,1}, n_{1,2}, \ldots, n_{1,\ell}, n_{2,1}, n_{2,2}, \ldots, n_{w,\ell-1}, n_{w,\ell})$$

denotes the maximum number of happy edges that can be obtained in G_Φ simultaneously with respect to a coloring such that the number of vertices with a label i in G_Φ that are colored with a color a in G_Φ is exactly $n_{i,a}$. Formally,

$$OPT(\Phi, n_{1,1}, \ldots, n_{w,\ell}) = \max \left\{ |\mathcal{E}(G_\Phi, c)| \;\middle|\; \begin{array}{r} c : V(G_\Phi) \to [\ell], \\ \forall i \in [w], a \in [\ell] : \\ |c^{-1}(a) \cap V_i(\Phi)| = n_{i,a} \end{array} \right\},$$

where $\mathcal{E}(G_\Phi, c)$ is the set of edges that are happy in G_Φ with respect to c. If there are no colorings corresponding to a cell $OPT(\Phi, n_{1,1}, \ldots, n_{w,\ell})$, we put its value equal to $-\infty$.

The algorithm computes the values of OPT in a bottom-up approach, starting from the simplest subexpressions of Ψ up to Ψ itself. Thus, when the algorithm starts computing the values of $OPT(\Phi, \cdot)$ for a subexpression Φ of Ψ, it has all values of OPT computed for each subexpression of Φ. There are four possible cases of computing values of $OPT(\Phi, \cdot)$ depending on the topmost operator in Φ.

1. $\Phi = i(v)$. Since G_Φ contains a single vertex with label i and no edges, it is enough to iterate over all possible colors of this vertex. If v is not precolored, for each color $a \in [\ell]$ put $OPT(\Phi, 0, \ldots, 0, n_{i,a} = 1, 0, \ldots, 0) = 0$. Otherwise, the color of v can only be $p(v)$, so do this only for $a = p(v)$. Thus, exactly ℓ values (or exactly one value) of $OPT(\Phi, \cdot)$ are put equal to 0, and all other values should equal $-\infty$ by the definition of OPT.
2. $\Phi = \Phi' \oplus \Phi''$. Consider a cell $OPT(\Phi, n_{1,1}, \ldots, n_{w,\ell})$. Any coloring c corresponding to this cell is split uniquely in the two colorings $c' = c|_{V(G_{\Phi'})}$ and $c'' = c|_{V(G_{\Phi''})}$ of $G_{\Phi'}$ and $G_{\Phi''}$ respectively. In its order, these colorings correspond to cells $OPT(\Phi', n'_{1,1}, \ldots, n'_{w,\ell})$ and $OPT(\Phi'', n''_{1,1}, \ldots, n''_{w,\ell})$, where $n'_{i,a}$ and $n''_{i,a}$ are unique for each choice of $i \in [w], a \in [\ell]$. As G_Φ is the disjoint union of $G_{\Phi'}$ and $G_{\Phi''}$, the number of happy edges in G_Φ with respect

to c can be found as a sum of happy edges with respect to c' and c'' in the corresponding graphs. Hence,

$$OPT(\Phi, n_{1,1}, \ldots, n_{w,\ell}) = \max_{n'_{i,a}+n''_{i,a}=n_{i,a}} \{OPT(\Phi', n'_{1,1}, \ldots, n'_{w,\ell}) + OPT(\Phi'', n''_{1,1}, \ldots, n''_{w,\ell})\}. \quad (1)$$

3. $\Phi = \rho_{i \to j}\Phi'$. Consider again a coloring c corresponding to a cell $OPT(\Phi, n_{1,1}, \ldots, n_{w,\ell})$. Note that Φ contains no vertices with label i, so $n_{i,a} = 0$ for each $a \in [\ell]$. Moreover, c is a coloring of $G_{\Phi'}$, thus it corresponds to the unique cell $OPT(\Phi, n'_{1,1}, \ldots, n'_{w,\ell})$, where $n'_{i,a} + n'_{j,a} = n_{j,a}$ for each $a \in [\ell]$ and $n'_{k,a} = n_{k,a}$ for each choice of label k distinct from i and j, and for each color $a \in [\ell]$. The number of happy edges in G_Φ with respect to c is the same as that in $G_{\Phi'}$. Hence,

$$OPT(\Phi, n_{1,1}, \ldots, n_{w,\ell}) = \begin{cases} -\infty, \text{ if } \exists a \in [\ell] : n_{i,a} \neq 0, \\ \max\left\{OPT(\Phi', n'_{1,1}, \ldots, n'_{w,\ell}) \middle| \begin{matrix} \forall a \in [\ell] : n'_{i,a} + n'_{j,a} = n_{j,a} \\ \forall k \in [w] \setminus \{i,j\}, a \in [\ell] : n'_{k,a} = n_{k,a} \end{matrix} \right\}. \end{cases} \quad (2)$$

4. $\Phi = \eta_{i,j}\Phi'$. This is the only case where edges are introduced. Any coloring c of G_Φ is a coloring of G'_Φ. Moreover, if c corresponds to $OPT(\Phi, n_{1,1}, \ldots, n_{w,\ell})$ then, clearly, c corresponds to $OPT(\Phi', n_{1,1}, \ldots, n_{w,\ell})$ as well. Thus, one shall only compute the number of newly-introduced edges that are happy with respect to c. As Ψ is a nice w-expression, each edge between vertices with label i and vertices with label j is newly-introduced. Each of such happy edge should connect a vertex with the label i and a vertex with the label j that are colored with the same color a for some a. The number of such edges for a fixed a is $n_{i,a} \cdot n_{j,a}$. Hence, $OPT(\Phi, n_{1,1}, \ldots, n_{w,\ell}) = OPT(\Phi', n_{1,1}, \ldots, n_{w,\ell}) + \sum_{a=1}^{\ell} n_{i,a} \cdot n_{j,a}$.

The description of all possible cases for Φ and corresponding recurrence relations is finished. Note that there are at most $|\Psi| \cdot n^{\ell \cdot w}$ cells in the OPT table, and each of them is computed in $\mathcal{O}(n^{\ell \cdot w})$ time (computation in the case of disjoint union and the case of relabelling takes the most time) by the algorithm. Thus, the whole computation of OPT takes $\mathcal{O}(|\Psi| \cdot n^{2\ell \cdot w})$ running time. Clearly, the maximum number of happy edges that can be obtained in G simultaneously equals $\max_{n_{1,1},\ldots,n_{w,\ell}} OPT(\Psi, n_{1,1}, \ldots, n_{w,\ell})$, which is found in $\mathcal{O}(n^{\ell \cdot w})$ time. This finishes the proof. □

Corollary 2. Maximum Happy Edges parameterized by cw + ℓ admits an XP-algorithm.

The fixed-parameter tractability of MHE with respect to cw $+\ell$ remains unknown though. We note that Theorem 2 does not imply that no FPT-algorithm exists for MHE parameterized by cw + ℓ under P $\neq$ NP. But it at least implies that no algorithm with running time $\mathcal{O}^*(\mathsf{poly}(\ell)^{f(\mathrm{cw})})$ exists for MHE, unless P = NP. We leave the FPT-membership of MHE parameterized by cw + ℓ as an open question.

4 Maximum Happy Vertices

We start this section by answering the complexity of MAXIMUM HAPPY VERTICES parameterized by cw + ℓ. We note that MHV is W[2]-hard when parameterized by the clique-width of the input graph alone [5]. In contrast to this, we show that MHV is in FPT if the clique-width parameter is extended by the number of colors ℓ.

Theorem 3 (⋆). MAXIMUM HAPPY VERTICES *can be solved in* $(\ell+1)^{\mathcal{O}(\mathrm{w})} \cdot n^{\mathcal{O}(1)}$ *running time, if a* w*-expression of* G *is given.*

Corollary 3. MAXIMUM HAPPY VERTICES parameterized by cw +ℓ admits an FPT-algorithm.

In the rest of this section we show that MAXIMUM HAPPY VERTICES is polynomially solvable on the class of interval graphs, that is related to clique-width in the following sense. Interval graphs have unbounded clique-width, moreover, unit interval graphs are minimal hereditary graph class of unbounded clique-width [20]. Since threshold graphs are a subclass of interval graphs, this result also covers the result of Choudhari and Reddy in [6], where they showed that MHV is polynomially solvable on the class of threshold graphs. We also note that MHE, in contrast to MHV, is NP-hard on the class of interval graphs, which is a corollary of Theorem 1.

Theorem 7 (⋆). *There is* $\mathcal{O}(\ell n^2)$ *running time algorithm for* MAXIMUM HAPPY VERTICES *on interval graphs.*

References

1. Agrawal, A.: On the parameterized complexity of happy vertex coloring. In: Brankovic, L., Ryan, J., Smyth, W.F. (eds.) IWOCA 2017. LNCS, vol. 10765, pp. 103–115. Springer, Cham (2018). https://doi.org/10.1007/978-3-319-78825-8_9
2. Aravind, N.R., Kalyanasundaram, S., Kare, A.S.: Linear time algorithms for happy vertex coloring problems for trees. In: Mäkinen, V., Puglisi, S.J., Salmela, L. (eds.) IWOCA 2016. LNCS, vol. 9843, pp. 281–292. Springer, Cham (2016). https://doi.org/10.1007/978-3-319-44543-4_22
3. Aravind, N., Kalyanasundaram, S., Kare, A.S., Lauri, J.: Algorithms and hardness results for happy coloring problems. arXiv preprint arXiv:1705.08282 (2017)
4. Bliznets, I., Sagunov, D.: Lower bounds for the happy coloring problems. In: Du, D.-Z., Duan, Z., Tian, C. (eds.) COCOON 2019. LNCS, vol. 11653, pp. 490–502. Springer, Cham (2019). https://doi.org/10.1007/978-3-030-26176-4_41
5. Bliznets, I., Sagunov, D.: On happy colorings, cuts, and structural parameterizations. In: Sau, I., Thilikos, D.M. (eds.) WG 2019. LNCS, vol. 11789, pp. 148–161. Springer, Cham (2019). https://doi.org/10.1007/978-3-030-30786-8_12

6. Choudhari, J., Reddy, I.V.: On structural parameterizations of happy coloring, empire coloring and boxicity. In: Rahman, M.S., Sung, W.-K., Uehara, R. (eds.) WALCOM 2018. LNCS, vol. 10755, pp. 228–239. Springer, Cham (2018). https://doi.org/10.1007/978-3-319-75172-6_20
7. Courcelle, B., Olariu, S.: Upper bounds to the clique width of graphs. Discrete Appl. Math. **101**(1–3), 77–114 (2000)
8. Cygan, M., et al.: Parameterized Algorithms. Springer, Cham (2015). https://doi.org/10.1007/978-3-319-21275-3
9. Diestel, R.: Graph Theory. Springer, Heidelberg (2018). https://doi.org/10.1007/978-3-662-53622-3
10. Espelage, W., Gurski, F., Wanke, E.: How to solve NP-hard graph problems on clique-width bounded graphs in polynomial time. In: Brandstädt, A., Le, V.B. (eds.) WG 2001. LNCS, vol. 2204, pp. 117–128. Springer, Heidelberg (2001). https://doi.org/10.1007/3-540-45477-2_12
11. Fomin, F.V., Golovach, P.A., Lokshtanov, D., Saurabh, S.: Almost optimal lower bounds for problems parameterized by clique-width. SIAM J. Comput. **43**(5), 1541–1563 (2014)
12. Gerber, M.U., Kobler, D.: Algorithms for vertex-partitioning problems on graphs with fixed clique-width. Theor. Comput. Sci. **299**(1–3), 719–734 (2003)
13. Giménez, O., Hliněný, P., Noy, M.: Computing the tutte polynomial on graphs of bounded clique-width. In: Kratsch, D. (ed.) WG 2005. LNCS, vol. 3787, pp. 59–68. Springer, Heidelberg (2005). https://doi.org/10.1007/11604686_6
14. Hartmann, T.A.: Target set selection parameterized by clique-width and maximum threshold. In: Tjoa, A.M., Bellatreche, L., Biffl, S., van Leeuwen, J., Wiedermann, J. (eds.) SOFSEM 2018. LNCS, vol. 10706, pp. 137–149. Springer, Cham (2018). https://doi.org/10.1007/978-3-319-73117-9_10
15. Hliněǹy, P., Oum, S., Seese, D., Gottlob, G.: Width parameters beyond tree-width and their applications. Comput. J. **51**(3), 326–362 (2007)
16. Kobler, D., Rotics, U.: Polynomial algorithms for partitioning problems on graphs with fixed clique-width. In: Proceedings of the 12th Annual ACM-SIAM Symposium on Discrete algorithms, SODA 2001, pp. 468–476. SIAM (2001)
17. Kobler, D., Rotics, U.: Edge dominating set and colorings on graphs with fixed clique-width. Discrete Appl. Math. **126**(2–3), 197–221 (2003)
18. Lackner, M., Pichler, R., Rümmele, S., Woltran, S.: Multicut on graphs of bounded clique-width. In: Lin, G. (ed.) COCOA 2012. LNCS, vol. 7402, pp. 115–126. Springer, Heidelberg (2012). https://doi.org/10.1007/978-3-642-31770-5_11
19. Lewis, R., Thiruvady, D., Morgan, K.: Finding happiness: an analysis of the maximum happy vertices problem. Comput. Oper. Res. **103**, 265–276 (2019)
20. Lozin, V.V.: Clique-width of unit interval graphs. arXiv:0709.1935 preprint (2007)
21. Mahadev, N.V.R., Peled, U.N.: Threshold Graphs and Related Topics. Elsevier, Amsterdam (1995)
22. Mahadev, N., Peled, U.: Threshold Graphs and Related Topics. In: Annals of Discrete Mathematics, vol. 56. North Holland (1995)
23. Makowsky, J.A., Rotics, U., Averbouch, I., Godlin, B.: Computing graph polynomials on graphs of bounded clique-width. In: Fomin, F.V. (ed.) WG 2006. LNCS, vol. 4271, pp. 191–204. Springer, Heidelberg (2006). https://doi.org/10.1007/11917496_18
24. Misra, N., Reddy, I.V.: The parameterized complexity of happy colorings. In: Brankovic, L., Ryan, J., Smyth, W.F. (eds.) IWOCA 2017. LNCS, vol. 10765, pp. 142–153. Springer, Cham (2018). https://doi.org/10.1007/978-3-319-78825-8_12

25. Xu, Y., Goebel, R., Lin, G.: Submodular and supermodular multi-labeling, and vertex happiness. arXiv e-prints p. 1606.03185 (2016)
26. Zhang, P., Jiang, T., Li, A.: Improved approximation algorithms for the maximum happy vertices and edges problems. In: Xu, D., Du, D., Du, D. (eds.) COCOON 2015. LNCS, vol. 9198, pp. 159–170. Springer, Cham (2015). https://doi.org/10.1007/978-3-319-21398-9_13
27. Zhang, P., Li, A.: Algorithmic aspects of homophyly of networks. Theor. Comput. Sci. **593**, 117–131 (2015)
28. Zhang, P., Xu, Y., Jiang, T., Li, A., Lin, G., Miyano, E.: Improved approximation algorithms for the maximum happy vertices and edges problems. Algorithmica **80**(5), 1412–1438 (2018)

Graph Hamiltonicity Parameterized by Proper Interval Deletion Set

Petr A. Golovach[1], R. Krithika[2(✉)], Abhishek Sahu[3], Saket Saurabh[3], and Meirav Zehavi[4]

[1] University of Bergen, Bergen, Norway
Petr.Golovach@uib.no
[2] Indian Institute of Technology Palakkad, Palakkad, India
krithika@iitpkd.ac.in
[3] The Institute of Mathematical Sciences, HBNI, Chennai, India
{asahu,saket}@imsc.res.in
[4] Ben-Gurion University, Beersheba, Israel
meiravze@bgu.ac.il

Abstract. The Path Cover and Cycle Cover problems are well-known generalizations of the classical Hamiltonian Path and Hamiltonian Cycle problems. Here, we are given an undirected graph on n vertices and a positive integer r and the task is to check if there are r vertex-disjoint paths (cycles) that together visit all the vertices of the graph exactly once. Path Cover and Cycle Cover remain NP-hard even when restricted to chordal graphs (Information Processing Letters 1986) but are polynomial-time solvable on proper interval graphs (Discrete Mathematics 1993 and Proceedings of WADS 2019). In this paper, we study the complexity of Path Cover and Cycle Cover with respect to a structural parameter, namely, distance to proper interval graphs. In particular, we show that Path Cover and Cycle Cover are fixed-parameter tractable (FPT) when parameterized by k, the size of a proper interval deletion set (a set of vertices whose deletion results in a proper interval graph). For this purpose, we design an algorithm with $\mathcal{O}(2^{\mathcal{O}(k \log k)} n^{\mathcal{O}(1)})$ running time for each of these problems. Our algorithms use several interesting properties of proper interval graphs and a dynamic programming procedure over clique partitions to solve these problems in the mentioned time. As a consequence, we get the same fixed-parameter tractability results for Hamiltonian Cycle and Hamiltonian Path problems with the same parameterization. Recently, Chaplick et al. (Proceedings of WADS 2019) obtained polynomial kernels and compression algorithms for Path Cover and Cycle Cover parameterized by a different measure of similarity with proper interval graphs. Our FPT algorithms also adds to this study of structural parameterizations for these classical problems.

The paper received support from the Research Council of Norway via the project "MULTIVAL" (grant no. 263317). This project received funding from the European Research Council (ERC) under the European Union's Horizon 2020 research and innovation programme (grant agreement No 819416). erc .

Y. Kohayakawa and F. K. Miyazawa (Eds.): LATIN 2020, LNCS 12118, pp. 104–115, 2020.
https://doi.org/10.1007/978-3-030-61792-9_9

1 Introduction

Given an undirected graph, a Hamiltonian path (cycle) is a path (cycle) that visits every vertex exactly once. Determining whether such paths and cycles exist in a graph are the classical NP-complete Hamiltonian Path and Hamiltonian Cycle problems [11]. A path (cycle) cover of a graph is a set of vertex-disjoint paths (cycles) such that every vertex of the graph appears in one path (cycle) in the set. A minimum path (cycle) cover is a path (cycle) cover of minimum size. By definition, the size of a minimum path (cycle) cover of a graph with a Hamiltonian path (cycle) is 1. In the Path Cover and Cycle Cover problems, we are given an undirected graph on n vertices and a positive integer r and the task is to determine if there exist a path cover and cycle cover, respectively, of size at most r. As both these problems are generalizations of Hamiltonian Path and Hamiltonian Cycle, they are NP-hard.

In algorithm theory, an active research area is to efficiently find solutions to otherwise NP-hard problems by restricting the class of admissible inputs. To this end, over the years, many special graph classes have been described in an attempt to deepen the understanding of fundamental properties of graphs and to improve the ability to solve practical problems efficiently. Examples include perfect graphs, bipartite graphs, chordal graphs, split graphs, interval graphs and proper interval graphs. The set of perfect graphs includes bipartite graphs and chordal graphs and the set of chordal graphs includes interval graphs (which in turn includes proper interval graphs) and split graphs. Many problems NP-hard on general graphs like Independent Set, Clique, Coloring and Clique Cover [11] are polynomial-time solvable on these special graph classes [12]. However, Hamiltonian Path and Hamiltonian Cycle are known to be NP-complete on chordal graphs (and even on split graphs) [19].

Another approach to solving NP-hard problems is to design efficient exponential-time algorithms. Parameterized algorithms is one class of such algorithms. In the parameterized algorithms and complexity framework, each problem instance is associated with a non-negative integer k called *parameter*, and a problem is said to be *fixed-parameter tractable* (FPT) if it can be solved in $f(k)n^{\mathcal{O}(1)}$ time for some computable function f, where n is the input size. Algorithms with such running times are called FPT algorithms or parameterized algorithms. Note that any non-polynomial factor in the running time of a parameterized algorithm is required to be a function of only the parameter. Informally, a problem is FPT if it can be solved efficiently (in polynomial-time) on instances with small parameter values. For convenience, the running time $f(k)n^{\mathcal{O}(1)}$ where f grows superpolynomially with k is denoted as $\mathcal{O}^*(f(k))$. Further details on parameterized algorithms can be found in [7].

Parameterizations for Path Cover and Cycle Cover. A natural parameter for Path Cover and Cycle Cover is the solution size, i.e., the size r of the path cover or cycle cover that we are looking for. In order for an FPT algorithm to exist for a parameterized problem, it is necessary that it must be solvable in polynomial time when the parameter is a constant. However, for $r = 1$, Path

Cover is Hamiltonian Path and Cycle Cover is Hamiltonian Cycle. Therefore, an $\mathcal{O}^*(f(r))$-time algorithm (or even an $\mathcal{O}^*(n^{f(r)})$-time algorithm) for either of the problems will imply P=NP. In other words, Path Cover and Cycle Cover are paraNP-hard when parameterized by solution size.

In the early years of parameterized complexity and algorithms, problems were almost always parameterized by the solution size. However, recent research has focused on other parameterizations based on structural parameters in the input. A parameter that has gained significant attention in this context is the size of a *modulator* to a family of graphs. Let $\mathcal{F}$ be a family of graphs. Given a graph H and a set $S \subseteq V(H)$, we say that S is an $\mathcal{F}$*-modulator* if $H - S$ is in $\mathcal{F}$. For example, if $\mathcal{F}$ is the family of independent sets, forests, bipartite graphs, interval graphs and chordal graphs, then the modulator corresponds to a vertex cover, feedback vertex set, odd cycle transversal, interval deletion set and chordal deletion set, respectively. The size of S is also called the *vertex-deletion distance* to $\mathcal{F}$. One of the earliest works in the realm of alternate parameterizations is by Cai [4] who studied Coloring problems parameterized by the vertex-deletion distance to various graph classes including bipartite graphs and split graphs. Fellows et al. [10] studied alternate parameterizations for problems that were proven to be intractable with respect to the standard parameterization. This led to a whole new ecology program and opened up a floodgate of new and exciting research. We refer to [15] for a detailed introduction to the whole program as well as the thesis of Jansen [14]. There has also been an extensive study of structural parameters for problems related to Path Cover and Cycle Cover such as Cycle Packing, Longest Path and Longest Cycle [3,17].

Our Choice of Parameter and Our Result. Focusing on structural parameters for Path Cover and Cycle Cover, the main topic of this paper, as these problems are NP-complete on chordal graphs [2], we cannot hope to have an algorithm with running time $\mathcal{O}^*(n^{f(k)})$, where k is the size of the given chordal deletion set, unless P=NP. Therefore, a natural parameter in this context is the size of the given modulator to a subclass of chordal graphs on which these problems are polynomial-time solvable. This is the starting point of our work. A popular graph class that is a subset of the class of chordal graphs (and even interval graphs) is the one of *proper interval graphs* (also known as *indifference graphs* and *unit interval graphs* in the literature). A graph is a proper interval graph if its vertices can be assigned to intervals such that there is an edge between two vertices if and only if their corresponding intervals have non-empty intersection. Further, this set of intervals should satisfy the property that no interval properly contains another. Proper interval graphs have a rich geometric structure which makes them amenable to efficient algorithms for most classical problems [12]. They have applications in several fields like scheduling, archaeology, developmental psychology and DNA sequencing [12]. Also, proper interval graphs are well studied in the framework of parameterized algorithms in the context of vertex-deletion problems [5,13]. It is also known that Cycle Cover and Path Cover can be solved in polynomial time on proper interval graphs [6,8]. In this paper, we study the complexity of Path Cover and Cycle Cover parameterized by the size of a proper interval deletion set. A set of vertices is called a *proper interval deletion set* if its deletion results in a proper interval graph. We show that PATH COVER and CYCLE COVER parameterized by the size of a proper interval deletion set are FPT and this is the main result of the paper.

Theorem 1. PATH COVER *and* CYCLE COVER *parameterized by the size k of a proper interval deletion set can be solved in $\mathcal{O}^*(2^{\mathcal{O}(k \log k)})$ time.*

We assume that the proper interval deletion set T is part of the input. This assumption is reasonable as given a graph H and an integer k, there is an algorithm that, in $\mathcal{O}^*(6^k)$ time, outputs a proper interval deletion set of size at most k (if one exists) [5,13]. Our algorithms use several interesting properties of proper interval graphs and a dynamic programming procedure over clique partitions to solve these problems in the mentioned time. As a consequence, we get the same fixed-parameter tractability results for HAMILTONIAN CYCLE and HAMILTONIAN PATH problems with the same parameterization. By parameterizing PATH COVER and CYCLE COVER with respect to the size of a proper interval deletion set as parameter, we attempt to understand the complexity of the problem on *almost proper interval graphs.* Recently, Chaplick et al. [6] obtained polynomial kernels and compression algorithms for PATH COVER and CYCLE COVER parameterized by a different measure of similarity with proper interval graphs. Our FPT algorithms also add to this study of structural parameterizations for these classical problems.

Overview of Our Algorithm and Techniques. Consider an instance $\mathcal{I} = (H, T, r)$ of PATH COVER or CYCLE COVER. Let $\mathcal{P}$ be a minimum path cover of H that we are looking for. We first guess the following properties of $\mathcal{P}$. Intialize the set of variables $\mathcal{S}$ to be the empty set. Let G denote $H - T$.

- We guess the number ℓ of paths in $\mathcal{P}$ that have a vertex from T. Clearly, $\ell \leq k$ and the number of choices for ℓ is k. Let $\mathcal{P}_m = \{P_1, \ldots, P_\ell\}$ denote the set of these paths.
- For each $P_i \in \mathcal{P}_m$, we guess if P_i has zero, one or two endpoints in T. The number of possible choices in this step is $2^{\mathcal{O}(k)}$.
- For each $P_i \in \mathcal{P}_m$, we guess the order $\lambda(P_i)$ of the vertices of $V(P_i) \cap T$ on P_i. The number of possible choices in this step is $2^{\mathcal{O}(k \log k)}$.
- For each $P_i \in \mathcal{P}_m$ that starts at a vertex in G, we add $S^i(\$, t)$ to $\mathcal{S}$ where t is the first vertex according to $\lambda(P_i)$. The variable $S^i(\$, t)$ indicates that we need to assign it a path in G that ends in a neighbour of t.
- For each $P_i \in \mathcal{P}_m$ that ends at a vertex in G, we add $S^i(t', \$)$ to $\mathcal{S}$ where t' is the last vertex according to $\lambda(P_i)$. The variable $S^i(t', \$)$ indicates that we need to assign it a path in G that starts in a neighbour of t'.
- For each $P_i \in \mathcal{P}_m$, for each pair of vertices $t \in T$ and $t' \in T$ that are consecutive according to $\lambda(P_i)$, we add $S^i(t, t')$ to $\mathcal{S}$ indicating that $S^i(t, t')$ should be assigned a path in G that is between a neighbour of t and a neighbour of t'.

Clearly, $|\mathcal{S}| = \mathcal{O}(k)$ and the task of finding a minimum path cover of H reduces to the problem of finding an assignment of vertex-disjoint paths in G to the variables in $\mathcal{S}$ satisfying the appropriate endpoint constraints while minimizing the size of a minimum path cover of $G[X]$ where $V(G) \setminus X$ is the set of vertices that are on some path assigned to some variable in $\mathcal{S}$. Similarly, the task of finding a minimum cycle cover of H also boils down to the problem of finding certain constrained paths in G which is a proper interval graph. We first show that these paths are very structured due to the properties given by a proper interval ordering. Then, we describe a dynamic programming procedure to find such structured paths.

Preliminaries. For graph theoretic terms not defined here, refer to [9]. A *path* $P = (v_1, \ldots, v_\ell)$ is a sequence of distinct vertices where every consecutive pair of vertices are adjacent. We call P a (v_1, v_ℓ)-path and say that P *starts* at v_1 and *ends* at v_ℓ. The *length* of P is defined as $|V(P)|$. A *cycle* $(v_1, \ldots, v_\ell)$ is a sequence of vertices such that $v_1, \ldots, v_\ell$ is a path and $v_\ell v_1$ is an edge. For a collection $\mathcal{P}$ of paths (or cycles), $V(\mathcal{P})$ denotes the set $\bigcup_{P \in \mathcal{P}} V(P)$. The concatenation of paths $P_1 = (v_1, \cdots, v_{j-1}, v_j)$ and $P_2 = (v_j, v_{j+1}, \cdots, v_\ell)$ such that $V(P_1) \cap V(P_2) = \{v_j\}$ is defined as the path $P_3 = (v_1, \cdots, v_{j-1}, v_j, v_{j+1}, \cdots, v_\ell)$. The vertices of a proper interval graph G can be ordered by a permutation $\pi : V(G) \to [|V(G)|]$ having the following property.

Proposition 1 ([18]). *Let G be a proper interval graph with proper interval ordering π. For every pair u, v of vertices with $\pi(u) < \pi(v)$, if $uv \in E(G)$, then $\{w \in V(G) \mid \pi(u) \leq \pi(w) \leq \pi(v)\}$ is a clique in G.*

Given a subset S of vertices of a proper interval graph G with proper interval ordering π, the leftmost vertex in S is the vertex v in S with least $\pi(v)$ and the rightmost vertex in S is the vertex u in S with greatest $\pi(u)$. Given a proper interval representation, the following result states that the vertex set of the proper interval graph can be organized into a sequence of cliques satisfying certain properties.

Proposition 2 ([16]). *Given a proper interval graph G with proper interval ordering π, there is a linear-time algorithm that outputs a partition of $V(G)$ into a sequence $Q_1, \cdots, Q_q$ of (pairwise vertex-disjoint) cliques satisfying the following properties. (1) For each pair of vertices $u \in Q_i$, $v \in Q_j$ with $1 \leq i < j \leq q$, $\pi(v) > \pi(u)$. (2) For every edge $uv \in E(G)$, there exists $1 \leq i \leq q$ such that either $u, v \in Q_i$ or $u \in Q_i$ and $v \in Q_{i+1}$.*

Observe that this partition is different from the classical clique path decomposition of (proper) interval graphs. We refer to the ordered set of cliques $\mathcal{Q} = \{Q_1, \cdots, Q_q\}$ as *clique partition* of G.

2 Structure of Path and Cycle Covers

Recall that a path cover $\mathcal{P}$ (cycle cover $\mathcal{C}$) of a graph H is a set of vertex-disjoint paths (cycles) in H such that $V(\mathcal{P}) = V(H)$ ($V(\mathcal{C}) = V(H)$). Consider an instance $\mathcal{I} = (H, T, r)$ of PATH COVER or CYCLE COVER. Let π and $\mathcal{Q} = \{Q_1, \cdots, Q_q\}$ denote the proper interval ordering and clique partition of $G = H - T$, respectively, obtained in polynomial time [12,16]. Let $T = \{t_1, \ldots, t_k\}$.

2.1 Paths and Cycles in Proper Interval Graphs

In this section, we list some fundamental properties of paths and cycles in proper interval graphs.

Proposition 3 ([1]). *Every connected proper interval graph has a Hamiltonian path, and a proper interval graph has a Hamiltonian cycle if and only if it is 2-connected with at least three vertices.*

Definition 1 ***(Monotone path).*** *Let G be a proper interval graph with proper interval ordering π. A path $P = (v_1, \ldots, v_r)$ in G is called* monotone *if $\pi(v_1) < \pi(v_2) < \cdots < \pi(v_r)$ or $\pi(v_1) > \pi(v_2) > \cdots > \pi(v_r)$.*

Observation 2. *If P is a path in a proper interval graph G with proper interval ordering π, then there is a monotone path P' in G with $V(P) = V(P')$.*

Definition 2 ***(2-monotone cycle).*** *Let G be a proper interval graph with proper interval ordering π. A cycle $C = (v_1, v_2, \ldots, v_i, v_{i+1}, \ldots, v_r)$ in G is called* 2-monotone *if there is an integer $i \in [r]$ such that $(v_1, v_2, \ldots, v_i)$ and $(v_1, v_r, \ldots, v_i)$ are monotone paths that are internally vertex-disjoint and start and end at the same vertices.*

Observation 3. *If C is a cycle in a proper interval graph G with proper interval ordering π, then there is a 2-monotone cycle C' in G such that $V(C) = V(C')$.*

Next, we define the notion of i-monotone paths in proper interval graphs.

Definition 3 ***(i-Monotone path).*** *Let G be a proper interval graph with proper interval ordering π. For a positive integer i, a path P is called* i-monotone *if P is the concatenation of i monotone paths.*

Proposition 4. *If P is a path from a vertex s in a proper interval graph G with proper interval ordering π, then there is an i-monotone path P' in G from s with $V(P) = V(P')$ for some $i \in [2]$.*

Proposition 5. *If P is a (s,t)-path in a proper interval graph G with proper interval ordering π, then there is an i-monotone (s,t)-path P' in G with $V(P) = V(P')$ for some $i \in [3]$.*

2.2 Canonical Minimum Path and Cycle Covers

For a path cover $\mathcal{P}$ of H, define the following sets.

- $\mathcal{P}_o = \{P_i \in \mathcal{P} : V(P_i) \cap T = \emptyset\}$, the set of paths in $\mathcal{P}$ that are completely contained in G.

- $\mathcal{P}_m = \mathcal{P} \setminus \mathcal{P}_o$, the set of paths in $\mathcal{P}$ that have at least one vertex from T.
- $M(\mathcal{P})$ is the set of maximal subpaths of paths in $\mathcal{P}_m$ that are contained in G. That is, for each P in $\mathcal{P}_m$, a subpath S of P with $V(S) \subseteq V(G)$ is in $M(\mathcal{P})$ if and only if there is no subpath S' of P such that $V(S') \subseteq V(G)$ and $V(S) \subset V(S')$.
- $S(\mathcal{P})$ is the set of maximal subpaths of paths in $M(\mathcal{P})$ that are monotone. That is, for each P in $M(\mathcal{P})$, a subpath S of P is in $S(\mathcal{P})$ if and only if S is monotone and there is no monotone subpath S' of P such that $V(S) \subset V(S')$.

We refer to elements of $S(\mathcal{P})$ as segments of $\mathcal{P}$. In the example shown below, $\mathcal{P}_m$ contains a path P from t_1 to t_2 and $M(\mathcal{P}) = \{S\}$ where $S = (a, b, c, d, e, f, g, h, i, j, k, l)$. The set of segments is $S(\mathcal{P}) = \{S_1, S_2, S_3\}$ where $S_1 = (c, b, a)$, $S_2 = (c, d, e, f, g, h)$ and $S_3 = (l, k, j, i, h)$.

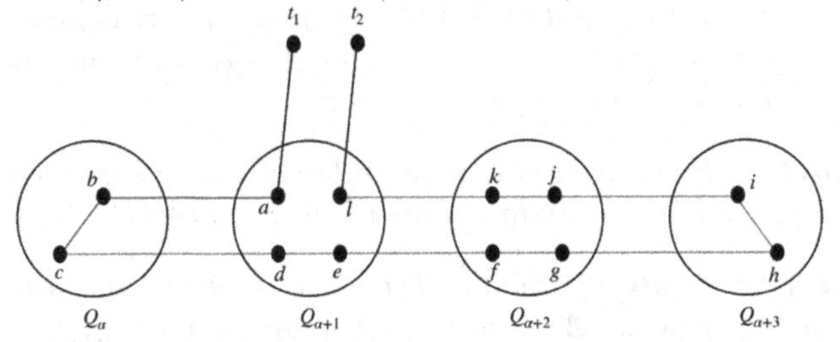

Definition 4 ***(Pseudo-consecutive vertices).*** *Two vertices $u, v \in T$ are said to be pseudo-consecutive if u and v are in the same path P in $\mathcal{P}$ and there is no other vertex of T that is in the subpath of P between u and v.*

In the example shown above, t_1 are t_2 are pseudo-consecutive.

Definition 5 ***(Pseudo-adjacent vertices).*** *Let y be a vertex in G that is an endpoint of some path in $\mathcal{P}$. A vertex $x \in T$ is said to be pseudo-adjacent to y if x and y are in the same path P in $\mathcal{P}$ and there is no other vertex of T that is in the subpath of P between y and x.*

Definition 6 **(Relevant and irrelevant vertices in Q_i).** *For a path S in $S(\mathcal{P}) \cup \mathcal{P}_o$ that contains at least one vertex from Q_i, the set $R_i(\mathcal{P}, S)$ of relevant vertices is $V(S) \cap Q_i$ if $|V(S) \cap Q_i| \leq 2$, otherwise $R_i(\mathcal{P}, S)$ consists of the leftmost and the rightmost vertices of $V(S) \cap Q_i$. The collection $\mathcal{R}_i(\mathcal{P})$ of relevant vertices contains the set $R_i(\mathcal{P}, S)$ of every path S in $S(\mathcal{P}) \cup \mathcal{P}_o$ that contains at least one vertex from Q_i. A vertex in Q_i that is not in $\mathcal{R}_i(\mathcal{P})$ called irrelevant.*

An example is shown in the following figure.

Definition 7 ***(Nice path cover).*** *A path cover $\mathcal{P}$ is said to be nice if the following properties hold.*

- *Every path in $\mathcal{P}_o$ is monotone.*
- *For any $i \in [q]$, there is at most one path P in $\mathcal{P}_o$ such that $Q_i \cap V(P) \neq \emptyset$.*

- *For every path P in $\mathcal{P}_m$, for every pair of pseudo-consecutive modulator vertices t, t' in P that are not consecutive in P, the maximal subpath of P between t and t' that is contained in G is i-monotone for some $i \in [3]$.*
- *For every path P in $\mathcal{P}_m$ starting (or ending) at a vertex s in G whose pseudo-adjacent modulator vertex is t, the maximal subpath of P contained in G that is between s and the neighbour of t in P is i-monotone for some $i \in [2]$.*
- *For any $i \in [q]$, if $|Q_i| > 10k$, then each segment $S \in \mathcal{S}(\mathcal{P})$ with $V(S) \cap Q_i \neq \emptyset$ that neither starts nor ends at a vertex in Q_i satisfies $|V(S) \cap Q_i| \geq 2$.*

Lemma 1. *If $\mathcal{P}$ is a nice minimum path cover of H, then $|\mathcal{S}(\mathcal{P})| \leq 4k$ and for any $i \in [q]$, $|\mathcal{R}_i(\mathcal{P})| \leq 8k + 2$.*

Definition 8 ***(Leftmost and rightmost set of vertices).*** *Consider a subset S of vertices of G. If $|S| > 10k$, then let $LM(S)$ denote the $10k$ leftmost vertices of S and $RM(S)$ denote the $10k$ rightmost vertices of S. Otherwise, $LM(S) = RM(S) = S$.*

Definition 9 ***(Boundary vertices of Q_i).*** *Consider the following sets.*

- *$L^i = LM(Q_i)$ and $R^i = RM(Q_i)$.*
- *For each $x \in T$, $L^i_x = LM(Q_i \cap N(x))$ and $R^i_x = RM(Q_i \cap N(x))$.*
- *For each $x, y \in T$, $L^i_{xy} = LM(Q_i \cap N(x) \cap N(y))$ and $R^i_{xy} = RM(Q_i \cap N(x) \cap N(y))$.*

The set $B(Q_i) = L^i \cup R^i \cup_{x \in T} (L^i_x \cup R^i_x) \cup_{y,z \in T} (L^i_{yz} \cup R^i_{yz})$ is called the boundary vertices of Q_i.

Notice that for any $i \in [q]$, the size of $B(Q_i)$ is $\mathcal{O}(k^3)$.

Definition 10. *For every vertex $v \in V(G)$, let $\rho(v)$ denote the tuple (m_1, m_2) defined as follows: m_1 is the vertex in T that preceeds v in the path in $\mathcal{P}$ that contains v. If no such vertex exists, then $m_1 = \$$. Similarly, m_2 is the vertex in T that succeeds v in the path in $\mathcal{P}$ that contains v. If no such vertex exists, then $m_2 = \$$.*

Definition 11 ***(Canonical path cover).*** *A nice path cover $\mathcal{P}$ is said to be canonical if for each $i \in [q]$ and each $S \in \mathcal{S}(\mathcal{P})$, $\mathcal{R}_i(\mathcal{P}, S) \subseteq B(Q_i)$ and the following properties are satisfied.*

- *For a segment $S \in \mathcal{S}(\mathcal{P})$ passing through Q_i, its first and last vertex in Q_i are in $L^i \cup R^i$.*
- *For a segment $S \in \mathcal{S}(\mathcal{P})$ that ends in Q_i, the following hold.*
  - *If $S \cap Q_i = (a)$ and $\rho(a) = (\$, \$)$, then $a \in L^i$. If $S \cap Q_i = (a)$ and $\rho(a) = (m_1, m_2)$, then $a \in L^i_{m_1 m_2} \cup R^i_{m_1 m_2}$. If $S \cap Q_i = (a)$ and $\rho(a) = (m_1, \$)$, then $a \in L^i_{m_1} \cup R^i_{m_1}$. If $S \cap Q_i = (a)$ and $\rho(a) = (\$, m_2)$, then $a \in L^i_{m_2} \cup R^i_{m_2}$.*
  - *If $\mathcal{R}_i(\mathcal{P}, S) = (a, b)$ and $\rho(a) = (m_1, \$)$, $\rho(b) = (\$, m_2)$, then $a \in L^i_{m_1} \cup R^i_{m_1}, b \in R^i_{m_2} \cup L^i_{m_2}$. If $\mathcal{R}_i(\mathcal{P}, S) = (a, b)$ and $\rho(a) = (\$, \$)$, $\rho(b) = (\$, m_2)$, then $a \in L^i, b \in R^i_{m_2} \cup L^i_{m_2}$. If $\mathcal{R}_i(\mathcal{P}, S) = (a, b)$ and $\rho(a) = (m_1, \$)$, $\rho(b) = (\$, \$)$, then $a \in L^i_{m_1} \cup R^i_{m_1}, b \in R^i \cup L^i$. If $\mathcal{R}_i(\mathcal{P}, S) = (a, b)$ and $\rho(a) = (\$, \$)$, $\rho(b) = (\$, \$)$, then $a \in L^i, b \in R^i$.*

- *For a segment $S \in S(\mathcal{P})$ that starts in Q_i, the following hold.*
  - *If $S \cap Q_i = (a)$ and $\rho(a) = (\$, \$)$, then $a \in R^i$. If $S \cap Q_i = (a)$ and $\rho(a) = (m_1, \$)$, then $a \in L^i_{m_1} \cup R^i_{m_1}$. If $S \cap Q_i = (a)$ and $\rho(a) = (\$, m_2)$, then $a \in R^i_{m_2} \cup L^i_{m_2}$.*
  - *If $\mathcal{R}_i(\mathcal{P}, S) = (a, b)$ and $\rho(a) = (\$, \$)$, $\rho(b) = (\$, m_2)$, then $a \in L^i \cup R_i, b \in R^i_{m_2} \cup L^i_{m_2}$. If $\mathcal{R}_i(\mathcal{P}, S) = (a, b)$ and $\rho(a) = (m_1, \$)$, $\rho(b) = (\$, \$)$, then $a \in L^i_{m_1} \cup R^i_{m_1}, b \in R^i$. If $\mathcal{R}_i(\mathcal{P}, S) = (a, b)$ and $\rho(a) = (\$, \$)$, $\rho(b) = (\$, \$)$, then $a \in L^i, b \in R^i$.*

Lemma 2. *Given a path cover $\mathcal{P}$ of H, a canonical path cover $\mathcal{P}^*$ of H with $|\mathcal{P}^*| \leq |\mathcal{P}|$ can be obtained in polynomial time.*

We define a similar notion of canonical cycle covers and show that there exists a minimum cycle cover of H that is canonical. The details are deferred to the full version of the paper.

3 Finding Canonical Minimum Path and Cycle Covers

Let $\mathcal{P}$ denote a minimum canonical path cover of H. We define the following functions to understand the relationship between the segments of a canonical path cover $\mathcal{P}$. The functions $\mathcal{F} : S(\mathcal{P}) \times S(\mathcal{P}) \to \{0, 1\}$ and $\mathcal{L} : S(\mathcal{P}) \times S(\mathcal{P}) \to \{0, 1\}$ are defined such that $\mathcal{F}(S, S') = 1$ ($\mathcal{L}(S, S') = 1$) if and only if S and S' start (end) at the same vertex. The functions $\mathcal{F}_1 : S(\mathcal{P}) \to T \cup \{0\}$ and $\mathcal{L}_1 : S(\mathcal{P}) \to T \cup \{0\}$ are defined such that $\mathcal{F}_1(S) = t$ if S starts immediately after t, otherwise $\mathcal{F}_1(S) = 0$. Similarly, $\mathcal{L}_1(S) = t$ if S ends just before t, otherwise $\mathcal{L}_1(S) = 0$.

Given $\mathcal{P}$, determining $S(\mathcal{P})$ is easy and in turn given $S(\mathcal{P})$, determining $\mathcal{F}$, $\mathcal{F}_1$, $\mathcal{L}$ and $\mathcal{L}_1$ is easy. It is now natural to ask what choices of $(\mathcal{F}, \mathcal{L}, \mathcal{F}_1, \mathcal{L}_1)$ lead to a set $S(\mathcal{P})$ that in turn leads to a minimum canonical path cover $\mathcal{P}$. Let us first guess the size of $S(\mathcal{P})$. From Lemma 1, it is at most $4k$. For a correct choice of this number, the choice $(\mathcal{F}, \mathcal{L}, \mathcal{F}_1, \mathcal{L}_1)$ that minimizes the size of a minimum path cover of $G[X]$ where $V(G) \setminus X$ is the set of vertices that are in some segment assigned to a variable in $\mathcal{S}$ is the one the results in that minimum path cover.

The Guessing Phase. With this information, we proceed as follows. Let $\mathcal{P}$ be a minimum canonical path cover that we are looking for. We first guess the following properties of $\mathcal{P}$. Intialize $\mathcal{S}$ to be the empty set. We guess the number ℓ of paths in $\mathcal{P}_m$. Clearly, $\ell \leq k$ and let $P_1, \ldots, P_\ell$ denote the paths in $\mathcal{P}_m$. For each $P_i \in \mathcal{P}_m$, we guess if P_i has zero, one or two endpoints in T. For each $P_i \in \mathcal{P}_m$, we guess the order of vertices of $V(P_i) \cap T$. For each $P_i \in \mathcal{P}_m$, for each pair of pseudo-consecutive vertices t and t' in P_i, we guess if t and t' are consecutive in P_i (in which case t and t' must be adjacent) or not. It t and t' are not consecutive in P_i, then we guess if the maximal subpath P of the (t, t')-path that is contained in G is 1-monotone or 2-monotone or 3-monotone.

- If P is 1-monotone, then add S^i to $\mathcal{S}$ and set $\mathcal{F}_1(S^i) = t$, $\mathcal{L}_1(S^i) = t'$.
- If P is 2-monotone, then add S_1^i and S_2^i to $\mathcal{S}$ and either set $\mathcal{L}_1(S_1^i) = t$, $\mathcal{F}_1(S_1^i) = 0$, $\mathcal{F}_1(S_2^i) = 0$, $\mathcal{L}_1(S_2^i) = t'$, $\mathcal{F}(S_1^i, S_2^i) = 1$, $\mathcal{L}(S_1^i, S_2^i) = 0$ or set $\mathcal{F}_1(S_1^i) = t$, $\mathcal{L}_1(S_1^i) = 0$, $\mathcal{F}_1(S_2^i) = t'$, $\mathcal{L}_1(S_2^i) = 0$, $\mathcal{L}(S_1^i, S_2^i) = 1$, $\mathcal{F}(S_1^i, S_2^i) = 0$.
- If P is 3-monotone, then we add the variables S_1^i, S_2^i and S_3^i to $\mathcal{S}$. We set $\mathcal{L}_1(S_1^i) = t$, $\mathcal{F}_1(S_1^i) = 0$, $\mathcal{F}_1(S_2^i) = 0$, $\mathcal{L}_1(S_2^i) = 0$, $\mathcal{F}_1(S_3^i) = t'$, $\mathcal{L}_1(S_3^i) = 0$, $\mathcal{F}(S_1^i, S_2^i) = 1$, $\mathcal{L}(S_2^i, S_3^i) = 1$, $\mathcal{F}(S_1^i, S_3^i) = 0$, $\mathcal{F}(S_2^i, S_3^i) = 0$, $\mathcal{L}(S_1^i, S_2^i) = 0$, $\mathcal{L}(S_1^i, S_3^i) = 0$.

For each $P_i \in \mathcal{P}_m$, for each ordered pair of pseudo-adjacent vertices $x \in T$ and $y \in V(G)$, we guess if the maximal subpath P of the (x, y)-path that is contained in G is 1-monotone or 2-monotone. We also add variables to $\mathcal{S}$ as mentioned earlier. For each pair S and S' of variables in $\mathcal{S}$ such that $\mathcal{F}(S, S')$ (or $\mathcal{L}(S, S')$) is not yet set is set to 0. Similarly, for each variable S in $\mathcal{S}$ such that $\mathcal{F}_1(S)$ (or $\mathcal{L}_1(S)$) is not yet set is set to 0.

The total number of choices is $2^{\mathcal{O}(k \log k)}$. Once a choice is fixed, the problem of finding a mnimum path cover $\mathcal{P}$ now reduces to the problem of finding an assignment of segments to variables in $\mathcal{S}$ that satisfy the relationships given by $(\mathcal{S}, \mathcal{F}, \mathcal{L}, \mathcal{F}_1, \mathcal{L}_1)$ while minimizing the size of a minimum path cover of $G - X$ where X is the set of vertices of H that are in a segment assigned to some variable in $\mathcal{S}$. In other words, we find an assignment of segments to variables in $\mathcal{S}$ that satisfy the relationships given by $(\mathcal{S}, \mathcal{F}, \mathcal{L}, \mathcal{F}_1, \mathcal{L}_1)$ resulting in a set of paths $\mathcal{P}_m$ while minimizing the number of paths in $\mathcal{P}_o$. Note that not all choices of $(\mathcal{S}, \mathcal{F}, \mathcal{L}, \mathcal{F}_1, \mathcal{L}_1)$ may necessarily lead to a minimum path cover $\mathcal{P}_o \cup \mathcal{P}_m$ of H. However, at least one of the choices that we generate leads to one.

Consider a particular choice of $(\mathcal{S}, \mathcal{F}, \mathcal{L}, \mathcal{F}_1, \mathcal{L}_1)$. This fixes how the paths in $\mathcal{P}_m$ interact with T. That is, for any path P in $\mathcal{P}_m$, the vertices of T that are in P and their order in P are fixed. Furthernore, the paths between any two pseudo-consecutive vertices and the paths between any two pseudo-adjacent vertices are also fixed. This also fixes the number of segments and the relationship among the segments. We will describe a dynamic programming algorithm that finds a minimum canonical path cover respecting this choice $\vartheta = (\mathcal{S}, \mathcal{F}, \mathcal{L}, \mathcal{F}_1, \mathcal{L}_1)$.

Finding an Assignment of Segments for $\vartheta = (\mathcal{S}, \mathcal{F}, \mathcal{L}, \mathcal{F}_1, \mathcal{L}_1)$. Let $Q_0 = \emptyset$. For each $i \in [q]$, let G_i denote the graph $G[Q_1 \cup \cdots \cup Q_i]$. Let us first understand the interaction of the solution (minimum canonical path cover with the properties given by $(\mathcal{S}, \mathcal{F}, \mathcal{L}, \mathcal{F}_1, \mathcal{L}_1)$) with G_i. Subsequently, we refer to $\mathcal{S}$ as segments instead of variables that have to be assigned segments.

Index of an Entry: An entry in the table T_i is indexed by a tuple $(\mathcal{S}_f, \mathcal{X}, X_o, \mathcal{A}, B)$ with the following interpretation.

- $\mathcal{S}_f \subseteq \mathcal{S}$ denotes the segments that have no vertex from $Q_{i+1} \cup \ldots \cup Q_q$. That is, these segments are completely contained in G_i.
- $\mathcal{X}$ denotes the set of relevant vertices of all segments from $\mathcal{S}$ in Q_i. That is, for every $S \in \mathcal{S}$, X_S in $\mathcal{X}$ is the set of relevant vertices of S in Q_i. If X_S is the empty set, then the segment corresponding to S has no vertex from Q_i. Otherwise, X_S has a single vertex or an ordered pair of vertices. If X_S is an

ordered pair of vertices (v_1, v_2), we call v_1 the first relevant vertex (denoted by $X_S(1)$) and v_2 the last relevant vertex (denoted by $X_S(2)$) of S in Q_i. If X_S has a single vertex v, we call v first and last relevant vertex of S in Q_i.
- X_o denotes the set of relevant vertices of the unique monotone path in $\mathcal{P}_o$ that has a vertex from Q_i.
- $\mathcal{A} \in [0, 1, 2, 3, 4]^{|\mathcal{S}|}$ represents the interactions of segments from $\mathcal{S}$ with Q_i.
  - $a_S = 0$ iff the segment S does not intersect Q_i and $a_S = 1$ iff the segment S has at least one vertex from Q_i but neither starts nor ends in Q_i.
  - $a_S = 2$ ($a_S = 3$) iff the segment S starts (ends) but does not end (start) in Q_i and $a_S = 4$ iff the segment S starts and ends in Q_i.
- Similarly, $B \in \{0, 1, 2, 3, 4\}$ represents the interaction of the monotone path P in $\mathcal{P}_o$ with Q_i.

We define the notion of valid indices and show that for each $i \in [q]$, the maximum number of valid indices is $\mathcal{O}^*(2^{\mathcal{O}(k \log k)})$.

(Optimum) Partial Solutions: For $\sigma = (\mathcal{S}_f, \mathcal{X}, X_o, \mathcal{A}, B)$, a collection of paths $\mathcal{P}_d \uplus \mathcal{P}_u$ is a partial solution of $T_i(\sigma)$ if the following conditions hold. Let $h : \mathcal{S}_f \uplus \{S \in \mathcal{S} : a_S \in \{1, 2\}\} \to \mathcal{P}_u$ denote the assignment of paths in $\mathcal{P}_u$ to variables in $\mathcal{S}_f \uplus \{S \in \mathcal{S} : a_S \in \{1, 2\}\}$.

- $|\mathcal{P}_u| = |\mathcal{S}_f| + |\{S \in \mathcal{S} : a_S \in \{1, 2\}\}|$ and h is injective. Further, every path P in $\mathcal{P}_d \cup \mathcal{P}_u$ is monotone and satisfies $V(P) \subseteq Q_1 \cup \ldots Q_i$.
- Every vertex in $Q_1 \cup \cdots \cup Q_i$ is in a path in $\mathcal{P}_d \cup \mathcal{P}_u$. Further, the paths in $\mathcal{P}_u$ take their respective relevant vertices in Q_i according to the assignment $\mathcal{X}$ i.e. the first and last vertices of X_S are the first and last vertices $V(h(S)) \cap Q_i$.
- Every pair P_i, P_j of distinct paths in $\mathcal{P}_d \cup \mathcal{P}_u$ are internally vertex-disjoint. Further, they are vertex-disjoint except when $\mathcal{F}(P_i, P_j) = 1$ or $\mathcal{L}(P_i, P_j) = 1$.
- For each $S \in \text{dom}(h)$, $h(S)$ starts at a vertex in $N(m_1)$ if $\mathcal{F}_1(S) = m_1$ and for each $S \in \mathcal{S}_f$, $h(S)$ ends at a vertex in $N(m_2)$ if $\mathcal{L}_1(S) = m_2$.
- If $\mathcal{F}(S_i, S_j) = 1$ and $S_i \in \text{dom}(h)$, then $S_j \in \text{dom}(h)$ and $h(S_i)$ and $h(S_j)$ start at the same vertex. Also, if $\mathcal{L}(S_i, S_j) = 1$ and $S_i \in \mathcal{S}_f$, then $S_j \in \mathcal{S}_f$ and $h(S_i)$ and $h(S_j)$ end at the same vertex.
- At most one path from $\mathcal{P}_d$ has relevant vertices in Q_i and these vertices are given by X_o. Any path $P = h(S)$ with $S \in \text{dom}(h)$ and at most one path from $\mathcal{P}_d$ start and end in Q_i iff a_S (and/or B) is in $\{2, 4\}$. Any path $P = h(S)$ with $S \in \text{dom}(h)$ and at most one path from $\mathcal{P}_d$ do not start but end in Q_i iff (and/or B) is in $\{1, 3\}$.

Let $B^* = 1$ if $B \in \{3, 4\}$, 0 otherwise. Over all possible partial solutions $\mathcal{P}_d \cup \mathcal{P}_u$, $T_i(\sigma)$ stores the one that minimizes the value of $|\mathcal{P}_d| - (1 - B^*)$. Such a partial solution is called an optimum partial solution. In other words, $T_i(\sigma)$ stores a partial solution that minimizes the number of paths contained inside $\mathcal{P}_d$ that end in G_i. If there is a path in $\mathcal{P}_d$ with B value either 1 or 2 in Q_i, it has a vertex in Q_{i+1} and hence not counted. We also store the size of an optimum solution denoted by $|T_i(\sigma)|$. An optimum solution for an entry in T_q where every path in the solution has ended in $G_q = G$ gives the required answer. We show

how to compute all the entries in T in $\mathcal{O}^*(2^{\mathcal{O}(k \log k)})$ time. This results in an $\mathcal{O}^*(2^{\mathcal{O}(k \log k)})$-time algorithm for Path Cover.

A similar algorithm for Cycle Cover based on finding canonical minimum cycle covers is described in the full version of the paper.

References

1. Bertossi, A.A.: Finding hamiltonian circuits in proper interval graphs. Inf. Process. Lett. **17**(2), 97–101 (1983)
2. Bertossi, A.A., Bonuccelli, M.A.: Hamiltonian circuits in interval graph generalizations. Inf. Process. Lett. **23**(4), 195–200 (1986)
3. Bodlaender, H.L., Jansen, B.M.P., Kratsch, S.: Kernel bounds for path and cycle problems. Theoret. Comput. Sci. **511**, 117–136 (2013)
4. Cai, L.: Parameterized complexity of vertex colouring. Discrete Appl. Math. **127**(3), 415–429 (2003)
5. Cao, Y.: Unit interval editing is fixed-parameter tractable. Inf. Comput. **253**(Part 1), 109–126 (2017)
6. Chaplick, S., Fomin, F.V., Golovach, P.A., Knop, D., Zeman, P.: Kernelization of Graph Hamiltonicity: Proper H-Graphs. In: Friggstad, Z., Sack, J.-R., Salavatipour, M.R. (eds.) WADS 2019. LNCS, vol. 11646, pp. 296–310. Springer, Cham (2019). https://doi.org/10.1007/978-3-030-24766-9_22
7. Cygan, M., et al.: Parameterized Algorithms. Springer, Cham (2015). https://doi.org/10.1007/978-3-319-21275-3
8. Damaschke, P.: Paths in interval graphs and circular arc graphs. Discrete Math. **112**(1), 49–64 (1993)
9. Diestel, R.: Graph Theory. GTM, vol. 173. Springer, Heidelberg (2017). https://doi.org/10.1007/978-3-662-53622-3
10. Fellows, M.R., Lokshtanov, D., Misra, N., Mnich, M., Rosamond, F.A., Saurabh, S.: The complexity ecology of parameters: an illustration using bounded max leaf number. Theory Comput. Syst. **45**(4), 822–848 (2009)
11. Garey, M.R., Johnson, D.S.: Computers and Intractability: A Guide to the Theory of NP-Completeness. W.H.Freeman and Company (1979)
12. Golumbic, M.C.: Algorithmic Graph Theory and Perfect Graphs, Second Edition. Elsevier Science B.V. (2004)
13. van't Hof, P., Villanger, Y.: Proper interval vertex deletion. Algorithmica **65**(4), 845–867 (2013)
14. Jansen, B.M.P.: The Power of Data Reduction: Kernels for Fundamental Graph Problems. Ph.D. thesis, Utrecht University, The Netherlands (2013)
15. Jansen, B.M.P., Fellows, M.R., Rosamond, F.A.: Towards fully multivariate algorithmics: parameter ecology and the deconstruction of computational complexity. Eur. J. Combinat. **34**(3), 541–566 (2013)
16. Ke, Y., Cao, Y., Ouyang, X., Wang, J.: Unit Interval Vertex Deletion: Fewer Vertices are Relevant. arXiv e-prints p. 1607.01162 (2016)
17. Krithika, R., Sahu, A., Saurabh, S., Zehavi, M.: The parameterized complexity of cycle packing: indifference is not an issue. Algorithmica **81**(9), 3803–3841 (2019)
18. Looges, P.J., Olariu, S.: Optimal greedy algorithms for indifference graphs. Comput. Math. Appl. **25**(7), 15–25 (1993)
19. Müller, H.: Hamiltonian circuits in chordal bipartite graphs. Discrete Math. **156**(1), 291–298 (1996)

Graph Square Roots of Small Distance from Degree One Graphs

Petr A. Golovach[1(✉)], Paloma T. Lima[1], and Charis Papadopoulos[2]

[1] Department of Informatics, University of Bergen, Bergen, Norway
{petr.golovach,paloma.lima}@uib.no
[2] Department of Mathematics, University of Ioannina, Ioannina, Greece
charis@uoi.gr

Abstract. Given a graph class $\mathcal{H}$, the task of the $\mathcal{H}$-Square Root problem is to decide, whether an input graph G has a square root H that belongs to $\mathcal{H}$. We are interested in the parameterized complexity of the problem for classes $\mathcal{H}$ that are composed by the graphs at vertex deletion distance at most k from graphs of maximum degree at most one, that is, we are looking for a square root H such that there is a modulator S of size k such that $H - S$ is the disjoint union of isolated vertices and disjoint edges. We show that different variants of the problems with constraints on the number of isolated vertices and edges in $H - S$ are FPT when parameterized by k by providing algorithms with running time $2^{2^{\mathcal{O}(k)}} \cdot n^{\mathcal{O}(1)}$. We further show that the running time of our algorithms is asymptotically optimal and it is unlikely that the double-exponential dependence on k could be avoided. In particular, we prove that the VC-k Root problem, that asks whether an input graph has a square root with vertex cover of size at most k, cannot be solved in time $2^{2^{o(k)}} \cdot n^{\mathcal{O}(1)}$ unless the Exponential Time Hypothesis fails. Moreover, we point out that VC-k Root parameterized by k does not admit a subexponential kernel unless P=NP.

Keywords: Graph square root · Parameterized complexity · Structural parameterization.

1 Introduction

Squares of graphs and square roots constitute widely studied concepts in graph theory, both from a structural perspective as well as from an algorithmic point of view. A graph G is the *square* of a graph H if G can be obtained from H by the addition of an edge between any two vertices of H that are at distance two. In this case, the graph H is called a *square root* of G. It is interesting to notice that there are graphs that admit different square roots, graphs that have a unique square root and graphs that do not have a square root at all. In 1994, Motwani

The paper received support from the Research Council of Norway via the projects "CLASSIS" (grant no. 249994) and "MULTIVAL" (grant no. 263317).

Y. Kohayakawa and F. K. Miyazawa (Eds.): LATIN 2020, LNCS 12118, pp. 116–128, 2020.
https://doi.org/10.1007/978-3-030-61792-9_10

and Sudan [24] proved that the problem of determining if a given graph G has a square root is NP-complete. This problem is known as the SQUARE ROOT problem. The intractability of SQUARE ROOT has been attacked in two different ways. The first one is by imposing some restrictions on the input graph G. In this vein, the SQUARE ROOT problem has been studied in the setting in which G belongs to a specific class of graphs [4,10,11,18,22,23,25].

Another way of coping with the hardness of the SQUARE ROOT problem is by imposing some additional structure on the square root H. That is, given the input graph G, the task is to determine whether G has a square root H that belongs to a specific graph class $\mathcal{H}$. This setting is known as the $\mathcal{H}$-SQUARE ROOT problem and it is the focus of this work. The $\mathcal{H}$-SQUARE ROOT problem has been shown to be polynomial-time solvable for specific graph classes $\mathcal{H}$ [6,9,11,14–19,21]. It is interesting to notice that the fact that $\mathcal{H}$-SQUARE ROOT can be efficiently (say, polynomially) solved for some class $\mathcal{H}$ does not automatically imply that $\mathcal{H}'$-SQUARE ROOT is efficiently solvable for every subclass $\mathcal{H}'$ of $\mathcal{H}$. On the negative side, $\mathcal{H}$-SQUARE ROOT remains NP-complete on graphs of girth at least 4 [7], split graphs [15], and chordal graphs [15]. The fact that all known NP-hardness constructions involve dense graphs [7,15,24] and dense square roots, raised the question of whether $\mathcal{H}$-SQUARE ROOT is polynomial-time solvable for every sparse graph class $\mathcal{H}$.

We consider this question from the Parameterized Complexity viewpoint for structural parameterizations of $\mathcal{H}$ (we refer to the recent book of Cygan et al. [5] for an introduction to Parameterized Complexity). More precisely, we are interested in graph classes $\mathcal{H}$ that are at *small distance* from a (sparse) graph class for which $\mathcal{H}$-SQUARE ROOT can be solved in polynomial time. Within this scope, the distance is usually measured either by the number of edge deletions, edge additions or vertex deletions. This approach for the problem was first applied by Cochefert et al. in [3], who considered $\mathcal{H}$-SQUARE ROOT, where $\mathcal{H}$ is the class of graphs that have a feedback edge set of size at most k, that is, for graphs that can be made forests by at most k edge deletions. They proved that $\mathcal{H}$-SQUARE ROOT admits a compression to a special variant of the problem with $\mathcal{O}(k^2)$ vertices, implying that the problem can be solved in $2^{\mathcal{O}(k^4)} + \mathcal{O}(n^4 m)$ time, i.e., is fixed-parameter tractable (FPT) when parameterized by k. Herein, we study whether the same complexity behavior occurs if we measure the distance by the number of *vertex deletions* instead of edge deletions.

Towards such an approach, the most natural consideration for $\mathcal{H}$-SQUARE ROOT is to ask for a square root of feedback *vertex* set of size at most k. The approach used by Cochefert et al. [3] fails if $\mathcal{H}$ is the class of graphs that can be made forests by at most k vertex deletions and the question of the parameterized complexity of our problem for this case is open. In this context, we consider herein the $\mathcal{H}$-SQUARE ROOT problem when $\mathcal{H}$ is the class of graphs of bounded vertex deletion distance to a disjoint union of isolated vertices and edges. Our main result is that the problem is FPT when parameterized by the vertex deletion distance. Surprisingly, however, we conclude a notable difference on the running time compared to the edge deletion case even on such a relaxed variation: a double-exponential dependency on the vertex deletion distance is highly unavoidable. Therefore, despite the fact that both problems are FPT, the vertex deletion distance parameterization for the $\mathcal{H}$-SQUARE ROOT problem requires substantial effort. More formally, we are interested in the following problem.

DISTANCE-k-TO-$(pK_1 + qK_2)$ SQUARE ROOT
Input: A graph G and nonnegative integers p, q, k such that $p+2q+k = |V(G)|$.
Task: Decide whether there is a square root H of G such that $H - S$ is a graph isomorphic to $pK_1 + qK_2$, for a set S on k vertices.

Note that when $q = 0$, the problem asks whether G has a square root with a vertex cover of size (at most) k and we refer to the problem as VC-k ROOT. If $p = 0$, we obtain DISTANCE-k-TO-MATCHING ROOT. Observe also that, given an algorithm solving DISTANCE-k-TO-(pK_1+qK_2) SQUARE ROOT, then by testing all possible values of p and q such that $p + 2q = |V(G)| - k$, we can solve the DISTANCE-k-TO-DEGREE-ONE ROOT problem, whose task is to decide whether there is a square root H such that the maximum degree of $H - S$ is at most one for a set S on k vertices.

We show that DISTANCE-k-TO-(pK_1+qK_2) SQUARE ROOT can be solved in $2^{2^{\mathcal{O}(k)}} \cdot n^{\mathcal{O}(1)}$ time, that is, the problem is FPT when parameterized by k, the size of the deletion set. We complement this result by showing that the running time of our algorithm is asymptotically optimal in the sense that VC-k ROOT, i.e., the special case of DISTANCE-k-TO-$(pK_1 + qK_2)$ SQUARE ROOT when $q = 0$, cannot be solved in $2^{2^{o(k)}} \cdot n^{\mathcal{O}(1)}$ time unless the *Exponential Time Hypothesis* (*ETH*) of Impagliazzo, Paturi and Zane [12] (see also [5] for the introduction to the algorithmic lower bounds based on ETH) fails. We also prove that VC-k ROOT does not admit a kernel of subexponential in k size unless P=NP.

Motivated by the above results, we further show that the problem of testing whether a given graph has a square root of bounded deletion distance to a clique is also FPT parameterized by the size of the deletion set.

2 Preliminaries

We refer to the recent book of [5] for an introduction to Parameterized Complexity and the textbook by Bondy and Murty [1] for any undefined graph terminology. We denote by K_r the complete graph on r vertices. Given two graphs G and G', we denote by $G+G'$ the disjoint union of them. For a positive integer p, pG denotes the disjoint union of p copies of G. The *square* of a graph H is the graph $G = H^2$ such that $V(G) = V(H)$ and every two distinct vertices u and v are adjacent in G if and only if they are at distance at most two in H. If $G = H^2$, then H is a *square root* of G. Two vertices u, v are said to be *true twins* if $N_G[u] = N_G[v]$. A *true twin class* of G is a maximal set of vertices that are pairwise true twins. Note that the set of true twin classes of G constitutes a partition of $V(G)$.

We will use integer linear programming as a subroutine in the proof of our main result by translating part of our problem as an instance of the following:

p-VARIABLE INTEGER LINEAR PROGRAMMING FEASIBILITY
Input: An $m \times p$ matrix A over $\mathbb{Z}$ and a vector $b \in \mathbb{Z}^m$.
Task: Decide whether there is a vector $x \in \mathbb{Z}^p$ such that $Ax \leq b$.

Lenstra [20] and Kannan [13] showed that the above problem is FPT parameterized by p, while Frank and Tardos [8] showed that this algorithm can be made to run also in polynomial space. We will make use of these results:

Theorem 2.1 ([8,13,20]). p-VARIABLE INTEGER LINEAR PROGRAMMING FEASIBILITY *can be solved using* $\mathcal{O}(p^{2.5p+o(p)} \cdot L)$ *arithmetic operations and space polynomial in* L*, where* L *is the number of bits in the input.*

3 FPT Algorithm for Distance-k-to-$(pK_1 + qK_2)$ Square Root

In this section we give an FPT algorithm for the DISTANCE-k-TO-$(pK_1 + qK_2)$ SQUARE ROOT problem, parameterized by k. We use (G, p, q, k) to denote an instance of the problem. Suppose that (G, p, q, k) is a YES-instance and H is a square root of G such that there is $S \subseteq V(G)$ of size k and $H - S$ is isomorphic to $pK_1 + qK_2$. We say that S is a *modulator*, the p vertices of $H - S$ that belong to pK_1 are called *S-isolated* vertices and the q edges that belong to qK_2 are called *S-matching* edges. Slightly abusing notation, we also use these notions when H is not necessarily a square root of G but any graph such that $H - S$ has maximum degree one.

3.1 Structural Lemmas

We start by defining the following two equivalence relations on the set of ordered pairs of vertices of G. Two pairs of adjacent vertices (x, y) and (z, w) are called *matched twins*, denoted by $(x, y) \sim_{\text{mt}} (z, w)$, if the following conditions hold:

$$N_G[x] \setminus \{y\} = N_G[z] \setminus \{w\} \quad \text{and} \quad N_G[y] \setminus \{x\} = N_G[w] \setminus \{z\}.$$

A pair of vertices (x, y) is called *comparable* if $N_G[x] \subseteq N_G[y]$. Two comparable pairs of vertices (x, y) and (z, w) are *nested twins*, denoted by $(x, y) \sim_{\text{nt}} (z, w)$, if the following conditions hold:

$$N_G(x) \setminus \{y\} = N_G(z) \setminus \{w\} \quad \text{and} \quad N_G[y] \setminus \{x\} = N_G[w] \setminus \{z\}.$$

We use the following properties of matched and nested twins.

Lemma 3.1. *Let* (x, y) *and* (z, w) *be two distinct vertex pairs (resp. comparable pairs) of* G *that are matched twins (resp. nested twins). Then, the following holds:* (i) $\{x, y\} \cap \{z, w\} = \emptyset$, (ii) $xw, zy \notin E(G)$, (iii) $yw \in E(G)$, (iv) *if* $(x, y) \sim_{mt} (z, w)$ *then* $xz \in E(G)$, (v) *if* $(x, y) \sim_{nt} (z, w)$ *then* $xz \notin E(G)$, (vi) $G - \{x, y\}$ *and* $G - \{z, w\}$ *are isomorphic.*

In particular, these properties allow us to classify pairs of vertices with respect to $\sim_{\rm mt}$ and $\sim_{\rm nt}$.

Observation 3.1. *The relations $\sim_{\rm mt}$ and $\sim_{\rm nt}$ are equivalence relations on pairs of adjacent vertices and comparable pairs of vertices, respectively.*

Let H be a square root of a connected graph G with at least three vertices, such that H is at distance k from $pK_1 + qK_2$, and let S be a modulator. Note that $S \neq \emptyset$, because G is connected and $|V(G)| \geq 3$. Then an S-matching edge ab of H satisfies exactly one of the following conditions:

1. $N_H(a) \cap S = \emptyset$ and $N_H(b) \cap S \neq \emptyset$,
2. $N_H(a) \cap S, N_H(b) \cap S \neq \emptyset$ and $N_H(a) \cap N_H(b) \cap S = \emptyset$,
3. $N_H(a) \cap S, N_H(b) \cap S \neq \emptyset$ and $N_H(a) \cap N_H(b) \cap S \neq \emptyset$.

We refer to them as type 1, 2 and 3 edges, respectively. We use the same notation for every graph F that has a set of vertices S such that $F - S$ has maximum degree at most one.

In the following three lemmas, we show the properties of the S-matching edges of types 1, 2 and 3 respectively that are crucial for our algorithm.

Lemma 3.2. *Let H be a square root of a connected graph G with at least three vertices such that $H - S$ is isomorphic to $pK_1 + qK_2$ for $S \subseteq V(G)$. If a_1b_1 and a_2b_2 are two type 1 distinct edges such that $N_H(b_1) \cap S = N_H(b_2) \cap S \neq \emptyset$, then the following holds:* (i) *(a_1, b_1) and (a_2, b_2) are comparable pairs,* (ii) *$(a_1, b_1) \sim_{nt} (a_2, b_2)$,* (iii) *$(a_1, b_1) \nsim_{mt} (a_2, b_2)$.*

Lemma 3.3. *Let H be a square root of a connected graph G with at least three vertices such that $H - S$ is isomorphic to $pK_1 + qK_2$ for $S \subseteq V(G)$. If a_1b_1 and a_2b_2 are two distinct type 2 edges such that $N_H(a_1) \cap S = N_H(a_2) \cap S$ and $N_H(b_1) \cap S = N_H(b_2) \cap S$, then the following holds:* (i) *$(a_1, b_1) \sim_{mt} (a_2, b_2)$,* (ii) *$(a_1, b_1) \nsim_{nt} (a_2, b_2)$.*

Lemma 3.4. *Let H be a square root of a connected graph G with at least three vertices such that $H - S$ is isomorphic to $pK_1 + qK_2$ for $S \subseteq V(G)$. If a_1b_1 and a_2b_2 are two distinct type 3 edges such that $N_H(a_1) \cap S = N_H(a_2) \cap S$ and $N_H(b_1) \cap S = N_H(b_2) \cap S$, then the following holds:* (i) *$(a_1, b_1) \nsim_{mt} (a_2, b_2)$,* (ii) *$(a_1, b_1) \nsim_{nt} (a_2, b_2)$,* (iii) *$a_1$ and a_2 (resp. b_1 and b_2) are true twins in G.*

We need the following straightforward observation about S-isolated vertices.

Observation 3.2. *Let H be a square root of a connected graph G with at least three vertices such that $H - S$ is isomorphic to $pK_1 + qK_2$ for $S \subseteq V(G)$. Then every two distinct S-isolated vertices of H with the same neighbors in S are true twins in G.*

The next lemma is used to construct reduction rules that allow to bound the size of equivalence classes of pairs of vertices with respect to $\sim_{\rm nt}$ and $\sim_{\rm mt}$. The proof of the lemma is based on the properties of matched and nested twins given in Lemma 3.1 and Lemmas 3.2–3.4.

Lemma 3.5. *Let H be a square root of a connected graph G with at least three vertices such that $H - S$ is isomorphic to $pK_1 + qK_2$ for a modulator $S \subseteq V(G)$ of size k. Let Q be an equivalence class in the set of pairs of comparable pairs of vertices with respect to $\sim_{nt}$ (an equivalence class in the set of pairs of adjacent vertices with respect to $\sim_{mt}$, respectively). If $|Q| \geq 2k + 2^{2k} + 1$, then Q contains two distinct pairs (a_1, b_1) and (a_2, b_2) such that a_1b_1 and a_2b_2 are S-matching edges of type 1 in H satisfying $N_H(b_1) \cap S = N_H(b_2) \cap S \neq \emptyset$ (S-matching edges of type 2 in H satisfying $N_H(a_1) \cap S = N_H(a_2) \cap S$ and $N_H(b_1) \cap S = N_H(b_2) \cap S$, respectively).*

3.2 The Algorithm for Distance-k-to-$(pK_1 + qK_2)$ Square Root

In this section we prove our main result. First, we consider connected graphs. For this, observe that if a connected graph G has a square root H then H is connected as well.

Theorem 3.1. Distance-k-to-$(pK_1 + qK_2)$ Square Root *can be solved in time $2^{2^{\mathcal{O}(k)}} \cdot n^{\mathcal{O}(1)}$ on connected graphs.*

Proof (sketch). Let (G, p, q, k) be an instance of Distance-k-to-$(pK_1 + qK_2)$ Square Root with G being a connected graph. Recall that we want to determine if G has a square root H such that $H - S$ is isomorphic to $pK_1 + qK_2$, for a modulator $S \subset V(G)$ with $|S| = k$, where $p + 2q + k = n$. If G has at most two vertices, then the problem is trivial. Notice also that if $k = 0$, then (G, p, q, k) may be a Yes-instance only if G has at most two vertices, because G is connected. Hence, from now we assume that $n \geq 3$ and $k \geq 1$.

We exhaustively apply the following rule to reduce the number of type 1 edges in a potential solution. For this, we consider the set $\mathcal{A}$ of comparable pairs of vertices of G and find its partition into equivalence classes with respect to $\sim_{\text{nt}}$. Note that $\mathcal{A}$ contains at most $2m$ elements and can be constructed in time $\mathcal{O}(mn)$. Then the partition of $\mathcal{A}$ into equivalence classes can be found in time $\mathcal{O}(m^2n)$ by checking the neighborhoods of the vertices of each pair.

Rule 3.1. If there is an equivalence class $Q \subseteq \mathcal{A}$ with respect to $\sim_{\text{nt}}$ such that $|Q| \geq 2k + 2^{2k} + 2$, delete two vertices of G that form a pair of Q and set $q := q - 1$.

The safeness of the rule is proved by making use Lemmas 3.1 and 3.5, that is, we show that the rule constructs an equivalent instance of the problem and the obtained graph is connected.

We also want to reduce the number of type 2 edges in a potential solution. Let $\mathcal{B}$ be the set of pairs of adjacent vertices. We construct the partition of $\mathcal{B}$ into equivalence classes with respect to $\sim_{\text{mt}}$. We have that $|\mathcal{B}| = 2m$ and, therefore, the partition of $\mathcal{B}$ into equivalence classes can be found in time $\mathcal{O}(m^2n)$ by checking the neighborhoods of the vertices of each pair. We exhaustively apply the following rule.

Rule 3.2. If there is an equivalence class $Q \subseteq \mathcal{B}$ with respect to $\sim_{\rm mt}$ such that $|Q| \geq 2k + 2^{2k} + 2$, delete two vertices of G that form a pair of Q and set $q := q - 1$.

Similarly to Rule 3.2, Rule 3.2 is safe by Lemmas 3.1 and 3.5.

After exhaustive application of Rules 3.2 and 3.2 we obtain the following bounds on the number of edges of types 1 and 2 in a potential solution using Lemmas 3.2 and 3.3.

Claim 3.1. *Let* (G', p, q', k) *be the instance of* Distance-k-to-$(pK_1 + qK_2)$ Square Root *after exhaustive applications of Rules* 3.2 *and* 3.2. *Then* G' *is a connected graph and a potential solution* H *to the instance has at most* $2^k(2k + 2^{2k} + 1)$ S*-matching edges of type 1 and* $2^{2k}(2k + 2^{2k} + 1)$ S*-matching edges of type 2.*

For simplicity, we call (G, p, q, k) the instance obtained after exhaustive applications of Rules 3.2 and 3.2. Note that G can be constructed in polynomial time.

By Claim 3.2, in a potential solution, the number of S-matching edges of types 1 and 2 is bounded by a function of k. We will make use of this fact to make further guesses about the structure of a potential solution. To do so, we first consider the classes of true twins of G and show the following claim using Observation 3.1 and Lemmas 3.2–3.4.

Claim 3.2. *Let* $\mathcal{T} = \{T_1, \ldots, T_r\}$ *be the partition of* $V(G)$ *into classes of true twins. If* (G, p, q, k) *is a* Yes*-instance to our problem, then* $r \leq 2(2^k + 2^{2k})(2k + 2^{2k} + 1) + k + 2^k + 2 \cdot 2^{2k}$.

Observe that the partition $\mathcal{T} = \{T_1, \ldots, T_r\}$ of $V(G)$ into classes of true twins can be constructed in time $\mathcal{O}(n^2)$ by comparing the neighbors of vertices. Using Claim 3.2, we apply the following rule.

Rule 3.3. If $|\mathcal{T}| > 2(2^k + 2^{2k})(2k + 2^{2k} + 1) + k + 2^k + 2 \cdot 2^{2k}$, then return No and stop.

From now on, we assume that we do not stop by Rule 3.2. This means that $|\mathcal{T}| = \mathcal{O}(2^{4k})$.

Suppose that (G, p, q, k) is a Yes-instance to Distance-k-to-$(pK_1 + qK_2)$ Square Root and let H be a square root of G that is a solution to this instance with a modulator S. We say that F is the *skeleton* of H with respect to S if F is obtained from H be the exhaustive application of the following rules:

(i) if H has two distinct type 3 S-matching edges xy and $x'y'$ with $N_H(x) \cap S = N_H(x') \cap S$ and $N_H(y) \cap S = N_H(y') \cap S$, then delete x and y,
(ii) if H has two distinct S-isolated vertices x and y with $N_H(x) = N_H(y)$, then delete x.

In other words, we replace the set of S-matching edges of type 3 with the same neighborhoods on the end-vertices in S by one representative and we replace the set of S-isolated vertices with the same neighborhoods by one representative.

We say that a graph F is a *potential solution skeleton* with respect to a set $S \subseteq V(F)$ of size k for (G, p, q, k) if the following holds:

(i) $F - S$ has maximum degree one, that is, $F - S$ is isomorphic to $sK_1 + tK_2$ for some nonnegative integers s and t,
(ii) for every two distinct S-isolated vertices x and y of F, $N_F(x) \neq N_F(y)$,
(iii) for every two distinct S-matching edges xy and $x'y'$ of type 3, either $N_F(x) \cap S \neq N_H(x') \cap S$ or $N_F(y) \cap S \neq N_H(y') \cap S$,
(iv) for every $A, B \subseteq S$ such that $A \cap B = \emptyset$ and at least one of A and B is nonempty, $\{xy \in E(F - S) \mid N_F(x) \cap S = A \text{ and } N_F(y) \cap S = B\}$ has size at most $2k + 2^{2k} + 1$.

Note that (iv) means that the number of type 1 and type 2 S-matched edges with the same neighbors in S is upper bounded by $2k+2^{2k}+1$. Since Rules 3.2 and 3.2 cannot be applied to (G, p, q, k), we obtain the following claim by Lemmas 3.2(ii) and 3.3(ii).

Claim 3.3. *Every skeleton of a solution to (G, p, q, k) is a potential solution skeleton for this instance with respect to the modulator S.*

We observe that each potential solution skeleton has bounded size.

Claim 3.4. *For every potential solution skeleton F for (G, p, q, k),*

$$|V(F)| \leq k + 2^k + 2 \cdot 2^{2k} + 2 \cdot 2^{2k}(2k + 2^{2k} + 1).$$

Moreover, we can construct the family $\mathcal{F}$ of all potential solution skeletons together with their modulators.

Claim 3.5. *The family $\mathcal{F}$ of all pairs (F, S), where F is a potential solution skeleton and $S \subseteq V(F)$ is a modulator of size k, has size at most $2^{\binom{k}{2}} + 2^{2^k} + 2^{2^{2k}} + (2k + 2^{2k} + 2)^{2^{2k}}$ and can be constructed in time $2^{2^{\mathcal{O}(k)}}$.*

Using Claim 3.2, we construct $\mathcal{F}$, and for every $(F, S) \in \mathcal{F}$, we check whether there is a solution H to (G, p, q, k) with a modulator S', whose skeleton is isomorphic to F with an isomorphism that maps S to S'. If we find such a solution, then (G, p, q, k) is a YES-instance. Otherwise, Claims 3.2 guarantees that (G, p, q, k) is a NO-instance.

Assume that we are given $(F, S) \in \mathcal{F}$ for the instance (G, p, q, k). Recall that we have the partition $\mathcal{T} = \{T_1, \ldots, T_r\}$ of $V(G)$ into true twin classes of size at most $2(2^k + 2^{2k})(2k + 2^{2k} + 1) + k + 2^k + 2 \cdot 2^{2k}$ by Rule 3.2. We define the *prime-twin* graph $\mathcal{G}$ of G as the graph with the vertex set $\mathcal{T}$ such that two distinct vertices T_i and T_j of $\mathcal{G}$ are adjacent if and only if $uv \in E(G)$ for $u \in T_i$ and $v \in T_j$. Clearly, given G and $\mathcal{T}$, $\mathcal{G}$ can be constructed in linear time. For an induced subgraph R of G, we define $\tau_R \colon V(R) \to \mathcal{T}$ to be a mapping such that $\tau_R(v) = T_i$ if $v \in T_i$ for $T_i \in \mathcal{T}$.

Let $\varphi \colon V(F) \to \mathcal{T}$ be a surjective mapping. We say that φ is *$\mathcal{G}$-compatible* if every two distinct vertices u and v of F are adjacent in F^2 if and only if $\varphi(u)$ and $\varphi(v)$ are adjacent in $\mathcal{G}$. Then by the definition of F and Lemma 3.4 we obtain the following.

Claim 3.6. *Let F be the skeleton of a solution H to (G, p, q, k). Then $\tau_F \colon V(F) \to \mathcal{T}$ is a $\mathcal{G}$-compatible surjection.*

Our next step is to reduce our problem to solving a system of linear integer inequalities. Let $\varphi\colon V(F) \to \mathcal{T}$ be a $\mathcal{G}$-compatible surjective mapping. Let X_1, X_2 and X_3 be the sets of end-vertices of the S-matching edges of type 1, type 2 and type 3 respectively in F. Let Y be the set of S-isolated vertices of F. For every vertex $v \in V(F)$, we introduce an integer variable x_v. Informally, x_v is the number of vertices of a potential solution H that correspond to a vertex v.

$$\begin{cases} x_v = 1 & \text{for } v \in S \cup X_1 \cup X_2, \\ x_v \geq 1 & \text{for } v \in Y \cup X_3, \\ x_u - x_v = 0 & \text{for every type 3 edge } uv, \\ \sum_{v \in Y} x_v = p, & \\ \sum_{v \in X_1 \cup X_2 \cup X_3} x_v = 2q, & \\ \sum_{v \in \varphi^{-1}(T_i)} x_v = |T_i| & \text{for } T_i \in \mathcal{T}. \end{cases} \tag{1}$$

Using Claim 3.2 together with Lemma 3.4 and Observation 3.1 we show the following crucial claim.

Claim 3.7. *The instance (G, p, q, k) has a solution$H$ with a modulator S' such that there is an isomorphism $\Psi\colon V(F) \to V(F')$ for the skeleton F' of H mapping S to S' if and only if there is a $\mathcal{G}$-compatible surjective mapping $\varphi\colon V(F) \to \mathcal{T}$ such that the system* (1) *has a solution.*

By Claim 3.2, we can state our task as follows: verify whether there is a $\mathcal{G}$-compatible surjection $\varphi\colon V(F) \to \mathcal{T}$ such that (1) has a solution. For this, we consider all at most $|V(F)|^{|\mathcal{T}|} = 2^{2^{\mathcal{O}(k)}}$ surjections $\varphi\colon V(F) \to \mathcal{T}$. For each φ, we verify whether it is $\mathcal{G}$-compatible. Clearly, it can be done in time $\mathcal{O}(|V(F)|^3)$. If φ is $\mathcal{G}$-compatible, we construct the system (1) with $|V(F)| = 2^{\mathcal{O}(k)}$ variables in time $\mathcal{O}(|V(F)|^2)$. Then we solve it by applying Theorem 2.1 in $2^{2^{\mathcal{O}(k)}} \log n$ time. This completes the description of the algorithm and its correctness proof.

To evaluate the total running time, notice that the preprocessing step, that is, the exhaustive application of Rules 3.2 and 3.2 is done in polynomial time. Then the construction of $\mathcal{T}$, $\mathcal{G}$ and the application of Rule 3.2 is polynomial as well. By Claim 3.2, $\mathcal{F}$ is constructed in time $2^{2^{\mathcal{O}(k)}}$. The final steps, that is, constructing φ and systems (1) and solving the systems, can be done in time $2^{2^{\mathcal{O}(k)}} \log n$. Therefore, the total running time is $2^{2^{\mathcal{O}(k)}} \cdot n^{\mathcal{O}(1)}$. □

For simplicity, in Theorem 3.1, we assumed that the input graph is connected but it is not difficult to extend the result for general case.

Corollary 3.1. Distance-k-to-$(pK_1 + qK_2)$ Square Root *can be solved in time $2^{2^{\mathcal{O}(k)}} \cdot n^{\mathcal{O}(1)}$.*

Corollary 3.1 gives the following statement for the related problems.

Corollary 3.2. VC-k Root, Distance-k-to-Matching Root *and* Distance-k-to-Degree-One Root *can be solved in time $2^{2^{\mathcal{O}(k)}} \cdot n^{\mathcal{O}(1)}$.*

4 Lower Bounds for Distance-k-to-$(pK_1 + qK_2)$ Square Root

In this section, we show that the running time of our algorithm for Distance-k-to-$(pK_1 + qK_2)$ Square Root given in Sect. 3 (see Theorem 3.1) cannot be significantly improved and we cannot expect the existence of a polynomial kernel. In fact, we show lower bounds for $q = 0$, that is, for the case of VC-k Root. To provide our lower bounds, we will give a parameterized reduction from the Biclique Cover problem. This problem takes as input a bipartite graph G and a nonnegative integer k, and the task is to decide whether the edges of G can be covered by at most k complete bipartite subgraphs. Chandran et al. [2] showed the following result about the Biclique Cover problem that will be of interest to us.

Theorem 4.1 ([2]). Biclique Cover *cannot be solved in time* $2^{2^{o(k)}} \cdot n^{\mathcal{O}(1)}$ *unless* ETH *is false and does not admit a kernel of size* $2^{o(k)}$ *unless* $P = NP$.

Lemma 4.1. *There exists a polynomial time algorithm that, given an instance* (B, k) *for* Biclique Cover, *produces an equivalent instance* $(G, k+4)$ *for* VC-k Root, *with* $|V(G)| = |V(B)| + k + 6$.

Proof (Sketch). Let (B, k) be an instance of Biclique Cover where (X, Y) is the bipartition of $V(B)$. Let $X = \{x_1, \ldots, x_p\}$ and $Y = \{y_1, \ldots, y_q\}$. We construct the instance $(G, k+4)$ for VC-k Root such that $V(G) = X \cup Y \cup \{z_1, \ldots, z_k\} \cup \{u, v, w, u', v', w'\}$. Denote by Z the set $\{z_1, \ldots, z_k\}$. The edge set of G is defined in the following way: $G[X \cup Z \cup \{u\}]$, $G[X \cup \{v\}]$, $\{u, v, w\}$, $G[Y \cup Z \cup \{u'\}]$, $G[Y \cup \{v'\}]$ and $\{u', v', w'\}$ are cliques and $x_i y_j \in E(G)$ if and only if $x_i y_j \in E(B)$.

For the forward direction, suppose (B, k) is a Yes-instance for Biclique Cover. We will show that $(G, k+4)$ is a Yes-instance for VC-k Root. Note that if B has a biclique cover of size strictly less than k, we can add arbitrary bicliques to this cover and obtain a biclique cover for B of size exactly k. Let $\mathcal{C} = \{C_1, \ldots, C_k\}$ be such a biclique cover. We construct the following square root candidate H for G with $V(H) = V(G)$. Add the edges uv, vw, $u'v'$ and $v'w'$ to H, and also all the edges between u and X, all the edges between u' and Y and all the edges in $G[Z]$. Finally, for each $1 \leq i \leq k$, add to H all the edges between z_i and the vertices of C_i. We show that the constructed graph H is indeed a square root of G by checking the adjacencies in H^2. We conclude that $(G, k+4)$ is a Yes-instance for VC-k Root using the fact that $Z \cup \{u, v, u', v'\}$ is a vertex cover of H of size $k+4$.

For the reverse direction, we state the next three claims.

(i) The edges uv, vw, $u'v'$ and $v'w'$ belong to any square root of G,
(ii) The edges $\{ux_i, u'y_j \mid 1 \leq i \leq p, 1 \leq j \leq q\}$ belong to any square root of G.
(iii) The edges $\{x_i y_j \mid 1 \leq i \leq p, 1 \leq j \leq q\}$ do not belong to any square root of G.

Now assume that G has a square root H that has a vertex cover of size at most $k+4$. By (iii), for every edge of G of the form x_iy_j, it holds that $x_iy_j \notin E(H)$. This implies that, for every such edge, there exists an induced P_3 in H having x_i and y_j as its endpoints. Since $N_G(x_i) \cap N_G(y_j) = Z$, only vertices of Z can be the middle vertices of these paths. For $1 \leq \ell \leq k$, let $C_\ell = N_H(z_\ell) \cap (X \cup Y)$. We will now show that $\mathcal{C} = \{C_1, \ldots, C_k\}$ is a biclique cover of B. First, note that since for every edge x_iy_j, there exists $z_h \in Z$ such that $z_hx_i, z_hy_j \in E(H)$, we conclude that $x_iy_j \in C_h$, which implies that $\mathcal{C}$ is an edge cover of B. Furthermore, for a given ℓ, since every vertex of C_ℓ is adjacent to z_ℓ in H, $G[C_\ell]$ is a clique and, therefore, $B[C_\ell]$ is a biclique. This implies that $\mathcal{C}$ is indeed a biclique cover of B of size k, which concludes the proof of the theorem. □

Combining Theorem 4.1 and Lemma 4.1, we obtain the following.

Theorem 4.2. VC-k Root *cannot be solved in time* $2^{2^{o(k)}} \cdot n^{\mathcal{O}(1)}$ *unless ETH is false and does not admit a kernel of size* $2^{o(k)}$ *unless* $P = NP$.

5 Conclusion

We believe that it would be interesting to further investigate the parameterized complexity of $\mathcal{H}$-Square Root for sparse graph classes $\mathcal{H}$ under structural parameterizations. The natural candidates are the Distance-k-to-Linear-Forest Root and Feedback-Vertex Set-k Root problems, whose tasks are to decide whether the input graph has a square root that can be made a linear forest, that is, a union of paths, and a forest respectively by (at most) k vertex deletions. Recall that the existence of an FPT algorithm for $\mathcal{H}$-Square Root does not imply the same for subclasses of $\mathcal{H}$. However, it can be noted that our complexity lower bounds still hold and, therefore, we cannot expect that these problems would be easier.

Parameterized complexity of $\mathcal{H}$-Square Root is widely open for other, not necessarily sparse, graph classes. As a step to this direction, we consider the Distance-k-to-Clique Square Root problem in which the task is to decide whether for a given graph G and a nonnegative integer k, there is a square root H of G such that $H - S$ is a complete graph for a set S on k vertices. We prove that it is FPT when parameterized by k.

Theorem 5.1. Distance-k-to-Clique Square Root *can be solved in time* $2^{2^{\mathcal{O}(k)}} \cdot n^{\mathcal{O}(1)}$.

What can be said if we ask for a square root that is at deletion distance k from a *cluster graph* (disjoint union of cliques)? We believe that our techniques allows to show that this problem is FPT when parameterized by k if the number of cliques is a fixed constant. Is the problem FPT without this constraint?

References

1. Bondy, J.A., Murty, U.S.R.: Graph Theory. Springer, London (2008)
2. Chandran, S., Issac, D., Karrenbauer, A.: On the parameterized complexity of biclique cover and partition. In: Proceedings of the 11th International Symposium on Parameterized and Exact Computation, IPEC 2016. LIPIcs, vol. 63, pp. 11:1–11:13 (2017)
3. Cochefert, M., Couturier, J., Golovach, P.A., Kratsch, D., Paulusma, D.: Parameterized algorithms for finding square roots. Algorithmica **74**(2), 602–629 (2016)
4. Cochefert, M., Couturier, J., Golovach, P.A., Kratsch, D., Paulusma, D., Stewart, A.: Computing square roots of graphs with low maximum degree. Discrete Appl. Math. **248**, 93–101 (2018)
5. Cygan, M., et al.: Lower bounds for kernelization. Parameterized Algorithms, pp. 523–555. Springer, Cham (2015). https://doi.org/10.1007/978-3-319-21275-3_15
6. Ducoffe, G.: Finding cut-vertices in the square roots of a graph. Discrete Appl. Math. **257**, 158–174 (2019)
7. Farzad, B., Lau, L.C., Le, V.B., Tuy, N.N.: Complexity of finding graph roots with girth conditions. Algorithmica **62**(1–2), 38–53 (2012)
8. Frank, A., Tardos, E.: An application of simultaneous diophantine approximation in combinatorial optimization. Combinatorica **7**, 49–65 (1987)
9. Golovach, P.A., Heggernes, P., Kratsch, D., Lima, P.T., Paulusma, D.: Algorithms for outerplanar graph roots and graph roots of pathwidth at most 2. Algorithmica **81**, 2795–2828 (2019)
10. Golovach, P.A., Kratsch, D., Paulusma, D., Stewart, A.: A linear kernel for finding square roots of almost planar graphs. Theor. Comput. Sci. **689**, 36–47 (2017)
11. Golovach, P.A., Kratsch, D., Paulusma, D., Stewart, A.: Finding cactus roots in polynomial time. Theory Comput. Syst. **62**, 1409–1426 (2018)
12. Impagliazzo, R., Paturi, R., Zane, F.: Which problems have strongly exponential complexity? J. Comput. Syst. Sci. **63**(4), 512–530 (2001)
13. Kannan, R.: Minkowski's convex body theorem and integer programming. Math. Oper. Res. **12**, 415–440 (1987)
14. Lau, L.C.: Bipartite roots of graphs. ACM Trans. Algor. **2**(2), 178–208 (2006)
15. Lau, L.C., Corneil, D.G.: Recognizing powers of proper interval, split, and chordal graph. SIAM J. Discrete Math. **18**(1), 83–102 (2004)
16. Le, V.B., Oversberg, A., Schaudt, O.: Polynomial time recognition of squares of ptolemaic graphs and 3-sun-free split graphs. Theor. Comput. Sc. **602**, 39–49 (2015)
17. Le, V.B., Oversberg, A., Schaudt, O.: A unified approach for recognizing squares of split graphs. Theor. Comput. Sc. **648**, 26–33 (2016)
18. Le, V.B., Tuy, N.N.: The square of a block graph. Discr. Math. **310**, 734–741 (2010)
19. Le, V.B., Tuy, N.N.: A good characterization of squares of strongly chordal split graphs. Inf. Process. Lett. **111**(3), 120–123 (2011)
20. Lenstra, H.W.: Integer programming with a fixed number of variables. Math. Oper. Res. **8**, 538–548 (1983)

21. Lin, Y.L., Skiena, S.: Algorithms for square roots of graphs. SIAM J. Discrete Math. **8**(1), 99–118 (1995)
22. Milanic, M., Oversberg, A., Schaudt, O.: A characterization of line graphs that are squares of graphs. Discrete Appl. Math. **173**, 83–91 (2014)
23. Milanic, M., Schaudt, O.: Computing square roots of trivially perfect and threshold graphs. Discrete Appl. Math. **161**, 1538–1545 (2013)
24. Motwani, R., Sudan, M.: Computing roots of graphs is hard. Discrete Appl. Math. **54**(1), 81–88 (1994)
25. Nestoridis, N.V., Thilikos, D.M.: Square roots of minor closed graph classes. Discrete Appl. Math. **168**, 34–39 (2014)

Structural Parameterizations for Equitable Coloring

Guilherme C. M. Gomes(✉), Matheus R. Guedes, and Vinicius F. dos Santos

Universidade Federal de Minas Gerais, Belo Horizonte, Minas Gerais, Brazil
{gcm.gomes,matheusresende,viniciussantos}@dcc.ufmg.br

Abstract. An n-vertex graph is equitably k-colorable if there is a proper coloring of its vertices such that each color is used either $\lfloor n/k \rfloor$ or $\lceil n/k \rceil$ times. While classic VERTEX COLORING is fixed parameter tractable under well established parameters such as pathwidth and feedback vertex set, equitable coloring is W[1]-hard. We prove that EQUITABLE COLORING is fixed parameter tractable when parameterized by distance to cluster or co-cluster graphs, improving on the FPT algorithm of Fiala et al. (2011) parameterized by vertex cover. In terms of intractability, we adapt the proof of Fellows et al. (2011) to show that EQUITABLE COLORING is W[1]-hard when simultaneously parameterized by distance to disjoint paths and number of colors. We also revisit the literature and derive other results on the parameterized complexity of the problem through minor reductions or other simple observations.

Keywords: Equitable coloring · Parameterized complexity · Distance to cluster · Distance to co-cluster · Distance to disjoint paths

1 Introduction

EQUITABLE COLORING is a variant of the classical VERTEX COLORING problem: we want to partition an n vertex graph into k independent sets such that each of these sets has either $\lfloor n/k \rfloor$ or $\lceil n/k \rceil$ vertices. The smallest integer k for which G admits an equitable k-coloring is called the *equitable chromatic number* of G. EQUITABLE COLORING was first discussed in [18], with an intended application for municipal garbage collection, and later in processor task scheduling [1], communication control [14], and server load balancing [20]. Lih [17] presented an extensive survey covering many of the results developed in the last 50 years. Its focus, however, is not algorithmic, and most of the presented results are bounds on the equitable chromatic number for various graph classes.

Many complexity results for EQUITABLE COLORING arise from a related problem, known as BOUNDED COLORING, as observed by Bodlaender and Fomin [2]. On BOUNDED COLORING, we ask that the size of the independent sets be bounded by an integer ℓ. Among the positive results for BOUNDED COLORING, the problem is known to be solvable in polynomial time for: split graphs [7],

Y. Kohayakawa and F. K. Miyazawa (Eds.): LATIN 2020, LNCS 12118, pp. 129–140, 2020.
https://doi.org/10.1007/978-3-030-61792-9_11

complements of interval graphs [3], complements of bipartite graphs [3], and forests [1]. Baker and Coffman [1] present the first algorithm for BOUNDED COLORING on trees, while Jarvis and Zhou [16] show how to compute an optimal ℓ-bounded coloring of a tree through a novel characterization. For cographs, bipartite and interval graphs, there are polynomial-time algorithms when the number of colors k is fixed. In terms of parameterized complexity, in [2] an XP algorithm is given for EQUITABLE COLORING parameterized by treewidth, while Fiala et al. [11] show that the problem is FPT parameterized by vertex cover. Recently, Gomes et al. [13] proved that, when parameterized by the treewidth of the complement graph, EQUITABLE COLORING is FPT, and Reddy [19] did the same when the parameter is the distance to threshold graphs.

The main contributions of this work are complexity results on EQUITABLE COLORING for parameterizations that are weaker than vertex cover, in the sense that the parameters are upper bounded by the vertex cover number. In particular, we show that EQUITABLE COLORING is fixed parameter tractable when parameterized by distance to cluster or by distance to co-cluster. Not only are the parameters weaker, but also in the first case, the algorithm is slightly faster than the one previously known for vertex cover, as it does not rely on INTEGER LINEAR PROGRAMMING; the running time, however, is still super exponential. On the negative side, we show that the combined parameterization distance to disjoint paths and number of colors is insufficient to guarantee tractability. Along with some of the works discussed here and in Sect. 2, our results cover many branches of the known graph parameter hierarchy [21]. The proofs can be found in the full version of the paper[1].

Notation and Terminology. We use standard graph theory notation and nomenclature for our parameters, following classical textbooks in the areas [4,9]. Define $[k] = \{1, \dots, k\}$ and 2^S the *powerset* of S. A *k-coloring* φ of a graph G is a function $\varphi : V(G) \mapsto [k]$. Alternatively, a k-coloring is a k-partition $V(G) \sim \{\varphi_1, \dots, \varphi_k\}$ such that $\varphi_i = \{u \in V(G) \mid \varphi(u) = i\}$. A k-coloring is said to be *equitable* if, for every $i \in [k]$, $\lfloor n/k \rfloor \leq |\varphi_i| \leq \lceil n/k \rceil$; it is *proper* if every φ_i is an independent set. Unless stated, all colorings are proper. The EQUITABLE COLORING problem asks whether or not G can be equitably k-colored. A graph is a *cluster graph* if each of its connected components is a clique; a *co-cluster graph* is the complement of a cluster graph. The *distance to cluster* (*co-cluster*) of a graph G, denoted by $dc(G)$ ($d\overline{c}(G)$), is the size of the smallest set $U \subseteq V(G)$ such that $G - U$ is a cluster (co-cluster) graph. Using the terminology of [6], a set $U \subseteq V(G)$ is an *$\mathcal{F}$-modulator* of G if the graph $G - U$ belongs to the graph class $\mathcal{F}$. When the context is clear, we omit the qualifier $\mathcal{F}$. For cluster and co-cluster graphs, one can decide if G admits a modulator of size k in time FPT on k [5].

2 Literature Corollaries and Minor Observations

The original `NP-complete` results of Bodlaender and Jansen [3], despite being initially regarded as polynomial reductions for BOUNDED COLORING, are a nice

[1] Permanently available at https://arxiv.org/abs/1911.03297.

source of parameterized hardness. To adapt their proofs to show that Equitable Coloring parameterized by the number of colors is W[1]-hard on cographs and paraNP-hard on bipartite graphs, it suffices to consider the version of Bin-Packing where each bin must be completely filled for the first case, while the latter follows immediately since they prove that Bounded 3-Coloring is NP-hard on bipartite graphs; these imply that adding the distance to theses classes in the parameterization yields no additional power whatsoever. Fellows et al. [10] show that Equitable Coloring parameterized by treewidth and number of colors is W[1]-hard, while an XP algorithm parameterized by treewidth is given for both Equitable Coloring and Bounded Coloring by Bodlaender and Fomin [2]. In fact, the reduction shown in [10] prove that, even when simultaneously parameterized by feedback vertex set, treedepth, and number of colors, Equitable Coloring remains W[1]-hard. Gomes et al. [13] show that the problem parameterized by number of colors, maximum degree and treewidth is W[1]-hard on interval graphs. However, their intractability statement can be strengthened to number of colors and *bandwidth*, with no changes to the reduction. In [6], Cai proves that Vertex Coloring is W[1]-hard parameterized by distance to split.

We also reviewed results on parameters weaker than distance to clique, since both distance to cluster and co-cluster fall under this category. For minimum clique cover, we resort to the classic result of Garey and Johnson [12] that Partition into Triangles is NP-hard. By definition, a graph G can be partitioned into vertex-disjoint triangles if and only if its complement graph can be equitably $(n/3)$-colored. The reduction given in [12] is from Exact Cover by 3-Sets, and their gadget (which we reproduce in Fig. 1) has the nice property that the complement graph $\overline{G}$ has a trivial clique cover of size nine: it suffices to pick one gadget i and one clique for each a_i^j. Thus, we have that Equitable Coloring is paraNP-hard parameterized by minimum clique cover. To see that when also parameterizing by the number of colors there is an FPT algorithm, we first look at the parameterization maximum independent set α and number of colors k, both of which we assume to be given on the input. First, if $k\alpha < n$, the instance is trivially negative, so we may assume $k\alpha \geq n$; but, in this case, we can spend exponential time on the number of vertices and still run in FPT time. Finally, we reduce from Equitable Coloring parameterized by the number of colors k to Equitable Coloring parameterized by k and minimum dominating set. If we take the source graph G and add $\frac{n}{k}$ vertices $D = v_i, \dots, v_{\frac{n}{k}}$ with $N(v_i) = V(G)$ for all $v_i \in D$, the set $\{v_1, u\}$, with $u \in V(G)$, is a dominating set of the resulting graph G'; moreover, G has an equitable k-coloring if and only if G' is equitably $(k+1)$-colorable, thus proving that Equitable Coloring parameterized by k and minimum dominating set is paraNP-hard. A summary of the results discussed in this work is displayed in Fig. 2.

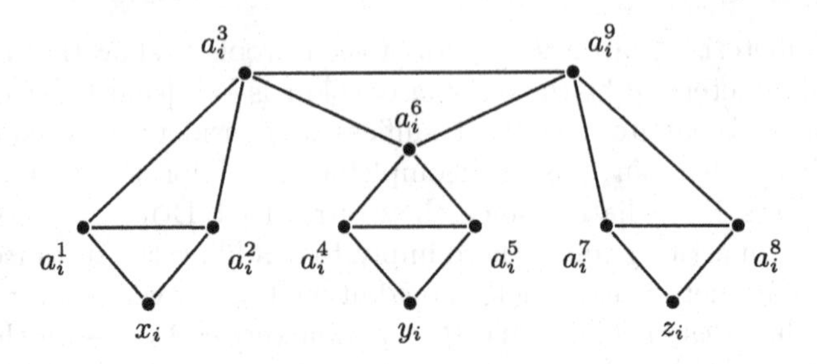

Fig. 1. EXACT COVER BY 3-SETS to PARTITION INTO TRIANGLES gadget of [12] representing the set $C_i = \{x_i, y_i, z_i\}$.

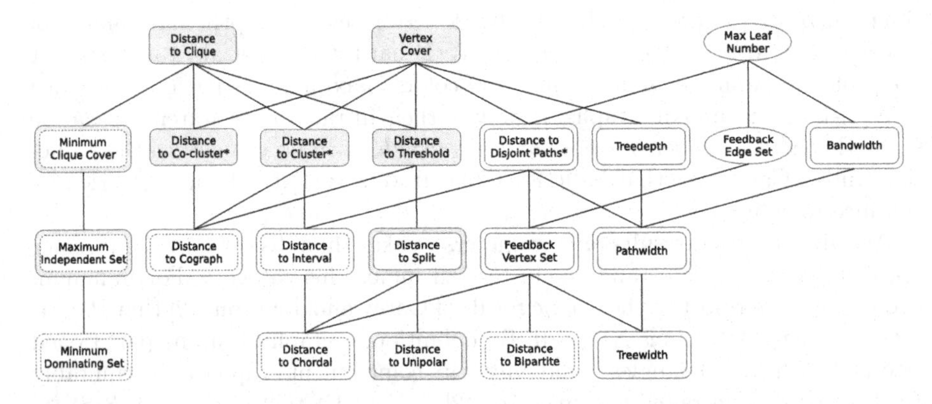

Fig. 2. Hasse diagram of the parameterizations of EQUITABLE COLORING and their complexities. A single shaded box indicates that the problem is FPT; two solid boxes represent W[1]-hard even if also parameterized by the number of colors; if the inner box is dashed, the problem is paraNP-hard; if the outer box is solid and shaded, additionally using the number of colors results in an FPT algorithm; if it is not shaded, it remains W[1]-hard. Entries with a * are our main contributions. Ellipses mark open cases.

3 Equitable Coloring Parameterized by Distance to Cluster

The goal of this section is to prove that EQUITABLE COLORING can be solved in FPT time when parameterized by the distance to cluster of the input graph. As a corollary of this result, we show that unipolar graphs – the class of graphs that have a clique as a modulator – can be equitably k-colored in polynomial time. Throughout this section, we denote the modulator by U, the connected components of $G - U$ by $\mathcal{C} = \{C_1, \dots, C_r\}$, and define $\ell = \lfloor \frac{n}{k} \rfloor$.

The central idea of our algorithm is to guess one of the possible $|U|^{|U|}$ colorings of the modulator and extend this guess to the clique vertices using max-flow. First, given U, $\mathcal{C}$, and a coloring φ' of U, we build an auxiliary graph H as follows: $V(H) = \{s, t\} \cup A \cup W \cup V(G) \setminus U$, where $A = \{a_1, \dots, a_k\}$ represents the colors we may assign to vertices, $W = \{w_{ij} \mid i \in [k], j \in [r]\}$ whose role is to maintain the property of the coloring, s is the source of the flow, t is the sink of the flow, and $V(G) = \{v_1, \ldots, v_n\}$ are the vertices of G. For the arcs, we have $E(H) = S \cup F_0 \cup F_1 \cup R \cup T$, where $S = \{(s, a_i) \mid i \in [k]\}$, $F_0 = \{(a_i, t) \mid i \in [k]\}$, $F_1 = \{(a_i, w_{ij}) \mid i \in [k], j \in [r]\}$, $R = \{(w_{ij}, v_p) \mid v_p \in C_j, N(v_p) \cap \varphi'_i = \emptyset\}$, and $T = \{(v_i, t) \mid v_i \in V(G) \setminus U\}$. As to the capacity of the arcs, we define $c : E(H) \to \mathbb{N}$, with $c(e \in S) = \ell$, $c((a_i, t)) = |\varphi'_i \cap U|$ and $c(e \in F_1 \cup R \cup T) = 1$. Semantically, the vertices of A correspond to the k colors, while each w_{ij} ensures that cluster C_j has at most one vertex of color i. Regarding the arcs, F_0 corresponds to the initial assignment of colors to the vertices of the modulator, and R encodes the adjacency between vertices of the clusters and colored vertices of the modulator. Note that the arcs in F_0 and R are the only ones affected by the pre-coloring φ'. An example of the constructed graph can be found in Fig. 3.

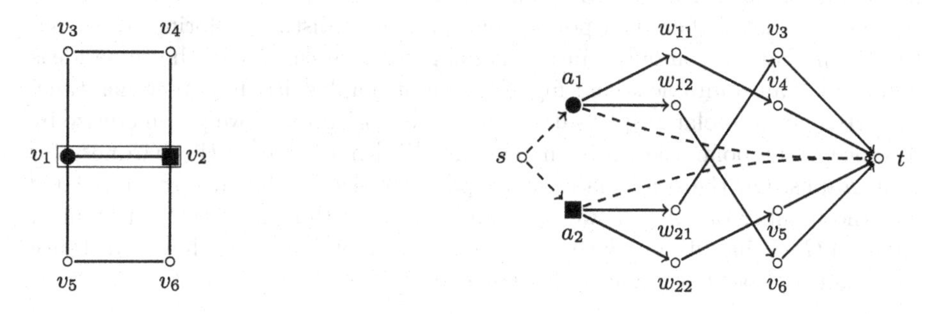

Fig. 3. (left) The input graph with $U = \{v_1, v_2\}$; (right) Auxiliary graph constructed from the precoloring of U. Solid arcs have unit capacity.

Now, let $f : E(H) \to \mathbb{N}$ be the function corresponding to the max-flow from s to t obtained using any of the algorithms available in the literature [8]. Our first observation, as given by the following lemma, is that, if no (s,t)-flow saturates the outbound arcs of s, then we cannot extend φ' to equitably k-color G.

Lemma 1. *If there is some $e \in S$ with $f(e) < \ell$, then G does not admit an equitable k-coloring that extends φ'.*

We may now assume that $f(e) = c(e)$ for every $e \in S$. Let $c'(e \notin S) = c(e)$, $c'(e \in S) = c(e) + 1$. We resume the search for augmenting paths on the network, replacing $c(\cdot)$ with $c'(\cdot)$, until it stops and returns the maximum (s,t)-flow g.

Lemma 2. *For every $e \in S$, $g(e) \geq f(e)$.*

Lemma 3. *The maximum (s,t)-flow F given by g is equal to the number of vertices of G if and only if there is an equitable k-coloring of G that extends φ'.*

At this point we are essentially done. Lemmas 1 and 3 guarantee that, if the max-flow algorithm fails to yield a large enough flow, a fixed pre-coloring of U cannot be extended; moreover, the latter also implies that, if an extension is possible, max-flow correctly finds it. Now, given U, for each of the $\mathcal{O}(|U|^{|U|})$ possible colorings of U, construct H and execute the above algorithm which, since max-flow can be solved in polynomial time [8], results in an FPT algorithm. If we are not given the modulator U, the same can be computed in FPT time.

Theorem 1. Equitable Coloring *parameterized by distance to cluster can be solved in FPT time.*

It is worthy to note here that there is nothing special about the capacities of the arcs in S; they act only as upper bounds to the number of vertices a color may be assigned to. Thus, not surprisingly, the same algorithm applies to problems where the size of each color class is only upper bounded. This will be particularly useful in the next session. Looking at the proof of Theorem 1, the only non-polynomial step is guessing the coloring of the modulator. A straightforward corollary is that if there is a polynomial number of distinct colorings of U and this family can be computed in polynomial time, we can apply the same ideas and check if an equitable k-coloring of the input graph exists in polynomial time. In particular, unipolar graphs satisfy the above condition. If we parameterize by distance to unipolar the problem remains W[1]-hard due to the hardness for split graphs. On the other hand, if we parameterize by distance to unipolar d and the number of colors k we have an FPT algorithm: note that the central clique of $G-U$ has at most k vertices, so we can treat G as a graph with distance to cluster at most $k+d$ and apply Theorem 1.

Corollary 1. Equitable Coloring *on unipolar graphs is in P. When parameterized by distance to unipolar, the problem remains $W[1]$-hard; if also parameterized by the number of colors, there is an FPT algorithm that solves it.*

4 Equitable Coloring Parameterized by Distance to Co-Cluster

Before proceeding to our hardness results, we discuss an FPT algorithm when parameterized by distance to co-cluster. Interestingly, the key ingredient to our approach is the algorithm presented in Sect. 3, which we use to compute the transitions between states of our dynamic programming table. Much like in the previous section, we denote by U the set of vertices such that $G-U$ is a co-cluster graph, and by $\mathcal{I} = \{I_1, \dots, I_r\}$ the independent sets of $G-U$. The following observation follows immediately from the fact that $G-U$ is a complete r-partite graph; it allows us to color the sets of $\mathcal{I}$ independently.

Observation 1. *In any k-coloring φ of G, for every color i, there is at most one $j \in [r]$ such that $\varphi_i \cap I_j \neq \emptyset$.*

Suppose we are already given U, a coloring ψ of U, and the additional restriction that colors $P \subseteq \psi(U)$ must be used on $\ell+1$ vertices. We index our dynamic programming table by (S, p, q, j), where $S \subseteq \psi(U)$ stores which colors of the modulator still need to be extended, p is the number of colors not in $\psi(U)$ that must still be used $\ell+1$ times, q the number of colors not in $\psi(U)$ that must still be used on ℓ vertices, and $j \in [r]$ indicates which of the independent sets we are trying to color. Our goal is to show that $f_{\psi,P}(S,p,q,j) = 1$ if and only if there is a coloring of $G_j = G[U \bigcup_{i=j}^{r} I_i]$ respecting the constraints given by (S,p,q,j). Intuitively, guessing P allows us to index the table by the number of colors not in $\psi(U)$ according only to the capacity of each color, otherwise it would be significantly harder to know, at any time in the algorithm, how many colors should be used on $\ell+1$ vertices. To compute $f_{\psi,P}(S,p,q,j)$, we essentially test every possibility of extension of the colors in S that respects the constraint imposed by P and allows the completion of the coloring of the j-th independent set of $G-U$. Because of Observation 1, the colors not in $\psi(U)$ are confined to a single independent set and, thus, it suffices to consider only how many colors of size $\ell+1$ we are going to use in I_j. We implement this transitioning according to Eq. 1:

$$f_{\psi,P}(S,p,q,j) = \max_{(R,x,y)\,\in\, ext(S,p,q,j)} f_{\psi,P\setminus R}(S \setminus R, p-x, q-y, j+1) \tag{1}$$

where $ext(S,p,q,j)$ is the set of all triples (R,x,y), with $R \subseteq S$, such that each color $i \in R$ can be extended to I_j, while x and y satisfy the system:

$$\begin{cases} x(\ell+1) + y\ell = |I_j| - \alpha_j \\ 0 \leq x \leq p \\ 0 \leq y \leq q \end{cases}$$

where α_j is the number of vertices of I_j used to extend the colors of R to I_j. Note that $|ext(S,p,q,j)| \leq 2^{|S|}n$, so it holds that, for each fixed ψ and P, our dynamic programming table can be computed in $\mathcal{O}^*(3^{|U|})$ time if and only if we can compute $ext(S,p,q,j)$ in $\mathcal{O}^*(|ext(S,p,q,j)|)$ time.

Lemma 4. *$ext(S,p,q,j)$ can be computed in $\mathcal{O}^*(|ext(S,p,q,j)|)$ time.*

Lemma 5. *$f_{\psi,P}(S,p,q,j) = 1$ if and only if ψ can be extended to a coloring φ of G_j using the colors of S, with each color in P used in $\ell+1$ vertices, p extra color classes of size $\ell+1$, and q color classes of size ℓ.*

Finally, all that is left is to show that the number of colorings of U and the constraint set P can both be computed in FPT time.

Theorem 2. EQUITABLE COLORING *can be solved in FPT time when parameterized by distance to co-cluster.*

It is important to note that the above algorithm does not contradict the *NP-hardness* of EQUITABLE COLORING on bipartite graphs, since solving the problem on *complete* bipartite graphs is in P. Moreover, if $U = \emptyset$, all steps of the algorithm are performed in polynomial time, yielding the following corollary.

Corollary 2. EQUITABLE COLORING *of complete multipartite graphs is in P.*

5 Distance to Disjoint Paths

The last parameterization we investigate for EQUITABLE COLORING is *distance to disjoint paths*, which is upper bounded by vertex cover and lower bounded by feedback vertex set. Contrary to our expectations, we show that the problem is $W[1]$-*hard* even if we also parameterize by the number of colors. To accomplish this, we make use of two intermediate problems, namely NUMBER LIST COLORING and EQUITABLE LIST COLORING parameterized by the number of colors. The latter is very similar to EQUITABLE COLORING but to each vertex v is assigned a list $L(v) \subseteq [k]$ of admissible colors. NUMBER LIST COLORING generalizes it in the sense that now we are given a function $h : [k] \mapsto \mathbb{N}$ and color i must be used *exactly* $h(i)$ times. As a first step, we show that NUMBER LIST COLORING parameterized by the number of colors is $W[1]$-*hard* on paths. By roughly doubling the number of colors and vertices used in the construction of [10], we are able to use, essentially, the same arguments. The source problem is MULTICOLORED CLIQUE parameterized by the solution size k: given a graph H such that $V(H)$ is partitioned in k color classes $\{V_1, \ldots, V_k\}$, we want to determine if there is a k-colored clique in H. We denote the edges between V_i and V_j by $E(i,j)$, $|V(H)|$ by n, and $|E(H)|$ by m. We may assume that $|V_i| = N$ and $|E(i,j)| = M$ for every i, j; to see why this is possible, we may take $k!$ disjoint copies of H, each corresponding to a permutation of the color classes and, for each edge $uv \in E(H)$, we connect each copy of u to each copy of v. In our reduction, we interpret a clique as a set of *oriented* edges between color classes, i.e. an edge $e \in E(i,j)$ is selected *twice*: once from V_i to V_j, and once from V_j to V_i. As such, we have *two* gadgets for each edge of H.

Construction. Due to the list nature of the problem, we assign semantic values to each set of colors. In our case, we separate them in four types:

Selection: The colors $\mathcal{S} = \{\sigma(i,j) \mid (i,j) \in [k]^2, i \neq j\}$ and $\mathcal{S}' = \{\sigma'(i,j) \mid (i,j) \in [k]^2, i \neq j\}$ are used to select which edges must belong to the clique.

Helper: $\mathcal{Y}$ and $\mathcal{X}$ satisfy $|\mathcal{Y}| = |\mathcal{X}| = |\mathcal{S}|$. These two sets of colors force the choice made at the root of the edge gadgets to be consistent across the gadget.

Symmetry: The colors $\mathcal{E} = \{\varepsilon(i,j) \mid (i,j) \in [k]^2, i < j\}$ and $\mathcal{E}' = \{\varepsilon'(i,j) \mid (i,j) \in [k]^2, i < j\}$ guarantee that, if edge $e \in E(i,j)$ is picked from V_i to V_j, it must also be picked from V_j to V_i.

Consistency: Colors $\mathcal{T} = \{\tau_i(r,s) \mid i \in [k],\ r,s \in [k] \setminus \{i\},\ r < s\}$ and $\mathcal{T}' = \{\tau_i'(r,s) \mid i \in [k],\ r,s \in [k] \setminus \{i\},\ r < s\}$ ensure that if the edge uv is chosen between V_i and V_j, the edge between V_i and V_r must also be incident to u.

Before detailing the gadgets themselves, we define what is, in our perception, one of the most important pieces of the proof. For each vertex $v \in V(H)$, choose an arbitrary but unique integer in the range $[n^2+1, n^2+n]$ and, for each edge e, a unique integer in the range $[2n^2+1, 2n^2+m]$. These are the *up-identification* numbers of vertex v and edge e, denoted by $v_\uparrow$ and $e_\uparrow$, respectively. Now, choose a suitably huge integer Z, say n^3, and define the *down-identification number* for

v as $v_\downarrow = Z - v_\uparrow$. These quantities play a key role on the numerical targets for the symmetry and consistency colors; since they are unique, these identification numbers tie together the choices between edge gadgets.

For each pair $i, j \in [k]$, with $i < j$, the input graph G of Number List Coloring has the groups of gadgets $\mathcal{G}(i,j)$ and $\mathcal{G}(j,i)$, each containing M edge gadgets corresponding to the edges of $E(i,j)$. We say that $\mathcal{G}(i,j)$ is the *forward group* and that $\mathcal{G}(j,i)$ is the *backward group*. For the description of the gadgets, we always assume $i < j$, $e \in E(i,j)$ with $u \in V_i$ and $v \in V_j$.

Forward Edge Gadget. The gadget $G(i,j,e)$ has a root vertex $r(i,j,e)$, with list $\{\sigma(i,j), \sigma'(i,j)\}$, and two neighbors, both with the list $\{\sigma(i,j), y(i,j)\}$, which for convenience we call $a(i,j,e)$ and $b(i,j,e)$. We equate membership of edge e in the solution to Multicolored Clique to the coloring of $r(i,j,e)$ with $\sigma(i,j)$. When discussing the vertices of the remaining vertices of the gadget, we say that a vertex is *even* if its distance to $r(i,j,e)$ is even, otherwise it is *odd*. To $a(i,j,e)$, we append a path with $2e_\downarrow + 2(k-1)u_\downarrow$ vertices. First, we choose $e_\downarrow$ even vertices to assign the list $\{y(i,j), \varepsilon'(i,j)\}$. Next, for each r in $j < r \leq k$, choose $u_\downarrow$ even vertices to assign the list $\{y(i,j), \tau_i'(j,r)\}$. Similarly, for each $s \neq i$ satisfying $s < j$, choose $u_\downarrow$ even vertices and assign the list $\{y(i,j), \tau_i(s,j)\}$. All the odd vertices - except $a(i,j,e)$ and $b(i,j,e)$ - are assigned the list $\{y(i,j), x(i,j)\}$. The path appended to $b(i,j,e)$ is similarly defined, except for two points: (i) the length and number of chosen vertices are proportional to $e_\uparrow$ and $u_\uparrow$; and (ii) when color $\varepsilon(i,j)$ (resp. $\tau_i(s,r)$) should be in the list, we add $\varepsilon'(i,j)$ (resp. $\tau_i'(s,r)$), and vice-versa. For an example of the edge gadget, please refer to Fig. 4.

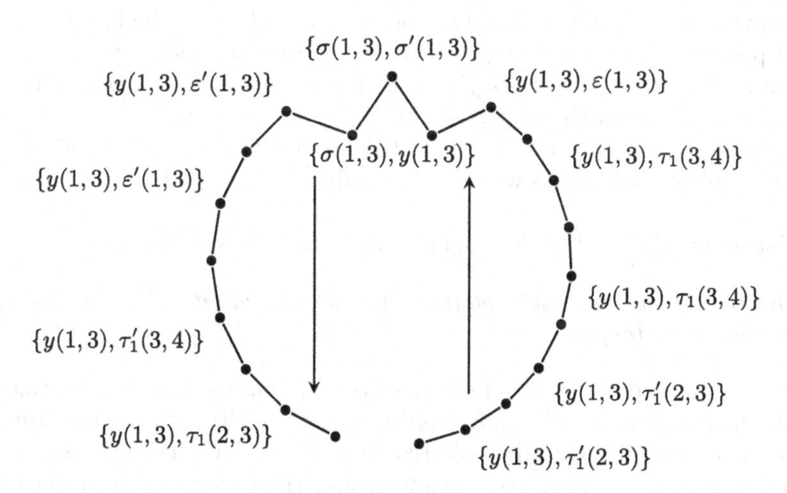

Fig. 4. Example of a forward edge gadget $G(1,3,e)$ of group $\mathcal{G}(1,3)$, with $k = 4$, $Z = 3$, $e_\downarrow = 2$, and $u_\downarrow = 1$. Vertices with no explicit list have list equal to $\{y(1,3), x(1,3)\}$.

Backward Edge Gadget. Gadget $G(j,i,e)$ has vertices $r(j,i,e)$, $a(j,i,e)$, and $b(j,i,e)$ defined similarly as to the forward gadget, with the root vertex having the list $\{\sigma(j,i), \sigma'(j,i)\}$, while the other two have the list $\{\sigma(j,i), y(j,i)\}$. To $a(j,i,e)$, we append a path with $2e_\downarrow + 2(k-1)v_\downarrow$ vertices. First, choose $e_\downarrow$ even vertices to assign the list $\{y(j,i), \varepsilon(i,j)\}$. Now, for each r in $j < r \leq k$, choose $v_\downarrow$ even vertices to assign the list $\{y(j,i), \tau'_j(i,r)\}$. Then, for each $s \neq i$ satisfying $s < j$, choose $v_\downarrow$ even vertices and assign the list $\{y(j,i), \tau_j(s,i)\}$. All the odd vertices are assigned the list $\{y(j,i), x(j,i)\}$. The path appended to $b(j,i,e)$ is similarly defined, except that: (i) the length and number of chosen vertices are proportional to $e_\uparrow$ and $v_\uparrow$; and (ii) when the color $\varepsilon(i,j)$ (resp. $\tau_j(s,r)$) is in the list, we replace it with $\varepsilon'(i,j)$ (resp. $\tau'_j(s,r)$), and vice-versa. Note that, for every edge gadget, either forward or backward, the number of vertices is equal to $3 + 2(e_\uparrow + e_\downarrow) + 2(k-1)(u_\uparrow + u_\downarrow) = 3 + 2kZ$. We say that $G(i,j,e)$ is *selected* if $r(i,j,e)$ is colored with $\sigma(i,j)$, otherwise it is *passed.*

Numerical Targets. Before defining the numerical targets, given by $h(\cdot)$, recall that $|E(i,j)| = M$ for every pair i,j and that, for every vertex u and edge e, the identification numbers satisfy the identity $v_\uparrow + v_\downarrow = Z$ and $e_\uparrow + e_\downarrow = Z$. We present the numerical targets of our instance - and some intuition - below.

Selection: $h(\sigma(i,j)) = 1 + 2(M-1)$ and $h(\sigma'(i,j)) = M - 1$. Since only one edge may be chosen from V_i to V_j, the non-selection color $\sigma'(i,j)$ must be used in $M-1$ edges of $\mathcal{G}(i,j)$. Thus, exactly one $G(i,j,e)$ is selected and, to achieve the target of $1 + 2(M-1)$, for every $f \in E(i,j) \setminus \{e\}$, both $a(i,j,f)$ and $b(i,j,f)$ must also be colored with $\sigma(i,j)$.

Helper: $h(y(i,j)) = 2 + kMZ$ and $h(x(i,j)) = kMZ - kZ$. The goal here is that, if $G(i,j,e)$ is selected, all the odd positions must be colored with $y(i,j)$, otherwise every even position must be colored with it. In the latter case, the odd positions of all but one gadget of $\mathcal{G}(i,j)$ must be colored with $x(i,j)$.

Symmetry: $h(\varepsilon(i,j)) = h(\varepsilon'(i,j)) = Z$. If the previous condition holds and $r(i,j,e)$ is colored with $\sigma(i,j)$, then $\varepsilon(i,j)$ appears in $e_\uparrow$ vertices of the gadget rooted at $r(i,j,e)$. To meet the target Z, $e_\downarrow$ vertices of another gadget must also be colored with it, as we show, the only way is if $r(j,i,e)$ is colored with $\sigma(j,i)$.

Consistency: $h(\tau_i(s,r)) = h(\tau'_i(s,r)) = Z$. Similar to symmetry colors.

Lemma 6. *If H has a k-multicolored clique, then G admits a list coloring meeting the numerical targets.*

We now proceed to the proof of the converse. Lemma 7 guarantees that the decision made at the root of a gadget propagates throughout the entire structure; Lemma 8 ensures that the edge selected from V_j to V_i is the same as the edge selected from V_i to V_j; finally, Lemma 9 equates the vertex of V_i incident to the edge between V_i and V_j to the vertex incident to the edge between V_i and V_s.

Lemma 7. *In every list coloring of G satisfying h, exactly one gadget of each $\mathcal{G}(i,j)$ is selected, each passed $G(i,j,e)$ has all of its kZ even vertices colored with $y(i,j)$, and the selected $G(i,j,f)$ has all of its $2 + kZ$ odd vertices colored with $y(i,j)$.*

Lemma 8. *In every list coloring φ of G, if $G(i, j, e)$ is selected, so is $G(j, i, e)$.*

Lemma 9. *In every list coloring φ of G, if $G(i, j, e)$ is selected and $e = uv$, then, for every $s \neq i$, the edge f of H corresponding to the selected gadget $G(i, s, f)$ must be incident to u.*

Theorem 3. Number List Coloring *on paths parameterized by the number of colors that appear on the lists is $W[1]$-hard.*

Corollary 3. Equitable List Coloring *on paths parameterized by the number of colors that appear on the lists is $W[1]$-hard.*

Corollary 4. Equitable Coloring *parameterized by the number of colors and distance to disjoint paths is $W[1]$-hard.*

Theorem 3 and its corollaries follow from the previous lemmas. For the first corollary, we add isolated vertices with lists of size one to the input of Number List Coloring so as to make all colors have the same numerical target. For the second, we add a clique of size r, the number of colors of the instance of Equitable List Coloring, and label them using the integers $[r]$; afterwards, for each vertex u of the input graph that does not have color i in its list, we add an edge between the i-th vertex of the clique and u.

6 Final Remarks

In this work we presented an extensive study of multiple parameterizations for the Equitable Coloring problem, obtaining both tractability and intractability results. Specifically, we proved that it is fixed parameter tractable when parameterized by distance to cluster and distance to co-cluster, and as corollaries that there is an FPT algorithm when parameterized by distance to unipolar and number of colors. Meanwhile, the problem remains W[1]-hard when simultaneously parameterized by distance to disjoint paths and number of colors. We also revisited previous works in the literature and restated them in terms of parameterized complexity. This review settled the complexity for some parameterizations weaker than distance to clique and show that our results are, in a sense, optimal: searching for parameters weaker than distance to (co-)cluster will most likely not yield FPT algorithms. Vertex Coloring is already notoriously hard to find polynomial kernels for, as shown by Jansen and Kratsch [15]; in fact, most of the parameterizations under which classical coloring admits a polynomial kernel do not make Equitable Coloring tractable, painting a quite bleak future for kernelization algorithms for Equitable Coloring. However, Reddy [19] does present a polynomial kernel parameterized by distance to threshold and number of colors, so not all hope is lost. All things considered, we believe that future work should be directed to the study of the parameterizations max leaf number or feedback edge set, as these are the two main cases of the graph parameter hierarchy we leave open. Another interesting direction would be to improve the running times of known tractable cases and determining lower bounds for parameters such as vertex cover and distance to clique, specially if the super exponential dependency is necessary.

References

1. Baker, B.S., Coffman, E.G.: Mutual exclusion scheduling. Theoret. Comput. Sci. **162**(2), 225–243 (1996)
2. Bodlaender, H.L., Fomin, F.V.: Equitable colorings of bounded treewidth graphs. Theoret. Comput. Sci. **349**(1), 22–30 (2005)
3. Bodlaender, H.L., Jansen, K.: Restrictions of graph partition problems. Part I. Theoret. Comput. Sci. **148**(1), 93–109 (1995)
4. Bondy, J.A., Murty, U.S.R.: Graph Theory with Applications, vol. 290. Macmillan, London (1976)
5. Boral, A., Cygan, M., Kociumaka, T., Pilipczuk, M.: Fast branching algorithm for cluster vertex deletion. arXiv e-prints p. 1306.3877 (2013)
6. Cai, L.: Parameterized complexity of vertex colouring. Discrete Appl. Math. **127**(3), 415–429 (2003)
7. Chen, B.-L., Ko, M.-T., Lih, K.-W.: Equitable and m-bounded coloring of split graphs. In: Deza, M., Euler, R., Manoussakis, I. (eds.) CCS 1995. LNCS, vol. 1120, pp. 1–5. Springer, Heidelberg (1996). https://doi.org/10.1007/3-540-61576-8_67
8. Cormen, T.H., Leiserson, C.E., Rivest, R.L., Stein, C.: Introduction to Algorithms, 3rd edn. The MIT Press, Cambridge (2009)
9. Cygan, M., et al.: Parameterized Algorithms, vol. 3. Springer, Cham (2015). https://doi.org/10.1007/978-3-319-21275-3
10. Fellows, M.R., et al.: On the complexity of some colorful problems parameterized by treewidth. Inf. Comput. **209**(2), 143–153 (2011)
11. Fiala, J., Golovach, P.A., Kratochvíl, J.: Parameterized complexity of coloring problems: treewidth versus vertex cover. Theoret. Comput. Sci. **412**(23), 2513–2523 (2011)
12. Garey, M.R., Johnson, D.S.: Computers and Intractability: A Guide to the Theory of NP-Completeness. W. H. Freeman & Co., New York (1979)
13. Gomes, G.d.C.M., Lima, C.V.G.C., Santos, V.F.D.: Parameterized Complexity of Equitable Coloring. Discrete Math. Theoret. Comput. Sci. **21**(1) (2019). ICGT 2018
14. Irani, S., Leung, V.: Scheduling with conflicts, and applications to traffic signal control. In: Proceedings of the 7th Annual ACM-SIAM Symposium on Discrete Algorithms, SODA 1996, pp. 85–94. SIAM, USA (1996)
15. Jansen, B.M., Kratsch, S.: Data reduction for graph coloring problems. Inf. Comput. **231**, 70–88 (2013). FCT 2011
16. Jarvis, M., Zhou, B.: Bounded vertex coloring of trees. Discrete Math. **232**(1–3), 145–151 (2001)
17. Lih, K.W.: Equitable coloring of graphs. In: Pardalos, P., Du, D.Z., Graham, R. (eds.) Handbook of Combinatorial Optimization, pp. 1199–1248. Springer, New York (2013). https://doi.org/10.1007/978-1-4419-7997-1_25
18. Meyer, W.: Equitable coloring. Am. Math. Monthly **80**(8), 920–922 (1973)
19. Reddy, I.V.: Parameterized coloring problems on threshold graphs. arXiv e-prints p. 1910.10364 (2019)
20. Smith, B., Bjorstad, P., Gropp, W.: Domain Decomposition: Parallel Multilevel Methods for Elliptic Partial Differential Equations. Cambridge University Press, Cambridge (2004)
21. Sorge, M., Weller, M.: The graph parameter hierarchy (2019, unpublished manuscript)

Algorithms and Data Structures

Dynamically Optimal Self-adjusting Single-Source Tree Networks

Chen Avin[1], Kaushik Mondal[2(✉)], and Stefan Schmid[3]

[1] School of Electrical and Computer Engineering,
Ben Gurion University of the Negev, Beersheba, Israel
avin@cse.bgu.ac.il

[2] Department of Mathematics,
Indian Institute of Technology Ropar, Rupnagar, India
kaushik.mondal@iitrpr.ac.in

[3] Faculty of Computer Science, University of Vienna, Vienna, Austria
schmiste@gmail.com

Abstract. This paper studies a fundamental algorithmic problem related to the design of demand-aware networks: networks whose topologies adjust toward the traffic patterns they serve, in an online manner. The goal is to strike a tradeoff between the benefits of such adjustments (shorter routes) and their costs (reconfigurations). In particular, we consider the problem of designing a self-adjusting tree network which serves single-source, multi-destination communication. The problem has interesting connections to self-adjusting datastructures. We present two constant-competitive online algorithms for this problem, one randomized and one deterministic. Our approach is based on a natural notion of *Most Recently Used (MRU)* tree, maintaining a *working set*. We prove that the working set is a cost lower bound for any online algorithm, and then present a randomized algorithm RANDOM-PUSH which *approximates* such an MRU tree at low cost, by pushing less recently used communication partners down the tree, along a random walk. Our deterministic algorithm MOVE-HALF does not directly maintain an MRU tree, but its cost is still proportional to the cost of an MRU tree, and also matches the working set lower bound.

1 Introduction

While datacenter networks traditionally rely on a *fixed* topology, recent optical technologies enable *reconfigurable* topologies which can adjust to the demand (i.e., traffic pattern) they serve *in an online manner*, e.g. [7,11,13,16]. Indeed, the physical topology is emerging as the next frontier in an ongoing effort to render networked systems more flexible.

In principle, such topological reconfigurations can be used to provide shorter routes between frequently communicating nodes, exploiting structure in traffic patterns [1,15,16], and hence to improve performance. However, the design of self-adjusting networks which dynamically optimize themselves toward the demand introduces an algorithmic challenge: an online algorithm needs to be devised which

Y. Kohayakawa and F. K. Miyazawa (Eds.): LATIN 2020, LNCS 12118, pp. 143–154, 2020.
https://doi.org/10.1007/978-3-030-61792-9_12

guarantees an efficient tradeoff between the benefits (i.e., shorter route lengths) and costs (in terms of reconfigurations) of topological optimizations.

This paper focuses on the design of a self-adjusting *complete tree* (CT) network: a network of nodes (e.g., servers or racks) that forms a complete tree, and we measure the routing cost in terms of the length of the shortest path between two nodes. Trees are not only a most fundamental topological structure of their own merit, but also a crucial building block for more general self-adjusting network designs: Avin et al. [2,3] recently showed that multiple tree networks (optimized individually for a single source node) can be combined to build general networks which provide low degree and low distortion. The design of a dynamic single-source multi-destination communication tree, as studied in this paper, is hence a stepping stone.

The focus on trees is further motivated by a relationship of our problem to problems arising in self-adjusting datastructures [5]: self-adjusting datastructures such as self-adjusting search trees [21] have the appealing property that they optimize themselves to the workload, leveraging temporal locality, but without knowing the future. Ideally, self-adjusting datastructures store items which will be accessed (frequently) *in the future*, in a way that they can be accessed quickly (e.g., close to the root, in case of a binary search tree), while also accounting for reconfiguration costs. However, in contrast to most datastructures, in a *network*, the search property is not required: the network supports *routing*. Accordingly our model can be seen as a novel flavor of such self-adjusting binary search trees where lookup is supported by a *map*, enabling shortest path routing (more details will follow).

We present a formal model for this problem later, but a few observations are easy to make. If we restrict ourselves to the special case of a *line* network (a "linear tree"), the problem of optimally arranging the destinations of a given single communication source is equivalent to the well-known *dynamic list update* problem: for such self-adjusting (unordered) lists, dynamically optimal online algorithms have been known for a long time [20]. In particular, the simple move-to-front algorithm which immediately promotes the accessed item to the front of the list, fulfills the *Most-Recently Used (MRU)* property: the ith furthest away item from the front of the list is the ith most recently used item. In the list (and hence on the line), this property is enough to guarantee optimality. The MRU property is related to the so called *working set property*: the cost of accessing item x at time t depends on the number of distinct items accessed since the last access of x prior to time t, including x. Naturally, we wonder whether the *MRU* property is enough to guarantee optimality also in our case. The answer turns out to be non-trivial.

A first contribution of this paper is the observation that if we count only *access* cost (ignoring any rearrangement cost, see Definition 1 for details), the answer is affirmative: the most-recently used tree is what is called *access optimal.* Furthermore, we show that the corresponding access cost is a lower bound for any algorithm which is dynamically optimal. But securing this property, i.e., maintaining the most-recently used items close to the root in the tree, introduces a new challenge: how to achieve this *at low cost*? In particular, assuming that *swapping* the locations of items comes at a *unit cost*, can the property be maintained at cost proportional to the *access* cost? As we show, *strictly* enforcing the most-recently used property in a tree is too costly to achieve optimality. But, as we will show, when turning to an *approximate* most-recently used property, we are able to show two important properties: *i)* such an approximation is good enough to guarantee access optimality; and *ii)* it can be maintained in expectation using a *randomized* algorithm: less recently used communication partners are pushed down the tree along a random walk.

While the most-recently used property is *sufficient*, it is not necessary: we provide a deterministic algorithm which is dynamically optimal but does not even maintain the MRU property approximately. However, its cost is still proportional to the cost of an MRU tree (Definition 4).

Succinctly, we make the following ***contributions***. First we show a working set lower bound for our problem. We do so by proving that an MRU tree is *access optimal*. In the following theorem, let $WS(\sigma)$ denote the working set of σ (a formal definition will follow later).

Theorem 1. *Consider a request sequence* σ*. Any algorithm* ALG *serving* σ *using a self-adjusting complete tree, has cost at least* $\text{cost}(ALG(\sigma)) \geq WS(\sigma)/4$, *where* $WS(\sigma)$ *is the working set of* σ*.*

Our main contribution is a deterministic online algorithm $MOVE-HALF$ which maintains a constant competitive self-adjusting Complete Tree (CT) network.

Theorem 2. $MOVE-HALF$ *algorithm is dynamically optimal.*

Interestingly, $MOVE-HALF$ does not require the MRU property and hence does not need to maintain MRU tree. This implies that maintaining a working set on CTs is not a necessary condition for dynamic optimality, although it is a sufficient one.

Furthermore, we present a dynamically optimal, i.e., constant competitive (on expectation) randomized algorithm for self-adjusting CTs called RANDOM-PUSH. RANDOM-PUSH relies on maintaining an approximate MRU tree.

Theorem 3. *The RANDOM-PUSH algorithm is dynamically optimal on expectation.*

Due to space constraints, proofs and longer discussions appear in a technical report [4].

2 Model and Preliminaries

Our problem can be formalized using the following simple model. We consider a single *source* that needs to communicate with a set of n nodes $V = \{v_1, \ldots, v_n\}$. The nodes are arranged in a complete binary tree and the source is connected the root of the tree. While the tree describes a reconfigurable *network*, we will use terminology from datastructures, to highlight this relationship and avoid the need to introduce new terms.

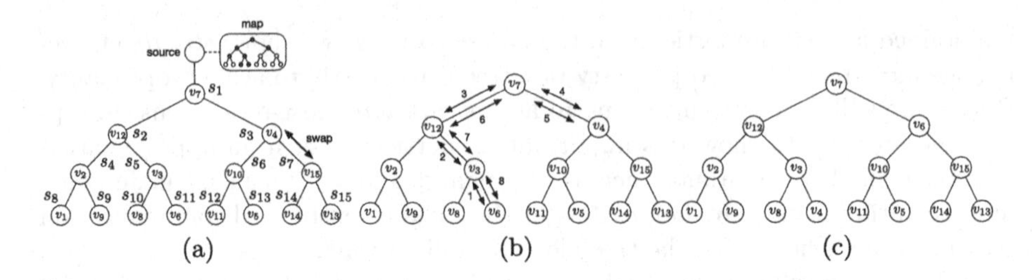

Fig. 1. (a) Our *complete tree* model: a source with a map, a tree of servers that host items (nodes) and a *swap* operation between neigboring items. (b) The node's tree network implied by the tree T from (a) and the set of swaps needed to interchange the location of v_6 and v_4. (c) The tree network after the interchange and swap operations of (b).

We consider a complete tree T connecting n servers $S = \{s_1, \ldots, s_n\}$. We will denote by $s_1(T)$ the root of the tree T, or s_1 when T is clear from the context, and by s_i.left (resp. s_i.right) the left (resp. right) child of server s_i. We assume that the n servers store n items (nodes) $V = \{v_1, \ldots, v_n\}$, one item per server. For any $i \in [1, n]$ and any time t, we will denote by $s_i.\text{guest}^{(t)} \in V$ the item mapped to s_i at time t. Similarly, $v_i.\text{host}^{(t)} \in S$ denotes the server hosting item v_i. Note that if $v_i.\text{host}^{(t)} = s_j$ then $s_j.\text{guest}^{(t)} = v_i$.

The *depth* of a server s_i is fixed and describes the distance from the root; it is denoted by s_i.dep, and $s_1.\text{dep} = 0$. The depth of an item v_i *at time* t is denoted by $v_i.\text{dep}^{(t)}$, and is given by the depth of the server to which v_i is mapped at time t. Note that $v_i.\text{dep}^{(t)} = v_i.\text{host.dep}^{(t)}$.

To this end, we interpret communication requests from the source as *accesses* to *items* stored in the (unordered) tree. All access requests (resp. communication requests) to items (resp. nodes) originate from the root s_1. If an item (resp. node) is frequently requested, it can make sense to move this item (node) closer to the root of T: this is achieved by *swapping* items which are neighboring in the tree (resp. by performing local topological swaps).

Access requests occur over time, forming a (finite or infinite) sequence $\sigma = (\sigma^{(1)}, \sigma^{(2)}, \ldots)$, where $\sigma^{(t)} = v_i \in V$ denotes that item v_i is requested, and needs to be accessed at time t. The sequence σ (henceforth also called the *workload*) is revealed one-by-one to an online algorithm ON. The *working set* of an item v_i at time t is the set of distinct elements accessed since the last access of v_i prior to time t, including v_i. We define the *rank* of item v_i at time t to be the size of the working set of v_i at time t and denote it as $v_i.\text{rank}^{(t)}$. When t is clear of context, we simply write v_i.rank. The working set bound of sequence σ of m requests is defined as $WS(\sigma) = \sum_{t=1}^{m} \log(\sigma^{(t)}.\text{rank})$.

Both serving (i.e., *routing*) the request and adjusting the configuration comes at a cost. We will discuss the two cost components in turn. Upon a request, i.e., whenever the source wants to communicate to a partner, it routes to it via the tree T. To this end, a message passed between nodes can include, for each node it passes, a bit indicating which child to forward the message next (requires $O(\log n)$ bits). Such a *source routing* header can be built based on a dynamic global *map* of the tree that is maintained at the source node. As mentioned, the source node is a direct neighbor of the root of the tree, aware of all requests, and therefore it can maintain the map. The *access cost* is hence given by the distance between the root and the requested item, which is basically the depth of the item in the tree.

The *reconfiguration cost* is due to the adjustments that an algorithm performs on the tree. We define the unit cost of reconfiguration as a *swap*: a swap means changing position of an item with its parent. Note that, any two items u, v in the tree can be *interchanged* using a number of swaps equal to twice the distance between them. This can be achieved by u first swapping along the path to v and then v swapping along the same path to initial location of u. This interchange operation results in the tree staying the same, but only u and v changing locations. We assume that to interchange items, we first need to access one of them. See Fig. 1 for an example of our model and interchange operation.

Definition 1 *(Cost). The cost incurred by an algorithm ALG to serve a request* $\sigma^{(t)} = v_i$ *is denoted by* $\text{cost}(ALG(\sigma^{(t)}))$, *short* $\text{cost}^{(t)}$. *It consists of two parts,* access *cost, denoted* $\text{acc-cost}^{(t)}$, *and* adjustment *cost, denoted* $\text{adj-cost}^{(t)}$. *We define access cost simply as* $\text{acc-cost}^{(t)} = v_i.\text{dep}^{(t)}$ *since ALG can maintain a* global map *and access* v_i *via the shortest path. Adjustment cost,* $\text{adj-cost}^{(t)}$, *is the total number of swaps, where a single swap means changing position of an item with its parent or a child. The total cost, incurred by ALG is then*

$$\text{cost}(ALG(\sigma)) = \sum_t \text{cost}(ALG(\sigma^{(t)})) = \sum_t \text{cost}^{(t)} = \sum_t (\text{acc-cost}^{(t)} + \text{adj-cost}^{(t)})$$

Our main objective is to design online algorithms that perform almost as well as optimal offline algorithms (which know σ ahead of time), even in the worst-case. In other words, we want to devise online algorithms which minimize the competitive ratio:

Definition 2 *(Competitive Ratio ρ). We consider the standard definition of (strict) competitive ratio ρ, i.e.,* $\rho = \max_\sigma \text{cost}(ON)/\text{cost}(OPT)$ *where σ is any input sequence and where OPT denotes the optimal offline algorithm.*

If an online algorithm is constant competitive, independently of the problem input, it is called *dynamically optimal.*

Definition 3 (Dynamic Optimality). *An (online) algorithm ON achieves* dynamic optimality *if it asymptotically matches the offline optimum on every access sequence. In other words, the algorithm ON is $O(1)$-competitive.*

We also consider a weaker form of competitivity (similarly to the notion of *search-optimality* in related work [6]), and say that *ON* is *access-competitive* if we consider only the access cost of *ON* (and ignore any adjustment cost) when comparing it to *OPT* (which needs to pay both for access and adjustment). For a randomized algorithm, we consider an oblivious online adversary which does not know the random bits of the online algorithm a priori.

The ***Self-adjusting Complete Tree Problem*** considered in this paper can then be formulated as follows: Find an online algorithm which serves any (finite or infinite) online request sequence σ with minimum cost (including both access and rearrangement costs), on a self-adjusting complete binary tree.

3 Access Optimality: A Working Set Lower Bound

For *fixed* trees, it is easy to see that keeping frequent items close to the root, i.e., using a *Most-Frequently Used* (MFU) policy, is optimal (cf. the technical report [4]). The design of online algorithms for *adjusting* trees is more involved. In particular, it is known that MFU is not optimal for lists [20]. A natural strategy could be to try and keep items close to the root which have been frequent "recently". However, this raises the question over which time interval to compute the frequencies. Moreover, changing from one MFU tree to another one may entail high adjustment costs.

This section introduces a natural *pendant* to the MFU tree for a dynamic setting: the *Most Recently Used (MRU) tree.* Intuitively, the MRU tree tries to keep the "working set" resp. *recently* accessed items close to the root. In this section we show a working set lower bound for any self-adjusting complete binary tree.

While the move-to-front algorithm, known to be dynamically optimal for self-adjusting lists [20], naturally provides such a "most recently used" property, generalizing move-to-front to the tree is non-trivial. We therefore first show that any algorithm that maintains an MRU tree is *access-competitive.* With this in mind, let us first formally define the MRU tree.

Definition 4 (MRU Tree). *For a given time t, a tree T is an* MRU *tree if and only if,*

$$v_i.\text{dep} = \lfloor \log v_i.\text{rank} \rfloor \tag{1}$$

Accordingly the root of the tree (level zero) will always host an item of rank one. More generally, servers in level i will host items that have a rank between $(2^i, 2^{i+1} - 1)$. Upon a request of an item, say v_j with rank r, the rank of v_j is updated to one, and only the ranks of items with rank smaller than r are increased, each by 1. Therefore, the rank of items with rank higher than r do not change and their level (i.e., depth) in the MRU tree remains the same (but they may switch location within the same level).

Definition 5 (MRU algorithm). *An online algorithm ON has the* MRU *property (or the working set property) if for each time t, the tree $T^{(t)}$ that ON maintains, is an* MRU *tree.*

The working set lower bound will follow from the following theorem which states that any algorithm that has the *MRU* property is *access competitive.*

Theorem 4. *Any online algorithm ON that has the* MRU *property is 4 access-competitive.*

Recall that an analogous statement of Theorem 4 is known to be true for a *list* [20]. As such, one would hope to find a simple proof that holds for complete trees, but it turns out that this is not trivial, since OPT has more freedom in trees. We therefore present a direct proof based on a potential function, similar in spirit to the list case.

Based on Theorem 4 we can now show our working set lower bound:

Theorem 5. *Consider a request sequence* σ*. Any algorithm* ALG *serving* σ *using a self-adjusting complete tree, has cost at least* $\text{cost}(ALG(\sigma)) \geq WS(\sigma)/4$*, where* $WS(\sigma)$ *is the working set of* σ*.*

Proof. The sum of the access costs of items from an MRU tree is exactly $WS(\sigma)$. For the sake of contradiction assume that there is an algorithm ALG with cost $\text{cost}(ALG(\sigma)) < WS(\sigma)/4$. If follows that Theorem 4 is not true. A contradiction. □

4 Deterministic Algorithm

4.1 Efficiently Maintaining an MRU Tree

It follows from the previous section that if we can maintain an MRU tree at the cost of *accessing* an MRU tree, we will have a dynamically optimal algorithm. So we now turn our attention to the problem of efficiently maintaining an MRU tree. To achieve optimality, we need that the tree adjustment cost will be proportional to the access cost. In particular, we aim to design a tree which on one hand achieves a good approximation of the MRU property to capture temporal locality, by providing fast *access* (resp. *routing*) to items; and on the other hand is also adjustable at low cost over time.

Let us now assume that a certain item $\sigma^{(t)} = u$ is accessed at some time t. In order to re-establish the (strict) MRU property, u needs to be promoted to the root. This however raises the question of where to move the item currently located at the root, let us call it v. A natural idea to make space for u at the root while preserving locality, is to *push down* items from the root, including item v. However, note that simply pushing items down along the path between u and v (as done in lists) will result in a poor performance in the tree. To see this, let us denote the sequence of items along the path from u to v by $P = (u, w_1, w_2, \ldots, w_\ell, v)$, where $\ell = u.\text{dep}$, *before* the adjustment. Now assume that the access sequence σ is such that it repeatedly cycles through the sequence P, in this order. The resulting cost per request is in the order of $\Theta(\ell)$, i.e., could reach $\Theta(\log n)$ for $\ell = \Theta(\log n)$. However, an algorithm which assigns (and then fixes) the items in P to the top $\log \ell$ levels of the tree, will converge to a cost of only $\Theta(\log \ell) \in O(\log \log n)$ per request: an exponential improvement.

Another basic idea is to try and keep the MRU property at every step. Let us call this strategy Max-Push. Consider a request to item u which is at depth $u.\text{dep} = k$. Initially u is moved to the root. Then the Max-Push strategy chooses for each depth $i < u.\text{dep}$, the *least* recently accessed (and with maximum rank)

Algorithm 1: Upon request to u in $MOVE-HALF$'s TREE

1: **access** $u = s$.guest along the tree branches (cost: u.dep)
2: let v be the item with the highest rank at depth $\lfloor u.\text{dep}/2 \rfloor$
3: **swap** u along tree branches to node v (cost: $\frac{3}{2}u$.dep)
4: **swap** v along tree branches to server s (cost: $\frac{3}{2}u$.dep)

item from level i: formally, $w_i = \arg\max_{v \in V: v.\text{dep}=i} v.\text{rank}$. We then push w_i to the host of w_{i+1}. It is not hard to see that this strategy will actually maintain a perfect MRU tree. However, items with the maximum rank in different levels, i.e., w_i.host and w_{i+1}.host, may not be in a parent-child relation. So to push w_i to w_{i+1}.host, we may need to travel all the way from w_i.host to the root and then from the root to w_{i+1}.host, resulting in a cost proportional to i per level i. This accumulates a rearrangement cost of $\sum_{i=1}^{k} i > k^2/2$ to push all the items with maximum rank at each layer, up to layer k. This is not proportional to the original access cost k of the requested item and therefore, leads to a non-constant competitive ratio as high as $\Omega(\log n)$.

Later, in Sect. 5, we will present a randomized algorithm that maintains a tree that approximates an MRU tree at a low cost. But first, we will present a simple deterministic algorithm that does not directly maintain an MRU tree, but has cost that is proportional to the MRU cost and is hence dynamically optimal.

4.2 The $MOVE-HALF$ Algorithm

In this section we propose a simple deterministic algorithm, $MOVE-HALF$, that is proven to be dynamically optimal. Interestingly $MOVE-HALF$ does not maintain the MRU property but its cost is shown to be competitive to the *access cost* on an MRU tree, and therefore, to the working set lower bound.

$MOVE-HALF$ is described in Algorithm 1. Initially, $MOVE-HALF$ and OPT start from the same tree (which is assumed w.l.o.g. to be an MRU tree). Then, upon a request to an item u, $MOVE-HALF$ first accesses u and then interchanges its position with node v that is the highest ranked item positioned at half of the depth of u in the tree. After the interchange the tree remains the same, only u and v changed locations. See Fig. 1 (b) for an example of $MOVE-HALF$ operation where v_6 at depth 3 is requested and is then interchanged with v_4 at depth 1 (assuming it is the highest rank node in level 1).

The *access cost* of $MOVE-HALF$ is proportional to the access cost of an MRU tree.

Theorem 6. *Algorithm $MOVE-HALF$ is 4 access-competitive to an MRU algorithm.*

Proof (Proof of Theorem 2). Using Theorem 4 and Theorem 6, $MOVE-HALF$ is 16-access competitive. It is easy to see from Algorithm 1 that total cost of $MOVE-HALF$'s tree is 4 times the access cost. Considering these, $MOVE-HALF$ is 64-competitive. □

In the coming section we show techniques to maintain MRU trees cheaply. This is another way to maintain dynamic optimality.

5 Randomized MRU Trees

The question of how, and if at all possible, to maintain an MRU tree deterministically (where for each request $\sigma^{(t)}$, $\sigma^{(t)}.depth = \lfloor \log \sigma^{(t)}.\text{rank} \rfloor$) at low cost is still an open problem. But, in this section we show that the answer is affirmative with two relaxations: namely by using randomization and approximation. We believe that the properties of the algorithm we describe next may also find applications in other settings, and in particular data structures like skip lists [9].

At the heart of our approach lies an algorithm to maintain a constant approximation of the MRU tree at any time. First we define MRU(β) trees for any constant β.

Definition 6 *(MRU(β) Tree). A tree T is called an* MRU(β) *tree if it holds for any item u and any time that,* $u.\text{dep} = \lfloor \log u.\text{rank} \rfloor + \beta$.

Note that, any MRU(0) tree is also an MRU tree. In particular, we prove in the following that a constant additive approximation is sufficient to obtain dynamic optimality.

Theorem 7. *Any online MRU(β) algorithm is* $4(1 + \lceil \frac{\beta}{2} \rceil)$ *access-competitive.*

To efficiently achieve an MRU(β) tree, we propose the RANDOM-PUSH strategy (see Algorithm 2). This is a simple randomized strategy which selects a random path starting at the root, and then steps down the tree to depth $k = u.\text{dep}$ (the accessed item depth), by choosing uniformly at random between the two children of each server at each step. This can be seen as a simple k-step random walk in a directed version of the tree, starting from the root of the tree. Clearly, the adjustment cost of RANDOM-PUSH is also proportional to k and its actions are independent of any oblivious online adversary. The main technical challenge of this section is proving the following theorem.

Theorem 8. *RANDOM-PUSH maintains an MRU*(4) *(Definition 6) tree in expectation, i.e., the expected depth of the item with rank r is less than* $\log r + 3 < \lfloor \log r \rfloor + 4$ *for any sequence σ and any time t.*

It now follows almost directly from Theorems 7 and 8 that RANDOM-PUSH is dynamically optimal.

Theorem 9. *The RANDOM-PUSH algorithm is dynamically optimal on expectation.*

Algorithm 2: Upon access to u in PUSH-DOWN TREE

1: **access** $s = u$.host along tree branches (cost: u.dep)
2: let $v = s_1$.guest be the item at the current root
3: **move** u to the root server s_1, setting s_1.guest $= u$ (cost: u.dep)
4: employ RANDOM-PUSH to **shift** down v to depth s.dep (cost: u.dep)
5: let w be the item at the end of the push-down path, where w.dep $= s$.dep
6: **move** w to s, i.e., setting s.guest $= w$ (cost: u.dep $\times$ 2)

Proof. Let the t-th requested item have rank r_t, then the access cost is $D(r_t)$. According to the RANDOM-PUSH (Algorithm 2), the total cost is $5D(r_t)$ which is five times the access cost on the MRU(4) tree. Formally, using Theorem 7 and Theorem 8, the expected total cost is:

$$\begin{aligned}\mathbb{E}\left[\text{cost}(RANDOM-PUSH)\right] &= \mathbb{E}\left[\sum_{i=1}^{t} 5D(r_i)\right] = 5\sum_{i=1}^{t}\mathbb{E}\left[D(r_i)\right] \leq 5\sum_{i=1}^{t}(\log(r_i)+3)\\ &\leq 5\sum_{i=1}^{t}(\lfloor\log(r_i)\rfloor+4) \leq 5\sum_{i=1}^{t}\text{cost}^{(t)}(MRU(4))\\ &\leq 5\cdot\text{cost}(MRU(4)) = 60\cdot\text{cost}(OPT)\end{aligned}$$

□

6 Related Work

The work most closely related to ours arises in the context of self-adjusting datastructures. However, while datastructures need to be *searchable*, networks come with *routing* protocols: the presence of a *map* allows us to trivially access a node (or item) at distance k from the front at a cost k. Interestingly, while we have shown in this paper that dynamically optimal algorithms for tree networks exist, the quest for constant competitive online algorithms for binary search trees remains a major open problem [21]. Nevertheless, there are self-adjusting binary search trees that are known to be *access optimal* [6], but their rearrangement cost it too high.

In the following, we first review most related work on datastructures and then discuss literature in the context of networks. A more detailed discussion appears in the technical report[4]. In contrast to CTs, self-adjustments in Binary Search Trees (BSTs) are based on *rotations* (which are assumed to have unit cost). While BSTs have the working set property, we are missing a matching lower bound: the *Dynamic Optimality Conjecture*, the question whether splay trees [21] are dynamically optimal, continues to puzzle researchers. We are also not the first to consider *Unordered* Trees (UTs) and it is known that existing lower bounds for (offline) algorithms on BSTs also apply to UTs that use rotations [12]. However, it is also known that this correspondance between ordered and unordered trees no longer holds under weaker measures such as *key independent processing costs* and in particular *Iacono's measure* [14]: the expected cost of the sequence which results from a random assignment of keys from the search tree to the items specified in an access request sequence. Iacono's work is also one example of prior work which shows that for specific scenarios, working set and dynamic optimality properties are equivalent. Regarding the current work, we note that the reconfiguration operations in UTs are more powerful than the swapping operations considered in our paper: a rotation allows to move entire subtrees at unit costs, while the corresponding cost in CTs is linear in the subtree size. We also note that in our model, we cannot move freely between levels, but moves can only occur between parent and child. In contrast to UTs, CTs are bound to be balanced.

Intriguingly, although Skip Lists (SLs) and BSTs can be transformed to each other, Bose et al. [8] were able to prove dynamic optimality for (a restricted kind of) SLs as well as B-Trees (BTs). However, the quest for proving dynamic optimality for general skip lists remains an open problem: two restricted types of models were considered in [8], bounded and weakly bounded. Due to the relationship between SLs and BSTs, a dynamically optimal SL would imply a working set lower bound for BST. Moreover, while both in their model and ours, proving the working set property is key, the problems turn out to be fundamentally different. In contrast to SLs, CTs revolve around *unordered* (and balanced) trees (that do not provide a simple search mechanism), rely on a different reconfiguration operation (i.e., swapping or *pushing* an item to its parent comes at unit cost), and, as we show in this paper, actually provide dynamic optimality for their general form. Finally, we note that [8] (somewhat implicitly) also showed that a random walk approach can achieve the working set property; in our paper, we show that the working set property can even be achieved deterministically and without maintaining MRU.

Finally, little is known about self-adjusting *networks*. While there exist several algorithms for the design of *static* demand-aware networks, e.g. [2,3,10,19], online algorithms which also minimize reconfiguration costs are less explored. The most closely related work to ours are *SplayNets* [17,18], which are also based on a tree topology (but a searchable one). However, *SplayNets* do not provide any formal guarantees over time, besides convergence properties in case of certain fixed demands.

Acknowledgments. Research supported by the ERC Consolidator grant *AdjustNet* (agreement no. 864228).

References

1. Avin, C., Ghobadi, M., Griner, C., Schmid, S.: On the complexity of traffic traces and implications. In: Proceedings of the International Conference on Measurement and Modeling of Computer Systems, ACM SIGMETRICS, pp. 47–48 (2020)
2. Avin, C., Mondal, K., Schmid, S.: Demand-aware network designs of bounded degree. Distrib. Comput., 311–325 (2019). https://doi.org/10.1007/s00446-019-00351-5

3. Avin, C., Mondal, K., Schmid, S.: Demand-aware network design with minimal congestion and route lengths. In: Proceedings of the 38th IEEE International Conference on Computer Communications, IEEE INFOCOM, pp. 1351–1359 (2019)
4. Avin, C., Mondal, K., Schmid, S.: Push-down trees: optimal self-adjusting complete trees. In: Technical Report arXiv 1807.04613 (2020)
5. Avin, C., Schmid, S.: Toward demand-aware networking: a theory for self-adjusting networks. ACM SIGCOMM Comput. Commun. Rev. **48**(5), 31–40 (2019)
6. Blum, A., Chawla, S., Kalai, A.: Static optimality and dynamic search-optimality in lists and trees. In: Proceedings of the 13th Annual ACM-SIAM Symposium on Discrete Algorithms (SODA) (2002)
7. Bojja Venkatakrishnan, S., Alizadeh, M., Viswanath, P.: Costly circuits, submodular schedules and approximate carathéodory theorems. In: Proceedings of the 2016 ACM SIGMETRICS International Conference on Measurement and Modeling of Computer Science, pp. 75–88 (2016)
8. Bose, P., Douïeb, K., Langerman, S.: Dynamic optimality for skip lists and b-trees. In: Proceedings of the 19th Annual ACM-SIAM Symposium on Discrete Algorithms (SODA), pp. 1106–1114 (2008)
9. Dean, B.C., Jones, Z.H.: Exploring the duality between skip lists and binary search trees. In: Proceedings of the 45th Annual Southeast Regional Conference, ACM-SE, New York, NY, USA, vol. 45, pp. 395–399. ACM (2007)
10. Foerster, K.T., Ghobadi, M., Schmid, S.: Characterizing the algorithmic complexity of reconfigurable data center architectures. In: Proceedings of ACM/IEEE Symposium on Architectures for Networking and Communications Systems (ANCS) (2018)
11. Foerster, K.T., Schmid, S.: Survey of reconfigurable data center networks: enablers, algorithms, complexity. SIGACT News **50**(2), 62–79 (2019)
12. Fredman, M.L.: Generalizing a theorem of Wilber on rotations in binary search trees to encompass unordered binary trees. Algorithmica **62**(3–4), 863–878 (2012)
13. Hamedazimi, N., et al.: Firefly: a reconfigurable wireless data center fabric using free-space optics. Proc. ACM SIGCOMM Comput. Commun. Rev. (CCR) **44**, 319–330 (2014)
14. Iacono, J.: Key-independent optimality. Algorithmica **42**(1), 3–10 (2005)
15. Kandula, S., Sengupta, S., Greenberg, A., Patel, P., Chaiken, R.: The nature of data center traffic: measurements and analysis. In: Proceedings of the 9th ACM Internet Measurement Conference (IMC), pp. 202–208 (2009)
16. M. Ghobadi et al.: Projector: Agile reconfigurable data center interconnect. In: Proceedings of the 2016 ACM SIGCOMM Conference, pp. 216–229 (2016)
17. Peres, B., Otavio, A.D.O., Goussevskaia, O., Avin, C., Schmid, S.: Distributed self-adjusting tree networks. In: Proceedings of the 38th IEEE International Conference on Computer Communications, IEEE INFOCOM, pp. 145–153 (2019)
18. Schmid, S., Avin, C., Scheideler, C., Borokhovich, M., Haeupler, B., Lotker, Z.: SplayNet: towards locally self-adjusting networks. IEEE/ACM Trans. Networks **24**(3), 1421–1433 (2016)
19. Singla, A., Singh, A., Ramachandran, K., Xu, L., Zhang, Y.: Proteus: a topology malleable data center network. In: Proceedings of the 9th ACM Workshop on Hot Topics in Networks (HotNets), pp. 1–6 (2010)
20. Sleator, D.D., Tarjan, R.E.: Amortized efficiency of list update and paging rules. Commun. ACM **28**(2), 202–208 (1985)
21. Sleator, D.D., Tarjan, R.E.: Self-adjusting binary search trees. J. ACM **32**(3), 652–686 (1985)

Batched Predecessor and Sorting with Size-Priced Information in External Memory

Michael A. Bender[1], Mayank Goswami[2(✉)], Dzejla Medjedovic[3], Pablo Montes[4], and Kostas Tsichlas[5]

[1] Department of Computer Science, Stony Brook University, Stony Brook, NY 11794, USA
bender@cs.stonybrook.edu
[2] Department of Computer Science, Queens College, CUNY, NY 11367, USA
mayank.goswami@qc.cuny.edu
[3] International University of Sarajevo, 71210 Sarajevo, Bosnia and Herzegovina
dzmedjedovic@ius.edu.ba
[4] Google Inc, Mountain View, CA 94043, USA
pabmont@gmail.com
[5] Aristotle University of Thessaloniki, 54124 Thessaloniki, Greece
tsichlas@csd.auth.gr

Abstract. The fundamental problems of sorting and searching, traditionally studied in the unit-cost comparison model, have been generalized to include priced information, where different pairs of items have different comparison costs. These costs can be arbitrary (Charikar et al. STOC 2000), structured (Gupta et al. FOCS 2001), or stochastic (Angelov et al. LATIN 2008). Motivated by the database setting where the comparison cost depends on the sizes of the records, we consider the problems of sorting and batched predecessor where two non-uniform sets of items A and B are given as input. In the RAM model, pairwise comparisons (A-A, A-B and B-B) have respective comparison costs a, b and c. We give upper and lower bounds for the case $a \leq b \leq c$, which serves as a warmup for the generalization to the external-memory model. In the Disk-Access Model (DAM), where transferring elements between disk and RAM is the main bottleneck, we consider the scenario where elements in B are larger than elements in A. All items are required in their entirety for comparisons in RAM. A key observation is that the complexity of sorting depends on the interleaving of the small and large items in the final sorted order, and with a high degree of interleaving, the lower bound is dominated by an associated batched predecessor problem. We give output-sensitive bounds on the batched predecessor and sorting; our bounds are tight in most cases. Our lower bounds require novel generalizations of lower bound techniques in external memory to accommodate non-uniform keys.

This work was supported in part by NSF grants CCF-1725543, CSR-1763680, CCF-1716252, CCF-1617618, CNS-1938709, and by Sandia National Laboratories. Supported by NSF grants CRII-1755791 and CCF-1910873.

Y. Kohayakawa and F. K. Miyazawa (Eds.): LATIN 2020, LNCS 12118, pp. 155–167, 2020.
https://doi.org/10.1007/978-3-030-61792-9_13

Keywords: Priced information · Sorting · Batched predecessor · External memory · Output-sensitive algorithms

1 Introduction

In most published literature on sorting and other comparison-based problems (e.g., searching and selection), the traditional assumption is that a comparison between any two elements costs one unit, and the efficiency of an algorithm depends on the total number of comparisons required to solve the problem. In this paper, we study a natural extension to sorting, where the cost of a comparison between a pair of elements can vary, and the comparison cost is the function of the elements being compared. We derive worst-case lower and upper bounds in the random-access-machine (RAM) and the disk-access-machine (DAM) [1] models for sorting and the batched predecessor problems.

In the RAM model, we assume that comparisons between a pair of keys have an associated cost that depends on the type of keys involved. As a toy problem, consider n red and n blue keys, where a comparison between a pair of red keys costs a, between a red key and a blue key costs b, and between a pair of blue keys costs c. Without loss of generality we can assume $a < c$, which gives rise to three cases to be considered, $a < b < c$, $a < c < b$, and $b < a < c$ (when $b = 1$ but $a = c = \infty$ corresponds to the well-known nuts and bolts problem [2].) In this paper we consider the setting of [16], where the comparison cost depends on the length of the keys being compared. However, our analysis considers the worst-case cost parameterized by the specific distribution (or the "interleaving") of the elements in the final sorted order.

Then we turn to the *disk access machine* (*DAM*) model (also called the *external-memory model* or the *I/O model*) [1], designed to capture the key aspect of data-intensive applications, where transferring data is the main bottleneck, as oppose to CPU computation. In this simplified model of modern memory hierarchy, data is transferred from an external disk of infinite capacity to the main memory of size M in blocks of size B, where $M > B$ and input size $N >> M$; the cost of the algorithm is measured by the number of block transfers (I/Os) that it needs; once data is in memory, all computation comes for free.

In the DAM model, the notion of the comparison cost naturally comes into play when elements have different *sizes* (or *lengths*) because the larger the elements are, the fewer of them can fit in a block transfer. Specifically, if a key has length w, where $w \leq B$, then up to B/w keys can be fetc.hed with one I/O; similarly, if $w \geq B$, then it takes w/B I/Os to bring that key into memory. Moreover, a long element, when brought into RAM, will displace a larger volume of keys than a short element, thus reducing the parallelism that many external-memory algorithms such as external merge sort benefit from.[1] In the DAM model, we are given two sets of keys, S keys of unit size (short keys) and L/w (long) keys of size w each (total volume L). We express our results parameterized by the interleaving of the elements in their final sorted order. Let the interleaving parameter k denote the number of consecutive runs of large keys (i.e., *stripes*) in the final sorted order. Our goal is to express the performance of the sorting algorithm, as a function of S, L, w, and k.

Sorting with two key lengths illustrates a special connection between sorting and the batched searching problem that we call *PLE* (Placement of Large Elements). Given S keys of unit size (short keys) in the sorted order, and L/w (long) keys of size w each, the objective in the PLE is to find the short key that is the immediate predecessor of each long key. The PLE problem is a lower bound on sorting because it starts off with more information than the original sorting and asks to do less; in many cases, it dominates the sorting cost.

However, obtaining lower bounds on the PLE presents several challenges. First, having records of different sizes implies that standard information-theoretic lower bounds on the unit-sized case do not apply, because now different types of I/Os can contribute different amounts of information. Second, depending on the interleaving of the small and large elements, expressed in our bounds with the interleaving parameter k, and the size of the large element w, the PLE bounds substantially change. And lastly, PLE is a batched searching problem with a nontrivial preprocessing-query tradeoff [9], an interesting problem on its own.

Related Work. In RAM, algorithms for inputs with priced information have been studied in the context of competitive analysis [3,11,12,16]. Another example of varying comparison costs is the well-known nuts-and-bolts problem [2]. Interleaving-sensitive lower bounds and batched searching are related to lower bounds for sorting multisets [19] and distribution-sensitive set-partitioning [14].

Aggarwal and Vitter [1] introduced the external-memory (DAM) model. The fundamental lower bounds were further generalized in [6] and [15] to the external algebraic decision tree model. Prominent examples studying lower bounds on batched and predecessor searching are found in [4,7,9].

Most previous work that considers variable-length keys does so in the context of B-trees [8,13,17,18,20], and string sorting [5], where in the latter, authors derive upper and lower bounds under different models of key divisibility. Strings are divisible, and that brings down the complexity of sorting, but the lower bounds in our paper also imply the worst-case bounds for the string scenario where the tie is broken at the last character. Indivisible keys are found in practical settings, where sorting and searching libraries such as GNU Sort [21] or

[1] DAM model can represent any two levels of memory, which is related to the record size in our problem. If the two levels are the disk and the main memory, elements could be larger than B but are much smaller than M. If the levels are cache and main memory, then the elements could have a length that is a nontrivial fraction of M.

Oracle Berkeley DB [10] allow the developer to pass in a comparison function as a parameter.

Organization. In Sect. 2 we present the RAM version of our problem. We present the sorting problem in external-memory in Sect. 3 and relate it to the batched predecessor problem. We then derive lower and upper bounds on the batched predecessor problem in Sect. 5 and end with open problems in Sect. 6

2 Warmup: The RAM Version

Two Types, RAM Version (2RAMSORT). The input is n red and n blue keys, and the output is the sorted sequence of all keys. A comparison between a pair of red keys costs a, between a red key and a blue key costs b, and between a pair of blue keys costs c. Without loss of generality we can assume that $a < c$.

The optimal sorting cost in RAM depends on the final interleaving of the elements in the final sorted order. If in the final sorted order all red keys come before all blue keys, then $\Theta(an \log n + cn \log n)$ is the optimal total comparison cost, and the algorithm that separately sorts the two sets, and uses only one red-blue comparison to concatenate is optimal. However, if the red and blue keys alternate in the final sorted order, then no blue-blue comparisons are ever required to sort, rendering the previous algorithm suboptimal.

Stripes and the Interleaving Parameter k. A consecutive run of red or blue keys in the final sorted order is called a *stripe*. For simplicity we assume we have as many red stripes as blue. Define k to be the number of stripes, and let ℓ_i (respectively s_i) be the number of blue (respectively red) keys in stripe i. The notation ℓ and s are chosen to correspond with the later sections when red elements will be small and blue elements will be large.

Below we prove the tight bounds for the case $a < b < c$: this is the most natural case to serve as a warmup for the I/O-model, because if we consider the red elements small, blue elements large, then the comparisons involving red elements cost less than those involving blue elements, thus $a < b < c$.

Theorem 1. 2RAMSORT *has the following comparison cost complexity for the comparison-cost case $a < b < c$:*

$$\Theta(an \log n + b(k \log n + n \log k) + c \sum_{i=1}^{k} \ell_i \log \ell_i)$$

Proof. The lower bounds follow by counting the number of permutations needed to solve the following subproblems of 2RAMSORT:

1. The total number of permutations any algorithm for 2RAMSORT must achieve is at least $n!$. A binary comparison reduces the number of permutations by a factor of at most 2, and the cheapest comparison cost is a, thus giving a lower bound $a \log n! = \Omega(an \log n)$, our first term.

2. Consider the following instance of the problem, where the red elements are given as sorted, and the locations and the contents of stripes of blue elements are also provided but stripes are unsorted inside. This gives us a lower bound of $\Omega(c\sum_{i=1}^{k} \ell_i \log \ell_i)$, our third term.

3. Proving the second term as a lower bound involves addressing the following instance: the red elements are given sorted, and the algorithm is just required to discover the contents and the locations of k blue stripes. Once the algorithm solves this batched predecessor problem, the stripes are individually sorted for free.

There are at least $P = \binom{n}{k} S(n,k)$ permutations to consider, first term corresponding to finding k positions for stripes, and the second term, $S(n,k)$ is the *Stirling number of the second kind*, i.e., the number of ways to partition a set of size n into k non-empty, disjoint subsets, which corresponds to distributing contents among the k blue stripes. Considering that $S(n,k) \geq k^{n-k}$, we obtain that $P \geq (n/k)^k k^{n-k}$. Since red elements are already sorted, comparisons of cost a are useless, and the cheapest available comparisons are those costing b. Thus we get a lower bound of $\Omega(b(k \log(n/k) + (n-k)\log k))$, which equals $\Omega(b(k \log n + n \log k - 2k \log k))$. One can then show by a case analysis that the last $-2k \log k$ term can be removed from the above expression, a detail we allocate to the full version.

Since we derived the lower bound on a constant number of instances of 2RAMSORT, we can claim a lower bound of the maximum complexity of these instances, that in turn equals the sum of the complexities in $\Omega(.)$ notation.

Due to space constraints, we refer the reader to the full version of the paper for the description of the upper bound. The steps in the upper bound match the respective lower-bound terms, by 1) sorting the red elements, 2) using two balanced binary trees, one for discovering new stripes of blue elements, and the other for assigning blue elements to existing stripes, and lastly, 3) sorting the stripes of blue elements.

3 Sorting and Batched Predecessor in External Memory with Size-Priced Information

The input to the two-sized sorting and batched predecessor problems are $\mathcal{S} = \{s_*\}$ (the small records) and $\mathcal{L} = \{\ell_*\}$ (the large records, each of size $1 < w \leq M/2$). A set of large elements forms a *stripe* if for each pair of large elements ℓ_i and ℓ_j in the stripe, there does not exist a small element between ℓ_i and ℓ_j in the final sorted order. Let k be the number of large-element stripes, and let the large-element stripes be $\mathcal{L}_1, \mathcal{L}_2, \ldots, \mathcal{L}_k$, as they are encountered in the ascending sorted order. The parameters in the complexity analysis of sorting and batched predecessor are thus $S = |\mathcal{S}|$, $L = |\mathcal{L}|$, w, k, and $\{L_i\}_{i=1}^{k}$, where $L_i = |\mathcal{L}_i|$.

Definition 1 (Two-Sized Sorting Sort (S, L)). *The input is an (unsorted) set of elements $N = S \cup L$. Set S consists of S unit-size elements, and L consists of*

L/w elements, each of size w.[2] *The output comprises the elements in N, sorted and stored contiguously in external memory.*

Definition 2 (PLE-Placement of Large Elements)*: The input is the* sorted *set of small elements $\mathcal{S} = \{s_1, s_2, \ldots, s_S\}$, and the* unsorted *set of large elements $\mathcal{L} = \{\ell_1, \ell_2, \ldots, \ell_{L/w}\}$. In the output, elements in $\mathcal{S}$ are sorted, and elements in L are sorted according to which stripe they belong to, but arbitrarily ordered within their stripe.*

Theorem 2 ***(Sorting complexity).*** *Denote by* $\mathrm{PLE}\,(S, L)$ *the complexity of the* PLE *problem. Then the I/O complexity of Two-Sized Sorting* $\mathrm{Sort}\,(S, L)$ *is*

$$\Theta\left(\frac{S}{B}\log_{M/B}\frac{S}{B} + \mathrm{PLE}\,(S, L) + \left(\sum_{i=1}^{k}\left(\frac{L_i}{B}\log_{M/w}\frac{L_i}{w}\right) + \frac{L}{B}\right)\right).$$

The first term refers to the sorting the short elements, and the third term refers to sorting the individual stripes of large elements. The first term is identical to the conventional $O(\frac{N}{B}\log_{M/B}\frac{N}{B})$ bound by Aggarwal and Vitter [1]. The third term requires a slight generalization of that bound, because large elements are larger than a block, a proof we include in our full version.

As in the RAM setting, since we have three subproblems, their maximum complexity, and hence the complexity of their sum, is a lower bound on $\mathrm{Sort}\,(S, L)$. We have thus reduced the sorting problem to the batched predecessor problem, which will occupy the rest of this article.

3.1 Main Challenges in the Batched Predecessor Problem

To solve the $\mathrm{PLE}\,(S, L)$, we need the notion of a **fan-out**, which measures the efficiency of an I/O. Before an I/O, a large element has a set of candidate positions where it might land, and this interval gets reduced by a certain factor (possibly 1) after an I/O. The fan-out of an I/O is defined to be the product of all such factors for all large elements that reduce their search intervals during this I/O. Some of the challenges involved in adapting the RAM results (and classical DAM results) to the DAM model include:

1. Non-uniformity of Record Sizes: In the unit-sized setting, the transfer of a block to main memory can decrease the number of permutations by a factor of at most[3] $B!\binom{M}{B}$. In our setting, the number of comparisons performed by an

[2] We overload notation for convenience of presentation. We assume $w \geq B$ also for the convenience of presentation. Our bounds hold for any $1 < w \leq M/2$ and are presented in the full version of the paper.

[3] The proof of the lower bound for sorting N unit-sized keys in [15] proceeds in the following fashion: assuming that all blocks are sorted (using a linear scan costing N/B I/Os), there are $N!/(B!)^{(N/B)}$ permutations required to achieve, and the transfer of a block of B sorted elements into the main memory containing $M - B$ sorted elements reduce the number of permutations by at most a factor $\binom{M}{B}$ (the "fan-out," since this is the degree of the node in the decision tree). Standard algebra gives a lower bound of $\Omega(\frac{N}{B}\log_{M/B}\frac{N}{B})$.

I/O varies depending on whether the block transfer carries large records or small records, and what the contents of RAM are at the time of the I/O, which yields following possibilities:

- The transfer of a large element into main memory full of large elements gives only $\binom{M/w}{1}$ per w/B I/Os as a large-element transfer costs w/B.
- The transfer of B small records into main memory filled with small records gives $\binom{M}{B}$.
- The transfer of a large element into a memory full of M small elements gives a fan-out of $M+1$.
- While the above three cases are tight, the main issue is in getting *an upper bound on how much a small block I/O can achieve.* The main memory can hold $p = (M-B)/w$ large elements, and an incoming small block has B small elements. Naively the maximum fan-out can be upper bounded by B^p, which is not tight. Our main aim is to get a better bound on this fan-out.

Both our upper and lower bounds are a minimum of two terms, where one dominates the other depending on how large the large elements are.

2. Requiring Output-Sensitive Lower Bound Limits Adversarial Arguments: Lower bounds on the unit-sized batched predecessor problem in external memory were recently obtained in [9]. In our setting, a more complicated adversarial analysis is required, that forms exactly k stripes at the end. Using this, we can argue a fan-out of at most 2^B on *most* small block I/Os.

4 Complexity of the Batched Predecessor Problem: Lower Bounds

The following theorem presents the PLE(S, L) lower bound:

Theorem 3 (PLE Lower Bound) .

$$\mathrm{PLE}\,(S,L) = \Omega\left(\min\left\{\frac{kw}{B}\log_M S + \frac{L}{B}\log_M k, \frac{k}{B}\log S + \frac{L}{wB}\log k + \frac{L}{B}\right\}\right).$$

In order to prove Theorem 3, we first divide the PLE(S, L) problem further into three subproblems for which we develop a common adversary argument framework:

1. S-k: An instance with only *one large element* in each large-element stripe.
   - Input: Set S of unit-sized elements $s_1, \ldots, s_S$ *(sorted)*, where $s_1 = -\infty$ and $s_S = \infty$, and large elements $\ell_1, \ldots, \ell_k$ (volume kw) *unsorted.*
   - Output: For each ℓ_i output s_j such that $s_j \le \ell_i \le s_{j+1}$. It is guaranteed that no other ℓ_k satisfies $s_j \le \ell_k \le s_{j+1}$ (one large element per stripe).
2. k-$\tilde{k}$: An instance with only *one small element* in each small-element stripe.
   - Input: Unit-sized elements $s_1, \ldots, s_{k+1}$ *sorted,* where $s_1 = -\infty$ and $s_{k+1} = \infty$, and large elements $\ell_1, \ldots, \ell_{\tilde{k}}$ (volume $\tilde{k}w$) *unsorted.*

  - Output: For each ℓ_i, output its predecessor and successor in S.
3. k-k: An instance with only *one element in each stripe*, large or small.
  - Input: Unit-sized elements $s_1, \ldots, s_{k+1}$ *sorted*, where $s_1 = -\infty$ and $s_{k+1} = \infty$, and large elements $\ell_1, \ldots, \ell_k$ (volume kw) *unsorted*.
  - Output: The entire set in the sorted order.

The format of lower bounds for S-k, k-$\tilde{k}$ and k-k is as follows: let X be the logarithm of the total number of permutations that an algorithm needs to achieve in order to solve the problem. As is easily observed, the values of X (modulo constant factors) for these three subproblems are $k \log(S/k)$, $\tilde{k} \log k$, and $k \log k$, respectively. Lemma 1 below is the most technical part of this paper, and it helps us quantify the behavior of the adversary during small-block and large-element inputs for all three subproblems. We use this lemma to prove the lower bounds for the individual three subproblems (found in Lemma 2, Lemma 3, and Lemma 4). Then we put the three lemmas together to obtain the expression from Theorem 3.

Lemma 1. *Consider any algorithm for the S-k, k-$\tilde{k}$, or the k-$k$problem. There exists an adversary such that:*

- *On the input of any block of B short elements, the adversary answers comparisons between all elements in main memory such that the fan-out of this I/O is at most 2^B. In other words, the number of permutations the algorithm needs to check is reduced by a factor at most 2^B.*
- *On the input of any large element (costing w/B I/Os), the adversary answers comparisons between all elements in main memory such that the fan-out of this I/O is at most $O(M)$. In other words, the number of permutations the algorithm needs to check is reduced by a factor at most $O(M)$.*

Proof of Lemma 1: We prove this lemma by describing the adversary. We capture the information learned at every point of the algorithm by assigning a *search interval* $R(\ell) = (s_i, s_j)$ for a large element ℓ at step t as the narrowest interval of small elements where ℓ can possibly land in the final sorted order, given the information the algorithm has learned so far. *Bits* of information is the logarithm of the search interval (e.g., halving the search interval means learning one bit of information.)

It will be useful to consider the binary tree $\mathcal{T}$ on the set $\mathcal{S}$. The search interval of any large element at any point during the execution of the algorithm is a contiguous collection of leaves in $\mathcal{T}$.

For simplicity we will assume that the size of $\mathcal{S}$ is a power of 2, and hence $\mathcal{T}$ is perfectly balanced. Also, if $R(\ell) = (s_i, s_j)$ is the range of a large element, we will make sure the adversary "rounds off" the search space so that the new range corresponds exactly to a subtree of some node in $\mathcal{T}$. This is accomplished by first finding the least common ancestor lca of s_i and s_j, and then shrinking the search space of ℓ to either the search space in the left subtree of lca or to the search space in the right subtree of lca, whichever is larger. Thus each large element ℓ at any time has an associated node in $\mathcal{T}$, which we denote by $v(\ell)$.

We also denote the interval corresponding to $v(\ell)$ (this is just the interval of its subtree) as $I(v(\ell))$.

Mechanics of the Adversary's Strategy: Our adversary will try to maintain the following invariant at all times during the execution of the algorithm.

Invariant: The search intervals of large elements in main memory are disjoint.

We denote by $\{\ell_i^{p-1}\}_{i=1}^{M/w}$ the set of at most M/w large elements in memory before the pth I/O. By hypothesis, the nodes in $\mathcal{T}$ belonging to the set $\{v(\ell_i^{p-1})\}_{i=1}^{M/w}$ have no ancestor-descendant relationships between them. We write S_i^{p-1} to denote $I(v(\ell_i^{p-1}))$, the search interval of large element ℓ_i^{p-1} at step $p-1$.

Small-Block Input. Consider the incoming block. We denote $n_{p,i}$ as the number of incoming small elements that belong to S_i^{p-1}. These elements divide S_i^{p-1} into $n_{p,i}+1$ parts $\{P_1, \ldots, P_{n_{pi}+1}\}$, some of them possibly empty. The largest of these parts (say P_j) is of size at least $1/(n_{p,i}+1)$ times the size of S_i^{p-1}. The new search interval of ℓ_i^p is defined to be the highest node in $\mathcal{T}$ such that $I(v) \subset P_j$.

Large Element Input. On an input of a large element ℓ_{new}^p (with search interval S_{new}^{p-1}), the adversary uses a strategy similar to that one on a small-block input to compare ℓ_{new}^p with the (at most) M small elements present in memory. These M small elements divide S_{new}^{p-1} into at most M parts, and the new search interval of ℓ_{new}^p corresponds to the highest node in $\mathcal{T}$ that contains the largest part. This is the temporary search interval S_{new}, with the corresponding node v_{new}. S_{new} can be related to the search intervals of large elements in memory in three ways:

- Case 1: The element ℓ_{new}^p shares a node with another large element ℓ_i^p. The conflict is resolved by sending ℓ_{new}^p and ℓ_i^p to the left and right children of v_{new}, respectively.
- Case 2: The element ℓ_{new}^p has an ancestor in memory. The ancestor is sent one level down, to the child that does not contain v_{new} in its subtree. Thus the conflict is resolved while giving at most $O(1)$ bit.
- Case 3: The element ℓ_{new}^p has descendants in memory. Denote the nodes that are descendants of v_{new} in $\mathcal{T}$ as $v_1, \ldots, v_{M/w}$. Let the corresponding search intervals be $S_1^{p-1}, \ldots, S_{M/w}^{p-1}$, respectively. Let $X = \cup_{i=1}^{M/w} S_i^{p-1}$ and $Y = S_{\text{new}} \backslash X$. The set Y is a union of at most $M/w+1$ intervals, each of which we denote by Y_i. Let Z be the largest interval from the set $\{S_1^{p-1}, \ldots, S_{M/w}^{p-1}, Y_1, \ldots, Y_{M/w}\}$. Hence, $|Z| \geq |S_{\text{new}}|/(2M/w)$.

  There are two cases to consider. The first case is when $Z = S_i^{p-1}$ for some i. In this case, $S_{\text{new}} = S_i^{p-1}$. In doing this we have given at most $O(\log M)$ bits. Now we proceed as in Case 1 to resolve the conflict with at most $O(1)$ extra bits. Otherwise, if $Z = Y_i$ for some i, then the adversary allots ℓ_{new}^p to the highest node v in $\mathcal{T}$ such that $I(v) \subseteq Z$.

We show that the above strategy produces the following guarantees (proof in full version):

1. On a small-block input, the adversary gives at most $O(\log(n_{p,i}+1))$ bits to ℓ_i^p.
2. On a small-block input, the adversary gives at most $O(B)$ bits.
3. On the input of a large element, the adversary gives at most $O(\log M)$ bits.

4.1 Putting It All Together: Lower Bounds for S-k, k-$\tilde{k}$ and k-k

1) S-kLower Bound. The proof rests on the following action of the adversary: in the very beginning, the adversary gives the algorithm the extra information that the ith largest large element lies somewhere between $s_{(i-1)\alpha}$ and $s_{i\alpha}$, where $\alpha = S/k$. In other words, the adversary tells the algorithm that the large elements are equally distributed across $\mathcal{S}$, one in each chunk of size S/k in $\mathcal{S}$. This deems the invariant of large elements in main memory having disjoint search intervals automatically satisfied. Since any algorithm that solves S-k must achieve $\Omega(k\log(S/k))$ bits of information, we have that

Lemma 2. $S{-}k = \Omega\left(\min\left(\frac{kw}{B}\log_M\frac{S}{k}, \frac{k}{B}\log\frac{S}{k} + \frac{kw}{B}\right)\right)$.

2) k-$\tilde{k}$Lower Bound. To solve k-$\tilde{k}$, an algorithm needs to learn $k\log\tilde{k}$ bits of information. Using the adversary strategy we described, we observe:

Lemma 3. k-$\tilde{k} = \Omega\left(\min\left(\frac{\tilde{k}w}{B}\log_M k, \frac{\tilde{k}}{B}\log k + \frac{\tilde{k}w}{B}\right)\right)$.

3) k-kLower Bound. To solve k-k, an algorithm needs to learn $k\log k$ bits of information. In the k-k problem, we expect to produce the perfect interleaving of the small and large elements in the final sorted order. That is, *each element lands in its own leaf of $\mathcal{T}$*. Therefore, the adversary does not posses the freedom to route elements down the tree at all times using the strategy we described. Instead, the strategy is used for a fraction of total bits the algorithm learns, and the remaining fraction is used to make up for the potential imbalance created by sending more elements to the left or to the right. We call these *type one* and *type two* bits, respectively. Type two bits are effectively given away for free by the adversary.

More formally, we define the *node capacity* ($c^T(v)$) as the number of large elements that pass through v during the execution of an algorithm. If the k-k algorithm runs in T I/Os, then the node capacity of v at a level h of $\mathcal{T}$ is designated by $c^T(v) = k/2^h$.

Definition 3 (type one *and* type two bits). *A bit gained by a large element ℓ is an* type one *bit if, when ℓ moves from v to one of v's children, at most $c^T(v)/4 - 1$ other large elements have already passed through v. The remainder of the bits are* type two *bits.*

Because a small-block input gives $O(B)$ bits and a large-element input gives $O(\log M)$ bits, and we need to achieve all type one bits to solve the problem (there are $(k\log k)/4$ of them), we obtain the following lower bound:

Lemma 4 $k\text{–}k = \Omega\left(\min\left(\frac{kw}{B}\log_M k, \frac{k}{B}\log k + \frac{kw}{B}\right)\right)$.

Proof of Theorem 3. The lower bounds for k-k, k-$\tilde{k}$ and S-k are each a minimum of two terms; it is safe to add the respective terms as the transition between which term dominates occurs at exactly the same value of w for each of the subproblems. Adding the terms for the lower bounds of k-k and S-k provides the $\frac{k}{B}\log S$ and $\frac{kw}{B}\log_M S$ terms in Theorem 3. Adding the terms for the lower bounds of k-k and k-$\tilde{k}$, and using that $k + \tilde{k} = L/w$ provides the $\frac{L}{wB}\log S$ and $\frac{L}{B}\log_M S$ terms in Theorem 3. The theorem also holds for other item sizes ($w_1 \leq w_2 \leq B$) and we prove the generalized result in the full version.

5 Upper Bounds

Our algorithm for Sort(S, L) works in three steps: 1) sort the short elements using traditional multi-way external memory merge-sort [1], 2) solve the associated PLE(S, L) problem, and 3) sort the long stripes obtained again using multi-way mergesort. The first and third steps give the first and third terms in the sorting complexity in Theorem 2.

We give two algorithms to solve PLE(S, L): PLE-DFS and PLE-BFS. The final upper bound is the minimum of the two terms, as presented in Theorem 4.

PLE-DFS: PLE-DFS builds a static B-tree T on S, and searches for large elements in T one by one. This approach is preferred in the case of really large elements, and it is better to input them fewer times.

We dynamically maintain a smaller B-tree T' that contains only *border elements* (the two small elements sandwiching each large element in the final sorted order) and has depth at most $\log_B k$. All large elements first travel down T' to locate their stripe. Only those elements for which their stripe has not yet been discovered need to travel down T. After a new stripe is discovered in T, it is then added to T'. The total cost becomes

$$O\left(\frac{L}{w}\log_B k + k\log_B S + \frac{L}{B} + \frac{S}{B}\right). \tag{1}$$

PLE-BFS: Our second algorithm for PLE uses a batch-searching tree with fanout $\Theta(M)$. When a node of the tree is brought into memory, we route all large elements via the node to the next level. We process the nodes of the M-tree level by level so all large elements proceed at an equal pace from the root to leaves. This technique is helpful when large elements are sufficiently small so that bringing them many times into memory does not hurt while they benefit from a large fanout.

The analysis is as follows: at each level of M-tree, the algorithm spends $\Theta(L/B)$ I/Os in large-element inputs. Every node of the tree is brought in at most once, which results in total $O(S/B)$ I/Os in small-element inputs. The total number of memory transfers for PLE-BFS then becomes $O\left(\frac{L}{B}\log_M S + \frac{S}{B}\right)$.

Our final upper bound is the better of the two algorithms:

Theorem 4 *(PLE Upper Bound).*

$$\mathrm{PLE}\,(S, L) = O\left(\min\left(\frac{L}{B}\log_M S + \frac{S}{B}, \frac{L}{w}\log_B k + k\log_B S + \frac{L}{B} + \frac{S}{B}\right)\right).$$

Substituting the lower and upper bounds of the batched predecessor problem (PLE (S, L)) derived in Theorems 3 and 4 into the complexity of sorting in Theorem 2 gives us lower and upper bounds on the sorting problem Sort (S, L).

Remark 1: One observes that in Theorem 4 (PLE (S, L) lower bound), the transition between the two terms in the minimum happens at $w = B\log M$. This is because when large elements are very large, the bound obtained by algorithms that do not input the large elements too often (PLE-DFS) is smaller than algorithms that input large elements multiple times (e.g., PLE-BFS).

Remark 2: The upper and lower bounds on PLE (S, L) are tight for a wide range of parameters. Moreover, if the first and third terms in the complexity of sorting (Theorem 2) dominate the complexity of the associated PLE (S, L) problem, our sorting algorithms are tight.

6 Conclusion and Open Problems

We derived upper and lower bounds on sorting and batched predecessor in the RAM and DAM models, when comparison or I/O costs depend on the length of the items being compared. In many settings, we show that the optimal sorting algorithm involves the optimal batched predecessor problem as a subroutine, and develop algorithms for the batched predecessor problem.

While our results are for the two-size setting, we would like to point out that our algorithms generalize to the multiple-sizes setting. However, generalizing our lower bound techniques to the multiple-size setting requires more ideas.

References

1. Aggarwal, A., Vitter, J.: The input/output complexity of sorting and related problems. Commun. ACM **31**, 1116–1127 (1988)
2. Alon, N., Blum, M., Fiat, A., Kannan, S., Naor, M., Ostrovsky, R.: Matching nuts and bolts. In: Proceedings of the 5th Annual ACM-SIAM Symposium on Discrete Algorithms, SODA, pp. 690–696 (1994)
3. Angelov, S., Kunal, K., McGregor, A.: Sorting and selection with random costs. In: Laber, E.S., Bornstein, C., Nogueira, L.T., Faria, L. (eds.) LATIN 2008. LNCS, vol. 4957, pp. 48–59. Springer, Heidelberg (2008). https://doi.org/10.1007/978-3-540-78773-0_5
4. Arge, L.: The buffer tree: a technique for designing batched external data structures. Algorithmica **37**(1), 1–24 (2003)
5. Arge, L., Ferragina, P., Grossi, R., Vitter, J.: On sorting strings in external memory (extended abstract). In: Proceedings of the 29th Annual ACM Symposium on Theory of Computing, STOC, pp. 540–548 (1997)

6. Arge, L., Knudsen, M., Larsen, K.: A general lower bound on the I/O-complexity of comparison-based algorithms. In: Proceedings of the 3rd Workshop on Algorithms and Data Structures, WADS, pp. 83–94 (1993)
7. Arge, L., Procopiuc, O., Ramaswamy, S., Suel, T., Vitter, J.: Theory and practice of I/O-efficient algorithms for multidimensional batched searching problems. In: Proceedings of the 9th Annual ACM-SIAM Symposium on Discrete Algorithms, SODA (1998)
8. Bender, M.A., Hu, H., Kuszmaul, B.C.: Performance guarantees for B-trees with different-sized atomic keys. In: Proceedings of the 29th ACM SIGMOD-SIGACT-SIGART Symposium on Principles of Database Systems, PODS, pp. 305–316 (2010)
9. Bender, M.A., Farach-Colton, M., Goswami, M., Medjedovic, D., Montes, P., Tsai, M.-T.: The batched predecessor problem in external memory. In: Schulz, A.S., Wagner, D. (eds.) ESA 2014. LNCS, vol. 8737, pp. 112–124. Springer, Heidelberg (2014). https://doi.org/10.1007/978-3-662-44777-2_10
10. Berkeley DB C API Reference. http://www.berkeleydb.com/
11. Charikar, M., Fagin, R., Guruswami, V., Kleinberg, J.P., Raghavan, A.S.: Query strategies for priced information. In: Proceedings of the 32nd Annual ACM Symposium on Theory of Computing, STOC, pp. 582–591 (2000)
12. Cicalese, F., Laber, E.: A new strategy for querying priced information. In: Proceedings of the 37th Annual ACM Symposium on Theory of Computing, STOC, pp. 674–683 (2005)
13. Diehrand, G., Faaland, B.: Optimal pagination of B-trees with variable-length items. Commun. ACM **27**(3), 241–247 (1984)
14. Elmasry, A.: Distribution-sensitive set multi-partitioning. In: 1st International Conference on the Analysis of Algorithms (2005)
15. Erickson, J.: Lower bounds for external algebraic decision trees. In: Proceedings of the 16th Annual ACM-SIAM Symposium on Discrete Algorithms, SODA, pp. 755–761 (2005)
16. Gupta, A., Kumar, A.: Sorting and selection with structured costs. In: Proceedings of the 42nd IEEE Symposium on Foundations of Computer Science, FOCS, pp. 416–425. IEEE (2001)
17. Larmore, L., Hirschberg, D.: Efficient optimal pagination of scrolls. Commun. ACM **28**(8), 854–856 (1985)
18. McCreight, E.: Pagination of B*-trees with variable-length records. Commun. ACM **20**(9), 670–674 (1977)
19. Munro, J., Spira, P.: Sorting and searching in multisets. SIAM J. Comput. **5**(1), 1–8 (1976)
20. Pinchuk, A.P., Shvachko, K.V.: Maintaining dictionaries: space-saving modifications of B-trees. In: Biskup, J., Hull, R. (eds.) ICDT 1992. LNCS, vol. 646, pp. 421–435. Springer, Heidelberg (1992). https://doi.org/10.1007/3-540-56039-4_57
21. The GNU C Library: qsort. http://www.gnu.org/software/libc/manual/

Probabilistically Faulty Searching on a Half-Line (Extended Abstract)

Anthony Bonato[1], Konstantinos Georgiou[1](✉), Calum MacRury[2], and Paweł Prałat[1]

[1] Department of Mathematics, Ryerson University, Toronto, ON M5B 2K3, Canada
{abonato,konstantinos,pralat}@ryerson.ca
[2] Department of Computer Science, University of Toronto, Toronto, ON M5S 2E4, Canada
cmacrury@cs.toronto.edu

Abstract. We study *p-Faulty Search*, a variant of the classic cow-path optimization problem, where a unit speed robot searches the half-line (or 1-ray) for a hidden item. The searcher is probabilistically faulty, and detection of the item with each visitation is an independent Bernoulli trial whose probability of success p is known. The objective is to minimize the worst case expected detection time, relative to the distance of the hidden item to the origin. A variation of the same problem was first proposed by Gal [29] in 1980. Alpern and Gal [3] proposed a so-called monotone solution for searching the line (2-rays); that is, a trajectory in which the newly searched space increases monotonically in each ray and in each iteration. Moreover, they conjectured that an optimal trajectory for the 2-rays problem must be monotone. We disprove this conjecture when the search domain is the half-line (1-ray). We provide a lower bound for all monotone algorithms, which we also match with an upper bound. Our main contribution is the design and analysis of a sequence of refined search strategies, outside the family of monotone algorithms, which we call *t-sub-monotone algorithms*. Such algorithms induce performance that is strictly decreasing with t, and for all $p \in (0, 1)$. The value of t quantifies, in a certain sense, how much our algorithms deviate from being monotone, demonstrating that monotone algorithms are sub-optimal when searching the half-line.

Keywords: Linear search · Online algorithms · Competitive analysis · Faulty robot · Probabilistic faults

1 Introduction

The problem of searching for a hidden item in a specified continuous domain dates back to the early 1960's and to the early works of Beck [8] and Bellman [9].

A. Bonato, K. Georgiou and P. Prałat—Research supported in part by NSERC.
C. MacRury—Research supported by a NSERC USRA, held at Ryerson University, Dept. of Mathematics.

Y. Kohayakawa and F. K. Miyazawa (Eds.): LATIN 2020, LNCS 12118, pp. 168–180, 2020.
https://doi.org/10.1007/978-3-030-61792-9_14

In its simplest form, a unit speed robot (that is, a mobile agent) starts at a known location, the origin, in a known search-domain. An item, sometimes called the *treasure* or the *exit*, is located (hidden) at an unknown distance d away from the origin, and it can be located by the robot only if it walks over it. What is the robot's trajectory that minimizes the worst case relative time that the treasure is located, compared to d? This worst case measure of efficiency is known as the competitive ratio of the trajectory. Interestingly, numerous variations of the problem admit trajectories inducing constant competitive ratios. In certain cases, for example, in the so-called linear-search problem where the domain is the line, tight lower bounds are known that require elaborate arguments.

We consider *p-Faulty Search* (FS_p), a probabilistic version of the classic linear-search problem in which the hidden item lies in a half-line (or 1-ray), and the item is detected with constant probability p (with independent Bernoulli trials) every time the robot walks over the item. This is a special case of a problem first proposed by Gal [29], where the search-domain is the line (or 2-rays). Natural solutions to the problem are so-called cyclic and monotone search patterns; that is, trajectories that process each direction periodically and where the searched space in each direction expands monotonically. In [3], Alpern and Gal proposed such a solution for searching 2-rays and they conjectured that an optimal trajectory must be cyclic and monotone. Angelopoulos in [5] extended the upper bound results using cyclic and monotone trajectories for searching m-rays. We prove that monotone trajectories are sub-optimal for searching a 1-ray. We do so first by establishing a lower bound for all monotone algorithms to the problem (which we also match with an upper bound), and second by designing a sequence of non-monotone trajectories inducing increasingly better performance (and deviating increasingly from being monotone).

Related Work: Search-type problems are concerned with finding a specific type of information placed within a well specified discrete or continuous domain. As a topic, it spans various sub-fields of Theoretical Computer Science and has given rise to a number of book-length treatments [1,3,21,41]. Applications range from data structures and mobile agent computing, to foraging and evolution, among others, for example, see [2,16,34,36,40].

The problem of searching for a hidden item in one-dimensional domains was first proposed more than 50 years ago by Beck [8] and Bellman [9] in a Bayesian context. In the 1990's, solutions to basic problem's variations were rediscovered, for example, see [7,35]. Since then, several studies of various search-type problems have resulted in an extensive literature. Below we give representative and selective examples, with an attempt to cite relatively recent results. Variations of search-type problems that share many similarities range from the type of search domain (for example, 1 or 2-dimensional [27,33], d-dimensional grid [18], cycle [38], polygons [23], graphs [6], grid [15], m-rays [13]), to the number of searchers (1 or more [37]), to the criterion for termination (for example, search, evacuation [14], priority evacuation [20], fetching [31]) to the communication model (for example, wireless or face-to-face [19]) to the type of the objective (for example, minimize worst case or average case [17]) to cost specs (for example, turning costs [26], cost for revisiting [11]), to the measure of efficiency (for example, time, energy [24]) to the knowledge of the input (none or partial [12]) and to other robots' specs (for example, speeds [22], faults [32], memory [39]), just to name a few. More recently, Fraigniaud et al. considered in [28] a Bayesian search problem in a discrete space, where a set of searchers are trying to locate a treasure placed, according to some distribution, in one of the boxes indexed by positive integers. Since it is outside the scope of this work to provide a comprehensive list of the large related literature, we further refer the interested reader to [3,4,25,30].

The version of linear search that we study, where the searcher is probabilistically faulty, was presented as an open problem by Gal in [29]. Later in [3] (see Chap. 8.6.2), Alpern and Gal provided a search strategy when the search domain is a line. In particular, they considered *cyclic* search trajectories where the robot alternates between searching each of the two directions, and each time monotonically increasing the searched space. Among the same family of algorithms that moreover expand the searched space in each direction geometrically, the authors provided the optimal trajectory. In addition, they conjectured that cyclic and monotone trajectories are in fact optimal. Along the same lines, [5] studied cyclic and monotone trajectories for searching m-rays. In a variation of the problem where the hidden item detections are not Bernoulli trials, [5] showed also that cyclic trajectories are in fact sub-optimal. For this and many other variations of probabilistically searching, where the probability of success is not known, optimal strategies remain open.

Main Contributions: We introduce and study p-Faulty Search (FS_p), a variation of the classic linear-search (cow-path) problem, in which the search space is the half-line, and detection of the hidden item (treasure) happens with known probability p. We are interested in designing search strategies that induce small competitive ratio, as a function of p; that is, that minimize the worst case expected detection time of the hidden item, with respect to its placement d, relative to the optimal performance of an algorithm that knows in advance the location of the item (so we normalize the expected performance both by d and p).

We focus on two families of search algorithms, which indicate that optimal solutions to FS_p may be particularly challenging to find. First, we study a natural family of algorithms, that we call monotone algorithms, which intuitively are determined by non-decreasing turning points x_i where searcher returns to the origin before expanding the searched space. Given that turning points increase geometrically; that is, when $x_i = b^i$, relatively straightforward calculations determine the optimal expansion factor $b = b(p)$. In fact, a simplified argument shows that in the cow-path problem (that is, when the search space consists of 2-rays and $p = 1$) the optimal expansion factor is $b = 2$. A more tedious argument (and one of our technical contributions), as in the cow-path problem, shows that the aforementioned choice of geometrically increasing x_i's for FS_p is in fact optimal among the family of monotone algorithms. Our main technical contribution pertains to the design and analysis of a family of algorithms that we call t-sub-monotone, which provide a sequence of refined search strategies which induce competitive ratios that strictly decrease with t, for every $p \in (0,1)$. Somehow surprisingly, our findings show that plain-vanilla, and previously considered, algorithms for FS_p are sub-optimal. All omitted proofs from this extended abstract can be found in the full version of the paper [10].

2 Problem Definition and Preliminary Observations

In *p-Faulty Searching on a Halfline* (FS_p) a speed-1 searcher (or robot) is located at the origin of the infinite half-line. At unknown distance d bounded away from the origin, which bound we set arbitrarily to 1, there is an item (or treasure) which is located/detected by the robot with constant and known probability p every time the robot passes over it (that is, detection trials are mutually independent and each has probability of success p). Also, for the sake of simplifying the analysis, we assume that the probability of detection becomes 1 if the treasure is placed exactly at a point where the robot changes direction. As we will see later, the worst placements of the treasure will be proven to be arbitrarily close to the turning points.

Given a robot's trajectory T, probability p and distance d, the *termination time* $\mathcal{E}_T(d)$ is defined as the expected time that the robot detects the treasure for the first time. Feasible solution to FS_p are robot's trajectories that induce bounded termination time (as a function of p, d) for all $p \in (0,1)$ and for all $d \geq 1$.

Note that p is part of the input to an algorithm for FS_p, while d is unknown. Hence, trajectories may depend on p but not on d. It is also evident that for a robot's trajectory to induce bounded termination time for all treasure placements, the robot needs to visit every point of the half-line, past point 1, infinitely many times. As it is also common in competitive analysis, we measure the performance of a search strategy relative to the optimal offline algorithm; that is, an algorithm that knows where the treasure is. Since such an algorithm needs to travel for time d to reach the treasure, as well as one would need $1/p$ trials, in expectation, before detecting it, we are motivated to introduce the following measure of efficiency for search trajectories.

Definition 1. *The competitive ratio of search strategy T for* FS_p *is defined as* $\mathcal{C}_p^T := \sup_{d \geq 1} \left\{ \frac{p\mathcal{E}_T(d)}{d} \right\}$.

Trajectory solutions (or search strategies) to problem FS_p are in correspondence with infinite sequences $\{t_i\}_{i \geq 0}$ of turning points, satisfying $t_0 = 0$, $t_i \geq 0$, $t_{2i+1} > t_{2i}$ and $t_{2i} < t_{2i-1}$, for all $i \geq 0$. Indeed such a sequence $\{t_i\}_{i \geq 0}$ corresponds to the trajectory in which robot moves from t_{2i} to t_{2i+1} (moving away from the origin), and from t_{2i-1} to t_{2i} (moving toward the origin), each time changing direction of movement, where $i = 1, 2, \ldots$.

For search strategy T and treasure location d (except from the turning points of T), let f_i denote the time till the robot passes over the treasure for the i'th time. Since the probability of successfully detecting the treasure is p, we have $\mathcal{E}_T(d) = \sum_{i=1}^{\infty} p(1-p)^{i-1} f_i$. In what follows, we express the expected termination time with respect to the additional time between two visitations of the treasure.

Lemma 1. *Let* $f_0 = 0$, *and let* $g_i = f_i - f_{i-1}$. *We then have that* $\mathcal{E}_T(d) = \sum_{i=1}^{\infty}(1-p)^{i-1}g_i$.

3 Monotone Trajectories

We explore the simplest possible trajectories for FS_p in which the searcher repeatedly returns to the origin every time she changes direction during exploration and before exploring new points in the half-line. More formally, *monotone trajectories* for FS_p are search algorithms $T = \{t_i\}_{i\geq 1}$, defined as[1] $t_{2i} = 0,\ t_{2i+1} = x_i,\ i = 1, 2, \ldots$, where $\{x_i\}_{i\geq 1}$ is a strictly increasing sequence with $x_i \to \infty$. Note that, in particular, we allow $x_i = x_i(p)$. The present section is devoted into determining the best monotone algorithm for FS_p. More specifically, we prove the following.

Theorem 1. *The optimal monotone algorithm for* FS_p *has competitive ratio* $\frac{4+4\sqrt{1-p}}{2-p} - p$.

3.1 An Upper Bound Using Monotone Trajectories

In this section we propose a specific monotone algorithm with the performance promised by Theorem 1. In particular, we consider "restricted" trajectories determined by increasing sequences $\{x_i\}_{i\geq 1}$, where $x_i = b^i$ and $b = b(p) > 1$. Within this sub-family, we determine the optimal choice of b that induces the smallest competitive ratio. For this, we first determine the placements of the treasure that induce the worst competitive ratio, given a search trajectory. As stated before, in the following analysis we make the assumption that the treasure is not placed at any turning point.

Lemma 2. *Consider a monotone algorithm* T, *determined by the strictly increasing sequence* $\{x_i\}_{i\geq 1}$. *If the treasure appears in interval* (x_r, x_{r+1}), *then the competitive ratio is no more than* $2\frac{p}{x_r}\sum_{i=1}^{r} x_i + 2\frac{p}{x_r}\sum_{i\geq 1}(1-p)^{2i-1}x_{r+i} + \frac{p^2}{2-p}$.

We are now ready to prove the promised upper bound.

Lemma 3. *The monotone trajectory* $T = \{x_i\}_{\geq 1}$, *where* $x_i = b^i$ *and* $b := \frac{1}{\sqrt{1-p}(2-p-\sqrt{1-p})}$ *has competitive ratio* $\frac{4+4\sqrt{1-p}}{2-p} - p$.

3.2 Lower Bounds for Monotone Trajectories

This section is devoted to proving the following lemma.

Lemma 4. *Every monotone trajectory has competitive ratio at least* $\frac{4+4\sqrt{1-p}}{2-p} - p$.

[1] Alternatively, we could have defined monotone trajectories so as to return to location 1, instead of the origin, since we know that $d \geq 1$. Our analysis next shows that such a modification would not improve the competitive ratio.

Consider an arbitrary monotone algorithm $T = \{f_i\}_{i\geq 0}$, where f_i is a monotone sequence tending to infinity, and which determines the turning points of the algorithm. Without loss of generality, we set $f_0 = 1$, as otherwise we may scale all turning points by f_0. Our lower bound will be obtained by restricting the placement of the treasure arbitrary close to (and $\epsilon > 0$ away after) turning points f_k (this may only result in a weaker lower bound). Taking $\epsilon \to 0$, we obtain that $g_1^k = 2\sum_{i=0}^{k} f_i + f_k, g_{2i}^k = 2(f_{k+i} - f_k), g_{2i+1}^k = 2f_k$, where the superscript k of g_i^k indicates exactly the placement of the treasure at f_k. In what follows, and for a fixed integer ℓ, we define

$$\alpha := \frac{1}{2} + \frac{1}{2-p} - \frac{c}{2p}, \quad \beta_{i,k} := (1-p)^{2(i-k)-1}, \text{for } k+1 \leq i \leq \ell,$$

$$\gamma_{\ell,k} := \frac{(1-p)^{2(\ell-k)+1}}{p(2-p)}.$$

We have the following lemma.

Lemma 5. *Let c be the optimal competitive ratio that can be achieved by monotone trajectory T. For every integer ℓ and for every $0 \leq k \leq \ell$ we have that*

$$\sum_{i=0}^{k-1} f_i + \alpha f_k + \sum_{i=k+1}^{\ell} \beta_{i,k} f_i + \gamma_{\ell,k} f_\ell \leq 0. \tag{1}$$

Recall that $f_0 = 1$. Our lower bound derived in the proof of Lemma 4 is obtained by finding the smallest c satisfying constraints (1), and in particular, inducing a strictly increasing sequence of f_i in i. Note that minimizing c subject to constraints (1) in variables $f_1, \ldots, f_\ell, c$ is a non-linear program. To obtain a lower bound for c, we observe that the only negative coefficients of variables f_i are those on the diagonal; that is, the coefficient of f_k in the k'th constraint. This allows us to apply repeatedly back substitution to obtain a lower bound for all f_i and hence, c as well, assuming that the visiting points f_i are increasing in i. Equivalently, for the optimal c that an algorithm can achieve, we may treat (for the sake of the analysis) all inequalities (1) as being tight, giving rise to the linear system $A_\ell f = a$ in variables $f^T = (f_1, \ldots, f_\ell)$, where

$$A_\ell := \begin{bmatrix} \beta_{1,0} & \beta_{2,0} & \beta_{3,0} & \cdots & \gamma_{\ell,0} + \beta_{\ell,0} \\ \alpha & \beta_{2,1} & \beta_{3,1} & \cdots & \gamma_{\ell,1} + \beta_{\ell,1} \\ 1 & \alpha & \beta_{3,2} & \cdots & \gamma_{\ell,2} + \beta_{\ell,2} \\ 1 & 1 & \alpha & \cdots & \gamma_{\ell,3} + \beta_{\ell,3} \\ \vdots & \vdots & \vdots & \cdots & \vdots \\ 1 & 1 & 1 & \cdots & \gamma_{\ell,\ell-1} + \beta_{\ell,\ell-1} \end{bmatrix}, \quad a := \begin{bmatrix} -\alpha \\ -1 \\ -1 \\ -1 \\ \vdots \\ -1 \end{bmatrix}.$$

Constraints $A_\ell f = a$ may be thought as the defining linear system on f_i's that give the optimal turning strategies, assuming that the treasure can only be placed arbitrarily close and after any of the ℓ first turning points of a search trajectory. In other words, given that any monotone algorithm is defined by a sequence of turning points, these points can be chosen so as to minimize the competitive ratio with the assumption that the hidden item will be nearly missed after each turning point. Having the competitive ratio be independent of the treasure's placement gives a lower bound to the competitive ratio of the algorithm. The proof of Lemma 4 follows directly from the following technical lemma.

Lemma 6. *Linear system $A_\ell f = a$, in variables f_i, defines a monotone sequence of turning points only if $c \geq \frac{4+4\sqrt{1-p}}{2-p} - p$.*

4 Sub-monotone Trajectories

For a fixed integer t, we consider a *t-sub-monotone trajectory* that is defined by a strictly increasing sequence $\{x_i\}_{i\geq 1}$, where $x_i = \beta^i$ for some $\beta = \beta(p) > 1$, and $\{\gamma_i\}_{i=1,\ldots,t}$ (where $\gamma_i = \gamma_i(p)$) satisfying $1 < \gamma_1 < \gamma_2 < \ldots < \gamma_t < \beta$. For convenience, we introduce abbreviations $\gamma_0 = 1$ and $\gamma_{t+1} = \beta$. For the formal description of the trajectory, we introduce the notion of a *t-hop between consecutive points* x_r, x_{r+1}, see Algorithm 1, which is a sub-trajectory of the robot starting from x_r and finishing at x_{r+1}. Given parameters γ_i and β, the t-suborigin trajectory is defined in Algorithm 2.

Algorithm 1. t-Hop between x_r, x_{r+1}

1: **for** $j = 1, \ldots, t$ **do**
2: Move from $\gamma_{j-1}x_r$ to $\gamma_j x_r$
3: Move from $\gamma_j x_r$ to $\gamma_{j-1}x_r$
4: Move from $\gamma_{j-1}x_r$ to $\gamma_j x_r$
5: **end for**
6: Move from $\gamma_t x_r$ to x_{r+1}

Algorithm 2. t-Sub-Monotone Trajectory

1: Move from the origin to x_1, then to the origin and then to x_1.
2: **for** $r = 1, \ldots, \infty$ **do**
3: Perform a t-hop between x_r, x_{r+1}.
4: Move from x_{r+1} to the origin
5: Move from the origin to x_{r+1}
6: **end for**

4.1 Performance Analysis of t-Sub-monotone Trajectories

For the remainder of the paper, we introduce the following expressions: $A = 2(1-p), B = \frac{2}{\beta-1} + \frac{2(1-p)^3}{1-\beta(1-p)^2}, C = \frac{2p(1-p)^3(2-p)\beta}{1-\beta(1-p)^2}, D = \frac{-2p^4+12p^3-26p^2+23p-4}{2-p}, E = \frac{2p(1-p)(2-p)\beta}{1-\beta(1-p)^2}, F = p\left(2\left(\frac{\beta(1-p)+1}{(\beta-1)(1-\beta(p-1)^2)}\right) + \frac{5-2p}{2-p}\right)$, where, in particular, $A = A(p), B = B(\beta, p), C = C(\beta, p), D = D(p), E = E(\beta, p), F = F(\beta, p)$.

The purpose of this section is to prove the following theorem.

Theorem 2. *For any $i = 1, \ldots, t+1$ and given that the treasure lies in interval $A_i := (\gamma_{i-1}x_r, \gamma_i x_r)$, the worst case induced competitive ratio R_i is given by the formula*

$$R_i = \begin{cases} p\left(\frac{A\gamma_i + B\gamma_t + C}{\gamma_{i-1}} + D\right), & \text{if } i = 1, \ldots, t \\ p\left(\frac{E}{\gamma_t} + F\right), & \text{if } i = t \end{cases}$$

An immediate consequence of Theorem 2 is that the best t-sub-monotone algorithm with expansion factor β within consecutive t-hops and intermediate turning points $\gamma_1, \gamma_2, \ldots, \gamma_t$ is the solution (if it exists) to optimization problem

$$\begin{aligned} \min_{\beta,\gamma_1,\ldots,\gamma_t} \quad & \max\{R_1, R_2, \ldots, R_t, R_{t+1}\} \\ \text{s.t.} \quad & 1 < \gamma_1 < \ldots < \gamma_t < \beta < \tfrac{1}{(1-p)^2}. \end{aligned} \tag{2}$$

Alternatively, any solution $\beta, \gamma_1, \ldots, \gamma_t$ which is feasible to (2) has competitive ratio $\max_{i=1,\ldots,t+1} R_i$.

4.2 Choosing Efficient t-Sub-Monotone Trajectories

The purpose of this section is to propose a method for choosing parameters $\beta, \gamma_1, \ldots, \gamma_t$ of a t-sub-monotone algorithm which are feasible to (2), hence, inducing competitive ratio $\max_{i=1,\ldots,t+1} R_i$. The main idea of our approach is to treat the induced competitive ratio as an unknown R, and then impose, for all $i = 1, \ldots, t+1$, that $R_i = R$. The choices of γ_i are solutions to a recurrence relation. From numerical calculations, we know that our method proposes *optimal* solutions to (2), where in particular, all strict inequality constraints are satisfied with slack. However, a proof of optimality is not evident.

For the values of $A(p), B(p,\beta), C(p,\beta), D(p), E(p,\beta), F(p,\beta)$, we provide a way of obtaining t-sub-monotone algorithms by solving one non-linear equation. To this end, we also introduce abbreviations: $x := \frac{R/p-D}{A}$, $y := \frac{\frac{B\ E}{R/p-F}+C}{A}$, where in particular $x = x(p,R)$ and $y = (p,\beta,R)$ (the fact that x is independent of β will be used later). Moreover, we introduce the concept of the *t-characteristic polynomial* of a pair (p,R), which is the degree-2 polynomial $q_0 + q_1\beta + q_2\beta^2$ where $q_0 = q_0(p,R,t), q_1 = q_1(p,R,t), q_2 = q_2(p,R,t)$ are defined as

$$q_0 = \left(p^2(2p((p-6)p+12)-17)-(p-2)R\right)\left(p^2+(p-2)R\right)x^t \tag{3}$$

$$\begin{aligned} q_1 =& 2(p-2)^4(p-1)p^3(R-p) + x^t \times \\ & \left((p(p(2p(p(2p-19)+74)-297)+308)-134)p^4\right. \\ & \left.-2(p-2)(p(p((p-8)p+25)-35)+20)p^2R-(p-2)^2((p-2)p+2)R^2\right) \end{aligned} \tag{4}$$

$$\begin{aligned} q_2 =& (p-1)\left(2(p-2)^4p^3(3p-R)\right. \\ & \left.-(p-1)\left(p^2(2p-5)-(p-2)R\right)\left((2(p-4)p+9)p^2+(p-2)R\right)x^t\right) \end{aligned} \tag{5}$$

Note that the discriminant of the t-characteristic polynomial of a pair (p,R) is a rational function of p, R (where the numerator and denominator are polynomials of degree $\Theta(t)$), and hence, a function exclusively of R, for every fixed p.

Given $p \in (0,1)$, we say that pair (β, R) is *feasible* if

$$x - y - 1 > 0, \tag{6}$$

$$\beta - \frac{E}{R/p - F} > 0. \tag{7}$$

As we shall see, constraints above guarantee that β is a valid expansion factor, and that the last turning point of a sub-monotone algorithm happens before a t-hop is completed. We will also require that

$$\left(1-\frac{y}{x-1}\right)x^t+\frac{y}{x-1}-\frac{E}{R/p-F}=0. \tag{8}$$

As the treasure could be located in any of the $t+1$ sub-intervals associated with a t-hop, constraint (8) will guarantee that the competitive ratio is independent of that placement. Our main theorem is the following.

Theorem 3. *Fix $p \in (0,1)$, and let $R \geq 3$ be such that the discriminant of the t-characteristic polynomial of pair (p,R) is equal to 0. Let $\beta = -q_1/2q_2$ and suppose that pair (β, R) is feasible. We also set $\gamma_i = \left(1-\frac{y}{x-1}\right)x^i+\frac{y}{x-1}$, $i = 1\ldots,t$. We then have that $\beta, \gamma_1, \ldots, \gamma_t$ is a t-sub-monotone algorithm with competitive ratio R for problem* FS_p.

4.3 Numerical Computation of t-Sub-Monotone Trajectories, $t \leq 10$

We summarize the numerical results we obtain by invoking Theorem 3 for $t = 1,\ldots,10$, obtaining t-sub-monotone algorithms that induce better and better competitive ratios. For each t and (enough many) $p \in (0,1)$ we compute the smallest root $R = R(p,t)$ at least 3 of the t-characteristic polynomial, and the associated value of the expansion factor $\beta = \beta(p,t)$. For every pair (β, R) we verify that the induced values of γ_i do define a feasible search trajectory by showing that pair (β, R) is feasible. Note that constraints (6) and (7) guarantee that β is a valid expansion factor, and that the intermediate turning points of a t-hop are well defined, assuming that the worst case competitive ratio is the same in all subintervals of a t-hop, as required by constraint (8).

The improvement in the competitive ratio, when $t = 1,\ldots,4$ is apparent from a plot of the competitive ratio as a function of p, see Fig. 1. Finally, it can be shown that the proposed solution is valid (by checking constraints (6) and (7)), or in other words that the reported competitive ratio of Fig. 1 is correct. The horizontal axis in all figures is probability p. The vertical axis is explained in detail in each of the captions.

For values $t = 5,\ldots,10$ we need to deploy heuristic comparisons in order to display the behavior of the achieved competitive ratio, along with the corresponding expansion factor (this is due to that improvements are negligible, even though strictly positive). Figure 2 compares the achieved competitive ratios.

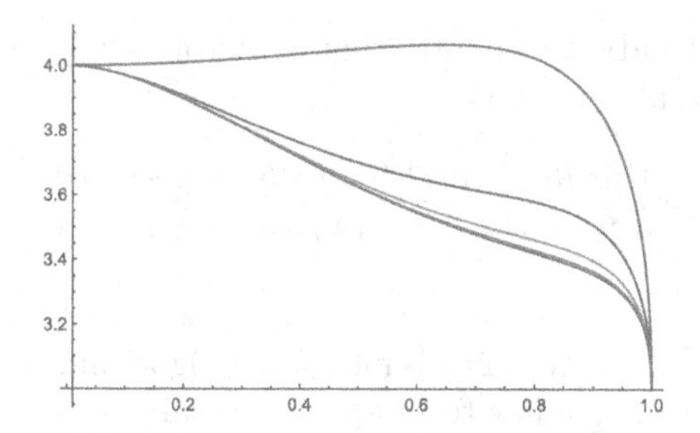

Fig. 1. The vertical axis shows the behavior of the achieved competitive ratio $R_t = R_t(p)$ of various t-sub-monotone algorithms. Purple line corresponds to the monotone algorithm of Lemma 3; that is, when $t = 0$. The subsequent improvements for $t = 1, 2, 3, 4$ are shown in colors blue, yellow, green and red, respectively. (Color figure online)

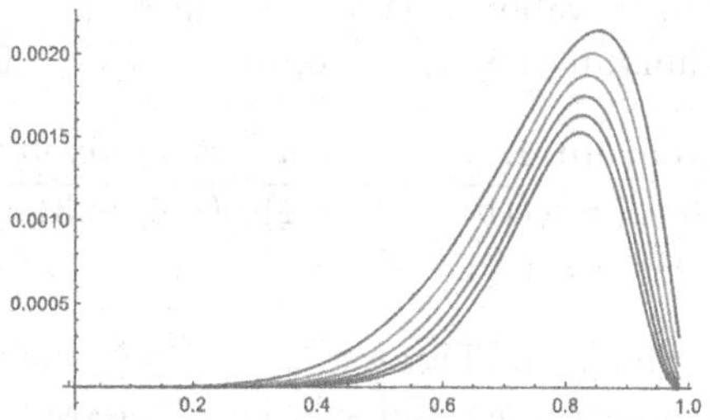

Fig. 2. Figure summarizes the behavior of t-sub-monotone algorithms for $t = 5, 6, 7, 8, 9, 10$, see colors blue, yellow, green, red, purple and brown, respectively. For each $t = 5, \ldots, 10$, the vertical axis corresponds to the scaled marginal improvements $4^{t-5}(R_{t-1} - R_t)$ between two consecutive values of t, which show that the competitive ratio does improve with t, still the improvement is increasingly negligible. (Color figure online)

4.4 Some Closed Formulae

As already discussed, we conjecture that the t-sub-monotone algorithms derived by Theorem 3 are optimal solutions to optimization problem (2), even though our conjecture does not compromise the correctness of our algorithms for problem FS_p. Nevertheless, a disadvantage of our approach, and in general of t-sub-monotone algorithms, is that our choices of parameters $\beta, \gamma_1, \ldots, \gamma_t$ do not admit closed form descriptions as functions of p. In this section, we deviate from our goal to determine the best possible t-sub-monotone algorithms, and we present specific choices of parameters $\beta, \gamma_1, \ldots, \gamma_t$ with closed formulas which induce nearly optimal competitive ratios.

Apart from our monotone trajectories, all our positive results were summarized in Sect. 4.3 and were based on numerical, and computer assisted, calculations. In light of Theorem 3, it is immediate that closed formulas for the achieved competitive ratios of t-sub-monotone algorithms do not exist. An exception, apart from the degenerate case $t = 0$, is the case $t = 1$. In particular, the discriminant of the 1-characteristic polynomial of pair (p, R) can be factored in two polynomials in R of degree 4 and of degree 2. One of the roots to the degree-4 polynomial is the competitive ratio of the 1-sub-monotone algorithm (as also per Theorem 3). Hence, the achieved competitive ratio R of the 1-sub-monotone algorithm, along with the corresponding expansion factor β admit closed formulas, even though they are enormous. Nevertheless, we show in the next theorem how to obtain an 1-sub-monotone and nearly optimal algorithm with performance and expansion factor that admit elegant closed formulas. Note that Theorem 1 combined with Theorem 4 below show provably, and not (computer-assisted and) numerically, that monotone algorithms are strictly sub-optimal for FS_p, for all $p \in (0,1)$.

Theorem 4. *There is a 1-sub-monotone algorithm for* FS_p *with competitive ratio* $R = \sqrt{(p-2)(p-1)(p(p(4p-3)+5)+2)} + \frac{4}{2-p} - (2-p)p$, *and expansion factor* $\beta = 1/(1-p)$.

Similar to Theorem 4, it is possible to identify a 2-sub-monotone algorithm with nearly optimal solution. Moreover, using similar techniques, one can compute the best competitive ratio possible by t-sub-monotone algorithms if we allow t to grow. We omit the details of this extended abstract.

5 Discussion and Open Problems

We studied *p-Faulty Search* (FS_p), a search problem on a 1-ray, where the searcher is probabilistically faulty with known probability $1-p$. Our main contribution pertains to the disproof of a conjecture that optimal trajectories for such problems are monotone. Whether the same conjecture is wrong for searching m-rays, and in particular, the line ($m = 2$) remains an open problem. When it comes to searching the half-line, all our algorithms have competitive ratio at least 4 when $p \to 0$ and at least 3 when $p \to 1$. The value of 3 is provably a lower bound to any search strategy since the searcher has to return at least once close to the origin before attempting for a second time an expansion of the searched space. No other general lower bound is known for the problem, whereas all our algorithms have competitive ratio at least $4-p$. Is $4-p$ a lower bound to any algorithm for FS_p, and if yes can this be matched by an upper bound? We conjecture that the lower bound is valid, as well as that our t-sub-monotone algorithms are sub-optimal.

Acknowledgements. The authors would like to thank Huda Chuangpishit, Sophia Park, Bhargav Parsi and Benjamin Reiniger for many fruitful discussions.

References

1. Ahlswede, R., Wegener, I.: Search problems. Wiley-Interscience (1987)
2. Albers, S., Henzinger, M.R.: Exploring unknown environments. SIAM J. Comput. **29**(4), 1164–1188 (2000)
3. Alpern, S., Gal, S.: The Theory of Search Games and Rendezvous. Springer, Berlin (2003)
4. Alpern, S., Fokkink, R., Gasieniec, L., Lindelauf, R., Subrahmanian, V. (eds.): Search Theory: A Game Theoretic Perspective, pp. 223–230. Springer NY, New York (2013)
5. Angelopoulos, S.: Further connections between contract-scheduling and ray-searching problems. In: 0001, Q.Y., Wooldridge, M.J. (eds.) Proceedings of the 24th International Joint Conference on Artificial Intelligence, IJCAI 2015, Buenos Aires, Argentina, 25–31 July 2015, pp. 1516–1522. AAAI Press (2015)

6. Angelopoulos, S., Dürr, C., Lidbetter, T.: The expanding search ratio of a graph. Discrete Appl. Math. **260**, 51–65 (2019)
7. Baeza Yates, R., Culberson, J., Rawlins, G.: Searching in the plane. Inf. Comput. **106**(2), 234–252 (1993)
8. Beck, A.: On the linear search problem. Israel J. Math. **2**(4), 221–228 (1964)
9. Bellman, R.: An optimal search. SIAM Rev. **5**(3), 274–274 (1963)
10. Bonato, A., Georgiou, K., MacRury, C., Pralat, P.: Probabilistically faulty searching on a half-line (2020). arXiv e-prints p. 2002.07797
11. Bose, P., De Carufel, J.L.: A general framework for searching on a line. Theor. Comput. Sci. **703**, 1–17 (2017)
12. Bose, P., De Carufel, J.L., Durocher, S.: Searching on a line: a complete characterization of the optimal solution. Theor. Comput. Sci. **569**, 24–42 (2015)
13. Brandt, S., Foerster, K.T., Richner, B., Wattenhofer, R.: Wireless evacuation on m rays with k searchers. In: Proceedings of the 24th International Colloquium on Structural Information and Communication Complexity, SIROCCO, pp. 140–157 (2017)
14. Brandt, S., Laufenberg, F., Lv, Y., Stolz, D., Wattenhofer, R.: Collaboration without communication: evacuating two robots from a disk. In: Fotakis, D., Pagourtzis, A., Paschos, V.T. (eds.) CIAC 2017. LNCS, vol. 10236, pp. 104–115. Springer, Cham (2017). https://doi.org/10.1007/978-3-319-57586-5_10
15. Brandt, S., Uitto, J., Wattenhofer, R.: A tight lower bound for semi-synchronous collaborative grid exploration. In: Proceedings of the 32nd International Symposium on Distributed Computing, DISC, pp. 13:1–13:17 (2018)
16. Burgard, W., Moors, M., Stachniss, C., Schneider, F.E.: Coordinated multi-robot exploration. IEEE Trans. Robot. **21**(3), 376–386 (2005)
17. Chuangpishit, H., Georgiou, K., Sharma, P.: Average case - worst case tradeoffs for evacuating 2 robots from the disk in the face-to-face model. In: Gilbert, S., Hughes, D., Krishnamachari, B. (eds.) ALGOSENSORS 2018. LNCS, vol. 11410, pp. 62–82. Springer, Cham (2019). https://doi.org/10.1007/978-3-030-14094-6_5
18. Cohen, L., Emek, Y., Louidor, O., Uitto, J.: Exploring an infinite space with finite memory scouts. In: Proceedings of the Twenty-Eighth Annual ACM-SIAM Symposium on Discrete Algorithms, pp. 207–224. SIAM (2017)
19. Czyzowicz, J., et al.: Evacuating robots via unknown exit in a disk. In: Kuhn, F. (ed.) DISC 2014. LNCS, vol. 8784, pp. 122–136. Springer, Heidelberg (2014). https://doi.org/10.1007/978-3-662-45174-8_9
20. Czyzowicz, J., et al.: Priority Evacuation from a Disk Using Mobile Robots. In: Lotker, Z., Patt-Shamir, B. (eds.) SIROCCO 2018. LNCS, vol. 11085, pp. 392–407. Springer, Cham (2018). https://doi.org/10.1007/978-3-030-01325-7_32
21. Czyzowicz, J., Georgiou, K., Kranakis, E.: Group search and Evacuation. Distrib. Comput. Mob. Entities Curr. Res. Moving Comput. **11340**, 335–370 (2019)
22. Czyzowicz, J., Kranakis, E., Krizanc, D., Narayanan, L., Opatrny, J., Shende, S.: Linear search with terrain-dependent speeds. In: Fotakis, D., Pagourtzis, A., Paschos, V.T. (eds.) CIAC 2017. LNCS, vol. 10236, pp. 430–441. Springer, Cham (2017). https://doi.org/10.1007/978-3-319-57586-5_36
23. Czyzowicz, J., Kranakis, E., Krizanc, D., Narayanan, L., Opatrny, J., Shende, S.: Wireless autonomous robot evacuation from equilateral triangles and squares. In: Papavassiliou, S., Ruehrup, S. (eds.) ADHOC-NOW 2015. LNCS, vol. 9143, pp. 181–194. Springer, Cham (2015). https://doi.org/10.1007/978-3-319-19662-6_13
24. Czyzowicz, J., et al.: Energy consumption of group search ona line. In: Proceedings of the 46th International Colloquium on Automata, Languages, and Programming, ICALP. LIPIcs, vol. 132, pp. 1–15 (2019)

25. Czyzowicz, J., Georgiou, K., Kranakis, E.: Group search and evacuation. In: Flocchini, P., Prencipe, G., Santoro, N. (eds.) Distributed Computing by Mobile Entities; Current Research in Moving and Computing, pp. 335–370. Springer (2019)
26. Demaine, E.D., Fekete, S.P., Gal, S.: Online searching with turn cost. Theor. Comput. Sci. **361**(2), 342–355 (2006)
27. Feinerman, O., Korman, A., Lotker, Z., Sereni, J.S.: Collaborative search on the plane without communication. In: Proceedings of the 2012 ACM Symposium on Principles of Distributed Computing, pp. 77–86 (2012)
28. Fraigniaud, P., Korman, A., Rodeh, Y.: Parallel Bayesian search with no coordination. J. ACM (JACM) **66**(3), 1–28 (2019)
29. Gal, S.: Search Games. Academic Press, New York (1980)
30. Gal, S.: Search Games. Wiley Encyclopedia for Operations Research and Management Science (2011)
31. Georgiou, K., Karakostas, G., Kranakis, E.: Search-and-fetch with 2 robots on a disk: Wireless and face-to-face communication models. Discrete Math. Theor. Comput. Sci. **21**(3) (2019)
32. Georgiou, K., Kranakis, E., Leonardos, N., Pagourtzis, A., Papaioannou, I.: Optimal circle search despite the presence of faulty robots. In: Dressler, F., Scheideler, C. (eds.) ALGOSENSORS 2019. LNCS, vol. 11931, pp. 192–205. Springer, Cham (2019). https://doi.org/10.1007/978-3-030-34405-4_11
33. Jeż, A., Łopuszański, J.: On the two-dimensional cow search problem. Inf. Process. Lett. **109**(11), 543–547 (2009)
34. Kagan, E., Ben-Gal, I.: Search and Foraging: Individual Motion and Swarmdynamics. CRC Press, Boca Raton (2015)
35. Kao, M.Y., Reif, J.H., Tate, S.R.: Searching in an unknown environment: an optimal randomized algorithm for the cow-path problem. Inf. Comput. **131**(1), 63–79 (1996)
36. Koutsoupias, E., Papadimitriou, C., Yannakakis, M.: Searching a fixed graph. In: Proc. of the 23rd International Colloquium on Automata, Languages, and Programming, ICALP, pp. 280–289. Springer (1996)
37. Lamprou, I., Martin, R., Schewe, S.: Fast two-robot disk evacuation with wireless communication. In: DISC, pp. 1–15 (2016)
38. Pattanayak, D., Ramesh, H., Mandal, P., Schmid, S.: Evacuating two robots from two unknown exits on the perimeter of a disk with wireless communication. In: Proceedings of the 19th International Conference on Distributed Computing and Networking, ICDCN, pp. 20:1–20:4 (2018)
39. Reingold, O.: Undirected st-connectivity in log-space. In: Proceedings of the 37th Annual ACM Symposium on Theory of Computing, STOC, pp. 376–385 (2005)
40. Schwefel, H.P.P.: Evolution and Optimum Seeking: The Sixth Generation. John Wiley & Sons Inc., Hoboken (1993)
41. Stone, L.: Theory of Optimal Search. Academic Press, New York (1975)

Query Minimization Under Stochastic Uncertainty

Steven Chaplick[1,2], Magnús M. Halldórsson[3], Murilo S. de Lima[4(✉)], and Tigran Tonoyan[5]

[1] Lehrstuhl für Informatik I, Universität Würzburg, Würzburg, Germany
[2] Department of Data Science and Knowledge Engineering, Maastricht University, Maastricht, The Netherlands
s.chaplick@maastrichtuniversity.nl
[3] ICE-TCS, Department of Computer Science, Reykjavik University, Reykjavik, Iceland
mmh@ru.is
[4] School of Informatics, University of Leicester, Leicester, UK
mslima@ic.unicamp.br
[5] Computer Science Department, Technion Institute of Technology, Haifa, Israel
ttonoyan@gmail.com

Abstract. We study problems with stochastic uncertainty data on intervals for which the precise value can be queried by paying a cost. The goal is to devise an adaptive decision tree to find a correct solution to the problem in consideration while minimizing the expected total query cost. We show that sorting in this scenario can be performed in polynomial time, while finding the data item with minimum value seems to be hard. This contradicts intuition, since the minimum problem is easier both in the online setting with adversarial inputs and in the offline verification setting. However, the stochastic assumption can be leveraged to beat both deterministic and randomized approximation lower bounds for the online setting. Although some literature has been devoted to minimizing query/probing costs when solving uncertainty problems with stochastic input, none of them have considered the setting we describe. Our approach is closer to the study of query-competitive algorithms, and it gives a better perspective on the impact of the stochastic assumption.

Keywords: Stochastic optimization · Query minimization · Sorting · Selection · Online algorithms

1 Introduction

Consider the problem of sorting n data items that are updated concurrently by different processes in a distributed system. Traditionally, one ensures that the data is strictly consistent, e.g., by assigning a master database that is queried by

Partially supported by Icelandic Research Fund grant 174484-051 and by EPSRC grant EP/S033483/1. This work started while M.S.L. and T.T. were at Reykjavik University, during a research visit by S.C.

Y. Kohayakawa and F. K. Miyazawa (Eds.): LATIN 2020, LNCS 12118, pp. 181–193, 2020.
https://doi.org/10.1007/978-3-030-61792-9_15

the other processes, or by running a distributed consensus algorithm. However, those operations are expensive, and we wonder if we could somehow avoid them. One different approach has been proposed for the TRAPP distributed database by Olston and Widom [15], and is outlined as follows. Every update is sent to the other processes, and each process maintains an interval on which each data item may lie. Whenever the precise value is necessary, a query on the master database can be performed. Some computations (e.g., sorting) can be performed without knowing the precise value of all data items, so one question that arises is how to perform these while minimizing the total query cost. Another scenario in which this type of problem arises is when market research is required to estimate the data input: a coarser estimation can be performed for a low cost, and more precise information can be obtained by spending more effort in research. The problem of sorting under such conditions, called the **uncertainty sorting problem with query minimization**, was recently studied by Halldórsson and de Lima [12].

The study of uncertainty problems with query minimization dates back to the seminal work of Kahan [13] and the TRAPP distributed database system by Olston and Widom [15], which dealt with simple problems such as computing the minimum and the sum of numerical data with uncertainty intervals. More recently, more sophisticated problems have been studied in this framework, such as geometric problems [1], shortest paths [6], minimum spanning tree and minimum matroid base [5,14], linear programming [16,19], and NP-hard problems such as the knapsack [8] and scheduling problems [3]. See [4] for a survey.

The literature describes two kinds of algorithms for this setting. Though the nomenclature varies, we adopt the following one. An **adaptive** algorithm may decide which queries to perform based on results from previous queries. An **oblivious** algorithm, however, must choose the whole set of queries to perform in advance; i.e., it must choose a set of queries that certainly allow the problem to be solved without any knowledge of the actual values.

Two main approaches have been proposed to analyze both types of algorithms. In the first, an oblivious (adaptive) algorithm is compared to a hypothetical optimal oblivious (adaptive) strategy; this is the approach in [6,13,15]. However, for more complex problems, and in particular for adaptive algorithms, it usually becomes more difficult to understand the optimal adaptive strategy. A second (more robust) approach is competitive analysis, which is a standardized metric for online optimization. In this setting, both oblivious and adaptive algorithms are compared to an **optimum query set**, a minimum-cost set of queries that a clairvoyant adversary, who knows the actual values but cannot disclose them without performing a query, can use to prove the obtained solution to be correct. An algorithm (either adaptive or oblivious) is **α-query-competitive** if it performs a total query cost of at most α times the cost of an offline optimum query set. This type of analysis is performed in [1,5,11–14]. For NP-hard problems, since we do not expect to find the "correct" solution in polynomial time, there are two approaches in the literature: either we have an objective function which combines query and solution costs (this is how the scheduling problem is addressed in [3]), or we have a fixed query budget and the objective function is based only on the solution cost (as for the knapsack problem in [8]).

Competitive analysis is, however, rather pessimistic. In particular, many problems such as minimum, sorting and spanning tree have a deterministic lower bound of 2 and a randomized lower bound of 1.5 for adaptive algorithms, and a simple 2-competitive deterministic adaptive algorithm, even if queries are allowed to return intervals [5,11,12,14]. For the sorting problem, e.g., Halldórsson and de Lima [12] showed that there is essentially one structure preventing a deterministic adaptive algorithm from performing better than 2.

One natural alternative to competitive analysis is to assume stochastic inputs, i.e., that the precise value in each interval follows a known probability distribution, and we want to build a decision tree specifying a priority ordering for querying the intervals until the correct solution is found, so that the expected total query cost is minimized.[1] In this paper, we show that the adaptive sorting problem in this setting can be solved exactly in polynomial time. Very surprisingly, however, we have evidence that the problem of finding the data item with minimum value is hard, though it can be approximated very well.

Some literature is devoted to a similar goal of this paper, but we argue that there are some essential differences. One first line of work consists of the **stochastic probing problem** [7,9,10,17], which is a general stochastic optimization problem with queries. Even though those works presented results for wide classes of constraints (such as matroid and submodular), they differ in two ways from our work. First, they assume that a solution can only contain elements that are queried, or that the objective function is based on the expectation of the non-queried elements. Second, the objective function is either a combination of the solution and query costs, or there is a fixed budget for performing queries. Since most of these variants are NP-hard [7], some papers [9,17] focused on devising approximation algorithms, while others [7,10] on bounding the ratio between an oblivious algorithm and an optimal adaptive algorithm (the **adaptive gap**). Another very close work is that of Welz [18, Section 5.3] and Yamaguchi and Maehara [19], who, like us, assume that a solution may contain non-queried items. Welz presented some results for the minimum spanning tree and traveling salesman problems, but they make strong assumptions on the probability distributions, while Yamaguchi and Maehara devised algorithms for a wide class of problems, which also yield improved approximation algorithms for some classical stochastic optimization problems. However, both works focus on obtaining approximate solutions, while we wish to obtain an exact one, and they only give asymptotic bounds on the number of queries performed, but do not compare this to the expected cost of an optimum query set. To sum up, our work gives a better understanding on how the stochastic assumption differs from the competitive analysis, since other assumptions are preserved and we use the same metric to analyze the algorithms: minimizing query cost while finding the correct solution.

Our Results. We prove that, for the sorting problem with stochastic uncertainty, we can construct an adaptive decision tree with minimum expected query cost

---

[1] Note that, unless some sort of nondeterminism is allowed, the stochastic assumption cannot be used to improve the oblivious results, so we focus on adaptive algorithms.

in polynomial time. We devise a dynamic programming algorithm which runs in time $O(n^3d^3) = O(n^6)$, where d is the clique number of the interval graph induced by the uncertainty intervals. We then discuss why simpler strategies fail, such as greedy algorithms using only local information, or relying on witness sets, which is a standard technique for solving query-minimization problems with adversarial inputs [1,5]. We also discuss why we believe that the dynamic programming algorithm cannot be improved to something better than $O(n^3)$.

Surprisingly, on the other hand, we present evidence that finding an adaptive decision tree with minimum expected query cost for the problem of finding the data item with minimum value is hard, although the online version (with adversarial inputs) and the offline (verification) version of the problem are rather simple. If the leftmost interval is the first one to be queried, we know how to compute the decision tree with minimum expected query cost easily. This also implies that, for any other decision tree, one branch can be calculated easily. However, if the leftmost interval is not the first to be queried, we prove that it should be the last one to be considered in the decision tree. The hard part, then, is to find the order in which the other intervals are considered in the "hard branch" of the decision tree. We discuss why various heuristics fail to this case. A simple approximation result with factor $1 + 1/d_1$ for uniform query costs, where d_1 is the degree of the leftmost interval in the interval graph, follows from the online version with adversarial inputs [13]. For arbitrary query costs, we show that the stochastic assumption can be used to beat both deterministic and randomized lower bounds for the online version with adversarial inputs.

Organization of the Paper. Section 2 is devoted to the sorting problem with stochastic uncertainty, and Sect. 3 to the problem of finding the minimum data item. We conclude the paper with future research questions in Sect. 4.

2 Sorting

The problem is to sort n numbers $v_1, \ldots, v_n \in \mathbb{R}$ whose actual values are unknown. We are given n intervals $I_1, \ldots, I_n$ such that $v_i \in I_i = [\ell_i, r_i]$. We can query interval I_i by paying a cost w_i, and after that we know the value of v_i. We want to find a permutation $\pi : [n] \rightarrow [n]$ such that $v_i \leq v_j$ if $\pi(i) < \pi(j)$ by performing a minimum-cost set of queries. We focus on adaptive algorithms, i.e., we can make decisions based on previous queries. We are interested in a stochastic variant of this problem in which v_i follows some known probability distribution on I_i. The only constraints are that (1) values in different intervals have independent probabilities, and (2) for any subinterval $(a, b) \subseteq I_i$, we can calculate $\mathbf{Pr}[v_i \in (a, b)]$ in constant time. The goal is to devise a strategy (i.e., a decision tree) to query the intervals so that the expected query cost is minimized. More precisely, this decision tree must tell us which interval to query first and, depending on where its value falls, which interval to query second, and so on.

Definition 1. *Two intervals I_i and I_j such that $r_i > \ell_j$ and $r_j > \ell_i$ are* **dependent**. *Two intervals that are not dependent are* **independent**.

The following lemma and proposition are proved in [12]. The lemma tells us that we have to remove all dependencies in order to be able to sort the numbers.

Lemma 1 ([12]). *The relative order between two intervals can be decided without querying either of them if and only if they are independent.*

Proposition 1 ([12]). *Let I_i and I_j be intervals with actual values v_i and v_j. If $v_i \in I_j$ (and, in particular, when $I_i \subseteq I_j$), then I_j is queried by every solution.*

Note that the dependency relation defines an interval graph. Proposition 1 implies that we can immediately query any interval containing another interval, hence we may assume a proper interval graph. We may also assume the graph is connected, since the problem is independent for each component, and that there are no single-point intervals, as they would give a non-proper or disconnected graph.

An Optimal Algorithm. We describe a dynamic programming algorithm to solve the sorting problem with stochastic uncertainty. Since we have a proper interval graph, we assume intervals are in the natural total order, with $\ell_1 \leq \cdots \leq \ell_n$ and $r_1 \leq \cdots \leq r_n$. We also pre-compute the **regions** $S_1, \ldots, S_t$ defined by the intervals, where $t \leq 2n - 1$. A region is the interval between two consecutive points in the set $\bigcup_{i=1}^{n} \{\ell_i, r_i\}$; we assume that the regions are ordered. We write $S_x = (a_x, b_x)$ with $a_x < b_x$, and we denote by $\mathcal{I}_x(y, z) := \{i : S_x \subseteq I_i \subseteq (a_y, b_z)\}$ the indices of the intervals contained in (a_y, b_z) that contain S_x. For simplicity we assume that, for any interval I_i and any region S_x, $\mathbf{Pr}[v_i = a_x] = \mathbf{Pr}[v_i = b_x] = 0$; this is natural for continuous probability distributions, and for discrete distributions we may slightly perturb the distribution support so that this is enforced. Since the dependency graph is a connected proper interval graph, we can also assume that each interval contains at least two regions.

Before explaining the recurrence, we first examine how Proposition 1 limits our choices with an example. In Fig. 1(a), suppose we first decide to query I_3 and its value falls in region S_5. Due to Proposition 1, all intervals that contain S_5, namely I_2 and I_4, have to be queried as well. In Fig. 1(b), we assume that v_2 falls in S_3 and v_4 falls in S_6. This forces us to query I_1 but also results in a solution without querying I_5. Therefore, each time we approach a subproblem by first querying an interval I_i whose value falls in region S_x, we are forced to query all other intervals that contain S_x, and so on in a cascading fashion, until we end up with subproblems that are independent of current queried values. To find the best solution, we must pick a first interval to query, and then recursively calculate the cost of the best solution, depending on the region in which its value falls. Here, the proper interval graph can be leveraged by having the cascading procedure follow the natural order of the intervals.

We solve the problem by computing three tables. The first table, M, is indexed by two regions $y, z \in \{1, \ldots, t\}$, and $M[y, z]$ is the minimum expected query cost for the subinstance defined by the intervals contained in (a_y, b_z). Thus, the value of the optimum solution for the whole problem is $M[1, t]$. To compute $M[y, z]$, we suppose the first interval in (a_y, b_z) that is queried by the optimum solution is I_i. Then, for each region $S_x \subseteq I_i$, when $v_i \in S_x$, we are

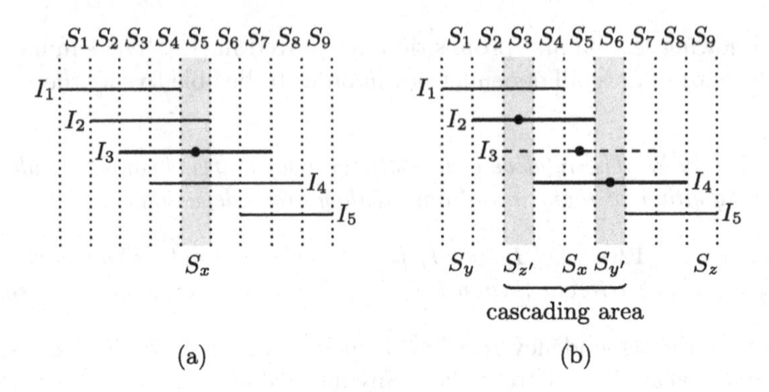

Fig. 1. A simulation of the querying process for a fixed realization of the values. (a) Querying I_3 first and assuming $v_3 \in S_5$. (b) Assuming $v_2 \in S_3$ and $v_4 \in S_6$.

forced to query every interval I_j with $j \in \mathcal{I}_x(y, z)$ and this cascades, forcing other intervals to be queried depending on where v_j falls. So we assume that, for all $j \in \mathcal{I}_x(y, z)$, v_j falls in the area defined by regions $z', z'+1, \ldots, y'-1, y'$, with $z' \leq x \leq y'$, and that this area is minimal (i.e., some point is in $S_{z'}$, and some point is in $S_{y'}$). We call this interval $(a_{z'}, b_{y'})$ the **cascading area** of I_i in $\mathcal{I}_x(y, z)$. In Fig. 1(b), we have $i = 3$, $x = 5$, $z' = 3$ and $y' = 6$. As the dependency graph is a proper interval graph, the remaining intervals (which do not contain S_x) are split in two independent parts, whose value is computed by two tables, L and R, which we describe next. So the recurrence for $M[y, z]$ is

$$\begin{cases} 0, & \text{if } (a_y, b_z) \text{ contains less than 2 intervals; otherwise,} \\ \underbrace{\min_{I_i \subseteq (a_y, b_z)}}_{\text{first interval to query}} \underbrace{\sum_{S_x \subseteq I_i} \mathbf{Pr}[v_i \in S_x] \cdot}_{\text{where point } v_i \text{ falls}} & \left(\begin{array}{c} \overbrace{\sum_{j \in \mathcal{I}_x(y,z)} w_j}^{\text{cost of cascading}} + \sum_{\substack{z' \leq x \\ y' \geq x}} \overbrace{p(y, z, i, x, z', y')}^{\text{cascading area}} \cdot \\ \underbrace{\cdot \left(\begin{array}{l} L[y, z', \min \mathcal{I}_x(y, z)] + \\ +R[y', z, \max \mathcal{I}_x(y, z)] \end{array} \right)}_{\text{cost of left/right subproblems}} \end{array} \right), \end{cases}$$

where $p(y, z, i, x, z', y')$ is the probability that $(a_{z'}, b_{y'})$ is the cascading area of I_i in $\mathcal{I}_x(y, z)$. We omit the description of how to calculate this probability.

The definitions of L and R are symmetric, so we focus on L. For region indices y, z, z' with $z \geq z'$, let $I_{j'}$ be the leftmost interval contained in (a_y, b_z). Now, $L[y, z', j]$ is the minimum expected query cost of solving the subproblem consisting of intervals $I_{j'}, I_{j'+1}, \ldots, I_{j-1}$, assuming that a previously queried point lies in the region $S_{z'}$. We ensure that z' is the leftmost region in (a_y, b_z) that contains a queried point so that we query all intervals that contain some point. For example, in Fig. 1(b), after querying I_2, I_3 and I_4, the left subproblem has $z' = 3$ and $j = 2$. It holds that L can be calculated in the following way. If no interval before I_j contains $S_{z'}$, then the cascading is finished and we can refer to table M for regions $y, y+1, \ldots, z'-1$. Otherwise I_{j-1} must contain $S_{z'}$,

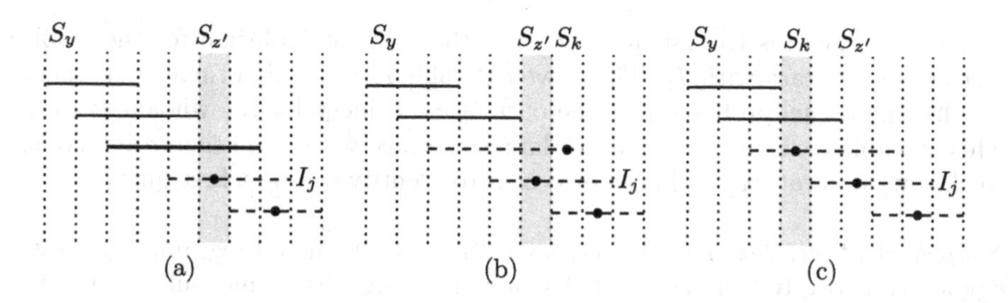

Fig. 2. An illustration of the definition of table L. (a) $L[y, z', j]$. (b) If $k \geq z'$, we recurse on $L[y, z', j-1]$. (c) If $k < z'$, we recurse on $L[y, k, j-1]$.

we query it, and either v_{j-1} falls to the right of $\ell_{z'}$ and we proceed to the next interval, or v_{j-1} falls in a region S_k with $k < z'$, and we proceed to the next interval with the leftmost queried point now being in S_k. Thus, we have

$$L[y, z', j] = \begin{cases} M[y, z'-1], & \text{if } j \leq 1 \text{ or } \ell_{j-1} < a_y \text{ or } I_{j-1} \not\supseteq S_{z'} \\ w_{j-1} + \sum_{S_k \subseteq I_{j-1}} \mathbf{Pr}[v_{j-1} \in S_k] \cdot L[y, \min(k, z'), j-1], & \text{otherwise.} \end{cases}$$

We illustrate this in Fig. 2. In Fig. 2(a), the subproblem contains $I_{j-1}, I_{j-2}, \ldots$, and the leftmost queried point is in $S_{z'}$. Since $S_{z'} \subseteq I_{j-1}$, we query I_{j-1} and assume v_{j-1} falls in a region S_k. In Fig. 2(b), we have that $k \geq z'$, so we recurse on $L[y, z', j-1]$; this will recurse on $M[y, z'-1]$ in its turn, since $S_{z'} \not\subseteq I_{j-2}$. In Fig. 2(c), we have that $k < z'$, so we recurse on $L[y, k, j-1]$, which in its turn will have to query I_{j-2}.

At this point it is not hard to see that the next theorem follows by a standard optimal substructure argument; we omit the proof.

Theorem 1. *$M[1, t]$ is the value of the minimum expected query cost to solve the stochastic sorting problem with uncertainty.*

The recurrences can be implemented in a bottom-up fashion that consumes time $O(n^6)$: if we precompute $p(y, z, i, x, z', y')$, then each entry of M is computed in $O(n^4)$, and each entry of L and R can be computed in linear time. It is possible to precompute $p(y, z, i, x, z', y')$ in time $O(n^4)$ (we omit how). A more careful analysis shows that the time consumption is $O(n^3 d^3)$, where d is the clique number of the interval graph. Note that, in a proper interval graph, an interval contains at most $2d - 1$ regions. Another simple fact is that $\mathcal{I}_x(y, z)$ contains at most d intervals, since every such interval contains S_x.

It seems difficult to improve this dynamic programming algorithm to something better than $O(n^3 \cdot \text{poly}(d))$. Note that the main information that the decision tree encodes is which interval should be queried first in a given independent subproblem (and there are $\Omega(n^2)$ such subproblems). We could hope to find an optimal substructure that would not need to test every interval as a first query, and that this information could somehow be inferred from smaller subproblems. However, consider $I_1 = (0, 100)$, $I_2 = (6, 105)$, and $I_3 = (95, 198)$, with uniform query costs and uniform probability distributions. The optimum solution for the first two intervals is to first query I_1, but the optimum solution for the whole instance is to start with I_2. Thus, even though I_2 is a suboptimal first query for the smaller subproblem, it is the optimal first query for the whole instance. This example could be adapted to a larger instance with more than d intervals, so that we need at least a linear pass in n to identify the best first query.

Simpler Strategies that Fail. It may seem that our dynamic programming strategy above is overly complex, and that a simpler algorithm may suffice to solve the problem. Below, we show sub-optimality of two such strategies.

We begin by showing that any greedy strategy that only takes into consideration local information (such as degree in the dependency graph or overlap area) fails. Consider a 5-path *abcde*, in which each interval has query cost 1 and an overlap of 1/3 with each of its neighbors, and the exact value is uniformly distributed in each interval. It can be shown by direct calculation that if we query I_b (or, equivalently, I_d) first, then we get an expected query cost of at most $29/9 = 3.2\bar{2}$, while querying I_c first yields an expected query cost of at least $11/3 = 3.6\bar{6}$. However, a greedy strategy that only takes into consideration local information cannot distinguish between I_b and I_c.

One technique that has been frequently applied in the literature of uncertainty problems with query minimization is the use of **witness sets**. A set of intervals W is a witness if a correct solution cannot be computed unless at least one interval in W is queried, even if all other intervals not in W are queried. Witness sets are broadly adopted because they simplify the design of query-competitive adaptive algorithms. If, at every step, an algorithm queries disjoint witness sets of size at most α, then this algorithm is α-query-competitive. This concept was proposed in [1]. For the sorting problem, by Lemma 1, any pair of dependant intervals constitute a witness set. However, we cannot take advantage of witness sets for the stochastic version of the problem, even for uniform query costs and uniform probability distributions, and even if we take advantage of the proper interval order. Consider the following intervals: $(0, 100), (95, 105), (98, 198)$. The witness set consisting of the first two intervals may lead us to think that either of them is a good choice as the first query. However, the unique optimum solution first queries the third interval. (The costs are $843/400 = 2.1075$ if we first query the first interval, $277/125 = 2.216$ if we first query the second interval, and $4182/2000 = 2.0915$ if we first query the third interval).

3 Finding the Minimum

We also consider the problem of finding the minimum (or, equivalently, the maximum) of n unknown values $v_1, \ldots, v_n$. Assume that the intervals are sorted by the left endpoint, i.e., $\ell_1 \leq \ell_2 \leq \cdots \leq \ell_n$. We may assume without loss of generality that $\ell_1 < \ell_2 < \cdots < \ell_n$. Let $\mathcal{I} = \{I_1, \ldots, I_n\}$. We begin by discussing some assumptions we can make. First, we can assume that the interval graph is a clique: with two independent intervals, we can remove the one on the right.

(However, we cannot assume a proper interval graph, as we did for sorting.) The second assumption is based on the following remark, whose proof we omit.

Remark 1. If I_1 contains some I_j, then I_1 is queried in every solution.

Thus we can assume that I_1 does not contain another interval; this implies that $r_1 = \min_i r_i$. It is also useful to understand how to find an optimum query set, i.e., to solve the problem assuming we know $v_1, \ldots, v_n$.

Lemma 2 ([13]). *The offline optimum solution either*

(a) queries interval I_i with minimum v_i and each interval I_j with $\ell_j < v_i$; or
(b) queries all intervals except for I_1, if v_1 is the minimum, $v_j > r_1$ for all $j > 1$, and this is better than option (a).

Option (b) can be better not only due to a particular non-uniform query cost configuration, but also with uniform query costs, when $v_1 \in I_2, \ldots, I_n$. Note also that I_1 is always queried in option (a). We omit the proof of this lemma.

We first discuss what happens if the first interval we query is I_1. In Fig. 3(a), we suppose that $v_1 \in S_3$. This makes I_2 become the leftmost interval, so it must be queried, since it contains v_1 and I_3. At this point we also know that we do not need to query I_4, since $v_1 < \ell_4$. After querying I_2, we have two possibilities. In Fig. 3(b), we suppose that $v_2 \in S_2$, so we already know that v_2 is the minimum and no other queries are necessary. In Fig. 3(c), we suppose that $v_2 \in S_6$, so we still need to query I_3 to decide if v_1 or v_3 is the minimum. Note that, once I_1 has been queried, we do not have to guess which interval to query next, since any interval that becomes the leftmost interval will either contain v_1 or will be to the right of v_1. Since this is an easy case of the problem, we formalize how to solve it. The following claim is clear: if we have already queried $I_1, \ldots, I_{i-1}$ and $v_1, \ldots, v_{i-1} \in I_i$, then we have to query I_i. (This relies on I_i having minimum ℓ_i among $I_i, \ldots, I_n$.) If we decide to first query I_1, then we are discarding option (b) in the offline solution, so all intervals containing the minimum value must be queried. The expected query cost is then $\sum_{i=1}^{n} w_i \cdot \mathbf{Pr}[I_i \text{ must be queried}]$. Given an interval I_i, it will not need to be queried if there is some I_j with $v_j < \ell_i$, thus the former probability is the probability that no value lies to the left of I_i. Since the probability distribution is independent for each interval, the expected query cost will be $\sum_{i=1}^{n} w_i \cdot \prod_{j<i} \mathbf{Pr}[v_j > \ell_i]$. This can be computed in $O(n^2)$ time.

Now let us consider what happens if the optimum solution does not start by querying I_1, but by querying some I_k with $k > 1$. When we query I_k, we have two cases: (1) if v_k falls in I_1, then we have to query I_1 and proceed as discussed above, querying I_2 if $v_1 > \ell_2$, then querying I_3 if $v_1, v_2 > \ell_3$ and so on; (2) if $v_k \notin I_1$, then v_k falls to the right of ℓ_i, for all $i \neq k$, so essentially the problem consists of finding the optimum solution for the remaining intervals, and this value will be independent of v_k. Therefore, the cost of first querying I_k is $w_k + \mathbf{Pr}[v_k \notin I_1] \cdot \text{opt}(\mathcal{I} \setminus \{I_k\}) + \mathbf{Pr}[v_k \in I_1] \cdot \sum_{i \neq k} w_i \cdot \prod_{j<i} \mathbf{Pr}[v_j > \ell_i | v_k \in I_1]$. Thus, we can see that a decision tree can be specified simply by a permutation of the intervals, since the last term in the last equation is fixed. More precisely, let $a(1), \ldots, a(n)$ be a permutation of the intervals, where $a(k) = i$ means that I_i

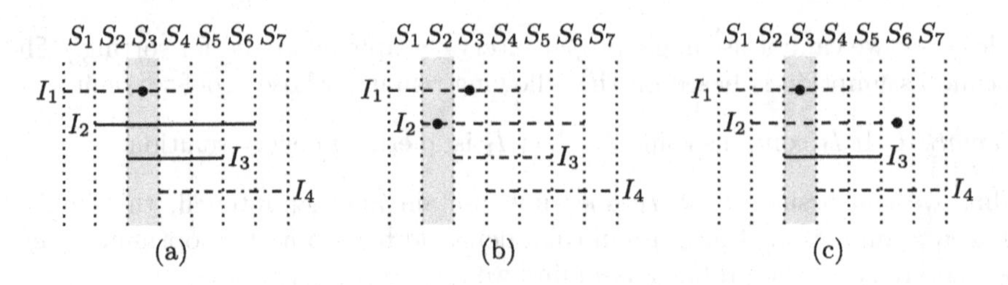

Fig. 3. A simulation of the querying process when we decide to first query I_1. (a) If $v_1 \in S_3$, I_2 must be queried, but not I_4. (b) If $v_2 \in S_2$, then v_2 is the minimum. (c) If $v_2 \in S_6$, then we still have to query I_3.

is the k-th interval in the permutation. We have two types of subtrees. Given a subset $X_k = \{a(k), \ldots, a(n)\}$ that contains 1, let $\hat{T}_k$ be the tree obtained by first querying I_1, then querying the next leftmost interval in X_k if it contains v_1 and so on. The second type of subtree T_k is defined by a suffix $a(k), \ldots, a(n)$ of the permutation. If $a(k) \neq 1$, then T_k is a decision tree with a root querying $I_{a(k)}$ and two branches. One branch, with probability $\mathbf{Pr}[v_{a(k)} \in I_1]$, consists of $\hat{T}_{k+1}$; the other branch, with probability $\mathbf{Pr}[v_{a(k)} \notin I_1]$, consists of T_{k+1}. If $a(k) = 1$, then $T_k = \hat{T}_k$, unless $k = n$, in which case T_n will be empty: I_1 does not need to be queried, because all other intervals have already been queried and their values fall to the right of I_1. We have that $\text{cost}(T_k)$

$$= \begin{cases} 0, & \text{if } a(k) = a(n) = 1 \\ \text{cost}(\hat{T}_k), & \text{if } a(k) = 1 \text{ but } k \neq n; \text{ otherwise,} \\ \mathbf{Pr}[v_{a(k)} \in I_1] \cdot \text{cost}(\hat{T}_k | v_{a(k)} \in I_1) + \mathbf{Pr}[v_{a(k)} \notin I_1] \cdot (w_{a(k)} + \text{cost}(T_{k+1})). \end{cases}$$

Note that in the last case we need to condition $\text{cost}(\hat{T}_k)$ to the fact that $v_{a(k)} \in I_1$. The cost of $\hat{T}_k$ conditioned to E is $\sum_{i=k}^{n} w_i \cdot \prod_{\substack{j \geq k \\ a(j) < a(i)}} \mathbf{Pr}[v_{a(j)} > \ell_{a(i)} | E]$.

It holds that, if I_1 is not the last interval in a decision tree permutation, then it is always better to move I_1 one step towards the beginning of the permutation. (We omit the proof due to space limitations.) Thus, by induction, the optimum solution either first queries I_1, or has I_1 at the end of the permutation. If I_1 is the last interval to be queried, then it does not have to be queried if all other values fall to its right. Thus it may be that, in expectation, having I_1 as the last interval is optimal.

We do not know, however, how to efficiently find the best permutation ending in I_1. Simply considering which interval begins or ends first, or ordering by $\mathbf{Pr}[v_i \in I_1]$ is not enough. To see this, consider the following two instances with uniform costs and uniform probabilities. In the first, $I_1 = (0, 100)$, $I_2 = (5, 305)$ and $I_3 = (6, 220)$; the best permutation is I_2, I_3, I_1 and has cost 2.594689. If we just extend I_2 a bit to the right, making $I_2 = (5, 405)$, then the best permutation is I_3, I_2, I_1, whose cost is 2.550467.

If there was a way to determine the relative order in the best permutation between two intervals $I_j, I_k \neq I_1$, simply by comparing some value not depending on the order of the remaining intervals (for example, by comparing the cost of $I_j I_k I_1 \cdots$ and $I_k I_j I_1 \cdots$), then we could find the best permutation easily. Unfortunately, the ordering of the permutations is not always consistent, i.e., given a permutation, consider what happens if we swap I_j and I_k: it is not always best to have I_j before I_k, or I_k before I_j. Consider intervals $I_1 = (0, 1000)$, $I_2 = (3, 94439)$, $I_3 = (8, 6924)$, and $I_4 = (9, 2493)$, with uniform query cost and uniform probability distributions. The best permutation is I_4, I_3, I_2, I_1, and the costs of the permutations ending in I_1 are as follows. Note that it is sometimes better that I_2 comes before I_3, and sometimes the opposite.

$$\begin{array}{lll} \mathrm{cost}(4,2,3,1) = 3.48611 & \mathrm{cost}(2,4,3,1) = 3.48715 & \mathrm{cost}(2,3,4,1) = 3.48889 \\ \mathrm{cost}(4,3,2,1) = 3.48593 & \mathrm{cost}(3,4,2,1) = 3.48770 & \mathrm{cost}(3,2,4,1) = 3.48859 \end{array}$$

This issue also seems to preclude greedy and dynamic programming algorithms from succeeding. It seems that it is not possible to find an optimal substructure, since the ordering is not always consistent among subproblems and the whole problem. We have implemented various heuristics and performed experiments on random instances, and could always find instances in which the optimum was missed, even for uniform query costs and uniform probabilities.

Another reason to expect hardness is that the following similar problem is NP-hard [7]. Given stochastic uncertainty intervals $I_1, \ldots, I_n$, costs $w_1, \ldots, w_n$, and a query budget C, find a set $S \subseteq \{1, \ldots, n\}$ with $w(S) \leq C$ that minimizes $\mathbf{E}[\min_{i \in S} v_i]$.

To conclude, we note that there are good approximation algorithms, which have been proposed for the online version [13]. If query costs are uniform, then first querying I_1 costs at most opt + 1, which yields a factor $1 + 1/d_1$, where d_1 is the degree of I_1 in the interval graph. For arbitrary costs, there is a randomized 1.5-approximation algorithm using weighted probabilities in the two solutions stated in Lemma 2. Those results apply to the stochastic version of the problem simply by linearity of expectation.

Theorem 2. *The minimum problem admits a* $(1+1/d_1)$*-approximation for uniform query costs, and a randomized* 1.5*-approximation for arbitrary costs.*

Those results have matching lower bounds for the online setting, and for arbitrary query costs there is a deterministic lower bound of 2. We show that the stochastic assumption can be used to beat those lower bounds for arbitrary costs. First, the randomized 1.5-approximation algorithm can be derandomized, simply by choosing which solution has smaller expected query cost: either first querying I_1, or first querying all other intervals and if necessary querying I_1. We know how to calculate both expected query costs; the latter is $\sum_{i>1} w_i + w_1 \cdot \left(1 - \prod_{i>1} \mathbf{Pr}[v_i > r_1]\right)$. We omit the proof of the following theorem and the description of the randomized algorithm.

Theorem 3. *There is a deterministic* 1.5*-approximation algorithm and a randomized* 1.45*-approximation algorithm for arbitrary query costs.*

4 Further Questions

Can we extend our approach for sorting so as to handle a dynamic setting, as in [2]? E.g., where some intervals can be inserted/deleted from the initial set. Updating the dynamic program should be faster than building it again from scratch.

Is the minimum problem NP-hard or can it be solved in polynomial time? If it is NP-hard, then so are the median and the minimum spanning tree problems. Can we devise polynomial time algorithms with better approximation guarantees than the best respective competitive online results?

References

1. Bruce, R., Hoffmann, M., Krizanc, D., Raman, R.: Efficient update strategies for geometric computing with uncertainty. Theory Comput. Syst. **38**(4), 411–423 (2005). https://doi.org/10.1007/s00224-004-1180-4
2. Busto, D., Evans, W., Kirkpatrick, D.: Minimizing interference potential among moving entities. In: Proceedings of the 30th Annual ACM-SIAM Symposium on Discrete Algorithms, SODA, pp. 2400–2418 (2019). http://dl.acm.org/citation.cfm?id=3310435.3310582
3. Dürr, C., Erlebach, T., Megow, N., Meißner, J.: Scheduling with explorable uncertainty. In: Proceedings of the 9th Innovations in Theoretical Computer Science Conference, ITCS, LIPIcs, vol. 94, pp. 30:1–30:14 (2018). https://doi.org/10.4230/LIPIcs.ITCS.2018.30
4. Erlebach, T., Hoffmann, M.: Query-competitive algorithms for computing with uncertainty. Bull. EATCS **116**, 22–39 (2015). http://bulletin.eatcs.org/index.php/beatcs/article/view/335
5. Erlebach, T., Hoffmann, M., Krizanc, D., Mihal'ák, M., Raman, R.: Computing minimum spanning trees with uncertainty. In: STACS, pp. 277–288 (2008). https://doi.org/10.4230/LIPIcs.STACS.2008.1358
6. Feder, T., Motwani, R., O'Callaghan, L., Olston, C., Panigrahy, R.: Computing shortest paths with uncertainty. J. Algorithms **62**(1), 1–18 (2007). https://doi.org/10.1016/j.jalgor.2004.07.005
7. Goel, A., Guha, S., Munagala, K.: Asking the right questions: model-driven optimization using probes. In: Proceedings of the 25th ACM SIGMOD-SIGACT-SIGART Symposium on Principles of Database Systems, PODS, pp. 203–212 (2006). https://doi.org/10.1145/1142351.1142380
8. Goerigk, M., Gupta, M., Ide, J., Schöbel, A., Sen, S.: The robust knapsack problem with queries. Comput. Oper. Res. **55**, 12–22 (2015). https://doi.org/10.1016/j.cor.2014.09.010
9. Gupta, A., Nagarajan, V.: A stochastic probing problem with applications. In: Goemans, M., Correa, J. (eds.) IPCO 2013. LNCS, vol. 7801, pp. 205–216. Springer, Heidelberg (2013). https://doi.org/10.1007/978-3-642-36694-9_18
10. Gupta, A., Nagarajan, V., Singla, S.: Algorithms and adaptivity gaps for stochastic probing. In: Proceedings of the 27th Annual ACM-SIAM Symposium on Discrete algorithms, SODA, pp. 1731–1747 (2016). https://doi.org/10.1137/1.9781611974331.ch120

11. Gupta, M., Sabharwal, Y., Sen, S.: The update complexity of selection and related problems. Theory Comput. Syst. **59**(1), 112–132 (2016). https://doi.org/10.1007/s00224-015-9664-y
12. Halldórsson, M.M., de Lima, M.S.: Query-competitive sorting with uncertainty. In: Proceedings of the 44th International Symposium on Mathematical Foundations of Computer Science, MFCS, LIPIcs, vol. 138, pp. 7:1–7:15 (2019). https://doi.org/10.4230/LIPIcs.MFCS.2019.7
13. Kahan, S.: A model for data in motion. In: Proceedings of the 23rd Annual ACM Symposium on Theory of Computing, STOC, pp. 265–277 (1991). https://doi.org/10.1145/103418.103449
14. Megow, N., Meißner, J., Skutella, M.: Randomization helps computing a minimum spanning tree under uncertainty. SIAM J. Comput. **46**(4), 1217–1240 (2017). https://doi.org/10.1137/16M1088375
15. Olston, C., Widom, J.: Offering a precision-performance tradeoff for aggregation queries over replicated data. In: Proceedings of the 26th International Conference on Very Large Data Bases, VLBD, pp. 144–155 (2000). http://ilpubs.stanford.edu:8090/437/
16. Ryzhov, I.O., Powell, W.B.: Information collection for linear programs with uncertain objective coefficients. SIAM J. Optim. **22**(4), 1344–1368 (2012). https://doi.org/10.1137/12086279X
17. Singla, S.: The price of information in combinatorial optimization. In: Proceedings of the 29th Annual ACM-SIAM Symposium on Discrete Algorithms, SODA, pp. 2523–2532 (2018). https://doi.org/10.1137/1.9781611975031.161
18. Welz, W.A.: Robot Tour Planning with High Determination Costs. Ph.D. thesis, Technischen Universität Berlin (2014). https://www.depositonce.tu-berlin.de/handle/11303/4597
19. Yamaguchi, Y., Maehara, T.: Stochastic packing integer programs with few queries. In: Proceedings of the 29th Annual ACM-SIAM Symposium on Discrete Algorithms, SODA, pp. 293–310 (2018). https://dl.acm.org/doi/abs/10.5555/3174304.3175288

Suffix Trees, DAWGs and CDAWGs for Forward and Backward Tries

Shunsuke Inenaga[1,2](✉)

[1] Department of Informatics, Kyushu University, Fukuoka, Japan
inenaga@inf.kyushu-u.ac.jp
[2] PRESTO, Japan Science and Technology Agency, Kawaguchi, Japan

Abstract. The suffix tree, DAWG, and CDAWG are fundamental indexing structures of a string, with a number of applications in bioinformatics, information retrieval, data mining, etc. An edge-labeled rooted tree (trie) is a natural generalization of a string, which can also be seen as a compact representation of a set of strings. Kosaraju [FOCS 1989] proposed the suffix tree for a backward trie, where the strings in the trie are read in the leaf-to-root direction. In contrast to a backward trie, we call a usual trie as a forward trie. Despite a few follow-up works after Kosaraju's paper, indexing forward/backward tries is not well understood yet. In this paper, we show a full perspective on the sizes of indexing structures such as suffix trees, DAWGs, and CDAWGs for forward and backward tries. In particular, we show that the size of the DAWG for a forward trie with n nodes is $\Omega(\sigma n)$, where σ is the number of distinct characters in the trie. This becomes $\Omega(n^2)$ for an alphabet of size $\sigma = \Theta(n)$. Still, we show that there is a compact $O(n)$-space implicit representation of the DAWG for a forward trie, whose space requirement is independent of the alphabet size. This compact representation allows for simulating each DAWG edge traversal in $O(\log \sigma)$ time, and can be constructed in $O(n)$ time and space over any integer alphabet of size $O(n)$.

1 Introduction

Text indexing is a fundamental problem in theoretical computer science that dates back to 1970's when suffix trees were invented [26]. Here the task is to preprocess a given text string S so that subsequent patten matching queries on S can be answered efficiently. Suffix trees have numerous other applications e.g. sequence comparisons [26], lossless data compression [2], data mining [23], and bioinformatics [15,21].

A trie is a rooted tree where each edge is labeled with a single character. A *backward* trie is an edge-reversed trie. Kosaraju [19] was the first to consider the trie indexing problem, and he proposed the suffix tree of a backward trie that takes $O(n)$ space, where n is the number of nodes in the backward trie. Kosaraju also claimed an $O(n \log n)$-time construction. Breslauer [7] showed how to build the suffix tree of a backward trie in $O(\sigma n)$ time and space, where σ is the

Y. Kohayakawa and F. K. Miyazawa (Eds.): LATIN 2020, LNCS 12118, pp. 194–206, 2020.
https://doi.org/10.1007/978-3-030-61792-9_16

Table 1. Summary of the numbers of nodes and edges of the suffix tree, DAWG, and CDAWG for a forward/backward trie with n nodes over an alphabet of size σ. The new bounds obtained in Sect. 5 of this paper are highlighted in bold. All the bounds here are valid with any alphabet size σ ranging from $\Theta(1)$ to $\Theta(n)$. Also, all these upper bounds are tight in the sense that there are matching lower bounds (see Sect. 5).

	Forward trie		Backward trie	
Indexing structure	# of nodes	# of edges	# of nodes	# of edges
Suffix tree	$O(n^2)$	$O(n^2)$	$O(n)$	$O(n)$
DAWG	$O(n)$	$O(\sigma n)$	$O(n^2)$	$O(n^2)$
CDAWG	$O(n)$	$O(\sigma n)$	$O(n)$	$O(n)$

alphabet size. Shibuya [25] presented an $O(n)$-time and space construction for the suffix tree of a backward trie over an integer alphabet of size $O(n)$. This line of research has been followed by the invention of XBWTs [11], suffix arrays [11], enhanced suffix arrays [18], and position heaps [24] for backward tries.

This paper considers the suffix trees, the *directed acyclic word graphs* (*DAWGs*) [5,9], and the *compact DAWGs* (*CDAWGs*) [6] built on a backward trie or on a forward (ordinary) trie. While all these indexing structures support linear-time pattern matching queries on tries, their sizes can significantly differ. We present *tight* lower and upper bounds on the sizes of all these indexing structures, as summarized in Table 1. Probably the most interesting result in our size bounds is the $\Omega(n^2)$ lower bound for the size of the DAWG for a forward trie with n nodes over an alphabet of size $\Theta(n)$ (Theorem 6), since this reveals that Mohri et al.'s algorithm [22] that constructs the DAWG for a forward trie with n nodes must take at least $\Omega(n^2)$ time and space in the worst case. We show that, somewhat surprisingly, there exists an *implicit compact representation* of the DAWG for a forward trie that occupies only $O(n)$ space independently of the alphabet size, and allows for simulating traversal of each DAWG edge in $O(\log \sigma)$ time. We also present an algorithm that builds this implicit representation of the DAWG for a forward trie in $O(n)$ time and space for any integer alphabet of size $O(n)$.

DAWGs for strings have important applications to pattern matching with don't cares [20], online Lempel-Ziv factorization in compact space [27], finding minimal absent words [13], etc. CDAWGs for strings can be regarded as *grammar compression* of input strings and can be stored in space linear in the number of right-extensions of maximal repeats [3]. It is known that the number of maximal repeats can be much smaller than the string length, particularly in highly repetitive strings. Hence, studying and understanding DAWGs/CDAWGs for tries are very important and are expected to lead to further research on efficient processing of tries.

Omitted proofs and supplemental figures can be found in a full version [16].

2 Preliminaries

Let Σ be an ordered alphabet. Any element of Σ^* is called a *string.* For any string S, let $|S|$ denote its length. Let ε be the empty string, namely, $|\varepsilon| = 0$. Let $\Sigma^+ = \Sigma^* \setminus \{\varepsilon\}$. If $S = XYZ$, then X, Y, and Z are called a *prefix*, a *substring*, and a *suffix* of S, respectively. For any $1 \leq i \leq j \leq |S|$, let $S[i..j]$ denote the substring of S that begins at position i and ends at position j in S. For convenience, let $S[i..j] = \varepsilon$ if $i > j$. For any $1 \leq i \leq |S|$, let $S[i]$ denote the ith character of S. For any string S, let $\overline{S}$ denote the reversed string of S, i.e., $\overline{S} = S[|S|] \cdots S[1]$. Also, for any set $\mathbf{S}$ of strings, let $\overline{\mathbf{S}}$ denote the set of the reversed strings of $\mathbf{S}$, namely, $\overline{\mathbf{S}} = \{\overline{S} \mid S \in \mathbf{S}\}$.

A *trie* T is a rooted tree (V, E) such that (1) each edge in E is labeled by a single character from Σ and (2) the character labels of the out-going edges of each node begin with mutually distinct characters. In this paper, a *forward trie* refers to an (ordinary) trie as defined above. On the other hand, a *backward trie* refers to an edge-reversed trie where each path label is read in the leaf-to-root direction. We will denote by $\mathsf{T_f} = (\mathsf{V_f}, \mathsf{E_f})$ a forward trie and by $\mathsf{T_b} = (\mathsf{V_b}, \mathsf{E_b})$ the backward trie that is obtained by reversing the edges of $\mathsf{T_f}$. We denote by a triple $(u, a, v)_\mathsf{f}$ an edge in a forward trie $\mathsf{T_f}$, where $u, v \in \mathsf{V}$ and $a \in \Sigma$. Each reversed edge in $\mathsf{T_b}$ is denoted by a triple $(v, a, u)_\mathsf{b}$. Namely, there is a directed labeled edge $(u, a, v)_\mathsf{f} \in \mathsf{E_f}$ iff there is a reversed directed labeled edge $(v, a, u)_\mathsf{b} \in \mathsf{E_b}$.

For a node u of a forward trie $\mathsf{T_f}$, let $anc(u, j)$ denote the jth ancestor of u in $\mathsf{T_f}$ if it exists. Alternatively, for a node v of a backward $\mathsf{T_b}$, let $des(v, j)$ denote the jth descendant of v in $\mathsf{T_b}$ if it exists. We use a *level ancestor* data structure [4] on $\mathsf{T_f}$ (resp. $\mathsf{T_b}$) so that $anc(u, j)$ (resp. $des(v, j)$) can be found in $O(1)$ time for any node and integer j, with linear space.

For nodes u, v in a forward trie $\mathsf{T_f}$ s.t. u is an ancestor of v, let $str_\mathsf{f}(u, v)$ denote the string spelled out by the path from u to v in $\mathsf{T_f}$. Let r denote the root of $\mathsf{T_f}$ and $\mathsf{L_f}$ the set of leaves in $\mathsf{T_f}$. The sets of substrings and suffixes of the forward trie $\mathsf{T_f}$ are respectively defined by $Substr(\mathsf{T_f}) = \{str_\mathsf{f}(u, v) \mid u, v \in \mathsf{V_f}, u \text{ is an ancestor of } v\}$ and $Suffix(\mathsf{T_f}) = \{str_\mathsf{f}(u, l) \mid u \in \mathsf{V_f}, l \in \mathsf{L_f}\}$.

For nodes v, u in a backward trie $\mathsf{T_b}$ s.t. v is a descendant of u, let $str_\mathsf{b}(v, u)$ denote the string spelled out by the reversed path from v to u in $\mathsf{T_b}$. The sets of substrings and suffixes of the backward trie $\mathsf{T_b}$ are respectively defined by $Substr(\mathsf{T_b}) = \{str_\mathsf{b}(v, u) \mid v, u \in \mathsf{V_b},\ v \text{ is a descendant of } u\}$ and $Suffix(\mathsf{T_b}) = \{str_\mathsf{b}(v, r) \mid v \in \mathsf{V_b},\ r \text{ is the root of } \mathsf{T_b}\}$.

In what follows, let n be the number of nodes in $\mathsf{T_f}$ (or equivalently in $\mathsf{T_b}$).

Fact 1. *(a) For any* $\mathsf{T_f}$ *and* $\mathsf{T_b}$, $Substr(\mathsf{T_f}) = \overline{Substr(\mathsf{T_b})}$. *(b) For any forward trie* $\mathsf{T_f}$, $|Suffix(\mathsf{T_f})| = O(n^2)$. *For some forward trie* $\mathsf{T_f}$, $|Suffix(\mathsf{T_f})| = \Omega(n^2)$. *(c)* $|Suffix(\mathsf{T_b})| \leq n - 1$ *for any backward trie* $\mathsf{T_b}$.

Fact 1-(a), Fact 1-(c) and the upper bound of Fact 1-(b) should be clear from the definitions. To see the lower bound of Fact 1-(b) in detail, consider a forward trie $\mathsf{T_f}$ with root r such that there is a single path of length k from r to a node v, and there is a complete binary tree rooted at v with k leaves. Then, for all nodes u in the path from r to v, the total number of strings in the set $\{str_f(u,l) \mid l \in \mathsf{L}_f\} \subset Suffix(\mathsf{T}_f)$ is at least $k(k+1)$, since each $str_f(u,l)$ is distinct for each path (u,l). By setting $k \approx n/3$ so that the number $|\mathsf{V}_f|$ of nodes in T_f equals n, we obtain Fact 1-(b). The lower bound is valid for alphabets of size σ ranging from 2 to $\Theta(k) = \Theta(n)$.

3 Maximal Substrings in Forward/Backward Tries

Blumer et al. [6] introduced the notions of right-maximal, left-maximal, and maximal substrings in a set $\mathbf{S}$ of strings, and presented clean relationships between the right-maximal/left-maximal/maximal substrings and the suffix trees/DAWGs/CDAWGs for $\mathbf{S}$. Here we give natural extensions of these notions to substrings in our forward and backward tries T_f and T_b, which will be the basis of our indexing structures for T_f and T_b.

Maximal Substrings on Forward Tries: For any substring X in a forward trie T_f, X is said to be *right-maximal* on T_f if (i) there are at least two distinct characters $a, b \in \Sigma$ such that $Xa, Xb \in Substr(\mathsf{T}_f)$, or (ii) X has an occurrence ending at a leaf of T_f. Also, X is said to be *left-maximal* on T_f if (i) there are at least two distinct characters $a, b \in \Sigma$ such that $aX, bX \in Substr(\mathsf{T}_f)$, or (ii) X has an occurrence beginning at the root of T_f. Finally, X is said to be *maximal* on T_f if X is both right-maximal and left-maximal in T_f. For any $X \in Substr(\mathsf{T}_f)$, let $r\text{-}mxml_f(X)$, $l\text{-}mxml_f(X)$, and $mxml_f(X)$ respectively denote the functions that map X to the shortest right-maximal substring $X\beta$, the shortest left-maximal substring αX, and the shortest maximal substring $\alpha X \beta$ that contain X in T_f, where $\alpha, \beta \in \Sigma^*$.

Maximal Substrings on Backward Tries: For any substring Y in a backward trie T_b, Y is said to be *left-maximal* on T_b if (i) there are at least two distinct characters $a, b \in \Sigma$ such that $aY, bY \in Substr(\mathsf{T}_b)$, or (ii) Y has an occurrence beginning at a leaf of T_b. Also, Y is said to be *right-maximal* on T_b if (i) there are at least two distinct characters $a, b \in \Sigma$ such that $Ya, Yb \in Substr(\mathsf{T}_b)$, or (ii) Y has an occurrence ending at the root of T_b. Finally, Y is said to be *maximal* on T_b if Y is both right-maximal and left-maximal in T_b. For any $Y \in Substr(\mathsf{T}_b)$, let $l\text{-}mxml_b(Y)$, $r\text{-}mxml_b(Y)$, and $mxml_b(Y)$ respectively denote the functions that map Y to the shortest left-maximal substring γY, the shortest right-maximal substring $Y\delta$, and the shortest maximal substring $\gamma Y \delta$ that contain Y in T_b, where $\gamma, \delta \in \Sigma^*$.

Clearly, the afore-mentioned notions are symmetric over T_f and T_b, namely:

Fact 2. *String X is right-maximal (resp. left-maximal) on T_f iff $\overline{X}$ is left-maximal (resp. right-maximal) on T_b. Also, X is maximal on T_f iff $\overline{X}$ is maximal on T_b.*

4 Indexing Forward/Backward Tries and Known Bounds

A compact tree for a set **S** of strings is a rooted tree such that (1) each edge is labeled by a non-empty substring of a string in **S**, (2) each internal node is branching, (3) the string labels of the out-going edges of each node begin with mutually distinct characters, and (4) there is a path from the root that spells out each string in **S**, which may end on an edge. Each edge of a compact tree is denoted by a triple (u, α, v) with $\alpha \in \Sigma^+$. We call internal nodes that are branching as *explicit nodes*, and we call loci that are on edges as *implicit nodes*. We will sometimes identify nodes with the substrings that the nodes represent.

In what follows, we will consider DAG or tree data structures built on a forward trie or backward trie. For any DAG or tree data structure D, let $|\mathsf{D}|_{\#Node}$ and $|\mathsf{D}|_{\#Edge}$ denote the numbers of nodes and edges in D, respectively.

4.1 Suffix Trees for Forward Tries

The *suffix tree* of a forward trie $\mathsf{T_f}$, denoted $\mathsf{STree}(\mathsf{T_f})$, is a compact tree which represents $\mathit{Suffix}(\mathsf{T_f})$. All non-root nodes in $\mathsf{STree}(\mathsf{T_f})$ represent right-maximal substrings on $\mathsf{T_f}$. Since now all internal nodes are branching, and since there are at most $|\mathit{Suffix}(\mathsf{T_f})|$ leaves, the numbers of nodes and edges in $\mathsf{STree}(\mathsf{T_f})$ are proportional to the number of suffixes in $\mathit{Suffix}(\mathsf{T_f})$. The following (folklore) quadratic bounds hold due to Fact 1-(b).

Theorem 1. *For any forward trie $\mathsf{T_f}$ with n nodes, $|\mathsf{STree}(\mathsf{T_f})|_{\#Node} = O(n^2)$ and $|\mathsf{STree}(\mathsf{T_f})|_{\#Edge} = O(n^2)$. These upper bounds hold for any alphabet. For some forward trie $\mathsf{T_f}$ with n nodes, $|\mathsf{STree}(\mathsf{T_f})|_{\#Node} = \Omega(n^2)$ and $|\mathsf{STree}(\mathsf{T_f})|_{\#Edge} = \Omega(n^2)$. These lower bounds hold for a constant-size or larger alphabet.*

4.2 Suffix Trees for Backward Tries

The *suffix tree* of a backward trie $\mathsf{T_b}$, denoted $\mathsf{STree}(\mathsf{T_b})$, is a compact tree which represents $\mathit{Suffix}(\mathsf{T_b})$. Since $\mathsf{STree}(\mathsf{T_b})$ contains at most $n-1$ leaves by Fact 1-(c) and all internal nodes of $\mathit{Suffix}(\mathsf{T_b})$ are branching, the following precise bounds follow from Fact 1-(c), which were implicit in the literature [7,19].

Theorem 2. *For any backward trie $\mathsf{T_b}$ with $n \geq 3$ nodes, $|\mathsf{STree}(\mathsf{T_b})|_{\#Node} \leq 2n-3$ and $|\mathsf{STree}(\mathsf{T_b})|_{\#Edge} \leq 2n-4$, independently of the alphabet size.*

The above bounds are tight since the theorem translates to the suffix tree with $2m-1$ nodes and $2m-2$ edges for a string of length m (e.g., $a^{m-1}b$), which can be represented as a path tree with $n = m+1$ nodes. By representing each edge label α by a pair $\langle v, u \rangle$ of nodes in $\mathsf{T_b}$ such that $\alpha = \mathit{str}_\mathsf{b}(u, v)$, $\mathsf{STree}(\mathsf{T_b})$ can be stored with $O(n)$ space.

Suffix Links and Weiner Links: For each explicit node aU of the suffix tree $\mathsf{STree}(\mathsf{T_b})$ of a backward trie $\mathsf{T_b}$ with $a \in \Sigma$ and $U \in \Sigma^*$, let $\mathit{slink}(aU) = U$. This is called the *suffix link* of node aU. For each explicit node V and $a \in \Sigma$, we also define the *reversed suffix link* $\mathcal{W}_a(V) = aVX$ where $X \in \Sigma^*$ is the shortest string such that aVX is an explicit node of $\mathsf{STree}(\mathsf{T_b})$. $\mathcal{W}_a(V)$ is undefined if $aV \notin \mathit{Substr}(\mathsf{T_b})$. These reversed suffix links are also called as *Weiner links* (or *W-link* in short) [8]. A W-link $\mathcal{W}_a(V) = aVX$ is said to be *hard* if $X = \varepsilon$, and *soft* if $X \in \Sigma^+$. The suffix links, hard and soft W-links of nodes in the suffix tree $\mathsf{STree}(\mathsf{T_f})$ of a forward trie $\mathsf{T_f}$ are defined analogously.

4.3 DAWGs for Forward Tries

The *directed acyclic word graph* (*DAWG*) of a forward trie $\mathsf{T_f}$ is a (partial) DFA that recognizes all substrings in $\mathit{Substr}(\mathsf{T_f})$. Hence, the label of every edge of $\mathsf{DAWG}(\mathsf{T_f})$ is a single character from Σ. $\mathsf{DAWG}(\mathsf{T_f})$ is formally defined as follows: For any substring X from $\mathit{Substr}(\mathsf{T_f})$, let $[X]_{E,\mathsf{f}}$ denote the equivalence class w.r.t. $\mathit{l\text{-}mxml}_\mathsf{f}(X)$. There is a one-to-one correspondence between the nodes of $\mathsf{DAWG}(\mathsf{T_f})$ and the equivalence classes $[\cdot]_{E,\mathsf{f}}$, and hence we will identify the nodes of $\mathsf{DAWG}(\mathsf{T_f})$ with their corresponding equivalence classes $[\cdot]_{E,\mathsf{f}}$. By the definition of equivalence classes, every member of $[X]_{E,\mathsf{f}}$ is a suffix of $\mathit{l\text{-}mxml}_\mathsf{f}(X)$. If X, Xa are substrings in $\mathit{Substr}(\mathsf{T_f})$ and $a \in \Sigma$, then there exists an edge labeled with character $a \in \Sigma$ from node $[X]_{E,\mathsf{f}}$ to node $[Xa]_{E,\mathsf{f}}$ in $\mathsf{DAWG}(\mathsf{T_f})$. This edge is called *primary* if $|\mathit{l\text{-}mxml}_\mathsf{f}(X)| + 1 = |\mathit{l\text{-}mxml}_\mathsf{f}(Xa)|$, and is called *secondary* otherwise. For each node $[X]_{E,\mathsf{f}}$ of $\mathsf{DAWG}(\mathsf{T_f})$ with $|X| \geq 1$, let $\mathit{slink}([X]_{E,\mathsf{f}}) = Z$, where Z is the longest suffix of $\mathit{l\text{-}mxml}_\mathsf{f}(X)$ not belonging to $[X]_{E,\mathsf{f}}$. This is the *suffix link* of this node $[X]_{E,\mathsf{f}}$.

Mohri et al. [22] introduced the *suffix automaton* for an acyclic DFA G, which is a small DFA that represents all suffixes of strings accepted by G. They considered equivalence relation $\equiv$ of substrings X and Y in an acyclic DFA G such that $X \equiv Y$ iff the following paths of the occurrences of X and Y in G are equal. Mohri et al.'s equivalence class is identical to our equivalence class $[X]_{E,\mathsf{f}}$ when $\mathsf{G} = \mathsf{T_f}$. To see why, recall that $\mathit{l\text{-}mxml}_\mathsf{f}(X) = \alpha X$ is the shortest substring of $\mathsf{T_f}$ such that αX is left-maximal, where $\alpha \in \Sigma^*$. Therefore, X is a suffix of $\mathit{l\text{-}mxml}_\mathsf{f}(X)$ and the following paths of the occurrences of X in $\mathsf{T_f}$ are identical to the following paths of the occurrences of $\mathit{l\text{-}mxml}_\mathsf{f}(X)$ in $\mathsf{T_f}$. Hence, in case where the input DFA G is in form of a forward trie $\mathsf{T_f}$ such that its leaves are the accepting states, then Mohri et al.'s suffix automaton is identical to our DAWG for $\mathsf{T_f}$. Mohri et al. [22] showed the following:

Theorem 3 (Corollary 2 of [22]). *For any forward trie* $\mathsf{T_f}$ *with* $n \geq 3$ *nodes,* $|\mathsf{DAWG}(\mathsf{T_f})|_{\#Node} \leq 2n - 3$*, independently of the alphabet size.*

We remark that Theorem 3 is immediate from Theorem 2 and Fact 2. This is because there is a one-to-one correspondence between the nodes of $\mathsf{DAWG}(\mathsf{T_f})$ and the nodes of $\mathsf{STree}(\mathsf{T_b})$, which means that $|\mathsf{DAWG}(\mathsf{T_f})|_{\#Node} = |\mathsf{STree}(\mathsf{T_b})|_{\#Node}$. Recall that the bound in Theorem 3 is only on the number of *nodes* in $\mathsf{DAWG}(\mathsf{T_f})$. We shall show later that the number of *edges* in $\mathsf{DAWG}(\mathsf{T_f})$ is $\Omega(\sigma n)$ in the worst case, which can be $\Omega(n^2)$ for a large alphabet.

4.4 DAWGs for Backward Tries

The DAWG of a backward trie $\mathsf{T_b}$, denoted $\mathsf{DAWG}(\mathsf{T_b})$, is a (partial) DFA that recognizes all strings in $\mathit{Substr}(\mathsf{T_b})$. The label of every edge of $\mathsf{DAWG}(\mathsf{T_b})$ is a single character from Σ. $\mathsf{DAWG}(\mathsf{T_b})$ is formally defined as follows: For any substring Y from $\mathit{Substr}(\mathsf{T_b})$, let $[Y]_{E,\mathsf{b}}$ denote the equivalence class w.r.t. $\mathit{l\text{-}mxml}_{\mathsf{b}}(Y)$. There is a one-to-one correspondence between the nodes of $\mathsf{DAWG}(\mathsf{T_b})$ and the equivalence classes $[\cdot]_{E,\mathsf{b}}$, and hence we will identify the nodes of $\mathsf{DAWG}(\mathsf{T_b})$ with their corresponding equivalence classes $[\cdot]_{E,\mathsf{b}}$. The notions of primary edges, secondary edges, and the suffix links of $\mathsf{DAWG}(\mathsf{T_b})$ are defined in similar manners to $\mathsf{DAWG}(\mathsf{T_f})$, but using the equivalence classes $[Y]_{E,\mathsf{b}}$ for substrings Y in the backward trie $\mathsf{T_b}$.

Symmetries Between Suffix Trees and DAWGs: The well-known *symmetry* between the suffix trees and the DAWGs (refer to [5,6,10]) also holds in our case of forward and backward tries. Namely, the suffix links of $\mathsf{DAWG}(\mathsf{T_f})$ (resp. $\mathsf{DAWG}(\mathsf{T_b})$) are the (reversed) edges of $\mathsf{STree}(\mathsf{T_b})$ (resp. $\mathsf{STree}(\mathsf{T_f})$). Also, the hard W-links of $\mathsf{STree}(\mathsf{T_f})$ (resp. $\mathsf{STree}(\mathsf{T_b})$) are the primary edges of $\mathsf{DAWG}(\mathsf{T_b})$ (resp. $\mathsf{DAWG}(\mathsf{T_f})$), and the soft W-links of $\mathsf{STree}(\mathsf{T_f})$ (resp. $\mathsf{STree}(\mathsf{T_b})$) are the secondary edges of $\mathsf{DAWG}(\mathsf{T_b})$ (resp. $\mathsf{DAWG}(\mathsf{T_f})$).

4.5 CDAWGs for Forward Tries

The *compact directed acyclic word graph* (*CDAWG*) of a forward trie $\mathsf{T_f}$, denoted $\mathsf{CDAWG}(\mathsf{T_f})$, is the edge-labeled DAG where the nodes correspond to the equivalence class of $\mathit{Substr}(\mathsf{T_f})$ w.r.t. $\mathit{mxml}_{\mathsf{f}}(\cdot)$. In other words, $\mathsf{CDAWG}(\mathsf{T_f})$ can be obtained by merging isomorphic subtrees of $\mathsf{STree}(\mathsf{T_f})$ rooted at internal nodes and merging leaves that are equivalent under $\mathit{mxml}_{\mathsf{f}}(\cdot)$, or by contracting non-branching paths of $\mathsf{DAWG}(\mathsf{T_f})$.

Theorem 4 ([17]). *For any forward trie* $\mathsf{T_f}$ *with* n *nodes over a* constant-size alphabet, $|\mathsf{CDAWG}(\mathsf{T_f})|_{\#Node} = O(n)$ *and* $|\mathsf{CDAWG}(\mathsf{T_f})|_{\#Edge} = O(n)$.

We emphasize that the above result by Inenaga et al. [17] states size bounds of $\mathsf{CDAWG}(\mathsf{T_f})$ only in the case where $\sigma = O(1)$. We will later show that this bound does not hold for the number of edges, in the case of a large alphabet.

4.6 CDAWGs for Backward Tries

The *compact directed acyclic word graph* (*CDAWG*) of a backward trie $\mathsf{T_b}$, denoted $\mathsf{CDAWG}(\mathsf{T_b})$, is the edge-labeled DAG where the nodes correspond to the equivalence class of $\mathit{Substr}(\mathsf{T_b})$ w.r.t. $\mathit{mxml}_{\mathsf{b}}(\cdot)$. Similarly to its forward trie counterpart, $\mathsf{CDAWG}(\mathsf{T_b})$ can be obtained by merging isomorphic subtrees of $\mathsf{STree}(\mathsf{T_b})$ rooted at internal nodes and merging leaves that are equivalent under $\mathit{mxml}_{\mathsf{f}}(\cdot)$, or by contracting non-branching paths of $\mathsf{DAWG}(\mathsf{T_b})$.

5 New Size Bounds on Indexing Forward/Backward Tries

To make the analysis simpler, we assume each of the roots, the one of $\mathsf{T_f}$ and the corresponding one of $\mathsf{T_b}$, is connected to an auxiliary node $\perp$ with an edge labeled by a unique character \$ that does not appear elsewhere in $\mathsf{T_f}$ or in $\mathsf{T_b}$.

5.1 Size Bounds for DAWGs for Forward/Backward Tries

Theorem 5. *For any backward trie $\mathsf{T_b}$ with n nodes, $|\mathsf{DAWG}(\mathsf{T_b})|_{\#Node} = O(n^2)$ and $|\mathsf{DAWG}(\mathsf{T_b})|_{\#Edge} = O(n^2)$. These upper bounds hold for any alphabet. For some backward trie $\mathsf{T_b}$ with n nodes, $|\mathsf{DAWG}(\mathsf{T_b})|_{\#Node} = \Omega(n^2)$ and $|\mathsf{DAWG}(\mathsf{T_b})|_{\#Edge} = \Omega(n^2)$. These lower bounds hold for a constant-size or larger alphabet.*

Theorem 6. *For any forward trie $\mathsf{T_f}$ with n nodes, $|\mathsf{DAWG}(\mathsf{T_f})|_{\#Edge} = O(\sigma n)$. For some forward trie $\mathsf{T_f}$ with n nodes, $|\mathsf{DAWG}(\mathsf{T_f})|_{\#Edge} = \Omega(\sigma n)$ which is $\Omega(n^2)$ for a large alphabet of size $\sigma = \Theta(n)$.*

Proof. Since each node of $\mathsf{DAWG}(\mathsf{T_f})$ can have at most σ out-going edges, the upper bound $|\mathsf{DAWG}(\mathsf{T_f})|_{\#Edge} = O(\sigma n)$ follows from Theorem 3.

To obtain the lower bound $|\mathsf{DAWG}(\mathsf{T_f})|_{\#Edge} = \Omega(\sigma n)$, we consider $\mathsf{T_f}$ which has a broom-like shape such that there is a single path of length $n - \sigma - 1$ from the root to a node v which has out-going edges with σ distinct characters $b_1, \ldots, b_\sigma$. Since the root of $\mathsf{T_f}$ is connected with the auxiliary node $\perp$ with an edge labeled \$, each root-to-leaf path in $\mathsf{T_f}$ represents $\$a^{n-\sigma+1}b_i$ for $1 \leq i \leq \sigma$. Now a^k for each $1 \leq k \leq n - \sigma - 2$ is left-maximal since it is immediately preceded by a and \$. Thus $\mathsf{DAWG}(\mathsf{T_f})$ has at least $n - \sigma - 2$ internal nodes, each representing a^k for $1 \leq k \leq n - \sigma - 2$. On the other hand, each $a^k \in \mathit{Substr}(\mathsf{T_f})$ is immediately followed by b_i with all $1 \leq i \leq \sigma$. Hence, $\mathsf{DAWG}(\mathsf{T_f})$ contains $\sigma(n - \sigma - 2) = \Omega(\sigma n)$ edges when $n - \sigma - 2 = \Omega(n)$. By choosing e.g. $\sigma \approx n/2$, we obtain $\mathsf{DAWG}(\mathsf{T_f})$ that contains $\Omega(n^2)$ edges. □

Mohri et al. (Proposition 4 of [22]) claimed that one can construct $\mathsf{DAWG}(\mathsf{T_f})$ in time proportional to its size. The following corollary is immediate from Theorem 6:

Corollary 1. *The DAWG construction algorithm of [22] applied to a forward trie with n nodes must take at least $\Omega(n^2)$ time in the worst case for an alphabet of size $\sigma = \Theta(n)$.*

5.2 Size Bounds for CDAWGs for Forward/Backward Tries

Theorem 7. *For any backward trie $\mathsf{T_b}$ with n nodes, $|\mathsf{CDAWG}(\mathsf{T_b})|_{\#Node} \leq 2n - 3$ and $|\mathsf{CDAWG}(\mathsf{T_b})|_{\#Edge} \leq 2n - 4$. These bounds are independent of the alphabet size.*

Proof. Since any maximal substring in $Substr(\mathsf{T_b})$ is right-maximal in $Substr(\mathsf{T_b})$, by Theorem 2 we have $|\mathsf{CDAWG}(\mathsf{T_b})|_{\#Node} \leq |\mathsf{STree}(\mathsf{T_b})|_{\#Node} \leq 2n-3$ and $|\mathsf{CDAWG}(\mathsf{T_b})|_{\#Edge} \leq |\mathsf{STree}(\mathsf{T_b})|_{\#Edge} \leq 2n-4$. □

The bounds in Theorem 7 are tight: Consider an alphabet $\{a_1, \ldots, a_{\lceil \log_2 n \rceil}, b_1, \ldots, b_{\lceil \log_2 n \rceil}, \$\}$ of size $2\lceil \log_2 n \rceil + 1$ and a binary backward trie $\mathsf{T_b}$ with n nodes where the binary edges at each depth $d \geq 2$ are labeled by the sub-alphabet $\{a_d, b_d\}$ of size 2. Because every suffix $S \in Suffix(\mathsf{T_b})$ is maximal in $\mathsf{T_b}$, $\mathsf{CDAWG}(\mathsf{T_b})$ for this $\mathsf{T_b}$ contains $n-1$ sinks. Also, since for each suffix S in $\mathsf{T_b}$ there is a unique suffix $S' \neq S$ that shares the longest common prefix with S, $\mathsf{CDAWG}(\mathsf{T_b})$ for this $\mathsf{T_b}$ contains $n-2$ internal nodes (including the source). This also means $\mathsf{CDAWG}(\mathsf{T_b})$ is identical to $\mathsf{STree}(\mathsf{T_b})$ for this backward trie $\mathsf{T_b}$.

Theorem 8. *For any forward trie $\mathsf{T_f}$ with n nodes, $|\mathsf{CDAWG}(\mathsf{T_f})|_{\#Node} \leq 2n-3$ and $|\mathsf{CDAWG}(\mathsf{T_f})|_{\#Edge} = O(\sigma n)$. For some forward trie $\mathsf{T_f}$ with n nodes, $|\mathsf{CDAWG}(\mathsf{T_f})|_{\#Edge} = \Omega(\sigma n)$ which is $\Omega(n^2)$ for a large alphabet of size $\sigma = \Theta(n)$.*

Proof. It immediately follows from Fact 1-(a), Fact 2, and Theorem 7 that $|\mathsf{CDAWG}(\mathsf{T_f})|_{\#Node} = |\mathsf{CDAWG}(\mathsf{T_b})|_{\#Node} \leq 2n-3$. Since a node in $\mathsf{CDAWG}(\mathsf{T_f})$ can have at most σ out-going edges, the upper bound $|\mathsf{CDAWG}(\mathsf{T_f})|_{\#Edge} = O(\sigma n)$ of the number of edges trivially holds. To obtain the lower bound, we consider the same broom-like forward trie $\mathsf{T_f}$ as in Theorem 6. In this $\mathsf{T_f}$, a^k for each $1 \leq k \leq n-\sigma-2$ is maximal and thus $\mathsf{CDAWG}(\mathsf{T_f})$ has at least $n-\sigma-2$ internal nodes each representing a^k for $1 \leq k \leq n-\sigma-2$. By the same argument to Theorem 6, $\mathsf{CDAWG}(\mathsf{T_f})$ for this $\mathsf{T_f}$ contains at least $\sigma(n-\sigma-2) = \Omega(\sigma n)$ edges, which accounts to $\Omega(n^2)$ for a large alphabet of size e.g. $\sigma \approx n/2$. □

The upper bound of Theorem 8 generalizes the bound of Theorem 4 for constant-size alphabets. Remark that $\mathsf{CDAWG}(\mathsf{T_f})$ for the broom-like $\mathsf{T_f}$ is almost identical to $\mathsf{DAWG}(\mathsf{T_f})$, except for the unary path $\$a$ that is compacted in $\mathsf{CDAWG}(\mathsf{T_f})$.

6 Constructing $O(n)$-size Representation of $\mathsf{DAWG}(\mathsf{T_f})$ in $O(n)$ Time

We have seen that $\mathsf{DAWG}(\mathsf{T_f})$ for any forward trie $\mathsf{T_f}$ with n nodes contains only $O(n)$ nodes, but can have $\Omega(\sigma n)$ edges for some $\mathsf{T_f}$ over an alphabet of size σ ranging from $\Theta(1)$ to $\Theta(n)$. Thus some $\mathsf{DAWG}(\mathsf{T_f})$ can have $\Theta(n^2)$ edges for $\sigma = \Theta(n)$ (Theorem 3 and Theorem 6). Hence, in general it is impossible to build an *explicit* representation of $\mathsf{DAWG}(\mathsf{T_f})$ within linear $O(n)$-space. By an explicit representation we mean an implementation of $\mathsf{DAWG}(\mathsf{T_f})$ where each edge is represented by a pointer between two nodes.

We show that there exists an $O(n)$-space *implicit* representation of $\mathsf{DAWG}(\mathsf{T_f})$ for any alphabet of size σ ranging from $\Theta(1)$ to $\Theta(n)$, that allows us $O(\log \sigma)$-time access to each edge of $\mathsf{DAWG}(\mathsf{T_f})$. This is trivial in case $\sigma = O(1)$, and hence in what follows we consider an alphabet of size σ such that σ ranges from $\omega(1)$

to $\Theta(n)$. Also, we suppose that our alphabet is an integer alphabet $\Sigma = [1..\sigma]$ of size σ. Then, we show that such an implicit representation of $\mathsf{DAWG}(\mathsf{T_f})$ can be built in $O(n)$ time and working space.

Based on the property stated in Sect. 4, constructing $\mathsf{DAWG}(\mathsf{T_f})$ reduces to maintaining hard and soft W-links over $\mathsf{STree}(\mathsf{T_b})$. Our data structure explicitly stores all $O(n)$ hard W-links, while it only stores carefully selected $O(n)$ soft W-links. The other soft W-links can be simulated by these explicitly stored W-links, in $O(\log \sigma)$ time each.

Our algorithm is built upon the following facts which are adapted from [12]:

Fact 3. *Let a be any character from Σ.*

(a) If there is a (hard or soft) W-link $\mathcal{W}_a(V)$ for a node V in $\mathsf{STree}(\mathsf{T_b})$, then there is a (hard or soft) W-link $\mathcal{W}_a(U)$ for any ancestor U of V in $\mathsf{STree}(\mathsf{T_b})$.
(b) If two nodes U and V have hard W-links $\mathcal{W}_a(U)$ and $\mathcal{W}_a(V)$, then the LCA Z of U and V also has a hard W-link $\mathcal{W}_a(Z)$.

In the following statements (c), (d), and (e), let V be any node of $\mathsf{STree}(\mathsf{T_b})$ *such that V has a* soft *W-link $\mathcal{W}_a(V)$ for $a \in \Sigma$.*

(c) There is a descendant U of V s.t. $U \neq V$ and U has a hard W-link $\mathcal{W}_a(V)$.
(d) The highest descendant of V that has a hard W-link for character a is unique. This fact follows from (b).
(e) Let U be the unique highest descendant of V that has a hard W-link $\mathcal{W}_a(U)$. For every node Z in the path from V to U, $\mathcal{W}_a(Z) = \mathcal{W}_a(U)$, i.e. the W-links of all nodes in this path for character a point to the same node in $\mathsf{STree}(\mathsf{T_b})$.

We construct a micro-macro tree decomposition [1] of $\mathsf{STree}(\mathsf{T_b})$ in a similar manner to [14], such that the nodes of $\mathsf{STree}(\mathsf{T_b})$ are partitioned into $O(n/\sigma)$ connected components (called *micro-trees*), each of which contains $O(\sigma)$ nodes. Such a decomposition always exists and can be computed in $O(n)$ time. The *macro tree* is the induced tree from the roots of the micro trees, and thus the macro tree contains $O(n/\sigma)$ nodes.

In every node V of the macro tree, we explicitly store all soft and hard W-links from V. Since there can be at most σ W-links from V, this requires $O(n)$ total space for all nodes in the macro tree. Let mt denote any micro tree. We compute the ranks of all nodes in a pre-order traversal in mt. Let $a \in \Sigma$ be any character such that there is a node V in mt that has a hard W-link $\mathcal{W}_a(V)$. Let $\mathsf{P}_a^{\mathsf{mt}}$ denote an array that stores a sorted list of pre-order ranks of nodes V in mt that have hard W-links for character a. Hence the size of $\mathsf{P}_a^{\mathsf{mt}}$ is equal to the number of nodes in mt that have hard W-links for character a. For all such characters a, we store $\mathsf{P}_a^{\mathsf{mt}}$ in mt. The total size of these arrays for all the micro trees is $O(n)$.

Let $a \in \Sigma$ be any character, and V any node in $\mathsf{STree}(\mathsf{T_b})$ which does not have a hard W-link for a. We wish to know if V has a soft W-link for a, and if so, we want to retrieve the target node of this link. Let mt denote the micro-tree that V belongs to. Consider the case where V is not the root R of mt, since otherwise $\mathcal{W}_a(V)$ is explicitly stored. If $\mathcal{W}_a(R)$ is nil, then by Fact 3-(a) no nodes in the micro tree has W-links for character a. Otherwise (if $\mathcal{W}_a(R)$ exists), then we can find $\mathcal{W}_a(W)$ as follows:

(A) If the predecessor P of V exists in $\mathsf{P}_a^{\mathsf{mt}}$ and P is an ancestor of V, then we follow the hard W-link $\mathcal{W}_a(P)$ from P. Let $Q = \mathcal{W}_a(P)$, and c be the first character in the path from P to V.
   (i) If Q has an out-going edge whose label begins with c, the child of Q below this edge is the destination of the soft W-link $\mathcal{W}_a(V)$ from V for a.
   (ii) Otherwise, then there is no W-link from V for a.
(B) Otherwise, $\mathcal{W}_a(R)$ from the root R of mt is a soft W-link, which is explicitly stored. We follow it and let $U = \mathcal{W}_a(R)$.
   (i) If $Z = \mathit{slink}(U)$ is a descendant of V, then U is the destination of the soft W-link $\mathcal{W}_a(V)$ from V for a.
   (ii) Otherwise, then there is no W-link from V for a.

The correctness of this algorithm follows from Fact 3-(e). Since each micro-tree contains $O(\sigma)$ nodes, the size of $\mathsf{P}_a^{\mathsf{mt}}$ is $O(\sigma)$ and thus the predecessor P of V in $\mathsf{P}_a^{\mathsf{mt}}$ can be found in $O(\log \sigma)$ time by binary search. We can check if one node is an ancestor of the other node (or vice versa) in $O(1)$ time, after standard $O(n)$-time preprocessing over the whole suffix tree. Hence, this algorithm simulates soft W-link $\mathcal{W}_a(V)$ in $O(\log \sigma)$ time.

Lemma 1. *Given a backward trie* $\mathsf{T_b}$ *with* n *nodes, we can compute* $\mathsf{STree}(\mathsf{T_b})$ *with all hard W-links in* $O(n)$ *time and space.*

Lemma 2. *We can compute, in* $O(n)$ *time and space, all W-links of the macro tree nodes and the arrays* $\mathsf{P}_a^{\mathsf{mt}}$ *for all the micro trees* mt *and characters* $a \in \Sigma$.

Proof. We perform a pre-order traversal on each micro tree mt. At each node V visited during the traversal, we append the pre-order rank of V to array $\mathsf{P}_a^{\mathsf{mt}}$ iff V has a hard W-link $\mathcal{W}_a(V)$ for character a. Since the size of mt is $O(\sigma)$ and since we have assumed an integer alphabet $[1..\sigma]$, we can compute $\mathsf{P}_a^{\mathsf{mt}}$ for all characters a in $O(\sigma)$ time. It takes $O(\frac{n}{\sigma} \cdot \sigma) = O(n)$ time for all micro trees.

The preprocessing for the macro tree consists of two steps. Firstly, we need to compute soft W-links from the macro tree nodes (recall that we have already computed hard W-links from the macro tree nodes by Lemma 1). For this sake, in the above preprocessing for micro trees, we additionally pre-compute the successor of the root R of each micro tree mt in each non-empty array $\mathsf{P}_a^{\mathsf{mt}}$. By Fact 3-(d), this successor corresponds to the unique descendant of R that has a hard W-link for character a. As above, this preprocessing also takes $O(\sigma)$ time for each micro tree, resulting in $O(n)$ total time. Secondly, we perform a bottom-up traversal on the macro tree. Our basic strategy is to "propagate" the soft W-links in a bottom up fashion from lower nodes to upper nodes in the macro tree (recall that these macro tree nodes are the roots of micro trees). In so doing, we first compute the soft W-links of the macro tree leaves. By Fact 3-(c) and -(e), this can be done in $O(\sigma)$ time for each leaf using the successors computed above. Then we propagate the soft W-links to the macro tree internal nodes. The existence of soft W-links of internal nodes computed in this way is justified by Fact 3-(a), however, the destinations of some soft W-links of some macro tree internal nodes may not be correct. This can happen when the corresponding micro trees contain hard W-links (due to Fact 3-(e)). These destinations can be modified by using the successors of the roots computed in the first step, again due to Fact 3-(e). Both of our propagation and modification steps take $O(\sigma)$ time for each macro tree node of size $O(\sigma)$, and hence, it takes a total of $O(n)$ time. □

Theorem 9. *Given a forward trie* $\mathsf{T_f}$ *of size* n *over an integer alphabet* $\Sigma = [1..\sigma]$ *with* $\sigma = O(n)$*, we can construct an* $O(n)$*-space representation of* $\mathsf{DAWG}(\mathsf{T_f})$ *in* $O(n)$ *time and working space.*

References

1. Alstrup, S., Secher, J.P., Spork, M.: Optimal on-line decremental connectivity in trees. Inf. Process. Lett. **64**(4), 161–164 (1997)
2. Apostolico, A., Lonardi, S.: Off-line compression by greedy textual substitution. Proc. IEEE **88**(11), 1733–1744 (2000)
3. Belazzougui, D., Cunial, F.: Fast label extraction in the CDAWG. In: Fici, G., Sciortino, M., Venturini, R. (eds.) SPIRE 2017. LNCS, vol. 10508, pp. 161–175. Springer, Cham (2017). https://doi.org/10.1007/978-3-319-67428-5_14
4. Bender, M.A., Farach-Colton, M.: The level ancestor problem simplified. Theor. Comput. Sci. **321**(1), 5–12 (2004)
5. Blumer, A., et al.: The smallest automaton recognizing the subwords of a text. Theor. Comput. Sci. **40**, 31–55 (1985)
6. Blumer, A., Blumer, J., Haussler, D., Mcconnell, R., Ehrenfeucht, A.: Complete inverted files for efficient text retrieval and analysis. J. ACM **34**(3), 578–595 (1987)
7. Breslauer, D.: The suffix tree of a tree and minimizing sequential transducers. Theor. Comput. Sci. **191**(1–2), 131–144 (1998)
8. Breslauer, D., Italiano, G.F.: Near real-time suffix tree construction via the fringe marked ancestor problem. J. Discrete Algorithms **18**, 32–48 (2013)
9. Crochemore, M.: Transducers and repetitions. Theor. Comput. Sci. **45**(1), 63–86 (1986)
10. Crochemore, M., Rytter, W.: Text Algorithms. Oxford University Press, Cambridge (1994)
11. Ferragina, P., Luccio, F., Manzini, G., Muthukrishnan, S.: Compressing and indexing labeled trees, with applications. J. ACM **57**(1), 4:1–4:33 (2009)
12. Fischer, J., Gawrychowski, P.: Alphabet-dependent string searching with wexponential search trees. In: Proceedings of the 26th Annual Symposium on Combinatorial Pattern Matching, CPM, pp. 160–171 (2015). Full version: https://arxiv.org/abs/1302.3347
13. Fujishige, Y., Tsujimaru, Y., Inenaga, S., Bannai, H., Takeda, M.: Computing DAWGs and minimal absent words in linear time for integer alphabets. In: Proceedings of the 41st International Symposium on Mathematical Foundations of Computer Science, MFCS, pp. 38:1–38:14 (2016)
14. Gawrychowski, P.: Simple and efficient LZW-compressed multiple pattern matching. J. Discrete Algorithms **25**, 34–41 (2014)

15. Gusfield, D.: Algorithms on Strings, Trees, and Sequences. Cambridge University Press (1997)
16. Inenaga, S.: Suffix trees, DAWGs and CDAWGs for forward and backward tries. arXiv e-prints p. 1904.04513 (2019). http://arxiv.org/abs/1904.04513
17. Inenaga, S., Hoshino, H., Shinohara, A., Takeda, M., Arikawa, S.: Construction of the CDAWG for a trie. In: Proceedings of the Prague Stringology Conference, PSC 2001, pp. 37–48 (2001)
18. Kimura, D., Kashima, H.: Fast computation of subpath kernel for trees. In: Proceedings of the 29th International Conference on Machine Learning, ICML (2012)
19. Kosaraju, S.R.: Efficient tree pattern matching (preliminary version). In: Proceedings of the 30th Annual Symposium on Foundations of Computer Science, FOCS, pp. 178–183 (1989)
20. Kucherov, G., Rusinowitch, M.: Matching a set of strings with variable length don't cares. Theor. Comput. Sci. **178**(1–2), 129–154 (1997)
21. Mäkinen, V., Belazzougui, D., Cunial, F., Tomescu, A.I.: Genome-Scale Algorithm Design: Biological Sequence Analysis in the Era of High-Throughput Sequencing. Cambridge University Press, Cambridge (2015)
22. Mohri, M., Moreno, P.J., Weinstein, E.: General suffix automaton construction algorithm and space bounds. Theor. Comput. Sci. **410**(37), 3553–3562 (2009)
23. Muthukrishnan, S.: Efficient algorithms for document retrieval problems. In: Proceedings of the 13th Annual ACM-SIAM Symposium on Discrete Algorithms, SODA, pp. 657–666 (2002)
24. Nakashima, Y., I, T., Inenaga, S., Bannai, H., Takeda, M.: The position heap of a Trie. In: Calderón-Benavides, L., González-Caro, C., Chávez, E., Ziviani, N. (eds.) SPIRE 2012. LNCS, vol. 7608, pp. 360–371. Springer, Heidelberg (2012). https://doi.org/10.1007/978-3-642-34109-0_38
25. Shibuya, T.: Constructing the suffix tree of a tree with a large alphabet. IEICE Trans. **E86-A(5)**, 1061–1066 (2003)
26. Weiner, P.: Linear pattern-matching algorithms. In: Proceedings of the 14th IEEE Annual Symposium on Switching and Automata Theory, pp. 1–11 (1973)
27. Yamamoto, J., Tomohiro, I., Bannai, H., Inenaga, S., Takeda, M.: Faster compact on-line Lempel-Ziv factorization. In: Proceedings of the 31st International Symposium on Theoretical Aspects of Computer Science, STACS, pp. 675–686 (2014)

Towards a Definitive Measure of Repetitiveness

Tomasz Kociumaka[1], Gonzalo Navarro[2,3], and Nicola Prezza[4(✉)]

[1] Department of Computer Science, Bar-Ilan University, Ramat Gan, Israel
kociumaka@mimuw.edu.pl

[2] Millennium Institute for Foundational Research on Data (IMFD), Santiago, Chile

[3] Department of Computer Science, University of Chile, Santiago, Chile

[4] Department of Business and Management, Luiss Guido Carli, Rome, Italy
nprezza@luiss.it

Abstract. Unlike in statistical compression, where Shannon's entropy is a definitive lower bound, no such clear measure exists for the compressibility of repetitive sequences. Since statistical entropy does not capture repetitiveness, ad-hoc measures like the size z of the Lempel–Ziv parse are frequently used to estimate repetitiveness. Recently, a more principled measure, the size γ of the smallest string *attractor*, was introduced. The measure γ lower bounds all the previous relevant ones (including z), yet length-n strings can be represented and efficiently indexed within space $O(\gamma \log \frac{n}{\gamma})$, which also upper bounds most measures (including z). While γ is certainly a better measure of repetitiveness than z, it is NP-complete to compute, and no $o(\gamma \log n)$-space representation of strings is known. In this paper, we study a smaller measure, $\delta \leq \gamma$, which can be computed in linear time. We show that δ better captures the compressibility of repetitive strings. For every length n and every value $\delta \geq 2$, we construct a string such that $\gamma = \Omega(\delta \log \frac{n}{\delta})$. Still, we show a representation of any string S in $O(\delta \log \frac{n}{\delta})$ space that supports direct access to any character $S[i]$ in time $O(\log \frac{n}{\delta})$ and finds the *occ* occurrences of any pattern $P[1..m]$ in time $O(m \log n + occ \log^{\varepsilon} n)$ for any constant $\varepsilon > 0$. Further, we prove that no $o(\delta \log n)$-space representation exists: for every length n and every value $2 \leq \delta \leq n^{1-\varepsilon}$, we exhibit a string family whose elements can only be encoded in $\Omega(\delta \log \frac{n}{\delta})$ space. We complete our characterization of δ by showing that, although γ, z, and other repetitiveness measures are always $O(\delta \log \frac{n}{\delta})$, for strings of any length n, the smallest context-free grammar can be of size $\Omega(\delta \log^2 n / \log \log n)$. No such separation is known for γ.

Part of this work was carried out during the Dagstuhl Seminar 19241, "25 Years of the Burrows–Wheeler Transform".

T. Kociumaka—Supported by ISF grants no. 1278/16, 824/17, and 1926/19, a BSF grant no. 2018364, and an ERC grant MPM (no. 683064) under the EU's Horizon 2020 Research and Innovation Programme.

G. Navarro—Supported in part by Fondecyt grant 1-170048, Chile; Millennium Institute for Foundational Research on Data (IMFD), Chile.

Y. Kohayakawa and F. K. Miyazawa (Eds.): LATIN 2020, LNCS 12118, pp. 207–219, 2020.
https://doi.org/10.1007/978-3-030-61792-9_17

Keywords: Data compression · Lempel–Ziv parse · Repetitive sequences

1 Introduction

The recent rise in the amount of data we aim to handle [41] is driving research into compressed data representations that can be used directly in compressed form [32]. Interestingly, much of today's fastest-growing data is highly repetitive, which enables space reductions of orders of magnitude [19]: genome collections, versioned text and software repositories, periodic sky surveys, and other sources produce data where each element in the collection is very similar to others.

Since a significant fraction of the data of interest consists of sequences, text indexes are important actors in this research. These are data structures that offer fast pattern matching (and possibly other more sophisticated capabilities) over a collection of strings. Though compressed text indexes are already mature [33] and offer fast pattern searching within space close to the statistical entropy of the string collection, such kind of entropy is unable to capture repetitiveness [28,32]. Achieving orders-of-magnitude space reductions requires instead to resort to other kinds of compressors, such as Lempel–Ziv [29], grammar compression [26], run-length compressed Burrows–Wheeler transform [19], and others. Various compressed indexes build on those methods; see a thorough review [19].

Unlike statistical compression, where Shannon's notion of entropy [40] is a clear lower bound to what compressors can achieve, a similar notion capturing repetitiveness has been elusive. Beyond Kolmogorov's complexity [27], which is uncomputable, repetitiveness is measured in ad-hoc terms, as the results of what specific compressors achieve. A list of such measures on a string $S[1..n]$ follows:

Lempel–Ziv compression [29] parses S into a sequence of *phrases*, with each phrase defined as the longest string that has appeared previously in S. The associated measure is the number z of phrases produced. The measure can be computed in $O(n)$ time [38].

Bidirectional macro schemes [42] extend Lempel–Ziv so that the source of each phrase may precede or follow it, as long as no circular dependencies are introduced. The associated measure b is the number of phrases of the smallest parsing. It holds $b \le z = O(b \log \frac{n}{b})$ [18], but computing b is NP-complete [20].

Grammar-based compression [26] builds a context-free grammar that generates S and only S. The associated measure is the size g of the smallest grammar (i.e., the total length of the right-hand sides of the rules). It holds $z \le g = O(z \log \frac{n}{z})$ and, while it is NP-complete to compute g, grammars of size $O(z \log \frac{n}{z})$ can be constructed in linear time [11,21,39].

Run-length grammar compression [35] allows in addition rules $A \rightarrow B^t$ (t repetitions of B) of constant size. The measure is the size g_{rl} of the smallest run-length grammar, and it holds $\frac{z}{2} \le g_{rl} \le g$ and $g_{rl} = O(b \log \frac{n}{b})$ [18].

Collage systems [25] extend run-length grammars by allowing truncation: in constant space we can refer to a prefix or a suffix of another nonterminal. The associated measure c satisfies $c \le g_{rl}$ and $c = O(z)$ [31].

Burrows–Wheeler transform (BWT) [10] is a permutation of S that tends to have long runs of equal letters if S is repetitive. The number r of maximal equal-letter runs in the BWT can be found in linear time. It is known that $g_{rl} = O(r \log \frac{n}{r})$ [19] and $\frac{b}{2} \le r = O(b \log^2 n) = O(z \log^2 n)$ [18,23].

CDAWGs [9] are automata that recognize every substring of S. The associated measure of repetitiveness is e, the size of the smallest such automaton (compressed by dissolving states of in-degree and out-degree one), which is built in linear time [9]. The measure e is always larger than r, g, and z [3,4].

An improvement to this situation is the recent introduction of the concept of string *attractor* [24]. An attractor Γ is a set of positions in S such that any substring of S has an occurrence covering a position in Γ. The size γ of the smallest attractor asymptotically lower bounds all the repetitiveness measures listed above. Recent results [13,24,34,36] show that efficient queries can be supported within $O(\gamma \log \frac{n}{\gamma})$ space[1] and that $g_{rl} = O(\gamma \log \frac{n}{\gamma})$. Previous solutions support random access to S, or indexed searches on S, within space $O(z \log \frac{n}{z})$ [5,6,12,17], $O(g)$ [1,8,14–16], $O(g_{rl})$ [19], $O(r)$ or $O(r \log \frac{n}{r})$ [4,19,30], and $O(e)$ [2,3], none improving in general upon the space $O(\gamma \log \frac{n}{\gamma})$ within which one can offer efficient access [24] and indexing [13,34]. Using indexes based on γ is not exempt of problems, however. Computing γ is NP-hard [24], and therefore one has to resort to approximations like z, in which case the representation is only guaranteed to be of size $O(z \log \frac{n}{z})$. While this problem has been recently sidestepped [13], it is still unclear whether γ is the definitive measure of repetitiveness. In particular, it is unknown whether one can always represent S within $O(\gamma)$ space (while this is possible in $O(b)$ space) or even within $o(\gamma \log n)$ space.

Our Contributions. In this paper, we study a new measure of repetitiveness, δ, which arguably captures better the concept of compressibility in repetitive strings and is more convenient to deal with. Although this measure was already introduced in a stringology context [37] and used to build indexes of size $O(\gamma \log \frac{n}{\gamma})$ without knowing γ [13], its properties and full potential have not been explored. It always holds that $\delta \le \gamma$, and δ can be computed in $O(n)$ time [13]. First, we show that δ can be asymptotically strictly smaller than γ: for every length n and every value $\delta \ge 2$, there exist a string such that $\gamma = \Omega(\delta \log \frac{n}{\delta})$. Still, we develop a representation of S of size $O(\delta \log \frac{n}{\delta})$ that allows accessing any character $S[i]$ in time $O(\log \frac{n}{\delta})$ and finds the *occ* occurrences of any pattern $P[1..m]$ in time $O(m \log n + occ \log^{\varepsilon} n)$ for any constant $\varepsilon > 0$. For this, we reduce the size of block trees [5] to $O(\delta \log \frac{n}{\delta})$. Therefore, we obtain improved space and the same time performance compared to previous results based on γ [24,34,36].[2] Further, we show that, for every length n and every value $2 \le \delta \le n^{1-\varepsilon}$ (where $\varepsilon > 0$ is an arbitrary constant), there exists a string family whose elements can only be represented in $\Omega(\delta \log \frac{n}{\delta})$ space. Thus, $o(\delta \log n)$ space is unreachable in general; no such limit is known for γ. We complete our characterization of

[1] Throughout the paper, the size of data structures is measured in machine words.

[2] The most recent index [13] locates patterns in $O(m + (occ + 1) \log^{\epsilon} n)$ time and $O(\gamma \log \frac{n}{\gamma})$ space (being thus faster but still using more space).

δ by proving that, although γ, b, z, and c are always $O(\delta \log \frac{n}{\delta})$, the smallest context-free grammar can be of size $g = \Omega(\delta \log^2 n / \log \log n)$ for strings of any length n. Again, no such lower bound is known to hold on γ.

2 Measure δ

The measure δ has recently been defined by Christiansen et al. [13, Section 5.1], though it is based on the expression $d_k(S)/k$, introduced by Raskhodnikova et al. [37] to approximate z. Below we summarize what is known about it.

Definition 1. *Let $d_k(S)$ be the number of distinct length-k substrings in S. Then*

$$\delta = \max\{d_k(S)/k : k \in [1..n]\}.$$

Lemma 1 (Based on [[37], Lemma 3]). *It always holds that $z = O(\delta \log \frac{n}{\delta})$.*

Proof. Raskhodnikova et al. [37] prove that if $d_\ell(S) \leq m \cdot \ell$ for every $\ell \leq \ell_0$, then $z \leq 4(m \log \ell_0 + \frac{n}{\ell_0})$. Plugging $\ell_0 = \frac{n}{\delta}$ and $m = \delta$, we conclude that $z \leq 4(\delta \log \frac{n}{\delta} + \delta) = O(\delta \log \frac{n}{\delta})$. □

Since b, c, and γ are $O(z)$, these three measures are all upper bounded by $O(\delta \log \frac{n}{\delta})$. Additionally, we conclude that $g_{rl} \leq g = O(z \log \frac{n}{z}) = O(\delta \log^2 \frac{n}{\delta})$, and note that $r = O(\delta \log^2 n)$ has been proved recently [23].

Before we proceed, let us recall the concept of an attractor.

Definition 2 (Kempa and Prezza [24]). *An* attractor *of a string $S[1..n]$ is a set of positions $\Gamma \subseteq [1..n]$ such that every substring $S[i..j]$ has at least one occurrence $S[i'..j'] = S[i..j]$ that covers an attractor position $p \in \Gamma \cap [i'..j']$.*

Lemma 2 ([[13], Lemma 5.6]). *Every string S satisfies $\delta \leq \gamma$.*

Proof. Every length-k substring has an occurrence covering an attractor position, so there can be at most $k\gamma$ distinct substrings, i.e., $d_k(S)/k \leq \gamma$ for all $k \leq n$. □

Lemma 3. ([[13], Lemma 5.7]). *The measure δ can be computed in $O(n)$ time and space given $S[1..n]$.*

Proof One can use the suffix tree or the LCP table of S to retrieve $d_k(S)$ for all $k \in [1..n]$ in $O(n)$ time, and then compute δ from this information. □

3 Lower Bounds in Terms of δ

In this section, we prove lower bounds in terms of the measure δ. First, we show that there exist string families where $\delta = o(\gamma)$; in fact, δ can be smaller by up to a logarithmic factor. Second, we prove that there are string families that cannot be encoded in $o(\delta \log n)$ space: for every length n and every value $2 \leq \delta \leq n^{1-\varepsilon}$ (where $\varepsilon > 0$ is an arbitrary constant), there is a string family whose elements require $\Omega(\delta \log \frac{n}{\delta})$ space to represent. Third, although in the next section we give an $O(\delta \log \frac{n}{\delta})$-space representation, below we construct a family of strings which cannot be represented using context-free grammars of size $O(\delta \log \frac{n}{\delta})$; a nearly logarithmic-factor separation exists.

3.1 Lower Bounds on Attractors

Consider an infinite string $S_\infty[1..]$, where $S_\infty[i] = \mathtt{b}$ if $i = 2^j$ for some integer $j \geq 0$, and $S_\infty[i] = \mathtt{a}$ otherwise. For $n \geq 1$, let S_n be the length-n prefix of S. We shall prove that the strings in this family satisfy $\delta = O(1)$ and $\gamma = \Omega(\log n)$.

Lemma 4. *For every $n \geq 1$, the string S_n satisfies $\delta \leq 2$ and $\gamma \geq \frac{1}{2}\lfloor \log n \rfloor$.*

Proof. For each $j \geq 1$, every pair of consecutive $\mathtt{b}$s in $S_\infty[2^{j-1}+1..]$ is at distance at least 2^j. Therefore, the only distinct substrings of length $k \leq 2^j$ in $S_\infty[2^{j-1}+1..]$ are of the form $\mathtt{a}^k$ or $\mathtt{a}^i\mathtt{b}\mathtt{a}^{k-i-1}$ for $i \in [0..k-1]$. Hence, the distinct length-k substrings of S_∞ are those starting up to position 2^{j-1}, $S_\infty[i..i+k-1]$ for $i \in [1..2^{j-1}]$, and the $k+1$ already mentioned strings, for a total of $d_k(S_\infty) \leq 2^{j-1} + k + 1$. Plugging $j = \lceil \log k \rceil$, we get $d_k(S_\infty) \leq 2^{\lceil \log k \rceil - 1} + k + 1 \leq 2^{\log k} + k \leq 2k$, concluding that $\delta(S_n) \leq 2$ holds for every n.

Next, observe that for each $j \geq 0$, the substring $\mathtt{b}\mathtt{a}^{2^j-1}\mathtt{b}$ has its unique occurrence in S_∞ at $S_\infty[2^j..2^{j+1}]$. The covered regions are disjoint across *even* integers j, so each one requires a distinct attractor element. Consequently, $\gamma(S_n) \geq \frac{j}{2}$ for $n \geq 2^j$. Plugging $j = \lfloor \log n \rfloor$, we get $\gamma(S_n) \geq \frac{1}{2}\lfloor \log n \rfloor$. □

We can also show that there are strings with $\delta = o(\gamma)$ as long as $2 \leq \delta \leq o(n)$.

Theorem 1. *For every length n and value $\delta \in [2..n]$, there is a string $S[1..n]$ with $\gamma = \Omega(\delta \log \frac{n}{\delta})$.*

Proof. Let us first fix an integer $m \geq 1$ such that $n \geq 4m - 1$ and decompose $n - m + 1 = \sum_{i=1}^{m} n_i$ roughly equally (so that $n_i \geq 3$ and $n_i = \Omega(\frac{n}{m})$). We shall build a string S over an alphabet consisting of $3m-1$ characters: $\mathtt{a}_i$ and $\mathtt{b}_i$ for $i \in [1..m]$ and $\$_i$ for $i \in [1..m-1]$. For this, we take $S^{(i)}$ to be the string S_{n_i} built for Lemma 4, with alphabet $\{\mathtt{a}, \mathtt{b}\}$ replaced by $\{\mathtt{a}_i, \mathtt{b}_i\}$, and we define S to be the concatenation of the strings $S^{(i)}$ interleaved with sentinels $\$_i$.

Notice that, for each k, we have $d_k(S) \leq (m-1)k + \sum_{i=1}^{m} d_k(S^{(i)})$ because every substring contains $\$_i$ or is contained in $S^{(i)}$ for some i. Hence, $\delta(S) \leq 3m - 1$. (In fact, $\delta(S) = 3m - 1$ because $d_1(S) = 3m - 1$.) Furthermore, $\gamma(S) \geq \sum_{i=1}^{m} \gamma(S^{(i)}) = \Omega(m \log \frac{n}{m}) = \Omega(\delta \log \frac{n}{\delta})$ since the alphabets of $S^{(i)}$ are disjoint.

This construction proves the theorem for $\delta = 3m - 1$ and $n \geq 4m - 1$. If $\delta \bmod 3 \neq 2$, we pad the string with $O(1)$ additional sentinels. Each one increases $\delta(S)$, $\gamma(S)$, and n by 1. Finally, we note that the claim for $\delta = \Omega(n)$ reduces to $\gamma = \Omega(\delta)$, and the latter relation follows directly from Lemma 2. □

3.2 Lower Bounds on Text Entropy and Grammar Size

We now show that there are string families that cannot be encoded in $o(\delta \log n)$ space, that is, $o(\delta \log^2 n)$ bits. It is not known if the same occurs with γ.

Consider a family $\mathcal{S}^*$ consisting of variants of the infinite string S_∞ constructed in the previous section, where the positions of $\mathtt{b}$s are further apart and slightly perturbed. More specifically, for each $S \in \mathcal{S}^*$, the first b is placed at position $S[1]$ and then, for $j \geq 2$, the jth b is placed anywhere in $S[2\cdot 4^{j-2}+1..4^{j-1}]$. The family $\mathcal{S}^*_n$ consists of length-n prefixes of the infinite strings of the family $\mathcal{S}^*$.

Lemma 5. *For every $n \geq 1$, the family $\mathcal{S}^*_n$ needs $\Omega(\log^2 n)$ bits to be encoded.*

Proof. In our definition of $\mathcal{S}^*$, the location of the jth b can be chosen among $2 \cdot 4^{j-2}$ positions, and each combination of these choices generates a different string in $\mathcal{S}^*_n$ as long as $n \geq 4^{j-1}$. Hence, $|\mathcal{S}^*_n| = \prod_{j=2}^{i+1} 2\cdot 4^{j-2} = 2^{\Omega(i^2)}$ for $n \geq 4^i$. To distinguish strings in $\mathcal{S}^*_n$, any encoding needs $\log|\mathcal{S}^*_n| = \Omega(\log^2 n)$ bits. □

Theorem 2. *For every length n and value $\delta \in [2..n]$, there exists a family of length-n strings of common measure δ that needs $\Omega(\delta \log^2 \frac{n}{\delta})$ bits to be encoded.*

Proof. By Lemma 5, encoding $\mathcal{S}^*_n$ requires $\Omega(\log^2 n)$ bits. Below, we prove that the measure δ for any string in $\mathcal{S}^*_n$ is at most 2. Starting from position $4^{j-1}+1$, the distances between two consecutive bs are at least 4^j. Therefore, the distinct substrings of length $k \leq 4^j$ are either those that start at position $i \in [1..4^{j-1}]$ or those of the form $\mathtt{a}^k$ or $\mathtt{a}^i\mathtt{b}\mathtt{a}^{k-i-1}$ for $i \in [0..k-1]$, which yields a total of $d_k(S) \leq 4^{j-1}+k+1$. Plugging $j = \lceil \frac{1}{2}\log k \rceil$, we get $d_k(S) \leq 4^{\lceil \frac{1}{2}\log k\rceil -1}+k+1 \leq 4^{\frac{1}{2}\log k}+k \leq 2k$. By definition of δ, we conclude that $\delta(S) \leq 2$ for every $S \in \mathcal{S}^*_n$.

As in the proof of Theorem 1, one can generalize this result to larger δ. □

The family $\mathcal{S}^*_n$ also gives strings that do not satisfy $g = O(\delta \log n)$.

Theorem 3. *For every length n, there is a string with $g = \Omega(\delta \log^2 n/\log\log n)$.*

Proof. Consider the same family $\mathcal{S}^*_n$, which needs $\Omega(\log^2 n)$ bits to be represented. If we could encode it with a grammar of size g, each grammar element would be a nonterminal that could be encoded with $O(\log g)$ bits. Therefore, our grammar representation would require $O(g \log g)$ bits. Since this must be $\Omega(\log^2 n)$, it follows that $g = \Omega(\log^2 n/\log\log n)$ for any grammar of size g encoding $\mathcal{S}^*_n$. Since $\delta = O(1)$ for every string $S \in \mathcal{S}^*_n$, it follows that $g = \Omega(\delta \log^2 n/\log\log n)$. □

4 Block Trees in δ-Bounded Space

The block tree [5] is a data structure designed to represent repetitive strings $S[1..n]$ in $O(z \log \frac{n}{z})$ space while offering efficient access. In this section, we show that the block tree is easily tuned to use $O(\delta \log \frac{n}{\delta})$ space while retaining its functionality. Note that, given the lower bounds of Sect. 3, we cannot hope for a representation of size $o(\delta \log \frac{n}{\delta})$.

4.1 Block Trees

Given integer parameters r and s, the root of the block tree divides S into s equal-sized (that is, with the same number of characters) blocks (assume for simplicity that $n = s \cdot r^t$ for some integer t).[3] Blocks are then classified into *marked* and *unmarked*. If two adjacent blocks B_1, B_2 form the leftmost occurrence of the underlying substring B_1B_2, then both B_1 and B_2 are marked. Blocks B that remain unmarked are replaced by a pointer to the pair of adjacent blocks B_1, B_2 that contains the leftmost occurrence of B, and the offset $\epsilon \geq 0$ where B starts inside B_1. Marked blocks are divided into r equal-sized sub-blocks, which form the children of the current block tree's level, and processed similarly in a recursive fashion. Let σ be the alphabet size. The level where the blocks become of length below $\log_\sigma n$ corresponds to the leaves of the block tree, and its blocks store their plain string content using $O(\log n)$ bits. The height of the block tree is then $h = O(\log_r \frac{n/s}{\log_\sigma n}) = O(\log_r \frac{n \log \sigma}{s \log n}) \subseteq O(\log \frac{n}{s})$.

The block tree construction guarantees that the blocks B_1 and B_2 to which any unmarked block points exist and are marked. Therefore, any access to a position $S[i]$ can be carried out in $O(h)$ time, by descending from the root to a leaf and spending $O(1)$ time in each level: To obtain $B[i]$ from a marked block B, we simply compute to which sub-block $B[i]$ belongs among the children of B. To obtain $B[i]$ from an unmarked block B pointing to B_1, B_2 with offset ϵ, we switch either to $B_1[\epsilon + i]$ or to $B_2[\epsilon + i - |B_1|]$, which are marked blocks.

By storing further data associated with marked and unmarked blocks, the block tree offers the following functionality [5]:

Access: any substring $S[i..i + \ell - 1]$ is extracted in time $O(h\lceil \ell / \log_\sigma n \rceil)$.
Rank: $rank_a(S, i)$ is the number of times symbol a occurs in $S[1..i]$. It is computed in time $O(h)$ by multiplying the space by $O(\sigma)$.
Select: $select_a(S, j)$ is the position of the jth occurrence of symbol a in S. It is computed in time $O(\log \log \frac{n}{s} + h \log \log r)$ by multiplying the space by $O(\sigma)$.

It is shown that there are only $O(zr)$ blocks in each level of the block tree (except the first, which has s); therefore its size is $O(s + zr \log_r \frac{n \log \sigma}{s \log n})$.

4.2 Bounding the Space in Terms of δ

We now prove that there are only $O(\delta r)$ blocks in each level of the block tree, and therefore, choosing $s = \delta$ yields a structure of size $O(\delta r \log_r \frac{n \log \sigma}{\delta \log n})$ with height $O(\log_r \frac{n \log \sigma}{\delta \log n})$. For $r = O(1)$, the space is $O(\delta \log \frac{n}{\delta})$ and the height is $O(\log \frac{n}{\delta})$.

Let us call level k of the block tree the one where blocks are of length r^k. In level k, then, S is covered regularly with blocks $B = S[r^k(i-1)+1..r^k i]$ of length r^k (though not all of them are present in the block tree). Note that

[3] If not, we simply pad S with spurious symbols at the end; whole spurious blocks are not represented. The extra space incurred is only $O(rh)$ for a block tree of height h. The actual construction [5] uses instead blocks of sizes $\lfloor n/s \rfloor$ and $\lceil n/s \rceil$.

k reaches its maximum in the root (where we have the largest blocks) and the minimum in the leaves of the block tree.

Lemma 6. *The number of marked blocks of length r^k in the block tree is $O(\delta)$.*

Proof. Any marked block B must belong to a sequence of three blocks, $B^- \cdot B \cdot B^+$, such that B is inside the leftmost occurrence of $B^- \cdot B$ or $B \cdot B^+$, or both (B^- and B^+ do not exist for the first and last block, respectively).

For the sake of computing our bound, let # be a symbol not appearing in S and let us add $2 \cdot r^k$ characters equal to # at the beginning of S and r^k characters equal to # at the end of S. We index the added prefix in negative positions (up to index 0), so that $S[-2 \cdot r^k + 1..0] = \#^{2 \cdot r^k}$. Now consider all the r^k text positions p belonging to a marked block B. The long substring $E = S[p - 2 \cdot r^k..p + 2 \cdot r^k - 1]$ centered at p, of length $4r^k$, contains $B^- \cdot B \cdot B^+$, and thus E contains the leftmost occurrence L of $B^- \cdot B$ or $B \cdot B^+$. All those long substrings E must then be distinct: if two long substrings E and E' are equal, and E' appears after E in S, then E' does not contain the leftmost occurrence of any substring L.

Since we added a prefix of length $2 \cdot r^k$ and a suffix of length r^k consisting of character # to S, the number of distinct substrings of length $4r^k$ is at most $d_{4r^k}(S) + 3r^k$. Therefore, there can be at most $d_{4r^k}(S) + 3r^k$ long substrings E as well, because they must all be distinct. Since each position p inside a block B induces a distinct long substring E, and each marked block B contributes r^k distinct positions p, there are at most $(d_{4r^k}(S) + 3r^k)/r^k$ marked blocks B of length r^k. The total number of marked blocks of length r^k is thus at most $(d_{4r^k}(S) + 3r^k)/r^k = 4 \cdot d_{4r^k}(S)/(4r^k) + 3r^k/r^k \leq 4\delta + 3$. □

Since the block tree has at most $4\delta + 3$ marked blocks per level, it has $O(\delta r)$ blocks across all the levels except the first. This yields the following result.

Theorem 4. *Let $S[1..n]$, over alphabet $[1..\sigma]$, have compressibility measure δ. Then the block tree of S, with parameters r and s, is of size $O(s + \delta r \log_r \frac{n \log \sigma}{s \log n})$ words and height $h = O(\log_r \frac{n \log \sigma}{s \log n})$.*

Note that $\frac{n}{\delta} = O(\frac{n}{z} \log \frac{n}{\delta}) = O(\frac{n}{z}\sqrt{\frac{n}{\delta}})$ due to Lemma 1, so $\log \frac{n}{\delta} = O(\log \frac{n^2}{z^2}) = O(\log \frac{n}{z}) = O(\log \frac{n}{\gamma})$. Hence, the query time we obtain using $O(\delta \log \frac{n}{\delta})$ space is asymptotically the same as the $O(\log \frac{n}{\gamma})$ time obtained in $O(\gamma \log \frac{n}{\gamma})$ space [34, 36] or the $O(\log \frac{n}{z})$ time obtained in $O(z \log \frac{n}{z})$ space [5].

5 Text Indexing in δ-Bounded Space

We now show that not only efficient access of S can be supported within $O(\delta \log \frac{n}{\delta})$ space, but also text indexing, that is, efficiently listing all the positions in S where a pattern $P[1..m]$ appears. For consistency with previous works, in this section we speak of a text $T[1..n]$ instead of a string $S[1..n]$.

Our index builds on top of a slight variant of the block tree of the previous sections, with $r = 2$, $s = \delta$, and stopping only when the leaves are of length 1. This block tree is of size $O(\delta \log \frac{n}{\delta})$ and of height $O(\log \frac{n}{\delta})$.

To build the index, we follow the same ideas of the "universal index" [34], whose space will be improved without affecting its search time complexities. That index builds on a variant of block trees designed for attractors: the Γ-tree has a first level with γ equal-sized blocks, and at any other level k, it marks the blocks that are at distance $< 2^k$ from an attractor position. Unmarked blocks B then point to some copy of B that crosses an attractor position (the blocks overlapping that copy are marked by definition). In the Γ-tree pointers can go leftward or rightward, not necessarily to a leftmost occurrence. The space of the Γ-tree is $\Theta(\gamma \log \frac{n}{\gamma})$, which we now know, by Theorem 4, that is never asymptotically smaller than that of block trees with parameters $r = 2$ and $s = \delta$.

Karp–Rabin fingerprinting [22] assigns a string $S[1..\ell]$ the signature $\kappa(S) = (\sum_{i=1}^{\ell} S[i] \cdot c^{i-1}) \bmod \mu$ for suitable integers $c > 1$ and prime μ. It is possible to build a signature formed by a pair of functions $\langle \kappa_1, \kappa_2 \rangle$ guaranteeing no collisions between substrings of $S[1..n]$, in $O(n \log n)$ expected time [7]. Our index will need to compute Karp–Rabin fingerprints $\kappa(T[i..j])$ in time $O(\log \frac{n}{\delta})$. This is done on block trees by using the same algorithm described for the Γ-tree.

Lemma 7. *Let $T[1..n]$ have compressibility measure δ, and let κ be a Karp–Rabin function. Then we can store a data structure of size $O(\delta \log \frac{n}{\delta})$ supporting the computation of κ on any substring of T in $O(\log \frac{n}{\delta})$ time.*

Proof. The structure is the described block tree variant, with some further fields. We store $\kappa(T[1..2^k i])$ at the ith top-level block, for all i and $k = \lceil \log \frac{n}{\delta} \rceil$. We also store $\kappa(B)$ for each block B stored in the tree and, for the unmarked blocks B pointing to B_1, B_2 with offset ϵ, we also store $\kappa(B_1[1+\epsilon..])$. Navarro and Prezza [34, Lem. 1] show that this suffices to compute $\kappa(T[i..j])$ within $O(1)$ time per level of the Γ-tree; their proof holds verbatim for the block tree. $\square$

Let us say that a block is *explicit* if it is stored in the block tree. Thus, a block is explicit if and only if it is marked or it is the child of a marked block.

Lemma 8. (See [[34], Lem. 2]) *Any substring $T[i..j]$ of length at least 2 either overlaps two consecutive explicit blocks or is completely inside an unmarked block.*

Proof. The leaves of the block tree, read left to right, partition T into a sequence of explicit blocks. The leaves are either unmarked blocks or blocks of length 1. Since $|T[i..j]| \geq 2$, if it is not completely inside an unmarked block, it cannot be contained in a leaf, so it must cross a boundary between two explicit blocks. $\square$

We now divide the possible occurrences of $P[1..m]$ in T into *primary* (those overlapping two consecutive explicit blocks) and *secondary* (those inside an unmarked block). The technique used on Γ-trees [34, Sec. 3] applies verbatim here: Primary occurrences are found using a grid of $(s-1) \times (s-1)$, where

$s = O(\delta \log \frac{n}{\delta})$ is the number of leaves in the block tree, which finds the occ_p primary occurrences in time $O((m + occ_p)\log^\varepsilon s)$, for any constant $\varepsilon > 0$. The ranges to search in the grid are obtained using their following result [34, Lem. 3].

Lemma 9. *Let $\mathcal{X}$ be a sorted set of suffixes of T, and κ a Karp–Rabin function. If one can extract a substring of length ℓ from T in time $f_e(\ell)$ and compute κ on it in time $f_h(\ell)$, then one can build a data structure of size $O(|\mathcal{X}|)$ that obtains the lexicographic ranges in $\mathcal{X}$ of the $m-1$ suffixes of a given pattern P in worst-case time $O(m(f_h(m) + \log m) + f_e(m))$, provided that κ is collision-free among substrings of T whose lengths are powers of two.*

Since in our case $f_e(m) = O(m \log \frac{n}{\delta})$ and $f_h(m) = O(\log \frac{n}{\delta})$, we can find all the ranges to search for in time $O(m \log \frac{nm}{\delta})$. The occ_s secondary occurrences are obtained as on Γ-trees [34, Sec. 3.2], within $O((occ_p + occ_s)\log\log \frac{n}{\delta})$ time.

Theorem 5. *Let $T[1..n]$ have measure δ. Then there exists a data structure of size $O(\delta \log \frac{n}{\delta})$ such that the occurrences of any pattern $P[1..m]$ in T can be located in time $O(m \log n + occ \log^\varepsilon n)$, for any constant $\varepsilon > 0$.*

6 Conclusions

We have made a step towards establishing the right measure of repetitiveness for a string $S[1..n]$. Compared with the most principled prior measure, the size γ of the smallest attractor, the proposed measure δ has several important advantages:

1. It lower bounds the previous measure, $\delta \le \gamma$, and can be computed in linear time, while finding γ is NP-hard.
2. We can always encode S in $O(\delta \log \frac{n}{\delta})$ space, and this is worst-case optimal in terms of δ: for any length n and any value $2 \le \delta \le n^{1-\varepsilon}$ (where $\varepsilon > 0$ is an arbitrary constant), there are text families needing $\Omega(\delta \log \frac{n}{\delta})$ space. Thus, $o(\delta \log n)$ space is unreachable. Instead, no text family is known to require $\omega(\gamma)$ space, nor it is known if $o(\gamma \log n)$ space can be reached.
3. Measures γ, b, c, and z are upper bounded by $O(\delta \log \frac{n}{\delta})$, and $g = O(\delta \log^2 \frac{n}{\delta})$, but there are text families where the smallest context-free grammar is of size $g = \Omega(\delta \log^2 n / \log\log n)$. This lower bound is not known to hold on γ.
4. The encodings using $O(\delta \log \frac{n}{\delta})$ space support direct access and indexed searches, with the same complexities obtained within attractor-bounded space, $O(\gamma \log \frac{n}{\gamma})$. An exception is a very recent faster index [13].

An ideal compressibility measure for repetitive sequences should be always reachable and string-wise optimal, apart from being practical to compute. Measure $\delta \log \frac{n}{\delta}$ is reachable and fast to compute, though optimal only in a coarse sense (i.e., not string-wise but within the class of all the strings with the same δ value).

Note that we do not know if one can always encode a string within $O(\gamma)$ space. If this was the case, then γ would be a better measure than $\delta \log \frac{n}{\delta}$, except for being hard to compute. Otherwise, a good alternative could be b, which is always reachable and might be string-wise optimal within some broad class of representations that exploit repetitiveness, yet NP-hard to compute. It is not known, however, if b or γ are monotone, that is, smaller on T than on TT', whereas δ clearly is. This fascinating quest is then still open.

On the more practical side, it would be interesting to obtain faster indexes of size $O(\delta \log \frac{n}{\delta})$. Our index requires $O(m \log n + occ \log^{\epsilon} n)$ search time, while in $O(\gamma \log \frac{n}{\gamma})$ space, it is possible to search in $O(m + (occ + 1) \log^{\epsilon} n)$ time [13].

References

1. Belazzougui, D., Cording, P.H., Puglisi, S.J., Tabei, Y.: Access, rank, and select in grammar-compressed strings. In: Bansal, N., Finocchi, I. (eds.) ESA 2015. LNCS, vol. 9294, pp. 142–154. Springer, Heidelberg (2015). https://doi.org/10.1007/978-3-662-48350-3_13
2. Belazzougui, D., Cunial, F.: Fast label extraction in the CDAWG. In: Fici, G., Sciortino, M., Venturini, R. (eds.) SPIRE 2017. LNCS, vol. 10508, pp. 161–175. Springer, Cham (2017). https://doi.org/10.1007/978-3-319-67428-5_14
3. Belazzougui, D., Cunial, F.: Representing the suffix tree with the CDAWG. In: Proceedings of 28th Annual Symposium on Combinatorial Pattern Matching, CPM, pp. 7:1–7:13 (2017). https://doi.org/10.4230/LIPIcs.CPM.2017.7
4. Belazzougui, D., Cunial, F., Gagie, T., Prezza, N., Raffinot, M.: Composite repetition-aware data structures. In: Cicalese, F., Porat, E., Vaccaro, U. (eds.) CPM 2015. LNCS, vol. 9133, pp. 26–39. Springer, Cham (2015). https://doi.org/10.1007/978-3-319-19929-0_3
5. Belazzougui, D., et al.: Queries on LZ-bounded encodings. In: Proceedings of the 2015 Data Compression Conference, DCC, pp. 83–92 (2015). https://doi.org/10.1109/DCC.2015.69
6. Bille, P., Ettienne, M.B., Gørtz, I.L., Vildhøj, H.W.: Time-space trade-offs for Lempel-Ziv compressed indexing. Theor. Comput. Sci. **713**, 66–77 (2018). https://doi.org/10.1016/j.tcs.2017.12.021
7. Bille, P., Gørtz, I.L., Sach, B., Vildhøj, H.W.: Time-space trade-offs for longest common extensions. J. Discrete Algorithms **25**, 42–50 (2014). https://doi.org/10.1016/j.jda.2013.06.003
8. Bille, P., Landau, G.M., Raman, R., Sadakane, K., Satti, S.R., Weimann, O.: Random access to grammar-compressed strings and trees. SIAM J. Comput. **44**(3), 513–539 (2015). https://doi.org/10.1137/130936889
9. Blumer, A., Blumer, J., Haussler, D., McConnell, R.M., Ehrenfeucht, A.: Complete inverted files for efficient text retrieval and analysis. J. ACM **34**(3), 578–595 (1987). https://doi.org/10.1145/28869.28873
10. Burrows, M., Wheeler, D.J.: A block-sorting lossless data compression algorithm. Technical report 124, Digital Equipment Corporation (1994). https://www.hpl.hp.com/techreports/Compaq-DEC/SRC-RR-124.pdf
11. Charikar, M., et al.: The smallest grammar problem. IEEE Trans. Inf. Theory **51**(7), 2554–2576 (2005). https://doi.org/10.1109/TIT.2005.850116
12. Christiansen, A.R., Ettienne, M.B.: Compressed indexing with signature grammars. In: Bender, M.A., Farach-Colton, M., Mosteiro, M.A. (eds.) LATIN 2018. LNCS, vol. 10807, pp. 331–345. Springer, Cham (2018). https://doi.org/10.1007/978-3-319-77404-6_25

13. Christiansen, A.R., Ettienne, M.B., Kociumaka, T., Navarro, G., Prezza, N.: Optimal-time dictionary-compressed indexes. arXiv e-prints p. 1811.12779 (2019). https://arxiv.org/abs/1811.12779
14. Claude, F., Navarro, G.: Self-indexed grammar-based compression. Fundam. Inform. **111**(3), 313–337 (2011). https://doi.org/10.3233/FI-2011-565
15. Claude, F., Navarro, G.: Improved grammar-based compressed indexes. In: Calderón-Benavides, L., González-Caro, C., Chávez, E., Ziviani, N. (eds.) SPIRE 2012. LNCS, vol. 7608, pp. 180–192. Springer, Heidelberg (2012). https://doi.org/10.1007/978-3-642-34109-0_19
16. Gagie, T., Gawrychowski, P., Kärkkäinen, J., Nekrich, Y., Puglisi, S.J.: A faster grammar-based self-index. In: Dediu, A.-H., Martín-Vide, C. (eds.) LATA 2012. LNCS, vol. 7183, pp. 240–251. Springer, Heidelberg (2012). https://doi.org/10.1007/978-3-642-28332-1_21
17. Gagie, T., Gawrychowski, P., Kärkkäinen, J., Nekrich, Y., Puglisi, S.J.: LZ77-based self-indexing with faster pattern matching. In: Pardo, A., Viola, A. (eds.) LATIN 2014. LNCS, vol. 8392, pp. 731–742. Springer, Heidelberg (2014). https://doi.org/10.1007/978-3-642-54423-1_63
18. Gagie, T., Navarro, G., Prezza, N.: On the approximation ratio of Lempel-Ziv parsing. In: Bender, M.A., Farach-Colton, M., Mosteiro, M.A. (eds.) LATIN 2018. LNCS, vol. 10807, pp. 490–503. Springer, Cham (2018). https://doi.org/10.1007/978-3-319-77404-6_36
19. Gagie, T., Navarro, G., Prezza, N.: Fully-functional suffix trees and optimal text searching in BWT-runs bounded space. J. ACM **67**(1), 1–54 (2020). https://doi.org/10.1145/3375890
20. Gallant, J.K.: String Compression Algorithms. Ph.D. thesis, Princeton Univ. (1982)
21. Jez, A.: A really simple approximation of smallest grammar. Theor. Comput. Sci. **616**, 141–150 (2016). https://doi.org/10.1016/j.tcs.2015.12.032
22. Karp, R.M., Rabin, M.O.: Efficient randomized pattern-matching algorithms. IBM J. Res. Dev. **31**(2), 249–260 (1987). https://doi.org/10.1147/rd.312.0249
23. Kempa, D., Kociumaka, T.: Resolution of the Burrows-Wheeler transform conjecture. arXiv e-prints p. 1910.10631 (2019). https://arxiv.org/abs/1910.10631
24. Kempa, D., Prezza, N.: At the roots of dictionary compression: String attractors. In: Proceedings of the 50th Annual ACM Symposium on the Theory of Computing, STOC, pp. 827–840 (2018). https://doi.org/10.1145/3188745.3188814
25. Kida, T., Matsumoto, T., Shibata, Y., Takeda, M., Shinohara, A., Arikawa, S.: Collage system: a unifying framework for compressed pattern matching. Theor. Comput. Sci. **298**(1), 253–272 (2003). https://doi.org/10.1016/S0304-3975(02)00426-7
26. Kieffer, J.C., Yang, E.: Grammar-based codes: a new class of universal lossless source codes. IEEE Trans. Inf. Theory **46**(3), 737–754 (2000). https://doi.org/10.1109/18.841160
27. Kolmogorov, A.N.: Three approaches to the quantitative definition of information. Int. J. Comput. Math. **2**(1–4), 157–168 (1968). https://doi.org/10.1080/00207166808803030
28. Kreft, S., Navarro, G.: On compressing and indexing repetitive sequences. Theor. Comput. Sci. **483**, 115–133 (2013). https://doi.org/10.1016/j.tcs.2012.02.006
29. Lempel, A., Ziv, J.: On the complexity of finite sequences. IEEE Trans. Inf. Theory **22**(1), 75–81 (1976). https://doi.org/10.1109/TIT.1976.1055501
30. Mäkinen, V., Navarro, G., Sirén, J., Välimäki, N.: Storage and retrieval of highly repetitive sequence collections. J. Comput. Biol. **17**(3), 281–308 (2010). https://doi.org/10.1089/cmb.2009.0169

31. Navarro, G., Ochoa, C., Prezza, N.: On the approximation ratio of ordered parsings. arXiv e-prints p. 1803.09517 (2019). https://arxiv.org/abs/1803.09517
32. Navarro, G.: Compact Data Structures - A Practical Approach. Cambridge University Press, New York (2016). https://doi.org/10.1017/cbo9781316588284
33. Navarro, G., Mäkinen, V.: Compressed full-text indexes. ACM Comput. Surv. **39**, 1 (2007). https://doi.org/10.1145/1216370.1216372
34. Navarro, G., Prezza, N.: Universal compressed text indexing. Theor. Comput. Sci. **762**, 41–50 (2019). https://doi.org/10.1016/j.tcs.2018.09.007
35. Nishimoto, T., Tomohiro, I., Inenaga, S., Bannai, H., Takeda, M.: Fully dynamic data structure for LCE queries in compressed space. In: Proceedings of the 41st International Symposium on Mathematical Foundations of Computer Science, MFCS, pp. 72:1–72:15 (2016). https://doi.org/10.4230/LIPIcs.MFCS.2016.72
36. Prezza, N.: Optimal rank and select queries on dictionary-compressed text. In: Proceedings of the 30th Annual Symposium on Combinatorial Pattern Matching, CPM, pp. 4:1–4:12 (2019). https://doi.org/10.4230/LIPIcs.CPM.2019.4
37. Raskhodnikova, S., Ron, D., Rubinfeld, R., Smith, A.D.: Sublinear algorithms for approximating string compressibility. Algorithmica **65**(3), 685–709 (2013). https://doi.org/10.1007/s00453-012-9618-6
38. Rodeh, M., Pratt, V.R., Even, S.: Linear algorithm for data compression via string matching. J. ACM **28**(1), 16–24 (1981). https://doi.org/10.1145/322234.322237
39. Rytter, W.: Application of Lempel-Ziv factorization to the approximation of grammar-based compression. Theor. Comput. Sci. **302**(1–3), 211–222 (2003). https://doi.org/10.1016/S0304-3975(02)00777-6
40. Shannon, C.E.: A mathematical theory of communication. Bell Syst. Tech. J. **27**, 398–403 (1948). https://doi.org/10.1002/j.1538-7305.1948.tb01338.x
41. Stephens, Z.D., et al.: Big data: astronomical or genomical? PLOS Biol. **13**(7), e1002195 (2015). https://doi.org/10.1371/journal.pbio.1002195
42. Storer, J.A., Szymanski, T.G.: Data compression via textual substitution. J. ACM **29**(4), 928–951 (1982). https://doi.org/10.1145/322344.322346

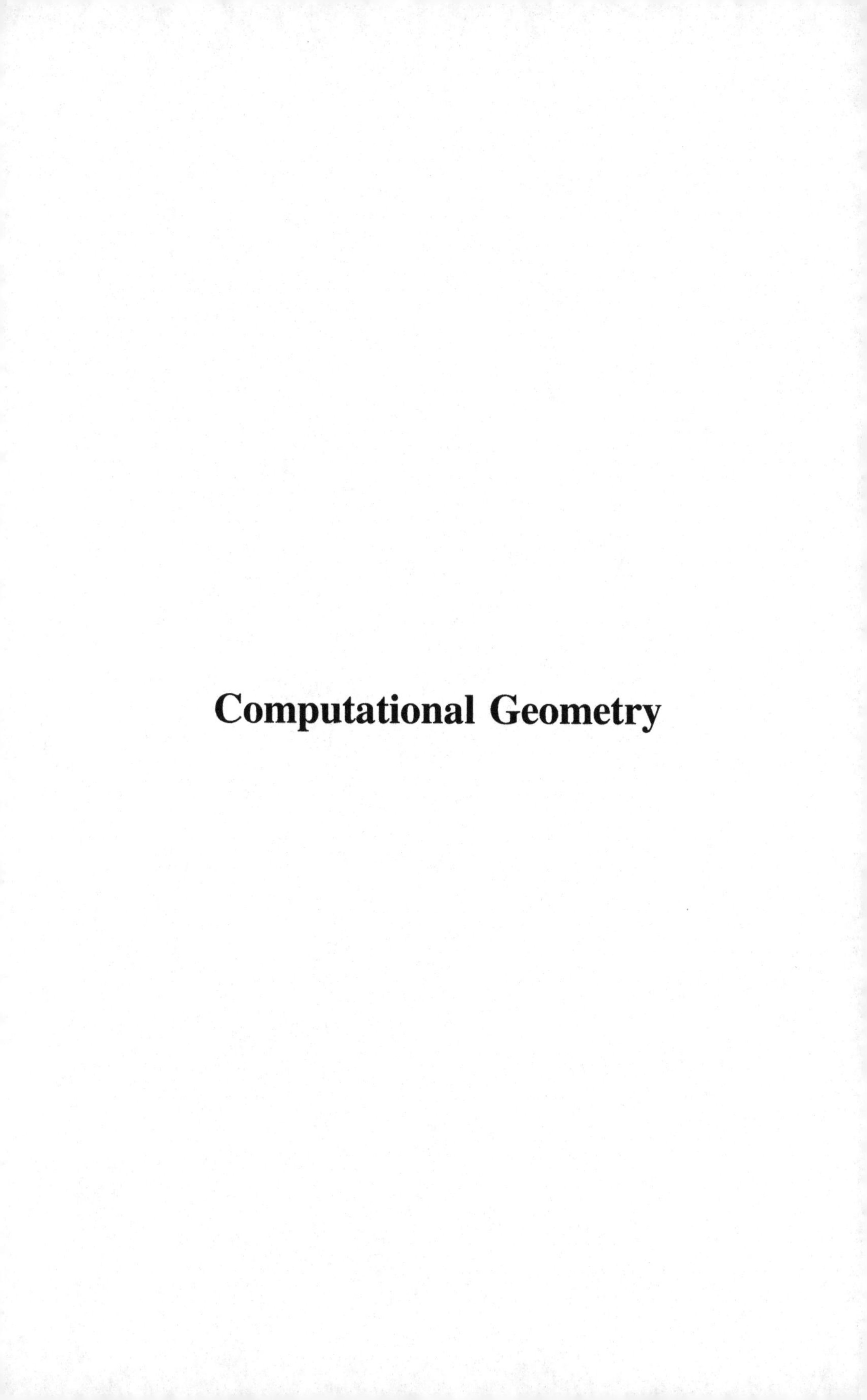

Computational Geometry

Flips in Higher Order Delaunay Triangulations

Elena Arseneva[1], Prosenjit Bose[2], Pilar Cano[2,3](✉), and Rodrigo I. Silveira[3]

[1] St. Petersburg State University, St. Petersburg, Russia
e.arseneva@spbu.ru
[2] Carleton University, Ottawa, Canada
jit@scs.carleton.ca
[3] Universitat Politècnica de Catalunya, Barcelona, Spain
m.pilar.cano@upc.edu, rodrigo.silveira@upc.edu

Abstract. We study the flip graph of higher order Delaunay triangulations. A triangulation of a set S of n points in the plane is order-k Delaunay if for every triangle its circumcircle encloses at most k points of S. The *flip graph* of S has one vertex for each possible triangulation of S, and an edge connecting two vertices when the two corresponding triangulations can be transformed into each other by a *flip* (i.e., exchanging the diagonal of a convex quadrilateral by the other one). The flip graph is an essential structure in the study of triangulations, but until now it had been barely studied for order-k Delaunay triangulations. In this work we show that, even though the order-k flip graph might be disconnected for $k \geq 3$, any order-k triangulation can be transformed into some other order-k triangulation by at most $k-1$ flips, such that the intermediate triangulations are of order at most $2k-2$, in the following settings: (1) for any $k \geq 0$ when S is in convex position, and (2) for any $k \leq 5$ and any point set S. Our results imply that the flip distance between two order-k triangulations is $O(kn)$, as well as an efficient enumeration algorithm.

1 Introduction

Triangulations are one of the most important geometric objects studied in discrete and computational geometry. Given a set S of points in the plane, a *triangulation* of S is a decomposition of the convex hull of S into triangles, such that each triangle has its three vertices in S. Triangulations are important not only from a theoretical point of view, but also due to their many applications in areas like mesh generation, computer aided geometric design, and geographic information science [3, 7, 10].

E.A. was supported by RFBR, project number 20-01-00488. P.B. was partially supported by NSERC. P.C. was supported by CONACYT, MX. R.S. was supported by MINECO through the Ramón y Cajal program. P.C. and R.S. were also supported by projects MINECO MTM2015-63791-R and Gen. Cat. 2017SGR1640. This project has received funding from the European Union's Horizon 2020 research and innovation programme under the Marie Skłodowska-Curie grant agreement No 734922.

Y. Kohayakawa and F. K. Miyazawa (Eds.): LATIN 2020, LNCS 12118, pp. 223–234, 2020.
https://doi.org/10.1007/978-3-030-61792-9_18

A point set S in the plane can have many different triangulations. More precisely, there are point sets with n points that have $\Omega(8.65^n)$ triangulations [12], while the best upper bound is currently 30^n [24]. Despite all the different possible triangulations, in practice, most of the time the *Delaunay triangulation* is used. A Delaunay triangulation of S, denoted $DT(S)$, is a triangulation where each triangle satisfies the *empty circle property*: the circumcircle of each triangle does not enclose points of S. When no four points of S are co-circular, $DT(S)$ is unique. The Delaunay triangulation has many good properties, including maximizing the minimum interior angle in its triangles among all triangulations of any fixed point set. For this reason its triangles are considered "well-shaped", and this is one of the main explanations for its popularity. For a survey see [4,13].

However, the fact that the DT is unique becomes an issue in certain application domains where extra flexibility is needed. For instance, triangulations are often used to model terrains. In this case, the points in S are samples of a 3D surface, thus have also an elevation. But the Delaunay triangulation neglects the elevation information, potentially resulting in poor models where important terrain features, such as valley or ridge lines, are ignored [10,15]. This motivated Gudmundsson et al. [14] to propose *higher order Delaunay triangulations*, a generalization of the DT that intends to provide well-shaped triangles, while giving flexibility to choose from a larger set of triangulations. A triangulation T of S is an *order-k Delaunay triangulation*—or, simply, *order-k*—if the circumcircle[1] of each triangle of T contains at most k points of S in its interior. Note that an order-0 triangulation is a standard Delaunay triangulation. As soon as $k > 0$, one obtains a class of triangulations that for small values of k are expected to have well-shaped triangles, but with potentially many triangulations to choose from. Order-k triangulations have been used for terrain modeling, minimum interference networks and triangulation of polygons [6,11,23,26].

A similar generalization, but based on edges instead of triangles, was proposed by Abellanas et al. [2]. In their work, an *order-k (Delaunay) edge* e is defined as an edge for which there exists a circle through the endpoints of e that encloses at most k points of S. All edges in an order-k triangulation are order-k. But the converse is not true: a triangle composed of three order-k edges can have order greater than k. In fact, the lowest order triangulation containing an order-k edge can have order up to $2k-2$ [15].

A fundamental operation to work with triangulations is the *edge flip*. It consists in removing an edge, shared by two triangles that form a convex quadrilateral, and inserting the other diagonal of the quadrilateral. A flip transforms a triangulation T into another triangulation T' that differs by exactly one edge and two triangles. The flip operation leads naturally to the definition of the *flip graph* of S. Each triangulation of S is represented by a vertex in this graph, and two vertices are adjacent if their corresponding triangulations differ by exactly one flip.

[1] We refer to the *interior of a circumcircle* as the interior of the disk defined by such a circle, in order to comply with terminology of earlier papers on the subject.

The importance of flips in triangulations comes from the fact, first proved by Lawson [18], that the flip graph is connected. Moreover, the sequence of edge flips connecting any two triangulations has length $O(n^2)$. In fact, it was later shown [19,25] that any triangulation of S can be converted into $DT(S)$ by performing $O(n^2)$ flips. Each flip in this transformation also results in an increase of the *angle vector* of the triangulation, i.e., the vector of all angles of each of its triangles in increasing order. It is also known that the quadratic upper bound on the diameter of the flip graph is tight in general [13,16], although it goes down to $\Theta(n)$ if the points in S are in convex position [28]. In general, computing the distance in the flip graph between two given triangulations is a difficult problem, whose complexity was open until recently, when it was shown to be APX-hard [20,22]. This has drawn considerable attention to the study of certain subgraphs of the flip graph, which define the flip graph of certain classes of triangulations. We refer to [9] for a survey.

In this paper we study the flip graph of higher order Delaunay triangulations. Most previous work on such triangulations focused on algorithmic questions related to finding order-k triangulations that are optimal with respect to extra criteria [17,26,27], or on evaluating their effectiveness in practical settings [8,11]. One of the few theoretical aspects studied is the asymptotic number of order-k triangulations [21]. In that work, the authors showed that for points drawn uniformly at random, already for $k = 1$ one can expect an exponential number of different order-k triangulations. However, almost nothing is known about the flip graph of order-k triangulations, except that it is connected only for $k \leq 2$ [1]. Similarly, Abellanas et al. [2] showed that the flip graph of triangulations of point sets with edges of order k is connected for $k \leq 1$, but might be disconnected for $k \geq 2$. On the other hand, they proved that for point sets in convex position the flip graph is always connected [2]. However, their proof implies an exponential bound in the diameter of the flip graph. The only previous work on the flip graph of order-k triangulations is by Abe and Okamoto [1], in the context of enumeration algorithms. They observed that for $k \leq 2$, the fact that the flip graph is connected implies that the reverse enumeration framework by Avis and Fukuda [5] can be applied to enumerate all order-k triangulations, spending polynomial time on each of them.

Our Work. We present several structural properties of the flip graph of order-k triangulations. For points in convex position, we show that for any $k > 2$ there exists a point set in convex position for which the flip graph is not connected. However, we prove that no order-k triangulation is too far from all the other order-k triangulations, in the sense that for any order-k triangulation T there exists another order-k triangulation T' at distance at most $k - 1$ in the flip graph of order-$(2k - 2)$ triangulations. It is noteworthy that each flip on the path from T to T' increases the angle vector of the triangulation. The bottom line is that while order-k triangulations are not connected via the flip operation, they become connected if a slightly relaxed condition on the order is considered. For points in generic (non-convex) position, we prove the same result for up to $k \leq 5$, although we conjecture that it holds for all k. Our results have several implications on the flip distance between order-k triangulations , as well as on their efficient algorithmic enumeration. Due to space limitations, most of the proofs are omitted.

2 Preliminaries and General Observations

In this section we give some definitions and observations that will be useful for the rest of the paper.

Let S be a point set in the plane. We assume that set S is in *general position*, i.e., that no three points of S lie on a line and no four points of S lie on a circle. Let T be a triangulation of S, and let $\triangle uyv$ be a triangle in T with vertices u,y,v. We will denote by $\bigcirc uyv$ the open disk defined by the enclosed area of the *circumcircle* of $\triangle uyv$ (i.e., the unique circle through u, y, and v). Thus, $\partial \bigcirc uyv$ denotes the circumcircle of $\triangle uyv$. Triangle $\triangle uyv$ is a triangle *of order* k, also called an *order-k* triangle, if $\bigcirc uyv$ contains at most k points of S. A triangulation where all triangles are order-k is an *order-k (Delaunay) triangulation*. Hence, a triangulation T is not order-k if $\bigcirc uvy$ contains more than k points of S for some $\triangle uyv$ in T. The set of all order-k triangulations of S will be denoted $\mathcal{T}_k(S)$.

Let $e = uv$ be an edge in T. Edge e is *flippable* if e is incident to two triangles $\triangle uxv$ and $\triangle uyv$ of T and $uxvy$ is a convex quadrilateral. The edge e is called *illegal* if $\bigcirc uxv$ contains y. Note that this happens if and only if $\bigcirc uyv$ contains x. See Fig. 1b. Otherwise, the edge uv is called *legal*. It is easy to see that an illegal edge is flippable. The *angle vector* $\alpha(T)$ of a triangulation T is the vector whose components are the angles of each triangle in T ordered in increasing order. Let $T' \neq T$ be another triangulation of S. We say that $\alpha(T) > \alpha(T')$ if $\alpha(T) > \alpha(T')$ in lexicographic order. It is well-known that if T' is the triangulation obtained by flipping an illegal edge of T, then $\alpha(T') > \alpha(T)$ [13]. Moreover, since $DT(S)$ maximizes the minimum angle, it follows that $DT(S)$ is the only triangulation where all the edges are legal [25]. This property leads to a simple algorithm for computing the $DT(S)$: start from an arbitrary triangulation of S, and flip illegal

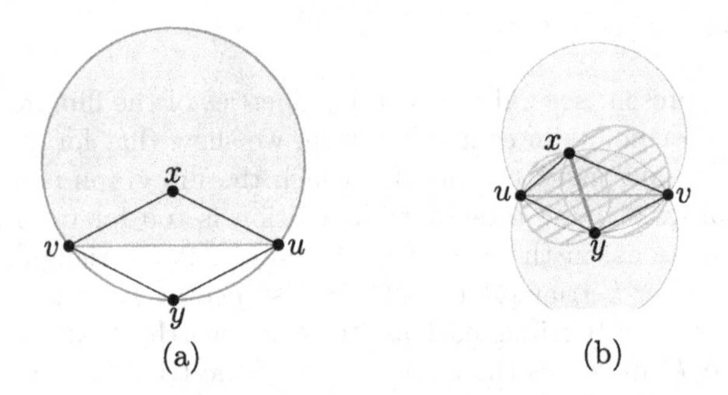

Fig. 1. (a) An illegal edge uv, with region $\bigcirc^{uv}_{y}$ in gray. (b) The union of the dashed filled disks $\bigcirc uxy$ and $\bigcirc xvy$ is contained in the union of the gray disks $\bigcirc uxv$ and $\bigcirc uyv$.

edges until none is left. This also implies that the flip graph of all triangulations of S is connected, since any triangulation can be transformed into $DT(S)$ by a finite number of flips. Let $G(\mathcal{T}_k(S))$ denote the flip graph of $\mathcal{T}_k(S)$.

Next we present several important facts. We start with a well-known observation, illustrated in Fig. 1b.

Observation 1. *Let $\triangle uxv$ and $\triangle uyv$ be two adjacent triangles in a triangulation of S. If edge uv is illegal, then $(\bigcirc uxy \cup \bigcirc xyv) \subset (\bigcirc uxv \cup \bigcirc uvy)$.*

In the context of order-k triangulations, Observation 1 implies the following.

Lemma 1 (Abe and Okamoto [1]**).** *Let T be a triangulation of S, let uv be an illegal edge of T, and let $\triangle uvx$ and $\triangle uyv$ be the triangles incident to uv in T. If $\triangle uxv$ is of order k, and $\triangle uyv$ is of order l, then triangles $\triangle uxy$ and $\triangle xyv$ are of order k' and l', respectively, for some k', l' with $k' + l' \leq k + l - 2$.*

Now we need an extra piece of notation, which we will use extensively. For a triangle $\triangle uyv$, we let $\ominus_y^{uv}$ denote the open region bounded by edge uv and the arc of $\partial \bigcirc uyv$ that does not contain y. See Fig 1a.

In what remains of this section we will consider a triangulation T of order $k \geq 3$, and an illegal edge uv adjacent to triangles $\triangle uxv$ and $\triangle uyv$ of T.

Throughout this work we will often refer to points of S that are contained—or not—in a certain region. For brevity, we will sometimes omit "of S", and simply refer to points in a certain region, as we do in the next observation.

Observation 2. *If $\ominus_y^{uv}$ does not contain points, then $\bigcirc uxy \cap \bigcirc xyv \cap \ominus_y^{uv}) \subset (\bigcirc uxv \cap \ominus_y^{uv}$ does not contain points.*

Consider the triangulation T' resulting from flipping uv in T. For the sake of simplicity, for any region R in the plane, we denote by $|R|$ the number of points of S in the interior of R. Using that $|\bigcirc uxv \setminus \{y\}| \leq k-1, |\bigcirc uyv \setminus \{x\}| \leq k-1$ and $|\bigcirc uxy| \geq k+1$, a rather simple counting argument implies the following.

Observation 3. *Each of $\ominus_y^{ux} \setminus \bigcirc uxv$ and $\ominus_x^{uy} \setminus \bigcirc uyv$ contains at least two points.*

The next two observations concern the case where the region $\bigcirc uxy \setminus \ominus_y^{ux}$ contains the maximum possible number of points, i.e., $k-1$. As shown next, in this case the intersection $\ominus_y^{ux} \cap \bigcirc uxv$ does not contain any points of S.

Observation 4. *If $\bigcirc uxy \setminus \ominus_y^{ux}$ contains $k-1$ points, then $\ominus_y^{ux} \cap \bigcirc uxv$ does not contain any point.*

Let $p_1 \neq y$ in S be such that $\triangle up_1x$ is in T. The next lemma implies that ux is an illegal edge.

Lemma 2. *If $\bigcirc uxy \setminus \ominus_y^{ux}$ contains $k-1$ points, then $\bigcirc up_1x$ contains point y, but does not contain any point of $\ominus_y^{ux}$.*

3 Points in Convex Position

In this section we show that $k-1$ flips are sufficient to transform any order-k triangulation of a convex point set into some other order-k triangulation, such that all the intermediate triangulations are of order $2k-2$.

Before that, we show that our result is tight in how large the flip distance between two order-k triangulations can be.

Theorem 5. *For any $k > 2$ there is a set S_k of $2k+2$ points in convex position such that $G(\mathcal{T}_k(S_k))$ is not connected. Moreover, there is a triangulation T_k in $\mathcal{T}_k(S_k)$ such that in order to transform T_k into any other triangulation in $\mathcal{T}_k(S_k)$ one needs to perform at least $k-1$ flips.*

Proof (sketc.h). We construct set S_k and the triangulation T_k as follows, see Fig. 2a. Start with a horizontal line segment uv and add points $S' = p_1, \ldots, p_k$ above it and points $S'' = q_1, \ldots, q_k$ below it, such that each point q_i is the reflection of the point p_i with respect to the line through uv. Point p_1 is placed close enough to uv, and each next point p_{i+1} for $i = 2, \ldots, k-1$ is: (1) inside $\bigcirc(uq_ip_i)$, (2) below the line through $p_{i-1}p_i$, (3) above the line through uv, and (4) outside $\bigcirc(up_{i-1}p_i)$ (we set $p_0 = v$). The set S_k is $\{u, v\} \cup S' \cup S''$. Triangulation T_k of S_k is formed by all the triangles $\triangle up_ip_{i+1}$ and $\triangle uq_iq_{i+1}$ (where $p_0 = q_0 = v$). It turns out that any $\bigcirc(up_ip_{i+1})$ (resp., $\bigcirc uq_iq_{i+1}$) contains exactly the k points of S'' (resp., of S') and no other point of S_k. Thus, T_k is in $\mathcal{T}_k(S_k)$. We observe that a triangulation of S_k containing edge p_ip_t with $k \geq i > t+1$ cannot be order-k (the case for q_iq_t is symmetric). Consider $T' \neq T_k$ in $\mathcal{T}_k(S_k)$. Each edge in $T' \setminus T_k$ must have one endpoint in S' and one in S''. Thus, edge uv has to be flipped in order to transform T_k to T'. Triangle $\triangle uq_1p_1$ is of order $2k-2$. The second part of the statement follows from the observation that for any i, j with $k \geq i > 0$ and $k \geq j > 0$, the triangle $\triangle up_iq_j$ is of order $2k-i-j$. Thus, $k-1$ flips are needed to get T', since some $\triangle up_iq_j$ with $i+j = k$ has to be in T'. Otherwise uv is in T', a contradiction. □

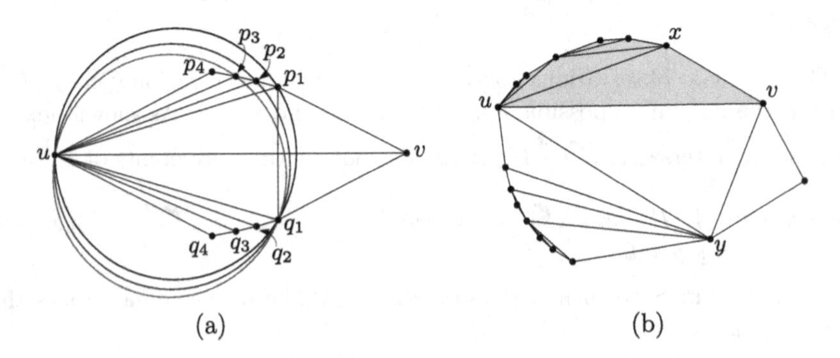

Fig. 2. (a) An order-k triangulation at distance at least $k-1$ from any other order-k triangulation (k=4). (b) The gray area corresponds to T_{uv}^x.

We now proceed to prove the upper bound. Let S be a point set in convex position. Let T be an order-k triangulation of S. We say that T is *minimal* if flipping any illegal edge in T results in a triangulation that is not order-k. Let uv be a diagonal in T and let $\triangle uxv$ and $\triangle uyv$ be the triangles incident to it. Since S is in convex position, the diagonal uv partitions T into two sub-triangulations that only share edge uv. Let T^x_{uv} (respectively, T^y_{uv}) denote the sub-triangulation that contains triangle $\triangle uxv$ (respectively, $\triangle uyv$). See Fig. 2b.

Theorem 6. *For a point set S in convex position and $k \geq 2$, let $T \neq DT(S)$ be a triangulation in $\mathcal{T}_k(S)$. Then, there exists T' in $\mathcal{T}_k(S)$ such that there is a path between T and T' in $G(\mathcal{T}_{2k-2}(S))$ of length at most $k-1$, where each edge of the path corresponds to flipping an illegal edge.*

Proof (sketch). For $k = 2$ the statement is trivial, since $G(\mathcal{T}_2(S))$ is connected. Assume $k \geq 3$. If T is not minimal, then T contains an illegal edge e such that flipping e results in an order-k triangulation. Thus, we assume that triangulation T is minimal. Observe first that there must be an illegal edge uv in T incident to triangle $\triangle uxv$ such that all edges of T^x_{uv} are legal. Indeed, since T is not an order-0 triangulation, T contains an illegal edge. Any triangulation of S has at least two ears, i. e., triangles with two edges in the convex hull of S, and for any ear in T all three of its edges are legal, otherwise T is not minimal.

Let uv be an illegal edge incident to triangle $\triangle uxv$ such that all edges in T^x_{uv} are legal. Let $\triangle uyv$ be the other triangle in T incident to uv. Consider triangulation $T_1 = (T \setminus \{uv\}) \cup \{xy\}$. Since T is minimal, T_1 is not order-k. The only triangles in T_1 that could be not order-k are the new triangles $\triangle uxy$ and $\triangle xyv$. Without loss of generality assume that $\triangle uxy$ is not order-k. By Lemma 1, it follows that $\triangle uxy$ is the only one that is not order-k. In addition, $\triangle uxy$ is order-$(2k-2)$. By Observation 3 it follows that $2 \leq |\bigcirc^{ux}_{y}| \leq k-1$.

By induction on the number of points in $\bigcirc^{ux}_{y}$ we show that T_1 can be transformed into an order-k triangulation T' by flipping at most $k-2$ illegal edges. Moreover, in the sequence of triangulations from T_1 to T' every triangulation is of order $2k-2$. The claim follows. □

The above theorem has two implications. First, due to the known property of illegal edges, each flip in the path from T to T' increases the angle vector of the triangulation. The second implication is as follows. Each edge of an order-k triangulation T is also order-k; there are $O(kn)$ such edges [2,14]. Theorem 6 implies that T can be transformed into $DT(S)$ by a sequence of $O((2k-2)n) = O(kn)$ flips, since only illegal edges are flipped and thus no edge can be flipped twice.

4 General Point Sets

In this section we consider a general point set S. We show that a triangulation of order $k = 3, 4$ or 5 of S can be transformed into some other order-k triangulation of S by flipping at most $k-1$ illegal edges, and that the intermediate triangulations are of order $2k-2$. Moreover, since we flip only illegal edges, after each flip that transforms a triangulation T to T', we have that $\alpha(T') > \alpha(T)$. Thus, if we keep applying this procedure, we will eventually reach $DT(S)$.

Theorem 7. *Let S be a point set in general position and let T be a triangulation in $\mathcal{T}_k(S)$ for $2 \leq k \leq 5$. There exists T' in $\mathcal{T}_k(S)$ such that there is a path from T to T' in $G(\mathcal{T}_{2k-2}(S))$ of length at most $k-1$, where each edge of the path corresponds to flipping an illegal edge.*

In order to prove Theorem 7, we consider whether T is minimal. If not, the statement follows trivially. If T is minimal, then $k > 2$. Also, for any illegal edge uv in T, flipping uv produces a new and unique triangle $\triangle uxy$ that is not order-k. Since $k = 3, 4, 5$, we notice that there are only two cases to consider for the number of points in each region of $\bigcirc uxy$. For $k = 3, 4$, since $\triangle uxy$ is not of order k, by Observation 3 it follows that one of $\bigcirc uxy \setminus \ominus^{ux}_{y}$ and $\bigcirc uxy \setminus \ominus^{uy}_{x}$ contains $k-1$ points. For $k = 5$, if none of $\bigcirc uxy \setminus \ominus^{ux}_{y}$ and $\bigcirc uxy \setminus \ominus^{uy}_{x}$ contains $k-1$ points of S then by Observation 3 and the fact that T is of order k, it follows that each of $\ominus^{ux}_{y}$ and $\ominus^{uy}_{x}$ contains 3 points of S.

Therefore, in order to show that with at most $k-1$ flips the triangulation T can be transformed into some other triangulation T' of order k such that only illegal edges are flipped, for $k = 3, 4, 5$, we consider two cases:

A) [$k = 3, 4, 5$] There are exactly $k-1$ points in $\bigcirc uxy \setminus \ominus^{ux}_{y}$ or exactly $k-1$ points in $\bigcirc uxy \setminus \ominus^{uy}_{x}$. See Fig 3a.

B) [$k = 5$] There are exactly $k-2 = 3$ points in $\ominus^{ux}_{y}$ and exactly $k-2 = 3$ points in $\ominus^{uy}_{x}$. See Fig 3b.

Case A Since the case when there are exactly $k-1$ points in $\bigcirc uxy \setminus \ominus^{ux}_{y}$ is symmetric to the one in $\bigcirc uxy \setminus \ominus^{uy}_{x}$, without loss of generality we assume that there are exactly $k-1$ points in $\bigcirc uxy \setminus \ominus^{ux}_{y}$.

We consider the resulting triangulation T_1 after flipping the illegal edge uv with the following properties: (1) $\triangle uxy$ is not of order k, (2) there are $k-1$ points in $\bigcirc uxy \setminus \ominus^{ux}_{y}$.

We show that there exists an order-k triangulation T' that can be reached from T in $G(\mathcal{T}_{2k-2}(S))$ after flipping at most $k-1$ illegal edges.

Next we summarize the main ideas behind the proof. By Observation 3 there are at least two points in $\ominus^{ux}_{y}$. Consider $p_1 \neq y$ such that $\triangle up_1x \in T$. By Lemma 2, ux is an illegal edge and $\bigcirc up_1x$ does not contain points of $\ominus^{ux}_{y}$. We consider the triangulation $T_2 = (T_1 \setminus \{ux\}) \cup \{p_1y\}$. By Observation 1,

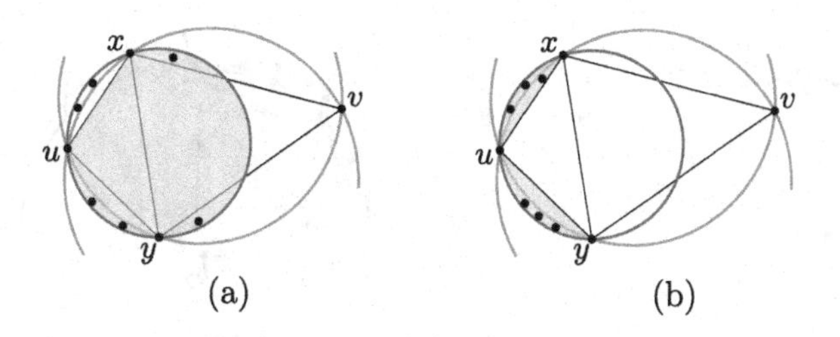

Fig. 3. In both cases, $k = 5$. (a) There are four points in the gray region $\bigcirc uxy \setminus \oslash_{y}^{ux}$. (b) There are exactly three points in $\oslash_{y}^{ux}$ and exactly three points in $\oslash_{x}^{uy}$.

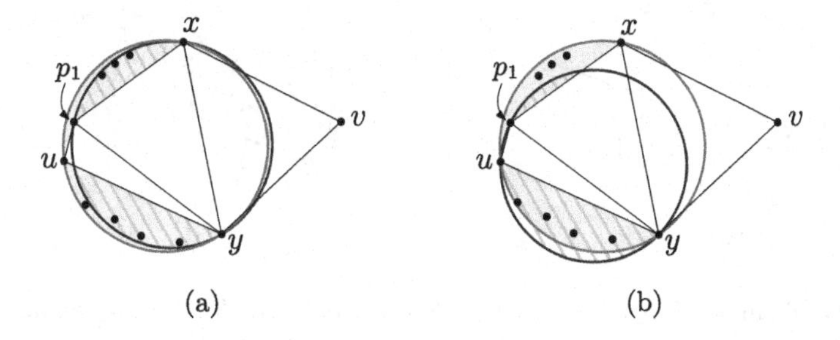

Fig. 4. In both cases: $k = 5$, the gray filled disk is $\oslash_{y}^{ux}$ and $\oslash_{x}^{uy}$ contain 4 points. (a) Disk $\bigcirc p_1xy$ does not contain the 4 points in $\bigcirc uxy \setminus \oslash_{y}^{ux}$. (b) The blue disk $\bigcirc up_1y$ contains the 4 points in $\bigcirc uxy \setminus \oslash_{y}^{ux}$.

$(\bigcirc up_1y \cup \bigcirc p_1xy) \subset (\bigcirc uxy \cup \bigcirc up_1x)$. Then, $\triangle up_1y$ and $\triangle p_1xy$ are order-$(2k-3)$ triangles. Thus, for $k = 3$ the statement follows. For $k = 4, 5$, we consider T_2 that is not of order k. Since $\bigcirc up_1x \cap \oslash_{y}^{ux}$ does not contain points, it follows from Observation 2 that exactly one triangle $\blacktriangle$ of $\triangle up_1y$ and $\triangle p_1xy$ is of order greater than k. So, for $k = 4$ such triangle is an instance of Case A. So, using analogous arguments, we show that with 3 flips T is transformed into another order-k triangulation T', by flipping only illegal edges. Similarly, the statement follows for $k = 5$, when is an instance of Case A. Otherwise, we show that $|\oslash_{y}^{ux}| = |\oslash_{x}^{uy}| = k - 1$ and that the triangle $\blacktriangle$ with order 6 (since $|\bigcirc \blacktriangle \setminus \oslash_{y}^{ux}| < k - 1 = 4)$) is $\triangle p_1xy$. See Fig 4. Moreover, we show that p_1x is an illegal edge and that the triangulation T_3 obtained after flipping p_1x is of order 6. In addition, if T_3 is not of order 5, then we get again a Case A and we apply the same arguments as before and show that with 4 flips, the statement follows for $k = 5$.

Case B We consider the resulting triangulation T_1 after flipping uv such that Case B occurs: (1) $\triangle uxy$ is of order greater than 5, (2) there are exactly three points in both $\oslash_{y}^{ux}$ and $\oslash_{x}^{uy}$.

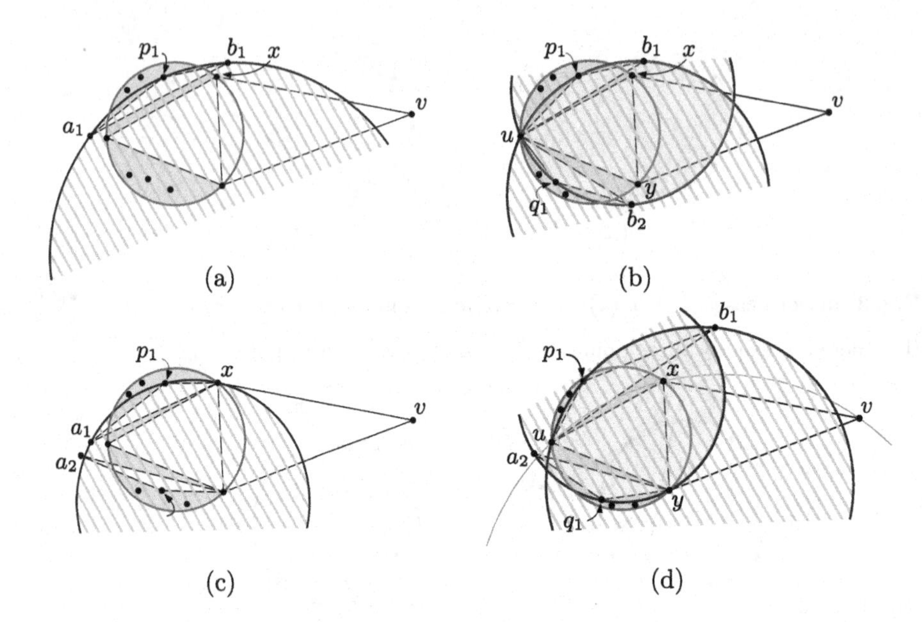

Fig. 5. (a) Triangle $\triangle a_1p_1b_1$ is not order-5. (b) Triangle $\triangle up_1b_2$ is not order-5. (c) Triangle $\triangle a_2q_1y$ is not order-5. (d) Triangles $\triangle up_1b_1$ and $\triangle a_2q_1y$ are order-5.

We show that there exists an order-5 triangulation T' that can be reached from T in the flip graph of $\mathcal{T}_{2k-2}(S)$ by flipping at most four illegal edges.

Next we summarize the main ideas behind the proof. We consider two cases depending on whether one of ux or uy are illegal, or not. If one of ux or uy is an illegal edge, say ux, then we consider the triangulation $T_2 = (T_1 \setminus \{ux\}) \cup \{p_1x\}$ where $p_1 \in \bigcirc^{ux}_{y}$. Using Observation 2 and counting arguments, we conclude that T_2 is of order $2k - 4 = 6$. Moreover, exactly one of the triangles $\triangle up_1y$ or $\triangle p_1xy$ is of order 6 and that it is an instance of Case A. Thus, using analogous arguments as in Case A, the statement follows. If none of ux and uy are illegal edges, then we show that there exists an illegal edge ub_1 in $\triangle ub_1x$ and $\triangle up_1b_1$ such that b_1 is in either $\bigcirc uxv \setminus \bigcirc uxy$ or $\bigcirc uyv \setminus \bigcirc uxy$, say $\bigcirc uxv \setminus \bigcirc uxy$, and $\in \bigcirc^{ux}_{y}$. See Fig. 5d. Otherwise T is not of order 5, a contradiction. See Fig. 5 for all possible cases. Hence, we consider $T_2 = (T_1 \setminus \{ub_1\}) \cup \{p_1x\}$ and show that T_2 is of order 6. If T_2 is not of order 5, then we consider $T_3 = (T_2 \setminus \{ux\}) \cup \{p_1y\}$ since $\in \bigcirc^{ux}_{y}$. Therefore, T_3 is again an order-6 triangulation and if T_3 is not of order 5 then we are again in an instance of Case A. Using similar arguments as in Case A, the statement follows. Therefore, Theorem 7 holds.

Finally, since flipping illegal edges always increases the angle vector and there are $O(kn)$ order-k edges (see [2,14]), it follows from Theorem 7 that for $k \leq 5$, any order-k triangulation can be transformed into $DT(S)$ by a sequence of at most $O(kn)$ triangulations of order $2k - 2$.

5 Conclusions

In this paper we presented the first general results on the flip graph of order-k Delaunay triangulations. We showed that already for points in convex position, the flip graph may not be connected. This is in contrast to the flip graph of triangulations that consist of order-k *edges*, for which the flip graph is always connected [2]. Our main result is that $k-1$ flips are sometimes necessary and always sufficient to transform an order-k triangulation into some other order-k triangulation, for any $k \geq 2$ if the points are in convex position. Moreover, we proved that these $k-1$ flips go through triangulations of order $2k-2$. This is a noteworthy result, and one of the first results on order-k Delaunay triangulations proven for any value of k. In the setting of general point sets , we also showed that for $k = 3, 4, 5$, the order-k triangulations are at flip distance at most $k-1$ from some other order-k triangulation within the flip graph of order-$(2k-2)$ triangulations. This result also implies that the flip distance between any two order-k triangulations is $O(kn)$, which is consistent with the fact that the diameter of the flip graph of all triangulations is $\Theta(n^2)$.

Our results imply an enumeration algorithm using the Avis and Fukuda framework [5], generalizing the results that Abe and Okamoto obtained for $k \leq 2$ [1]. For the case of convex position this is not of practical importance, as in that case one can obtain a more efficient method by first pre-computing all order-k triangles, and then recursively enumerating all order-k triangulations. However, our results imply the first non-trivial enumeration results for points in generic position for $3 \leq k \leq 5$. It should be mentioned that small values of k are the most important ones in practice, since for small orders the triangle shape is still close to Delaunay, but at the same time they are enough to obtain significantly better triangulations [23].

Clearly, the main question left open is what happens in general when $k \geq 6$. For larger orders our techniques present issues due to a large increase in the number of cases that need to be considered. However, we conjecture that the same results obtained for convex position hold in general, and in particular, that any order-k triangulation is at flip distance at most $k-1$ from another order-k triangulation.

References

1. Abe, Y., Okamoto, Y.: On algorithmic enumeration of higher-order Delaunay triangulations. In: Proceedings of the 11th Japan-Korea Joint Workshop on Algorithms and Computation, Seoul, Korea, pp. 19–20 (2008)
2. Abellanas, M., Bose, P., García, J., Hurtado, F., Nicolás, C.M., Ramos, P.: On structural and graph theoretic properties of higher order Delaunay graphs. Int. J. Comput. Geom. Appl. **19**(06), 595–615 (2009)
3. Aichholzer, O., et al.: Triangulations intersect nicely. Discrete Comput. Geom. **16**(4), 339–359 (1996). https://doi.org/10.1007/BF02712872
4. Aurenhammer, F., Klein, R., Lee, D.T.: Voronoi Diagrams and Delaunay Triangulations. World Scientific Publishing Company, River Edge (2013)

5. Avis, D., Fukuda, K.: Reverse search for enumeration. Discrete Appl. Math. **65**(1–3), 21–46 (1996)
6. Benkert, M., Gudmundsson, J., Haverkort, H., Wolff, A.: Constructing interference-minimal networks. In: Wiedermann, J., Tel, G., Pokorný, J., Bieliková, M., Štuller, J. (eds.) SOFSEM 2006. LNCS, vol. 3831, pp. 166–176. Springer, Heidelberg (2006). https://doi.org/10.1007/11611257_14
7. Bern, M., Eppstein, D.: Mesh generation and optimal triangulation. Comput. Euclid. Geom. **1**, 23–90 (1992)
8. Biniaz, A., Dastghaibyfard, G.: Slope fidelity in terrains with higher-order Delaunay triangulations. In: Proc. of the 16th International Conference in Central Europe on Computer Graphics, Visualization and Computer Vision, pp. 17–23 (2008)
9. Bose, P., Hurtado, F.: Flips in planar graphs. Comput. Geom. **42**(1), 60–80 (2009)
10. De Floriani, L.: Surface representations based on triangular grids. Vis. Comput. **3**(1), 27–50 (1987)
11. De Kok, T., Van Kreveld, M., Löffler, M.: Generating realistic terrains with higher-order Delaunay triangulations. Comput. Geom. **36**(1), 52–65 (2007)
12. Dumitrescu, A., Schulz, A., Sheffer, A., Tóth, C.D.: Bounds on the maximum multiplicity of some common geometric graphs. SIAM J. Discrete Math. **27**(2), 802–826 (2013)
13. Fortune, S.: Voronoi Diagrams and Delaunay Triangulations, pp. 225–265. World Scientific (1995)
14. Gudmundsson, J., Hammar, M., van Kreveld, M.: Higher order Delaunay triangulations. Comput. Geom. **23**(1), 85–98 (2002)
15. Gudmundsson, J., Haverkort, H.J., Van Kreveld, M.: Constrained higher order Delaunay triangulations. Comput. Geom. **30**(3), 271–277 (2005)
16. Hurtado, F., Noy, M., Urrutia, J.: Flipping edges in triangulations. Discrete Comput. Geom. **22**(3), 333–346 (1999)
17. van Kreveld, M.J., Löffler, M., Silveira, R.I.: Optimization for first order Delaunay triangulations. Comput. Geom. **43**(4), 377–394 (2010)
18. Lawson, C.L.: Transforming triangulations. Discrete Math. **3**(4), 365–372 (1972)
19. Lawson, C.L.: Software for C1 surface interpolation. In: Mathematical Software, pp. 161–194. Elsevier (1977)
20. Lubiw, A., Pathak, V.: Flip distance between two triangulations of a point set is NP-complete. Comput. Geom. **49**, 17–23 (2015)
21. Mitsche, D., Saumell, M., Silveira, R.I.: On the number of higher order Delaunay triangulations. Theor. Comput. Sci. **412**(29), 3589–3597 (2011)
22. Pilz, A.: Flip distance between triangulations of a planar point set is APX-hard. Comput. Geom. **47**(5), 589–604 (2014)
23. Rodríguez, N., Silveira, R.I.: Implementing data-dependent triangulations with higher order Delaunay triangulations. ISPRS Int. J. Geo-Inf. **6**(12), 390 (2017)
24. Sharir, M., Sheffer, A.: Counting triangulations of planar point sets. Electr. J. Comb. **18**(1) (2011)
25. Sibson, R.: Locally equiangular triangulations. Comput. J. **21**(3), 243–245 (1978)
26. Silveira, R.I., van Kreveld, M.: Optimal higher order Delaunay triangulations of polygons. Comput. Geom. **42**(8), 803–813 (2009)
27. Silveira, R.I., van Kreveld, M.J.: Towards a definition of higher order constrained Delaunay triangulations. Comput. Geom. **42**(4), 322–337 (2009)
28. Sleator, D.D., Tarjan, R.E., Thurston, W.P.: Rotation distance, triangulations, and hyperbolic geometry. J. Amer. Math. Soc. **1**(3), 647–681 (1988)

An $\Omega(n^3)$ Lower Bound on the Number of Cell Crossings for Weighted Shortest Paths in 3-Dimensional Polyhedral Structures

Frank Bauernöppel[1(✉)], Anil Maheshwari[2], and Jörg-Rüdiger Sack[2]

[1] Hochschule für Technik und Wirtschaft, Computer Engineering, Berlin, Germany
frank.bauernoeppel@htw-berlin.de
[2] Carleton University, School of Computer Science, Ottawa, Canada
{anil,sack}@scs.carleton.ca

Abstract. A new lower bound of $\Omega(n^3)$ on the maximum number of cell crossings for weighted shortest paths in 3-dimensional polyhedral structures consisting of a linear number of $\mathcal{O}(n)$ polyhedral cells and cell faces is derived. This is a generalization and sharpening of the formerly known $\Omega(n^2)$ lower bound on the maximum number of cell crossings for weighted shortest path in 2-dimensional polyhedral structures and has been a long-standing open problem for the 3-dimensional case.

Keywords: Computational geometry · Algorithms · Computational complexity · Weighted shortest path · Lower bound

1 Introduction

1.1 Motivation

Shortest path problems and their practical solutions are of significant interest not only in computer science but also in areas such as optimization, geo-sciences, and robotics. They are also omnipresent in today's routing appliances. Already in ancient times, Hero of Alexandria (Heron) formulated a "Principle of the Shortest Path of Light" stating that light follows a shortest path when reflected on flat mirrors [9]. This was later generalized to paths passing through different media (Snell's law of refraction).

Efficient shortest path algorithms for graph having weighted edges like Dijkstra's Algorithm [8] or the A^* algorithm [13] are well studied and practical implementations are readily available in many software libraries like the Boost C++ library [4].

In many geometric shortest path problems however, the path is not restricted to use the edges of a finite graph but may freely travel along 2-dimensional surfaces or pass through 3-dimensional bodies. In addition to dimensionality, the problem gets more complex when the domains are not homogeneous. In

Y. Kohayakawa and F. K. Miyazawa (Eds.): LATIN 2020, LNCS 12118, pp. 235–246, 2020.
https://doi.org/10.1007/978-3-030-61792-9_19

such applications, weighted shortest paths provide much more meaningful results than Euclidean distance. Such problems include e.g., path finding in terrains exhibiting different features (e.g. rock, water, forest), planning optimum routes for new streets or power lines in rural areas (2D, 2.5D), or for accessing new geological deposits and reservoirs (3D). Other examples include optimum path-planning in robotics and minimally invasive surgery. The class of weighted region shortest path problems has first been investigated by [10].

For the weighted region problem, a number of NP-hardness instances have been identified see e.g., [6,11]. More recently, an unsolvability result was established by Carufel et al. [7]. (For a survey article see e.g., Bose et al. [5].) Therefore, approximation algorithms have been devised, such as the $(1+\epsilon)$-approximation schemes developed by Aleksandrov et al. for the 2-dimensional case [3] or by Sun and Reif [17], in query mode [1], as well as for the 3-dimensional case [2].

Euclidean and weighted problems also exhibit very interesting structural differences; these are also of importance for the complexity analysis. In the Euclidean setting, paths may cross a convex cell (e.g., triangle) at most once. At first glance it might be surprising that there exist 2-dimensional polyhedral structures consisting of n convex cells and having weighted shortest paths consisting of as many as $\Theta(n^2)$ straight line segments. For these, $\Theta(n)$ convex cells will be crossed $\Theta(n)$ times each. Such polyhedral structures were first described by Mitchell and Papadimitriou [10].

This raises the question about the maximum possible number of straight line segments in the 3-dimensional case. This question remained open since 1991 until now. In the remainder of this paper we establish a lower bound for the 3-dimensional case by constructing novel 3-dimensional polyhedral structures $Q(n)$ consisting of $\mathcal{O}(n)$ convex cells in which a weighted shortest path has $\Theta(n^3)$ cell crossings and straight line path segments.

1.2 Previous Work

For the 2-dimensional case, a construction of weighted planar subdivisions with weighted shortest paths consisting of $\Theta(n^2)$ cell crossings has been presented in 1991 by Mitchell and Papadimitriou [10].

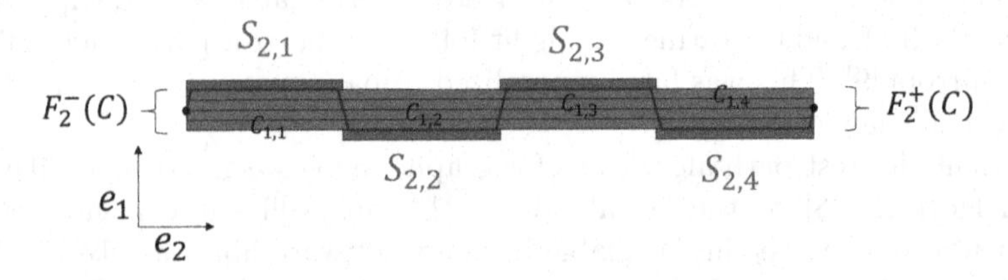

Fig. 1. A 2-dimensional polyhedral structure with a weighted shortest path (yellow) from a point of the left facet to a point of the right facet for $n = 4$. The polyhedral structure consists of n central polyhedral cells (red) and n satellite cells (green) of lesser weight. The weighted shortest path crosses the n red cells n^2 times in total.

Figure 1 resembles the construction by Mitchell and Papadimitriou [10] for the 2-dimensional case and $n = 4$. The weighted shorted path (yellow) starts on the facet $F_2^-(C)$ and terminates on the facet $F_2^+(C)$ of the central rectangular area C (red). The central red area is partitioned into n horizontal slices, each having a weight of 1. In contrast, the n green satellite areas have a lower weight of 1/4 only. If the width of the central area (along e_2) is sufficiently large and its height (along e_1) is sufficiently small, a weighted shortest path will use all the green satellite areas thereby crossing the red slices $\Theta(n^2)$ times in total. Detailed measures and proofs are derived for the 3-dimensional case in Sect. 4.

Although the $\Omega(n^2)$ lower bound trivially extends to higher dimensions, up to now no higher bound has been derived for the 3-dimensional case.

1.3 Example of the New 3-Dimensional Construction for $n = 6$

Figure 2 shows an approximate weighted shortest path in the novel 3-dimensional polyhedral structure $Q(n)$ for $n = 6$. An approximate weighted shortest path (yellow) from a point of the front facet to a point of the rear facet was computed using the wsp3dovm [19] software. The weighted shortest path crosses each of the n red slices at least n^2 times which makes a total of $\Theta(n^3)$ cell crossings. The wsp3dovm [19] software has been made available by the authors under an open source software license and uses the OpenVolumeMesh [12] library for representing 3D polyhedral structures. For calculating the approximate weighted shortest path, a tetrahedralization of $Q(6)$ was used. The tetrahedralization was generated automatically from an initial concise description of the cells of $Q(6)$ by using the tetgen software [15], [16]. The results were visualized by using ParaView [14] from the Visualization Toolkit [18]. Again, detailed measures and proofs are derived in Sect. 4.

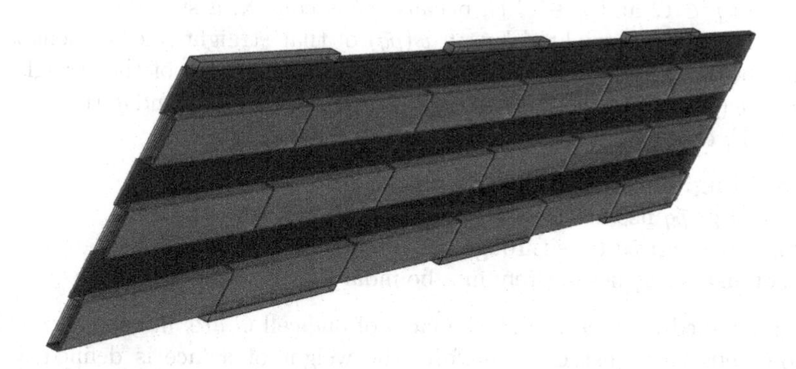

Fig. 2. Polyhedral structure $Q(n)$ for $n = 6$ with a zig-zagging weighted shortest path (yellow) from a point of the front facet (shown left) to a point of the rear facet (shown right). The front and rear facets resemble the 2-dimensional case shown in Fig. 1. Six new blue cuboids were attached to the top and bottom facets, having the least weight and a sufficient length to attract the weighted shortest path. Thus, this path has $\Theta(n^3)$ cell crossings of the $\mathcal{O}(n)$ red cells in total. (Color figure online)

2 Preliminaries

2.1 Polyhedral Cells and Polyhedral Structures

The basic definitions and concepts are introduced in this subsection which loosely follow those of [3] and [2]. Let $\mathbb{R}^3$ denote the 3-dimensional Euclidean vector space and let $E = \{e_1, e_2, e_3\}$ be its standard basis. A *polyhedral cell* $C \subset \mathbb{R}^3$ is the convex hull of a finite number of points in $\mathbb{R}^3$. The points which are not contained in the relative interior of C are called the *vertices* of C. The 1-dimensional boundary elements of C are called the *edges* of C, and the 2-dimensional boundary elements of C are called the *facets* of C. The facets, edges, and vertices of a polyhedral cell C are commonly called the *faces* of C.

A 3-dimensional *polyhedral structure* $P \subset \mathbb{R}^3$ is the union of a finite number of polyhedral cells $P = \cup_{i=1}^{k} C_i$ where the intersection of any pair of different polyhedral cells is either empty or a common face. All cells and the polyhedral structure P itself are assumed to be bounded, closed, and connected.

An axis-parallel *cuboid* C is a special polyhedral cell which has for each dimension $i = 1, 2, 3$ exactly one pair of parallel facets $F_i^-(C)$ and $F_i^+(C)$ orthogonal to basis vector e_i. The six facets are commonly called the left/right, bottom/top, and front/rear facet of the cuboid cell C. An axis-parallel cuboid can be defined by one point $v = F_1^-(C) \cap F_2^-(C) \cap F_3^-(C)$, the *origin* of cuboid C, and a size triple (s_1, s_2, s_3), where s_i denotes the distance between $F_i^-(C)$ and $F_i^+(C)$.

2.2 Weighted Shortest Paths in Polyhedral Structures

A positive real number, the *weight*, $w(C)$ is associated with each cell C of a polyhedral structure P. For a single polyhedral cell C, the shortest path between two points $p \in C$ and $q \in C$ is, because C is convex, a straight line segment denoted by $\overline{pq}$. The weighted length $w(\overline{pq})$ of that straight line segment is the product of the weight of the cell and the Euclidean length of the straight line segment: $w(\overline{pq}) = w(C)\|pq\|$. Depending on the location of p and q, the following cases for $\overline{pq}$ can be identified:

cell avoiding: $\overline{pq}$ has no point in common with the cell
cell visiting: $\overline{pq}$ has at least one point in common with the cell
  cell crossing: $\overline{pq}$ runs through the interior of the cell
  facet using: $\overline{pq}$ lies entirely in a boundary facet of the cell

In a polyhedral structure P, the faces of one cell C are, in general, incident to more cells than just C. Therefore, the weight of a face is defined as the minimum weight among all its incident cells. When a straight line segment $\overline{pq} \subseteq C$ runs along such a face, the minimum weight of all cells incident to $\overline{pq}$ is used to calculate the weight contribution of that straight line segment $w(\overline{pq}) = min\{w(C_i)\|pq\| : \overline{pq} \subseteq C_i\}$. Commonly, the straight line segment $\|pq\|$ *inherits* the weight from an incident cell C_i of minimum weight.

A *piece-wise linear* path $\overline{p_1 p_2 ... p_k}$ in a polyhedral structure P is a sequence of points $p_1, p_2, ..., p_k \in P$ where each consecutive pair of points p_i and p_{i+1} lies on the boundary of a common cell C_i of P. The weighted length of a piece-wise linear path is defined as the sum of the weighted lengths of the straight line segments involved: $w(\overline{p_1 p_2 ... p_k}) = \sum_{i=1}^{k-1} w(\overline{p_i p_{i+1}})$.

A *general* path $\tilde{p} = \widetilde{vw}$ in polyhedral structure P is a rectifiable (finite length) curve $\widetilde{vw} \subseteq P$ starting at some point $v \in P$ and terminating at point $w \in P$. The intersection of a general path $\tilde{p}$ with all boundaries of the cells of polyhedral structure P partitions $\tilde{p}$ into a finite sequence of consecutive subpaths $\tilde{p} = \tilde{p}_1 \tilde{p}_2 ... \tilde{p}_m$. The weighted length (cost) of $\tilde{p}$ is now defined by $w(\tilde{p}) = \sum_{i=1}^{m} w_i \|\tilde{p}_i\|$, where w_i is the weight of a cell of minimum weight among all cells containing subpath $\tilde{p}_i$ and $\|\tilde{p}_i\|$ denotes its Euclidean length.

Finally, a *weighted shortest path* in P from a point $p \in P$ to a point $q \in P$ is defined as a general path starting at p and terminating at q having minimum weighted length among all such paths.

2.3 Properties of Weighted Shortest Paths in Polyhedral Structures

A few useful properties of weighted shortest paths in polyhedral structures can be observed which follow from the convexity of the polyhedral cells. The proofs are omitted here.

Lemma 1. *Every weighted shortest general path in a polyhedral structure is a piece-wise linear path.*

Because of Lemma 1, we can restrict ourselves to searching for piece-wise linear paths having minimum weighted length.

Lemma 2. *Let $\widetilde{vw}$ be a weighted shortest path in a polyhedral structure P. Let C be a polyhedral cell of weight $w(C)$ in polyhedral structure P that is visited by $\widetilde{vw}$ more than once. I.e., the path $\widetilde{vw}$ leaves C at some point p and re-enters the cell C later at some other point $q \neq p$. Then, the subpath $\widetilde{pq}$ visits at least one cell of P with lower weight.*

Lemma 3. *Let C be a cell of minimum weight $w(C)$ among all cells of the polyhedral structure P. Then, a weighted shortest path $\widetilde{vw}$ in P visits C at most once.*

3 A Property of Weighted Shortest Paths Crossing a Cuboid

In this section, a sufficient condition for the structure of certain weighted shortest paths crossing a cuboid is derived in Theorem 1. This condition is the foundation for the construction of the novel 3-dimensional polyhedral structures $Q(n)$ presented in Sect. 4. We illustrate the following Theorem 1 in Fig. 3.

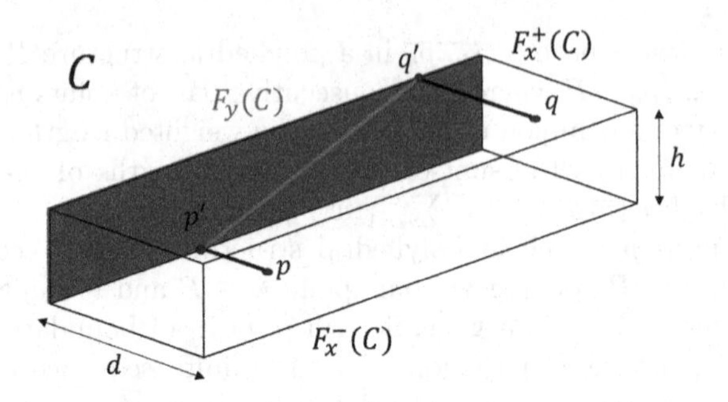

Fig. 3. Illustration of Theorem 1.

Theorem 1. *Let C be a 3-dimensional axis-parallel cuboid, let $F_x^-(C)$ and $F_x^+(C)$ be a pair of parallel facets of C, and let $F_y(C)$ be another facet of C. Let h denote the Euclidean distance between the facets $F_x^-(C)$ and $F_x^+(C)$, and let d denote the maximum Euclidean distance from a point in C to the facet $F_y(C)$. Let w_{min}, respectively $w_{max} = 1$, be the minimum, respectively maximum, weight in C including its borders, except for the facet $F_y(C)$ having the smallest weight, w_y. If the condition*

$$d/h \leq w_y < w_{min}/3 \tag{1}$$

is met, then every weighted shortest path $\widetilde{pq} \subset C$ between a point $p \in F_x^-(C)$ and a point $q \in F_x^+(C)$ is using the facet $F_y(C)$.

Proof. Suppose, by contradiction, that the statement is not true. Then, $\widetilde{pq}$ entirely avoids $F_y(C)$ and the following lower bound holds for the weighted length $w(\widetilde{pq})$ of path $\widetilde{pq}$

$$w_{min}\|pq\| \leq w(\widetilde{pq}). \tag{2}$$

Now, a better upper bound for the weighted length of a certain piece-wise linear path $w(\overline{pp'q'q})$ using $F_y(C)$ can be derived, which yields a contradiction.

Let p' be the projection of point p onto the facet $F_y(C)$, and let q' be the projection of point q onto the same facet. The Euclidean distances $\|pp'\|$ and $\|q'q\|$ are both bounded from above by d

$$\|pp'\| \leq d \tag{3}$$

$$\|q'q\| \leq d. \tag{4}$$

Since straight line segments $\overline{pp'}$ and $\overline{qq'}$ run entirely in C and have a weight of at most $w_{max} = 1$ and because of the left side of Eq. 1

$$w(\overline{pp'}) \leq w_y h. \tag{5}$$

$$w(\overline{q'q}) \leq w_y h. \tag{6}$$

The Euclidean distance of the straight line segment $\|p'q'\|$ is bounded from above by

$$\|p'q'\| \leq \|pq\|. \tag{7}$$

Because of $\overline{p'q'} \subset F_y(C)$ having weight w_y

$$w(\overline{p'q'}) \leq w_y \|pq\|. \tag{8}$$

Summing up and considering $h \leq \|pq\|$, one gets

$$w(\overline{pp'q'q}) \leq 3w_y \|pq\|. \tag{9}$$

Now, because of the right side of Eq. 1, we get the contradiction

$$w(\overline{pp'q'q}) < w(\widetilde{pq}) \tag{10}$$

which completes the proof. ∎

Note, that for the special case that p and q lie on the same facet, say $F_z(C)$, orthogonal to $F_x^-(C)$, $F_x^+(C)$, and $F_y(C)$, Theorem 1 yields a sufficient condition for the 2-dimensional case as well. Therefore, this construction applies to the $2-d$ and $3-d$ problem settings.

4 Construction of the Polyhedral Structure $Q(n)$

In this section, the novel polyhedral structure $Q(n)$ is constructed as the union of several axis-parallel cuboids:

- n central cuboids $C_{1,1}, C_{1,2}, ..., C_{1,n}$
- n satellite cuboids of rank 2 $S_{2,1}, S_{2,2}, ..., S_{2,n}$
- n satellite cuboids of rank 3 $S_{3,1}, S_{3,2}, ..., S_{3,n}$
- two shims T_2^- and T_2^+

We first observe that the total number of cells in $Q(n)$ is $\mathcal{O}(n)$. The construction is illustrated for $n = 4$ in Fig. 4.

For the construction, a weight value w_i and a length value l_i are defined for each dimension $i = 1, 2, 3$ by

$$w_i = \begin{cases} 1 & i = 1 \\ w_{i-1}/4 & i = 2, 3 \end{cases} \tag{11}$$

$$l_i = \begin{cases} 1 & i = 1 \\ n \cdot l_{i-1}/w_i & i = 2, 3. \end{cases} \tag{12}$$

4.1 Construction of the Central Cuboids

For $j = 1, 2, ..., n$, the central cuboid $C_{1,j}$ has its origin at point $((j-1)l_1, 0, 0)$ and sizes (l_1, nl_2, nl_3). Therefore, each consecutive pair of central cuboids $C_{1,j}$ and $C_{1,j+1}$ is adjacent, sharing one common facet: $F^+(C_{1,j}) = F^-(C_{1,j+1})$. The weight, w_1, of $C_{1,j}$ is set to 1.

The union of all central cuboids is denoted by $C = \cup_{j=1}^n C_{1,j}$. C has a cuboid like shape by itself with a size triple of (nl_1, nl_2, nl_3).

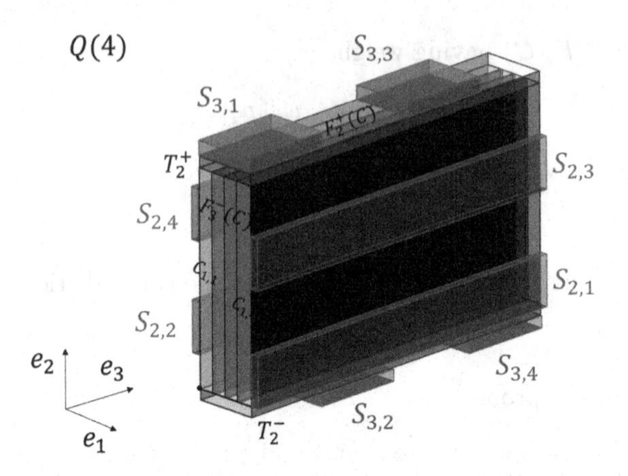

Fig. 4. Illustration of Q(4) showing the four central cuboids (red), the four satellite cuboids of rank 2 (green) and rank 3 (blue), and the two shims (orange). (Color figure online)

4.2 Adding the Satellite Cuboids of Rank 2

The n satellite cuboids of rank 2 will be attached in zig-zag order to the facets $F^-(C_{1,1})$ and $F^+(C_{1,n})$ as shown in Fig. 4.

More formally, for $j = 1, 2, ..., n$, if j is odd, $S_{2,j}$ has its origin at point $(nl_1, (j-1)l_2, 0)$, or, if j is even, at point $(-1, (j-1)l_2, 0)$. The size of each $S_{2,j}$ is $(1, l_2, nl_3)$ and the weight of each $S_{2,j}$ is set to $w_2 = 1/4$.

By construction, the $S_{2,j}$ are adjacent to $C_{1,1}$ if j even, and adjacent to $C_{1,n}$ if j odd. Because the height of the satellite cuboids is by a factor of n smaller than the height of the central cuboids, the facets $F^-(C_{1,1})$, respectively $F^+(C_{1,n})$, need to be partitioned into n congruent pieces of height l_2 each, in order to match the $F^+(S_{2,j})$, respectively $F^-(S_{2,j})$.

Strictly speaking, $C_{1,1}$ and $C_{1,n}$ are no cuboids but cuboid-like shaped polyhedral cells. But, since only two central cuboids are affected, the total number of facets, edges, and vertices remains $\mathcal{O}(n)$.

The union of all central cuboids and satellite cuboids of rank 2 is denoted by $R = C \cup S_{2,1} \cup S_{2,2} \cup ... \cup S_{2,n}$. R will be a subset of Q.

4.3 Weighted Shortest Paths in Polyhedral Structure R

The front facet $F_3^-(C)$ of the 3-dimensional polyhedral structure R resembles the 2-dimensional construction of Mitchell and Papadimitriou [10], cf. Fig. 1. Now, weighted shortest paths starting on the top facet $F_2^+(C)$ and terminating on the bottom facet $F_2^-(C)$, or vice versa, are considered. It will be shown that such paths visit all the satellite cuboids of rank 2. As a consequence, such paths will consist of $\Theta(n^2)$ cell crossings.

In the following, let v be a point on $F_2^-(C)$, let w be a point on $F_2^+(C)$, and let $\widetilde{vw}$ be a weighted shortest path from v to w in the polyhedral structure R.

Lemma 4. *Let v, w, and $\widetilde{vw}$ be as defined above. Then, the weighted shortest path $\widetilde{vw}$ is visiting all n satellite cuboids $S_{2,j}$ of rank 2.*

Proof. Let $S_{2,j}$ be a satellite cuboid of rank 2. Let $C_{2,j}$ denote the intersection of the central cuboid C with a cuboid with its origin at point $(0, (j-1)l_2, 0)$ and sizes (nl_1, l_2, nl_3), i.e. $C_{2,j}$ and $S_{2,j}$ share a common facet $C_{2,j} \cap S_{2,j}$, which is either $F_1^+(S_{2,j})$ or $F_1^-(S_{2,j})$, depending on j.

Since $\widetilde{vw}$ runs from $v \in F_2^-(C)$ and terminates in $w \in F_2^+(C)$, it must cross $C_{2,j}$. Let p be the point where $\widetilde{vw}$ enters $C_{2,j}$, and let q be the point where $\widetilde{vw}$ exits $C_{2,j}$. As one cannot exclude yet that there are more than one such p and q, the last possible entry point is chosen for p and the first exit point after p is chosen for q.

Now, Theorem 1 can be applied to cuboid $C_{2,j}$ and facet $C_{2,j} \cap S_{2,j}$, because the sufficient condition is met with $d = nl_1$, $h = l_2 = nl_1/w_2$, $w_{min} = w_{max} = w_1 = 1$ and $w_y = w_2 = 1/4$. ∎

Lemma 5. *Let v, w, and $\widetilde{vw}$ be as defined above. Then, $\widetilde{vw}$ is visiting every satellite cuboid $S_{2,j}$ of rank 2 exactly once.*

Proof. Every satellite cuboid $S_{2,j}$ is visited because of Lemma 4. The same satellite cuboid $S_{2,j}$ cannot be visited more than once in R because it is a cell of minimum weight in R, see Lemma 3. ∎

Lemma 6. *Let v, w, and $\widetilde{vw}$ be as defined above. Then, $\widetilde{vw}$ is visiting all n satellite cuboids $S_{2,j}$ of rank 2 in R in their natural order $1, 2, ..., n$.*

Proof. Suppose, by contradiction, that the Lemma does not hold. Then, there exist two satellite cuboids $S_{2,j}$ and $S_{2,k}$ such that $j < k$ but $S_{2,k}$ is visited before $S_{2,j}$. Because $\widetilde{vw}$ ends on facet $F_2^+(C)$, it then must re-visit satellite cuboid $S_{2,k}$, which contradicts Lemma 5. ∎

Because each pair of consecutive satellite cuboids of rank 2 is situated on opposite facets of C, the central cuboid C is crossed by $\widetilde{vw}$ $\Theta(n)$ times and, since $C = \cup_{j=1}^{n} C_{1,j}$, each crossing of C implies $\Theta(n)$ cell crossings; this is captured in the following Theorem.

Theorem 2. *Let v, w, and $\widetilde{vw}$ be as defined above. Then, the weighted shortest path $\widetilde{vw}$ has a total of $\Theta(n^2)$ cell crossings.*

4.4 Adding the Satellite Cuboids of Rank 3

Similarly to the satellite cuboids $S_{2,j}$, n satellite cuboids $S_{3,j}$ will be attached in zig-zag order to the facets $F_2^-(C)$ and $F_2^+(C)$ of R; this is illustrated in Fig. 5.

More formally, for $j = 1, 2, ..., n$, if j is odd, at point $(0, nl_2, (j-1)l_3)$, or, if j is even, $S_{3,j}$ has its origin at point $(0, -1, (j-1)l_3)$. The size of each $S_{3,j}$ is $(nl_1, 1, l_3)$ and the weight of each $S_{3,j}$ is set to $w_3 = 1/16$.

The union of R and all the satellite cuboids of rank 3 is denoted by $Q' = R \cup S_{3,1} \cup S_{3,2} \cup ... \cup S_{3,n}$.

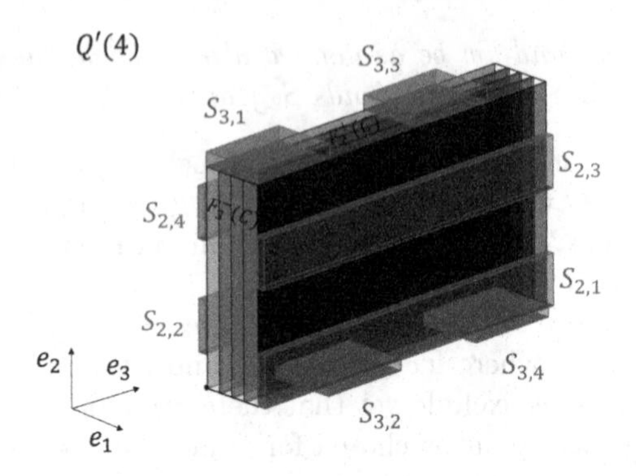

Fig. 5. Illustration of Q'(4) showing the central cuboid C (red) and the four satellite cuboids $S_{2,j}$ of rank 2 (green) and the four satellite cuboids $S_{3,j}$ of rank 3 (blue). (Color figure online)

4.5 Weighted Shortest Paths in Polyhedral Structure Q'

Now, a weighted shortest path $\widetilde{vw}$ starting at a point v on the front facet $F_3^-(C)$ and terminating at a point w on the rear facet $F_3^+(C)$ is considered. Analogously to Subsect. 4.3, it can be shown that this path visits all satellite cuboids of rank 3 exactly once and in their natural order. This is, because Theorem 1 can be applied to cuboid $C_{3,j}$ and facet $S_{3,j} \cap C_{3,j}$ with $d = nl_2$, $h = l_3 = nl_2/w_3$, $w_{min} = w_2 = 1/4$, $w_{max} = w_1 = 1$, and $w_y = w_3 = 1/16$. Here $C_{3,j}$ denotes a cuboid with its origin at point $(0, 0, (j-1)l_3)$ and sizes (nl_1, nl_2, l_3).

Now, let $S_{3,j}$ and $S_{3,j+1}$ be a pair of consecutive satellite cuboids of rank 3, let p be the last point on the weighted shortest path $\widetilde{vw}$ visiting $S_{3,j}$ and let q be the first point on the weighted shortest path $\widetilde{vw}$ visiting $S_{3,j+1}$. Then, from Subsect. 4.3 it follows that the subpath $\widetilde{pq}$ of the weighted shortest path $\widetilde{vw}$ has $\Theta(n^2)$ cell crossings.

Because there are $n-1$ pairs of consecutive satellite cuboids of rank 3, the weighted shortest path $\widetilde{vw}$ has $\Theta(n^3)$ cell crossings in total.

4.6 Adding the Shims

The drawback of Q' is the prohibitively large number of boundary elements, which is $\mathcal{O}(n^2)$. This is, because the top and bottom facets $F_2^+(C)$ and $F_2^-(C)$ were already split into n parts $F_2^+(C_{1,j})$ and $F_2^-(C_{1,j})$ in the direction of e_1 and are split again in the direction of e_3 when adding the satellite cuboids of rank 3.

In order to avoid this, the satellite cuboids of rank 3 will now be shifted away from the central cuboid C by a sufficiently small value $\delta > 0$ to their final positions in $Q(n)$. For $j = 1, 2, ..., n$, if j is even, the origin of $S_{3,j}$ is set to $(0, -1-\delta, jl_3)$, or, if j is odd, to $(0, nl_2+\delta, jl_3)$, see Fig. 4. The sizes and weights of the satellite cuboids of rank 3 remain unchanged.

The gaps between C and the satellite cuboids of rank 3 are now filled with two cuboid shaped polyhedral cells T_2^- and T_2^+, called the *shims*. The shims T_2^-

and T_2^+ have their origins at the points $(0, -\delta, 0)$ and $(0, nl_2, 0)$ and have sizes of (nl_1, δ, nl_3) and a weight of $w_1 = 1$.

The union of R, all the shifted satellite cuboids of rank 3, and the shims is denoted by Q.

By adding the shims, the number of boundary elements in Q remains $\mathcal{O}(n)$. Furthermore, if $\delta > 0$ is sufficiently small, Theorem 1 still holds, and the topology of the above weighted shortest path will not be altered. Thus, the main result of the paper can be stated as follows:

Theorem 3. *Let $n \in \mathbb{N}$ and $Q(n)$ the polyhedral structure constructed above. Then, $Q(n)$ consists of $\mathcal{O}(n)$ polyhedral cells and faces and a weighted shortest path in $Q(n)$ starting at a point p on the front facet $F_3^-(C)$ and terminating at a point q on the rear facet $F_3^+(C)$ has $\Theta(n^3)$ cell crossings.*

5 Conclusions

A new lower bound of $\Omega(n^3)$ for the number of cell crossings of a weighted shortest path in a 3-dimensional polyhedral structure consisting of $\mathcal{O}(n)$ cells has been devised. This generalizes and sharpens the lower bound know for the 2-dimensional case. This generalization had been open.

When using a cell weight of $w_1' = 1 - \epsilon$ for all odd numbered $C_{1,j}$ and a sufficiently small $\epsilon > 0$, Theorem 1 still holds and the topology of the considered weighted shortest paths is not altered. But, by Snell's law, each crossing of C is then refracted into n individual straight line segments. Therefore, Theorem 2, respectively Theorem 3, can be reformulated, stating that the weighted shortest paths consist of $\Theta(n^2)$, respectively $\Theta(n^3)$, straight line segments.

Finally, polyhedral structure Q can be made convex, without altering the complexity of the construction, by adding $\mathcal{O}(n)$ additional cuboid cells filling the gaps.

We believe that our techniques generalize to d dimensions, but, due to space consideration, the extension falls outside the scope of this work.

Acknowledgement. The authors would like to thank the referees for valuable comments made. The third author would like to thank Erik van Leeuwen, Ioana Bercea, Karl Bringmann, and Michael Sagraloff for fruitful initial discussions which took place while that author was visiting Max-Planck Institute, Algorithms, Saarbrücken.

References

1. Aleksandrov, L., Djidjev, H.N., Guo, H., Maheshwari, A., Nussbaum, D., Sack, J.-R.: Algorithms for approximate shortest path queries on weighted polyhedral surfaces. Discrete Comput. Geom. **44**(4), 762–801 (2009). https://doi.org/10.1007/s00454-009-9204-0
2. Aleksandrov, L., Djidjev, H., Maheshwari, A., Sack, J.-R.: An approximation algorithm for computing shortest paths in weighted 3-D domains. Discrete Comput. Geom. **50**(1), 124–184 (2013). https://doi.org/10.1007/s00454-013-9486-0

3. Aleksandrov, L., Maheshwari, A., Sack, J.: Determining approximate shortest paths on weighted polyhedral surfaces. J. ACM **52**(1), 25–53 (2005). https://doi.org/10.1145/1044731.1044733
4. The Boost C++ Library. http://www.boost.org/
5. Bose, P., Maheshwari, A., Shu, C., Wuhrer, S.: A survey of geodesic paths on 3D surfaces. Comput. Geom. **44**(9), 486–498 (2011). https://doi.org/10.1016/j.comgeo.2011.05.006
6. Canny, J.F., Reif, J.H.: New lower bound techniques for robot motion planning problems. In: Proceedings of the 28th Annual Symposium on Foundations of Computer Science, Los Angeles, California, USA, 27–29 October 1987, pp. 49–60. IEEE Computer Society (1987). https://doi.org/10.1109/SFCS.1987.42
7. Carufel, J.D., Grimm, C., Maheshwari, A., Owen, M., Smid, M.H.M.: A note on the unsolvability of the weighted region shortest path problem. Comput. Geom. **47**(7), 724–727 (2014). https://doi.org/10.1016/j.comgeo.2014.02.004
8. Dijkstra, E.: A note on two problems in connexion with graphs. Numerische Mathematik **1**(1), 269–271 (1959). https://doi.org/10.1007/BF01386390
9. Leonhardt, U., Philbin, T.: Geometry and Light: The Science of Invisibility. Courier Dover Publications (2012)
10. Mitchell, J.S.B., Papadimitriou, C.H.: The weighted region problem: finding shortest paths through a weighted planar subdivision. J. ACM **38**(1), 18–73 (1991). https://doi.org/10.1145/102782.102784
11. Mitchell, J.S.B., Sharir, M.: New results on shortest paths in three dimensions. In: Proceedings of the 20th Annual Symposium on Computational Geometry, SCG 2004, pp. 124–133. ACM, New York (2004). https://doi.org/10.1145/997817.997839
12. OpenVolumeMesh - A Generic and Versatile Index-Based Data Structure for Polytopal Meshes. https://www.openvolumemesh.org/
13. P. E. Hart, N.J.N., Raphael, B.: A formal basis for the heuristic determination of minimum cost paths. IEEE Trans. Syst. Sci. Cybern. SSC **4**(2), 100–107 (1968)
14. ParaView - An Open-Source, Multi-Platform Data Analysis and Visualization Application. http://www.paraview.org/
15. Si, H.: TetGen - A Quality Tetrahedral Mesh Generator and a 3D Delaunay Triangulator. http://www.tetgen.org/
16. Si, H.: TetGen - A Quality Tetrahedral Mesh Generator and a 3D Delaunay Triangulator. Technical report, Karl-Weierstraß-Institut für Angewandte Analysis und Stochastik, Berlin, Germany (2013)
17. Sun, Z., Reif, J.H.: On finding approximate optimal paths in weighted regions. J. Algorithms **58**(1), 1–32 (2006). https://doi.org/10.1016/j.jalgor.2004.07.004
18. Visualization Toolkit Website. http://www.vtk.org/
19. wsp3dovm - Weighted Shortest Paths with OpenVolumeMesh. https://github.com/FrankBau/wsp3dovm/

Computing Balanced Convex Partitions of Lines

Sergey Bereg(✉)

University of Texas at Dallas, Richardson, TX, USA
besp@utdallas.edu

Abstract. Dujmović and Langerman (2013) proved a ham-sandwich cut theorem for an arrangement of lines in the plane. Recently, Xue and Soberón (2019) generalized it to balanced convex partitions of lines in the plane. In this paper, we study the computational problems of computing a ham-sandwich cut balanced convex partitions for an arrangement of lines in the plane. We show that both problems can be solved in polynomial time.

Keywords: Ham-sandwich theorem · Arrangement of lines · Balanced convex partitions.

1 Introduction

Dujmović and Langerman [5] proved a ham-sandwich cut theorem for an arrangement of lines in the plane. For a set L of lines in the plane, we denote by $I(L)$ the set of pairwise intersection points of L.

Theorem 1 (Dujmović and Langerman [5]**).** *For any arrangements A_1 and A_2 of lines in $\mathbb{R}^2$, there exists a line ℓ bounding closed halfplanes ℓ^+ and ℓ^- and sets A_i^σ, $i \in 1, 2$, $\sigma \in +, -$ such that $A_i^\sigma \subseteq A_i, |A_i^\sigma| \geq |A_i|^{1/2}$, and $I(A_i^\sigma) \in \ell^\sigma$.*

We show that the ham-sandwich line can be computed in polynomial time. The ham sandwich theorem has been generalized to convex partitions of the plane. The following theorem was proven independently by Bespamyatnikh, Kirkpatrick, and Snoeyink [1], by Ito, Uehara, and Yokoyama [9] and by Sakai [14].

Theorem 2 ([1, Theorem 10]). *Given rn red and rm blue points in the plane in general position, there exists a subdivision of the plane into r convex regions each of which contains n red and m blue points.*

A subdivision of the plane satisfying Theorem 2 is called *equitable* [1]. The main tool in the proof of Theorem 2 is equitable k-cuttings for $k = 2, 3$. A 2-cutting is simply a partition of the plane by a line. A 3-cutting is a partition of the plane into 3 wedges by 3 rays starting from the same point.

Recently, Xue and Soberón [16] generalized Theorem 1 as follows. We use the notation $[k] = \{1, 2, \dots, k\}$.

The research is supported in part by NSF award CCF-1718994.

Y. Kohayakawa and F. K. Miyazawa (Eds.): LATIN 2020, LNCS 12118, pp. 247–257, 2020.
https://doi.org/10.1007/978-3-030-61792-9_20

Theorem 3 (Xue and Soberón [16]). *Let A, B be two finite sets of lines in $\mathbb{R}^2$ such that $A \cup B$ is in general position, and let r be a fixed positive integer. Then, there is a convex partition $(C_1, \ldots, C_r)$ of $\mathbb{R}^2$ into r parts such that for all $j \in [r]$ there exist sets $A_j \subset A$, $B_j \subset B$ such that $I(A_j) \subset C_j, I(B_j) \subset C_j$ and*

$$|A_j| \geq r^{\ln(2/3)}|A|^{1/r} - 2r, \qquad |B_j| \geq r^{\ln(2/3)}|B|^{1/r} - 2r.$$

In this paper a convex partition satisfying the conditions of Theorem 3 is called an *equitable partition* of sets A and B.

The proof of Theorem 3 is similar to the proof of Theorem 2 [1]. For this, we need a measure μ defined as follows. For a set of lines L in the plane and a set $K \subseteq \mathbb{R}^2$, we define

$$\mu_L(K) = \max\{|L'| : L' \subseteq L, I(L') \subseteq K\}.$$

The main idea is to apply equitable k-cuttings for $k = 2, 3$ to obtain the desired partition.

Definition 1. *Let A, B be two finite sets of lines in $\mathbb{R}^2$ such that $A \cup B$ is in general position, and let r be a fixed positive integer. A 2-cutting of the plane into two parts (C_1, C_2) is called* equitable *if there exist two positive integers r_1, r_2 such that $r_1 + r_2 = r$ and*

$$\mu_A(C_i) \geq \left(\frac{2\,A|}{3}\right)^{r_i/r} - 2, \quad \mu_B(C_i) \geq \left(\frac{2|B|}{3}\right)^{r_i/r} - 2 \qquad \text{for } i = 1, 2.$$

Definition 2. *Let A, B be two finite sets of lines in $\mathbb{R}^2$ such that $A \cup B$ is in general position, and let r be a fixed positive integer. A 3-cutting of the plane into three parts (C_1, C_2, C_3) is called* equitable *if there exist three positive integers r_1, r_2, r_3 such that $r_1 + r_2 + r_3 = r$ and*

$$\mu_A(C_i) \geq \left(\frac{2\,A|}{3}\right)^{r_i/r} - 2, \quad \mu_B(C_i) \geq \left(\frac{2|B|}{3}\right)^{r_i/r} - 2 \qquad \text{for } i = 1, 2, 3.$$

A k-cutting is called *convex* if its parts are convex. Theorem 3 follows from the following lemma [16].

Lemma 1. *Let A, B be two finite sets of lines in the plane, each in general position, and $r \geq 2$ be a positive integer. Then, there exists an equitable k-cutting for some $k \in \{2, 3\}$.*

Our results are the following.

- We show that the ham-sandwich line for two sets of lines in the plane can be computed in $O(n^2 \log^2 n)$ time (Theorem 4 in Sect. 3).
- An *equitable partition* of two sets of lines in the plane into r convex regions can be computed in $O(n^6 \log^2 n \log r)$ time (Sects. 4–7).

2 Preliminaries

The key lemma in the proof of Theorem 1 (and Theorem 3) is the following lemma [5].

Lemma 2. *For any two open halfplanes H_1 and H_2 in the plane and any finite set L of lines*

$$\mu_L(H_1 \cup H_2) \leq \mu_L(H_1) \cdot \mu_L(H_2).$$

Xue and Soberón [16] proved a key lemma for 3-cuttings.

Lemma 3. *For any convex partition of the plane into three wedges C_1, C_2, C_3 and any set A of n lines in general position in the plane*

$$\mu_A(C_1)\mu_A(C_2)\mu_A(C_3) \geq \frac{2n}{3}.$$

3 Computing Ham-Sandwich Cuts

Let $h^-(t)$ denote the set $\{(x, y) \mid x \leq t\}$ and let $h^+(t) = \{(x, y) \mid x \geq t\}$. The following problem is the basis of our algorithms.

Problem P1. Given a set L of lines in the plane and an integer $n_0 \leq n$, find smallest x_0 such that $\mu_L(h^-(x_0)) \geq n_0$.

Lemma 4. *Problem P1 can be solved in $O(n \log^2 n)$ time.*

Proof. For a given t, we can compute $\mu_L(h^-(t))$ in $O(n \log n)$ time as follows. Let $L = \{l_1, \ldots, l_n\}$. For each line l_i in L, compute its slope s_i and intercept r_i. Therefore, the equation of the line l_i is $y = s_i(x - t) + r_i$. We sort the line by the intercept and assume that $l_1, \ldots, l_n$ is the sorted order, i.e. $r_1 \leq r_2, \ldots, r_n$. Compute $l(t)$, the length of the longest increasing subsequence of $s_1, s_2, \ldots, s_n$ in $O(n \log n)$ time [8,15].

To compute x_0, we use an algorithm for slope selection [2–4,10,13]. In the dual setting this problem is the following. Given a set L of n non-vertical lines and an integer number $k \in [\binom{n}{2}]$, we want to find two lines from L such that their intersection point has the kth smallest x-coordinate. This problem can be solved in $O(n \log n)$ time. We apply binary search on the values of k and compute smallest x_0 such that $\mu_L(h^-(x_0)) \geq n_0$ using $\log \binom{n}{2} = O(\log n)$ tests. The total running time is $O(n \log^2 n)$. □

Theorem 4. *Let A and B be two finite sets of lines each in the plane such that no two lines of $A \cup B$ are parallel. A ham-sandwich line for the arrangement of $A \cup B$ can be computed in $O(n^2 \log^2 n)$ time where $n = |A| + |B|$.*

Proof. Using the algorithm for problem P1, we can compute

- x_0, the smallest t such that $\mu_A(h^-(t)) \geq \sqrt{|A|}$, and
- x_1, the largest t such that $\mu_A(h^+(t)) \geq \sqrt{|A|}$.

The value x_0 is computed by applying the algorithm to the set A and the value x_1 is computed by applying the algorithm to the set A' that is symmetric to A, i.e. if a line with the equation $y = ax + b$ is in A then the line with the equation $y = -ax - b$ is in A'. Then $x_1 = -t$ where t is the output value of the algorithm applied to set A'.

The interval $[x_0, x_1]$ is not empty because $\mu_A(h^-(t)) \cdot \mu_A(h^+(t)) \geq |A|$ for any value of $t \in \mathbb{R}$ by Lemma 2. In other words, $\mu_A(h^-(t)) \geq \sqrt{|A|}$ or $\mu_A(h^+(t)) \geq \sqrt{|A|}$. The running time for computing $[x_0, x_1]$ is $O(n \log^2 n)$.

We also compute an interval $[x'_0, x'_1]$ for set B using μ_B. If the intervals $[x_0, x_1]$ and $[x'_0, x'_1]$ intersect then the line with equation $x = t$ is a ham-sandwich line for any $t \in [x_0, x_1] \cap [x'_0, x'_1]$. Suppose that the intervals $[x_0, x_1]$ and $[x'_0, x'_1]$ do not intersect. Wlog the interval for A is to the left of the interval for B, i.e. $x_1 < x'_0$.

Let $A(\phi)$ (resp. $B(\phi)$) be the set of lines A rotated clockwise by an angle ϕ about the origin. Let $X(\phi) = [x_0(\phi), x_1(\phi)]$ and $X'(\phi) = [x'_0(\phi), x'_1(\phi)]$ be the corresponding intervals. We want to find an angle ϕ such that the intervals $X(\phi)$ and $X'(\phi)$ intersect.

For a set of lines L, we denote by $\mathcal{A}(L)$ be the arrangement of lines L. Consider the arrangement $\mathcal{A}(A \cup B)$. Let V be the set of $\binom{|A|+|B|}{2}$ vertices of the arrangement $\mathcal{A}(A \cup B)$. Let L be the set of all lines that contain at least two points of V. Let $\phi_1, \phi_2, \ldots, \phi_{|L|}$ be the sorted sequence of the slopes of lines in L. Consider an interval $I = (\phi_i, \phi_{i+1})$. For any $\phi \in I$, the numbers $x_0(\phi), x_1(\phi), x'_0(\phi), x'_1(\phi)$ preserve the order since each of them corresponds to a vertex of V rotated clockwise by angle ϕ. Therefore the intervals $X(\phi)$ and $X'(\phi)$ preserve the relation for all $\phi \in I$, i.e. either (i) they intersect or (ii) $X(\phi)$ is to the left of $X'(\phi)$ or (iii) $X(\phi)$ is to the right of $X'(\phi)$, for all $\phi \in I$.

We want to find an interval $I = (\phi_i, \phi_{i+1})$ such that $X(\phi)$ and $X'(\phi)$ intersect for all $\phi \in I$. Note that, for all ϕ in the first interval $\phi \in (-\infty, \phi_1)$ the interval $X(\phi)$ is to the left of the interval $X'(\phi)$ since $0 \in (-\infty, \phi_1)$. Also for all ϕ in the last interval $\phi \in (\phi_{|L|}, \infty)$ the interval $X(\phi)$ is to the right of the interval $X'(\phi)$ since $\pi \in (\phi_{|L|}, \infty)$. We apply binary search on the sequence $\phi_1, \phi_2, \ldots, \phi_{|L|}$. For any interval $I = (\phi_i, \phi_{i+1})$, we pick $\phi \in I$ and compute the intervals $X(\phi)$ and $X'(\phi)$ in $O(n \log^2 n)$ time (the rotation of the lines $A \cup B$ by ϕ can be done in linear time). The total time for computing these intervals is $O(n \log^3 n)$. For binary search we use the slope selection for the set of points V. Each slope selection takes $O(|V| \log |V|)$ time [2–4,10,13]. Then the total time is $O(n^2 \log^2 n)$. □

4 Computing an Equitable Partition

Our algorithm for computing an equitable partition of the plane is based on equitable 2- and 3-cuttings. In this section we show how 2-cuttings can be computed. In particular, we need to find the pair (r_1, r_2) for an equitable 2-cutting.

For convenience, let $M(X, i) = \left\lceil \left(\frac{2|X|}{3}\right)^{i/r} \right\rceil - 2$ for $i \in [r - 1]$.

We define the sign $\sigma(i)$ for $i \in [r-1]$ as follows. As in the proof of Theorem 4, we compute

- x_0, the smallest t such that $\mu_A(h^-(t)) \geq M(A,i)$, and
- x_1, the largest t such that $\mu_A(h^+(t)) \geq M(A,r-i)$.

We also compute an interval $[x'_0, x'_1]$ for set B using μ_B and the lower bounds $M(B,i), M(B,r-i)$. If the intervals $[x_0, x_1]$ and $[x'_0, x'_1]$ intersect then, for any $t \in [x_0, x_1] \cap [x'_0, x'_1]$, the line with equation $x = t$ is an equitable 2-cutting for $(r_1, r_2) = (i, r-i)$. We assign $\sigma(i) = 0$ in this case. Suppose that the intervals $[x_0, x_1]$ and $[x'_0, x'_1]$ do not intersect. If $x_1 < x'_0$, we set $\sigma(i) = 1$; otherwise $x_0 > x'_1$ and we set $\sigma(i) = -1$.

We apply the algorithm from Lemma 4 to compute the sign sequence $\sigma(1)$, $\sigma(2), \ldots, \sigma(r-1)$ in $O(rn\log^2 n)$ time. If there is a sign $\sigma(i) = 0$ then an equitable 2-cutting (by a vertical line) is found. Suppose that $\sigma(i) = \pm 1$ for all $i \in [r-1]$. We apply the following theorem from [1].

Theorem 5 **([1]).** *For any sequence of signs $\sigma(1), \sigma(2), \ldots, \sigma(r-1)$ with $\sigma(i) = \pm 1$, there is a pair (r_1, r_2) or a triple (r_1, r_2, r_3) with sum r and the same signs such that any $1 \leq r_i \leq 2r/3$.*

The proof of Theorem 5 [1] implies that a pair (r_1, r_2) or a triple (r_1, r_2, r_3) can be computed in $O(r)$ time if the sequence of signs is known. If it is a pair (r_1, r_2) then we can compute an equitable 2-cutting for (r_1, r_2) as follows.

As in the proof of Theorem 4, we use the rotated sets $A(\phi)$ and $B(\phi)$). We can also define, for the sets $A(\phi)$ and $B(\phi)$) and any $i \in [r-1]$,

(i) the corresponding intervals $X(\phi, i) = [x_0(\phi, i), x_1(\phi, i)]$ and $X'(\phi) = [x'_0(\phi, i), x'_1(\phi, i)]$, and
(ii) the signs $\sigma(\phi, i)$.

Suppose that $\sigma(\phi, i) \neq \sigma(0, i)$ for some angle ϕ and $i \in [r-1]$. Then an equitable 2-cutting for $(i, r-i)$ can be found using binary search in the set of slopes $\phi_1, \phi_2, \ldots$ as in the proof of Theorem 4.

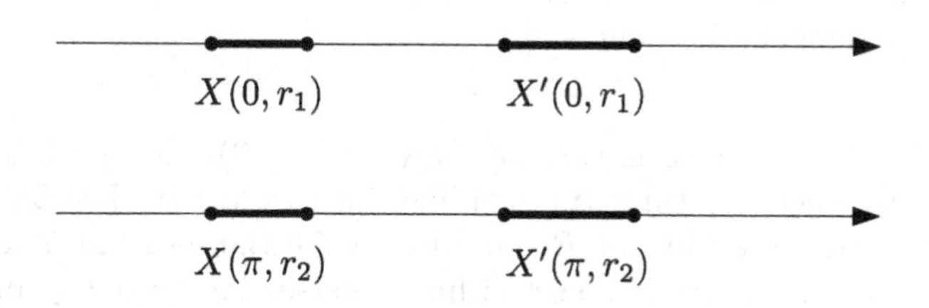

Fig. 1. The intervals for $\phi = 0, i = r_1$ and $\phi = \pi, i = r_2$ projected on the x-axis.

Note that $X(0,i) = X(\pi, r-i)$ and $X'(0,i) = X'(\pi, r-i)$ for $i \in [r-1]$. Then $\sigma(\pi, r-i) = -\sigma(0,i)$. Then $\sigma(\pi, r_2) = -\sigma(0, r_1) = -\sigma(0, r_2)$, see Fig. 1. Thus, we can find an equitable 2-cutting for (r_1, r_2).

Lemma 5. *If there is a pair* (r_1, r_2) *with sum* r *such that* $\sigma(r_1) = \sigma(r_2)$, *then an equitable 2-cutting for* (r_1, r_2) *can be computed in* $O(n^2 \log^2 n)$ *time.*

Note that a ham-sandwich cut for two point sets in the plane can be computed in linear time [12] and an equitable subdivision of two point sets in the plane can be computed in subquadratic time [1].

In the subsequent sections we will deal with 3-cuttings. We also may assume that, for any $i \in [r-1]$, the sign function $\sigma(\pi, i)$ is invariant for all θ as an equitable 2-cutting can be found otherwise.

5 Computing the Measure of a Wedge

In this section we present an algorithm for computing $\mu_L(W)$ for any set of lines L and a wedge W in the plane. We can assume that $W = \{(x, y) \mid x, y \geq 0\}$ by using affine transformations. Lines in L intersect both the x- and the y-axis. Let $x_1, \ldots, x_k$ and $y_1, \ldots, y_m$ be the sorted coordinates of the intersection points, see Fig. 2.

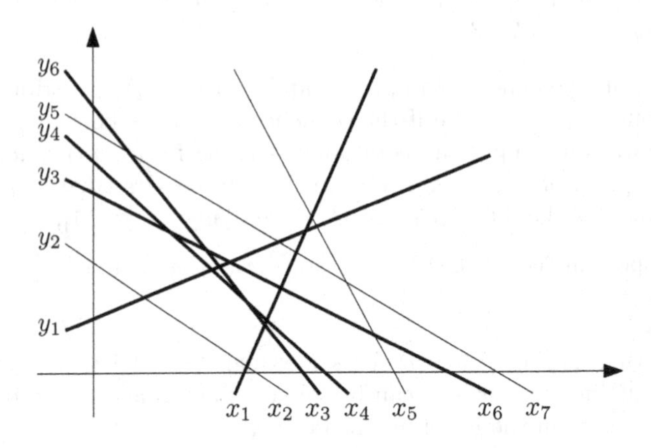

Fig. 2. 8 lines crossing the wedge $W = \{(x, y) \mid x, y \geq 0\}$. $\mu_L(W) = 5$ and the corresponding 5 lines are shown in bold.

The wedge W is between two rays $R_X = \{(x, 0) \mid x > 0\}$ and $R_X = \{(0, y) \mid y > 0\}$. There are three types of lines intersecting W. Let L_X be the set of lines intersecting R_X but not R_Y and let L_Y be the set of lines intersecting R_Y but not R_X. Let L_{XY} be the set of lines intersecting both R_X and R_Y.

Lemma 6. *Let* L' *be a subset of* L *such that* $I(L') \subset W$. *There exists a pair* (i, j) *such that*
(i) any line in $L' \cap L_{XY}$ *intersects the* x*-axis at* $x_{i'} \geq x_i$ *and the* y*-axis at* $y_{j'} \geq y_j$,
(ii) any line in $L' \cap L_X$ *intersects the* x*-axis at* $x_{i'} < x_i$ *and*
(iii) any line in $L' \cap L_Y$ *intersects the* y*-axis at* $y_{j'} < y_j$.

Proof. Let x_i be the smallest x-intercept of a line in $L' \cap L_{XY}$ and let l_1 be this line. Let y_j be the smallest y-intercept of a line in $L' \cap L_{XY}$ and let l_2 be this line. Clearly, the condition (i) holds. The condition (ii) holds too; otherwise line l_1 does not intersect all the lines in L_X. Similarly, the condition (iii) holds; otherwise line l_2 does not intersect all the lines in L_Y. □

We call a set of lines L' satisfying the conditions (i)-(iii) of Lemma 6, *(i,j)-set of lines*. For every (i,j), we compute $L^*_{i,j}$, a largest (i,j)-set of lines. By Lemma 6, $\mu_L(W) = |L^*_{i,j}|$.

Let $\mathbf{x}(l)$ and $\mathbf{y}(l)$ denote the x- and the y-intercept of a line, i.e. the equation of l can be expressed as $y = ax + \mathbf{y}(l)$ and $y = a(x - \mathbf{x}(l))$. For every pair (i,j) with $1 \leq i \leq k$ and $1 \leq j \leq m$, we show how to compute the largest set $L' \subseteq L_{XY}$ such that $I(L') \subset W$. Let $L' = \{l_1, \ldots, l_s\}$ be a set of lines in L_{XY} with $\mathbf{x}(l_1) \geq \mathbf{x}(l_2) \geq \ldots \mathbf{x}(l_s) \geq x_i$ and $\mathbf{y}(l_1), \mathbf{y}(l_2), \ldots, \mathbf{y}(l_s) \geq y_j$. Then $I(L') \subset W$ if and only if $\mathbf{y}(l_1) \leq \mathbf{y}(l_2) \leq \cdots \leq \mathbf{y}(l_s)$, for example three lines with x-intercepts x_6, x_4, and x_3 in Fig. 2 have y-intercepts y_3, y_4, and y_6, respectively. Therefore, we can use an algorithm for computing the longest increasing subsequence of $\mathbf{y}(l_1), \mathbf{y}(l_2), \ldots, \mathbf{y}(l_s)$ in $O(n \log n)$ time [8,15].

For every pair (i,j) with $1 \leq i \leq k$ and $1 \leq j \leq m$, we show how to compute the largest set $L' \subseteq L_X \cup L_Y$ such that $I(L') \subset W$. Let $L' = \{l_1, \ldots, l_s\}$ be a set of lines in $L_X \cup L_Y$ satisfying (i) and (ii). Suppose that they are sorted as follows. The lines from L_X first, then the lines from L_Y. The lines from L_X are sorted by x-intercept in decreasing order. The lines from L_Y are sorted by y-intercept in increasing order. Then $I(L') \subset W$ if and only if the slopes of he lines in L' are decreasing. Therefore, we can use an algorithm for computing the longest decreasing subsequence in $O(n \log n)$ time.

Therefore $|L^*_{i,j}| = |A^*_{i,j}| + |B^*_{i,j}|$ where $|A^*_{i,j}|$ is the maximum size of $L' \subseteq L_{XY}$ for (i,j) and $|B^*_{i,j}|$ is the maximum size of $L' \subseteq L_X \cup L_Y$ for (i,j). If we use the longest increasing/decreasing subsequence for all pairs (i,j), the total running time will be $O(n^3 \log n)$. We show that it can be reduced to $O(n^2 \log n)$.

We compute $|A^*_{i,j}|$ using $|A^*_{i+1,j}|$. Consider line l_t with $\mathbf{x}(l_t) = x_i$. If $\mathbf{y}(l_t) < y_j$, line l_t can be ignored. If $\mathbf{y}(l_t) \geq y_j$, we add it to the sequence $\mathbf{y}(l_1), \mathbf{y}(l_2), \ldots, \mathbf{y}(l_s)$ and compute the length of the longest increasing subsequence (LIS) of the new sequence. Since we add a new element to it, the length of LIS can be updated in $O(\log n)$ time. Similarly, the values of $|B^*_{i,j}|$ can be computed. Then the total running time is $O(n^2 \log n)$.

Theorem 6. *For any set L of n lines in the plane and any wedge W, the measure $\mu_L(W)$ can be computed in $O(n^2 \log n)$ time.*

6 Computing Canonical Cuttings

Let (r_1, r_2, r_3) be the triple provided by Theorem 5 (we assume that a pair (r_1, r_2) does not exist).

Similar to [1,16] we define a *canonical cutting.* For a point p, construct three rays r_0, r_1 and r_2 starting from p. The first ray r_0 is pointing downwards. Let $C_i, i = 1, 2$ be the region defined by rays r_0 and r_i as shown in Fig. 3. Let $\alpha_i, i = 1, 2$ be the angle rays r_0 and r_i. The canonical cutting is defined by choosing $\alpha_i, i = 1, 2$ to be the smallest angle such that $\mu_A(C_i) \geq M(A, r_i)$. We also denote these angles $\alpha_i(p), i = 1, 2$. Let C_3 be the region between rays r_1 and r_2, see Fig. 3. By Lemma 3, $\mu_A(C_3) \geq M(A, r_3)$.

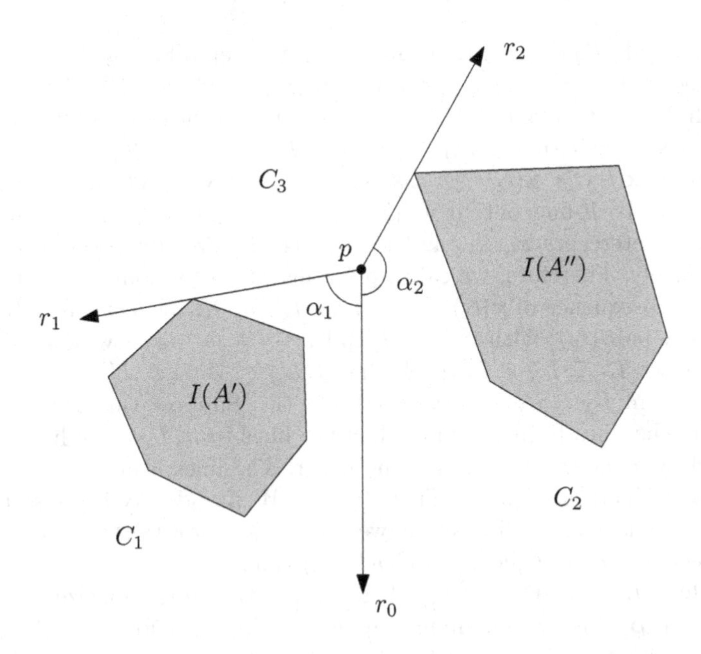

Fig. 3. Canonical 3-cutting.

Lemma 7. *For any point p in the plane, the canonical 3-cutting can be computed in $O(n^2 \log^2 n)$ time.*

Proof. First, we compute $I(A)$ in $O(n^2)$ time. For every point $q \in I(A)$, compute the slope of the vector pq. Sort $I(A)$ by slope. Compute ray r_1 of the canonical 3-cutting at p by using binary search in the sorted set $I(A)$. For any slope s in the binary search, compute the measure $\mu_A(C_1)$ using the algorithm from Theorem 6. Similarly the ray r_2 can be computed. The total time is $O(n^2 \log^2 n)$. □

The locus of all points p defining a convex canonical 3-cutting is

$$R = \{p \in \mathbb{R}^2 \mid x_0 \leq p_x \leq x_1 \text{ and } \alpha_1(p) + \alpha_2(p) \geq \pi\}.$$

The region R contains an apex of an equitable 3-cutting [16]. It can be proven using a coloring: a point $p \in R$ has color $i \in [3]$ if $\mu_B(C_i) \geq M(B, r_i)$. Note that a point may have more than one color and a point with three colors is an apex of the equitable 3-cutting. It exists by the following theorem which is applied to region R. The sides of the triangle correspond to the left/right and top sides of R, see Fig. 4.

Theorem 7 (Knaster, Kuratowski, Mazurkiewicz [11]). *Let Δ be a triangle with vertices $1, 2, 3$. Suppose that Δ is colored with colors $\{1, 2, 3\}$ such that every vertex i has color i, and every point on a side ij has at least one of the colors i or j. If every color class is a closed set, then there is a point with all three colors.*

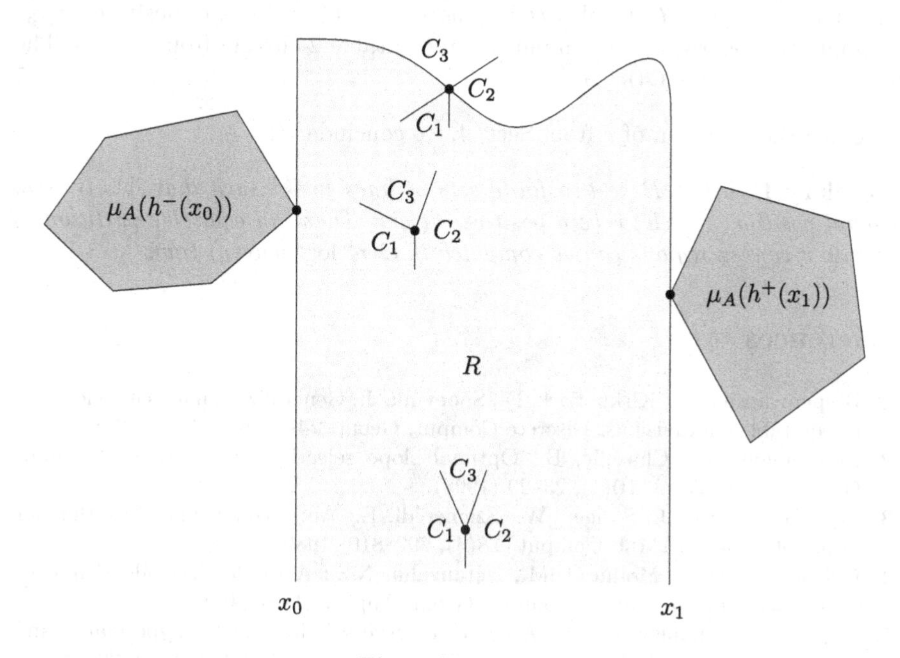

Fig. 4. Region R.

7 Computing an Equitable 3-Cutting

In this section we show that an equitable 2- or 3-cutting can be computed efficiently.

Theorem 8. *Let A, B be two finite sets of points in the plane, each in general position, and $r \geq 2$ be a positive integer. An equitable k-cutting for some $k \in \{2, 3\}$ can be computed in $O(n^6 \log^2 n)$ time.*

Proof. Let V be the set of vertices of the arrangement of lines $L = A \cup B$. Let $\mathcal{L}$ be the union of
(i) the set lines passing through two points of V, and
(ii) the vertical lines passing through V.
Then the arrangement $\mathcal{L}$ contains $O(n^2)$ lines and $O(n^4)$ faces.

We apply the topological sweep method of Edelsbrunner and Guibas [6,7] to traverse the faces of the arrangement of $\mathcal{L}$. For each face F of the arrangement, we can check the boundary conditions of the region R using angles $\alpha_i, i \in [3]$ computed for some point $p \in F$. If the canonical 3-cutting is convex, we compute the coloring of the face. The algorithm stops if all three colors are used for p. Note that all the points in F have the same coloring. When we reach the top boundary of region R, we also check the sign of C_3. If it is opposite of $\sigma(r_3)$ we apply the algorithm for computing an equitable 2-cutting from Sect. 4. The total running time is $O(n^6 \log^2 n)$. □

Using the partition of r from Sect. 4, we conclude

Corollary 1. *Let A, B be two finite sets of lines in $\mathbb{R}^2$ such that $A \cup B$ is in general position, and let r be a positive integer. Then, an equitable partition of $\mathbb{R}^2$ into r convex regions can be computed in $O(n^6 \log^2 n \log r)$ time.*

References

1. Bespamyatnikh, S., Kirkpatrick, D., Snoeyink, J.: Generalizing ham sandwich cuts to equitable subdivisions. Discrete Comput. Geom. **24**(4), 605–622 (2000)
2. Brönnimann, H., Chazelle, B.: Optimal slope selection via cuttings. Comput. Geom. Theory Appl. **10**(1), 23–29 (1998)
3. Cole, R., Salowe, J., Steiger, W., Szemerédi, E.: An optimal-time algorithm for slope selection. SIAM J. Comput. **18**(4), 792–810 (1989)
4. Dillencourt, M.B., Mount, D.M., Netanyahu, N.S.: A randomized algorithm for slope selection. Internat. J. Comput. Geom. Appl. **2**, 1–27 (1992)
5. Dujmovic, V., Langerman, S.: A center transversal theorem for hyperplanes and applications to graph drawing. Discrete Comput. Geom. **49**(1), 74–88 (2013)
6. Edelsbrunner, H., Guibas, L.J.: Topologically sweeping an arrangement. J. Comput. Syst. Sci. **38**, 165–194 (1989). Corrigendum in 42 (1991), 249–251
7. Edelsbrunner, H., Souvaine, D.L.: Computing median-of-squares regression lines and guided topological sweep. J. Am. Statist. Assoc. **85**, 115–119 (1990)
8. Fredman, M.L.: On computing the length of longest increasing subsequences. Discrete Math. **11**(1), 29–35 (1975)
9. Ito, H., Uehara, H., Yokoyama, M.: 2-dimension ham sandwich theorem for partitioning into three convex pieces. In: Akiyama, J., Kano, M., Urabe, M. (eds.) JCDCG 1998. LNCS, vol. 1763, pp. 129–157. Springer, Heidelberg (2000). https://doi.org/10.1007/978-3-540-46515-7_11
10. Katz, M.J., Sharir, M.: Optimal slope selection via expanders. Inform. Process. Lett. **47**, 115–122 (1993)
11. Knaster, B., Kuratowski, C., Mazurkiewicz, S.: Ein beweis des fixpunktsatzes für n-dimensionale simplexe. Fundamenta Mathematicae **14**(1), 132–137 (1929)

12. Lo, C.Y., Matoušek, J., Steiger, W.L.: Algorithms for ham-sandwich cuts. Discrete Comput. Geom. **11**, 433–452 (1994)
13. Matoušek, J.: Randomized optimal algorithm for slope selection. Inform. Process. Lett. **39**, 183–187 (1991)
14. Sakai, T.: Balanced convex partitions of measures in R^2. Graphs Combinat. **18**(1), 169–192 (2002)
15. Schensted, C.: Longest increasing and decreasing subsequences. Can. J. Math. **13**, 179–191 (1961)
16. Xue, A., Soberón, P.: Balanced convex partitions of lines in the plane. arXiv e-prints p. 1910.06231 (2019). https://arxiv.org/abs/1910.06231

Ordered Strip Packing

K. Buchin[1], D. Kosolobov[2], W. Sonke[1(✉)], B. Speckmann[1], and K. Verbeek[1]

[1] TU Eindhoven, Eindhoven, The Netherlands
{k.a.buchin,w.m.sonke,b.speckmann,k.a.b.verbeek}@tue.nl
[2] Ural Federal University, Yekaterinburg, Russia
dkosolobov@mail.ru

Abstract. We study an ordered variant of the well-known strip packing problem, which is motivated by applications in visualization and typography. Our input consists of a maximum width W and an ordered list of n blocks (rectangles). The goal is to pack the blocks into rows (not exceeding W) while obeying the given order and minimizing either the number of rows or the total height of the drawing. We consider two variants: (1) non-overlapping row drawing (NORD), where distinct rows cannot share y-coordinates, and (2) overlapping row drawing (ORD), where consecutive rows may overlap vertically. We present an algorithm that computes the minimum-height NORD in $O(n)$ time. Further, we study the worst-case tradeoffs between the two optimization criteria—number of rows and total height—for both NORD and ORD. Surprisingly, we show that the minimum-height ORD may require $\Omega(\log n/\log\log n)$ times as many rows as the minimum-row ORD. The proof of the matching upper bound employs a novel application of information entropy.

Keywords: Linear layouts · Packing

1 Introduction

Packing problems arise in many practical applications and have hence been studied extensively in the literature. Two well-known classes of 2D packing problems are *bin packing* and *strip packing*. In both problems, the aim is to pack (without rotation) a set of *blocks* (rectangles) into a shape such that they are internally disjoint. The goal of bin packing is to pack the blocks into rectangular *bins* of the same (given) size while minimizing the number k of bins used. The goal of strip packing is to pack the blocks into a *strip* of fixed width W and infinite height while minimizing the height H of the strip used.

In this paper we introduce a new *ordered* variant of strip packing where we are also given an order on the blocks which the packing must obey. Such orders arise naturally in applications such as typography or visualization: When dividing the

W. Sonke, B. Speckmann and K. Verbeek—Partially supported by the Netherlands Organisation for Scientific Research (NWO) under project no. 639.023.208 (W.S. and B.S.) and no. 639.021.541 (K.V.).

Y. Kohayakawa and F. K. Miyazawa (Eds.): LATIN 2020, LNCS 12118, pp. 258–270, 2020.
https://doi.org/10.1007/978-3-030-61792-9_21

text in a paragraph into lines, it is obviously necessary to retain the word order (see, for example, the word wrapping algorithm by Knuth and Plass still used in TEX [4]). Or consider the visual layout of linear sequences such as time lines or industrial processes: to make better use of the available screen space the linear layouts need to be "folded" into a strip (the screen) while keeping the sequence intact (see Fig. 1).

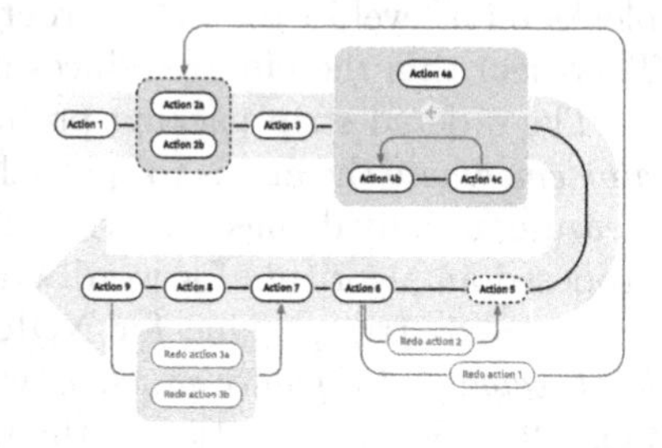

Fig. 1. Process visualization [9].

Problem Statement. Our input consists of an ordered list B of n blocks and a maximum width W. Each block $b = (w_b, h_b^\uparrow, h_b^\downarrow)$ is specified by its width w_b, its top height $h_b^\uparrow$, and its bottom height $h_b^\downarrow$. The goal is to place the blocks on rows. Each row consists of a horizontal line called the *spine*, on which blocks are aligned vertically (see Fig. 2a). The top and bottom heights $h_b^\uparrow$ and $h_b^\downarrow$ specify how much block b sticks out above and below the spine, respectively. A solution consists of a set of y-coordinates for the spines, and, for every block, a row and an x-coordinate. More precisely, the placement (*drawing*) of the blocks must follow these rules:

1. All blocks are interior disjoint.
2. All blocks lie completely in the x-range $[0, W]$.
3. The height of a block b above (below) the assigned spine is $h_b^\uparrow$ ($h_b^\downarrow$).
4. The interior of a block may not overlap with the spine of another row.
5. The order of the blocks from top to bottom, and then from left to right, must coincide with the specified order in the input (see Fig. 2b).

We consider two natural optimization questions: (1) minimizing the total height of the drawing, and (2) minimizing the number of rows. Furthermore, we distinguish two versions: In a *non-overlapping rows drawing (NORD)*, blocks on different rows may not share the same y-coordinates (see Fig. 2b). In an *overlapping rows drawing (ORD)* this restriction is omitted (see Fig. 2c). Minimizing the total height of an ORD is NP-hard, while a NORD of minimum total height can be computed in $O(n^2)$ time [9]. Furthermore, NORDs and ORDs that minimize the number of rows can easily be computed greedily in polynomial time.

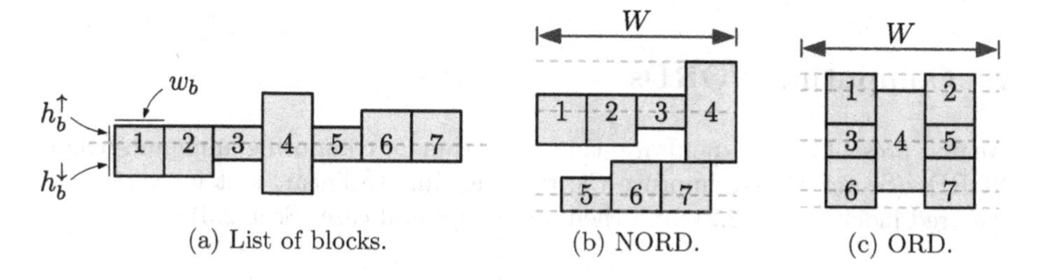

Fig. 2. Minimum-height NORD and ORD of a list of blocks.

Related Work. 2D bin packing and strip packing have been widely studied. Both problems are NP-hard, since 1D bin packing is already NP-hard [5]. Hence, research has focused on approximation algorithms. See, for example, the surveys by Coffman *et al.* [1] and Lodi *et al.* [7]. Some online approximation algorithms for strip packing are *level-based*, that is, they layout blocks on horizontal *levels*, akin to rows in our problem. For example, the *next-fit* algorithm [3] greedily puts blocks on a level in the input order, until the next block does not fit anymore. This algorithm therefore produces minimum-row NORDs.

The ordered strip packing problem for NORDs is related to the *word wrap problem*, where the aim is to split a list of words into lines such that some quality measure, usually defined in terms of line lengths, is maximized. Knuth and Plass proposed an algorithm for word wrapping, which is still in use in TEX [4].

Ordered strip packing for NORDs can be considered a special case of the *least-weight subsequence problem*, which was introduced by Hirschberg and Larmore [2] as a generalization of the word wrap problem. Given a weight function $f : \{1, \ldots, n\} \times \{1, \ldots, n\} \rightarrow \mathbb{R}$, they ask for a minimum-weight subsequence S of $[1, 2, \ldots, n]$, starting with 1 and ending with n. Here the weight of a sequence $\{s_i\}_{i=0}^{n}$ is given by $\sum_{i=1}^{n} f(s_{i-1}, s_i)$. Hirschberg and Larmore present a straightforward $O(n^2)$-time algorithm, which can be improved to $O(n \log n)$ time if f is concave (that is, $f(i, k) + f(j, l) \leq f(i, l) + f(j, k)$ for all $i \leq j < k \leq l$), and to $O(n)$ time if the weight function satisfies an additional condition. Wilber [10] improved this to an $O(n)$-time algorithm without requiring the additional condition. However, the weight function for minimum-height NORDs is not concave, and this method thus cannot be used.

Results and Organization. In Sect. 2 we present a new algorithm that can compute a minimum-height NORD with n blocks in $O(n)$ time, improving upon the best known existing algorithm [9] that runs in $O(n^2)$ time. In Sect. 3 we study the tradeoffs between the two optimization criteria: how is the height of the drawing affected if we minimize the number of rows, and how is the number of rows affected if we minimize the height. We present upper and lower bounds for these tradeoffs. The most interesting and most surprising bound shows that a minimum-height ORD may require $\Omega(\log n / \log \log n)$ times as many rows as the minimum-row ORD. We prove that this bound is tight using a novel application of the information entropy function. We believe that this new way of using entropy is of independent interest, and may have further applications in packing problems. Omitted proofs can be found in the full version of the paper.

2 Computing NORDs

We first present a basic quadratic-time algorithm for computing minimum-height NORDs (Sect. 2.1). We improve the running time to linear, first for vertically-centered blocks (Sect. 2.2), and then for the general case (Sect. 2.3).

2.1 Basic Algorithm

We need to determine for each block whether it starts a new row or not. To do this efficiently, we use the following observation. In a k-row drawing, we call the last block on row $k-1$ the *separating block.*

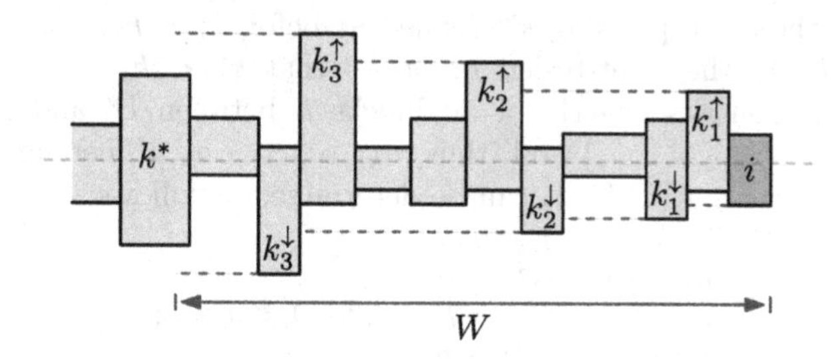

Fig. 3. Illustration of Lemma 1, which allows us to consider only a subset of the blocks to end row $k-1$. Blocks in $K_i^\uparrow$ and $K_i^\downarrow$ are marked here as $k_j^\uparrow$ and $k_j^\downarrow$, respectively.

Lemma 1. *Let B be a list of blocks. A minimum-height NORD for B exists such that the separating block either (1) has larger top height than all blocks on the last row, or (2) has larger bottom height than all blocks on the last row, or (3) together with the blocks on the last row, has total width larger than W.*

We process the blocks in order, incrementally creating minimum-height NORDs for the first $1, \dots, n$ blocks. Assume that we are constructing a minimum-height NORD for the first i blocks. By Lemma 1 we only consider drawings in which the separating block satisfies conditions (1), (2), or (3). Let k^* be the smallest integer such that blocks $k^*+1, \dots, i$ have total width at most W. Blocks $1, \dots, k^*-1$ cannot serve as the separating block, because that would overfill the last row. The blocks k after k^* that satisfy condition (1) are those with successively larger top heights (starting at block i going backwards to k^*); they form a "staircase" pattern. We call the set of these blocks the *top candidate set* $K_i^\uparrow$ for blocks $1, \dots, i$. Similarly, the blocks k after k^* that satisfy condition (2) form a staircase of increasing bottom heights; we call this the *bottom candidate set* $K_i^\downarrow$ (see Fig. 3). After computing $K_i^\uparrow$, $K_i^\downarrow$, and k^*, we compute

$$T[i] = \min_{k \in K_i^\uparrow \cup K_i^\downarrow \cup \{k^*\}} \big(T[k] + h_{\text{row}}(k+1, i)\big),$$

where $h_{\text{row}}(k+1,i) := h_{\text{row}}^\uparrow(k+1,i) + h_{\text{row}}^\downarrow(k+1,i)$, $h_{\text{row}}^\uparrow(k+1,i) = \max\{h_j^\uparrow \mid j = k+1, \dots, i\}$, and $h_{\text{row}}^\downarrow(k+1,i) = \max\{h_j^\downarrow \mid j = k+1, \dots, i\}$ are the height, top height, and bottom height of a row with blocks $k+1, \dots, i$. It follows from Lemma 1 that $T[i]$, for $i = 1, \dots, n$, is the height of a minimum-height NORD for the blocks $1, \dots, i$, provided $T[0] = 0$. As sets $K_i^\uparrow$ and $K_i^\downarrow$ may have linear size, computing $T[i]$ takes $O(i)$ time, resulting in a total running time of $O(n^2)$.

2.2 Algorithm for Vertically Centered Blocks

Next we consider the special case when all blocks are vertically centered, i.e., $h_i^{\uparrow} = h_i^{\downarrow}$ for all blocks i. We denote the total height of block i by h_i $(= h_i^{\uparrow} + h_i^{\downarrow})$. Since the top and bottom candidate sets $K_i^{\uparrow}$ and $K_i^{\downarrow}$ coincide in this case, we denote them simply as K_i.

Fix i and the corresponding k^* defined as before. Let $k_1, \ldots, k_s$ denote all blocks from K_i in the right-to-left order, so that $k^* < k_s < \cdots < k_1 < i$. Recall that K_i consists exactly of all blocks k between k^* and i such that $h_k > \max\{h_{k+1}, h_{k+2}, \ldots, h_i\}$ and, thus, $h_{k_s} > \ldots > h_{k_1}$. Therefore, the value $h_{\text{row}}(k+1, i)$, for $k \in K_i \cup \{k^*\}$, can be determined as follows:

$$
\begin{aligned}
h_{\text{row}}(k_1+1, i) &= h_i; \\
h_{\text{row}}(k_j+1, i) &= h_{k_{j-1}}, \text{ for } 1 < j \le s; \\
h_{\text{row}}(k^*+1, i) &= h_{k_s}.
\end{aligned}
$$

Instead of a naïve linear computation of the minimum as in the quadratic algorithm, we store K_i in a so-called *mindeque* [6]: a deque that supports the standard insertions and deletions in constant amortized time and that can compute the minimum of the values assigned to its elements in constant time. We assign to each $k_j \in K_i$ the value $T[k_j] + h_{k_{j-1}}$, for $j > 1$, and $T[k_1] + h_i$, for $j = 1$. Therefore, one can calculate $T[i] = \min_{k \in K_i \cup \{k^*\}}(T[k] + h_{\text{row}}(k+1, i))$ as the minimum of $T[k^*] + h_{k_s}$ and all values assigned to the deque blocks. Thus, $T[i]$ can be calculated in $O(1)$ time.

It remains to show that the mindeque storing K_i can be maintained with $O(n)$ insertions and deletions. For this, we describe how to modify k^* and transform K_i into K_{i+1}. First, k^* is updated by consecutive increments until $h_{k^*+1} + \cdots + h_{i+1} \le W$. From K_i we have to remove the blocks that are to the left of the new k^*. Thus, we dequeue the leftmost blocks $k_s, k_{s+1}, \ldots, k_t$ with $k_s < \cdots < k_t \le k^* < k_{t+1}$ for the new k^* (or $k_t = k_1$ if $k_1 \le k^*$). Denote by K_i' the updated set K_i. As k^* does not decrease, in total at most $O(n)$ such deletions are performed.

A block k belongs to K_{i+1} iff $h_k > \max_{k<j\le i+1} h_j$ and $k^* < k < i+1$. However, K_i' contains exactly all k such that $h_k > \max_{k<j\le i} h_j$ and $k^* < k < i$. To obtain K_{i+1} from K_i' we still need to remove all $k \in K_i'$ with $h_k \le h_{i+1}$ and then insert the block i if $h_i > h_{i+1}$. Now since $h_{k_s} > \cdots > h_{k_1}$, the blocks to be removed are rightmost in K_i'. Thus, we can dequeue blocks $k_p, \ldots, k_1$ until $h_{k_{p+1}} > h_{i+1} \ge h_{k_p}$. Thus, all modifications can be performed by deque operations.

As described above, any block is added at most once to the mindeque, and therefore any block is removed at most once. Thus, we perform $O(n)$ mindeque operations in total and the overall running time is linear. Since the mindeque data structure can identify an element on which the minimum is attained, a minimum-height NORD can be reconstructed via standard backtracking.

Theorem 1. *A minimum-height NORD of n vertically-centered blocks can be computed in $O(n)$ time using a mindeque.*

2.3 General Linear Algorithm

Consider the general case where $h_i^{\uparrow}$ and $h_i^{\downarrow}$ can differ. Fix i and k^* and denote by $k_1^{\uparrow}, \ldots, k_s^{\uparrow}$ and $k_1^{\downarrow}, \ldots, k_t^{\downarrow}$ all blocks of $K_i^{\uparrow}$ and $K_i^{\downarrow}$ in the right-to-left order. By definition, $K_i^{\uparrow}$ contains exactly all blocks k such that $k^* < k < i$ and $h_k^{\uparrow} > \max_{k<j\le i} h_j^{\uparrow}$; similarly, $k \in K_i^{\downarrow}$ iff $k \in (k^*, i)$ and $h_k^{\downarrow} > \max_{k<j\le i} h_j^{\downarrow}$. Therefore, as in Sect. 2.2, $K_i^{\uparrow}$ can be maintained in a deque and transformed into $K_{i+1}^{\uparrow}$ by removing the leftmost blocks 'swept' by k^* and the rightmost blocks k with $h_k^{\uparrow} \le h_{i+1}^{\uparrow}$, and by inserting i if $h_i^{\uparrow} > h_{i+1}^{\uparrow}$. In total, this requires $O(n)$ operations when i passes from 1 to n; $K_i^{\downarrow}$ can be processed analogously. It is unclear, however, how to efficiently use these two deques, since maintaining the row heights associated with the blocks in $K_i^{\uparrow}$ and $K_i^{\downarrow}$ requires a more subtle approach.

We split $K_i^{\uparrow}$ into maximal contiguous subsequences that do not interleave $K_i^{\downarrow}$: $(k_{j_\ell+1}^{\uparrow}, k_{j_\ell+2}^{\uparrow}, \ldots, k_{j_{\ell+1}}^{\uparrow})_{\ell=0}^{p}$, where for all $k \in K_i^{\downarrow}$ either $k > k_{j_\ell+1}^{\uparrow}$ or $k_{j_{\ell+1}}^{\uparrow} \ge k$ and $0 = j_0 < \cdots < j_p = s$. Likewise, $K_i^{\downarrow}$ is split into $(k_{j'_\ell+1}^{\downarrow}, k_{j'_\ell+2}^{\downarrow}, \ldots, k_{j'_{\ell+1}}^{\uparrow})_{\ell=0}^{q}$ non-interleaving with $K_i^{\uparrow}$, where $0 = j'_0 < \cdots < j'_q = t$. The subsequences are arranged from right to left and each of them is stored in a separate mindeque d_ℓ (the values assigned to the blocks for d_ℓ are defined below) in this order:

$$\underbrace{k_1^{\uparrow}, k_2^{\uparrow}, \ldots, k_{j_1}^{\uparrow}}_{d_1}, \underbrace{k_1^{\downarrow}, k_2^{\downarrow}, \ldots, k_{j'_1}^{\downarrow}}_{d_2}, \underbrace{k_{j_1+1}^{\uparrow}, k_{j_1+2}^{\uparrow}, \ldots, k_{j_2}^{\uparrow}}_{d_3}, \underbrace{k_{j'_1+1}^{\downarrow}, k_{j'_1+2}^{\downarrow}, \ldots, k_{j'_2}^{\downarrow}}_{d_4}, \ldots$$

The sequence is non-increasing and only adjacent blocks $k_{j_\ell}^{\uparrow}, k_{j'_{\ell-1}+1}^{\downarrow}$ or $k_{j'_\ell}^{\downarrow}, k_{j_\ell+1}^{\uparrow}$ can coincide in it (such blocks belong to both top and bottom candidate sets). When i increases, some rightmost and leftmost blocks from $K_i^{\uparrow}$ and $K_i^{\downarrow}$ are removed and, in the process, some d_ℓ might be deleted entirely. To retain maximality of the subsequences, we then have to join some d_ℓ. In order to do this efficiently, we store d_ℓ as *catenable mindeques* [6] that can be concatenated in $O(1)$ amortized time. Thus, all the d_ℓ are maintained in $O(n)$ overall time.

For any block k, we have $\max_{k<j\le i} h_j^{\downarrow} = h_b^{\downarrow}$ for $b = \min\{b \in K_i^{\downarrow} \cup \{i\} : k < b\}$. Hence, given a deque d_ℓ whose blocks are from $K_i^{\uparrow}$, all its blocks k yield the same value $\max_{k<j\le i} h_j^{\downarrow}$, which we denote a_ℓ: if $\ell = 1$, $a_\ell = h_i^{\downarrow}$; otherwise, $a_\ell = h_m^{\downarrow}$, where m is the leftmost block in $d_{\ell-1}$ that is not in d_ℓ (observe that d_ℓ and $d_{\ell-1}$ can share a block only if d_ℓ contains only one block). Analogously, for $d_\ell \subseteq K_i^{\downarrow}$ the maximum $\max_{k<j\le i} h_j^{\uparrow}$ is the same for all $k \in d_\ell$ and we denote it a_ℓ.

Each mindeque d_ℓ can compute the minimum, denoted $\min d_\ell$, of the values assigned to its blocks. We assign to each $k_r^{\uparrow} \in K_i^{\uparrow}$ (resp., $k_r^{\downarrow} \in K_i^{\downarrow}$) the value $T[k_r^{\uparrow}] + h_{k_{r-1}^{\uparrow}}^{\uparrow}$ (resp., $T[k_r^{\downarrow}] + h_{k_{r-1}^{\downarrow}}^{\downarrow}$), for $r > 1$, and $T[k_1^{\uparrow}] + h_i^{\uparrow}$ (resp., $T[k_1^{\downarrow}] + h_i^{\downarrow}$), for $r = 1$. We store pointers to $d_1, d_2, \ldots$ in a mindeque D and assign $a_\ell + \min d_\ell$ to the pointer to d_ℓ; D is easy to maintain along with the deques d_ℓ in $O(n)$ overall time. Recall that $T[i] = \min\{T[k] + h_{\text{row}}(k+1, i) : k \in K_i^{\uparrow} \cup K_i^{\downarrow} \cup \{k^*\}\}$. For each $k_r^{\uparrow}$ with $r > 1$, we have $T[k_r^{\uparrow}] + h_{\text{row}}(k_r^{\uparrow} + 1, i) = T[k_r^{\uparrow}] + h_{r-1}^{\uparrow} + \max_{k_r^{\uparrow}<j\le i} h_j$.

Observe that for the deque d_ℓ containing $k_r^\uparrow$, $\max_{k_r^\uparrow < j \le i} h_j = a_\ell$ and the value $T[k_r^\uparrow] + h_{r-1}^\uparrow$ is assigned to $k_r^\uparrow$; we analogously analyze $T[k] + h_{\text{row}}(k+1, i)$ for k equal to $k_1^\uparrow$, $k_1^\downarrow$, and $k_r^\downarrow$ with $r > 1$. Therefore, one can compute $T[i]$ as the minimum of $\min D$ and $T[k^*] + h_{\text{row}}(k^*+1, i)$, where $h_{\text{row}}(k^*+1, i) = h_s^\uparrow + h_t^\downarrow$.

Theorem 2. *A minimum-height NORD of n blocks can be found in $O(n)$ time.*

Table 1. All tradeoffs between minimizing rows and height for NORDs and ORDs.

Type	$\alpha(n)$		$\beta(n)$	
	Lower	Upper	Lower	Upper
NORD	≥ 2	≤ 2	$\geq 3/2$	≤ 2
ORD	≥ 4	≤ 4	$\Omega(\log n / \log\log n)$	$O(\log n / \log\log n)$

3 Optimization Tradeoffs

Depending on the application, we may want to either minimize the number of rows of the drawing, minimize the total height, or a combination. To study the effect of the two optimization criteria, we analyze the worst-case tradeoffs between them. More precisely, we consider the worst-case ratio $\alpha(n)$ between the heights of minimum-row and minimum-height drawings (ORD and NORD):

$$\alpha(n) := \sup_{\text{list } B \text{ of } n \text{ blocks}} \frac{\text{height of minimum-row drawing of } B}{\text{height of minimum-height drawing of } B}.$$

Secondly we consider the worst-case ratio $\beta(n)$ between the number of rows of minimum-height and minimum-row drawings:

$$\beta(n) := \sup_{\text{list } B \text{ of } n \text{ blocks}} \frac{\#\text{ rows of minimum-height drawing of } B}{\#\text{ rows of minimum-row drawing of } B}.$$

We assume that the two criteria are optimized lexicographically, for example the minimum-height drawing has the fewest rows among all drawings with the minimum height. All bounds are summarized in Table 1. Proofs for the simple cases can be found in the full version of the paper. Next we analyze the most interesting case, namely the number of rows of a minimum-height ORD.

3.1 Lower Bound for Minimum-Height ORDs

In this section we prove a lower bound on the number of rows of a minimum-height ORD compared to the minimum-row ORD. For any integer $k > 2$, we construct a list of $n = \Theta(k^k)$ blocks such that a minimum-row ORD uses 2 rows, and a minimum-height ORD requires at least $k+1$ rows (see Fig. 4). Let the

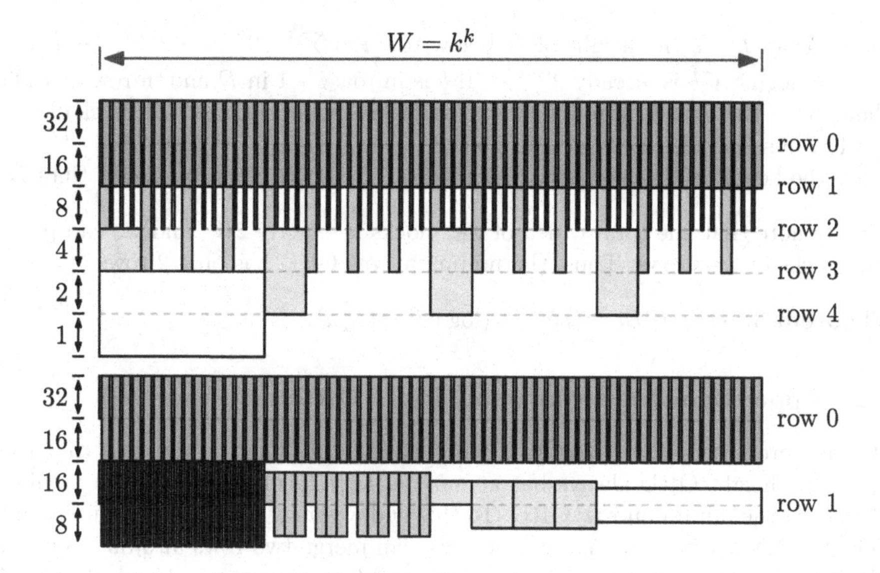

Fig. 4. Construction (here for $k = 4$) of a minimum-height ORD with $k + 1$ rows (above), whose minimum-row ORD (below) has 2 rows. (Not to scale vertically; distances between rows given on the left.)

width of the drawing be $W = k^k$. Row k contains one block of width k^{k-1}. Row $k-1$ contains k blocks, each of width k^{k-2}. Generally, row i ($1 \leq i \leq k$) contains k^{k-i} blocks of width k^{i-1}. Hence, for each row, the sum of its block widths is exactly k^{k-1}, and the sum of all block widths on rows $1, \ldots, k$ is k^k.

The top height of the blocks on row i is 2^{k-i+1}. The bottom height of the blocks on row i is 2^{k-i}, that is, equal to the top height of the blocks on the next row. However, every k-th block (starting with the first) on each row is *truncated*: it has bottom height 0. This ensures that the top and bottom heights of adjacent rows can fully overlap, so the vertical distance between the spines of rows i and $i+1$ in the minimum-height ORD D is exactly 2^{k-i}. We add an additional row of $2 \cdot k^{k-1}$ blocks on top of the drawing (the blue area in Fig. 4) that perfectly surround the k^{k-1} blocks on row 1. These blocks have top height 2^{k+1} and occupy the entire width of the row. Specifically, for every block on row 1 there is a corresponding block on row 0 with width 1 and bottom-height 0, and between two blocks of row 1 (and at the end) there is a block on row 0 with width $k-1$ and bottom-height 2^k. Therefore the total width of all blocks in the drawing is $2k^k$, and the height of the drawing is $2^{k+2} - 1$.

Lemma 2. *The ORD D constructed above has minimum height, and any other ORD D' with fewer rows is higher.*

Proof (sketch). It is easy to see that D has optimal height if the assignment of blocks to rows is fixed. Consider the first block b in D' that is assigned to a different row than in D. If b is in row i in D and in row $i+1$ in D', then D' is higher than D, as the height of D below row i is $\sum_{j=i}^{k} 2^{k-j} = 2^{k-i+1} - 1$ and the top height of b is already 2^{k-i+1}. If b is in row $i+1$ in D and in row i in D', then one of the blocks on row i cannot be placed below a truncated block of row $i-1$. Thus, the distance between row $i-1$ and i is 2^{k+2-i} instead of 2^{k+1-i}. Since the height of D below row i is only $2^{k-i+1} - 1$, D' must be higher than D.

Finally note that the total width of the blocks is exactly $2W$, and we can place the blocks on two rows. Thus, the minimum-row ORD has only 2 rows.

Theorem 3. *For ORDs,* $\beta(n) = \Omega(\log n / \log\log n)$.

3.2 Upper Bound for Minimum-Height ORDs

We now prove that the bound in Theorem 3 is tight. That is, we show that the minimum-height ORD always has at most $O(\log n / \log\log n)$ times the number of rows of the minimum-row ORD. To this end we show that, given a minimum-height ORD D with "too many" rows, we can merge two rows into one. Since a block on one row influences where blocks on adjacent rows can be placed, local modification may not allow us to merge two rows, unless we can guarantee that the merged rows have enough flexibility to move blocks horizontally.

The flexibility we have to move blocks on a given row depends on how the blocks on adjacent rows are arranged. If all blocks on a row r are placed consecutively (they form one *megablock*), then on the next row there is enough flexibility to move blocks to merge (assuming the rows are not too full). However, if all blocks on r are regularly spaced, then on the next row there may not be enough space to move blocks at all. Hence, we move blocks on a sequence of consecutive rows until we create a row with a single megablock. This cannot always be done in a constant number of rows: we might need $\Omega(\log n / \log\log n)$ rows. To measure how close we are to a single megablock, we use an entropy-like function H on the free space between the blocks. We say a row r is δ*-dense* if the total width of the blocks on r is at most δW, where δ is the *density* of the row. We aim to show that H is always reduced by some term depending on δ.

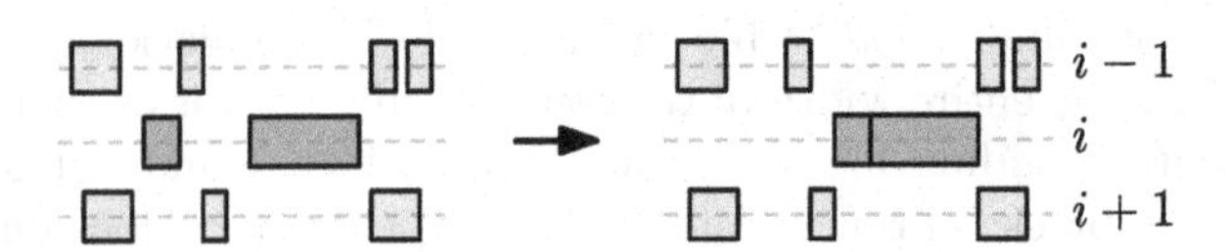

Fig. 5. Merge step: the two blue blocks are merged into a megablock.

Consider a set of k rows of the minimum-height ORD D, numbered row $1, \ldots, k$. Let δ_i be the density of row i. We first move and merge the blocks to obtain a *canonical placement* of the ORD. The blocks on row 1 are fixed. We then repeat the following operations until the ORD does not change anymore:

1. *Merge steps:* merge two consecutive megablocks b_1 and b_2 on the same row i into a larger megablock (see Fig. 5). A merge step is possible if and only if on rows $i-1$ and $i+1$, between b_1 and b_2, there is a large enough gap for the newly created megablock. There are *virtual* megablocks of width 0 on the left and right side of each row. That is, the leftmost megablock will jump into the leftmost available gap, and, if the row has more than one megablock, the rightmost megablock will jump into the rightmost available gap.
2. *Move steps:* move all blocks on row 2 to the left as much as possible. If a block starts hitting a block on row 3 below it, this block (and any other blocks on rows $i > 3$ further down) are moved along until a fixed block on a higher row is hit (possibly indirectly). The rightmost block on a row is moved to the right instead of to the left (unless there is only one megablock). After all blocks on row 2 are stuck, we fix the blocks on row 2 (only for this iteration of move steps) and repeat the process for the next row, and so on.

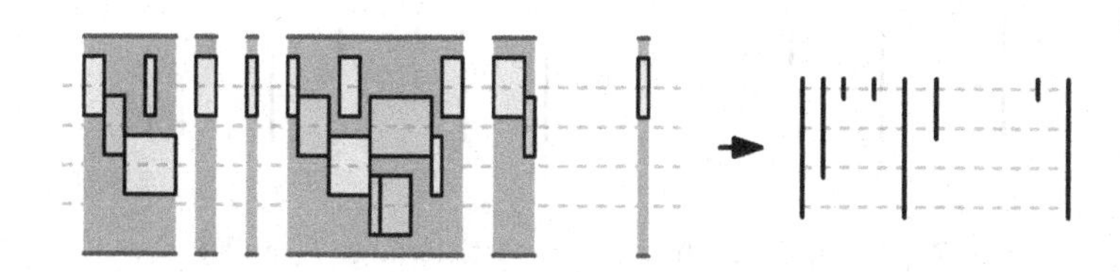

Fig. 6. Removing x-coordinates with blocks in them to obtain the free space partitioning. The rows in the drawing on the right represent $F_1, \ldots, F_4$.

As each merge step decreases the number of megablocks by one, we reach a canonical placement after a finite number of steps. In the remainder of this section we assume that we have a canonical placement of k rows. We say that a sequence of k rows is *Δ-dense* if the total width of all blocks is at most ΔW (typically, we use $\Delta < 1$). Clearly, for a set of Δ-dense rows, $\delta_i \leq \Delta$ for all i. In the following we assume w.l.o.g. that $W = 1$. Starting with the interval $[0, 1]$, we can obtain the *free space partitioning* of a set of rows by removing all x-coordinates occupied by any block (see Fig. 6). The free space partitioning is essentially a set of contiguous intervals of x-coordinates not used by any block. Note that the total length of the free space partitioning of a set of Δ-dense rows is at least $1 - \Delta$. Between two intervals of the free space partitioning there is an interval of x-coordinates occupied by blocks. We refer to such a set of blocks as a *separator*, and define the *index* of a separator S by the largest index of a row with a block in S (the leftmost and rightmost separator always have index k).

Lemma 3. *For a canonical placement, every internal separator S of index i contains a block for all rows j with $j \leq i$.*

We define the *i-th free space partitioning* F_i as the partitioning that includes only the separators with index at least i (see the right of Fig. 6), so F_i denotes gaps on row i. Further note that F_i is always a refinement of F_{i+1} by definition.

We define the *entropy* [8] of F_i as $H_i = -\sum_j x_j \log x_j$, where x_j is the length of the j-th interval of F_i. We aim to show via the entropy that, if the number of rows k is large enough, then F_k contains at most two intervals.

Lemma 4. *For a canonical placement of a Δ-dense set of rows ($\Delta \leq 1/2$) with width $W = 1$, either F_{i+1} consists of at most two intervals, or $H_i - H_{i+1} \geq (1-\Delta)\log((1-\Delta)/(2\delta_{i+1}))$, where δ_{i+1} is the density of row $i+1$.*

Proof. Consider an interval I of length y in F_{i+1}, where F_{i+1} has more than two intervals. Since F_i is a refinement of F_{i+1}, I is covered by r intervals with lengths $x_1, \ldots, x_r$ in F_i (see Fig. 7). Because the drawing is in canonical placement, $x_1, \ldots, x_r$ each must be smaller than the sum of the widths of the two megablocks in the separators surrounding I in F_{i+1}, which exist due to Lemma 3. (Otherwise these megablocks could have merged into that gap.) We denote the sum of the widths of the megablocks around I by z.

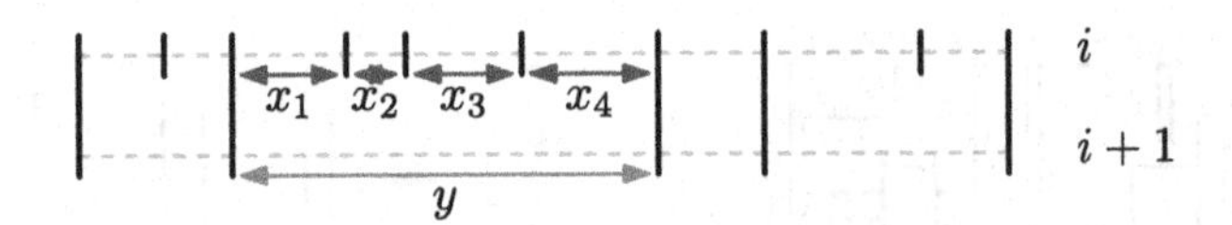

Fig. 7. Sketch of the free space partitionings F_i and F_{i+1}.

For this interval, the contribution to H_{i+1} is $-y \log y$, and the contribution to H_i is $\sum_j -x_j \log x_j$. Since $x_j < z$ for all j and $\sum_j x_j = y$, the contribution to H_i is at least $-y \log z$. Thus, the contribution to the difference $H_i - H_{i+1}$ in this interval is at least $y \log(y/z)$.

Next, we sum up these differences over all intervals of F_{i+1}. Let Z be the sum of all values z, and let Y be the sum of all values y. Y is the total length of the free space, which is at least $1 - \Delta$, while Z counts every block on row $i+1$ at most twice, thus $Z \leq 2\delta_{i+1}$. Using the log sum inequality we obtain that $H_i - H_{i+1} \geq Y \log Y/Z$. The claim follows from the bounds on Y and Z.

The next step is to use Lemma 4 to show that we can merge two rows without increasing height if we have sufficiently many rows. We first establish the conditions under which we can merge two consecutive rows.

Lemma 5. *Let D be an ORD with width W, 3 rows, and blocks of total width at most W. Then an ORD D' with only two rows exists which is not higher than D.*

Lemma 6. *Consider a Δ-dense set R of $2k+7$ consecutive rows of an ORD D with n blocks where $\Delta \leq 1/5$. If $(k/2)^{k/2} > 2n$, then we can obtain another ORD D' for which all rows not in R and the first and last row of R are the same as in D, D' has one fewer rows, and the height of D' is at most the height of D.*

Proof (sketch). First, we compute a canonical placement for the first $k+1$ rows, and, rotated by 180°, also for the last $k+1$ rows (see Fig. 8). Using the log sum inequality, the entropy of row 1 is bounded by $H_1 \leq (1-\Delta)\log(n/(1-\Delta)) \leq \log(2n)$, since $\Delta \leq 1/2$. By applying Lemma 4 repeatedly, we obtain that both F_{k+1} and F_{k+7} have only two intervals. Place the blocks on row $k+2$ together (as one megablock) as far to the left as possible; note that there is at most Δ free space to the left of this megablock. Similarly, place the blocks on row $k+6$ as far to the right as possible. For rows $k+3$ to $k+5$, the interval between $[2\Delta, 1-2\Delta]$ is free. Since the remaining width is $1-4\Delta$, and the total width of the blocks on these 3 rows is at most Δ, we can apply Lemma 5 to the restricted interval $[2\Delta, 1-2\Delta]$ if $\Delta \leq 1/5$. This reduces the number of rows without increasing the height and thus completes the proof.

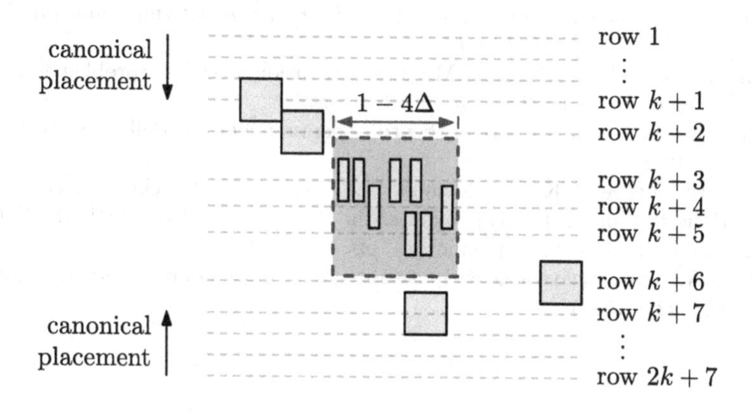

Fig. 8. Sketch of the proof of Lemma 6.

Lemma 7. *The minimum-height ORD D of an instance with n blocks has at most $O(r \log n/\log\log n)$ rows, where r is the number of rows of the minimum-row ORD.*

Proof. Assume w.l.o.g. that $W = 1$. Then the total width of all blocks is at most r. Now let k be the smallest integer such that $(k/2)^{k/2} > 2n$. Assume for the sake of contradiction that D has more than Crk rows for some large constant C. We partition the rows of D into consecutive groups of $(2k+7)$ rows. If none of these groups are Δ-dense for $\Delta = 1/5$, then we get that $r(Ck/(2k+7))/5 < r$ or $Ck/(10k+35) < 1$. We can easily choose C large enough (e.g. $C > 45$) such that this does not hold. Thus there must exist a Δ-dense group of $(2k+7)$ rows for $\Delta = 1/5$. We then apply Lemma 6 to obtain a contradiction. Thus, D has at most $O(rk)$ rows. To complete the proof, observe that $k = O(\log n/\log\log n)$.

Theorem 4. *For ORDs, $\beta(n) = O(\log n/\log\log n)$.*

References

1. Coffman, E.G., Csirik, J., Galambos, G., Martello, S., Vigo, D.: Bin packing approximation algorithms: survey and classification. In: Pardalos, P.M., Du, D.-Z., Graham, R.L. (eds.) Handbook of Combinatorial Optimization, pp. 455–531. Springer, New York (2013). https://doi.org/10.1007/978-1-4419-7997-1_35
2. Hirschberg, D.S., Larmore, L.L.: The least weight subsequence problem. SIAM J. Comput. **16**(4), 628–638 (1987)
3. Hofri, M.: Two-dimensional packing: expected performance of simple level algorithms. Inf. Control **45**, 1–17 (1980)
4. Knuth, D.E., Plass, M.F.: Breaking paragraphs into lines. Softw.-Pract. Exp. **11**, 1119–1184 (1981)
5. Korte, B., Vygen, J.: Bin-Packing. In: Combinatorial Optimization, pp. 426–441. Springer, Heidelberg (2005)
6. Kosaraju, S.R.: An optimal RAM implementation of catenable min double-ended queues. In: Proceedings of the 5th Annual ACM-SIAM Symposium on Discrete Algorithms, pp. 195–203 (1994)
7. Lodi, A., Martello, S., Monaci, M.: Two-dimensional packing problems: a survey. Eur. J. Oper. Res. **141**, 241–252 (2002)
8. Shannon, C.E.: A mathematical theory of communication. Bell Syst. Tech. J. **27**, 379–423 (1948)
9. Sonke, W., Verbeek, K., Meulemans, W., Verbeek, E., Speckmann, B.: Optimal algorithms for compact linear layouts. In: Proceedings of the 11th IEEE Pacific Visualization Symposium (PacificVis), pp. 1–10 (2018)
10. Wilber, R.: The concave least-weight subsequence problem revisited. J. Algorithms **9**(3), 418–425 (1988)

Shortest Rectilinear Path Queries to Rectangles in a Rectangular Domain

Mincheol Kim[1], Sang Duk Yoon[2], and Hee-Kap Ahn[1](✉)

[1] Department of Computer Science and Engineering, Graduate School of Artificial Intelligence, Pohang University of Science and Technology, Pohang, Korea
{rucatia,heekap}@postech.ac.kr
[2] Department of Service and Design Engineering, SungShin Women's University, Seoul, Korea
sangduk.yoon@sungshin.ac.kr

Abstract. Given a set of open axis-aligned disjoint rectangles in the plane, each of which behaves as both an obstacle and a target, we seek to find shortest obstacle-avoiding rectilinear paths from a query to the nearest target and the farthest target. In our problem, the distance to a target is determined by the point on the target achieving the minimum or maximum geodesic distance among all points on the boundary of the target. This problem arises in facility location and robot motion planning problems. We show how to construct a data structure for such shortest path queries to the nearest and farthest neighbors efficiently.

Keywords: Shortest path query · Rectangular domain · Nearest neighbor · Farthest neighbor

1 Introduction

Computing the nearest and farthest neighbors has been studied extensively in computational geometry and has applications in machine learning, data mining, computer vision, and many more fields. Given a set P of points in a space, we preprocess the points in P and construct a data structure such that given any query point the point in P nearest or farthest to the query point can be reported efficiently. One natural and efficient technique for points in the plane is to decompose the plane into regions with respect to the distances to the points and construct a search structure on the decomposition for point location queries. For instance, the nearest-site and farthest-site Voronoi diagrams of points in the plane serve as such a decomposition [5].

M. Kim and H.-K. Ahn were supported by the Institute of Information & communications Technology Planning & Evaluation(IITP) grant funded by the Korea government(MSIT) (No. 2017-0-00905, Software Star Lab (Optimal Data Structure and Algorithmic Applications in Dynamic Geometric Environment)) and (No. 2019-0-01906, Artificial Intelligence Graduate School Progra (POSTECH)). S.D. Yoon was supported by "Cooperative Research Program for Agriculture Science & Technology Development (Project No. PJ01526903)" Rural Development Administration, Republic of Korea.

Y. Kohayakawa and F. K. Miyazawa (Eds.): LATIN 2020, LNCS 12118, pp. 271–282, 2020.
https://doi.org/10.1007/978-3-030-61792-9_22

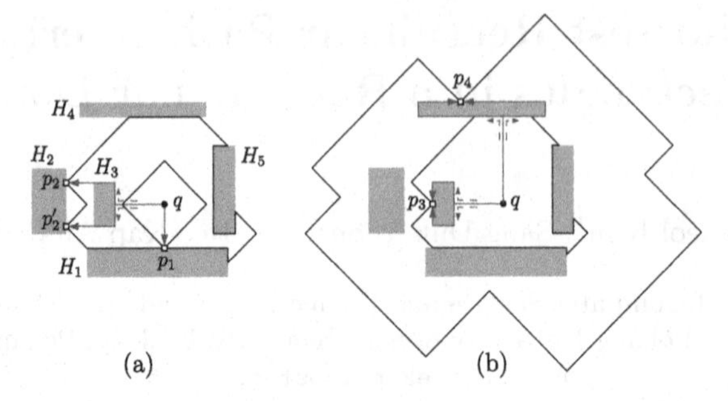

Fig. 1. Five gray rectangles $H_1, \ldots, H_5$, each behaves as both an obstacle and a site. (a) The nearest rectangle of q under minimum distance is determined by p_1. The farthest rectangle of q under minimum distance is determined by p_2 (and p_2'). (b) The nearest rectangle of q under maximum distance is determined by p_3. The farthest rectangle of q under maximum distance is determined by p_4.

The problem has also been considered in the presence of obstacles in the plane. In general, the obstacles are pairwise-disjoint polygons and the plane minus the interior of the obstacles is called *the polygonal domain.* The nearest or farthest point of a given query point can still be computed efficiently once we construct the geodesic nearest-site or geodesic farthest-site Voronoi diagram. However, there can be an exponential number of locally shortest simple paths between two points in the polygonal domain. It requires some additional work, possibly using data structures, to report a shortest path among them efficiently.

The problem becomes much more challenging when the sites are objects other than points. The distance from a query point q to an object X in the polygonal domain is then measured to the point in X whose geodesic distance is minimum or maximum among all points in X. Since such a point in X may vary with respect to query points, it is nontrivial for a given query to find a shortest path to such a point in X in the polygonal domain. The problem becomes even more complicated if each object behaves as an obstacle. Then the point in X whose geodesic distance is minimum or maximum from a query point always lies on the boundary of X and a shortest path to the point must avoid the interior of X.

In this paper, we consider nearest and farthest queries under the L_1 distance in the polygonal domain where each polygonal obstacle is an axis-aligned rectangle and it is also regarded as a site. We seek to find a shortest obstacle-avoiding rectilinear path to the nearest rectangle or to the farthest rectangle, where the distance to a rectangle is determined by a point on the rectangle achieving the minimum or the maximum geodesic L_1 distances among all points on its boundary. A *rectilinear path* is a path that consists only of horizontal and vertical segments. Figure 1 illustrates the problems we consider in the polygonal domain.

The problem arises typically in facility location and motion planning applications. The wire interconnects between chips in a VLSI circuit and PCB is a disjoint packing of rectilinear paths that avoid the interiors of chips. Since the pinouts of a chip are splayed out to its boundary, a path to a chip is connected to one of the pinouts. A challenge in the design is to locate VIAs or power connectors in the board such that the maximum path (wire) length is minimized to maintain low clock skew [8].

Related Works. There are two types of approaches for solving the shortest path queries, wavefront approaches and path-preserving structures. In the wavefront approach, we use a curve consisting of points at equal geodesic distance from the query, which is called *the geodesic circle*, that propagates continuously, taking the boundaries of the obstacles into account. Mitchell [14] constructs a Shortest Path Map with $O(n \log n)$ preprocessing time and $O(n)$ space, supporting $O(\log n)$ query time for the length of a shortest path.

In the path-preserving approach, we construct a structure (or a graph) that represents some geometric information and then solve the shortest path problem using the structure. Under the Euclidean distance, one-point and two-point shortest path queries in a polygonal domain with total n vertices have been studied extensively. Hershberger and Suri [12] gave a data structure with $O(n \log n)$ preprocessing time and $O(n \log n)$ space, supporting $O(\log n)$ query time for one-point shortest path queries. Chiang and Mitchell [4] gave a data structure with $O(n^{11})$ preprocessing time and $O(n^{11})$ space, supporting $O(\log n)$ query time for two-point queries. Later Guo et al. [10] gave a data structure using $O(n^2)$ space, supporting $O(h \log n)$ query time, where h is the number of holes.

Under the L_1 distance, Chen et al. [3] gave a data structure supporting $O(\log n)$ query time for two-point queries. Later Wang [16] reduced the preprocessing time to $O(n + (h^2 \log^4 h / \log \log h))$ and the space to $O(n + (h^2 \log^3 h / \log \log h))$ with $O(\log n)$ query time, where h is the number of holes in the polygonal domain. When all obstacles are rectangles, Elgindy and Mitra [7] gave a data structure with $O(n^2)$ preprocessing time and $O(n^2)$ space, supporting two-point L_1 shortest path query in $O(\log n)$ time. They used a planar subdivision and graph structure to store the distances and the shortest rectilinear geodesic paths between a finite number of junctions efficiently.

The nearest or farthest neighbor among m points from a query point in a polygonal domain can be found efficiently using the Voronoi diagrams [12,13]. Ben-Moshe et al. [2] gave a data structure with $O(nm \log(n+m))$ preprocessing time and $O(nm)$ space that supports $O(\log(n+m))$ query time. Later, Ben-Moshe et al. [1] proposed a time-space trade-off data structure.

However, we are not aware of any result on shortest path queries when the polygonal obstacles behave as both obstacles and targets.

Our Results. Our main result is to give data structures with efficient query algorithms for the shortest path queries to the nearest and farthest rectangles under minimum and maximum distances in a polygonal domain.

Given n axis-aligned disjoint rectangular obstacles in the plane, we construct a data structure in $O(n^2)$ time using $O(n^2)$ space such that for a query consisting of a point q and an obstacle Q in the domain, the minimum or the maximum L_1 geodesic distance from q to Q can be computed in $O(\log n)$ time. A shortest rectilinear geodesic path from q to Q achieving the distance can be reported in $O(\log n + K)$ time, where K is the number of line segments of the path.

We then present data structures for the nearest-site and farthest-site L_1 shortest path queries. We consider four versions of the problem depending on the nearest and farthest neighbors and on the minimum and maximum distance points of an object. We construct a data structure in $O(n \log n)$ time using $O(n)$ space such that for a query point q, the shortest rectilinear path from q to the nearest-min point can be computed in $O(\log n)$ time. For the nearest-max query, we construct a data structure in $O(n^2)$ time using $O(n^2)$ space such that for a query point q, the shortest rectilinear path from q to the nearest-max point can be computed in $O(\log n + K)$ time. For the farthest-min/farthest-max queries, we construct a data structure in $O(n^2 \log n)$ time using $O(n^2)$ space such that for a query point q, the shortest rectilinear path to the corresponding point in $O(\log n + K)$ time. Again, K is the number of line segments of the path.

Finally, we present data structures for a query consisting of a horizontal segment and an obstacle in the domain. The data structures can be constructed in $O(n^2 \log n)$ time and $O(n^2 \log n)$ space such that given a query of a horizontal segment h and an obstacle Q, the minimum (or the maximum) L_1 geodesic distance from h to Q can be computed in $O(\log n)$ time. A shortest rectilinear geodesic path from the point in h to the point in Q achieving the minimum (or the maximum) distance can also be reported in $O(\log n + K)$ time, where K is the number of line segments of the path.

The details and the omitted proofs will be found in the full version.

1.1 Notation and Preliminaries

Let P be a set of n disjoint axis-aligned (open) rectangles in $\mathbb{R}^2$. Each rectangle $H \in \mathsf{P}$ is open and plays as both an obstacle and a target in computing shortest paths in the plane. Let $\mathsf{D} := \mathbb{R}^2 - \cup_{H \in \mathsf{P}} H$ and call it a *rectangular domain.* A point p in D is represented by its x-coordinate $x(p)$ and y-coordinate $y(p)$. We use ∂H to denote the boundary of H.

The L_1 distance between two points $a \in \mathbb{R}^2$ and $b \in \mathbb{R}^2$ is defined to be $|x(a) - x(b)| + |y(a) - y(b)|$. A rectilinear path π consisting of k segments can be represented by a sequence of $k+1$ points, called *bends*, at which the path switches between vertical and horizontal, except at the endpoints of the path. The length of π is the sum of the segment lengths of π. The geodesic L_1 distance between any two points $p, q \in \mathsf{D}$, denoted by $d(p,q)$, is the length of a shortest rectilinear geodesic path connecting p and q. A path π is *x-monotone* (and *y-monotone*) if the intersection of π with any line perpendicular to the x-axis (and to the y-axis) is connected or empty. If π is x-monotone and y-monotone, π is *xy-monotone.*

We use $d_{\min}(p, H) := \min_{x \in \partial H} d(p, x)$ to denote *the minimum distance* from a point $p \in \mathsf{D}$ to a rectangle $H \in \mathsf{P}$. Likewise, we use $d_{\max}(p, H) := \max_{x \in \partial H} d(p, x)$ to denote *the maximum distance* from p to H. Let $p_{\min}(p, H) := \arg\min_{x \in \partial H} d(p, x)$ and $p_{\max}(p, H) := \arg\max_{x \in \partial H} d(p, x)$.

Carrier Graphs. *A carrier graph* is a directed acyclic graph constructed from P [7]. The carrier graph of n axis-aligned disjoint rectangles has $O(n)$ vertices and edges. Since the carrier graph is directed and acyclic, the distances of all pairs of vertices can be computed in $O(n^2)$ time. Moreover, the graph can encode information on shortest paths between vertices using $O(n^2)$ space. Using the carrier graph together with the information on the distances and shortest paths in $O(n^2)$ space, the distance between any two query points can be computed in $O(\log n)$ time and a shortest rectilinear path between them can be reported in $O(\log n + K)$ time, where K is the number of line segments of the path.

2 Point-to-Rectangle Shortest Path Queries

We construct a data structure that given a query consisting of a point q and a rectangle $Q \in \mathsf{P}$ finds the shortest paths and its distance from q to Q in D. We report two shortest paths from q to Q, one under minimum distance and one under maximum distance.

2.1 Wake of a Side of a Rectangle

We define the wake of a side of a rectangle $H \in \mathsf{P}$. Let $\ell = v_t v_b$ be the left side of H with $y(v_t) > y(v_b)$. We construct two unbounded xy-monotone paths $\pi(v_t)$ and $\pi(v_b)$ as the boundary chains of the wake of ℓ. Consider the horizontal ray from $v_t = \pi_0$ going rightwards. The ray stops when it hits another rectangle $H' \in \mathsf{P}$. Let π_1 be the point where the ray stops and π_2 be the top-left corner of H'. We repeat this process by taking the horizontal ray from π_2 going rightwards and so on until the ray goes to infinity. Then we obtain an xy-monotone path $\pi(v_t) = (\pi_0, \pi_1, \pi_2, \ldots)$. Likewise, we can obtain another xy-monotone path $\pi(v_b)$ from the bottom-left corner of H, by repeating the process with the bottom-left corners of the rectangles hit by the rays. We call the region bounded by $\pi(v_t)$, $\pi(v_b)$, and ℓ in D that is incident to the side of H opposite to ℓ the *wake* of ℓ and denote it by $W_\ell(H)$. $W_\ell(H)$ contains $\pi(v_t), \pi(v_b)$ and ℓ. See Fig. 2(a). The following lemma states the properties of shortest paths.

Lemma 1 (Rezende et al. [6]). *Every shortest path between any two points in a rectangular domain is x-monotone, y-monotone, or xy-monotone. Also, a shortest path from a point $s \in \ell$ to any point $t \in W_\ell(H)$ is x-monotone.*

The wakes of the other sides of H can be defined similarly, using rays going vertically upwards for the bottom side, rays going vertically downwards for the top side, and rays going horizontally leftwards for the right side of H.

We use the wakes in computing the distance and a shortest path from a query point q to a target point t in D as follows. We first identify the rectangles hit first by each of the two vertical rays (one upwards and one downwards) and two horizontal rays (one leftwards and one rightwards) from q. See Fig. 2(b). For each such rectangle, we use the wake of the side hit first by the ray. If t is contained

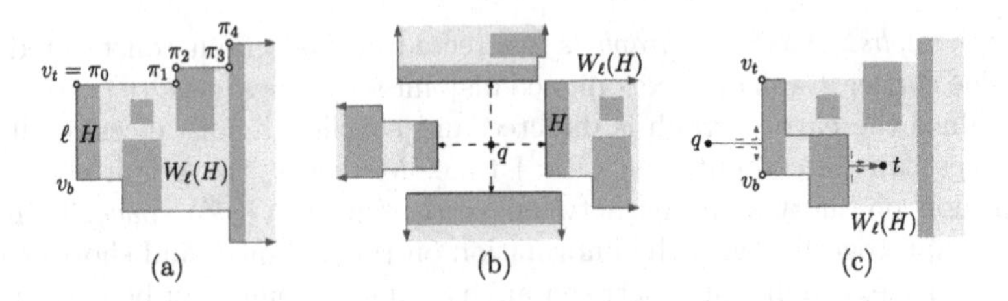

Fig. 2. (a) The wake $W_\ell(H)$ of the left side ℓ of H. (b) Four sides of rectangles hit by they rays from q, and their wakes. (c) There is a shortest path from q to $t \in W_\ell(H)$ containing either v_t or v_b as the first bend. They are x-monotone.

in none of the four wakes, every shortest path from q to t is xy-monotone and $d(q,t)$ is simply the sum of the differences of their coordinates.

If t is contained in one of the wakes, $d(q,t) = \min\{d(q,v_t)+d(v_t,t), d(q,v_b)+d(v_b,t)\}$, where v_t and v_b are the endpoints of the side defining the wake. Without loss of generality, assume that the ray from q going rightwards hits a rectangle $H \in \mathsf{P}$ first its left side ℓ, and that t is contained in $W_\ell(H)$. If $t \notin \ell$, there is a shortest path from q to t that contains either v_t or v_b as a bend. See Fig. 2(c) for an illustration. Thus, we use the carrier graph G for the positive x direction to compute the distance and a shortest path from v_t or v_b to t. This is the basically same method applied to find a shortest path between two points in [7].

2.2 Minimum Distance Between a Point and a Rectangle

A shortest rectilinear geodesic path π from q to $p_{\min}(q,Q)$ consists of one or more line segments. We determine whether π is a line segment or not in $O(\log n)$ time using ray shooting query structures with $O(n \log n)$ preprocessing time and $O(n)$ space [9]. If the horizontal or vertical ray from q hits Q first among the rectangles in P, π is the line segment qp, where p is the point the ray hits Q.

Thus, we consider the case that π consists of more than one line segment. We denote by H the rectangle in P hit first by the horizontal ray from q going rightwards and by ℓ the left side of H. Note that $H \neq Q$. We show how to deal with the case that Q is incident to $W_\ell(H)$. The other cases that Q is incident to one of three remaining wakes can be handled analogously.

Lemma 2. *For any two points $a, b \in W_\ell(H)$ satisfying $x(a) < x(b)$ and $y(a) = y(b)$, we have $d(q,a) < d(q,b)$.*

From Lemma 2, $p_{\min}(q,Q)$ lies on the left side of Q. We can specify the location of $p_{\min}(q,Q)$ by the following lemma.

Lemma 3. *Let $q \in \mathsf{D}$ be a point and $Q \in \mathsf{P}$ be a rectangle. If a shortest path from q to $p_{\min}(q,Q)$ consists of more than one line segment, then $p_{\min}(q,Q)$ must be a vertex of the carrier graph lying on the boundary of Q.*

By Lemma 2 and Lemma 3, $p_{\min}(q, Q)$ is a vertex of the carrier graph G for the positive x direction that lies on the left side (and its endpoints) of Q. Let ℓ be the left side of a rectangle $H \in \mathsf{P}$ which a rightward horizontal ray from q hits first. Let v_t and v_b be the top and bottom endpoints of ℓ, respectively.

To achieve $O(\log n)$ query time, we compute the minimum distance from any point $a \in \ell$ to each rectangle incident to $W_\ell(H)$ in advance. In specific, for each rectangle $H' \in \mathsf{P}\setminus\{H\}$ such that $\partial H' \in W_\ell(H)$, we compute $d_t = \min_{\eta\in F} d(v_t, \eta)$ and $d_b = \min_{\eta\in F} d(v_b, \eta)$ in $W_\ell(H)$, where F is the set of the vertices of G lying on the left side of H'. Let $f(y) = (y - b) + d_b$ and $g(y) = (t - y) + d_t$, where t, b are the y-coordinates of the top and bottom sides of H and y denotes $y(a)$. Observe that $f(y)$ denotes the length of a shortest path from a to H' that goes through v_b and $g(y)$ denotes the length of a shortest path from a to H' that goes through v_t. Thus, $d_{\min}(a, H') = \min\{f(y), g(y)\}$. That is, the lower envelope L of $f(y)$ and $g(y)$ for $y \in [b, t]$ shows the length of a shortest path from $a \in \ell$ to H' together with the information of whether the path goes through v_b or v_t. Note that the complexity of L is $O(1)$. If the horizontal ray from q going rightwards hits a rectangle first at the point a on its left side, $d_{\min}(q, Q) = |qa| + d_{\min}(a, Q)$. We compute such envelopes for each side of all rectangles in total $O(n^2)$ time.

Given point $q \in \mathsf{D}$ and rectangle $Q \in \mathsf{P}$, we use ray shooting query from q in $O(\log n)$ time to identify the four sides and their wakes. We check if Q is incident to one of the four wakes in constant time by maintaining for each wake the indices of the rectangles incident to the wake in a table. If Q is incident to one of the four wakes, we can compute the distance from the lower envelope for Q in constant time. If Q is incident to none of the four wakes, π is xy-monotone. If Q is contained in the first quadrant of q, $p_{\min}(q, Q)$ is the bottom-left corner of Q. Thus, we just return the sum of differences of coordinates between q and $p_{\min}(q, Q)$ as the distance in constant time. A shortest path from q to $p_{\min}(q, Q)$ can be computed using the carrier graph [7].

Theorem 1. *We can construct a data structure for the rectangular domain* D *induced by a set* P *of* n *axis-aligned disjoint rectangles in the plane in* $O(n^2)$ *time and* $O(n^2)$ *space such that given a pair of a point* $q \in \mathsf{D}$ *and a rectangle* $Q \in \mathsf{P}$*, the minimum* L_1 *geodesic distance from* q *to* Q *can be computed in* $O(\log n)$ *time. A shortest rectilinear geodesic path from* q *to the point in* Q *achieving the distance can be reported in* $O(\log n + K)$ *time, where* K *is the number of line segments of the path.*

2.3 Maximum Distance Between a Point and a Rectangle

Unlike the minimum distance case, a shortest rectilinear geodesic path π from q to $p_{\max}(q, Q)$ always consists of more than one line segment.

We denote by H the rectangle in P hit first by the horizontal ray from q going rightwards and by ℓ the left side of H. We show how to deal with the case that Q is incident to $W_\ell(H)$. It is possible that $H = Q$. The other cases that Q is incident to one of three remaining wakes can be handled analogously.

Let w_t and w_b be the top and bottom endpoints of the right side of Q, respectively. We can get the following Corollary from Lemma 2.

Corollary 1. *$p_{\max}(q, Q)$ lies on the right side $w_t w_b$ of Q. Moreover, there is a shortest rectilinear geodesic path from q to $p_{\max}(q, Q)$ going through w_t as a bend and a shortest rectilinear geodesic path from q to $p_{\max}(q, Q)$ going through w_b as a bend.*

We compute $p_{\max}(q, Q)$ when Q is incident to $W_\ell(H)$ as follows. Let v_t and v_b be the top and bottom endpoints of ℓ, respectively. By Corollary 1, $p_{\max}(q, Q)$ is determined by $d(q, w_t)$ and $d(q, w_b)$. w_t and w_b are in $W_\ell(H)$, so $d(q, w_i) = \min\{d(q, v_j) + d(v_j, w_i)\}$ for $i = t, b$ and $j = t, b$. The horizontal ray from q going rightwards hits ∂H at the left side of H, and thus $d(q, v_j) = |x(q) - x(v_j)| + |y(q) - y(v_j)|$. Since $d(u, v)$ for every pair (u, v) of vertices of the carrier graph is computed during the preprocessing phase, $d(v_j, w_i)$ can be reported in constant time. Thus, we have $d(q, w_t)$ and $d(q, w_b)$. We can obtain $p_{\max}(q, Q)$ on the right side of Q.

Given point $q \in \mathsf{D}$ and rectangle $Q \in \mathsf{P}$, we use ray shooting query from q in $O(\log n)$ time to identify the four sides and their wakes. If Q is incident to one of the four wakes, we compute $d_{\max}(q, Q)$ using $d(q, w_t)$ and $d(q, w_b)$ in constant time. Otherwise, π is xy-monotone. If Q is contained in the first quadrant of q, $p_{\max}(q, Q)$ is the top-right corner of Q. We just return the sum of differences of coordinates between q and $p_{\max}(q, Q)$ as $d_{\max}(q, Q)$. A shortest path from q to $p_{\max}(q, Q)$ can be computed using the carrier graph.

Theorem 2. *We can construct a data structure for the rectangular domain D induced by a set P of n axis-aligned disjoint rectangles in the plane in $O(n^2)$ time and $O(n^2)$ space such that given a pair of a point $q \in \mathsf{D}$ and a rectangle $Q \in \mathsf{P}$, the maximum L_1 geodesic distance from q to Q can be computed in $O(\log n)$ time. A shortest rectilinear geodesic path from q to the point in Q achieving the distance can be reported in $O(\log n + K)$ time, where K is the number of line segments of the path.*

3 Queries to the Nearest and Farthest Rectangles

We construct data structures such that given a point query q in D a rectangle in P satisfying the following criteria can be computed in $O(\log n)$ time: $\min_{H \in \mathsf{P}} d_{\min}(q, H)$ (Nearest-min query), $\max_{H \in \mathsf{P}} d_{\min}(q, H)$ (Farthest-min query), $\min_{H \in \mathsf{P}} d_{\max}(q, H)$ (Nearest-max query), and $\max_{H \in \mathsf{P}} d_{\max}(q, H)$ (Farthest-max query). A shortest rectilinear path from q to the rectangle can also be returned in $O(\log n + K)$ time, where K is the number of line segments of the path.

3.1 Nearest-Min Query

Given a query point q, we find a rectangle $H^* = \arg\min_{H \in \mathsf{P}} d_{\min}(q, H)$, and report $p_{\min}(q, H^*)$ together with $d_{\min}(q, H^*)$ and a shortest rectilinear geodesic path from q to $p_{\min}(q, H^*)$. Observe that an L_1 geodesic circle centered at q with radius $d_{\min}(q, H^*)$ in D forms a rhombus with all internal angles 90° at corners and that the boundary of H^* is incident to the rhombus. See the rhombus incident to p_1 in Fig. 1(a).

So we can find H^* from the L_1 nearest-site Voronoi diagram of the rectangles in P. Papadopoulou and Lee [15] show that the L_∞ Voronoi diagram of polygonal objects with total complexity n can be constructed in $O(n \log n)$ time. Thus, we can construct the L_1 Voronoi diagram of the rectangles in the same time. The resulting L_1 Voronoi diagram has $O(n)$ complexity, and a point location query for q can be answered in $O(\log n)$ time. Since a shortest rectilinear geodesic path from q to $p_{\min}(q, H^*)$ consists of at most two segments, it can be computed in constant time.

Theorem 3. *We can construct a data structure for the rectangular domain* D *induced by a set* P *of* n *axis-aligned disjoint rectangles in the plane in* $O(n \log n)$ *time and* $O(n)$ *space such that given a query point* $q \in \mathsf{D}$*, the nearest rectangle* H *of* P *from* q *under minimum* L_1 *geodesic distance and its distance from* q *can be found in* $O(\log n)$ *time. A shortest rectilinear geodesic path from* q *to the point in* H *achieving the distance can be reported in* $O(\log n)$ *time.*

3.2 Farthest-Min Query

Given a query point q, we find a rectangle $H^* = \arg\max_{H \in \mathsf{P}} d_{\min}(q, H)$, and report $p_{\min}(q, H^*)$ together with $d_{\min}(q, H^*)$ and a shortest rectilinear geodesic path from q to $p_{\min}(q, H^*)$. See the geodesic circle incident to p_2 in Fig. 1(a).

Consider $W_\ell(H)$ of the left side ℓ of a rectangle $H \in \mathsf{P}$. Let v_t and v_b denote the top and bottom endpoints of ℓ, respectively. Lemma 3 implies that for any point $q' \in \ell$ and any rectangle $H' \in \mathsf{P}$ incident to $W_\ell(H)$ such that $H \neq H'$, $p_{\min}(q', H')$ is a vertex of the carrier graph for the positive x direction lying on H. We can compute the lower envelope for ℓ defined in Sect. 2.2 in time linear to the number of vertices of the carrier graph lying on the rectangle, that represents the minimum geodesic distance from any point $a \in \ell$ to a rectangle in P incident to $W_\ell(H)$. Thus, the lower envelope for ℓ for each rectangle in P incident to $W_\ell(H)$ can be computed in total $O(n)$ time since the number of vertices of the carrier graph is $O(n)$. Then we compute the upper envelope of the lower envelopes for ℓ, which represents the farthest rectangle from any point $a \in \ell$ among all rectangles of P incident to $W_\ell(H)$. For each rectangle in P, we compute such an upper envelope for each of its sides. Each upper envelope can be computed in $O(n \log n)$ time and $O(n)$ space [11]. So, we compute and maintain $O(n)$ upper envelopes with $O(n^2 \log n)$ preprocessing time and $O(n^2)$ space. In case that there is no rectangle of $\mathsf{P} \setminus \{H\}$ incident to $W_\ell(H)$, we simply do not compute the envelope because H is obviously the farthest rectangle.

Given a query point q, we use ray shooting query from q in $O(\log n)$ time to identify the four sides and their wakes. Using the upper envelopes corresponding to the sides, we can find a rectangle $H^* = \arg\max_{H \in \mathsf{P}} d_{\min}(q, H)$ for rectangles H incident to one of the four wakes, and report $p_{\min}(q, H^*)$ together with $d_{\min}(q, H^*)$ in $O(\log n)$ time. A shortest rectilinear path from q to $p_{\min}(q, H^*)$

can also be reported in $O(\log n+K)$ time, where K is the number of line segments of the path.

Now we handle the case that H^* is incident to none of the four wakes of q. Without loss of generality, assume that H^* is contained in the first quadrant $\mathcal{Q}_1$ of q. We observe that the bottom left corner c of H^* should have the largest sum $x(c)+y(c)$ among all bottom-left corners of the rectangles in P. The rectangle H' with the largest sum can be obtained in $O(n)$ time in preprocessing and whether H' is incident to none of the four wakes and also contained in $\mathcal{Q}_1$ of q can be determined in $O(1)$ time. Again, a shortest path from q to the bottom left corner of H^* can be computed using the carrier graph.

Theorem 4. *We can construct a data structure for the rectangular domain* D *induced by a set* P *of* n *axis-aligned disjoint rectangles in the plane in* $O(n^2 \log n)$ *time and* $O(n^2)$ *space such that given a query point* $q \in \mathsf{D}$*, the farthest rectangle* H *from* q *under minimum* L_1 *geodesic distance and its distance from* q *can be found in* $O(\log n)$ *time. A shortest rectilinear geodesic path from* q *to the point in* H *achieving the distance can be reported in* $O(\log n + K)$ *time, where* K *is the number of line segments of the path.*

3.3 Nearest-Max Query

Given a query point q, we find a rectangle $H^* = \arg\min_{H\in\mathsf{P}} d_{\max}(q,H)$, and report $p_{\max}(q,H^*)$ together with $d_{\max}(q,H^*)$ and a shortest rectilinear geodesic path from q to $p_{\max}(q,H^*)$. See the geodesic circle incident to p_3 in Fig. 1(b).

Consider $W_\ell(H)$ of the left side $\ell = v_t v_b$ of a rectangle $H \in \mathsf{P}$ with $y(v_t) > y(v_b)$. By Corollary 1, $p_{\max}(q,H^*)$ lies on the right side of H^*. Thus, we can define a function for ℓ that represents the geodesic distance from any point $a \in \ell$ to the top-right corner, denoted by w_t, of a rectangle $H' \in \mathsf{P}$ incident to $W_\ell(H)$. More precisely, $f(y) = (y-b)+d_b$ and $g(y) = (t-y)+d_t$, where $d_b = d(v_b, w_t)$ and $d_t = d(v_t, w_t)$, where t, b are the y-coordinates of the top and bottom sides of H and y denotes $y(a)$ for point $a \in \ell$. Then the lower envelope L_t of $f(y)$ and $g(y)$ represents the geodesic distance from any point $a \in \ell$ to w_t. Similarly, let L_b be the lower envelope for ℓ that represents the geodesic distance from any point $a \in \ell$ to the bottom-right corner, denoted by w_b, of H'.

Using L_t and L_b, we can define a function $j(\ell, H')$ that represents $p_{\max}(a, H')$ for points $a \in \ell$. Observe that $j(\ell, H')$ a piecewise linear function whose graph consists of at most three line segments, one of slope $+1$, one of slope 0, and one of slope -1, in the order. There is a (possibly empty) range r in ℓ such that $d_{\max}(a, H')$ is constant for every point $a \in r$ (corresponding to the segment of slope 0). Moreover, a shortest rectilinear path from a point $a \in r$ to $p_{\max}(a, H')$ through v_t and a shortest rectilinear path from a to $p_{\max}(a, H')$ through v_b have the same length $d_{\max}(a, H')$. Observe that the lower envelope L_ℓ of the $j(\ell, H')$ functions for every rectangle $H' \in \mathsf{P}$ incident to $W_\ell(H)$ represents the nearest rectangle from any point $a \in \ell$ among all rectangles of P incident to $W_\ell(H)$.

In the preprocessing phase, we compute for each rectangle in P the lower envelope L_s for each side s of the rectangle. Since there are $O(n)$ rectangles in the wake of s and each of the distance functions (lower envelopes) defined on s consists of at most three segments of slopes $\{+1, 0, -1\}$ in the order, we can compute L_s in $O(n)$ time. Therefore, we compute and maintain $O(n)$ lower envelopes with $O(n^2)$ preprocessing time and $O(n)$ space.

Given a query point q, we use ray shooting query from q to identify the four sides and their wakes. Using the lower envelope corresponding to each of the sides, we can find H^* and report $p_{\max}(q, H^*)$ together with $d_{\max}(q, H^*)$ in $O(\log n)$ time. A shortest rectilinear geodesic path from q to $p_{\max}(q, H^*)$ can also be reported in $O(\log n + K)$ time, where K is the number of line segments of the path.

Theorem 5. *We can construct a data structure for the rectangular domain* D *induced by a set* P *of* n *axis-aligned disjoint rectangles in the plane in* $O(n^2)$ *time and* $O(n^2)$ *space such that given a query point* $q \in \mathsf{D}$*, the nearest rectangle* H *from* q *under maximum* L_1 *geodesic distance and its distance from* q *can be found in* $O(\log n)$ *time. A shortest rectilinear geodesic path from* q *to the point in* H *achieving the distance can be reported in* $O(\log n + K)$ *time, where* K *is the number of line segments of the path.*

3.4 Farthest-Max Query

Given a query point q, we find a rectangle $H^* = \arg\max_{H \in \mathsf{P}} d_{\max}(q, H)$, and report $p_{\max}(q, H^*)$ together with $d_{\max}(q, H^*)$ and a shortest rectilinear geodesic path from q to $p_{\max}(q, H^*)$. See the geodesic circle incident to p_4 in Fig. 1(b).

Theorem 6. *We can construct a data structure for the rectangular domain* D *induced by a set* P *of* n *axis-aligned disjoint rectangles in the plane in* $O(n^2 \log n)$ *time and* $O(n^2)$ *space such that given a query point* $q \in \mathsf{D}$*, the farthest rectangle* H *from* q *under maximum* L_1 *geodesic distance and its distance from* q *can be found in* $O(\log n)$ *time. A shortest rectilinear geodesic path from* q *to the point in* H *achieving the distance can be reported in* $O(\log n + K)$ *time, where* K *is the number of line segments of the path.*

4 Line Segment Queries

We construct a data structure such that given a horizontal line segment query $h \subset \mathsf{D}$ and a rectangle $Q \in \mathsf{P}$ the shortest rectilinear geodesic path from h to Q and its length can be found efficiently.

Theorem 7. *We can construct a data structure for the rectangular domain* D *induced by a set* P *of* n *axis-aligned disjoint rectangles in the plane in* $O(n^2 \log n)$ *time and* $O(n^2 \log n)$ *space such that given a pair of a horizontal line segment* $h \in \mathsf{D}$ *and a rectangle* $Q \in \mathsf{P}$*, the minimum (or maximum)* L_1 *geodesic distance from* h *to* Q *(and the corresponding points on* h *and* Q*) can be computed in* $O(\log n)$ *time. A shortest rectilinear geodesic path from the point in* h *to the point in* Q *achieving the minimum (or maximum) distance can also be reported in* $O(\log n + K)$ *time, where* K *is the number of line segments of the path.*

References

1. Ben-Moshe, B., Bhattacharya, B., Shi, Q.: Farthest neighbor Voronoi diagram in the presence of rectangular obstacles. In: Proceedings of the 13th Canadian Conference on Computational Geometry (CCCG), pp. 243–246 (2005)
2. Ben-Moshe, B., Katz, M., Mitchell, J.: Farthest neighbors and center points in the presence of rectangular obstacles. In: Proceedings of the 17th Annual Symposium on Computational Geometry (SoCG), pp. 164–171 (2001)
3. Chen, D., Inkulu, R., Wang, H.: Two-point L_1 shortest path queries in the plane. In: Proceedings of the 30th Annual Symposium on Computational Geometry (SoCG), pp. 406–415 (2014)
4. Chiang, Y.J., Mitchell, J.: Two-point Euclidean shortest path queries in the plane. In: Proceedings of the 10th Annual ACM-SIAM Symposium on Discrete Algorithms (SODA), pp. 215–224 (1999)
5. De Berg, M., Cheong, O., Van Kreveld, M., Overmars, M.: Computational Geometry: Algorithms and Applications, 3rd edn. Springer, Santa Clara (2008). https://doi.org/10.1007/978-3-540-77974-2
6. De Rezende, P., Lee, D.T., Wu, Y.F.: Rectilinear shortest paths with rectangular barriers. In: Proceedings of the 1st Annual Symposium on Computational Geometry (SoCG), pp. 204–213 (1985)
7. Elgindy, H., Mitra, P.: Orthogonal shortest route queries among axes parallel rectangular obstacles. Int. J. Comput. Geom. Appl. **4**(1), 3–24 (1994)
8. Gester, M., Müller, D., Nieberg, T., Panten, C., Schulte, C., Vygen, J.: BonnRoute: Algorithms and data structures for fast and good VLSI routing. ACM Trans. Des. Autom. Electron. Syst. **18**(2), 32:1–32:24 (2013)
9. Giora, Y., Kaplan, H.: Optimal dynamic vertical ray shooting in rectilinear planar subdivisions. ACM Trans. Algorithms **5**(3), 28:1–28:51 (2009)
10. Guo, H., Maheshwari, A., Sack, J.-R.: Shortest path queries in polygonal domains. In: Fleischer, R., Xu, J. (eds.) AAIM 2008. LNCS, vol. 5034, pp. 200–211. Springer, Heidelberg (2008). https://doi.org/10.1007/978-3-540-68880-8_20
11. Hershberger, J.: Finding the upper envelope of n line segments in $O(n \log n)$ time. Inf. Process. Lett. **33**(4), 169–174 (1989)
12. Hershberger, J., Suri, S.: An optimal algorithm for Euclidean shortest paths in the plane. SIAM J. Comput. **28**(6), 2215–2256 (1999)
13. Liu, C.H., Lee, D.: Higher-order geodesic Voronoi diagrams in a polygonal domain with holes. In: Proceedings of the 24th Annual ACM-SIAM Symposium on Discrete Algorithms (SODA), pp. 1633–1645 (2013)
14. Mitchell, J.: L_1 shortest paths among polygonal obstacles in the plane. Algorithmica **8**(1–6), 55–88 (1992). https://doi.org/10.1007/BF01758836
15. Papadopoulou, E., Lee, D.: The L_∞ Voronoi diagram of segments and VLSI applications. Int. J. Comput. Geom. Appl. **11**(05), 503–528 (2001)
16. Wang, H.: A divide-and-conquer algorithm for two-point L_1 shortest path queries in polygonal domains. In: Proceedings of the 35th International Symposium on Computational Geometry (SoCG), pp. 59:1–59:14 (2019)

Farthest Color Voronoi Diagrams: Complexity and Algorithms

Ioannis Mantas[1], Evanthia Papadopoulou[1(✉)], Vera Sacristán[2], and Rodrigo I. Silveira[2]

[1] Faculty of Informatics, Università della Svizzera Italiana, Lugano, Switzerland
{ioannis.mantas,evanthia.papadopoulou}@usi.ch
[2] Department de Matemàtiques, Universitat Politècnica de Catalunya, Barcelona, Spain
{vera.sacristan,rodrigo.silveira}@upc.edu

Abstract. The *farthest-color Voronoi diagram* (FCVD) is a *farthest-site* Voronoi structure defined on a family $\mathcal{P}$ of m point-clusters in the plane, where the total number of points is n. The FCVD finds applications in problems related to color spanning objects and facility location. We identify structural properties of the FCVD, refine its combinatorial complexity bounds, and list conditions under which the diagram has $O(n)$ complexity. We show that the diagram may have complexity $\Omega(n+m^2)$ even if clusters have disjoint convex hulls. We present construction algorithms with running times ranging from $O(n \log n)$, when certain conditions are met, to $O((n+s(\mathcal{P})) \log^3 n)$ in general, where $s(P)$ is a parameter reflecting the number of *straddles* between pairs of clusters in $\mathcal{P}$ ($s(P) \in O(mn)$). A pair of points $q_1, q_2 \in Q$ is said to *straddle* $p_1, p_2 \in P$ if the line segment q_1q_2 intersects (straddles) the line through p_1, p_2 and the disks through (p_1, p_2, q_1) and (p_1, p_2, q_2) contain no points of P, Q. The complexity of the diagram is shown to be $O(n+s(\mathcal{P}))$.

Keywords: Farthest color · MaxMin · Voronoi diagram · Point clusters

1 Introduction

The *Voronoi diagram* is a versatile and well-known geometric partitioning structure defined by a set of simple geometric objects in a space, called *sites*. The ordinary (*nearest-neighbor*) Voronoi diagram of a set of points in the plane is a subdivision into maximal regions such that all points in one region share the same

A preliminary version of this work was presented at EuroCG 2019. I. M. and E. P. were supported in part by the Swiss National Science Foundation, project SNF 200021E-154387. V. S. and R. S. were supported by projects MINECO MTM2015-63791-R and Gen. Cat. 2017SGR1640. This project has received funding from the European Union's Horizon 2020 research and innovation programme under the Marie Skłodowska-Curie grant agreement No 734922.

Y. Kohayakawa and F. K. Miyazawa (Eds.): LATIN 2020, LNCS 12118, pp. 283–295, 2020.
https://doi.org/10.1007/978-3-030-61792-9_23

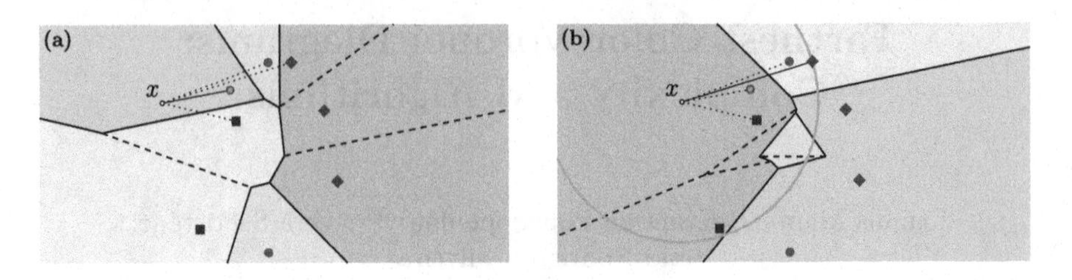

Fig. 1. (a) NCVD($\mathcal{P}$) and (b) FCVD($\mathcal{P}$) of a family $\mathcal{P}$, with $d_c(x, P)$ illustrated.

nearest site. In the *farthest-site Voronoi diagram* points in a single region have the same farthest site. Many generalizations of this simple concept have been considered, including generalized sites, metrics and spaces. For a comprehensive list of results and some of the many applications of Voronoi diagrams see [5,22].

We are interested in *color Voronoi diagrams*, where each site is a set of points in $\mathbb{R}^2$, referred to as a *cluster*. Conceptually, we identify a distinct color with each cluster and all points in a cluster have the same color. The distance between a point $x \in \mathbb{R}^2$ and a cluster P is realized by the nearest point in P, i.e., $d_c(x, P) = \min_{p \in P} d(x, p)$. The *nearest color Voronoi diagram* (NCVD) of a family $\mathcal{P}$ of clusters is a *min-min* diagram that can be easily derived from the ordinary Voronoi diagram of all points in $\mathcal{P}$: the region of a cluster P is the union of the Voronoi regions of points belonging to P (see Fig. 1a). In this paper we focus on the farthest counterpart, the *farthest color Voronoi diagram* (FCVD), a *max-min* diagram generalizing both the nearest and farthest Voronoi diagrams of points, and whose properties are still not well understood (see Fig. 1b).

Motivation. Farthest problems involving point clusters appear in several different situations. Clusters may represent locations of facilities of the same type that can be accessed interchangeably, while the farthest distance allows to give worst-case scenario bounds on the distance to reach an object of each type. For instance, consider the following typical *facility location* problem: given locations of multiple types of facilities (e.g., hospitals, schools, etc.), each type represented by a cluster, find a location such that the distance to all services is minimized. This can be found using the *minimum color spanning disk* [1], which can be extracted efficiently from the FCVD. Such problems arise also in spatial databases [26]. Many other *minimum color spanning objects* have been considered in the literature, see e.g., [2,14,18]. Similar problems also appear when considering imprecision in geometric data, as point clusters are a natural way to represent the possible locations of an object, whose exact location is unknown [17]. In this setting, the FCVD encodes proximity information, allowing to efficiently solve problems involving pairs of points [3] or larger clusters [12]. The diagram is also useful in *shape matching* [25], finding the translation that minimizes the Hausdorff distance between two point sets [15]. Finally, the FCVD has been used to solve variants of the *Steiner tree* problem [7], *sensor deployment* problems in wireless sensor networks [20] and to find *stabbing circles* for line segments [10].

Related Work. The FCVD was first studied by Huttenlocher et al. [15], showing that the combinatorial complexity of the diagram is $\Omega(mn)$ in the worst case and $O(mn\alpha(mn))$, where m is the number of clusters and n is the overall number of points. The worst case complexity was latter settled to $\Theta(mn)$ by Abellanas et al. [1]. Using a geometric transformation in 3 dimensions, the diagram can be computed in $O(mn \log n)$ time by computing the upper envelope of m *Voronoi surfaces* [15], one for each input cluster.

Closely related to the FCVD is the *Hausdorff Voronoi diagram* (HVD) of point clusters. The HVD is a *min-max* diagram: the distance from a point $x \in \mathbb{R}^2$ to a cluster P is the farthest distance, $d_f = \max_{p \in P} d(x,p)$, and the plane is subdivided into regions with the same nearest cluster. The HVD has been extensively studied [13,23], and many algorithmic paradigms have been considered for its construction, see e.g. [4,9,11,13,23,24]. Interestingly, the algorithm presented in [13] for the HVD can also be used to obtain an $O(n^2)$-time algorithm for the FCVD. This has already been remarked in [10] for 2-point clusters. In the worst case, this algorithm is optimal as the diagram may have complexity $\Theta(n^2)$. However, it remains $\Theta(n^2)$ even if the complexity of the diagram is $O(n)$.

A central question in the study of the FCVD is under what conditions the diagram has $O(n)$ complexity, and when it can be computed in subquadratic time. Some restricted instances of linear-size diagrams have already been considered by Bae [6], Claverol et al. [10] and Iacono et al. [16].

Contribution. We present structural properties of the FCVD and refine its combinatorial complexity bounds. We show that the complexity of the diagram is $O(n + s(\mathcal{P}))$, where $s(\mathcal{P})$ is a parameter reflecting the number of *straddles* between clusters ($s(\mathcal{P}) = O(mn)$, see Definition 5). Based on this, we list conditions under which the FCVD has $O(n)$ structural complexity and show that *linear separability* is not such a condition. Indeed, we establish that FCVD($\mathcal{P}$) may have complexity $\Omega(n + m^2)$ for a family $\mathcal{P}$ of linearly separable clusters. Finally, we present an $O((n + s(\mathcal{P})) \log^3 n)$-time construction algorithm for the FCVD, which is considerably more efficient than existing approaches when the straddling number is small. Due to lack of space, some proofs have been omitted.

2 Preliminaries

Let $\mathcal{P} = \{P_1, ..., P_m\}$ be a family of m *clusters of points* in $\mathbb{R}^2$, $m > 1$, where no two clusters share a common point. Let the set of all points be $\mathcal{P}^* = \bigcup_{P_i \in \mathcal{P}} P_i$, with $|\mathcal{P}^*| = n$. We assume that $\mathcal{P}^*$ is in general position, i.e., no three points are collinear and no four points are cocircular.

Let Vor(P) denote the ordinary (nearest-neighbor) Voronoi diagram of a set of points P in $\mathbb{R}^2$ and let $vreg(p, P)$ denote the Voronoi region of a point $p \in P$ in Vor(P), i.e., $vreg(p, P) = \{x \in \mathbb{R}^2 \mid d(x,p) < d(x,q)\ \forall q \in P,\ q \neq p\}$.

We use the following additional notation. The line through two points p and q is denoted by $L(p,q)$. Let $C(p,q,r)$ denote the circle through points p, q and r and let $D(p,q,r)$ denote the corresponding disk. The bisector of two points p

and q is denoted by $b(p,q)$. The line segment with endpoints p and q is denoted by $\overline{pq}$. The convex hull of a set of points P is denoted by $CH(P)$.

The distance of a point $x \in \mathbb{R}^2$ to a cluster P_i is $d_c(x,P_i) = min_{p \in P_i} d(x,p)$. We define the following two *color Voronoi diagrams.*

Definition 1. *The* nearest color Voronoi diagram *of* $\mathcal{P}$, *denoted* NCVD($\mathcal{P}$), *is the subdivision of* $\mathbb{R}^2$ *into nearest color regions. The* nearest color region of a cluster $P_i \in \mathcal{P}$ *is* $n_c reg(P_i, \mathcal{P}) = \{x \in \mathbb{R}^2 \mid d_c(x,P_i) < d_c(x,P_j)\ \forall P_j \in \mathcal{P},\ j \neq i\}$.

The NCVD($\mathcal{P}$) can be directly derived from the ordinary Voronoi diagram Vor($\mathcal{P}^*$), as $n_c reg(P_i,\mathcal{P}) = \bigcup_{p \in P_i} vreg(p, \mathcal{P}^*)$. Thus, it immediately follows that it has complexity $O(n)$ and it can be computed in $O(n \log n)$ time.

Definition 2. *The* farthest color Voronoi diagram *of* $\mathcal{P}$, *denoted* FCVD($\mathcal{P}$), *is the subdivision of* $\mathbb{R}^2$ *into farthest color regions. The* farthest color region of a cluster $P_i \in \mathcal{P}$ *is* $f_c reg(P_i, \mathcal{P}) = \{x \in \mathbb{R}^2 \mid d_c(x,P_i) > d_c(x,P_j)\ \forall P_j \in \mathcal{P},\ j \neq i\}$.

Region $f_c reg(P_i, \mathcal{P})$ may consist of several connected components, called *faces.* The faces of $f_c reg(P_i, \mathcal{P})$ are further subdivided into finer parts by the skeleton of Vor(P_i), which is called the *internal skeleton* of $f_c reg(P_i, \mathcal{P})$. For $p \in P_i$, let $f_c reg(p, \mathcal{P}) = \{x \in f_c reg(P_i, \mathcal{P}) \mid d(x,p) < d(x,q),\ \forall q \in P_i \backslash \{p\}\}$. The FCVD($\mathcal{P}$), augmented with the internal skeletons of its faces, is denoted by FCVD$_a$($\mathcal{P}$).

Definition 3. *Given two clusters* P *and* Q, *their* color bisector, *denoted* $b_c(P,Q)$, *is the locus of points equidistant from* P *and* Q, *i.e.,* $b_c(P,Q) = \{x \in \mathbb{R}^2 \mid d_c(x,P) = d_c(x,Q)\}$.

The color bisector $b_c(P,Q)$ is a subgraph of Vor($P \cup Q$). It is a collection of disjoint unbounded chains and cycles of total complexity $O(|P| + |Q|)$. By ordinary Voronoi diagram it easily follows that if P and Q have disjoint convex hulls then $b_c(P,Q)$ is a single unbounded, monotone chain. Moreover, $b_c(P,Q)$ contains only bounded chains if and only if $CH(P) \subset CH(Q)$ or $CH(Q) \subset CH(P)$.

The augmented FCVD contains different types of Voronoi edges and vertices. To distinguish them we use the following conventions (see Fig. 2a): Voronoi edges that are subsets of color bisectors are called *color Voronoi edges* , while the edges of the internal skeletons are called *internal.* Voronoi vertices that are incident to three color bisectors are called *color Voronoi vertices*, while vertices incident to two color bisectors and one internal edge are called *mixed vertices.* Vertices of the internal skeletons are called *internal vertices.*

The *farthest color disk* of a point $x \in \mathbb{R}^2$, such that x lies in the closure of $f_c reg(p, \mathcal{P})$ and $p \in P_i$, is the disk centered at x of radius $d(x,p)$ (see Fig. 1b). Such a disk contains no point of P_i in its interior and its closure contains at least one point from every cluster in $\mathcal{P}$.

In our algorithms, we use a refinement of FCVD derived from the *visibility decomposition*, similarly to [24]: For each region $f_c reg(p, \mathcal{P})$, and for each color or mixed vertex u on $\partial f_c reg(p, \mathcal{P})$, draw $L(p,u) \cap f_c reg(p, \mathcal{P})$ (see Fig. 2a). The intersection $L(p,u) \cap f_c reg(p, \mathcal{P})$ is connected due to the following *visibility*

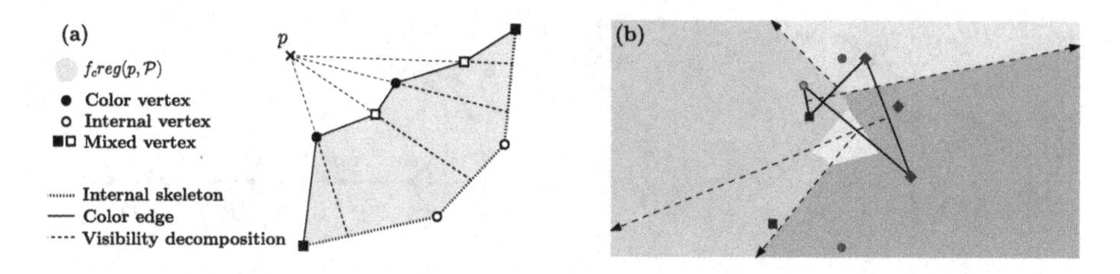

Fig. 2. (a) Features of the FCVD illustrated on a bounded face. (b) $CLH(\mathcal{P})$ of the family of Fig. 1. The rays at infinity coincide with the unbounded faces of $\text{FCVD}_\text{a}(\mathcal{P})$.

property [6]: Given points $p \in P_i$, $x \in f_c reg(p, \mathcal{P})$, and a ray r emanating from x in the direction away from p, the intersection $r \cap vreg(p, P_i)$ lies entirely in $f_c reg(p, \mathcal{P})$.

The *cluster hull* of $\mathcal{P}$ is a closed (non-simple) polygonal chain which characterizes the unbounded faces of the HVD. As shown in the sequel it also characterizes the unbounded faces of the FCVD (see Fig. 2b). We review its definition:

Definition 4 [24]. *Given a family of clusters $\mathcal{P}$, a point $p \in P_i$ is a* hull vertex *if p admits a supporting line ℓ such that P_i lies entirely in one halfplane defined by ℓ while every other cluster $P_j \neq P_i$ in $\mathcal{P}$ intersects the other halfplane. A* hull edge $e = (p, q)$ *connects two hull vertices $p \in P_i, q \in P_j$ if the line ℓ through p, q leaves P_i and P_j entirely on one side and every other cluster in $\mathcal{P}$ intersects the halfplane at the opposite side of ℓ. The edge e is associated with a unit vector that is normal to ℓ, pointing in the direction away from P_i, P_j. The hull edges sorted by the circular ordering of their normal vectors define a closed polygonal chain, called the* cluster hull *of $\mathcal{P}$ and denoted $CLH(\mathcal{P})$.*

3 Structural Properties and Complexity

Lemma 1. *A face f of $f_c reg(P_i, \mathcal{P})$ satisfies:*

1. *If f is bounded, it must contain a non-empty internal skeleton, which is a tree incident to the mixed vertices of ∂f.*
2. *If f is unbounded, its internal skeleton is a (possibly empty) forest; each tree of the internal forest has exactly one unbounded edge.*

The following lemma shows that the cluster hull $CLH(\mathcal{P})$ characterizes the unbounded faces of the FCVD($\mathcal{P}$). It is derived by establishing a one-to-one correspondence between the unbounded faces of the FCVD and the HVD. This also implies that the unbounded faces are $O(n)$, since the cluster hull has complexity $O(n)$ as shown in [24].

Lemma 2. *A region $f_c reg(p, \mathcal{P})$ is unbounded if and only if p is a vertex of $CLH(\mathcal{P})$. A counterclockwise traversal of $CLH(\mathcal{P})$ derives the clockwise ordering of the unbounded edges of* $\text{FCVD}_\text{a}(\mathcal{P})$.

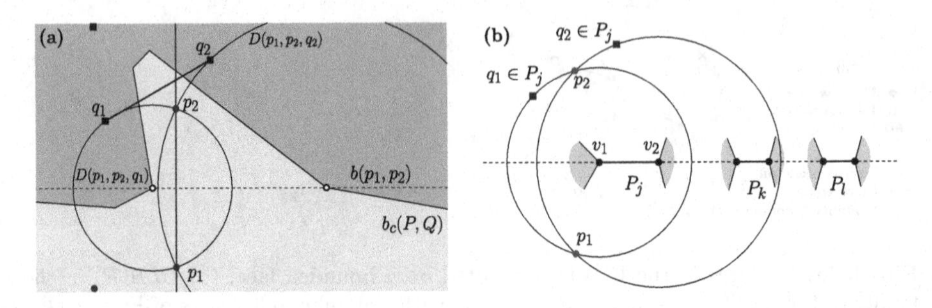

Fig. 3. Illustration of (a) a straddle and the proof of Lemma 4 and (b) the proof of Lemma 6 with the sequence of consecutive pairs of mixed vertices along $b(p_1, p_2)$.

Lemma 3. *FCVD($\mathcal{P}$) has $O(n)$ unbounded faces.*

To derive the total complexity of the diagram we need to consider its bounded faces, which in turn are determined by the mixed vertices incident to their non-empty internal skeleton. To this aim we define the notion of *straddles*.

Definition 5. *A cluster Q, and in particular a pair of points $q_1, q_2 \in Q$, is said to* straddle *points $p_1, p_2 \in P$ if the disks $D(p_1, p_2, q_1)$ and $D(p_1, p_2, q_2)$ contain no points of P and Q in their interior. Let $s(p_1, p_2)$ denote the number of clusters that straddle p_1, p_2. The* straddling number *of a family $\mathcal{P}$ is $s(\mathcal{P}) = \sum_{P_i \in \mathcal{P}} \sum_{(p_j, p_k) \in P_i} s(p_j, p_k)$.*

The condition of Definition 5 implies that the line segment $\overline{q_1 q_2}$ must intersect (straddle) the line $L(p_1, p_2)$ (see Fig. 3a and Lemma 4), hence, the name of the parameter $s(\mathcal{P})$. Note that segments $\overline{q_1 q_2}$ and $\overline{p_1 p_2}$ may or may not intersect.

Lemma 4. *Suppose that disks $D(p_1, p_2, q_1)$ and $D(p_1, p_2, q_2)$ contain no points of P and Q in their interior, i.e., their centers are mixed vertices of $b_c(P, Q)$. Then the segment $\overline{q_1 q_2}$ intersects (straddles) the line $L(p_1, p_2)$.*

Proof. The disks $D(p_1, p_2, q_1)$ and $D(p_1, p_2, q_2)$ are empty. Thus, point q_1 lies on $C(p_1, p_2, q_1) \backslash D(p_1, p_2, q_2)$ and point q_2 lies on $C(p_1, p_2, q_1) \backslash D(p_1, p_2, q_2)$. Hence, line $L(p_1, p_2)$ separates them and so $\overline{q_1 q_2} \cap L(p_1, p_2) \neq \emptyset$. Refer to Fig. 3a. □

Lemma 5. *The straddling number $s(\mathcal{P})$ is $O(mn)$.*

Proof. A pair (p_1, p_2) inducing vertices on bisector $b_c(P, Q)$, as in Definition 5, is incident to an edge in $\text{Vor}(P \cup Q)$, which is subset of the bisector $b(p_1, p_2)$. Bisector $b(p_1, p_2)$ can have only one occurrence as a Voronoi edge in $\text{Vor}(P \cup Q)$, so Q straddles (p_1, p_2) at most once, and hence $s(p_j, p_k) \leq m - 1$. Only pairs (p_j, p_k) inducing Voronoi edges in $\text{Vor}(P)$ can be straddled. Overall, there are $\sum_{P_i \in \mathcal{P}} O(|P_i|) = O(n)$ Voronoi edges, so $O(n)$ pairs may be straddled. □

In the following lemma, we show a property of consecutive mixed vertices. We then use it to bound the total number of mixed vertices in Lemma 7.

Lemma 6. *Let v_1, v_2 be two mixed vertices on the same bisector, incident to $f_c reg(P_i, \mathcal{P})$, such that the segment $\overline{v_1 v_2} \cap f_c reg(P_i, \mathcal{P}) = \emptyset$ (see Fig. 3b). Then v_1 and v_2 are induced by the same cluster Q.*

Lemma 7. FCVD($\mathcal{P}$) *has $O(n + s(\mathcal{P}))$ bounded faces.*

Proof. For any $P_i \in \mathcal{P}$ and for any pair (p_j, p_k) inducing an edge e in Vor(P_i), we count the number of mixed vertices appearing along e. There are at most two vertices not resulting from a straddle, the outermost from each side. Any other pair of vertices, by Lemma 6, is the result of a straddle. Thus, there at most $2 \cdot s(p_j, p_k) + 2$ mixed vertices incident to e. Such vertices appear consecutively along e (see Fig. 3b). Each pair of consecutive mixed vertices may create one bounded face incident to e, so there are at most $s(p_j, p_k) + 1$ bounded faces incident to e. Overall, there are $O(n)$ pairs, inducing Voronoi edges, concluding the proof. □

We conclude in Theorem 1 refining the $O(mn)$ upper bound for FCVD($\mathcal{P}$). This implies a sufficient condition for FCVD($\mathcal{P}$) to have $O(n)$ size, which is $s(\mathcal{P}) = O(n)$.

Theorem 1. FCVD($\mathcal{P}$) *has $O(n + s(\mathcal{P}))$ combinatorial complexity.*

4 Conditions for Linear-Size Diagrams

We derive conditions for the FCVD to have linear complexity. To this aim we first consider its relation to *abstract Voronoi diagrams*. These diagrams are defined in terms of bisecting curves satisfying some simple combinatorial properties, called axioms [19]. In the context of color Voronoi diagrams, these axioms can be interpreted as follows. For every subset $\mathcal{P}' \subseteq \mathcal{P}$:

(A1) Each region $n_c reg(P_i, \mathcal{P}')$ is non-empty and connected.
(A2) Each point in the plane belongs to the closure of a region $n_c reg(P_i, \mathcal{P}')$.
(A3) Each color bisector is an unbounded Jordan curve.
(A4) Any two color bisectors intersect transversally in a finite number of points.

A family of clusters is called *admissible* if the underlying system of bisectors satisfies axioms (A1)–(A4). By the structural properties of farthest abstract Voronoi diagrams [8,21] we derive the following.

Lemma 8. *If $\mathcal{P}$ is admissible, then the skeleton of* FCVD($\mathcal{P}$) *is a tree of combinatorial complexity $O(n)$.*

Next we give a necessary and sufficient condition under which $\mathcal{P}$ is admissible.

Theorem 2. *A family $\mathcal{P}$ is admissible if and only if the following two conditions hold: (1) No cluster is entirely enclosed within the convex hull of any other cluster in $\mathcal{P}$; and (2) for every $P_i \in \mathcal{P}$, region $n_c reg(P_i, \mathcal{P})$ is connected.*

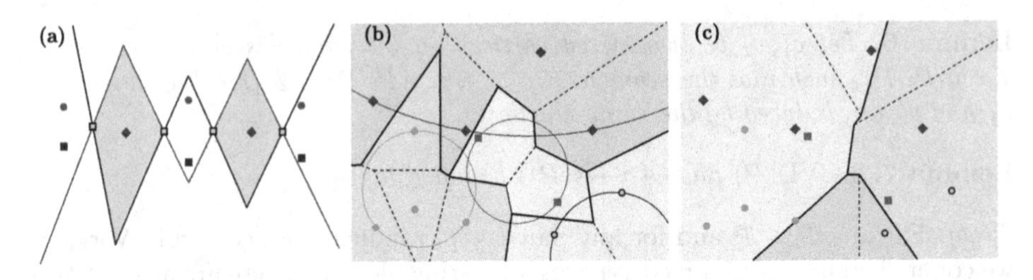

Fig. 4. (a) Two color bisectors *(red/green & blue/green)* intersecting linearly many times. (b) A disk-separable family $\mathcal{P}$ with NCVD($\mathcal{P}$) and (c) FCVD($\mathcal{P}$).

Proof (sketch). The bisector $b_c(P_i, P_j)$ contains no unbounded components if and only if P_i is entirely enclosed within $CH(P_j)$, or vice versa. Thus, if $\mathcal{P}$ is admissible then the axioms imply that both conditions hold. Next, we show that if $b_c(P_i, P_j)$ consists of more than one connected component then $n_c reg(P_i, \mathcal{P})$ is also disconnected. Thus, if both conditions of the lemma hold, then each bisector $b_c(P_i, P_j)$ is an unbounded Jordan curve, satisfying axiom (A3). Finally, we show that it suffices to check connectivity of axiom (A1) only for $\mathcal{P}$ without examining every subset $\mathcal{P}' \subset \mathcal{P}$ separately. □

Two clusters are called *linearly separable* if they have disjoint convex hulls. A family of pairwise linearly separable clusters is called *linearly separable*. Linear separability alone does not imply an admissible family. In particular, the bisectors of three linearly separable clusters, $b_c(P_i, P_j)$ and $b_c(P_j, P_k)$, may intersect $\Theta(|P_i| + |P_j| + |P_k|)$ times, and thus (A1) need not be satisfied (see Fig. 4a).

A linearly separable family $\mathcal{P}$ satisfies condition (1) of Theorem 2. So, deciding if $\mathcal{P}$ is admissible can be done by checking region connectivity in NCVD($\mathcal{P}$).

Theorem 3. *For a linearly separable family of clusters $\mathcal{P}$, we can decide if $\mathcal{P}$ is admissible in $O(n \log n)$ time.*

Disk-separability is another sufficient condition for a family to be admissible. A family $\mathcal{P}$ is called *disk-separable* if for every cluster $P_i \in \mathcal{P}$ there exists a disk that contains P_i and no point from another cluster P_j (see Fig. 4b).

Theorem 4. *If a family $\mathcal{P}$ is disk-separable then it is admissible.*

5 A Lower Bound for Linearly Separable Clusters

Linear separability is a natural property to investigate when characterizing properties of the FCVD. In this section we show that the FCVD may have complexity $\Omega(n + m^2)$ even if the clusters are linearly separable. To this aim, we define a family $\mathcal{P}$ of m linearly separable 2-point clusters $\mathcal{P} = \{P_i = \{l_i, u_i\}, 1 \leq i \leq m\}$ whose FCVD contains $\Theta(m^2)$ mixed vertices.

We construct the family $\mathcal{P}$ as follows, refer to Fig. 5. Let $l_1 = (0, 0)$ and $u_1 = (0, 2^m)$. Let C_i, $2 \leq i \leq m$, be a family of concentric circles centered at u_1,

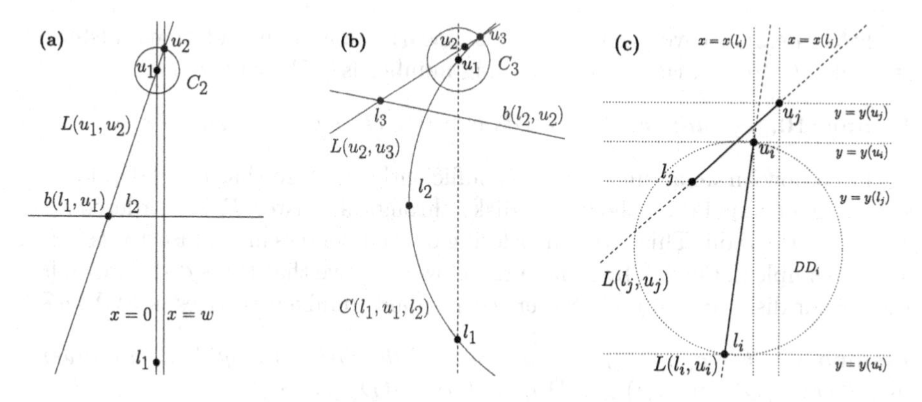

Fig. 5. (a) Placement of P_1, P_2 and (b) Placement of P_3. For any P_i, with $i \geq 3$, the placement is analogous. (c) The relation between two clusters P_i and P_j.

each of radius $2^{-(m-i+2)}$. Each *upper* point u_i is placed on circle C_i. Each *lower* point l_i is placed on bisector $b(l_{i-1}, u_{i-1})$. We control the placement of all points using a parameter w, $0 \leq w \ll 2^{-m}$. In particular, u_2 is placed at the upper intersection point of C_2 and the vertical line $x = w$. Each *lower* point l_i is placed at the intersection of line $L(u_{i-1}, u_i)$ and bisector $b(l_{i-1}, u_{i-1})$. Each *upper* point $u_i, i \geq 3$, is placed on the upper intersection of C_i and circle $C(l_{i-2}, u_{i-2}, l_{i-1})$.

As defined, the family $\mathcal{P}$ does not satisfy the general position assumption as every four points $(l_{i-2}, u_{i-2}, l_{i-1}, u_i)$ are cocircular and every three points (l_i, u_i, u_{i-1}) are collinear. However, general position can be easily enforced during the construction by infinitesimally translating u_i, for $i \geq 3$, on C_i towards the interior of $C(l_{i-2}, u_{i-2}, l_{i-1})$, and l_i, for $i \geq 2$, on $b(l_{i-1}, u_{i-1})$ towards the y-axis.

For $w = 0$, all points lie on the y-axis. As w increases lower points are translated up and left, while upper points are translated down and right. The quantity w needs to be sufficiently small so that l_j lies within the disk DD_i whose diameter is defined by l_i, u_i, for every $i < j$. Refer to our Geogebra applet[1] for an interactive visualization of $\mathcal{P}$ and the effect of changing w.

The following lemma points out properties of the family $\mathcal{P}$, see Fig. 5c. Let $x(p)$ (resp. $y(p)$) denote the x-coordinate (resp. y-coordinate) of a point p. We assume that line $L(a, b)$ is oriented from a to b.

Lemma 9. *Assuming that $l_j \in DD_i$, for any $i < j$, the following hold:*

(a) Point l_j is to the left of line $L(l_i, u_i)$ and point u_j is to its right.
(b) $0 < slope(L(l_j, u_j)) < slope(L(l_i, u_i))$.
(c) $y(l_i) < y(l_j) < y(u_i) < y(u_j)$ and $x(u_i) < x(u_j)$.
(d) Cluster P_i is below $L(l_j, u_j)$.

[1] http://compgeom.inf.usi.ch/FCVD/lowerbound.

From Lemma 9 we infer that $\mathcal{P}$ is linearly separable and that cluster P_j straddles P_j $\forall i < j$, thus, the straddling number is $s(\mathcal{P}) = \Theta(m^2)$.

Lemma 10. *For any m, there exists $w > 0$ such that $l_j \in DD_i$, $\forall i < j$.*

From now on we assume that w is sufficiently small so that $l_j \in \mathrm{DD}_i$, for any $i < j$. In Lemma 11 we show how disks through a cluster P_i are ordered with respect to the radii. This gives an ordering of their centers along bisector $b(l_i, u_i)$, as for example in Fig. 3b. In Lemma 12 below we prove that these disks are all farthest color disks and so their centers correspond to mixed vertices in FCVD($\mathcal{P}$).

Lemma 11. *For any $i < j < k$, the radii of the disks through l_i, u_i are ordered as follows: $r(D(l_i, u_i, u_j)) > r(D(l_i, u_i, l_j)) \geq r(D(l_i, u_i, u_k))$.*

Proof (sketch). We show separately that (1) $\mathrm{r}(D(l_i, u_i, u_j)) > \mathrm{r}(D(l_i, u_i, l_j))$ and (2) $\mathrm{r}(D(l_i, u_i, l_j)) \geq \mathrm{r}(D(l_i, u_i, u_k))$. For (1) we use the properties of Lemma 9. For (2) we first prove that $\mathrm{r}(D(l_i, u_i, l_j)) > \mathrm{r}(D(l_i, u_i, u_{j+1}))$. This is equivalent to showing that I_i, the upper intersection point of $C(l_i, u_i, l_j) \cap C(l_{j-1}, u_{j-1}, l_j)$, lies in the interior of C_{j+1}. We prove a simplified version of the statement using the coordinates of the points when $w = 0$, and we argue that the distance of I_i to C_{j+1} in that version, serves as an upper bound for the distance between I_i and C_{j+1}, as w increases. This statement combined with (1), then implies (2). □

Lemma 12. *Disks $D(l_i, u_i, l_j)$ and $D(l_i, u_i, u_j)$ are farthest color disks $\forall i < j$. They contain contain one point of cluster P_k, $\forall P_k \in \mathcal{P} \backslash \{P_i, P_j\}$. In particular:*

(i) if $k < i < j$, then $u_k \in D(l_i, u_i, l_j)$ *and* $u_k \in D(l_i, u_i, u_j)$,
(ii) if $i < k < j$, then $l_k \in D(l_i, u_i, l_j)$ *and* $l_k \in D(l_i, u_i, u_j)$,
(iii) if $i < j < k$, then $u_k \in D(l_i, u_i, l_j)$ *and* $u_k \in D(l_i, u_i, u_j)$.

Since the disks of Lemma 12 are farthest color disks, they induce mixed vertices in FCVD($\mathcal{P}$). Thus, FCVD($\mathcal{P}$) has $\Theta(m^2)$ mixed vertices appearing in pairs along bisectors. Each pair delimits a bounded face, hence, FCVD($\mathcal{P}$) has $\Theta(m^2)$ bounded faces.

Combining with the trivial $\Omega(n)$ bound, we conclude as follows.

Theorem 5. FCVD($\mathcal{P}$) *may have combinatorial complexity $\Omega(n + m^2)$, even if $\mathcal{P}$ is linearly separable.*

6 Construction Algorithms

We consider the divide & conquer paradigm. Split $\mathcal{P}$ into two sets $\mathcal{P}_L$ and $\mathcal{P}_R$ of roughly equal size. Recursively compute FCVD($\mathcal{P}_\mathrm{L}$) and FCVD($\mathcal{P}_\mathrm{R}$), and then merge them to obtain FCVD($\mathcal{P}$).

Merging FCVD($\mathcal{P}_\mathrm{L}$) and FCVD($\mathcal{P}_\mathrm{R}$) requires constructing the *merge curve* which is the set of color Voronoi edges in FCVD($\mathcal{P}_\mathrm{L} \cup \mathcal{P}_\mathrm{R}$) belonging to bisectors $b_c(P_i, P_j)$ with $P_i \in \mathcal{P}_L$ and $P_j \in \mathcal{P}_R$. The merge curve may consist of several connected components, both unbounded and bounded. To construct it, a *starting point* needs to be identified on each component, and then each component has to be *traced.*

Given a starting point on a component, we can efficiently trace it in $O(n)$ time by adapting standard tracing methods and exploiting the visibility decomposition, similarly to [24]. Moreover, due to Lemma 2, we can identify starting points on the unbounded components of the merge curve in $O(n)$ time, by merging the respective cluster hulls prior to the two diagrams, also similar to [24].

It remains to identify starting points on the bounded components of the merge curve. Bounded components enclose pieces of internal skeletons, due to Lemma 1. Hence, to identify starting points, we can search for the portions of the internal skeletons that are enclosed in such components. If an internal vertex is enclosed in a bounded component, it can be easily identified by point location in $O(\log n)$ time. However, no such vertex may be present and thus we need to identify an internal edge. To identify an internal edge, we use the data structure of Iacono et al. [16], which allows for efficient searches of intersections between two plane graphs. This has been also used in a similar way for the HVD [16]. Details are given in the following theorem.

Theorem 6. FCVD($\mathcal{P}$) *can be constructed in* $O((n + s(\mathcal{P})) \log^3 n)$ *time.*

Proof. We use the data structure of [16] as follows. At each recursive step, for each internal edge e of $\mathrm{FCVD_a}(\mathcal{P}_\mathrm{L})$ not entirely contained in the traced components we build a *search tree*, which is a balanced binary tree of depth $O(\log n)$, that implicitly stores the intersections of e with the skeleton of $\mathrm{FCVD_a}(\mathcal{P}_\mathrm{R})$. Identifying a portion of e requires a traversal of the search tree. An analogous procedure is done for every internal edge of $\mathrm{FCVD_a}(\mathcal{P}_\mathrm{R})$. The set of all search trees can be constructed in $O(n \log n)$.

Deciding which child to follow while navigating the search tree is done by point location at each node, which takes $O(\log n)$ time. Thus, a single traversal from the root to a leaf requires $O(\log^2 n)$ time. When traversing the tree, at some nodes the search might continue to both children, as portions of e may appear in more than one components, but this has to be a result of a straddle, from Lemma 6, Moreover, a single straddle occurs in at most one recursive step, since a pair of clusters can be considered at most once. By Lemma 7, there are $O(n + s(\mathcal{P}))$ bounded faces. This bounds the portions to be identified and the time all search trees are traversed, resulting in $O((n + s(\mathcal{P})) \log^3 n)$ time. □

If $\mathcal{P}$ is admissible, FCVD($\mathcal{P}$) can be computed using the randomized algorithm of Mehlhorn et al. [21]. Color bisectors, however, may have $\Theta(n)$ complexity, and a direct application would give time complexity $O(n^2 \log n)$. When $\mathcal{P}$ is admissible, by Lemma 8, all regions of FCVD($\mathcal{P}$) are unbounded, and thus, this is true for all components of the merge curve. So, each recursive step in our algorithm takes $O(n)$ time and we derive the following theorem.

Theorem 7. *If* $\mathcal{P}$ *is admissible,* FCVD($\mathcal{P}$) *can be constructed in* $O(n \log n)$ *time.*

Concluding, we remark that although the straddling number $s(\mathcal{P})$ is $\Theta(mn)$ in the worst case, this number could be small in ordinary instances of linearly separable clusters. In fact, configurations of linearly separable clusters, which can actually realize a quadratic straddling number are quite degenerate. In instances of clusters where the straddling number is small, our algorithm outperforms the existing $O(mn \log n)$-time [15] and $O(n^2)$-time [13] algorithms. This includes families of clusters that may or not be linearly separable.

References

1. Abellanas, M., et al.: The farthest color Voronoi diagram and related problems. In: Proceedings of the 17th European Workshop on Computational Geometry, pp. 113–116 (2001)
2. Acharyya, A., Nandy, S.C., Roy, S.: Minimum width color spanning annulus. Theor. Comput. Sci. **725**, 16–30 (2018)
3. Arkin, E.M., et al.: Bichromatic 2-center of pairs of points. Comput. Geom. **48**(2), 94–107 (2015)
4. Arseneva, E., Papadopoulou, E.: Randomized incremental construction for the Hausdorff Voronoi diagram revisited and extended. J. Comb. Optim. **37**(2), 579–600 (2018). https://doi.org/10.1007/s10878-018-0347-x
5. Aurenhammer, F., Klein, R., Lee, D.T.: Voronoi Diagrams and Delaunay Triangulations. World Scientific, Singapore (2013)
6. Bae, S.W.: On linear-sized farthest-color Voronoi diagrams. IEICE Trans. Inf. Syst. **95**(3), 731–736 (2012)
7. Bae, S.W., Lee, C., Choi, S.: On exact solutions to the Euclidean bottleneck Steiner tree problem. Inf. Process. Lett. **110**(16), 672–678 (2010)
8. Bohler, C., Cheilaris, P., Klein, R., Liu, C.H., Papadopoulou, E., Zavershynskyi, M.: On the complexity of higher order abstract Voronoi diagrams. Comput. Geom. **48**(8), 539–551 (2015)
9. Cheilaris, P., Khramtcova, E., Langerman, S., Papadopoulou, E.: A randomized incremental algorithm for the Hausdorff Voronoi diagram of non-crossing clusters. Algorithmica **76**(4), 935–960 (2016). https://doi.org/10.1007/s00453-016-0118-y
10. Claverol, M., Khramtcova, E., Papadopoulou, E., Saumell, M., Seara, C.: Stabbing circles for sets of segments in the plane. Algorithmica **80**(3), 849–884 (2018). https://doi.org/10.1007/s00453-017-0299-z
11. Dehne, F., Maheshwari, A., Taylor, R.: A coarse grained parallel algorithm for Hausdorff Voronoi diagrams. In: Proceedings of the 35th International Conference on Parallel Processing, ICPP, pp. 497–504. IEEE (2006)
12. Ding, H., Xu, J.: Solving the chromatic cone clustering problem via minimum spanning sphere. In: Aceto, L., Henzinger, M., Sgall, J. (eds.) ICALP 2011. LNCS, vol. 6755, pp. 773–784. Springer, Heidelberg (2011). https://doi.org/10.1007/978-3-642-22006-7_65
13. Edelsbrunner, H., Guibas, L.J., Sharir, M.: The upper envelope of piecewise linear functions: algorithms and applications. Discrete Comput. Geom. **4**(1), 311–336 (1989). https://doi.org/10.1007/BF02187733
14. Fleischer, R., Xu, X.: Computing minimum diameter color-spanning sets. In: Lee, D.-T., Chen, D.Z., Ying, S. (eds.) FAW 2010. LNCS, vol. 6213, pp. 285–292. Springer, Heidelberg (2010). https://doi.org/10.1007/978-3-642-14553-7_27
15. Huttenlocher, D.P., Kedem, K., Sharir, M.: The upper envelope of Voronoi surfaces and its applications. Discrete Comput. Geom. **9**(3), 267–291 (1993). https://doi.org/10.1007/BF02189323

16. Iacono, J., Khramtcova, E., Langerman, S.: Searching edges in the overlap of two plane graphs. WADS 2017. LNCS, vol. 10389, pp. 473–484. Springer, Cham (2017). https://doi.org/10.1007/978-3-319-62127-2_40
17. Jørgensen, A., Löffler, M., Phillips, J.M.: Geometric computations on indecisive points. In: Dehne, F., Iacono, J., Sack, J.-R. (eds.) WADS 2011. LNCS, vol. 6844, pp. 536–547. Springer, Heidelberg (2011). https://doi.org/10.1007/978-3-642-22300-6_45
18. Khanteimouri, P., Mohades, A., Abam, M.A., Kazemi, M.R.: Computing the smallest color-spanning axis-parallel square. In: Cai, L., Cheng, S.-W., Lam, T.-W. (eds.) ISAAC 2013. LNCS, vol. 8283, pp. 634–643. Springer, Heidelberg (2013). https://doi.org/10.1007/978-3-642-45030-3_59
19. Klein, R.: Concrete and Abstract Voronoi Diagrams. LNCS, vol. 400. Springer, Heidelberg (1989). https://doi.org/10.1007/3-540-52055-4
20. Lee, C., Shin, D., Bae, S.W., Choi, S.: Best and worst-case coverage problems for arbitrary paths in wireless sensor networks. Ad Hoc Netw. **11**(6), 1699–1714 (2013)
21. Mehlhorn, K., Meiser, S., Rasch, R.: Furthest site abstract Voronoi diagrams. Int. J. Comput. Geom. Appl. **11**(06), 583–616 (2001)
22. Okabe, A., Boots, B., Sugihara, K., Chiu, S.N.: Spatial Tessellations: Concepts and Applications of Voronoi Diagrams, vol. 501. Wiley, Hoboken (2009)
23. Papadopoulou, E.: The Hausdorff Voronoi diagram of point clusters in the plane. Algorithmica **40**(2), 63–82 (2004). https://doi.org/10.1007/s00453-004-1095-0
24. Papadopoulou, E., Lee, D.T.: The Hausdorff Voronoi diagram of polygonal objects: a divide and conquer approach. Int. J. Comput. Geom. Appl. **14**(06), 421–452 (2004)
25. Veltkamp, R.C., Hagedoorn, M.: State of the art in shape matching. In: Lew, M.S. (ed.) Principles of Visual Information Retrieval. ACVPR, pp. 87–119. Springer, London (2001). https://doi.org/10.1007/978-1-4471-3702-3_4
26. Zhang, D., Chee, Y.M., Mondal, A., Tung, A.K., Kitsuregawa, M.: Keyword search in spatial databases: towards searching by document. In: Proceedings of the 25th International Conference on Data Engineering, pp. 688–699. IEEE (2009)

Rectilinear Convex Hull of Points in 3D

Pablo Pérez-Lantero[1], Carlos Seara[2(✉)], and Jorge Urrutia[3]

[1] Departamento de Matemática y Ciencia de la Computación, USACH, Santiago, Chile
pablo.perez.l@usach.cl

[2] Departament de Matemàtiques, Universitat Politècnica de Catalunya, Barcelona, Spain
carlos.seara@upc.edu

[3] Instituto de Matemáticas, Universidad Nacional Autónoma de México, Mexico City, Mexico
urrutia@matem.unam.mx

Abstract. Let P be a set of n points in $\mathbb{R}^3$ in general position, and let $RCH(P)$ be the rectilinear convex hull of P. In this paper we obtain an optimal $O(n \log n)$-time and $O(n)$-space algorithm to compute $RCH(P)$. We also obtain an efficient $O(n \log^2 n)$-time and $O(n \log n)$-space algorithm to compute and maintain the set of vertices of the rectilinear convex hull of P as we rotate $\mathbb{R}^3$ around the z-axis. Finally we study some properties of the rectilinear convex hulls of point sets in $\mathbb{R}^3$.

1 Introduction

Let P be a set of n points in the plane. An *open quadrant* in the plane is the intersection of two open half-planes whose supporting lines are parallel to the x and y-axes. An open quadrant is called P-free if it contains no points of P. The rectilinear convex hull of P is the set

$$RCH(P) = \mathbb{R}^2 \setminus \bigcup_{W(P) \in \mathcal{W}} W(P),$$

where $\mathcal{W}$ denotes the set of all P-free open quadrants; see Fig. 1, left.

The rectilinear convex hull of point sets has been studied mostly in the plane; e.g., see Ottmann et al. [9], Alegría et al. [1], and Bae et al. [3].

An open θ-quadrant is the intersection of two open half-planes whose supporting lines are orthogonal, one of which when rotated clockwise by θ degrees becomes horizontal.

P. Perez-Lantero—Partially supported by projects CONICYT FONDECYT/Regular 1160543 (Chile), DICYT 041933PL Vicerrectoría de Investigación, Desarrollo e Innovación USACH (Chile), and Programa Regional STICAMSUD 19-STIC-02.

C. Seara—Research supported by projects MTM2015-63791-R MINECO/FEDER and Gen. Cat. DGR 2017SGR1640.

J. Urrutia—Research supported by PAPIIT IN105221 Programa de Apoyo a la Investigación e Innovación Tecnológica, UNAM.

Y. Kohayakawa and F. K. Miyazawa (Eds.): LATIN 2020, LNCS 12118, pp. 296–307, 2020.
https://doi.org/10.1007/978-3-030-61792-9_24

We define the θ-rectilinear convex hull $RCH_\theta(P)$ of a point set P as the set

$$RCH_\theta(P) = \mathbb{R}^2 \setminus \bigcup_{W(P)\in\mathcal{W}_\theta} W(P),$$

where $\mathcal{W}_\theta$ denotes the set of all P-free *open θ-quadrants.*

Note that $RCH_\theta(P)$ changes as θ changes. In fact, as θ changes from 0 to $\frac{\pi}{2}$ there are $O(n)$ combinatorially different rectilinear convex hulls; see [1,3]. Figure 1 right shows an example of a θ-rectilinear convex hull which happens to be disconnected.

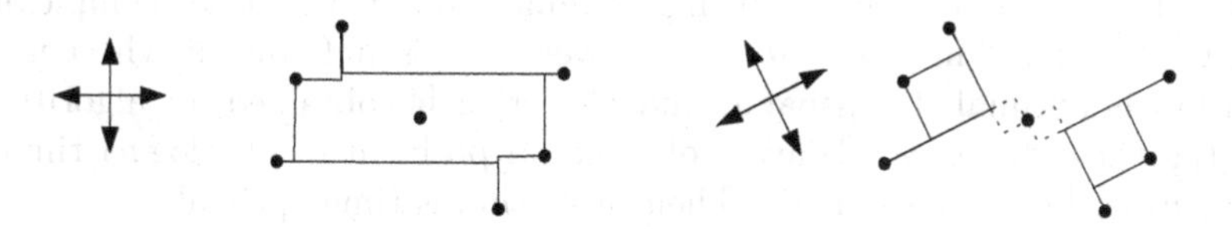

Fig. 1. Left: $RCH(P)$. Right: $RCH_{\pi/6}(P)$ of the same point set.

An open octant in $\mathbb{R}^3$ is the intersection of the three half-spaces, one perpendicular to the x-axis, one perpendicular to the y-axis, and another one perpendicular to the z-axis. As for the planar case, an octant is called P-free if it contains no elements of P. The rectilinear convex hull of a set of points in $\mathbb{R}^3$ is defined as

$$RCH^3(P) = \mathbb{R}^3 \setminus \bigcup_{W(P)\in\mathcal{W}} W(P),$$

where W denotes the set of all P-free open octants. In fact, in this paper and as an abuse of language, by $RCH^3(P)$ we will also denote the boundary of $RCH^3(P)$, and analogously for the similar definitions above. Thus, the rectilinear convex layers of a point set in $\mathbb{R}^3$ are defined recursively, as follows: calculate $RCH^3(P)$, and remove the elements of P in $RCH^3(P)$.

Results. In this paper we consider the rectilinear convex hull $RCH^3(P)$ of point sets in $\mathbb{R}^3$. We obtain an $O(n \log n)$ time and $O(n)$ space algorithm to calculate $RCH^3(P)$. We also give an $O(n \log^2 n)$ time and $O(n \log n)$ space algorithm to maintain the set of vertices of $RCH^3(P)$ as we rotate $\mathbb{R}^3$ around the z-axis. We present some results on the combinatorics of rectilinear convex hulls in $\mathbb{R}^3$ which are related to our algorithmic results, and interesting in their own right. In particular, we show that the rectilinear convex hull of a point set can change a quadratic number of times while its vertex set remains unchanged. Finally we present some open problems.

To avoid cumbersome terminology, from now on we will refer to $RCH^3(P)$ simply as $RCH(P)$.

1.1 Previous Work

The study of the rectilinear convex hull of point sets in the plane is closely related to that of finding the set of maximal points of point sets under vector dominance. This problem was introduced by Kung et al. [8], see also [11]. They obtained an optimal $O(n \log n)$ time and $O(n)$ space algorithm to solve this problem in the plane and in the three-dimensional space. They also found algorithms to solve the maxima problem in higher dimensions whose time complexity is $O(n \log^{d-2} n)$ for dimensions $d \geq 3$; however, it is not known whether their algorithm is optimal. Buchsbaum and Goodrich [4] obtained an algorithm to solve the three-dimensional layers of maxima problem in $O(n \log n)$ time and $O(n \log n / \log \log n)$ space in $\mathbb{R}^3$. Their algorithm is time optimal.

The rectilinear convex hull of a point set in the plane was first studied by Ottmann et al. [9] where they obtain an optimal $O(n \log n)$ time algorithm to calculate them. The reader may also consult the monograph Restricted-Orientation Geometry [5], where they study topics related to the problems we study here. Other variants of the problems studied here were also treated [5,6].

The rectilinear convex layers of a point set P in Euclidean spaces are defined recursively, as follows: calculate the rectilinear convex hull of P, remove its elements from P, and proceed recursively until P becomes empty. The rectilinear convex layers of point sets were first studied in Peláez et al. [10], where an optimal $O(n \log n)$ time algorithm to calculate them was obtained.

Variants of the $RCH(P)$ that were considered in the plane are: computing and maintaining $RCH_\theta(P)$ when we rotate the coordinate axes around the origin, or determining the angle of rotation θ such that the area of $RCH_\theta(P)$ is minimized or maximized [1,3].

A point set is a *rectilinear convex* set if all of its elements lie on the boundary of their rectilinear convex hull. Erdős-Szekeres type problems for finding rectilinear convex subsets of point sets were studied by González-Aguilar et al. [7]. They obtained algorithms to find the largest rectilinear convex subset of a point set, as well as finding their largest rectilinear convex hole, that is, subsets of points of P such that their rectilinear convex hull contains no element of P in the interior.

1.2 Notation and Preliminaries

For a point $p \in \mathbb{R}^3$, we refer to x_p, y_p, and z_p as the x-, y-, and z-coordinates of p, respectively. A point p satisfies a sign pattern; e.g., $(+,-,+)$ if $x_p \geq 0$, $y_p \leq 0$, and $z_p \geq 0$. There are eight possible sign patterns that a point can satisfy, namely: $(+,+,+)$, $(-,+,+)$, $(-,-,+)$, $(+,-,+)$, $(+,+,-)$, $(-,+,-)$, $(-,-,-)$, and $(+,-,-)$. The first sign pattern corresponds to all of the points in the first *octant* of $\mathbb{R}^3$. Similarly, we will say that the points satisfying the second pattern, $(-,+,+)$, correspond to points in the second octant, ..., and those satisfying $(+,-,-)$ correspond to the eighth octant of $\mathbb{R}^3$. The *open* octants are defined in a similar way, except that we require strict inequalities; e.g., the first open octant corresponds to points for which $x_p > 0$, $y_p > 0$, and $z_p > 0$. Given a point $p \in \mathbb{R}^3$, the octants with respect to p are the octants induced by a translation of the origin to p.

In addition, each of the eight sign patterns defines a partial order on the elements of $\mathbb{R}^3$ as follows: consider two points $p, q \in \mathbb{R}^3$. We say that p is dominates q according to the sign pattern $(+,+,+)$ if

$$(x_q \leq x_p) \wedge (y_q \leq y_p) \wedge (z_q \leq z_p).$$

We refer to this as $q \preceq_1 p$. In a similar way, we define the domination relations with respect to the other seven sign patterns, which we denote as $p \preceq_2 q$, $p \preceq_3 q$, $p \preceq_4 q$, $p \preceq_5 q$, $p \preceq_6 q$, $p \preceq_7 q$, and $p \preceq_8 q$. For example, the dominance relation with respect to the second sign pattern is:

$$q \preceq_2 p \iff (x_q \geq x_p) \wedge (y_q \leq y_p) \wedge (z_q \leq z_p).$$

These dominance relations define partial orders in P, and they can be extended to any dimension $d > 3$ in a straightforward way.

Definition 1. *A point* $p \in P$ *is a* maximal element *of* P *with respect to a partial order if there is no* $q \in P$ *that dominates* p*.*

The partial order $\preceq_1$ defined by $(+,+,+)$ is usually known as *vector dominance* [8].

2 Computing $RCH(P)$

In this section we show how to calculate the rectilinear convex hull of point sets in $\mathbb{R}^3$. The problem of finding the maximal elements of a point set in $\mathbb{R}^2$ and $\mathbb{R}^3$ with respect to vector dominance is called the maxima problem. The next result is well known.

Theorem 1 [8]. *The maxima problem in* $\mathbb{R}^d$*,* $d = 2, 3$*, can be solved in optimal* $O(n \log n)$ *time and* $O(n)$ *space.*

In fact, when solving the maxima problem, we obtain the set of faces and vertices of the *orthogonal polyhedron*, i.e., the faces meet at right angles and edges are parallel to the axes; call it $\mathcal{P}^1$ as shown in Fig. 2 left. In the right part of Fig. 2 we show the top view of $\mathcal{P}^1$. Observe that if we intersect a horizontal plane $\mathcal{Q}_c$ with equation $z = c$ with $\mathcal{P}^1$, we obtain an orthogonal polygon $\mathcal{C}_c^1$ that will change as we move $\mathcal{Q}_c$ up or down; i.e., as we increase or decrease c.

In Fig. 3 we show how $\mathcal{C}_c^1$ changes as we scan it from top to bottom starting at the top vertex of $\mathcal{P}^1$. We show the intersection of $\mathcal{P}^1$ with $\mathcal{Q}_c$ as $\mathcal{Q}_c$ sweeps through the first four top vertices of $\mathcal{P}^1$. It is easy to see that when $\mathcal{Q}_c$ moves from one vertex of $\mathcal{P}^1$ to the next, $\mathcal{C}_c^1$ changes in the following way: a new vertex

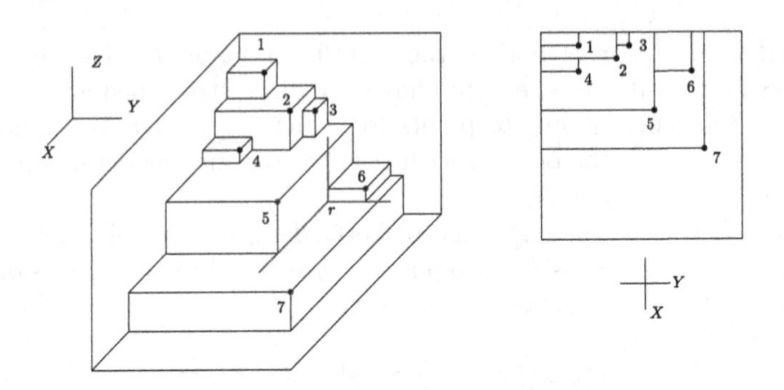

Fig. 2. Left: Maxima points in the first octant with the topmost antenna in p_1, and an extremal first open octant in blue. Right: The projection on the XY plane of the first octant maxima point set.

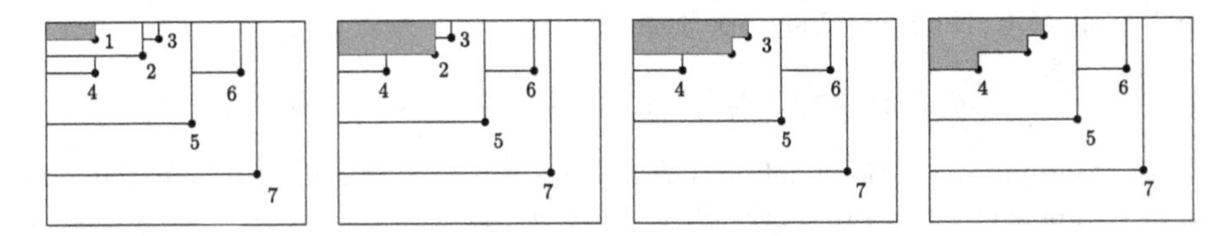

Fig. 3. How $\mathcal{C}_c^1$ changes as $\mathcal{Q}_c$ moves down from point 1 to point 4.

appears, which is the vertex of an *elbow* from which two rays emanate, one horizontal and one vertical, that extend until they hit $\mathcal{C}_c^1$ or go to infinity; see Fig. 3.

Analogously, we can define $\mathcal{P}^i$, $\mathcal{C}_c^i$, $i = 2, \ldots, 8$. Since $RCH(P) = \bigcap_{i=1,\ldots,8} \mathcal{P}^i$, we have the following.

Theorem 2. *For each constant c, the intersection of* $RCH(P)$ *with* $\mathcal{Q}_c$ *is the intersection of the orthogonal polygons* $\mathcal{C}_c^i$, $i = 1, \ldots, 8$.

To compute $RCH(P)$ we will sweep a plane $\mathcal{Q}_c$ from top to bottom, stopping at each point of P on $RCH(P)$. Each time we stop, we need to update $RCH(P) = \bigcap_{i=1,\ldots,8} \mathcal{P}^i$. We claim that we can do this in $O(\log n)$ time. To prove this, observe that when we move $\mathcal{Q}_c$ from a point p of P to the next point, say q, the only curves $\mathcal{C}_c^i$, $i = 2, \ldots, 8$ that change are those containing q, and these can be recomputed in $O(\log n)$ time.

As a consequence of the discussion above we have the following result.

Theorem 3. *Given a set* P *of* n *points in 3D, the rectilinear convex hull of* P*,* $RCH(P)$*, can be computed in optimal* $O(n \log n)$ *time and* $O(n)$ *space.*

3 Maintaining $RCH_\theta(P)$

In the plane, the problem of maintaining $RCH_\theta(P)$ as θ changes from 0 to 2π has been studied [1,3]. In this section we will study this problem in $\mathbb{R}^3$ restricted to rotations of $\mathbb{R}^3$ around the z-axis. Thus, in the rest of this section we will use octants defined as intersections of three mutually orthogonal semi-spaces whose supporting planes are orthogonal to three mutually orthogonal lines through the origin, one of which is the z-axis. Thus, two of these three lines lie on the XY-plane, and correspond to rotations of the x- and y-axis by an angle θ in the clockwise direction. We call such octants θ-octants, and the corresponding rectilinear convex hulls generated $RCH_\theta(P)$. In the rest of this section we will assume that elements of P are labeled $\{p_1, \ldots, p_n\}$ from top to bottom according to their z-coordinate.

For every $p \in \mathbb{R}^3$ there are eight θ-octants having p as their apex; we will call them p^θ-octants. Exactly four p^θ-octants contain points in $\mathbb{R}^3$ above the horizontal plane λ_p through p, and the other four have points below λ_p. We call the first four *up* p^θ*-octants*, and the other *down* p^θ*-octants.* Note that if an up p^θ-octant is no-P-free, it contains elements of P above λ_p, and that non-P-free down p^θ-octants contain points in P below λ_p.

Observe that a point $p \in P$ is a vertex of $RCH_\theta(P)$ if there is a P-free p^θ-octant. In this case we will say that p is a θ-active point, otherwise p is θ-inactive. Furthermore, if there is a P-free up p^θ-octant, we call p an up θ-active vertex. We define down θ-active vertices in a similar way.

We first analyze the set of angles for which points in P are up θ-active. Let p_i be a point of P, and consider the orthogonal projection onto λ_{p_i} of the points $p_1, \ldots, p_{i-1}$ (the points in P above λ_{p_i}), and let P'_i be the point set thus obtained; see Fig. 4 left. If for some θ, p_i is up θ-active, then there is a wedge of angular size θ_{p_i} at least $\frac{\pi}{2}$ on λ_{p_i} whose apex is p_i, that is P'_i-free; see Fig. 4 right. Clearly, no more than three such disjoint wedges can exist. In a similar way, we can prove that p_i is down active in at most three angular intervals. Thus, the following result, equivalent to a result in Avis et al. [2] for points on the plane, follows.

Theorem 4. *The set of angles θ for which a point $p_i \in P$ is active consists of at most six disjoint intervals in the set of directions* $[0, 2\pi]$.

Finding the angle intervals at which p_i is up-active is now reduced to finding, if they exist, P'_i-free wedges in λ_{p_i} whose apex is p_i of angular size at least $\frac{\pi}{2}$. We solve this as follows: note first that if one such wedge exists, it has to contain at least one of the four rays emanating from p_i parallel to the x- or y-axis. Let those rays be X_i^+, X_i^-, Y_i^+, and Y_i^-; see Fig. 4 right. For X_i^+, we will solve the following problem.

Rotate X_i^+ clockwise until it hits a point in P'_i. Next, rotate X_i^+ clockwise until it hits another point in P'. Measure the angle α formed by the two rays thus obtained. If $\alpha \geq \frac{\pi}{2}$ then we have found a set of intervals at which p_i is active, else discard X_i^+. Proceed in the same way with X_i^-, Y_i^+, and Y_i^-. We will have

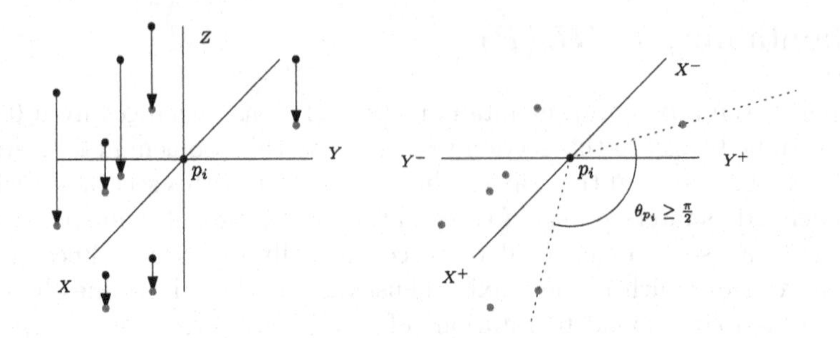

Fig. 4. Checking whether the point p_i is maximal with respect to the first octant for some angular interval: project the set of points $\{p_1, p_2, \ldots, p_{i-1}\}$ on the plane parallel to the XY plane passing through p_i, and obtain the projected points (red points in figure). (Color figure online)

to repeat this process for all of the points $p_i \in P$ from top to bottom. We now show how to process all the points in P in $O(n \log^2 n)$ time and $O(n \log n)$ space.

The main difficulty in finding the wedges in the above discussion is that as we process the points of P from top to bottom, the number of points in λ_{p_i} increases one by one, and thus, we need a dynamic data structure to solve the following problem.

Problem 1. Let $Q = \{q_1, \ldots, q_n\}$ be a set of points in the plane, and let $Q_{i-1} = \{q_1, \ldots, q_{i-1}\}$. For each i we want to solve the following problem: let r_i be the vertical ray that starts at q_i and points up. Find the first point a_i (respectively, b_i) of Q_{i-1} that r_i meets when we rotate it in the clockwise (respectively, counter-clockwise) direction around q_i.

One last point before proceeding with our results; instead of projecting the points of P on the planes λ_{p_i}, we will project them one by one on the XY-plane. Everything else remains unchanged. The next observation on binary trees will be useful.

Observation 1. *Let T be a balanced binary tree. For each node of $v \in T$ let S_v be the set of leaves of T that are descendants of v. Let u be a leaf of T, and consider the path p_u that joins u to the root of T, and let Δ_u be the set of nodes of T that are direct descendants of a node in p_u. Then, the sets S_v, $v \in \Delta_u$ induce a partition of the set of leaves of $T \setminus \{u\}$. Moreover if a node w in Δ_u is the right (left) descendant of a node in p_u, then the elements of S_w lie to the right (respectively, left) of v; see Fig. 5.*

Theorem 5. *Problem 1 can be solved in $O(n \log^2 n)$ time and $O(n \log n)$ space.*

Proof. Assume without loss of generality that no two points of Q lie on a horizontal line, and let $left_i$ and $right_i$ be the set of points in P lying to the left (respectively, right) of the vertical line through q_i. If we know the convex hulls

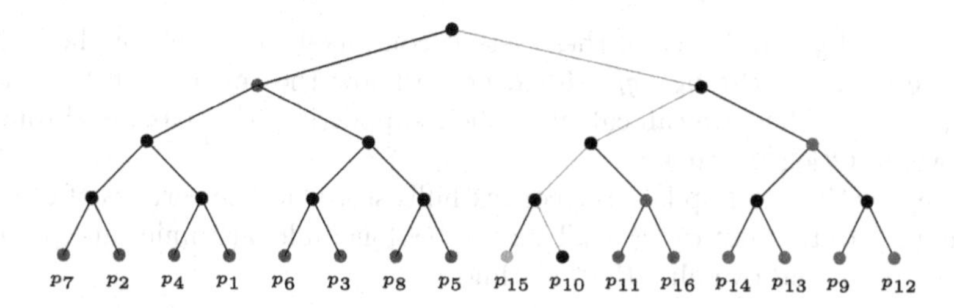

Fig. 5. Balanced binary tree $\mathcal{T}$.

of $left_i$ and $right_i$, then the points we are seeking, a_i and b_i, can be computed by calculating the supporting lines of the convex hull of $left_i$ and $right_i$ passing through q_i. It is well known that we can compute these lines in $O(\log n)$ time.

Furthermore, if $left_i$ and $right_i$ have been decomposed into k disjoint sets $W_i, \ldots, W_k$ where each W_i is contained in $left_i$ or in $right_i$, and we have the convex hull $conv(W_i)$ of all of these point sets, then we can find the supporting lines through q_i for all of them in overall $O(\log |W_1| + \cdots + \log |W_k|) = O(\log^2 n)$ time; see Fig. 6. This will now allow us to obtain a_i and b_i in $O(\log n)$ time. This is the main idea that will enable us to design a data structure to solve Problem 1 in $O(n \log^2 n)$ time.

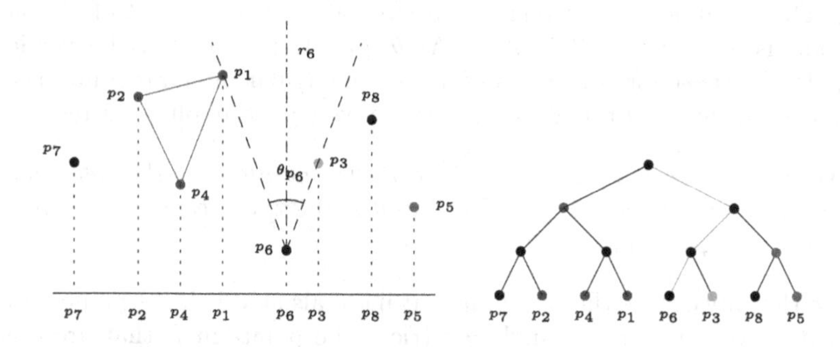

Fig. 6. Processing p_6. At this point, p_7 and p_8 are inactive, so they are in black.

Let $D(Q)$ be a balanced binary tree whose leaves are the elements of Q sorted in order from left to right according to their x-coordinate. Note that this order does not necessarily coincide with the labeling $q_1, \ldots, q_n$ of the elements of Q. Initially, every leaf of $D(Q)$ is considered *inactive*. For each vertex of $D(Q)$ we maintain the convex hull of $W(q)$, where $W(q)$ is the set of descendant leaves (points of Q) that are active.

For each i, from $i = 1, \ldots, n$ we execute the following algorithm:

- Consider the nodes of $D(Q)$ that are direct descendants of nodes in the path p_{q_i} connecting q_i to the root of $D(Q)$. By Observation 1, the active descendants of these nodes form a partition $W_i, \ldots, W_k$ of the set of active leaves

of $D(Q) \setminus \{q_i\}$, and each of these sets is contained to the left or the right of the vertical line through q_i. Moreover we know the convex hulls of each of $W_i, \ldots, W_k$. Thus, we can calculate their supporting lines passing through q_i in overall $O(\log^2 n)$ time.
- Make q_i active, and update the convex hulls stored at the vertices of the path joining q_i to the root of $D(Q)$. This can be done in logarithmic time per node of the path, and overall $O(\log^2 n)$ time.

Observe that the convex polygon associated with the root of $D(Q)$ can be of size n. For the vertices in the next level of $D(Q)$, the sum of the sizes of the convex polygons associated to them is n, and in general, the sum of the polygons associated to the vertices of $D(Q)$ is n. Since the number of levels of $D(Q)$ is $O(\log n)$, the space used is $O(n \log n)$. □

By using the results in Theorem 5 we can calculate the set of intervals at which all of the points in P are up-active. In a similar way we can determine the intervals for which the points in P are down-active; thus we have the following.

Theorem 6. *The set of intervals at which the points of P are θ-active can be computed in $O(n \log^2 n)$ time and $O(n \log n)$ space.*

Observe that the set of intervals at which two points of P are active define intervals in the unit circle C, where the points on C correspond to angles in $[0, 2\pi]$. Thus, if an angle θ belongs to an interval at which a point of P is active, this point is a vertex of $RCH_\theta(P)$. As θ goes from 0 to 2π, the vertices of $RCH_\theta(P)$ are those for which one of its active angular intervals contains θ.

As a consequence of the above discussion we have the following result.

Theorem 7. *Given a set P of n points in 3D, maintaining the elements of P that belong to the boundary of $RCH_\theta(P)$ as $\theta \in [0, 2\pi]$ can be done in $O(n \log^2 n)$ time and $O(n \log n)$ space.*

Note that if we store the set of angular intervals at which the points of P are active, then we can, for any angle θ retrieve the points in P that are θ active in linear time. In case that we want to compute the $RCH_\theta(P)$ for a particular value of θ, all we have to do is to retrieve that θ-active points of P in linear time, and use the algorithm presented in Theorem 3.

4 The Combinatorics of Rectilinear Convex Hulls in $\mathbb{R}^3$

In the previous section, we studied the problem of maintaining the set of points of P that are vertices of $RCH_\theta(P)$. The problem of maintaining $RCH_\theta(P)$ does not follow from our previous results. As we shall see, there are examples of point sets $P \subset \mathbb{R}^3$ such that the number of combinatorially different rectilinear convex hulls of P can be at least $\Omega(n^2)$ while the set of vertices of $RCH_\theta(P)$ remains unchanged.

To begin, we notice that $RCH_\theta(P)$ is not necessarily connected; this is easy to see, as for point sets in the plane this property does not hold; e.g., see Fig. 1 right. What is a bit more interesting is that even when $RCH_\theta(P)$ is connected and has non-empty interior, it is not necessarily simply connected. In Fig. 7 we show a rectilinear convex hull of a point set whose rectilinear convex hull is a torus. The elements of P are the vertices of the cubes glued together to obtain the figure. The reader will notice immediately that the points in P are not in general position. A slight perturbation of the elements of P, that would bring them to a point set in general position, will also yield a rectilinear convex hull whose interior is a torus. Evidently, we can construct similar examples to that shown in Fig. 7 in which we obtain oriented surfaces of arbitrarily large genus.

Fig. 7. The rectilinear convex hull of a point set that is not simply connected.

We now construct a point set such that for an angular interval $[\alpha, \beta]$ while $\theta \in [\alpha, \beta]$, $RCH_\theta(P)$ will maintain the same vertices while it changes a quadratic number of times.

Consider a circular cylinder $\mathcal{C}$ that is perpendicular to the XY-plane, and consider a geodesic curve $\mathcal{H}$ on $\mathcal{C}$ that joins two points p and q on $\mathcal{C}$. Choose a set of n points $P' = \{p'_1, \ldots, p'_n\}$ on $\mathcal{H}$ such that their projection on the XY-plane is a set of equidistant points on a small interval of the circle in which $\mathcal{C}$ and the XY-plane intersect, and such that if $i < j$ the z-coordinate of p_i is smaller than the z-coordinate of p_j; see Fig. 8.

Observe that there is a P'-free up octant $Q_{1,n}$ whose bottom, left, and back faces contain p'_1, p'_{n-1} and p'_n. Observe now that there is a translation of Q_1 up and to the left so as to produce a second octant $Q_{1,n-1}$ whose bottom, left and back faces contain p'_2, p'_{n-1}, and p'_{n-2}. We can iterate this process $\frac{n-1}{2}$ times to obtain a set of $\frac{n-1}{2}$ P'-free octants that have on their bottom, left and back faces p'_1, p'_i and p'_{i+1}, $i = \lfloor \frac{n-1}{2} \rfloor, \ldots, n-1$; see Fig. 8.

Rotating Q_2 slightly in the clockwise direction and moving it up we can obtain a P-free octant $Q_{2,n}$ containing on its bottom, left, and back faces p'_2, p'_n, and

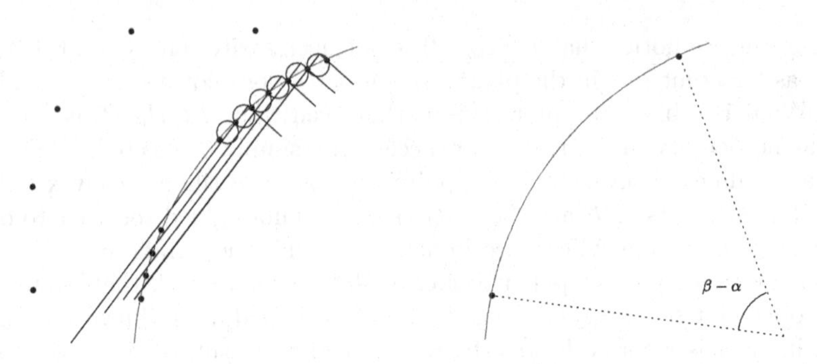

Fig. 8. A configuration of points that illustrates a point set for which its set of vertices remains unchanged while its rectilinear convex hull changes a quadratic number of times.

p'_{n-1}. We can now repeat the same process as we did with $\{p'_1, \ldots, p'_n\}$, and with $\{p'_2, \ldots, p'_n\}$, starting with p'_2 and $Q_{2,n}$ to obtain a new set of P'-free extremal octants. Repeat this process with $\{p'_3, \ldots, p'_n\}, \ldots, \{p'_{n-3}, \ldots, p'_n\}$, to obtain a quadratic number of P'-free *extremal* P'-octants. Let α and β be the angles of rotation of the XY-plane such that the x-axis becomes parallel to the line at which the back faces of $Q_{1,n}$ and $Q_{n-3,n}$ intersect the XY-plane. Observe that all of $p'_1, \ldots, p'_n$ are active points and on the boundary of RCH_θ for all $\theta \in [\alpha, \beta]$. In the meantime, all of the P-free octants we obtained above become active during an angular interval contained in $[\alpha, \beta]$. To complete our construction, we add a few points to the set $\{p'_1, \ldots, p'_n\}$. These points are located behind the circular cylinder $\mathcal{C}$, placed appropriately to ensure that the rectilinear convex hull of the point set P thus obtained has non-empty interior, and all of $\{p'_1, \ldots, p'_n\}$ are on its boundary. Thus, the rectilinear convex hull of P changes a quadratic number of times while its vertex set remains unchanged.

Theorem 8. *There are configurations of points in* $\mathbb{R}^3$ *such that for an angular interval* $[\alpha, \beta]$ *while* $\theta \in [\alpha, \beta]$, $RCH_\theta(P)$ *will maintain the same vertices while it changes a quadratic number of times.*

5 Final Remarks and Future Lines of Research

We have shown how to calculate the rectilinear convex hull of a point set in $O(n \log n)$ time. The rectilinear convex layers of a point set in $\mathbb{R}^3$ are defined in a recursive way as follows: calculate $RCH^3(P)$, and remove the elements of P in $RCH^3(P)$. It is clear that the rectilinear convex layers of P can be computed in a recursive way by removing the rectilinear convex hull of the point set until P becomes empty. If P has k rectilinear convex layers, this can be done in $O(kn \log n)$ time and $O(n)$ space. We conjecture that there exists an algorithm to find the rectilinear convex layers of P in better than $O(kn \log n)$ time and $O(n)$ space.

We proved that the number of times that the set of vertices of the rectilinear convex hull of a point set P changes while $\mathbb{R}^3$ is rotated around the z-axis is linear; however, the rectilinear convex hull of P may change a quadratic number of times. A future line of research is that of obtaining efficient algorithms to maintain the rectilinear convex hull of P in time proportional to the number of times it changes times a logarithmic factor.

Finally, we remark that obtaining the rectilinear convex hull of a point set, when we rotate $\mathbb{R}^3$ around any line through the origin, can be done trivially in $O(n \log n)$ time, as any such rotation can be achieved as a composition of a rotation around the z-axis followed by a rotation around the x-axis.

Acknowledgement. This work has received funding from the European Union's Horizon 2020 research and innovation programme under the Marie Skłodowska-Curie grant agreement No. 734922.

References

1. Alegría-Galicia, C., Orden, D., Seara, C., Urrutia, J.: On the $\mathcal{O}$-hull of planar point sets. In: Proceedings of the 30th European Workshop on Computational Geometry (2014)
2. Avis, D., et al.: Unoriented Θ-maxima in the plane: complexity and algorithms. SIAM J. Comput., **28**(1), 278–296 (1999)
3. Bae, S.W., Lee, Ch., Ahn, H.-K., Choi, S., Chwa, K.-Y.: Computing minimum-area rectilinear convex hull and L-shape. Comput. Geometry Theory Appl. **42**(9), 903–912 (2009)
4. Buchsbaum, A.L., Goodrich, M.T.: Three-dimensional layers of maxima. Algorithmica **39**(4), 275–286 (2004)
5. Fink, E., Wood, D.: Restricted-orientation Convexity. Monographs in Theoretical Computer Science (An EATCS Series), Springer-Verlag (2004). https://doi.org/10.1007/978-3-642-18849-7
6. Franěk, V., Matoušek, J.: Computing D-convex hulls in the plane. Comput. Geometry Theory Appl. **42**(1), 81–89 (2009)
7. González-Aguilar, H., Orden, D., Pérez-Lantero, P., Rappaport, D., Seara, C., Tejel, J., Urrutia, J.: Maximum rectilinear convex subsets. In: Gąsieniec, L.A., Jansson, J., Levcopoulos, C. (eds.) FCT 2019. LNCS, vol. 11651, pp. 274–291. Springer, Cham (2019). https://doi.org/10.1007/978-3-030-25027-0_19
8. Kung, H.-T., Luccio, F., Preparata, F.P.: On finding the maxima of a set of vectors. J. ACM **22**(4), 469–476 (1975)
9. Ottmann, T., Soisalon-Soininen, E., Wood, D.: On the definition and computation of rectilinear convex hulls. Inf. Sci. **33**(3), 157–171 (1984)
10. Peláez, C., Ramírez-Vigueras, A., Seara, C., and Urrutia, J.: On the rectilinear convex layers of a planar set. In: Proceedings of the Mexican Conference on Discrete Mathematics and Computational Geometry, 60th birthday of Jorge Urrutia (2013)
11. Preparata, F.P., Shamos, M.I.: Computational Geometry: An Introduction. Springer, New York (1985). https://doi.org/10.1007/978-1-4612-1098-6

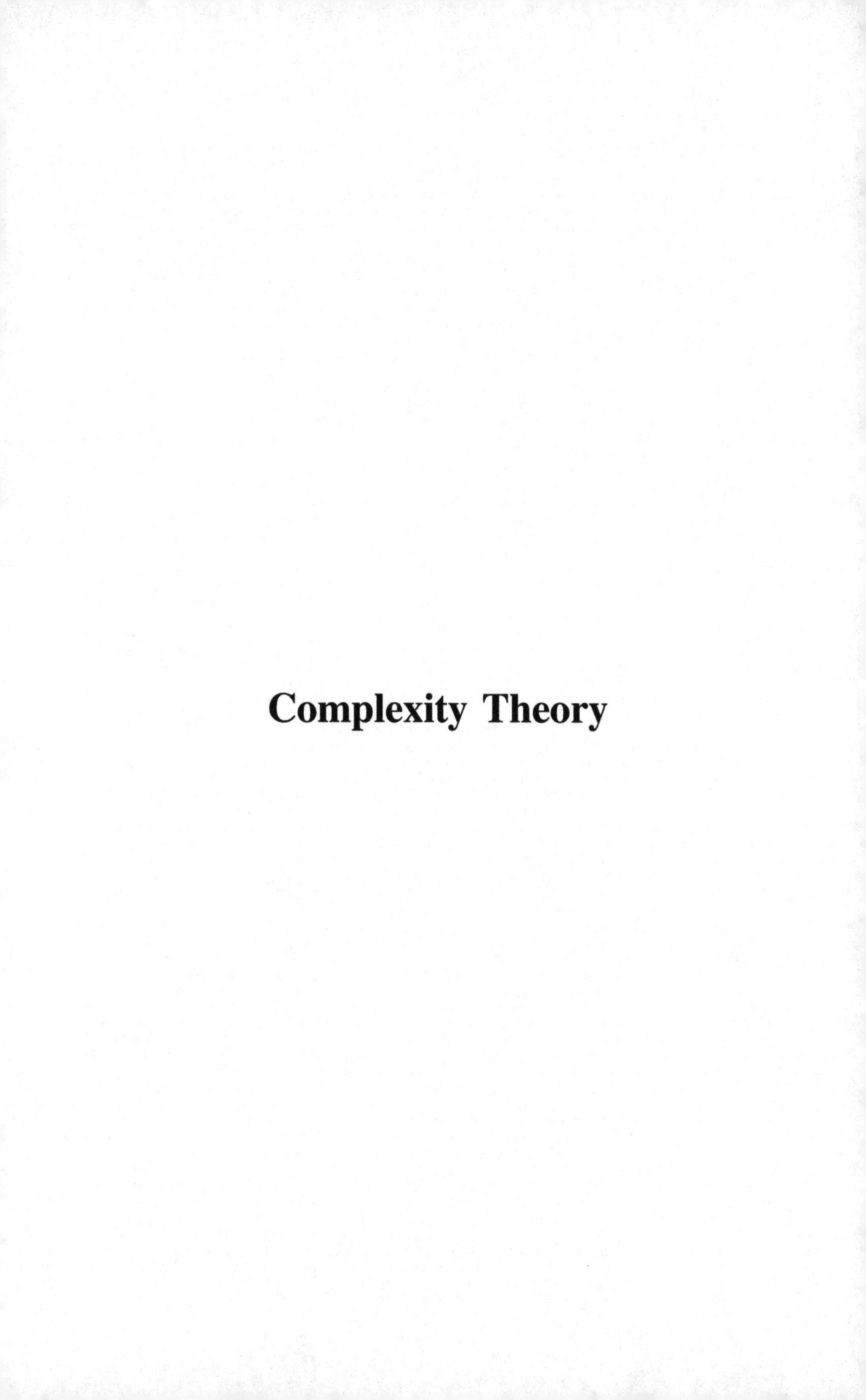

Complexity Theory

Monotone Circuit Lower Bounds from Robust Sunflowers

Bruno Pasqualotto Cavalar[1(✉)], Mrinal Kumar[2], and Benjamin Rossman[3]

[1] Institute of Mathematics and Statistics, University of São Paulo, São Paulo, Brazil
brunopc@ime.usp.br
[2] IIT Bombay, Mumbai, India
mrinalkumar08@gmail.com
[3] University of Toronto, Toronto, Canada
ben.rossman@utoronto.ca

Abstract. Robust sunflowers are a generalization of combinatorial sunflowers that have applications in monotone circuit complexity [14], DNF sparsification [6], randomness extractors [8], and recent advances on the Erdős-Rado sunflower conjecture [3,9,12]. The recent breakthrough of Alweiss, Lovett, Wu and Zhang [3] gives an improved bound on the maximum size of a w-set system that excludes a robust sunflower. In this paper, we use this result to obtain an $\exp(n^{1/2-o(1)})$ lower bound on the monotone circuit size of an explicit n-variate monotone function, improving the previous record $\exp(n^{1/3-o(1)})$ of Harnik and Raz [7]. We also show an $\exp(\Omega(n))$ lower bound on the monotone *arithmetic* circuit size of a related polynomial. Finally, we introduce a notion of robust clique-sunflowers and use this to prove an $n^{\Omega(k)}$ lower bound on the monotone circuit size of the CLIQUE function for all $k \leqslant n^{1/3-o(1)}$, strengthening the bound of Alon and Boppana [1].

1 Introduction

A monotone Boolean circuit is a Boolean circuit with AND and OR gates but no negations (NOT gates). Although a restricted model of computation, monotone Boolean circuits seem a very natural model to work with when computing *monotone* Boolean functions, i.e., Boolean functions $f : \{0,1\}^n \to \{0,1\}$ such that for all pairs of inputs $(a_1, a_2, \ldots, a_n), (b_1, b_2, \ldots, b_n) \in \{0,1\}^n$ where $a_i \leqslant b_i$ for every i, we have $f(a_1, a_2, \ldots, a_n) \leqslant f(b_1, b_2, \ldots, b_n)$. Many natural and well-studied Boolean functions such as Clique and Majority are monotone.

Monotone Boolean circuits have been very well studied in Computational Complexity over the years, and continue to be one of the few seemingly largest natural sub-classes of Boolean circuits for which we have exponential lower bounds. This line of work started with a very influential paper of Razborov [13] who proved a super-polynomial $n^{\Omega(\log n)}$ lower bound on the size of monotone circuits computing the $\mathsf{Clique}_{k,n}$ function for $k \leqslant \log n$. Prior to Razborov's result, we didn't even have super-linear lower bounds for monotone circuits, with the best bound being a lower bound of $4n$ due to Tiekenheinrich [15]. Further

Y. Kohayakawa and F. K. Miyazawa (Eds.): LATIN 2020, LNCS 12118, pp. 311–322, 2020.
https://doi.org/10.1007/978-3-030-61792-9_25

progress in this line of work included the results of Andreev [4] who proved an exponential lower bound for another explicit function. Alon and Boppana [1] extended Razborov's result by proving an $n^{\Omega(\sqrt{k})}$ lower bound for $\mathsf{Clique}_{k,n}$ for all $k \leqslant n^{2/3-o(1)}$. These state of art monotone circuit lower bounds saw a further quantitative improvement in a work of Harnik and Raz [7] who proved a lower bound of $2^{\Omega((n/\log n)^{1/3})}$ for an explicit n-variate function defined using a small probability space of random variables with bounded independence. However, to this day, the question of proving truly exponential lower bounds for monotone circuits (of the form $2^{\Omega(n)}$) for an explicit n-variate function) remains open! (Truly exponential lower bounds for monotone *formulas* were obtained only recently [11]).

In the present paper, we are able to improve the best known lower bound for monotone circuits by proving the first $2^{\Omega(n^{1/2}/(\log n)^{3/2})}$ lower bound for an explicit monotone Boolean function (Sect. 2). The function is based on the same construction first considered by Harnik and Raz [7], but our argument employs the approximation method of Razborov with recent improvements on robust sunflower bounds [3,12]. In the full paper, by applying the same technique with a variant of robust sunflowers that we call robust clique-sunflowers, we are able to prove an $n^{\Omega(k)}$ lower bound for the $\mathsf{Clique}_{k,n}$ function when $k \leqslant n^{1/3-o(1)}$, thus improving the result of Alon and Boppana when k is in this range. Finally, we are able to prove truly exponential lower bounds in the monotone arithmetic setting to a fairly general family of polynomials, which shares some similarities to the Harnik and Raz function (also in the full paper).

1.1 Monotone Circuit Lower Bounds and Sunflowers

The original lower bound for $\mathsf{Clique}_{k,n}$ due to Razborov employed a technique which came to be known as the *approximation method*. Given a monotone circuit C of "small size", it consists into constructing gate-by-gate, in a bottom-up fashion, another circuit $\widetilde{C}$ that approximates C on most inputs of interest. One then exploits the structure of this *approximator circuit* to prove that it differs from $\mathsf{Clique}_{k,n}$ on most inputs of interest, thus implying that no "small" circuit can compute this function. This technique was leveraged to obtain lower bounds for a host of other monotone problems [1].

A crucial step in Razborov's proof involved the sunflower lemma due to Erdős and Rado. A family $\mathcal{F}$ of subsets of $[n]$ is called a *sunflower* if there exists a set Y such that $F_1 \cap F_2 = Y$ for every $F_1, F_2 \in \mathcal{F}$. The sets of $\mathcal{F}$ are called *petals* and the set $Y = \bigcap \mathcal{F}$ is called the *core*. We say that the family $\mathcal{F}$ is ℓ-uniform if every set in the family has size ℓ.

Theorem 1 (Erdős and Rado [5]**).** *Let $\mathcal{F}$ be a ℓ-uniform family of subsets of $[n]$. If $|\mathcal{F}| \geqslant \ell!(r-1)^{\ell}$, then $\mathcal{F}$ contains a sunflower of r petals.*

Informally, the sunflower lemma allows one to prove that a monotone function can be approximated by one with fewer minterms by means of the "plucking" procedure: if the function has too many (more than $\ell!(r-1)^{\ell}$) minterms of size

ℓ, then it contains a sunflower with r petals; remove all the petals, replacing them with the core. One can then prove that this procedure does not introduce many errors.

The notion of *robust sunflowers* was introduced by the third author in [14], to achieve better bounds via the approximation method on the monotone circuit size of $\mathsf{Clique}_{k,n}$ when the negative instances are Erdős-Rényi random graphs $\boldsymbol{G}_{n,p}$ below the k-clique threshold.[1] A family $\mathcal{F} \subseteq 2^{[n]}$ is called a (p, ε)-*robust sunflower* if

$$\mathop{\mathbb{P}}_{\boldsymbol{W} \subseteq_p [n]} [\exists F \in \mathcal{F} : F \subseteq \boldsymbol{W} \cup Y] > 1 - \varepsilon,$$

where $Y := \bigcap \mathcal{F}$ and $\boldsymbol{W}$ is a p-random subset of $[n]$. Henceforth, we consistently write random objects using boldface symbols (such as $\boldsymbol{W}$, $\boldsymbol{G}_{n,p}$, etc.).

As remarked in [14], every ℓ-uniform sunflower of r petals is a $(p, e^{-rp^{\ell}})$-robust sunflower. Moreover, as observed in [9], every $(1/r, 1/r)$-robust sunflower contains a sunflower of r petals. A corresponding bound for the appearance of robust sunflowers in large families was also proved in [14].

Theorem 2 [14]. *Let $\mathcal{F}$ be a ℓ-uniform family such that $|\mathcal{F}| \geqslant \ell!(2\log(1/\varepsilon)/p)^{\ell}$. Then $\mathcal{F}$ contains a (p, ε)-robust sunflower.*

For many choice of parameters p and ε, this bound is better than the one by Erdős and Rado, thus leading to better approximation bounds. In a recent breakthrough, this result was significantly improved in [3].

Theorem 3 (Theorem 2.5 of [3]). *Let $\mathcal{F}$ be a ℓ-uniform family such that $|\mathcal{F}| \geqslant (\log \ell)^{\ell} \cdot (\log\log \ell \cdot \log(1/\varepsilon)/p)^{O(\ell)}$. Then $\mathcal{F}$ contains a (p, ε)-robust sunflower.*

Because of the connection between robust sunflowers and sunflowers explained above, this result was used by the authors to significantly improve the standard sunflower bounds of Erdős and Rado. Soon afterwards, Rao [12] provided an alternative proof which slightly improved the bound. It is this bound we are going to use, which we introduce in the next section.[2]

1.2 Slice Sunflowers

In what follows, let m be a positive integer such that $m < n$.

Definition 1. *Let $\mathcal{F}$ be a family of subsets of $[n]$ and let $Y := \bigcap \mathcal{F}$. Let al.so $\boldsymbol{W} \subseteq [n]$ be a set of size m chosen uniformly at random. The family $\mathcal{F}$ is called a (m, ε)-*slice-sunflower *if*

$$\mathop{\mathbb{P}}_{\boldsymbol{W}} [\exists F \in \mathcal{F} : F \subseteq \boldsymbol{W} \cup Y] > 1 - \varepsilon.$$

[1] Robust sunflowers were called *quasi-sunflowers* in [6,8,9,14] and *approximate sunflowers* in [10]. Following Alweiss *et al.* [3], we adopt the new name *robust sunflower*.

[2] Crucially for our application, the $O(\ell)$ exponent in the bound of Theorem 3 is only 2ℓ when $\varepsilon = 2^{-\Omega(\ell)}$. To get any improvement over the Harnik-Raz bound, we require $\ell + o(\ell)$, which is given by the result of Rao [12].

Theorem 4 [12]. *There exists an universal constant $B > 0$ such that the following holds. Let $p \in (0,1)$ and let $\mathcal{F} \subseteq \binom{[n]}{\ell}$ be such that $|\mathcal{F}| \geqslant (Bx \log x)^\ell$, where $x = \log(\ell/\varepsilon)/p$. Then $\mathcal{F}$ contains a (m, ε)-slice-sunflower, where $m = \lfloor np \rfloor$.*

The theorem above is implicit in Rao [12]. In the full paper, we include an explicit proof of this theorem, closely following the argument and notation of [12].

2 Harnik-Raz Function

The strongest lower bound known for monotone circuits computing an explicit n-variate monotone Boolean function is $\exp\left(\Omega((n/\log n)^{1/3})\right)$, and it was obtained by Harnik and Raz [7]. In this section, we will prove a lower bound of $\exp(\Omega(n^{1/2}/(\log n)^{3/2}))$ for the same Boolean function they considered. We apply the *method of approximations* [13] and the new *robust sunflower* bound [3,12]. We do not expect that a lower bound better than $\exp(n^{1/2-o(1)})$ can be obtained by this technique, even with better sunflower bounds.

We start by giving a high level outline of the proof. We define the Harnik-Raz function $f_{\mathrm{HR}} : \{0,1\}^n \to \{0,1\}$ and find two distributions $\boldsymbol{Y}$ and $\boldsymbol{N}$ with support in $\{0,1\}^n$ satisfying the following properties:

- f_{HR} outputs 1 on $\boldsymbol{Y}$ with high probability (Lemma 1);
- f_{HR} outputs 0 on $\boldsymbol{N}$ with high probability (Lemma 2).

Because of these properties, the distribution $\boldsymbol{Y}$ is called the *positive test distribution*, and $\boldsymbol{N}$ is called the *negative test distribution.* We also define a set of monotone Boolean functions called *approximators*, and we show that:

- every approximator commits many mistakes on either $\boldsymbol{Y}$ or $\boldsymbol{N}$ with high probability (Lemma 8);
- every Boolean function computed by a "small" monotone circuit agrees with an approximator on both $\boldsymbol{Y}$ and $\boldsymbol{N}$ with high probability (Lemma 9).

Together these suffice for proving that "small" circuits cannot compute f_{HR}. The crucial part where the robust sunflower result comes into play is in the second item.

2.1 Technical Preliminaries

For $A \subseteq [n]$, let $x_A \in \{0,1\}^n$ be the binary vector with support in A. For a set $A \in 2^{[n]}$, let $\lceil A \rceil$ be the indicator function satisfying

$$\lceil A \rceil(x) = 1 \iff x_A \leqslant x.$$

Define also $\{0,1\}^n_{=m} := \left\{x_A : A \in \binom{n}{m}\right\}$. For a monotone Boolean function $f : \{0,1\}^n \to \{0,1\}$, let $\mathcal{M}(f)$ denote the set of minterms of f, and let $\mathcal{M}_\ell(f) := \mathcal{M}(f) \cap \{0,1\}^n_{=\ell}$. Elements of $\mathcal{M}_\ell(f)$ are called ℓ-minterms of f. In what follows, we will mostly ignore ceilings and floors for the sake of convenience, since these do not make any substantial difference in the final calculations.

2.2 The Function

We now describe the construction of the function $f_{\mathrm{HR}} : \{0,1\}^n \to \{0,1\}$ considered by Harnik and Raz [7]. First observe that, for every n-bit monotone Boolean function f, there exists a family $\mathcal{S} \subseteq 2^{[n]}$ such that

$$f(x_1, \ldots, x_n) = f_{\mathcal{S}}(x_1, \ldots, x_n) := \bigvee_{S \in \mathcal{S}} \bigwedge_{j \in S} x_j.$$

Indeed, $\mathcal{S}$ can be chosen to be the family of the coordinate-sets of minterms of f. Now, in order to construct the Harnik-Raz function, we will suppose n is a prime power and let $\mathbb{F}_n$ be the field of n elements. Moreover, we fix two positive integers c and k with $c < k$. For a polynomial $P \in \mathbb{F}_n[x]$, we let S_P be the set of the valuations of P in each element of $\{1, 2, \ldots, k\}$ (in other words, $S_P = \{P(1), \ldots, P(k)\}$). Observe that it is not necessarily the case that $|S_P| = k$, since it may happen that $P(i) = P(j)$ for some i, j such that $i \neq j$. Finally, we consider the family $\mathcal{S}_{\mathrm{HR}}$ defined as

$$\mathcal{S}_{\mathrm{HR}} := \{S_P : P \in \mathbb{F}_n[x], P \text{ has degree at most } c-1 \text{ and } |S_P| \geqslant k/2\}.$$

We thus define f_{HR} as $f_{\mathrm{HR}} := f_{\mathcal{S}_{\mathrm{HR}}}$.

We now explain the choice of $\mathcal{S}_{\mathrm{HR}}$. First, the choice for valuations of polynomials with degree at most $c-1$ is explained by a fact observed in [2]. If a polynomial $\boldsymbol{P} \in \mathbb{F}_n[x]$ with degree $c-1$ is chosen uniformly at random, they observed that the random variables $\boldsymbol{P}(1), \ldots, \boldsymbol{P}(k)$ are c-wise independent, and are each uniform in $[n]$. This allows us to define a distribution on the inputs (the positive test distribution) that has high agreement with f_{HR} and is easy to analyze. Observe further that, since $|\mathcal{S}_{\mathrm{HR}}| \leqslant n^c$, the monotone complexity of f_{HR} is at most $2^{c \log n}$. Later we will chose c to be roughly $n^{1/2}$, and prove that the monotone complexity of f_{HR} is $2^{\Omega(c)}$.

Finally, the restriction $|S_P| \geqslant k/2$ is a truncation made to ensure that no minterm of f_{HR} is very small. Otherwise, if f_{HR} had small minterms, it might have been a function that almost always outputs 1. Such functions have very few maxterms and are therefore computed by a small CNF. Since we desire f_{HR} to have high complexity, this is an undesirable property. The fact that f_{HR} doesn't have small minterms is important in the proof that f_{HR} almost surely outputs 0 in the negative test distribution (Lemma 2).

We now define the positive and negative test distributions. Let $\boldsymbol{Y} \in \{0,1\}^n$ be the random variable which chooses a polynomial $\boldsymbol{P} \in \mathbb{F}_n[x]$ with degree at most $c-1$ uniformly at random, and maps it into the binary input $x_{S_{\boldsymbol{P}}} \in \{0,1\}^n$. Let

$$p := n^{-4c/k} \quad \text{and} \quad m := \lfloor np \rfloor.$$

Let also $\boldsymbol{N}$ be the distribution which chooses an input from $\{0,1\}^n_{=m}$ uniformly at random. For a Boolean function f and a probability distribution $\boldsymbol{\mu}$ on the inputs on f, we write $f(\boldsymbol{\mu})$ to denote the random variable which evaluates f on a random instance of $\boldsymbol{\mu}$. Harnik and Raz proved that f_{HR} outputs 1 on $\boldsymbol{Y}$ with high probability.

Lemma 1 (Claim 4.2 in [7]). *We have* $\mathbb{P}[f_{\mathrm{HR}}(\boldsymbol{Y}) = 1] \geqslant 1 - k/n$.

We now claim that f_{HR} also outputs 0 on $\boldsymbol{N}$ with high probability.

Lemma 2. *We have* $\mathbb{P}[f_{\mathrm{HR}}(\boldsymbol{N}) = 0] \geqslant 1 - n^{-c}$.

Proof. Let x_A be an input sampled from $\boldsymbol{N}$. Observe that $f_{\mathrm{HR}}(x_A) = 1$ only if there exists a minterm x of f_{HR} such that $x \leqslant x_A$. Since all minterms of f_{HR} have Hamming weight at least $k/2$ and f_{HR} has at most n^c minterms, we have

$$\mathbb{P}[f_{\mathrm{HR}}(\boldsymbol{N}) = 1] \leqslant n^c \cdot \frac{\binom{n-k/2}{m-k/2}}{\binom{n}{m}} \leqslant n^c \cdot \left(\frac{m}{n}\right)^{k/2} \leqslant n^{-c}.$$

As a consequence of Lemmas 1 and 2, we obtain the following result.

Lemma 3. *For large enough* n, *we have* $\mathbb{P}[f_{\mathrm{HR}}(\boldsymbol{Y}) = 1] + \mathbb{P}[f_{\mathrm{HR}}(\boldsymbol{N}) = 0] \geqslant 9/5$.

2.3 A Closure Operator

In this section, we describe a closure operator in the lattice of monotone Boolean functions. We prove that the closure of a monotone Boolean function f is a good approximation for f on the negative test distribution (Lemma 4), and we give a bound on the size of the set of minterms of *closed* monotone functions. This bound makes use of the robust sunflower lemma (Theorem 4), and is crucial to bounding errors of approximation (Lemma 7). Throughout this section, we let

$$\varepsilon := n^{-3c}.$$

Definition 2. *We say that a monotone function* $f : \{0,1\}^n \to \{0,1\}$ *is* ε-closed *if, for every* $A \in \binom{[n]}{\leqslant c}$, *we have*

$$\mathbb{P}[\ f(\boldsymbol{N} \vee x_A) = 1\ ] \geqslant 1 - \varepsilon \implies f(x_A) = 1.$$

This means that for, an ε-closed function, we always have $\mathbb{P}[f(\boldsymbol{N} \vee x_A) = 1] \notin [1 - \varepsilon, 1)$ when $|A| \leqslant c$. Note morever that if f, g are both ε-closed monotone Boolean functions, then so is $f \wedge g$. Therefore, there exists a unique minimum closed function $\mathrm{cl}(f)$ satisfying $f \leqslant \mathrm{cl}(f)$. We call $\mathrm{cl}(f)$ the *closure* of f. We now give a bound on the error of approximating f by $\mathrm{cl}(f)$ under the distribution $\boldsymbol{N}$.

Lemma 4. *For every monotone* $f : \{0,1\}^n \to \{0,1\}$, *we have*

$$\mathbb{P}\left[f(\boldsymbol{N}) = 0 \text{ and } \mathrm{cl}(f)(\boldsymbol{N}) = 1\right] \leqslant n^{-2c}.$$

Proof. We first prove that there exists a positive integer t and sets $A_1, \ldots, A_t$ and monotone functions $h_0, h_1, \ldots, h_t : \{0,1\}^n \to \{0,1\}$ such that

1. $h_0 = f$,
2. $h_i = h_{i-1} \vee \lceil A_i \rceil$,

3. $\mathbb{P}[h_{i-1}(\boldsymbol{N} \cup x_{A_i}) = 1] \geqslant 1 - \varepsilon$,
4. $h_t = \mathrm{cl}(f)$.

Indeed, if h_{i-1} is not closed, there exists $A_i \in \binom{[n]}{\leqslant c}$ such that $\mathbb{P}[h_{i-1}(\boldsymbol{N} \cup x_{A_i}) = 1] \geqslant 1 - \varepsilon$ but $h_{i-1}(x_{A_i}) = 0$. We let $h_i := h_{i-1} \vee \lceil A_i \rceil$. Clearly, we have that h_t is closed, and that the value of t is at most the number of subsets of $[n]$ of size at most c. Therefore, we get $t \leqslant \sum_{j=0}^{c} \binom{n}{j}$. Moreover, by induction we obtain that $h_i \leqslant \mathrm{cl}(f)$ for every $i \in [t]$. It follows that $h_t = \mathrm{cl}(f)$. Now, observe that

$$\begin{aligned}
\mathbb{P}\left[f(\boldsymbol{N}) = 0 \text{ and } \mathrm{cl}(f)(\boldsymbol{N}) = 1\right] &\leqslant \sum_{i=1}^{t} \mathbb{P}\left[f_{i-1}(\boldsymbol{N}) = 0 \text{ and } f_i(\boldsymbol{N}) = 1\right] \\
&= \sum_{i=1}^{t} \mathbb{P}\left[f_{i-1}(\boldsymbol{N}) = 0 \text{ and } x_{A_i} \subseteq \boldsymbol{N}\right] \\
&\leqslant \sum_{i=1}^{t} \mathbb{P}\left[f_{i-1}(\boldsymbol{N} \cup x_{A_i}) = 0\right] \\
&\leqslant \varepsilon \sum_{j=0}^{c} \binom{n}{j} \leqslant n^{-2c}.
\end{aligned}$$

We now bound the size of the set of ℓ-minterms of an ε-closed function. This bound is dependent on the robust sunflower theorem (Theorem 4).

Lemma 5. *Let $B > 0$ be as in Theorem 4. If a monotone function $f : \{0,1\}^n \to \{0,1\}$ is ε-closed, then, for all $\ell \in [c]$, we have*

$$|\mathcal{M}_\ell(f)| \leqslant \left(B \frac{\log(\ell/\varepsilon)}{p} \log\left(\frac{\log(\ell/\varepsilon)}{p}\right)\right)^{\ell}.$$

Proof. Fix $\ell \in [c]$. Suppose we have $|\mathcal{M}_\ell(f)| > (C\log(\ell/\varepsilon)/p \log(\log(\ell/\varepsilon)/p))^{\ell}$. Consider also the family $\mathcal{F} := \left\{A \in \binom{[n]}{\ell} : x_A \in \mathcal{M}_\ell(f)\right\}$. Observe that $|\mathcal{F}| = |\mathcal{M}_\ell(f)|$. By Theorem 4, there exists a (m, ε)-slice-sunflower $\mathcal{F}' \subseteq \mathcal{F}$. Let $Y := \bigcap \mathcal{F}'$ and let $W \in \binom{[n]}{m}$ be chosen uniformly at random. We have

$$\begin{aligned}
\mathbb{P}[f(\boldsymbol{N} \vee x_Y) = 1] &\geqslant \mathbb{P}[\exists x \in \mathcal{M}_\ell(f) : x \leqslant \boldsymbol{N} \vee x_Y] \\
&= \mathbb{P}[\exists F \in \mathcal{F} : F \subseteq W \cup Y] \\
&\geqslant \mathbb{P}[\exists F \in \mathcal{F}' : F \subseteq W \cup Y] \\
&\geqslant 1 - \varepsilon.
\end{aligned}$$

Therefore, since f is ε-closed, we get that $f(x_Y) = 1$. However, since $Y = \bigcap \mathcal{F}'$, there exists $F \in \mathcal{F}'$ such that $Y \subsetneq F$. This is a contradiction, because x_F is a minterm of f.

2.4 Trimmed Monotone Functions

In this section, we define a *trimming* operation for Boolean functions. We will bound the probability that a *trimmed* function gives the correct output on the distribution $\boldsymbol{Y}$, and we will give a bound on the error of approximating a Boolean function f by the trimming of f on that same distribution.

Definition 3. *We say that a monotone function* $f \in \{0,1\}^n \to \{0,1\}$ *is* trimmed *if all the minterms of* f *have size at most* $c/2$. *We define the trimming operation* $\mathrm{trim}(f)$ *as follows:*

$$\mathrm{trim}(f) := \bigvee_{\ell=1}^{c/2} \bigvee_{A \in \mathcal{M}_\ell(f)} \lceil A \rceil.$$

That is, the trim operation takes out from f all the minterms of size larger than $c/2$, yielding a trimmed function. We will first prove the following claim.

Claim. For every monotone function $f : \{0,1\}^n \to \{0,1\}$ and $\ell \leqslant c$, we have $\mathbb{P}[\exists x \in \mathcal{M}_\ell(f) : x \leqslant \boldsymbol{Y}] \leqslant (k/n)^\ell \, |\mathcal{M}_\ell(f)|$.

Proof. Recall (Sect. 2.2) that the distribution $\boldsymbol{Y}$ takes a polynomial $\boldsymbol{P} \in \mathbb{F}_n[x]$ with degree at most $c-1$ uniformly at random and returns the binary vector $x_{\{\boldsymbol{P}(1),\boldsymbol{P}(2),\ldots,\boldsymbol{P}(k)\}} \in \{0,1\}^n$. Let $A \in \binom{[n]}{\ell}$ for $\ell \leqslant c$. Observe that $x_A \leqslant \boldsymbol{Y}$ if and only if $A \subseteq \{\boldsymbol{P}(1), \boldsymbol{P}(2), \ldots, \boldsymbol{P}(k)\}$. Therefore, if $x_A \leqslant \boldsymbol{Y}$, then there exists indices $\{j_1, \ldots, j_\ell\}$ such that $\{\boldsymbol{P}(j_1), \boldsymbol{P}(j_2), \ldots, \boldsymbol{P}(j_\ell)\} = A$. Since $\ell \leqslant c$, we get by the c-wise independence of $\boldsymbol{P}(1), \ldots, \boldsymbol{P}(k)$ that the random variables $\boldsymbol{P}(j_1), \boldsymbol{P}(j_2), \ldots, \boldsymbol{P}(j_\ell)$ are independent. It follows that

$$\mathbb{P}[\{\boldsymbol{P}(j_1), \boldsymbol{P}(j_2), \ldots, \boldsymbol{P}(j_\ell)\} = A] = \frac{\ell!}{n^\ell}.$$

Therefore, we have

$$\mathbb{P}[x_A \leqslant \boldsymbol{Y}] = \mathbb{P}[A \subseteq \{\boldsymbol{P}(1), \boldsymbol{P}(2), \ldots, \boldsymbol{P}(k)\}] \leqslant \binom{k}{\ell} \frac{\ell!}{n^\ell} \leqslant \left(\frac{k}{n}\right)^\ell.$$

The claim now follows by an union bound.

Lemma 6. *If a monotone function* $f \in \{0,1\}^n \to \{0,1\}$ *is trimmed and* $f \neq \mathbb{1}$ *(i.e.,* f *is not identically* 1*), then*

$$\mathbb{P}[f(\boldsymbol{Y}) = 1] \leqslant \sum_{\ell=1}^{c/2} \left(\frac{k}{n}\right)^\ell |\mathcal{M}_\ell(f)|.$$

Proof. It suffices to see that, since f is trimmed, if $f(\boldsymbol{Y}) = 1$ and $f \neq \mathbb{1}$ then there exists a minterm x of f with Hamming weight between 1 and $c/2$ such that $x \leqslant \boldsymbol{Y}$. The result follows by the claim above.

Lemma 7. *Let $f \in \{0,1\}^n \to \{0,1\}$ be a monotone function, all of whose minterms have Hamming weight at most c. We have*

$$\mathbb{P}\left[f(\boldsymbol{Y}) = 1 \text{ and } \operatorname{trim}(f)(\boldsymbol{Y}) = 0\right] \leqslant \sum_{\ell=c/2}^{c} \left(\frac{k}{n}\right)^{\ell} |\mathcal{M}_\ell(f)|.$$

Proof. If we have $f(\boldsymbol{Y}) = 1$ and $\operatorname{trim}(f)(\boldsymbol{Y}) = 0$, then there was a minterm x of f with Hamming weight larger than $c/2$ that was removed by the trimming process. Therefore, since $|x| \leqslant c$ by assumption, the result follows by the claim.

2.5 The Approximators

Let $\mathcal{A} := \{\operatorname{trim}(\operatorname{cl}(f)) : f : \{0,1\}^n \to \{0,1\}$ is monotone$\}$. Functions in $\mathcal{A}$ will be called *approximators.* We define the *approximating* operations $\sqcup, \sqcap : \mathcal{A} \times \mathcal{A} \to \mathcal{A}$ as follows: for $f, g \in \mathcal{A}$, let

$$f \sqcup g := \operatorname{trim}(\operatorname{cl}(f \vee g)),$$
$$f \sqcap g := \operatorname{trim}(\operatorname{cl}(f \wedge g)).$$

Observe that every input function x_i is an approximator. Therefore, we can replace each gate of a monotone $\{\vee, \wedge\}$-circuit C by its corresponding approximating gate, thus obtaining a $\{\sqcup, \sqcap\}$-circuit $C^{\mathcal{A}}$ computing an approximator.

The rationale for choosing this set of approximators is as follows. By letting approximators be the trimming of a closed function, we are able to plug the bound on the set of ℓ-minterms given by the robust sunflower lemma (Lemma 5) on Lemmas 6 and 7, since the trimming operation can only *reduce* the set of minterms. Moreover, since trimmings can only help to get a negative answer on the negative test distribution, we can safely apply Lemma 4 when bounding the errors of approximation.

2.6 The Lower Bound

In this section, we will prove that the function f_{HR} requires monotone circuits of size $2^{\Omega(c)}$. By properly choosing c and k, this will imply the promised $\exp(\Omega(n^{1/2-o(1)}))$ lower bound for the Harnik-Raz function. First, we fix some parameters. Choose B as in Lemma 5. We also let

$$c := \frac{1}{6Be^{1/B}} \left(\frac{n}{(\log n)^3}\right)^{1/2}, \qquad k := \left(\frac{n}{\log n}\right)^{1/2}.$$

For simplicity, we assume these values are integers. We clearly have $c < k$. Moreover, observe that, because of this choice of parameters, we have $p = \Omega(1)$. Indeed, we have

$$p = n^{-4c/k} = n^{-2/(3Be^{1/B} \log n)} = e^{-2/(3Be^{1/B})} \geqslant e^{-1/B}.$$

We will now show that, when f is an approximator, the bound of Lemma 6 can be replaced by 1/2, and also that, when f is an ε-closed function, the bound of Lemma 7 can be replaced by $2^{-\Omega(c)}$. We will first need to bound the sequence s_ℓ, defined as follows. For every $1 \leqslant \ell \leqslant c$, let

$$s_\ell := \left(\frac{k}{n}\right)^\ell \cdot \left(B\frac{\log(c/\varepsilon)}{p} \log\left(\frac{\log(c/\varepsilon)}{p}\right)\right)^\ell.$$

Note that, when f is a n-bit ε-closed monotone function, we get by Lemma 5 that $\left(\frac{k}{n}\right)^\ell |\mathcal{M}_\ell(f)| \leqslant s_\ell$. In other words, the summands of Lemma 6 and Lemma 7 can be replaced by s_ℓ in some applications. Observe moreover that $s_\ell = (s_1)^\ell$. Now we are going to show that, for n sufficiently large, we have $s_1 \leqslant 1/3$, which implies $s_\ell \leqslant 3^{-\ell}$. First, observe that

$$\log(c/\varepsilon)/p = \log(n^{3c}c)/p \leqslant \log(n^{4c})/p = \frac{4c}{p}\log n.$$

Moreover, we have

$$\log\left(\log(c/\varepsilon)/p\right) = \log\left(\frac{4c}{p}\log n\right) = \frac{1}{2}\log n - \frac{1}{2}\log\log n + O(1) \leqslant \frac{1}{2}\log n,$$

for n sufficiently large. From the previous two inequalities, we obtain for n sufficiently large that

$$s_1 = B \cdot \frac{k}{n} \cdot \frac{\log(c/\varepsilon)}{p} \log\left(\frac{\log(c/\varepsilon)}{p}\right) \leqslant \frac{2B}{p} \cdot \frac{ck(\log n)^2}{n} \leqslant 1/3,$$

as desired.

Lemma 8 (Approximators make many errors). *For every approximator $f \in \mathcal{A}$, we have $\mathbb{P}[f(\boldsymbol{Y}) = 1] + \mathbb{P}[f(\boldsymbol{N}) = 0] \leqslant 3/2$.*

Proof. Let $f \in \mathcal{A}$. By definition, there exists an ε-closed function h such that $f = \mathrm{trim}(h)$. Observe that $\mathcal{M}_\ell(f) \subseteq \mathcal{M}_\ell(h)$ for every $\ell \in [c]$. Hence, applying Lemma 6 and the bounds for s_ℓ, we obtain that, if $f \neq \mathbb{1}$, we have

$$\mathbb{P}[f(\boldsymbol{Y}) = 1] \leqslant \sum_{\ell=1}^{c/2} \left(\frac{k}{n}\right)^\ell |\mathcal{M}_\ell(h)| \leqslant \sum_{\ell=1}^{c/2} s_\ell \leqslant \sum_{\ell=1}^{c/2} 3^{-\ell} \leqslant 1/2.$$

Therefore, for every $f \in \mathcal{A}$ we have $\mathbb{P}[f(\boldsymbol{Y}) = 1] + \mathbb{P}[f(\boldsymbol{N}) = 0] \leqslant 1 + 1/2 \leqslant 3/2$.

Lemma 9 (C is well-approximated by $C^{\mathcal{A}}$). *Let C be a monotone circuit. We have*

$$\mathbb{P}[C(\boldsymbol{Y}) = 1 \text{ and } C^{\mathcal{A}}(\boldsymbol{Y}) = 0] + \mathbb{P}[C(\boldsymbol{N}) = 0 \text{ and } C^{\mathcal{A}}(\boldsymbol{N}) = 1] \leqslant \mathrm{size}(C) \cdot 2^{-\Omega(c)}.$$

Proof. We begin by bounding the approximation errors under the distribution $\boldsymbol{Y}$. We will show that, for two approximators $f, g \in \mathcal{A}$, if $f \vee g$ accepts an input from $\boldsymbol{Y}$, then $f \sqcup g$ rejects that input with probability at most $2^{-\Omega(c)}$, and that the same holds for the approximation $f \sqcap g$.

First note that, if $f, g \in \mathcal{A}$, then all the minterms of both $f \vee g$ and $f \wedge g$ have Hamming weight at most c, since f and g are trimmed. Let now $h = \mathrm{cl}(f \vee g)$. We have $(f \sqcup g)(x) < (f \vee g)(x)$ only if $\mathrm{trim}(h)(x) < h(x)$. Since h is closed, we obtain the following inequality by Lemma 7 and the bounds on s_ℓ:

$$\mathbb{P}\left[(f \vee g)(\boldsymbol{Y}) = 1 \text{ and } (f \sqcup g)(\boldsymbol{Y}) = 0\right] \leqslant \sum_{\ell=c/2}^{c} \left(\frac{k}{n}\right)^{\ell} |\mathcal{M}_\ell(h)| \leqslant \sum_{\ell=c/2}^{c} s_\ell = 2^{-\Omega(c)}.$$

The same argument shows $\mathbb{P}\left[(f \wedge g)(\boldsymbol{Y}) = 1 \text{ and } (f \sqcap g)(\boldsymbol{Y}) = 0\right] = 2^{-\Omega(c)}$. Since there are $\mathrm{size}(C)$ gates in C, this implies that $\mathbb{P}[C(\boldsymbol{Y}) = 1 \text{ and } C^{\mathcal{A}}(\boldsymbol{Y}) = 0] \leqslant \mathrm{size}(C) \cdot 2^{-\Omega(c)}$.

To bound the approximation errors under $\boldsymbol{N}$, note that $(f \vee g)(x) = 0$ and $(f \sqcup g)(x) = 1$ only if $\mathrm{cl}(f \vee g)(x) \neq (f \vee g)(x)$, since trimming a Boolean function cannot decrease the probability that it rejects an input. Therefore, by Lemma 4 we obtain

$$\mathbb{P}\left[(f \vee g)(\boldsymbol{N}) = 0 \text{ and } (f \sqcup g)(\boldsymbol{N}) = 1\right] \leqslant n^{-2c} \leqslant 2^{-\Omega(c)}.$$

The same argument shows $\mathbb{P}\left[(f \wedge g)(\boldsymbol{N}) = 0 \text{ and } (f \sqcap g)(\boldsymbol{N}) = 1\right] = 2^{-\Omega(c)}$. Once again, doing this approximation for every gate in C implies $\mathbb{P}[C(\boldsymbol{N}) = 0 \text{ and } C^{\mathcal{A}}(\boldsymbol{N}) = 1] \leqslant \mathrm{size}(C) \cdot 2^{-\Omega(c)}$. This finishes the proof.

Theorem 5. *Any monotone circuit computing f_{HR} has size $2^{\Omega(c)} = 2^{\Omega(n^{1/2}/(\log n)^3)}$.*

Proof. Let C be a monotone circuit computing f_{HR}. For large n, we have

$$\begin{aligned} 9/5 &\leqslant \mathbb{P}[f_{\mathrm{HR}}(\boldsymbol{Y}) = 1] + \mathbb{P}[f_{\mathrm{HR}}(\boldsymbol{N}) = 0] \\ &\leqslant \mathbb{P}[C(\boldsymbol{Y}) = 1 \text{ and } C^{\mathcal{A}}(\boldsymbol{Y}) = 0] + \mathbb{P}[C^{\mathcal{A}}(\boldsymbol{Y}) = 1] \\ &\quad + \mathbb{P}[C(\boldsymbol{N}) = 0 \text{ and } C^{\mathcal{A}}(\boldsymbol{N}) = 1] + \mathbb{P}[C^{\mathcal{A}}(\boldsymbol{N}) = 0] \\ &= 3/2 + \mathrm{size}(C) 2^{-\Omega(c)}. \end{aligned}$$

This implies $\mathrm{size}(C) = 2^{\Omega(c)}$.

2.7 Discussion

In this application, we chose the values of c and k to be roughly $\sqrt{n}$. We expect that, if c were chosen to be closer to n, the implied Harnik-Raz function would still have $2^{\Omega(c)}$ complexity, and thus one would be able to improve our bound. However, we do not think that the present technique would work for any $c > \sqrt{n}$, as it seems to require that $ck \leqslant n$. Therefore, in order to obtain a stronger bound to the Harnik-Raz function, we think a different technique has to be considered.

Acknowledgements. Bruno Pasqualotto Cavalar was supported by São Paulo Research Foundation (FAPESP), grants #2018/22257-7 and #2018/05557-7, and he acknowledges CAPES (PROEX) for partial support of this work. A part of this work was done during a research internship of Bruno Pasqualotto Cavalar and a postdoctoral stay of Mrinal Kumar at the University of Toronto. Benjamin Rossman was supported by NSERC, Ontario Early Researcher Award and Sloan Research Fellowship.

References

1. Alon, N., Boppana, R.B.: The monotone circuit complexity of Boolean functions. Combinatorica **7**(1), 1–22 (1987). https://doi.org/10.1007/BF02579196
2. Alon, N., Babai, L., Itai, A.: A fast and simple randomized parallel algorithm for the maximal independent set problem. J. Algorithms **7**(4), 567–583 (1986). https://doi.org/10.1016/0196-6774(86)90019-2
3. Alweiss, R., Lovett, S., Wu, K., Zhang, J.: Improved bounds for the sunflower lemma. arXiv:1908.08483 (2019). https://arxiv.org/abs/1908.08483
4. Andreev, A.E.: A method for obtaining lower bounds on the complexity of individual monotone functions. Dokl. Akad. Nauk SSSR **282**(5), 1033–1037 (1985)
5. Erdős, P., Rado, R.: Intersection theorems for systems of sets. J. London Math. Soc. **35**, 85–90 (1960). https://doi.org/10.1112/jlms/s1-35.1.85
6. Gopalan, P., Meka, R., Reingold, O.: DNF sparsification and a faster deterministic counting algorithm. Comput. Complex. **22**(2), 275–310 (2013). https://doi.org/10.1007/s00037-013-0068-6
7. Harnik, D., Raz, R.: Higher lower bounds on monotone size. In: Proceedings of the 32nd Annual ACM Symposium on Theory of Computing, pp. 378–387. ACM, New York (2000). https://doi.org/10.1145/335305.335349
8. Li, X., Lovett, S., Zhang, J.: Sunflowers and quasi-sunflowers from randomness extractors. In: Approximation, Randomization, and Combinatorial Optimization. Algorithms and Techniques, APPROX/RANDOM. LIPIcs, vol. 116, pp. 51:1–51:13 (2018)
9. Lovett, S., Solomon, N., Zhang, J.: From DNF compression to sunflower theorems via regularity. arXiv preprint p. 1903.00580 (2019)
10. Lovett, S., Zhang, J.: DNF sparsification beyond sunflowers. In: Proceedings of the 51st Annual ACM SIGACT Symposium on Theory of Computing, pp. 454–460. ACM (2019)
11. Pitassi, T., Robere, R.: Strongly exponential lower bounds for monotone computation. In: Proceedings of the 49th Annual ACM SIGACT Symposium on Theory of Computing, pp. 1246–1255. ACM (2017)
12. Rao, A.: Coding for sunflowers (2019). arXiv preprint arXiv:1909.04774
13. Razborov, A.A.: Lower bounds on the monotone complexity of some Boolean functions. Dokl. Akad. Nauk SSSR **281**(4), 798–801 (1985)
14. Rossman, B.: The monotone complexity of k-clique on random graphs. SIAM J. Comput. **43**(1), 256–279 (2014). https://doi.org/10.1137/110839059
15. Tiekenheinrich, J.: A 4n-lower bound on the mononotone network complexity of a oneoutput Boolean function. Inf. Process. Lett. **18**, 201 (1984)

Tight Bounds on Sensitivity and Block Sensitivity of Some Classes of Transitive Functions

Siddhesh Chaubal(✉) and Anna Gál

Department of Computer Science, University of Texas, Austin, TX, USA
{siddhesh,panni}@cs.utexas.edu

Abstract. Nisan and Szegedy [16] conjectured that block sensitivity is at most polynomial in sensitivity for any Boolean function. Until a recent breakthrough of Huang [14], the conjecture had been wide open in the general case, and was proved only for a few special classes of Boolean functions. Huang's result [14] implies that block sensitivity is at most the 4th power of sensitivity for any Boolean function. It remains open if a tighter relationship between sensitivity and block sensitivity holds for arbitrary Boolean functions; the largest known gap between these measures is quadratic [3,8,9,11,18,21].

We prove tighter bounds showing that block sensitivity is at most 3rd power, and in some cases at most square of sensitivity for subclasses of transitive functions, defined by various properties of their DNF (or CNF) representation. Our results improve and extend previous results regarding transitive functions. We obtain these results by proving tight (up to constant factors) lower bounds on the smallest possible sensitivity of functions in these classes.

In another line of research, it has also been examined what is the smallest possible block sensitivity of transitive functions. Our results yield tight (up to constant factors) lower bounds on the block sensitivity of the classes we consider.

Keywords: Sensitivity · Block sensitivity · Transitive functions

1 Introduction

The *sensitivity* $s(f)$ of a Boolean function f is the maximum over all inputs x of the number of coordinate positions i such that changing the value of the i-th bit of x changes the value of the function. The *block sensitivity* $bs(f)$ of a Boolean function f is the maximum over all inputs x of the number of disjoint blocks of positions such that changing the value of all bits of x in any given block changes the value of the function. (See Sect. 2 for more formal definitions.) Nisan and Szegedy [16] conjectured that block sensitivity is at most polynomial in sensitivity for any Boolean function. A number of important complexity measures (such as CREW PRAM complexity, certificate complexity, decision tree depth

Y. Kohayakawa and F. K. Miyazawa (Eds.): LATIN 2020, LNCS 12118, pp. 323–335, 2020.
https://doi.org/10.1007/978-3-030-61792-9_26

in various models and degree) are polynomially related to block sensitivity, and therefore to each other. See [7,12] for a survey. Until a recent breakthrough by Huang [14], the best upper bounds on any of these measures were exponential in terms of sensitivity. The previous best upper bounds on block sensitivity in terms of sensitivity were by Ambainis et al. [2] giving $bs(f) \leq s(f)2^{s(f)-1}$, and by He et al. [13] who gave a constant factor improvement to this bound. Huang's result [14] implies that $bs(f) \leq s(f)^4$ for any Boolean function f. The best separation between sensitivity and block sensitivity remains quadratic [3,8,9,11,18,21].

Despite a lot of attention to the problem, until Huang's result, the conjecture was verified only for a few special classes of Boolean functions, including some special classes of *transitive functions*, such as symmetric functions, graph properties and minterm-transitive functions. The following questions have been raised in connection to sensitivity and block sensitivity of transitive functions.

1. An intriguing aspect of transitive functions is that no examples of transitive functions are known on n input bits with $o(n^{1/3})$ sensitivity. Chakraborty [8] constructed a transitive function on n variables with sensitivity $\Theta(n^{1/3})$. It is implicit in a paper by Sun [20] that for a transitive function f on n variables, $bs(f) \cdot s(f)^2 \geq n$. Together with Huang's result this gives that any transitive function f on n variables has $s(f) \geq \Omega(n^{1/6})$. Previously, Chakraborty [8] proved that every minterm-transitive function f on n variables has $s(f) \geq \Omega(n^{1/3})$. It remains open if the sensitivity of every transitive function is at least $\Omega(n^{1/3})$.
2. Another intriguing question is that considering transitive functions with $f(0^n) \neq f(1^n)$, we don't even have any examples with $o(n^{1/2})$ sensitivity. A remark in the survey [12] in combination with Huang's result [14] implies that any transitive function f on n variables where n is a prime power and $f(0^n) \neq f(1^n)$ has sensitivity $s(f) \geq \Omega(\sqrt{n})$. However, this does not seem to directly imply a similar consequence for transitive functions with $f(0^n) \neq f(1^n)$ when n is not a prime power, because for a transitive function, a subfunction obtained by fixing a subset of its bits is no longer necessarily transitive.
3. While it is still open if every transitive function has sensitivity $\Omega(n^{1/3})$, Sun [20] proved that every transitive function has *block sensitivity* at least $n^{1/3}$. This resulted in further studies of what is the smallest possible block sensitivity of transitive functions. Drucker [10] showed that minterm-transitive functions must have block sensitivity at least $\Omega(n^{3/7})$. This bound is tight for the class of minterm-transitive functions: Amano [1] constructed minterm-transitive functions with block sensitivity $O(n^{3/7})$, improving constructions of Sun [20] and Drucker [10] by logarithmic factors. It remains open if transitive functions with block sensitivity $o(n^{3/7})$ exist.

1.1 Our Results

In this paper we settle the above questions for some special classes of transitive functions, significantly extending previous results about subclasses of transitive functions. For the classes we consider, we show that for functions f on n variables $s(f) \geq \Omega(n^{1/3})$ which implies $bs(f) \leq O(s(f)^3)$. In addition, we prove that the block sensitivity of functions on n variables in all the classes we consider is at least $\Omega(n^{3/7})$. Furthermore, under the additional assumption that $f(0^n) \neq f(1^n)$, we show that $s(f) \geq \Omega(\sqrt{n})$ for transitive functions f represented by DNF (or CNF) such that the number of positive literals per term is the same up to constant factors. Previously this was not known to hold for arbitrary values of n, even for the special case of minterm-transitive functions.

Our lower bounds on both sensitivity and block sensitivity are tight up to constant factors for the corresponding classes.

We consider the following three subclasses of transitive functions.

Transitive Functions with Sparse DNF (or CNF). We consider transitive functions that can be represented by DNFs with up to $2^{n^{\frac{1}{2}-\epsilon}}$ terms, or by CNFs with up to $2^{n^{\frac{1}{2}-\epsilon}}$ clauses, for constant $\epsilon > 0$. For any non-constant function f of this form we prove that $s(f) \geq \Omega(\min\{n^{1/3}, n^{2\epsilon}\})$. In particular, setting $\epsilon = 1/6$ gives the bound $s(f) \geq \Omega(n^{1/3})$ for transitive functions represented by DNFs (or CNFs) of size up to $2^{n^{1/3}}$.

Comparing with previous results, we note that any DNF with at most t terms is also a read-t DNF. Thus, the results of [6] imply that non-constant functions represented by DNFs with at most $n^{\frac{1}{3}-\epsilon}$ terms have sensitivity $\Omega(n^{\epsilon})$. Our results significantly improve this to DNFs with up to an exponential $2^{n^{\frac{1}{2}-\epsilon}}$ number of terms, in the case of transitive functions.

Transitive Functions Represented by DNF (or CNF) with a Not-Too-Frequent Variable. We further extend these results to transitive functions represented by DNFs (or CNFs) of arbitrary sizes, as long as there exists a variable that appears at most $2^{n^{\frac{1}{2}-\epsilon}}$ times, for constant $\epsilon > 0$. As above, setting $\epsilon = 1/6$ gives $s(f) \geq \Omega(n^{1/3})$.

Transitive Functions Represented by DNF (or CNF) with Approximately the Same Number of Positive Literals per Term. Next we consider transitive functions represented by DNF (or CNF) where the number of terms as well as the size of the terms (i.e. the width of the DNF) are arbitrary, but the number of positive literals in each term is the same up to constant factors. We prove for transitive functions f on n variables with this property that $s(f) \geq \Omega(n^{1/3})$.

This class significantly extends the previously studied class of *minterm-transitive* functions. Roughly speaking, minterm-transitive functions have the property that all their 1-inputs are consistent with minterms that are equivalent to just one minterm, under permutations from the invariance group of the function. Chakraborty [8] proved that minterm-transitive functions f on n variables have $s(f) \geq \Omega(n^{1/3})$, and he noted that his argument extends to the case when the number of positive literals as well as the sizes of each term are the same up to constant factors. Our contribution is to further extend the argument without making any assumptions about the sizes of the terms.

Tightness of Our Bounds. As noted above, Chakraborty [8] gave an example of a transitive function on n variables with sensitivity $\Theta(n^{1/3})$, and Amano [1] gave an example of a transitive function on n variables with block sensitivity $\Theta(n^{3/7})$. Both functions are minterm-transitive, thus they can be represented by DNFs where each term has the same number of positive literals. On the other hand, both functions can be represented by DNFs with n terms, thus they also belong to the other two classes of transitive functions that we consider. This shows that our bounds $s(f) \geq \Omega(n^{1/3})$ and $bs(f) \geq \Omega(n^{3/7})$ are the best possible for these classes, up to constant factors.

We give a simple example of a minterm-transitive function f on n variables, with sensitivity $\Theta(\sqrt{n})$ such that $f(0^n) \neq f(1^n)$. This shows that our $\Omega(\sqrt{n})$ lower bound on sensitivity is tight up to constant factors for the corresponding class. We describe this example in the full version of the paper.

1.2 Our Techniques

First, we note that our arguments are independent of Huang's proof [14]. Instead, our results are based on new upper bounds on the *minimum certificate size*, that hold for *arbitrary* Boolean functions, not just transitive functions. We give two such bounds: one upper bounds the minimum certificate size by the sensitivity of the function and by the logarithm of the number of terms of the DNF (Lemma 5), the other relates the minimum certificate size to the number of occurrences of any given variable and the influence of that variable (Lemma 6). We note that relating the minimum certificate size to influence has been also used in [5] in a different context. These upper bounds allow us to take advantage of a result of Chakraborty [8] (see Corollary 1) which shows that for transitive functions, upper bounds on the minimum certificate size imply lower bounds on the sensitivity of the function.

We emphasize that our upper bounds on minimum certificate size hold for arbitrary Boolean functions, not just transitive functions. The part of our arguments that is specific to transitive functions, is using the fact that for transitive functions, upper bounds on minimum certificate size imply lower bounds on sensitivity, and the relationship between the influences of different variables of transitive functions.

We also provide a new, stronger tradeoff between sensitivity and the certificate size on two special inputs, (0^n and 1^n), that holds for arbitrary transitive functions (Lemma 8). This allows us to obtain tight, $\Omega(\sqrt{n})$ lower bounds on the sensitivity of functions f on n variables in our third class, when $f(0^n) \neq f(1^n)$.

Finally, we observe that upper bounds on the minimum certificate size also provide lower bounds on block sensitivity of arbitrary transitive functions, (Lemma 9), with a stronger tradeoff than what follows from tradeoffs between sensitivity and minimum certificate size. This allows us to obtain tight, $\Omega(n^{3/7})$ lower bounds on the block sensitivity of functions in all the classes we consider.

Due to page limits, some of our proofs are left to the full version of the paper.

2 Preliminaries

Let $f : \{0,1\}^n \to \{0,1\}$ be a Boolean function. For $x \in \{0,1\}^n$ and $i \in [n]$ we denote by x^i the input obtained by flipping the i-th bit of x. More generally, for $S \subseteq [n]$ we denote by x^S the input obtained by flipping the bits of x in all coordinates in the subset S.

Definition 1 Sensitivity. *The sensitivity $s(f,x)$ of a Boolean function f on input x is the number of coordinates $i \in [n]$ such that $f(x) \neq f(x^i)$. The 0-sensitivity and 1-sensitivity of f are defined as $s_0(f) = \max\{s(f,x) : f(x) = 0\}$ and $s_1(f) = \max\{s(f,x) : f(x) = 1\}$, respectively. The sensitivity of f is defined as $s(f) = \max\{s(f,x) : x \in \{0,1\}^n\} = \max\{s_0(f), s_1(f)\}$.*

Definition 2 Block Sensitivity. *The block sensitivity $bs(f,x)$ of a Boolean function f on input x is the maximum number of pairwise disjoint subsets $S_1, \ldots, S_k$ of $[n]$ such that for each $i \in [k]$ $f(x) \neq f(x^{S_i})$. The 0-block sensitivity and 1-block sensitivity of f are defined as $bs_0(f) = \max\{bs(f,x) : f(x) = 0\}$ and $bs_1(f) = \max\{bs(f,x) : f(x) = 1\}$, respectively. The block sensitivity of f is defined as $bs(f) = \max\{bs(f,x) : x \in \{0,1\}^n\} = \max\{bs_0(f), bs_1(f)\}$.*

It is convenient to refer to coordinates $i \in [n]$ such that $f(x) \neq f(x^i)$ as *sensitive bits* for f on x. Similarly, a subset $S \subseteq [n]$ is called a *sensitive block* for f on x if $f(x) \neq f(x^S)$.

Definition 3 Partial Assignment. *Given an integer $n > 0$, a partial assignment α is a function $\alpha\colon [n] \to \{0,1,\star\}$. A partial assignment α corresponds naturally to a setting of n variables $(x_1, x_2, \ldots, x_n)$ to $\{0,1,\star\}$ where x_i is set to $\alpha(i)$. The variables set to $\star$ are called unassigned or free, and we say that the variables set to 0 or 1 are fixed. We say that $x \in \{0,1\}^n$ agrees with α if $x_i = \alpha(i)$ for all i such that $\alpha(i) \neq \star$. The size of a partial assignment α is defined as the number of fixed variables of α.*

Definition 4 Certificate. *For a function $f\colon \{0,1\}^n \to \{0,1\}$ and input $x \in \{0,1\}^n$ a partial assignment α is a certificate of f on x if x agrees with α and any input y agreeing with α satisfies $f(y) = f(x)$. A certificate α is a 1-certificate (resp. 0-certificate) if $f(x) = 1$ (resp. $f(x) = 0$), on inputs x that agree with α.*

Definition 5 Minterms and Maxterms. *A certificate α is called minimal, if after changing any of its fixed variables to a free variable, the resulting partial assignment α' is not a certificate, that is the function is not constant on inputs agreeing with α'. A minimal 1-certificate is called a minterm, and a minimal 0-certificate is called a maxterm.*

Definition 6 Size and Weight of Certificates. *The size of a certificate α, denoted by $size(\alpha)$ is defined as the size of the partial assignment α. The weight of a certificate α, denoted by $wt(\alpha)$, is the number of bits fixed to 1 by α.*

Definition 7 Certificate Complexity. *The certificate complexity $C(f,x)$ of a Boolean function f on input x is the size of the smallest certificate of f on x. The 0-certificate complexity and 1-certificate complexity of f are defined as $C_0(f) = \max\{C(f,x) : f(x) = 0\}$ and $C_1(f) = \max\{C(f,x) : f(x) = 1\}$, respectively. The certificate complexity of f is defined as $C(f) = \max\{C(f,x) : x \in \{0,1\}^n\} = \max\{C_0(f), C_1(f)\}$.*

It is also useful to consider the following definition of the smallest certificate size over all inputs. Note that this can be rephrased as the co-dimension of the largest subcube of the Boolean cube $\{0,1\}^n$ where f is constant.

Definition 8 Minimum Certificate Size. *The minimum certificate size of a Boolean function $f\colon \{0,1\}^n \to \{0,1\}$ is defined as $C_{min}(f) = \min\{C(f,x) : x \in \{0,1\}^n\}$.*

We will use the following lemma of Simon [19].

Lemma 1. *[19] (see also [4]) Let $f : \{0,1\}^n \to \{0,1\}$ be a non-constant Boolean function. Then $|f^{-1}(1)| \geq 2^{n-s_1(f)}$ and $|f^{-1}(0)| \geq 2^{n-s_0(f)}$.*

2.1 Transitive Functions

Definition 9 Invariance Group. *A Boolean function $f\colon \{0,1\}^n \to \{0,1\}$ is invariant under a permutation $\sigma\colon [n] \to [n]$, if for any $x \in \{0,1\}^n$, $f(x_1, \dots, x_n) = f(x_{\sigma(1)}, \dots, x_{\sigma(n)})$. The set of all permutations under which f is invariant forms a group, called the invariance group of f.*

Definition 10 Transitive Function. *A Boolean function is transitive if its invariance group Γ is transitive, that is, for each $i, j \in [n]$, there is a $\sigma \in \Gamma$ such that $\sigma(i) = j$.*

For example, the set of all permutations on n bits, denoted by S_n is a transitive group of permutations. Another example of a transitive group of permutations is the set of all *cyclic shifts* on n bits, denoted by $Shift_n = \{\xi_0, \xi_1, \dots, \xi_{n-1}\}$, where the permutation ξ_j cyclically shifts the string by j positions.

We will use the following observations of Chakraborty about transitive functions. Recall that S_n denotes the group of all permutations on n bits.

We use the following notation: for a set $S \subseteq [n]$ and a permutation $\sigma \in S_n$ we denote by $\sigma(S)$ the set $\{\sigma(i) | i \in S\}$.

Lemma 2 (4.3 in [8]). *Let $\Gamma \subseteq S_n$ be a transitive group of permutations on n bits. Then, for any $\emptyset \neq S \subseteq [n]$ with $|S| = k$, there exists $\hat{\Gamma} \subseteq \Gamma$ with $|\hat{\Gamma}| \geq \frac{n}{k^2}$ such that for any two permutations $\sigma_1, \sigma_2 \in \hat{\Gamma}$ their images on S are disjoint, that is $\sigma_1(S) \cap \sigma_2(S) = \emptyset$.*

Lemma 3 (4.4 in [8]). *Let $f\colon \{0,1\}^n \to \{0,1\}$ be a non-constant transitive function. Let α be a 1-certificate (resp. 0-certificate) for some $x \in \{0,1\}^n$, with $size(\alpha) = k > 0$. Then, $s_0(f) \geq \frac{n}{k^2}$ (resp. $s_1(f) \geq \frac{n}{k^2}$).*

Corollary 1. *For a non-constant transitive function* $f\colon \{0,1\}^n \to \{0,1\}$ *we have:*

$$s(f)(C_{min}(f))^2 \geq n\,.$$

We will also use the following observation of Sun [20].

Lemma 4. *[20] Let* $\Gamma \subseteq S_n$ *be a transitive group of permutations on* n *bits. For any* $x, y \in \{0,1\}^n$*, if* $wt(x) \cdot wt(y) < n$*, then there exists some* $\sigma \in \Gamma$*, such that* $\sigma(x)$ *and* y *do not have any* 1*-s in the same position.*

3 Lower Bounds on Sensitivity of Transitive Functions

3.1 Sparse DNF (or CNF)

In this section we prove lower bounds on the sensitivity of transitive functions that can be represented by DNFs with up to $2^{n^{\frac{1}{2}-\epsilon}}$ terms, or by CNFs with up to $2^{n^{\frac{1}{2}-\epsilon}}$ clauses, for constant $\epsilon > 0$.

We start with a lemma that holds for any Boolean function, transitivity is not required.

Lemma 5. *Let* $f : \{0,1\}^n \to \{0,1\}$ *be a non-constant Boolean function. If* f *can be represented by a DNF with* t *terms, then* $C_{min}(f) \leq s_1(f) + \log t$ *, and if* f *can be represented by a CNF with* t *clauses, then* $C_{min}(f) \leq s_0(f) + \log t$ *.*

Proof. We prove the statement about DNFs, the proof for CNFs is analogous. Let $f : \{0,1\}^n \to \{0,1\}$ be a non-constant Boolean function that can be represented by a DNF with t terms. Notice that for each term of the DNF, we get a 1-certificate by fixing the variables that appear in the given term, to a value so that the term is satisfied, and leaving the remaining variables free. This means that the number of variables that participate in any given term must be at least $C_{min}(f)$. Thus, the number of different inputs that satisfy a given term is at most $2^{n-C_{min}(f)}$. This means that $|f^{-1}(1)| \leq t2^{n-C_{min}(f)}$. On the other hand, by Simon's Lemma (see Lemma 1 in Sect. 2) $|f^{-1}(1)| \geq 2^{n-s_1(f)}$. Combining these two inequalities implies the statement of the lemma. □

We obtain the following theorem.

Theorem 1. *Let* $\epsilon > 0$ *and let* $f : \{0,1\}^n \to \{0,1\}$ *be a non-constant transitive function that can be represented by a DNF with up to* $2^{n^{\frac{1}{2}-\epsilon}}$ *terms, or by a CNF with up to* $2^{n^{\frac{1}{2}-\epsilon}}$ *clauses. Then* $s(f) \geq \Omega(\min\{n^{1/3}, n^{2\epsilon}\})$.

Remark 1. Setting $\epsilon = 1/6$ gives $s(f) \geq \Omega(n^{1/3})$ for transitive functions represented by DNFs (or CNFs) of size up to $2^{n^{1/3}}$.

3.2 DNF (or CNF) with a Not-Too-Frequent Variable

In this section we further extend the results of the previous section. We show that the same lower bounds for sensitivity hold for transitive functions represented by DNFs with an arbitrary number of terms, as long as there exists a variable that appears in no more than $2^{n^{\frac{1}{2}-\epsilon}}$ terms, for constant $\epsilon > 0$. An analogous result holds considering CNFs.

We once again start with an observation that holds for arbitrary Boolean functions, not just transitive functions.

For a Boolean function $f\colon \{0,1\}^n \to \{0,1\}$, the influence of the i-th variable, denoted by $Inf_i(f)$ is defined as: $Inf_i(f) = \Pr_x[f(x) \neq f(x^i)]$ where the probability is taken over the uniform distribution on $\{0,1\}^n$.

Lemma 6. *Let $f : \{0,1\}^n \to \{0,1\}$ be a Boolean function that can be represented by a DNF (or CNF) such that its i-th variable appears in at most k terms (resp. clauses) of the formula, for some $i \in [n]$. Then we have: $C_{min}(f) \leq \log k + 1 - \log Inf_i(f)$*

Proof. We prove the statement about DNFs, the proof for CNFs is analogous.

As we noted in the proof of Lemma 5, for each term of the DNF, we get a 1-certificate by fixing the variables that appear in the given term, to a value so that the term is satisfied, and leaving the remaining variables free. This means that the number of variables that participate in any given term must be at least $C_{min}(f)$. Thus, the number of different inputs that satisfy a given term is at most $2^{n-C_{min}(f)}$.

Consider only those k terms that include the variable x_i. The number of inputs satisfying at least one of these terms is at most $k2^{n-C_{min}(f)}$. Also, notice that each of the 1-inputs that are sensitive to the i-th bit must satisfy one of the terms that include the variable x_i. (Each 1-input must satisfy at least one term, and an input that is sensitive to x_i cannot satisfy a term that does not depend on x_i.) Therefore, the number of 1-inputs that are sensitive to the i-th bit is at most $k2^{n-C_{min}(f)}$. On the other hand, the number of 1-inputs that are sensitive to the i-th bit equals $Inf_i(f) \cdot 2^{n-1}$. Thus, we get $Inf_i(f) \cdot 2^{n-1} \leq k2^{n-C_{min}(f)}$, and this gives the statement of the lemma. □

We are ready to prove the following theorem for transitive functions.

Theorem 2. *Let $\epsilon > 0$ and let $f : \{0,1\}^n \to \{0,1\}$ be a non-constant transitive function that can be represented by a DNF (or CNF) such that one of its variables appears in at most $2^{n^{\frac{1}{2}-\epsilon}}$ terms (resp. clauses) of the formula. Then $s(f) \geq \Omega(\min\{n^{1/3}, n^{2\epsilon}\})$.*

Proof. We prove the statement about DNFs, the proof for CNFs is analogous. As before, it is enough to prove the statement for $0 < \epsilon \leq 1/6$, since this will imply that $s(f) \geq \Omega(n^{1/3})$ whenever $\epsilon \geq 1/6$.

Let x_i be a variable that appears in at most $k = 2^{n^{\frac{1}{2}-\epsilon}}$ terms of the DNF. It is known (see e.g. [17]) that for transitive f, $Inf_i(f) = Inf_j(f)$ for any $j \in [n]$, and thus $Inf_i(f) = \max_{j \in [n]} Inf_j(f)$. By a theorem of Kahn, Kalai and Linial [15], $\max_{j \in [n]} Inf_j(f) \geq \Omega(p(1-p)\log n/n)$, where p is the probability that the function f equals 1. Then, by Lemma 6 we get $C_{min}(f) \leq \log k + 1 - \log Inf_i(f) \leq O(\log k + 1 + \log n + \log \frac{1}{p(1-p)})$.

Notice that $\frac{1}{p(1-p)} \geq 2^{n^{\frac{1}{2}-\epsilon}}$ implies that $\min\{|f^{-1}(1)|, |f^{-1}(0)|\} \leq 2^{n+1-n^{\frac{1}{2}-\epsilon}}$. Then by Lemma 1, $s(f) \geq \Omega(n^{\frac{1}{2}-\epsilon})$, which is at least $\Omega(n^{2\epsilon})$ when $\epsilon \leq 1/6$.

Otherwise, $\frac{1}{p(1-p)} < 2^{n^{\frac{1}{2}-\epsilon}}$, and we get $C_{min}(f) \leq O(n^{\frac{1}{2}-\epsilon})$. Then, Corollary 1 implies that $s(f) \geq \Omega(n^{2\epsilon})$. □

Remark 2. Setting $\epsilon = 1/6$ gives $s(f) \geq \Omega(n^{1/3})$ for transitive functions represented by DNFs (or CNFs) such that one of the variables appears no more than $2^{n^{1/3}}$ times.

3.3 DNF (or CNF) with Approximately the Same Number of Positive Literals per Term

In this section we consider transitive functions represented by DNFs where the number of terms as well as the size of the terms (i.e. the width of the DNF) are arbitrary, but the number of positive literals in each term is approximately the same. In other words, we consider transitive functions f such that the 1-inputs of f can be covered by subcubes that correspond to minterms with approximately equal weights.

Note that minterm-transitive functions have this property, since all their minterms have exactly the same weight. However, a minterm-transitive function f must have a single minterm α such that every 1-input of f agrees with either α or $\sigma(\alpha)$ for some σ in the invariance group of f. Our condition allows f to have a set $\Lambda = \{\alpha_1, \alpha_2, \ldots\}$ of an arbitrary number of different minterms, as long as they have approximately the same weight, and every 1-input of f agrees with some $\alpha_i \in \Lambda$.

Note also that we allow the different minterms in Λ to have different sizes, we only require that they have approximately equal weight. That is, we require that they each set approximately the same number of bits to 1 but they can set different numbers of bits to 0.

Remark 3. Our arguments would also work if we require the number of bits fixed to 0 to be approximately the same in each minterm. Analogous results hold for maxterms and CNFs as well.

First we prove a simple lemma that holds for arbitrary Boolean functions, not just for transitive functions.

Lemma 7. *Let $f : \{0,1\}^n \to \{0,1\}$ be a non-constant Boolean function. Let $\Lambda = \{\alpha_1, \alpha_2, \ldots\}$ be a set of minterms of f such that every 1-input of f agrees with some $\alpha_i \in \Lambda$. Let λ_1 denote the smallest number of 1-s fixed by any $\alpha_i \in \Lambda$, and let λ_0 denote the smallest number of 0-s fixed by any $\alpha_i \in \Lambda$. Then, $s_1(f) \geq \max\{\lambda_1, \lambda_0\}$.*

Note that the number of minterms in the set Λ can be arbitrarily large. An analogous statement considering sets of maxterms covering the 0-inputs of f gives a lower bound on $s_0(f)$. We obtain the following theorem.

Theorem 3. *Let $f : \{0,1\}^n \to \{0,1\}$ be a non-constant transitive function. Assume that there is a set $\Lambda = \{\alpha_1, \alpha_2, \ldots\}$ of minterms of f such that every 1-input of f agrees with some $\alpha_i \in \Lambda$. Let w be the weight of the smallest weight minterm in Λ, and assume that for some constant c, $wt(\alpha_i) \leq c \cdot w$ for all $\alpha_i \in \Lambda$. Then $s(f) = \Omega(n^{1/3})$.*

A Stronger Tradeoff between Certificate Size and Sensitivity

Our bounds in the previous sections are based on using the tradeoff between minimum certificate size and sensitivity proved by Chakraborty (see Corollary 1). Next we observe that considering the certificate size of a transitive function on either the all 0 or all 1 string, one can obtain a stronger tradeoff between certificate size and sensitivity. More precisely, we prove the following lemma.

Lemma 8. *For a non-constant transitive function $f : \{0,1\}^n \to \{0,1\}$,*

$$C(f, 0^n) \cdot s(f) \geq n \qquad \text{and} \qquad C(f, 1^n) \cdot s(f) \geq n\,.$$

Proof. We prove the first statement, the proof of the second statement is analogous. Let B be a minimal block such that $f(0^n) \neq f((0^n)^B)$. Since B is minimal, $s(f) \geq |B|$.

Let α be a certificate of f on the all zero input 0^n. If $size(\alpha) \cdot |B| < n$, then, we apply Lemma 4 to the characteristic vectors of the set of bits that α fixes and the set B. This gives that there is a $\sigma \in \Gamma$ where Γ is the invariance group of f, such that $\sigma(\alpha)$ and B do not have any indices in common. But this gives a contradiction, since every certificate of f on 0^n must intersect B. Thus, $size(\alpha) \cdot |B| \geq n$, which implies the statement. □

Lower Bound on Sensitivity when $f(0^n) \neq f(1^n)$

We use Lemma 8 to obtain stronger lower bounds on the sensitivity of transitive functions with approximately equal weight minterms in their DNF representation, under the additional condition that $f(0^n) \neq f(1^n)$.

Theorem 4. *Let $f : \{0,1\}^n \to \{0,1\}$ be a non-constant transitive function, such that $f(0^n) \neq f(1^n)$. Assume that there is a set $\Lambda = \{\alpha_1, \alpha_2, \ldots\}$ of minterms of f such that every 1-input of f agrees with some $\alpha_i \in \Lambda$. Let w be the weight of the smallest weight minterm in Λ, and assume that for some constant c, $wt(\alpha_i) \leq c \cdot w$ for all $\alpha_i \in \Lambda$. Then $s(f) = \Omega(\sqrt{n})$.*

Proof. First we consider the case when $f(0^n) = 0$. Then, $f(1^n) = 1$, and any DNF representing f must include a term with only positive literals. For a given DNF representing f, let w denote the smallest number of positive literals in any term. Then, by the condition of the Theorem, $C(f, 1^n) \leq c \cdot w$ for some constant c, and combining Lemma 8 with Lemma 7 we get that $s(f) \geq \Omega(\sqrt{n})$.

In the case when $f(0^n) = 1$, any DNF for f must include a term with only negative literals, and then our condition implies that the DNF uses only negative literals. That is, in this case the function must be anti-monotone, which implies that $s(f,x) = bs(f,x) = C(f,x)$ for every input x. Thus, $s(f) \geq C(f, 0^n)$. Since $f(1^n) \neq f(0^n)$, f is not constant, and we can apply Lemma 8, which now directly gives $s(f) \geq \sqrt{n}$. □

4 Lower Bounds on Block Sensitivity of Transitive Functions

We obtain the following tradeoff between the minimum certificate size and the block sensitivity of transitive functions.

Lemma 9. *For any non-constant transitive function $f : \{0,1\}^n \to \{0,1\}$, and an integer $5 \leq r \leq 15$, if $C_{min}(f) \leq O(n^{3/r})$, then $bs(f) \geq \Omega(n^{1-\frac{4}{r}})$.*

Combining Lemma 9 with our arguments in the previous sections, we prove $\Omega(n^{3/7})$ lower bounds on the block sensitivity of functions on n bits in each of the classes we considered.

Theorem 5. *Let $f : \{0,1\}^n \to \{0,1\}$ be a non-constant transitive function. If f can be represented by a DNF with at most $2^{n^{3/7}}$ terms, or with a CNF with at most $2^{n^{3/7}}$ clauses, then $bs(f) \geq \Omega(n^{3/7})$.*

As before, we can extend this theorem to DNFs (or CNFs) with an arbitrary number of terms (resp. clauses) as long as there is at least one variable that is not used too many times.

Theorem 6. *Let $f : \{0,1\}^n \to \{0,1\}$ be a transitive function that can be represented by a DNF (or CNF) such that its i-th variable appears in at most $2^{n^{3/7}}$ terms (resp. clauses) of the formula, for some $i \in [n]$. Then we have: $bs(f) \geq \Omega(n^{3/7})$*

Finally, we consider the class of transitive functions represented by DNFs (or CNFs) where the number of positive literals in each term (resp. clause) is approximately equal.

Theorem 7. *Let $f : \{0,1\}^n \to \{0,1\}$ be a non-constant transitive function. Assume that there is a set $\Lambda = \{\alpha_1, \alpha_2, \ldots\}$ of minterms of f such that every 1-input of f agrees with some $\alpha_i \in \Lambda$. Let w be the weight of the smallest weight minterm in Λ, and assume that for some constant c, $wt(\alpha_i) \leq c \cdot w$ for all $\alpha_i \in \Lambda$. Then $bs(f) = \Omega(n^{3/7})$.*

Acknowledgements. We thank the anonymous referees for helpful comments.

References

1. Amano, K.: Minterm-transitive functions with asymptotically smallest block sensitivity. Inf. Process. Lett. **111**(23–24), 1081–1084 (2011). https://doi.org/10.1016/j.ipl.2011.09.008
2. Ambainis, A., Bavarian, M., Gao, Y., Mao, J., Sun, X., Zuo, S.: Tighter relations between sensitivity and other complexity measures. In: Esparza, J., Fraigniaud, P., Husfeldt, T., Koutsoupias, E. (eds.) ICALP 2014. LNCS, vol. 8572, pp. 101–113. Springer, Heidelberg (2014). https://doi.org/10.1007/978-3-662-43948-7_9
3. Ambainis, A., Sun, X.: New separation between s(f) and bs(f). Electron. Colloquium Comput. Complex. (ECCC) **18**, 116 (2011). http://eccc.hpi-web.de/report/2011/116
4. Ambainis, A., Vihrovs, J.: Size of sets with small sensitivity: a generalization of Simon's Lemma. In: Jain, R., Jain, S., Stephan, F. (eds.) TAMC 2015. LNCS, vol. 9076, pp. 122–133. Springer, Cham (2015). https://doi.org/10.1007/978-3-319-17142-5_12
5. Arunachalam, S., Chakraborty, S., Koucký, M., Saurabh, N., de Wolf, R.: Improved bounds on Fourier entropy and Min-entropy. arXiv e-prints p. 1809.09819 (2018). http://arxiv.org/abs/1809.09819
6. Bafna, M., Lokam, S.V., Tavenas, S., Velingker, A.: On the sensitivity conjecture for read-k formulas. In: Proceedings of the 41st International Symposium on Mathematical Foundations of Computer Science (MFCS), pp. 16:1–16:14 (2016). https://doi.org/10.4230/LIPIcs.MFCS.2016.16
7. Buhrman, H., De Wolf, R.: Complexity measures and decision tree complexity: a survey. Theoret. Comput. Sci. **288**(1), 21–43 (2002). https://doi.org/10.1016/S0304-3975(01)00144-X
8. Chakraborty, S.: On the sensitivity of cyclically-invariant Boolean functions. Discr. Math. Theor. Comput. Sci. **13**(4), 51–60 (2011). http://dmtcs.episciences.org/552
9. Chaubal, S., Gál, A.: New constructions with quadratic separation between sensitivity and block sensitivity. In: Proceedings of the 38th IARCS Annual Conference on Foundations of Software Technology and Theoretical Computer Science (FSTTCS), pp. 13:1–13:16 (2018). https://doi.org/10.4230/LIPIcs.FSTTCS.2018.13
10. Drucker, A.: Block Sensitivity of minterm-transitive functions. Theor. Comput. Sci. **412**(41), 5796–5801 (2011). https://doi.org/10.1016/j.tcs.2011.06.025
11. Gopalan, P., Servedio, R.A., Tal, A., Wigderson, A.: Degree and sensitivity: tails of two distributions. arXiv e-prints p. 1604.07432 (2016), http://arxiv.org/abs/1604.07432
12. Hatami, P., Kulkarni, R., Pankratov, D.: Variations on the sensitivity conjecture. Theory Comput. Graduate Surv. **4**, 1–27 (2011). https://doi.org/10.4086/toc.gs.2011.004
13. He, K., Li, Q., Sun, X.: A tighter relation between sensitivity complexity and certificate complexity. Theoret. Comput. Sci. **762**, 1–12 (2019)
14. Huang, H.: Induced subgraphs of hypercubes and a proof of the sensitivity conjecture. arXiv e-prints arXiv:1907.00847, July 2019. https://arxiv.org/abs/1907.00847
15. Kahn, J., Kalai, G., Linial, N.: The influence of variables on Boolean functions. In: Proceedings of the 29th Annual Symposium on Foundations of Computer Science (FOCS), pp. 68–80 (1988). https://doi.org/10.1109/SFCS.1988.21923

16. Nisan, N., Szegedy, M.: On the degree of Boolean functions as real polynomials. Comput. Complex. **4**(4), 301–313 (1994). https://doi.org/10.1007/BF01263419
17. O'Donnell, R.: Analysis of Boolean functions. Cambridge University Press, Cambridge (2014)
18. Rubinstein, D.: Sensitivity vs block sensitivity of Boolean functions. Combinatorica **15**(2), 297–299 (1995). https://doi.org/10.1007/BF01200762
19. Simon, H.U.: A tight Ω(loglog n)-bound on the time for parallel RAM's to compute nondegenerated Boolean functions. Inf. Control **55**(1), 102–107 (1982). https://doi.org/10.1016/S0019-9958(82)90477-6
20. Sun, X.: Block sensitivity of weakly symmetric functions. In: Proceedings of the 3rd International Conference on Theory and Applications of Models of Computation (TAMC), pp. 339–344 (2006). https://doi.org/10.1007/11750321_32
21. Virza, M.: Sensitivity versus block sensitivity of Boolean functions. Inf. Process. Lett. **111**(9), 433–435 (2011). https://doi.org/10.1016/j.ipl.2011.02.001

Sherali-Adams and the Binary Encoding of Combinatorial Principles

Stefan Dantchev, Abdul Ghani, and Barnaby Martin(✉)

Department of Computer Science, Durham University, Durham, UK
barnabymartin@gmail.com

Abstract. We consider the Sherali-Adams (SA) refutation system together with the unusual *binary* encoding of certain combinatorial principles. For the unary encoding of the Pigeonhole Principle and the Least Number Principle, it is known that linear rank is required for refutations in SA, although both admit refutations of polynomial size. We prove that the binary encoding of the Pigeonhole Principle requires exponentially-sized SA refutations, whereas the binary encoding of the Least Number Principle admits logarithmic rank, polynomially-sized SA refutations. We continue by considering a refutation system between SA and Lasserre (Sum-of-Squares). In this system, the unary encoding of the Least Number Principle requires linear rank while the unary encoding of the Pigeonhole Principle becomes constant rank.

Keywords: Proof Complexity · Lift-and-project methods · Binary encoding

1 Introduction

It is well-known that questions on the satisfiability of propositional CNF formulae may be reduced to questions on feasible solutions for certain Integer Linear Programs (ILPs). In light of this, several ILP-based proof (more accurately, refutation) systems have been suggested for propositional CNF formulae, based on proving that the relevant ILP has no solutions. Typically, this is accomplished by relaxing an ILP to a continuous Linear Program (LP), which itself may have (non-integral) solutions, and then modifying this LP iteratively until it has a solution iff the original ILP had a solution (which happens at the point the LP has no solution). Among the most popular ILP-based refutation systems are Cutting Planes [6,11] and several proposed by Lovász and Schrijver [18].

Another method for solving ILPs was proposed by Sherali and Adams [22], and was introduced as a propositional refutation system in [7]. Since then it has been considered as a refutation system in the further works [1,9]. The Sherali-Adams system (SA) is of significant interest as a static variant of the Lovász-Schrijver system without semidefinite cuts (LS). It is proved in [15] that the SA rank of a polytope is less than or equal to its LS rank; hence we may claim

Y. Kohayakawa and F. K. Miyazawa (Eds.): LATIN 2020, LNCS 12118, pp. 336–347, 2020.
https://doi.org/10.1007/978-3-030-61792-9_27

that SA is at least as strong as LS (though it is unclear whether it is strictly stronger).

Various fundamental combinatorial principles used in Proof Complexity may be given in first-order logic as sentences φ with no finite models and in this article we will restrict attention to those in Π_2-form. Riis discusses in [21] how to generate from prenex φ a family of CNFs, the nth of which encodes that φ has a model of size n, which are hence contradictions. Following Riis, it is typical to encode the existence of the witnesses to an existentially quantified variable in longhand with a big disjunction, of the form $S_{\mathbf{a},1} \vee \ldots \vee S_{\mathbf{a},n}$, that we designate the *unary encoding*. Here the arity of $\mathbf{a}$ is the number of universally quantified variables preceding the existentially quantified variable, on which it might depend.

As recently investigated in the works [3,4,8,10,13,17], it may also be possible to encode the existence of such witnesses *succinctly* by the use of a *binary encoding*. Essentially, the existence of the witness is now given implicitly as any propositional assignment to the relevant variables $S_{\mathbf{a},1}, \ldots, S_{\mathbf{a},\log n}$, which we call S for Skolem, gives a witness; whereas in the unary encoding a solitary true literal tells us which is the witness. Combinatorial principles encoded in binary are interesting to study for Resolution-type systems since they still preserve the hardness of the combinatorial principle while giving a more succinct propositional representation. In certain cases this leads to obtain significant lower bounds in an easier way than for the unary case [4,8,10,17].

The binary encoding also implicitly enforces an at-most-one constraint at the same time as it does at-least-one. When some big disjunction $S_{\mathbf{a},1} \vee \ldots \vee S_{\mathbf{a},n}$ of the unary encoding is translated to constraints for an ILP it enforces $S_{\mathbf{a},1} + \ldots + S_{\mathbf{a},n} \geq 1$. Were we to insist that $S_{\mathbf{a},1} + \ldots + S_{\mathbf{a},n} = 1$ then we encode immediately also the at-most-one constraint. We paraphrase this variant as being (the unary) *encoding with equalities* or "SA-with-equalities".

The Pigeonhole Principle (PHP), which essentially asserts that n pigeons may not be assigned to $n-1$ holes such that no hole has more than one pigeon, and the Least Number Principle (LNP), which asserts that a partially-ordered n-set possesses a minimal element, are ubiquitous in Proof Complexity. Typically (and henceforth) we work under the same name with their negations, which are expressible in (Π_2) first-order logic as formulae with no finite models.

In [9] we have proved that the SA rank of (the polytopes associated with) (the unary encoding of) each of the Pigeonhole Principle and Least Number Principles is $n-2$ (where n is the number of pigeons and elements in the poset, respectively). It is known that SA polynomially simulates Resolution (see e.g. [9]) and it follows there is a polynomially-sized refutation in SA of the Least Number Principle. That there is a polynomially-sized refutation in SA of the Pigeonhole Principle is noted in [20].

In this paper we consider the binary encodings of the Pigeonhole Principle and the Least Number Principle as ILPs. We additionally consider their (unary) encoding with equalities. We first prove that the binary encoding of the Pigeonhole Principle requires exponential size in SA. We then prove that the (unary) encoding of the Least Number Principle with equalities has SA rank 2 and polynomial size. This allows us to prove that the binary encoding of the Least Number Principle has SA rank at most $2 \log n$ and polynomial size.

The divergent behaviour of these two combinatorial principles is tantalising – while the Least Number Principle becomes easier for SA in the binary encoding (in terms of rank), the Pigeonhole Principle becomes harder (in terms of size). Such variable behaviour has been observed for the Pigeonhole Principle in Resolution, where the binary encoding makes it easier for treelike Resolution (in terms of size) [8].

We continue by considering a refutation system SA + Squares which is between SA and Lasserre (Sum-of-Squares) [14] (see also [15] for comparison between these systems). SA+ Squares appears as Static LS_+ in [12]. In this system one can always assume the non-negativity of (the linearisation of) any squared polynomial. In contrast to our system SA-with-equalities, we see that the rank of the unary encoding of the Pigeonhole Principle is 2, while the rank of the Least Number Principle is linear. We prove this by showing a certain moment matrix in positive semidefinite. Our rank results for the unary encoding can be contrasted in Table 1. Owing to space restrictions, many of our proofs are omitted.

Table 1. Rank based complexity for the unary encoding in different systems (on the left) and size based complexity for the binary encoding (on the right). The lower table shows where the corresponding result is proved.

Unary case	SA	SA-with-equalities	SA + Squares
PHP	linear	linear	constant
LNP	linear	constant	linear

Binary case	SA
PHP	exponential
LNP	polynomial

Unary case	SA	SA-with-equalities	SA + Squares
PHP	[9]	[9]	Theorem 3 [12]
LNP	[9]	Theorem 2	Theorem 4

Binary case	SA
PHP	Theorem 1
LNP	Corollary 2

2 Preliminaries

Let $[m]$ be the set $\{1, \ldots, m\}$. Let us assume, without loss of much generality, that n is a power of 2. Cases where n is not a power of 2 are handled in the binary encoding by explicitly forbidding possibilities.

If P is a propositional variable, then $P^0 = \overline{P}$ indicates the negation of P, while P^1 indicates P. A *term* is a conjunction of propositional literals.

From a CNF formula $F := C_1 \wedge \ldots \wedge C_r$ in variables $v_1, \ldots, v_m$ we generate an ILP in $2m$ variables $Z_{v_\lambda}, Z_{\neg v_\lambda}$ ($\lambda \in [m]$). For literals $l_1, \ldots, l_t$ s.t. $(l_1 \vee \ldots \vee l_t)$ is a clause of F we have the constraining inequality

$$(2.1) \quad Z_{l_1} + \ldots + Z_{l_t} \geq 1.$$

We also have, for each $\lambda \in [m]$, the equalities of negation

$$(2.2) \quad Z_{v_\lambda} + Z_{\neg v_\lambda} = 1$$

together with the bounding inequalities

$$(2.3) \quad 0 \leq Z_{v_\lambda} \leq 1 \quad \text{and} \quad 0 \leq Z_{\neg v_\lambda} \leq 1.$$

Let $\mathcal{P}_0^F$ be the polytope specified by these constraints on the real numbers. It is clear that this polytope contains integral points iff the formula F is satisfiable. In general, $\mathcal{P}_0^F$ is non-empty; in fact, if F is a contradiction that does not admit refutation by unit clause propagation, this is the case (we may use unit clause propagation to assign $0-1$ values to some variables, thereafter assigning $1/2$ to those variables remaining). Note that it follows that any unsatisfiable Horn CNF F (i.e., where each clause contains at most one positive variable) has SA rank 0, since F must then admit refutation by unit clause propagation (which may be used to demonstrate $\mathcal{P}_0^F$ empty).

Sherali-Adams (SA) provides a static refutation method that takes the polytope $\mathcal{P}_0^F$ defined by (2.1)–(2.3) and r-*lifts* it to another polytope $\mathcal{P}_r^F$ in $\sum_{\lambda=0}^{r+1} \binom{2m}{\lambda}$ dimensions. Specifically, the variables involved in defining the polytope $\mathcal{P}_r^F$ are $Z_{l_1 \wedge \ldots \wedge l_{r+1}}$ ($l_1, \ldots, l_{r+1}$ literals of F) and $Z_\emptyset$. Let us say that the term $Z_{l_1 \wedge \ldots \wedge l_{r+1}}$ has *rank* r. Note that we accept commutativity and idempotence of the $\wedge$-operator, e.g. $Z_{l_1 \wedge l_2} = Z_{l_2 \wedge l_1}$ and $Z_{l_1 \wedge l_1} = Z_{l_1}$. Also $\emptyset$ represents the empty conjunct (boolean true); hence we set $Z_\emptyset := 1$. For literals $l_1, \ldots, l_t$, s.t. $(l_1 \vee \ldots \vee l_t)$ is a clause of F, we have the constraining inequalities

$$(2.1') \quad Z_{l_1 \wedge D} + \ldots + Z_{l_t \wedge D} \geq Z_D,$$

for D any conjunction of at most r literals of F. We also have, for each $\lambda \in [m]$ and D any conjunction of at most r literals, the equalities of negation

$$(2.2') \quad Z_{v_\lambda \wedge D} + Z_{\neg v_\lambda \wedge D} = Z_D$$

together with the bounding inequalities

$$(2.3') \quad 0 \leq Z_{v_\lambda \wedge D} \leq Z_D \quad \text{and} \quad 0 \leq Z_{\neg v_\lambda \wedge D} \leq Z_D.$$

For $r' \leq r$, the defining inequalities of $\mathcal{P}_{r'}^F$ are consequent on those of $\mathcal{P}_r^F$. Equivalently, any solution to the inequalities of $\mathcal{P}_r^F$ gives rise to solutions of the inequalities of $\mathcal{P}_{r'}^F$, when projected on to its variables. If D' is a conjunction of r' literals, then $Z_{D \wedge D'} \leq Z_D$ follows by transitivity from r' instances of (2.3′). We refer to the property $Z_{D \wedge D'} \leq Z_D$ as *monotonicity*. Finally, let us note that $Z_{v \wedge \neg v} = 0$ holds in $\mathcal{P}_1^F$ and follows from a single lift of an equality of negation.

The SA *rank* of the polytope $\mathcal{P}_0^F$ (formula F) is the minimal i such that $\mathcal{P}_i^F$ is empty. Thus, the notation rank is overloaded in a consistent way, since $\mathcal{P}_i^F$ is specified by inequalities in variables of rank at most i. The largest r for which $\mathcal{P}_r^F$ need be considered is $2m-1$, since beyond that there are no new literals to lift by. Even that is somewhat further than necessary, largely because, if the conjunction D contains both a variable and its negation, it may be seen from the equalities of negation that $Z_D = 0$. In fact, it follows from [15] that the SA rank of $\mathcal{P}_0^F$ is always $\leq m-1$ (for a contradiction F).

The number of defining inequalities of the polytope $\mathcal{P}_r^F$ is exponential in r; hence a naive measure of SA size would see it grow more than exponentially in rank. However, not all of the inequalities $(2.1') - (2.3')$ may be needed to specify the empty polytope. We therefore define the SA *size* of the polytope $\mathcal{P}_0^F$ (formula F) to be the size (of an encoding) of a minimal subset of the inequalities $(2.1') - (2.3')$ of $\mathcal{P}_{2m}^F$ that specifies the empty polytope.

Let us now consider principles which are expressible as first-order formulae, with no finite models, in Π_2-form, i.e. as $\forall \boldsymbol{x} \exists \boldsymbol{w} \varphi(\boldsymbol{x}, \boldsymbol{w})$ where $\varphi(\boldsymbol{x}, \boldsymbol{w})$ is a formula built on a family of relations $\boldsymbol{R}$. For example the *Least Number Principle*, which states that a finite partial order has a minimal element is one of such principles. Its negation can be expressed in Π_2-form as:

$$\forall x, y, z \exists w \neg R(x,x) \wedge (R(x,y) \wedge R(y,z) \rightarrow R(x,z)) \wedge R(x,w).$$

This can be translated into a unsatisfiable CNF using a unary encoding of the witness, as shown below alongside the binary encoding.

LNP$_n$: Unary encoding		LNP$_n$: Binary encoding	
$\overline{P}_{i,i}$	$\forall i \in [n]$	$\overline{P}_{i,i}$	$\forall x \in [n]$
$\overline{P}_{i,j} \vee \overline{P}_{j,k} \vee P_{i,k}$	$\forall i,j,k \in [n]$	$\overline{P}_{i,j} \vee \overline{P}_{j,k} \vee P_{i,k}$	$\forall i,j,k \in [n]$
$\overline{S}_{i,j} \vee P_{i,j}$	$\forall i,j \in [n]$	$\bigvee_{i \in [\log n]} S_{i,j}^{1-a_i} \vee P_{j,a}$	$\forall j, a \in [n]$
$\bigvee_{i \in [n]} S_{i,j}$	$\forall j \in [n]$	where $a_1 \ldots a_{\log n} = \text{bin}(a)$	

Note that we placed the witness in the Skolem variables $S_{i,x}$ as the first argument and not the second, as we had in the introduction. This is to be consistent with the $P_{i,j}$ and the standard formulation of LNP as the least, and not greatest, number principle. A more traditional form of the (unary encoding of the) LNP$_n$ has clauses $\bigvee_{i \in [n]} P_{i,j}$ which are consequent on $\bigvee_{i \in [n]} S_{i,j}$ and $\overline{S}_{i,j} \vee P_{i,j}$ (for all $i \in [n]$).

Indeed, one can see how to generate a binary encoding of C from any combinatorial principle C expressible as a first order formula in Π_2-form with no finite models. Exact details can be found in Definition 4 in [8].

As a second example we consider the *Pigeonhole Principle* which states that a total mapping from $[m]$ to $[n]$ has necessarily a collision when m and n are integers with $m > n$. The negation of its relational form for $m = n+1$ can be expressed as a Π_2-formula as

$$\forall x, y, z \exists w \neg R(x,0) \wedge (R(x,z) \wedge R(y,z) \rightarrow x = y) \wedge R(x,w)$$

where 0 represents the object that is among the $[n+1]$ but not among the $[n]$. Its usual unary and binary propositional encoding are:

PHP^m_n : Unary encoding

$$\bigvee_{j=1}^{n} P_{i,j} \qquad \forall i \in [m]$$
$$\overline{P}_{i,j} \vee \overline{P}_{i',j} \qquad \forall i \neq i' \in [m], j \in [n]$$

PHP^m_n : Binary encoding

$$\bigvee_{j=1}^{\log n} P_{i,j}^{1-a_j} \vee \bigvee_{j=1}^{\log n} P_{i',j}^{1-a_j}$$
$$\forall a \in [n], i \neq i' \in [m]$$
$$\text{where } a_1 \ldots a_{\log n} = \text{bin}(a)$$

where 0 no longer appears now m and n are explicit. Properly, the Pigeonhole Principle should also admit S variables (as with the LNP) but one notices that the existential witness w to the type *pigeon* is of the distinct type *hole*. Furthermore, pigeons only appear on the left-hand side of atoms $R(x,z)$ and holes only appear on the right-hand side. For the Least Number Principle instead, the transitivity axioms effectively enforce the type of y appears on both the left- and right-hand side of atoms $R(x,z)$. This accounts for why, in the case of the Pigeonhole Principle, we did not need to introduce any new variables to give the binary encoding, yet for the Least Number Principle a new variable S appears. However, our results would hold equally were we to have chosen the more complicated form of the Pigeonhole Principle. Note that our formulation of the Least Number Principle is symmetric in the elements and our formulation of the Pigeonhole Principle is symmetric in each of the pigeons and holes.

When we consider the Sherali-Adams r-lifts of, e.g., the Least Number Principle, we will identify terms of the form $Z_{P_{i,j} \wedge \overline{S}_{i',j'} \wedge \ldots}$ as $P_{i,j}\overline{S}_{i',j'} \ldots$. Thus, we take the subscript and use overline for negation and concatenation for conjunction. This prefigures the multilinear notation we will revert to in Sect. 5, but one should view for now $P_{i,j}\overline{S}_{i',j'} \ldots$ as a single variable and not a multilinear monomial.

Finally, we wish to discuss the encoding of the Least Number Principle and Pigeonhole Principle as ILPs *with equality*. For this, we take the unary encoding but instead of translating the wide clauses (e.g. from the LNP) from $\bigvee_{i \in [n]} S_{i,x}$ to $S_{1,x} + \ldots + S_{n,x} \geq 1$, we instead use $S_{1,x} + \ldots + S_{n,x} = 1$. This makes the constraint at-least-one into exactly-one (which is a priori enforced in the binary encoding). A reader who doesn't wish to consult the long version of this paper should consider the Least Number Principle as the combinatorial principle of the following lemma.

Lemma 1. *Let* C *be any combinatorial principle expressible as a first order formula in* Π_2*-form with no finite models. Suppose the unary encoding of* C *with equalities has an* SA *refutation of rank* r *and size* s. *Then the binary encoding of* C *has an* SA *refutation of rank at most* $r \log n$ *and size at most* s.

Proof. We take the SA refutation of the unary encoding of C with equalities of rank r, in the form of a set of inequalities, and build an SA refutation of the binary encoding of C of rank $r \log n$, by substituting terms $S_{x,a}$ in the former with $S^{a_1}_{x,1} \ldots S^{a_{\log n}}_{x,\log n}$, where $a_1 \ldots a_{\log n} = \text{bin}(a)$, in the latter. Note that the equalities of the form

$$\sum_{a_1 \ldots a_{\log n} = \text{bin}(a)} S^{a_1}_{x,1} \cdots S^{a_{\log n}}_{x,\log n} = 1$$

follow from the inequalities (2.2') and (2.3'). Further, inequalities of the form $S^{a_1}_{x,1} \ldots S^{a_{\log n}}_{x,\log n} \leq P_{x,a}$ follow since $S_{x,j}\overline{S}_{x,j} = 0$ for each $j \in [\log n]$.

3 The Lower Bound for the Binary Pigeonhole Principle

In this section we study the inequalities derived from the binary encoding of the Pigeonhole principle. We first prove a certain SA rank lower bound for a version of the binary PHP, in which only a subset of the holes is available.

Lemma 2. *Let $H \subseteq [n]$ be a subset of the holes and let us consider binary* $\text{PHP}^m_{|H|}$ *where each pigeon can go to a hole in H only. Any* SA *refutation of binary* $\text{PHP}^m_{|H|}$ *involves a term that mentions at least $|H|$ pigeons.*

The proof of the size lower bound for the binary PHP^{n+1}_n then is by a standard random-restriction argument combined with the rank lower bound above. Assume w.l.o.g that n is a perfect power of two. For the random restrictions $\mathcal{R}$, we consider the pigeons one by one and with probability $1/4$ we assign the pigeon uniformly at random to one of the holes still available. We first need to show that the restriction is "good" w.h.p., i.e. neither too big nor too small. The former is needed so that in the restricted version we have a good lower bound, while the latter will be needed to show that a good restriction coincides well any reasonably big term, in the sense that they have in common a sufficiency of pigeons. A simple application of a Chernoff bound gives the following

Fact 1. *If $|\mathcal{R}|$ is the number of pigeons (or holes) assigned by $\mathcal{R}$,*

1. *the probability that $|\mathcal{R}| < \frac{n}{8}$ is at most $e^{-n/32}$, and*
2. *the probability that $|\mathcal{R}| > \frac{3n}{8}$ is at most $e^{-n/48}$.*

So, from now on, we assume that $\frac{n}{8} \leq |\mathcal{R}| \leq \frac{3n}{8}$. We first prove that a given wide term, i.e. a term that mentions a constant fraction of the pigeons, survives the random restrictions with exponentially small probability.

Lemma 3. *Let T be a term that mentions at least $\frac{n}{2}$ pigeons. The probability that T does not evaluate to zero under the random restrictions is at most $\left(\frac{5}{6}\right)^{n/16}$.*

Proof. An application of a Chernoff bound gives the probability that fewer than $\frac{n}{16}$ pigeons mentioned by T are assigned by $\mathcal{R}$ is at most $e^{-n/64}$. For each of these pigeons the probability that a single bit-variable in T belonging to the pigeon is set by $\mathcal{R}$ to zero is at least $\frac{1}{5}$. This is because when $\mathcal{R}$ sets the pigeon, and thus the bit-variable, there were at least $\frac{5n}{8}$ holes available, while at most $\frac{n}{2}$ choices set the bit-variable to one. The difference is $\frac{n}{8}$ which divided by $\frac{5n}{8}$ gives $\frac{1}{5}$. Thus T survives under $\mathcal{R}$ with probability at most $e^{-n/64} + \left(\frac{4}{5}\right)^{n/16} < \left(\frac{5}{6}\right)^{n/16}$.

Finally, we can prove that

Theorem 1. *Any* SA *refutation of the binary* PHP_n^{n+1} *has to contain at least* $\left(\frac{6}{5}\right)^{n/16} - 1$ *terms.*

We now consider the so-called weak binary PHP, PHP_n^m, where m is potentially much larger than n. The weak unary PHP_n^m is interesting because it admits (significantly) subexponential-in-n refutations in Resolution when m is sufficiently large [5]. It follows that this size upper bound is mirrored in SA. However, as proved in [8], the weak binary PHP_n^m remains almost-exponential-in-n for minimal refutations in Resolution. We will see here that the weak binary PHP_n^m remains almost-exponential-in-n for minimally sized refutations in SA. In this weak binary case, the random restrictions $\mathcal{R}$ above do not work, so we apply quite different restrictions $\mathcal{R}'$ that are as follows: for each pigeon select independently a single bit uniformly at random and set it to 0 or 1 with probability of $1/2$ each.

We can easily prove the following

Lemma 4. *A term T that mentions n' pigeons does not evaluate to zero under $\mathcal{R}'$ with probability at most $e^{-n'/2\log n}$.*

Proof. For each pigeon mentioned in T, the probability that the bit-variable present in T is set by the random restriction is $\frac{1}{\log n}$, and if so, the probability that the bit-variable evaluates to zero is $\frac{1}{2}$. Since this happens independently for all n' mentioned pigeons, the probability that they all survive is at most $\left(1 - \frac{1}{2\log n}\right)^{n'}$.

Now, we only need to prove that in the restricted version of the pigeon-hole principle, there is always a big enough term.

Lemma 5. *The probability that an* SA *refutation of the binary* PHP_n^m, *for $m > n$, after $\mathcal{R}'$ does not contain a term mentioning $\frac{n}{2\log n}$ pigeons is at most $e^{-n/32\log^2 n}$.*

We now proceed as in the proof of Theorem 1 to deduce that any SA refutation of the binary PHP_n^m must have size exponential in n.

Corollary 1. *Any* SA *refutation of the binary* PHP_n^m, *$m > n$, has to contain at least $e^{n/32\log^2 n}$ terms.*

Proof. Assume for a contradiction, that there is a refutation with fewer terms of rank at most $\frac{n}{2\log n}$. By Lemma 4 and a union-bound, there is a specific restriction that evaluates all these terms to zero. However, this contradicts Lemma 5 .

4 The Least Number Principle with Equality

Recall that the unary *Least Number Principle* (LNP_n) *with equality* has the following set of SA axioms:

$$\begin{aligned} \mathit{self} : P_{i,i} = 0 \quad \forall\, i \in n \\ \mathit{trans} : P_{i,k} - P_{i,j} - P_{j,k} + 1 \geq 0 \quad \forall\, i, j, k \in [n] \\ \mathit{impl} : P_{i,j} - S_{i,j} \geq 0 \quad \forall\, i, j \in [n] \\ \mathit{lower} : \textstyle\sum_{i \in [n]} S_{i,j} - 1 = 0 \quad \forall\, j \in [n] \end{aligned}$$

Strictly speaking Sherali-Adams is defined for inequalities only. An equality axiom $a = 0$ is simulated by the two inequalities $a \geq 0, -a \geq 0$, which we refer to as the *positive* and *negative* instances of that axiom, respectively. Also, note that we have used $P_{i,j} + \overline{P}_{i,j} = 1$ to derive this formulation. We call two terms *isomorphic* if one term can be gotten from the other by relabelling the indices appearing in the subscripts by a permutation.

Theorem 2. *For n large enough, the* SA *rank of the* LNP_n *with equality is at most* 2 *and* SA *size at most polynomial in n.*

Corollary 2. *The binary encoding of* LNP_n *has* SA *rank at most* $2 \log n$ *and* SA *size at most polynomial in n.*

Proof. Immediate from Lemma 1.

5 SA + Squares

In this section we consider a proof system, SA + Squares, based on inequalities of multilinear polynomials. We now consider axioms as degree-1 polynomials in some set of variables and refutations as polynomials in those same variables. Then this system is gotten from SA by allowing addition of (linearised) squares of polynomials. In terms of strength this system will be strictly stronger than SA and at most as strong as Lasserre (also known as Sum-of-Squares), although we do not at this point see an exponential separation between SA + Squares and Lasserre. See [2,14,15] for more on the Lasserre proof system and [16] for tight degree lower bound results.

Consider the polynomial $S_{i,j}P_{i,j} - S_{i,j}P_{i,k}$. The square of this is

$$S_{i,j}P_{i,j}S_{i,j}P_{i,j} + S_{i,j}P_{i,k}S_{i,j}P_{i,k} - 2S_{i,j}P_{i,j}S_{i,j}P_{i,k}.$$

Using idempotence this linearises to $S_{i,j}P_{i,j} + S_{i,j}P_{i,k} - 2S_{i,j}P_{i,j}P_{i,k}$. Thus we know that this last polynomial is non-negative for all 0/1 settings of the variables. A *degree-d* SA + Squares refutation of a set of linear inequalities (over terms) $q_1 \geq 0, \ldots, q_x \geq 0$ is an equation of the form

$$-1 = \sum_{i=1}^{x} p_i q_i + \sum_{i=1}^{y} r_i^2 \tag{1}$$

where the p_i are polynomials with nonnegative coefficients and the degree of the polynomials $p_i q_i, r_i^2$ is at most d. We want to underline that we now consider a term like $S_{i,j}P_{i,j}P_{i,k}$ as a product of its constituent variables. This is opposed to the preceding sections in which we viewed it as a single variable $Z_{S_{i,j}P_{i,j}P_{i,k}}$. The translation from the degree discussed here to SA rank previously introduced may be paraphrased by "rank = degree – 1".

We show that the unary PHP becomes easy in this stronger proof system while the LNP remains hard. The following appears as Example 2.1 in [12].

Theorem 3 ([12]). *The unary* PHP_n^{n+1} *has an* SA + Squares *refutation of degree 2.*

We give our lower bound for the unary LNP_n by producing a linear function v (which we will call a *valuation*) from terms into $\mathbb{R}$ such that

1. for each axiom $p \geq 0$ and every term X with $deg(Xp) \leq d$ we have $v(Xp) \geq 0$, and
2. we have $v(r^2) \geq 0$ whenever $deg(r^2) \leq d$.
3. $v(1) = 1$.

The existence of such a valuation clearly implies that a degree-d SA + Squares refutation cannot exist, as it would result in a contradiction when applied to both sides of Eq. (1).

To verify that $v(r^2) \geq 0$ whenever $deg(r^2) \leq d$ we show that the so-called *moment-matrix* $\mathcal{M}_v$ is positive semidefinite. The degree-d moment matrix is defined to be the symmetric square matrix whose rows and columns are indexed by terms of size at most $d/2$ and each entry is the valuation of the product of the two terms indexing that entry. Given any polynomial σ of degree at most $d/2$ let c be its coefficient vector. Then if $\mathcal{M}_v$ is positive semidefinite:

$$v(\sigma^2) = \sum_{deg(T_1), deg(T_2) \leq d/2} c(T_1)c(T_2)v(T_1T_2) = c^\top \mathcal{M}_v c \geq 0.$$

(For more on this see e.g. [14], Sect. 2.)

Recall that the unary *Least Number Principle* (LNP_n) has the set of SA axioms *self, trans, impl* but where the last axiom *lower* now has the form $\sum_{i \in [n]} S_{i,j} - 1 \geq 0$, for all $j \in [n]$.

Theorem 4. *There is no* SA + Squares *refutation of the unary* LNP_n *with degree at most* $(n-3)/2$.

An alternative formulation of the Least Number Principle asks that the order be total, and this is enforced with axioms *anti-sym* of the form $P_{i,j} \vee P_{j,i}$, or $P_{i,j} + P_{j,i} \geq 1$, for $i \neq j \in [n]$. Let us call this alternative formulation TLNP. Ideally, lower bounds should be proved for TLNP, because they are potentially stronger. Conversely, upper bounds are stronger when they are proved on the ordinary LNP, without the total order. Looking into the last proof, one sees that the lifts of *anti-sym* are satisfied as we derive our valuation exclusively from total orders. This is interesting because an upper bound in Lasserre of order $\sqrt{n}$ is known for TLNP_n [19]. It is proved for a slightly different formulation of TLNP_n from ours, but we believe it is straightforward to translate it to our formulation. Thus, Theorem 4, together with [19], shows a quadratic rank separation between SA + Squares and Lasserre.

6 Conclusion

Our result that the unary encoding of the Least Number Principle with equalities has SA rank 2 contrasts strongly with the fact that the unary encoding of the Least Number Principle has SA rank $n-2$ [9]. Now we know the unary encoding of the Pigeonhole Principle has SA rank $n-2$ also. This leaves one wondering about the unary encoding of the Pigeonhole Principle with equalities, which does appear in Table 1. In fact, the valuation of [9] witnesses this still has SA rank $n-2$ (and we give the argument in the long version of this paper). That is, the Pigeonhole Principle does not drop complexity in the presence of equalities, whereas the Least Number Principle does.

Acknowlegements. We thank Nicola Galesi for collaboration on the binary encoding in Resolution-type systems. In particular, some of our definitions come from joint work with him in [8]. We thank several anonymous reviewers for careful reading of our paper and their insights.

References

1. Atserias, A., Lauria, M., Nordström, J.: Narrow proofs may be maximally long. ACM Trans. Comput. Log. **17**(3), 19:1–19:30 (2016). https://doi.org/10.1145/2898435
2. Barak, B., Steurer, D.: Sum-of-squares proofs and the quest toward optimal algorithms. In: Proceedings of the International Congress of Mathematicians (ICM), vol. IV, pp. 509–533 (2014)
3. Bonacina, I., Galesi, N.: A framework for space complexity in algebraic proof systems. J. ACM **62**(3), 23:1–23:20 (2015). https://doi.org/10.1145/2699438
4. Bonacina, I., Galesi, N., Thapen, N.: Total space in resolution. SIAM J. Comput. **45**(5), 1894–1909 (2016). https://doi.org/10.1137/15M1023269
5. Buss, S., Pitassi, T.: Resolution and the weak pigeonhole principle. In: Nielsen, M., Thomas, W. (eds.) CSL 1997. LNCS, vol. 1414, pp. 149–156. Springer, Heidelberg (1998). https://doi.org/10.1007/BFb0028012
6. Chvátal, V.: Edmonds polytopes and a hierarchy of combinatorial problems. Discrete Math. **4**, 305–337 (1973)
7. Dantchev, S.S.: Rank complexity gap for Lovász-Schrijver and Sherali-Adams proof systems. In: Proceedings of the 39th Annual ACM Symposium on Theory of Computing, New York, NY, USA, pp. 311–317. ACM Press (2007). https://doi.org/10.1145/1250790.1250837
8. Dantchev, S.S., Galesi, N., Martin, B.: Resolution and the binary encoding of combinatorial principles. In: Proceedings of the 34th Computational Complexity Conference, CCC 2019, New Brunswick, NJ, USA, 18–20 July 2019, pp. 6:1–6:25 (2019). https://doi.org/10.4230/LIPIcs.CCC.2019.6

9. Dantchev, S.S., Martin, B., Rhodes, M.N.C.: Tight rank lower bounds for the Sherali-Adams proof system. Theor. Comput. Sci. **410**(21-23), 2054–2063 (2009). https://doi.org/10.1016/j.tcs.2009.01.002
10. Filmus, Y., Lauria, M., Nordström, J., Ron-Zewi, N., Thapen, N.: Space complexity in polynomial calculus. SIAM J. Comput. **44**(4), 1119–1153 (2015). https://doi.org/10.1137/120895950
11. Gomory, R.E.: Solving linear programming problems in integers. In: Bellman, R., Hall, M. (eds.) Combinatorial Analysis, Proceedings of Symposia in Applied Mathematics. Providence, RI, vol. 10 (1960)
12. Grigoriev, D., Hirsch, E.A., Pasechnik, D.V.: Complexity of semi-algebraic proofs. In: Alt, H., Ferreira, A. (eds.) STACS 2002. LNCS, vol. 2285, pp. 419–430. Springer, Heidelberg (2002). https://doi.org/10.1007/3-540-45841-7_34
13. Hrubes, P., Pudlák, P.: Random formulas, monotone circuits, and interpolation. In: Proceedings of the 58th IEEE Annual Symposium on Foundations of Computer Science, FOCS 2017, Berkeley, CA, USA, 15–17 October 2017, pp. 121–131. IEEE Computer Society (2017). https://doi.org/10.1109/FOCS.2017.20
14. Lasserre, J.B.: An explicit exact SDP relaxation for nonlinear 0-1 programs. In: Aardal, K., Gerards, B. (eds.) IPCO 2001. LNCS, vol. 2081, pp. 293–303. Springer, Heidelberg (2001). https://doi.org/10.1007/3-540-45535-3_23
15. Laurent, M.: A comparison of the Sherali-Adams, Lovász-Schrijver and Lasserre relaxations for $0-1$ programming. Technical report PNA-R0108, Amsterdam (2001)
16. Lauria, M., Nordström, J.: Tight size-degree bounds for sums-of-squares proofs. Comput. Complex. **26**(4), 911–948 (2017). https://doi.org/10.1007/s00037-017-0152-4
17. Lauria, M., Pudlák, P., Rödl, V., Thapen, N.: The complexity of proving that a graph is Ramsey. Combinatorica **37**(2), 253–268 (2017). https://doi.org/10.1007/s00493-015-3193-9
18. Lovász, L., Schrijver, A.: Cones of matrices and set-functions and 0-1 optimization. SIAM J. Optim. **1**(2), 166–190 (1991)
19. Potechin, A.: Sum of squares bounds for the total ordering principle. arXiv e-prints, p. 1812.01163 (2018)
20. Rhodes, M.: Rank lower bounds for the Sherali-Adams operator. In: Cooper, S.B., Löwe, B., Sorbi, A. (eds.) CiE 2007. LNCS, vol. 4497, pp. 648–659. Springer, Heidelberg (2007). https://doi.org/10.1007/978-3-540-73001-9_67
21. Riis, S.: A complexity gap for tree resolution. Comput. Complex. **10**(3), 179–209 (2001). https://doi.org/10.1007/s00037-001-8194-y
22. Sherali, H.D., Adams, W.P.: A hierarchy of relaxations between the continuous and convex hull representations for zero-one programming problems. SIAM J. Discrete Math. **3**(3), 411–430 (1990)

Hardness of Variants of the Graph Coloring Game

Thiago Marcilon[1], Nicolas Martins[2], and Rudini Sampaio[3(✉)]

[1] Centro de Ciências e Tecnologia, University Federal do Cariri, Juazeiro do Norte, Brazil
thiago.marcilon@ufca.edu.br

[2] University Integração Internacional Lusofonia Afrobrasileira Unilab, Redenção, Brazil
nicolasam@unilab.edu.br

[3] Departamento de Computação, Universidade Federal do Ceará, Fortaleza, Brazil
rudini@dc.ufc.br

Abstract. Very recently, a long-standing open question proposed by Bodlaender in 1991 was answered: the graph coloring game is PSPACE-complete. In 2019, Andres and Lock proposed five variants of the graph coloring game and left open the question of PSPACE-hardness related to them. In this paper, we prove that these variants are PSPACE-complete for the graph coloring game and also for the greedy coloring game, even if the number of colors is the chromatic number. Finally, we also prove that a connected version of the graph coloring game, proposed by Charpentier et al. in 2019, is PSPACE-complete.

Keywords: Coloring game · Game chromatic number · PSPACE-hardness

1 Introduction

In the graph coloring game, given a graph G and a set C of integers (representing the color set), two players (Alice and Bob) alternate turns (starting with Alice) in choosing an uncolored vertex to be colored by an integer of C not already assigned to one of its colored neighbors. In the greedy coloring game, there is one additional constraint: the vertices must be colored by the least possible integer of C. Alice wins if all vertices are successfully colored. Otherwise, Bob wins the game. From the classical Zermelo-von Neumann theorem, one of the two players has a winning strategy, since it is a finite game without draw. Thus, the game chromatic number $\chi_g(G)$ and the game Grundy number $\Gamma_g(G)$ are defined as the least numbers of colors in the set C for which Alice has a winning strategy in the graph coloring game and the greedy coloring game, respectively.

R. Sampaio—Supported by CAPES [88887.143992/2017-00] DAAD Probral and [88881.197438/2018-01] STIC AmSud, CNPq Universal [401519/2016-3], [425297/2016-0] and [437841/2018-9], and FUNCAP [4543945/2016] Pronem.

Y. Kohayakawa and F. K. Miyazawa (Eds.): LATIN 2020, LNCS 12118, pp. 348–359, 2020.
https://doi.org/10.1007/978-3-030-61792-9_28

Clearly, $\chi(G) \leq \chi_g(G)$ and $\chi(G) \leq \Gamma_g(G) \leq \Gamma(G)$, where $\chi(G)$ is the chromatic number of G and $\Gamma(G)$ is the Grundy number of G (the maximum number of colors that can be used by a greedy coloring of G).

The graph coloring game was first considered by Brams about 38 years ago in the context of coloring maps and was described by Gardner in 1981 in his "*Mathematical Games*" column of Scientific American [8]. It remained unnoticed until Bodlaender [2] reinvented it in 1991.

Since then, the graph coloring game became a very active topic of research. In 1993, Faigle et al. [7] proved that $\chi_g(G) \leq 4$ in forests and, in 2007, Sidorowicz [16] proved that $\chi_g(G) \leq 5$ in cacti. In 1994, Kierstead and Trotter [11] proved that $\chi_g(G) \leq 7$ in outerplanar graphs. In 1999, Dinski and Zhu proved that $\chi_g(G) \leq k(k+1)$ for every graph with acyclic chromatic number k [6]. In 2000, Zhu proved that $\chi_g(G) \leq 3k+2$ in partial k-trees [18]. For planar graphs, Zhu [19] proved in 2008 that $\chi_g(G) \leq 17$, Sekiguchi [15] proved in 2014 that $\chi_g(G) \leq 13$ if the girth is at least 4 and Nakprasit et al. [13] proved in 2018 that $\chi_g(G) \leq 5$ if the girth is at least 7. In 2008, Bohman, Frieze and Sudakov [3] investigated the asymptotic behavior of $\chi_g(G_{n,p})$ for the random graph $G_{n,p}$.

In Bodlaender's 1991 paper, the complexity was left as "*an interesting open problem*". A point of difficulty to set the complexity is the definition of the decision problem. As pointed out by Zhu [17], the graph coloring game "*exhibits some strange properties*" and the following naive question is still open (Question 1 of [17]): Does Alice have a winning strategy for the coloring game with $k+1$ colors if she has a winning strategy with k colors? Thus it is possible to define two decision problems for the graph coloring game: given a graph G and an integer k: $\chi_g(G) \leq k$? Does Alice have a winning strategy with k colors? Both problems are equivalent if and only if Question 1 of [17] has an affirmative answer.

Nevertheless, it was proved in 2019 that both coloring game decision problems are PSPACE-complete [5], solving Bodlaender's 1991 question. Also in 2019, Andres and Lock [1] proposed five variants of the graph coloring game: g_B (Bob starts the game), $g_{A,A}$ (Alice starts and can pass turns), $g_{A,B}$ (Alice starts and Bob can pass turns), $g_{B,A}$ (Bob starts and Alice can pass turns) and $g_{B,B}$ (Bob starts and can pass turns). They left the following problem: "*the question of PSPACE-hardness remains open for all the game variants mentioned above*".

In 2019, Charpentier, Hocquard, Sopena and Zhu [4] proposed a connected version of the graph coloring game (starting with Alice): the subgraph induced by the set of colored vertices must be connected. They prove that Alice wins with 2 colors in bipartite graphs and with 5 colors in outerplanar graphs.

In 2013, Havet and Zhu [9] proposed the greedy coloring game and the game Grundy number $\Gamma_g(G)$. They proved that $\Gamma_g(G) \leq 3$ in forests and $\Gamma_g(G) \leq 7$ in partial 2-trees. They also posed two questions. Problem 5 of [9]: $\chi_g(G)$ can be bounded by a function of $\Gamma_g(G)$? Problem 6 of [9]: Is it true that $\Gamma_g(G) \leq \chi_g(G)$ for every graph G? In 2015, Krawczyk and Walczak [12] answered Problem 5 of [9] in the negative: $\chi_g(G)$ is not upper bounded by a function of $\Gamma_g(G)$. To the best of our knowledge, Problem 6 of [9] is still open. In 2019, it was proved that the greedy coloring game is PSPACE-complete [5]. It was also proved that the game Grundy number is equal to the chromatic number in split graphs and extended P_4-laden graphs, even if Bob starts and can pass any turn.

In this paper, we prove that all variants of the graph coloring game and the greedy coloring game are PSPACE-complete even if the number of colors is the chromatic number for any pair $Y \in \{Alice, Bob\}$ and $Z \in \{Alice, Bob, No\ one\}$, where Y starts the game and Z can pass turns, by reductions from POS-CNF-11 and POS-DNF-11. Finally, we also prove that the connected version of the graph coloring game is PSPACE-complete, by a reduction from the variant of the graph coloring game in which Bob starts the game.

2 PSPACE-Complete Variants of Graph Coloring Game

Firstly, let us consider the game variant g_B: Bob starts the game. Let $\chi_g^B(G)$ be the minimum number of colors in the set C for which Alice has a winning strategy in g_B. Clearly, $\chi_g^B(G) \geq \chi(G)$. With this, we can define two decision problems for g_B: given a graph G and an integer k,

- (Problem g_B-1) $\chi_g^B(G) \leq k$?
- (Problem g_B-2) Does Alice have a winning strategy in g_B with k colors?

In this section, we prove that the following more restricted problem is PSPACE-complete: given a graph G and its chromatic number $\chi(G)$,

- (Problem g_B-3) $\chi_g^B(G) = \chi(G)$?

Notice that $\chi(G)$ is part of the input of Problem g_B-3.

It is easy to see that Problems g_B-1 and g_B-2 are generalizations of Problem g_B-3, since both problems are equivalent to it for $k = \chi(G)$. For this, notice that $\chi_g^B(G) \leq k = \chi(G)$ if and only if $\chi_g^B(G) = \chi(G)$, which is true if and only if Alice has a winning strategy in g_B with $k = \chi(G)$ colors. Then the PSPACE-hardness of Problem g_B-3 implies the PSPACE-hardness of Problems g_B-1 and g_B-2. To the best of our knowledge, no paper have explicitly defined these decision problems or proved pertinence in PSPACE.

Lemma 1. *Problems g_B-1, g_B-2 and g_B-3 are in PSPACE.*

Proof (Sketch). Let G be a graph with n vertices and $k \leq n$ be an integer. Let us begin with Problem g_B-2. Since the number of turns is exactly n and, in each turn, the number of possible moves is at most $n \cdot k$ (there are at most n vertices to select and at most k colors to use), we have that Problem g_B-2 is a polynomially bounded two player game and then it is in PSPACE [10]. Consequently, Problem g_B-3 is also in PSPACE.

Finally, regarding Problem g_B-1, notice that it can be decided using Problem g_B-2 for all $k' = \chi(G), \ldots, k$. That is, if there is $k' \in \{\chi(G), \ldots, k\}$ such that Problem g_B-2 with k' colors is YES, then Problem g_B-1 is also YES. Otherwise, it is NO. Since Problem g_B-2 is in PSPACE, then Problem g_B-1 is also in PSPACE.

Now, we prove that Problem g_B-3 is PSPACE-complete. In [5], the PSPACE-hardness reduction of the graph coloring game used the POS-CNF problem, which is known to be log-complete in PSPACE [14]. In POS-CNF, we are given a set $\{X_1, \ldots, X_N\}$ of N variables and a CNF formula (conjunctive normal form: conjunction of disjunctions) with M clauses $C_1, \ldots, C_M$ (also called disjunctions), in which only positive variables appear (that is, no negations of variables). Players I and II alternate turns setting a previously unset variable True or False, starting with Player I. After all N variables are set, Player I wins if and only if the formula is True. Clearly, since there are only positive variables, we can assume that Players I and II always set variables True and False, respectively.

Unfortunately, by associating Player I with Alice and Player II with Bob, all our attempts to obtain a reduction for g_B similar to the one in [5] using POS-CNF have failed. However, another problem proved to be useful for g_B: POS-DNF, which is also PSPACE-complete [14]. In POS-DNF, we are given a DNF formula (disjunctive normal form: disjunction of conjunctions) instead of a CNF formula. In other words, Player I in POS-DNF has a similar role of Player II in POS-CNF: he wins if plays every variable of some conjunction. From now on, we will call Players I and II of POS-DNF as Bob and Alice, respectively. As an example, consider the DNF formula $(X_1 \wedge X_2 \wedge X_5) \vee (X_1 \wedge X_3 \wedge X_5) \vee (X_2 \wedge X_4 \wedge X_5) \vee (X_3 \wedge X_4 \wedge X_5)$. Note that Bob has a winning strategy for this formula firstly setting X_5 True, since it is in all conjunctions: if Alice sets X_1 False, Bob sets X_4 True; if Alice sets X_4 False, Bob sets X_1 True; if Alice sets X_2 False, Bob sets X_3 True; if Alice sets X_3 False, Bob sets X_2 True.

Lemma 2. *If Bob (resp. Alice) has a winning strategy in POS-CNF or POS-DNF, then he (resp. she) also has a winning strategy if Alice (resp. Bob) can pass any turn.*

Proof (Sketch). In both cases, if the opponent passed a turn, just assume that the opponent has selected some non-selected variable and keep playing following the winning strategy in the original game. If the opponent selects this assumed variable later in the game, just assume that other non-selected variable was selected and keep playing with the winning strategy. If all variables (including the assumed ones) were selected, then the formula is true and just select any assumed variable (in any order). With this, since the player have followed a winning strategy in POS-CNF or POS-DNF and the assumptions restricted only the player (and not the opponent), we are done.

If the disjunctions/conjunctions have at most 11 variables, we are in POS-CNF-11 and POS-DNF-11 problems, which are also PSPACE-complete [14].

One important ingredient of the reduction is the graph F_1 of Fig. 1, which has a clique K, an independent set Q of $|K| + 3$ vertices and three vertices s, w and y such that s and w are neighbors and are adjacent to all vertices in K and y is adjacent to all vertices in $K \cup Q$. We start proving that, in case of Bob firstly coloring s, Alice must color y with the same color of s in her first move.

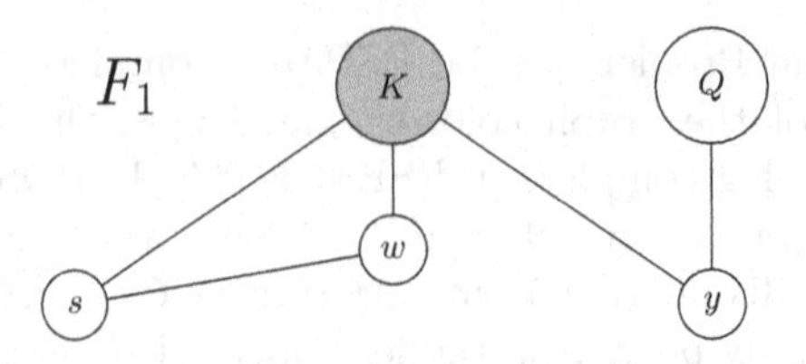

Fig. 1. Graph F_1: clique K with k vertices and independent set Q with $k+3$ vertices.

Lemma 3. *Consider the graph F_1 of Fig. 1 with $|K| = k$ and $|Q| = k+3$ and assume that Bob colored vertex s in the first move of g_B. Alice wins the game in F_1 with $k+2$ colors if and only if she colors vertex y with the same color of s in her first move.*

Proof (Sketch). Without loss of generality, assume s was colored with color 1. Note that F_1 can be colored with $k+2$ colors if and only if either y and s or y and w receive the same color. Then, she wins with $k+2$ colors if colors y with 1 in her first move. Thus assume Alice does not color y with 1 in her first move.

During the game, we say that a vertex v sees a color c if v has a neighbor colored c. Bob can win by coloring s, w and y with distinct colors in the following way. He firstly colors a vertex of Q with color 1, avoiding Alice to color y with color 1. Now, Bob has to guarantee that y and w receive different colors. For this, the following strategy holds: (i) if w is not colored and some color $c \neq 1$ appears in Q and does not appear in K, then he colors w with color c; (ii) If w is colored, y is not colored and does not see some color c distinct from the color of w, then he colors y with c; (iii) otherwise, he colors any vertex of F_1 preferring vertices of Q with any color not appearing in the neighborhood of y.

This strategy guarantees that every color seen by w is also seen by y. Moreover, after a Bob's move from (iii), he guarantees that some color c seen by y is not seen by w. Thus Alice must color a vertex of K with c, since otherwise Bob wins in his turn from (i) or (ii). Since $|Q| = |K| + 3$, Alice cannot do this indefinitely and Bob wins the game.

Theorem 1. *Given a graph G, deciding whether $\chi_g^B(G) = \chi(G)$ is PSPACE-complete. Thus, given k, deciding whether $\chi_g^B(G) \leq k$ or deciding if Alice has a winning strategy in g_B with k colors are PSPACE-complete problems.*

Proof (Sketch). From Lemma 1, the three decision problems are in PSPACE. Given a POS-DNF-11 formula with N variables $X_1, \ldots, X_N$ and M conjunctions $C_1, \ldots, C_M$, let p_j (for $j = 1, \ldots, M$) be the size of conjunction C_j ($p_j \leq 11$). We will construct a graph G such that $\chi(G) = M+3N+25$ and $\chi_g(G) = M+3N+25$ if and only if Alice has a winning strategy for the POS-DNF-11 formula.

Initially, the constructed graph G is the graph F_1 of Fig. 1 with $|K| = M + 3N + 23$ and $|Q| = |K| + 3$. See Fig. 2. For every variable X_i, create a vertex x_i in G. For every conjunction C_j, we create a *conjunction clique*. For this, first create a clique with vertices $\ell_{j,1}, \ldots, \ell_{j,p_j}$ and join $\ell_{j,k}$ to x_i with an edge if and only if both are associated to the same variable, for $k = 1, \ldots, p_j$. Also add the new vertex $\ell_{j,0}$ (which is not associated to variables) and join it with an edge to the vertex y. For every vertex $\ell_{j,k}$ ($j = 1, \ldots, M$ and $k = 0, \ldots, p_j$), replace it by two true-twin vertices $\ell'_{j,k}$ and $\ell''_{j,k}$, which are adjacent vertices with same neighborhood of $\ell_{j,k}$. Moreover, add to the conjunction clique of C_j a clique L_j with size $M + 3N + 25 - 2(p_j + 1) \geq 3N$ and join all vertices of L_j to s. With this, all conjunction cliques have exactly $M + 3N + 25$ vertices.

Figure 2 shows the constructed graph G for the formula $(X_1 \wedge X_2 \wedge X_5) \vee (X_1 \wedge X_3 \wedge X_5) \vee (X_2 \wedge X_4 \wedge X_5) \vee (X_3 \wedge X_4 \wedge X_5)$. Recall that Bob has a winning strategy in POS-DNF-11 firstly setting X_5 True: if Alice sets X_1 False, Bob sets X_4 True; if Alice sets X_4 False, Bob sets X_1 True; if Alice sets X_2 False, Bob sets X_3 True; if Alice sets X_3 False, Bob sets X_2 True. In the reduction of this example, we have $N = 5$ variables, $M = 4$ conjunctions, $p_j = 3$, $|K| = 42$, $|Q| = 45$, the cliques L_1 to L_4 have $M + 3N + 25 - 2(p_j + 1) = 36$ vertices each.

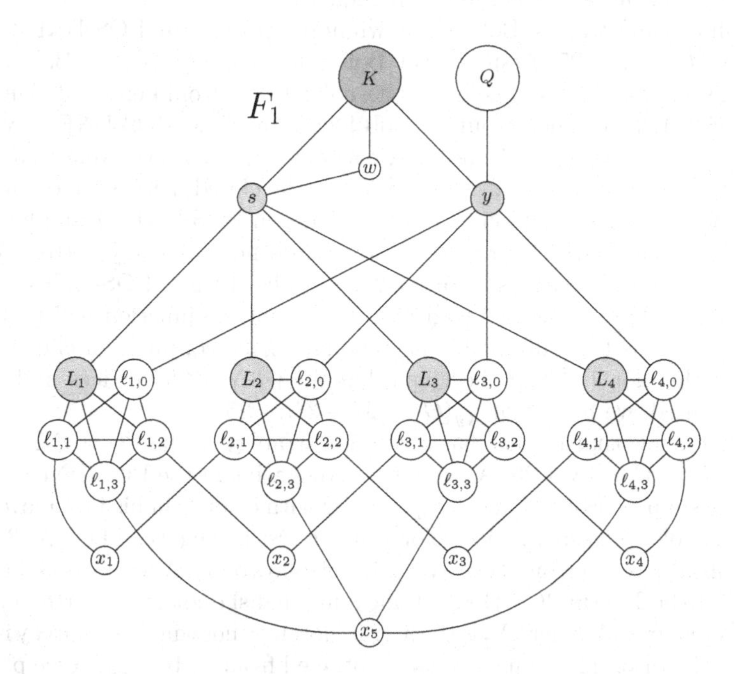

Fig. 2. Constructed graph G for the formula $(X_1 \wedge X_2 \wedge X_5) \vee (X_1 \wedge X_3 \wedge X_5) \vee (X_2 \wedge X_4 \wedge X_5) \vee (X_3 \wedge X_4 \wedge X_5)$. Recall that each vertex $\ell_{j,k}$ represents two true-twins $\ell'_{j,k}$ and $\ell''_{j,k}$; L_1, L_2, L_3, L_4 are cliques with 36 vertices; K is a clique with 42 vertices. Bob has a winning strategy avoiding 44 colors in the graph coloring game.

It is easy to check that $\chi(G) = M + 3N + 25$. For this, color s and all vertices in Q with color 1, the vertices of K with colors 2 to $M + 3N + 24$, color w, y and every vertex x_i ($i = 1, \ldots, n$) with color $M + 3N + 25$. For every $j = 1, \ldots, M$, color the vertices $\ell'_{j,k}$ and $\ell''_{j,k}$ with colors $2k+1$ and $2k+2$ ($k = 0, \ldots, p_j$). Finally, color the vertices of the clique L_j using the colors $2p_j+3,\ldots,M+3N+25$. Since the conjunction cliques contains $M+3N+25$ vertices, then $\chi(G) = M+3N+25$.

In the following, we show that Alice has a winning strategy in the graph coloring game if and only if she has a winning strategy in POS-DNF-11. From Lemma 3, in her first move, Alice must color vertex y of F_1 if Bob colored vertex s in his first move. Roughly speaking, we show that, in the best strategies, Bob colors vertex s first and Alice colors vertex y with the same color. Also notice that every vertex of a conjunction clique has degree exactly $M+3N+25$ (since it has exactly one neighbor outside the clique). In order to have the conjunction clique colored using the colors $1,\ldots,M+3N+25$, Alice must guarantee that all colors appearing in the outside neighbors of a conjunction clique also appears inside the clique. On the other hand, we show that Bob's strategy is making all outside neighbors of a conjunction clique to be colored with the same color of s and y (which will represent True in POS-DNF-11) and thus impeding Alice of using this color inside the conjunction clique.

We first show that if Bob has a winning strategy in POS-DNF-11, then $\chi_g(G) > M+3N+25$. Assume that Bob wins in POS-DNF-11. Bob uses the following strategy. He firstly colors s with color 1 and, from Lemma 3, Alice must color y with 1. In the next rounds, Bob follows his first POS-DNF-11 winning strategy: colors with color 1 the vertex associated to the variable that should receive True. If Alice colors a vertex in $N[x_i]$ (the closed neighborhood of x_i) for some i, Bob considers that she marked X_i False in POS-DNF-11 and he follows his winning POS-DNF-11 strategy; if Alice does not color any vertex in $N[x_i]$ for some i, then Bob plays as if Alice has passed her turn in POS-DNF-11 (recall Lemma 2). Then at some point all literals of some conjunction will be marked True. This means that all outside neighbors of some conjunction clique will be colored with color 1. Since the clique has $M+3N+25$ vertices and color 1 cannot be used, we have that $\chi_g(G) > M+3N+25$.

We now show that if Alice has a winning strategy in the POS-DNF-11 game then $\chi_g(G) = M+3N+25$. Assume that Alice wins in the POS-DNF-11 game.

Firstly suppose Bob colors s (resp. y), say with color 1, in his first move. Then Alice must color y (resp. s) with color 1 in her first move (recall Lemma 3). Alice can play using the following strategy: (1) if Bob plays on x_i, Alice plays as if Bob has chosen X_i to be True in POS-DNF-11, meaning that she colors the vertex x_j with a color different from 1, where X_j is the literal chosen by her winning strategy in POS-DNF-11; (2) if Bob plays on some twin obtained from vertex $\ell_{i,j}$, Alice plays the least available color in the other twin; (3) otherwise, Alice plays as if Bob has passed his turn in POS-DNF-11 (recall Lemma 2) if this game is not over yet; otherwise colors any non-colored vertex of G with the least available color. Following this strategy, every conjunction clique has a vertex colored 1. Since each clique L_j has at least $3N$ vertices, Alice and Bob can finish coloring every conjunction clique using the colors $1,\ldots,M+3N+25$.

Now assume Bob colored $v_1 \notin \{s,y\}$ in his first move (with some color c). Then Alice colors y firstly with a color $c' \neq c$, say $c' = 2$ w.l.g. Let v_2 be the 2nd vertex chosen by Bob. If $v_2 = s$ and its color is 2, we are done from the last paragraph (just replacing color 2 by color 1). Otherwise, Alice colors w with color 2 (and then s cannot be colored 2). We show that Alice has a winning strategy in this case. Assume w.l.g. that the color of s in the game will be 1 (otherwise we can relabel the colors). Thus no vertex of L_j is colored 1 ($j = 1, \ldots, M$).

With this, if Bob colored $\ell'_{i,0}$ or $\ell''_{i,0}$ for some i and the corresponding conjunction clique does not have a vertex colored 1, then Alice must color a vertex inside this conjunction clique with color 1. Otherwise, if there is a non-colored variable vertex, Alice colors it with a color distinct from 1. Since each clique L_j has at least $3N$ vertices, then Bob cannot color all vertices of some L_j before all variable vertices are colored. With this, Alice can guarantee that all colors of the variable vertices appear in the conjunction cliques and Alice wins.

Following the same path of g_B, we define $\chi_g^{A,A}(G)$, $\chi_g^{A,B}(G)$, $\chi_g^{B,A}(G)$ and $\chi_g^{B,B}(G)$: the minimum number of colors in C such that Alice has a winning strategy in $g_{A,A}$, $g_{A,B}$, $g_{B,A}$ and $g_{B,B}$, resp. We can also define three decision problems for each game $g_{Y,Z}$ ($Y, Z \in \{A, B\}$): given a graph G, its chromatic number $\chi(G)$ and an integer k,

- (Problem $g_{Y,Z}$-1) $\chi_g^{Y,Z}(G) \leq k$?
- (Problem $g_{Y,Z}$-2) Does Alice have a winning strategy in $g_{Y,Z}$ with k colors?
- (Problem $g_{Y,Z}$-3) $\chi_g^{Y,Z}(G) = \chi(G)$?

Corollary 1. *For every $Y, Z \in \{A, B\}$, the decision problems $g_{Y,Z}$-1, $g_{Y,Z}$-2 and $g_{Y,Z}$-3 are PSPACE-complete.*

Proof (Sketch). Let $Y, Z \in \{A, B\}$. Following similar arguments in Lemma 1, we obtain that they are PSPACE. The crucial argument to prove PSPCE-hardness is Lemma 2, which asserts that if Bob (resp. Alice) has a winning strategy in POS-CNF or POS-DNF-11, then he (resp. she) also has a winning strategy if Alice (resp. Bob) can pass any turn. Following the proof of Theorem 2.2 in [5] if $Y = A$ or the proof of Theorem 1 above if $Y = B$, we have that a winning strategy in g_Y is obtained from a winning strategy in POS-CNF/POS-DNF-11. If Alice (resp. Bob) has a winning strategy in the related logical game and $Z = B$ (resp. $Z = A$), she (resp. he) also has a winning strategy in $g_{Y,Z}$ by following the winning strategy in the logical game when the opponent passes a turn. Now if Alice (resp. Bob) has a winning strategy in the related logical game and $Z = A$ (resp. $Z = B$), she (resp. he) also has a winning strategy in $g_{Y,Z}$ (just not passing moves and simulating the obtained winning strategy in g_Y).

3 Connected Graph Coloring Game Is PSPACE-Complete

In this section, we prove that the connected version of the graph coloring game [4] is PSPACE-complete with a reduction from Problem g_B-3 of Sect. 2.

Theorem 2. *Given a graph G and an integer k, deciding whether Alice has an winning strategy with exactly k colors or at most k colors in the connected version of the graph coloring game (Alice starting) are PSPACE-complete problems.*

Proof (Sketch). As before, we first define a more restricted decision problem and prove that it is PSPACE-complete: given G and its chromatic number $\chi(G)$, Alice has a winning strategy with $\chi(G)$ colors?

We obtain a reduction from Problem g_B-3 of Sect. 2. Let $(G, \chi(G))$ be an instance of Problem g_B-3 with $|V(G)|$ odd. The reduction is as depicted in Fig. 3, where K is a clique with size $\chi(G)$ and s is connected to every vertex of G. Notice that $|V(G) \cup \{s\}|$ is even.

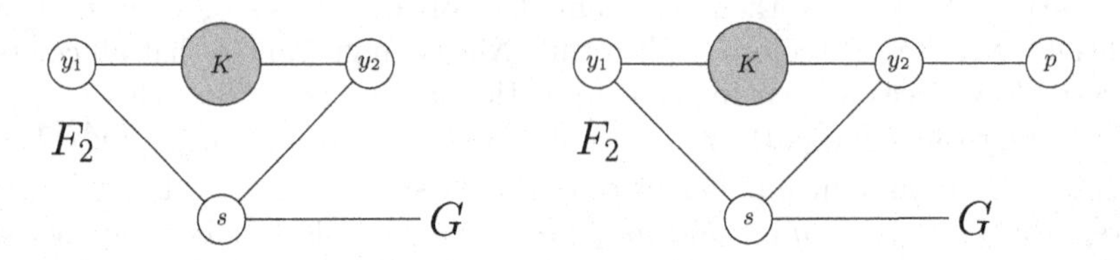

Fig. 3. The reduction from the graph G adding a gadget F_2 to it. The left if $\chi(G)$ is even or the right if $\chi(G)$ is odd.

The resulting graph G' has chromatic number $\chi(G') = \chi(G) + 1$. Note that, if y_1 and y_2 receive distinct colors, Bob wins since it is impossible to color G' with $\chi(G')$ colors.

Also, note that, if Alice does not play her first move in y_1, y_2 or K, she loses. This is because, since $|V(G) \cup \{s\}|$ is even, Bob can always guarantee that Alice will play on either y_1 or y_2 before him and consequently he can play a different color in the other, forcing distinct colors for y_1 and y_2. This is true even if she plays in p first or in y_2 first when p is a vertex of G'.

First, assume that Alice has a winning strategy for the variant g_B-3 of the graph coloring game (Bob starts the game). She has the following strategy: she begins playing y_1. Bob has to play in either s or K. She then plays in y_2 with the same color as y_1, ensuring her safety inside F_2. From there on, if Bob plays in G, she plays according to her strategy in g_B-3. If he plays in F_2, she also plays in F_2 (since $|V(F)|$ is odd, she can always do this).

Now, assume that Bob has a winning strategy in g_B-3. Assume, without loss of generality, that Alice plays either y_1 or K. If she plays in y_1, Bob can play in s and then she has to play in y_2. If she plays in K, he can play in y_1 and then she has to play in y_2. From now on, in either case, Bob can guarantee he is the first to make a move in G since $|V(F)|$ is odd. After this, if she plays in G, he also plays in G following his winning strategy. If she plays in F_2, he also plays in F_2 which is always possible.

4 PSPACE-Complete Variants of Greedy Coloring Game

As in Sect. 2, we define five variants of the greedy coloring game: g^*_B (Bob starts), $g^*_{A,A}$ (Alice starts and can pass any turn), $g^*_{A,B}$ (Alice starts and Bob can pass any turn), $g^*_{B,A}$ (Bob starts and Alice can pass any turn) and $g^*_{B,B}$ (Bob starts and can pass any turn). Unlike in the game coloring problem, the greedy game coloring problem satisfies the following:

Proposition 1. *If Alice (resp. Bob) has a winning strategy with k colors in $g^*_{Y,Z}$, then she (resp. he) also has a winning strategy with $k+1$ colors ($Y, Z \in \{A, B\}$).*

Proof. A winning strategy with k colors in the greedy coloring game is a strategy with $k+1$ colors that does not use the color $k+1$, since the coloring is greedy.

Let us start with g^*_B (Bob starts the greedy coloring game). Let $\Gamma_g^B(G)$ be the minimum number of colors in C such that Alice has a winning strategy in g^*_B. Clearly, $\chi(G) \leq \Gamma_g^B(G) \leq \Gamma(G)$. We can define two natural decision problem for g^*_B: given a graph G, its chromatic number $\chi(G)$ and an integer k,

- (Problem g^*_B-1) $\Gamma_g^B(G) \leq k$? Alice has winning strategy with k colors in g^*_B?
- (Problem g^*_B-2) $\Gamma_g^B(G) = \chi(G)$?

Clearly, Problem g^*_B-1 is a generalization of Problem g^*_B-2 (just set $k = \chi(G)$). Then the PSPACE-hardness of g^*_B-2 implies the PSPACE-hardness of g^*_B-1.

We obtain a reduction from POS-DNF-11 similar to the one of Sect. 2 for g_B. One important ingredient of the reduction is the graph F_3 of Fig. 4, which has a clique K with k vertices and three vertices s, w and y such that s and w are adjacent to all vertices in K and w is adjacent to y. We start proving that, in case of Bob firstly coloring s, Alice must color y in her first move.

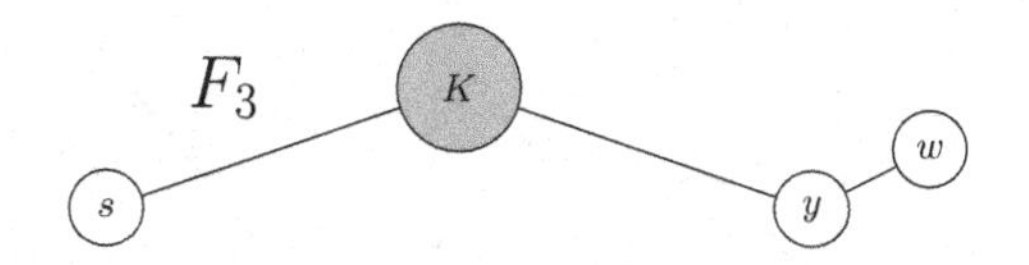

Fig. 4. Graph F_3: K is a clique.

Lemma 4. *Consider the graph F_3 of Fig. 4 with $|K| = k$ and assume that Bob colored vertex s in the first move of g^*_B. Alice wins the game in F_3 with $k+1$ colors if and only if she colors vertex y in her first move.*

Proof (Sketch). Clearly s is colored 1 and F_3 is colored with $k+1$ colors if and only if y and s receive the same color. If Alice colors y in her first move (color 1), she wins with $k+1$ colors. Thus assume Alice does not color y. Then Bob colors w, which receives color 1, forcing different colors for s and y.

Theorem 3. *g_B^*-1 and g_B^*-2 are PSPACE-complete.*

Proof (Sketch). Following similar arguments of Lemma 1, since the number of turns is exactly n and, in each turn, the number of possible moves is at most n, we have that both decision problems are PSPACE. We follow a very similar reduction of Theorem 1 (but from POS-CNF-11 instead of POS-DNF-11), including a neighbor $\overline{x_i}$ of degree 1 to each vertex x_i and replacing graph F_1 by graph F_3 with $|K| = M + 3N + 24$. Recall that, in POS-CNF, it is given a CNF formula (conjunctive normal form: conjunction of disjunctions). In POS-CNF-11, there is an additional constraint: the clauses have at most 11 variables. Figure 4 shows the constructed graph G for the formula $(X_1 \vee X_2) \wedge (X_1 \vee X_3) \wedge (X_2 \vee X_4) \wedge (X_3 \vee X_4)$. In the reduction of this example, we have $N = 4$ variables and $M = 4$ clauses. The cliques L_1 to L_M have $M + 3N + 19$ vertices each (Fig. 5).

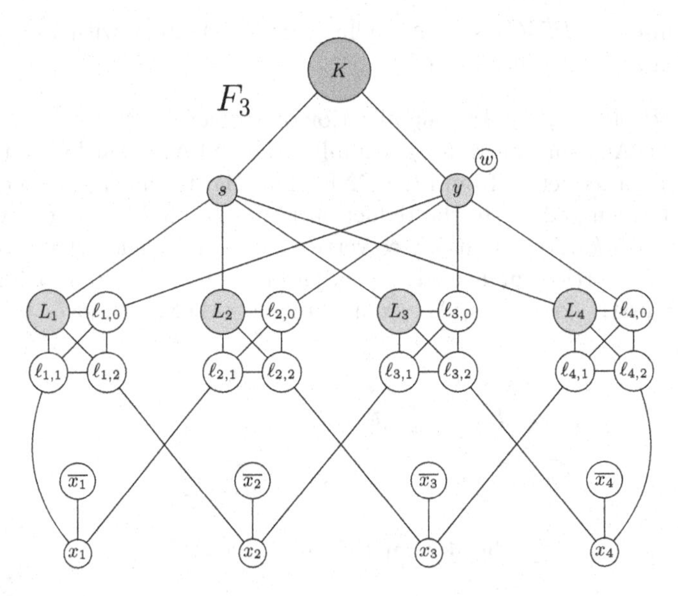

Fig. 5. Constructed graph G for the formula $(X_1 \vee X_2) \wedge (X_1 \vee X_3) \wedge (X_2 \vee X_4) \wedge (X_3 \vee X_4)$. Recall that each vertex $\ell_{j,k}$ represents two true-twins $\ell'_{j,k}$ and $\ell''_{j,k}$, L_1, L_2, L_3, L_4 are cliques with 35 vertices. Bob has a winning strategy avoiding 41 colors in the greedy coloring game.

As in Theorem 1, $\chi(G) = M + 3N + 25$. With similar arguments in Theorem 1, we obtain the result. The main difference is that, instead coloring a vertex x_i with a color distinct from 1, Alice colors the vertex $\overline{x_i}$ (with color 1).

As before, we define $\Gamma_g^{A,A}(G)$, $\Gamma_g^{A,B}(G)$, $\Gamma_g^{B,A}(G)$ and $\Gamma_g^{B,B}(G)$: the minimum number of colors in C such that Alice has a winning strategy in $g^*_{A,A}$, $g^*_{A,B}$, $g^*_{B,A}$ and $g^*_{B,B}$, resp. With this, we define two decision problem for each game $g^*_{Y,Z}$ with $Y, Z \in \{A, B\}$: given G, its chromatic number $\chi(G)$ and an integer k,

– (Problem $g^*_{Y,Z}$-1) $\Gamma_g^{Y,Z}(G) \leq k$? That is, does Alice have a winning strategy in $g_{Y,Z}$ with k colors?
– (Problem $g^*_{Y,Z}$-2) $\Gamma_g^{Y,Z}(G) = \chi(G)$?

Corollary 2. *For every $Y, Z \in \{A, B\}$, the decision problems $g^*_{Y,Z}$-1 and $g^*_{Y,Z}$-2 are PSPACE-complete.*

References

1. Andres, D., Lock, E.: Characterising and recognising game-perfect graphs. Discrete Math. Theoret. Comput. Sci. **21**(1) (2019). https://dmtcs.episciences.org/5499
2. Bodlaender, H.L.: On the complexity of some coloring games. In: Möhring, R.H. (ed.) WG 1990. LNCS, vol. 484, pp. 30–40. Springer, Heidelberg (1991). https://doi.org/10.1007/3-540-53832-1_29
3. Bohman, T., Frieze, A., Sudakov, B.: The game chromatic number of random graphs. Random Struct. Algorithms **32**(2), 223–235 (2008)
4. Charpentier, C., Hocquard, H., Sopena, E., Zhu, X.: A connected version of the graph coloring game. In: Proceedings of the 9th Slovenian International Conference on Graph Theory (Bled 2019) (2019). arXiv:1907.12276
5. Costa, E., Pessoa, V.L., Sampaio, R., Soares, R.: PSPACE-hardness of two graph coloring games. In: Electronic Notes in Theoretical Computer Science, vol. 346, pp. 333–344, Proceedings of the 10th Latin and American Algorithms, Graphs and Optimization Symposium, LAGOS 2019 (2019)
6. Dinski, T., Zhu, X.: A bound for the game chromatic number of graphs. Discrete Math. **196**(1), 109–115 (1999)
7. Faigle, U., Kern, U., Kierstead, H., Trotter, W.: On the game chromatic number of some classes of graphs. Ars Combinatoria **35**, 143–150 (1993)
8. Gardner, M.: Mathematical games. Sci. Am. **244**(4), 18–26 (1981)
9. Havet, F., Zhu, X.: The game Grundy number of graphs. J. Comb. Optim. **25**(4), 752–765 (2013)
10. Hearn, R.A., Demaine, E.D.: Games, Puzzles, and Computation. A. K. Peters Ltd., Natick (2009)
11. Kierstead, H.A., Trotter, W.T.: Planar graph coloring with an uncooperative partner. J. Graph Theory **18**(6), 569–584 (1994)
12. Krawczyk, T., Walczak, B.: Asymmetric coloring games on incomparability graphs. Electron. Notes Discrete Math. **49**, 803–811 (2015)
13. Nakprasit, K.M., Nakprasit, K.: The game coloring number of planar graphs with a specific girth. Graphs Comb. **34**(2), 349–354 (2018)
14. Schaefer, T.J.: On the complexity of some two-person perfect-information games. J. Comput. Syst. Sci. **16**(2), 185–225 (1978)
15. Sekiguchi, Y.: The game coloring number of planar graphs with a given girth. Discrete Math. **330**, 11–16 (2014)
16. Sidorowicz, E.: The game chromatic number and the game colouring number of cactuses. Inf. Process. Lett. **102**(4), 147–151 (2007)
17. Zhu, X.: The game coloring number of planar graphs. J. Comb. Theory, Ser. B **75**(2), 245–258 (1999)
18. Zhu, X.: The game coloring number of pseudo partial k-trees. Discrete Math. **215**(1), 245–262 (2000)
19. Zhu, X.: Refined activation strategy for the marking game. J. Comb. Theory, Ser. B **98**(1), 1–18 (2008)

Tractable Unordered 3-CNF Games

Md Lutfar Rahman(✉) and Thomas Watson

University of Memphis, Memphis, TN, USA
mrahman9@memphis.edu

Abstract. The classic TQBF problem can be viewed as a game in which two players alternate turns assigning truth values to a CNF formula's variables in a prescribed order, and the winner is determined by whether the CNF gets satisfied. The complexity of deciding which player has a winning strategy in this game is well-understood: it is NL-complete for 2-CNFs and PSPACE-complete for 3-CNFs.

We continue the study of the *unordered* variant of this game, in which each turn consists of picking any remaining variable and assigning it a truth value. The complexity of deciding who can win on a given CNF is less well-understood; prior work by the authors showed it is in L for 2-CNFs and PSPACE-complete for 5-CNFs. We conjecture it may be efficiently solvable on 3-CNFs, and we make progress in this direction by proving the problem is in P, indeed in L, for 3-CNFs with a certain restriction, namely that each width-3 clause has at least one variable that appears in no other clause. Another (incomparable) restriction of this problem was previously shown to be tractable by Kutz.

Keywords: 3-CNF · Games · Unordered · Logarithmic space

1 Introduction

Two-player games play an important role in complexity theory, particularly in the study of space-bounded computations. For example, the seminal PSPACE-complete problem TQBF—in which the goal is to determine whether a given quantified boolean formula $\exists x_1 \forall x_2 \exists x_3 \forall x_4 \cdots \varphi(x_1, \ldots, x_n)$ is true—can be viewed as deciding who has a winning strategy in the following two-player game: player 1 picks a bit value to assign to x_1, then player 2 assigns x_2, then player 1 assigns x_3, then player 2 assigns x_4, etc., with player 1 winning iff φ is satisfied.

Most commonly, φ is a conjunctive normal form (CNF) formula, which consists of a conjunction of clauses where each clause is a disjunction of literals. A w-CNF has at most w literals in each clause, and this *width* parameter w often governs the complexity of problems involving CNFs. For 2-CNFs, TQBF is NL-complete [2,4] (in particular, in P), while for 3-CNFs it is PSPACE-complete [12]. We call the corresponding game the *ordered CNF game* because the players are required to "play" the variables in a particular order prescribed in the input.

This work was supported by NSF grant CCF-1657377.

Y. Kohayakawa and F. K. Miyazawa (Eds.): LATIN 2020, LNCS 12118, pp. 360–372, 2020.
https://doi.org/10.1007/978-3-030-61792-9_29

Complexity of the Unordered CNF Game. In contrast, many real-world games have greater flexibility in terms of the set of moves available in each turn: the current player may be allowed to pick any of the remaining possible moves to do. We can define a variant of TQBF, called the *unordered CNF game*, which has this format: The input is again a CNF φ, and in each turn the current player picks a remaining (unassigned) variable and picks a bit value to assign it. The winner is determined by whether φ gets satisfied; we let T denote the player who wins when every clause of φ is true, and F denote the player who wins when some clause of φ is false. For 2-CNFs, deciding who has a winning strategy in this game is known to be in L [7], while PSPACE-completeness was shown for 11-CNFs [10,11], then for 6-CNFs [1], and then for 5-CNFs [7]. It remains a mystery what happens for widths 3 and 4.

We boldly conjecture that, in stark contrast to its ordered counterpart, *the unordered 3-CNF game may actually be tractable.* Progress toward confirming this conjecture can be made by considering certain restrictions on the input CNF, and showing that the game is tractable under these restrictions. The contribution of this paper is such a result. Before stating our result, for comparison we review other restrictions that have been studied.

One natural restriction is CNFs that are positive (a.k.a. monotone), meaning that all literal occurrences are unnegated variables; in this case, the unordered CNF game is equivalent to the so-called Maker–Breaker game (which is widely-studied in the combinatorics literature). In fact, [10,11] proved that the unordered CNF game is PSPACE-complete even for positive 11-CNFs (and a simplified proof for unbounded-width positive CNFs appears in [3]). Kutz [5,6] proved that for positive 3-CNFs, the unordered CNF game is tractable (in P) under an additional restriction on the hypergraph structure of the CNF, namely that no two clauses have more than one variable in common. This is the only previous result in the direction of confirming our conjecture.

It would be interesting to lift either the "positive" restriction or the "only one common variable" restriction in Kutz's result. We prove that *both* can be lifted if we instead impose a different (incomparable) restriction on the CNF's hypergraph structure. Specifically, we can view the variables in a clause as nodes, which are places where the clause can "connect" to other clauses (by sharing the variable). One difficulty in Kutz's analysis was handling width-3 clauses that use each of their 3 nodes to connect to other clauses. By restricting this difficulty away, we are able to address both limitations of Kutz's result, by handling general (not positive) CNFs that can have more than one common variable between pairs of clauses. (Our analysis does not end up resembling Kutz's very much, though.)

Thus our theorem can be stated as: the unordered 3-CNF game is in P, in fact in L, when each width-3 clause has at least one "spare" variable that appears in no other clauses. In the context of satisfiability, this restriction (each width-3 clause has a spare variable) is not very interesting since it would reduce to 2-SAT (the width-3 clauses could automatically be satisfied). Similarly, under this restriction, 3-TQBF would reduce to 2-TQBF since each clause with a spare variable belonging to T ($\exists$) would get satisfied (and thus disappear), and each clause whose spare variable belongs to F (∀) would shrink to a width-2 clause. However, for the unordered 3-CNF game there is no clear way to reduce this restricted version to a 2-CNF game, since both players can vie for any spare variable. As we show in this paper, combinatorially characterizing the winner of such a restricted unordered 3-CNF game turns out to be drastically more involved than for unordered 2-CNF games [7].

Proof Outline. To prove our theorem, there are multiple cases depending on who has the first move and who has the last move. The case where T goes first reduces to the case where F goes first (by trying all possibilities for T's opening move, and seeing whether any of them lead to a win for T in the residual game where F moves first), so we focus on the latter. Our proof separately handles the cases where F has both the first and last moves (Sect. 3) and where F has the first move and T has the last move (omitted due to space constraints).

The case where F has both the first and last moves (so the number of variables is odd) is somewhat simpler to analyze. We state and prove a characterization of who has a winning strategy in this case, in terms of certain features of the input formula; an efficient algorithm follows straightforwardly from this. To obtain the characterization, we begin by identifying various types of subformulas whose presence in the input formula would enable F to win. It is an elementary but non-trivial case analysis to verify that in any of these subformulas, F indeed has a strategy to ensure some clause gets falsified (Sect. 3.1). The more interesting part of the proof is to show that not only do these subformulas constitute "obstacles" to T winning, but in a sense they are the *only* obstacles (Sect. 3.2). Although it is not true that F can win iff at least one of those subformulas exists in the original formula, we prove something just as good: F can win iff he has an opening move that ensures at least one of those subformulas will exist in the residual formula at the end of the first round. (A round consists of an F move followed by a T move.)

In other words, if T can fend off all the obstacles for one round, then he will be able to fend them off for the entire game. This non-obvious fact is key to taming the combinatorial structure of the game. The proof of this fact involves a subtle induction that modifies the game rules to allow F to "pass" (forgo his turn) whenever he wishes—this can only make it harder for T to win, but it is needed for the induction to go through. After a round, we can prove that for each of the smaller components that were created in the residual formula: either we can design a direct winning strategy for T in that component by exploiting the absence of the obstacle subformulas, or T can fend off obstacles for one more round in that component, enabling us to apply the induction hypothesis. Finally, to combine the "sub-strategies" for the separate components into an overall strategy for T, we exploit the resilience of the sub-strategies against pass moves by F.

The case where F goes first and T goes last follows a similar structure but is more involved. Some of the above argument can be recycled, but the parts that relied on F moving last need to be changed. Now the "complete" set of obstacles is larger and more complicated. The inductive argument for T's winning strategy requires a more detailed analysis and uses a further modification of the game: the new rule says that a certain subformula gets immediately removed from the game (its variables become unplayable) whenever it is created in the residual formula. The deleted copies of this subformula are then dealt with "outside of" the induction, to recover a proof for the unmodified game.

Summary. One motivation for studying the unordered CNF game is that it is naturally analogous to a variety of real-world games where the same moves are available to both players. Indeed, the original result of Schaefer [10,11] has been used in many reductions to show PSPACE-completeness of other natural games with an unordered flavor (see [7] for a list). At a more fundamental level, the problem we study is very simple to define, and our result reveals new insights about CNFs, which are among the most ubiquitous representations of boolean functions.

A potential big payoff for this research direction is to show that the general unordered 3-CNF game is tractable. That may sound outlandish since arbitrary 3-CNFs are typically thought of as "too unstructured" to admit efficient algorithms for interesting problems. Our result together with the complementary result by Kutz [5,6] provides a glimpse into why the bold conjecture may be true, and a plausible roadmap for proving it: by combining our techniques, which handle negated literals and clauses that share two variables, with Kutz's techniques, which handle clauses without spare variables. Short of handling the general game, there are other open and interesting special cases to which our techniques may be germane, such as the Maker–Breaker game on general 3-uniform hypergraphs.

The proof of our result reveals a novel structural property: it is impossible for F to mount a "long-range" attack for creating a simple "obstacle" after a super-constant number of rounds—it is a "now or never" situation for F. We conjecture the same phenomenon holds for the game on unrestricted 3-CNFs, since we are unaware of any counterexamples. If a counterexample is found, it might be turned into a gadget for proving hardness of the general game. Even NL-hardness would be fundamentally interesting since our algorithm—based on detecting a simple obstacle after constantly many rounds—only uses logarithmic space. (As a side result—not included in this paper—we can show that the unordered 4-CNF game is NL-hard.)

Although our requirement that every width-3 clause has a spare variable seems to be a very strong restriction, and may not naturally show up in other contexts, we feel it is an important stepping stone for understanding more general games. It already adds a very significant layer of complexity over the unordered 2-CNF game, and it represents a reasonable way of suppressing some of the difficulties posed by the hypergraph structure of 3-CNFs (which Kutz's proof works hard to address), en route to a more general result.

Furthermore, our proof contributes some innovative techniques for analyzing games, including: modifying the game to facilitate an induction; our framework for showing how T can extend his good fortune from one round to all subsequent rounds; and a method for simplifying gameplay analysis by imagining that the moves happened in a different order.

2 Preliminaries

We define a **formula** as a pair (φ, X) where φ is a CNF and $X = \{x_1, \ldots, x_n\}$ contains all the variables that appear in φ (and possibly more). In the unordered CNF game there are two players, denoted T (for "true") and F (for "false"), who alternate turns. Each turn consists of picking a remaining (unassigned) variable from X and assigning it a value 0 or 1. The game ends when all variables of X have been assigned, and T wins if φ is satisfied, and F wins if it is not. We let G (for "game") denote the problem of deciding which player has a winning strategy, given the formula (φ, X) and a specification of which player goes first. We let G_w denote the restriction of G to instances where each clause has at most w literals (φ has width w). We define a **spare variable** as occurring in only one clause, and we assume without loss of generality that a spare variable appears as a positive literal. Then we let G_3^* denote the restriction of G_3 to instances where each width-3 clause in φ has at least one spare variable.

Theorem 1. G_3^* *is in polynomial time, in fact, in logarithmic space.*

We introduce subscripts to distinguish the different patterns for "who goes first" and "who goes last". For $a, b \in \{\mathrm{T}, \mathrm{F}\}$, the subscript $a \cdots b$ means player a goes first and player b goes last, $a \cdots$ means a goes first, and $\cdots b$ means b goes last. Thus $\mathrm{G}^*_{3,\mathrm{T}\cdots}$ corresponds to the game where T goes first, which (as noted in Sect. 1) reduces to $\mathrm{G}^*_{3,\mathrm{F}\cdots}$ by brute-forcing T's first move. So, we just prove Theorem 1 for $\mathrm{G}^*_{3,\mathrm{F}\cdots}$, which is split into the cases $\mathrm{G}^*_{3,\mathrm{F}\cdots\mathrm{F}}$ (F goes first and last, so $n = |X|$ must be odd) and $\mathrm{G}^*_{3,\mathrm{F}\cdots\mathrm{T}}$ (F goes first and T goes last, so $n = |X|$ must be even). We use the terms **move**, **turn**, or **play** interchangeably to mean T or F assigning a bit value to one variable. A **round** consists of two consecutive moves, and since we only need to consider F having the first move, each round will consist of one F move followed by one T move (except in $\mathrm{G}^*_{3,\mathrm{F}\cdots\mathrm{F}}$, the last round will have only one move).

A **subformula** (φ', X') of a formula (φ, X) is defined as φ' having a subset of clauses from φ and $X' \subseteq X$ containing all the variables that appear in φ' (and possibly more). After a move, the formula changes to a **residual formula** where the variable that got played is removed from X, and each clause containing the variable either disappears (since it is satisfied by a true literal) or shrinks (since a false literal might as well not be there). F wins if the residual formula has a width-0 clause, and T wins if it has no clauses. The residual formula after a move may or may not be a subformula of the formula before the move.

When we say F **can ensure** some property **within** k rounds, we formally mean that either

- the original formula has the property, or
- ($\exists$ F move) ($\forall$ T move) the residual formula has the property, or

- (∃ F move) (∀ T move) (∃ F move) (∀ T move) the residual formula has the property, or ⋯⋯
- (∃ F move) (∀ T move) ⋯ (∃ F move in k^{th} round) (∀ T move in k^{th} round) the residual formula has the property.

Note that the property is only checked at the boundary between rounds (and not after F's move but before T's move inside of a round).

A positive CNF is equivalent to a hypergraph where nodes are variables and hyperedges are clauses. In this paper, we use a hypergraph representation of general (not necessarily positive) CNFs. As shown in Fig. 1, a clause is a hyperedge where nodes represent variables, and signs are annotations representing variables' literal appearances. When we omit the sign of a variable on a diagram, it could be either + or − but it is not relevant.

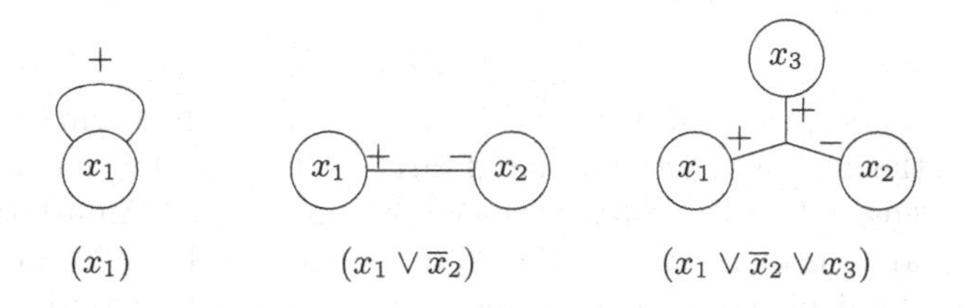

Fig. 1. Example clauses and their hypergraph representations

Two clauses in a general CNF can share any number of same signed or opposite signed literals. We think of a shared variable as a connection between two clauses, and we define two types of connections:

- **Pure Connection:** A variable that appears with the same sign in two clauses. For example, in $(x_1 \vee x_2 \vee x_3) \wedge (x_2 \vee x_4 \vee x_5)$ there is a pure connection at x_2. See Fig. 2 on the left. Another example: in $(x_1 \vee \overline{x}_2 \vee x_3) \wedge (\overline{x}_2 \vee x_4 \vee x_5)$ there is again a pure connection at x_2.
- **Mixed Connection:** A variable that appears with the opposite sign in two clauses. For example, in $(x_1 \vee x_2 \vee x_3) \wedge (\overline{x}_2 \vee x_4 \vee x_5)$ there is a mixed connection at x_2. See Fig. 2 on the right. Another example: in $(x_1 \vee \overline{x}_2 \vee x_3) \wedge (x_2 \vee x_4 \vee x_5)$ there is again a mixed connection at x_2.

A formula (φ, X) is called **connected** if the associated hypergraph is connected (with the signs being irrelevant); i.e., it is possible to get from any variable to any other variable by a sequence of clauses, each having a connection to the next. A formula is thus naturally partitioned into connected components, each of which is a subformula. An **isolated variable** is one that is in X but not in any clause of φ, and thus forms a connected component by itself since the associated node is incident to no hyperedges. A variable in a width-1 clause is not considered isolated.

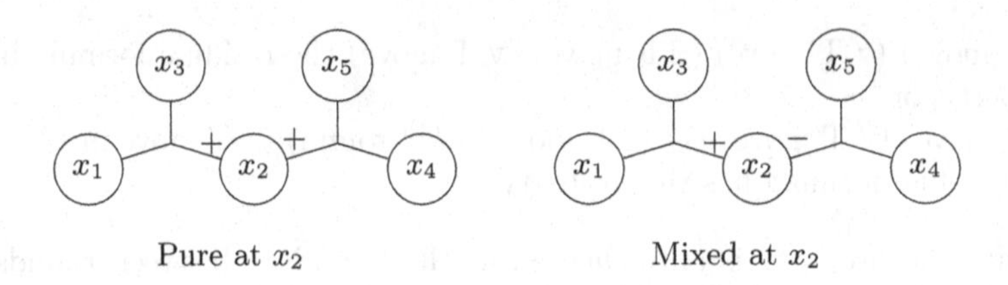

Fig. 2. Clause connections

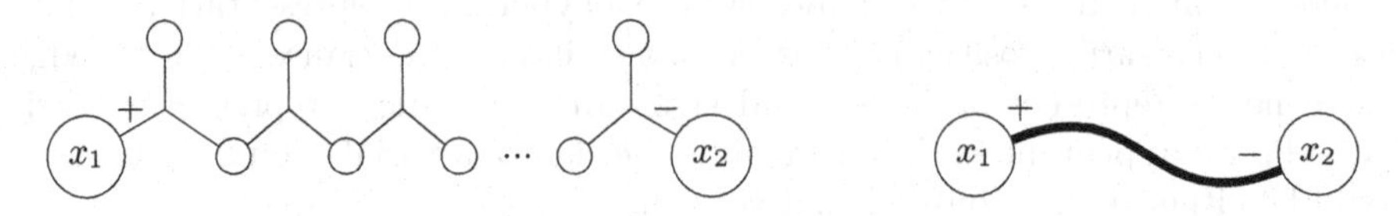

Fig. 3. A chain between x_1 and x_2

A **chain** is a sequence of distinct width-3 clauses each sharing exactly one variable with the next, and with no shared variables between two non-consecutive clauses. The length L of the chain is the number of clauses. An arbitrary chain between x_1 and x_2 is illustrated in Fig. 3 on the left. On the right, we show how the chain can be depicted by a thick line. If $L = 0$ then $x_1 = x_2$. If $L = 1$ then the only clause in the chain contains both x_1 and x_2.

A **cycle** is like a chain with $L > 2$ and $x_1 = x_2$. A **diamond** happens when two width-3 clauses share exactly two variables. Intuitively, a diamond is like the smallest case of a cycle, with $L = 2$.

3 $\mathrm{G}^*_{3,\mathrm{F}\cdots\mathrm{F}}$

We henceforth assume that in a formula (φ, X), φ is always a 3-CNF where each width-3 clause has at least one spare variable.

Lemma 1. F *has a winning strategy in a* $\mathrm{G}^*_{3,\mathrm{F}\cdots\mathrm{F}}$ *game iff* F *can ensure within one round at least one of the following subformulas exists.*

*(1) A width-*0 *or width-*1 *clause.*
*(2) Two width-*2 *clauses sharing both variables.*
*(3) Two width-*2 *clauses and a chain (of length* ≥ 0*) between them.*
*(4) A width-*2 *clause and a chain (of length* ≥ 1*) between its two variables with at least one mixed connection between the chain and the width-*2 *clause.*
*(5) A width-*2 *clause, a cycle or diamond containing at most one width-*2 *clause variable, and a chain (of length* ≥ 0*) between them.*

Moreover, if subformula (4) or (5) exists at the beginning of a round then F *can ensure subformula (1) or (2) or (3) exists within two more rounds.*

The proof of Lemma 1 is in Sect. 3.1 and Sect. 3.2.

Corollary 1. *F has a winning strategy in a $G^*_{3,F\cdots F}$ game iff F can ensure subformula (1) or (2) or (3) exists within the first three rounds.*

Corollary 1 yields a direct approach to devise an algorithm for $G^*_{3,F\cdots F}$:

> Try all possible sequences of 6 moves for the first 3 rounds. Check whether ($\exists$ F move) ($\forall$ T move) ($\exists$ F move) ($\forall$ T move) ($\exists$ F move) ($\forall$ T move): subformula *(1)* or *(2)* or *(3)* exists in the residual formula.

This can be implemented in log space, because keeping track of a sequence of the first six moves takes log space, searching for subformula *(1)* or *(2)* takes log space, and searching for subformula *(3)* also takes log space since it can be expressed as an undirected s–t connectivity problem [8,9]: for each pair of width-2 clauses, check whether there exists a chain between them.

We conjecture the same algorithm (possibly with a different number of brute-force rounds) actually solves $G_{3,F\cdots F}$; we are not aware of any counterexamples.

3.1 Right-to-left Implication of Lemma 1

Suppose at least one of the subformulas *(1–5)* exists when it is F's turn to play. We will handle each subformula in separate claims. For concreteness, we illustrate the arguments using literals with particular signs, but all the arguments work even if we negate all occurrences of any variable.

Claim 1. *If subformula (1) exists, F has a winning strategy.*

Proof. If a width-0 clause exists then T has no chance to satisfy it, so F wins. If a width-1 clause exists, say (x_1), then F can play $x_1 = 0$ and win. □

Claim 2. *If subformula (2) exists, F has a winning strategy.*

Proof. There are two possible ways that can happen:

- Case 1: *The clauses have opposite signs for one variable (mixed connection).* For example, in $(x_1 \vee x_2) \wedge (\overline{x}_1 \vee x_2)$ only x_1 has opposite signs. Then F can play $x_2 = 0$, and whatever the value of x_1, F will win.
- Case 2: *The clauses have opposite signs for both variables.* For example, in $(x_1 \vee x_2) \wedge (\overline{x}_1 \vee \overline{x}_2)$ both x_1 and x_2 have opposite signs. Since F moves last, F can wait by playing other variables until T has to play x_1 or x_2. Then F makes $x_1 = x_2$ and wins. □

Claim 3. *If subformula (3) exists, F has a winning strategy.*

Proof. We call this situation a manriki (a Japanese ninja weapon). The two width-2 clauses are like two handles and the chain in the middle can be arbitrarily long. We prove this claim by induction on the length of the chain.

Base case: The length of the chain is zero, i.e., the two handles directly share a variable. We can assume the two handles do not share both variables since otherwise that falls under Claim 2. There are two possible ways the handles can have one common variable:

- Case 1: *Pure Connection.* For example, in $(x_1 \vee x_2) \wedge (x_2 \vee x_3)$, x_2 forms a pure connection. F can play $x_2 = 0$. Then whatever T does, F plays $x_1 = 0$ or $x_3 = 0$ and wins.
- Case 2: *Mixed Connection.* For example, in $(x_1 \vee x_2) \wedge (\overline{x}_2 \vee x_3)$, x_2 forms a mixed connection. F can play $x_1 = 0$. If T plays $x_2 = 1$ then F plays $x_3 = 0$ and wins. If T plays $x_2 = 0$ then F wins. If T does not play x_2, F wins by playing $x_2 = 0$.

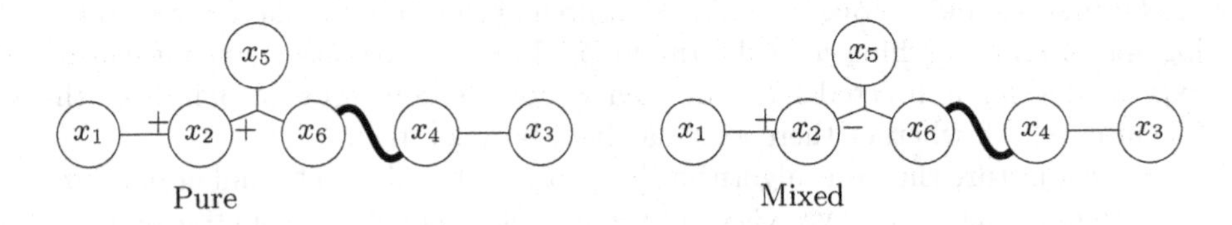

Fig. 4. Subformula *(3)* (Claim 3)

Induction Step: There are two cases depending on the type of connection at the common variable between one of the handles and the chain:

- Case 1: *Pure Connection.* For example, in Fig. 4 on the left, x_2 forms a pure connection between handle $(x_1 \vee x_2)$ and the chain. F can play $x_2 = 0$. If T plays $x_1 = 1$ then we have a smaller manriki from $(x_5 \vee x_6)$ to $(x_4 \vee x_3)$ where F can win by the induction hypothesis. If T plays $x_1 = 0$ then F wins. If T does not play x_1 then F wins by playing $x_1 = 0$.
- Case 2: *Mixed Connection.* For example, in Fig. 4 on the right, x_2 forms a mixed connection between handle $(x_1 \vee x_2)$ and the chain. F can play $x_1 = 0$. If T plays $x_2 = 1$ then we have a smaller manriki from $(x_5 \vee x_6)$ to $(x_4 \vee x_3)$ where F can win by the induction hypothesis. If T plays $x_2 = 0$ then F wins. If T does not play x_2 then F wins by playing $x_2 = 0$. □

Claim 4. *If subformula (4) exists,* F *has a winning strategy.*

Proof. There are three cases depending on how the width-2 clause is connected to the chain. For example, in Fig. 5, $(x_1 \vee x_2)$ is the width-2 clause and x_2 is a mixed connection. In the smallest versions, the chain (the bold line illustrated in the general versions) has length 1 for cases 1 and 2 and length 0 for case 3.

- Case 1: *Pure at* x_1. F can play $x_1 = 0$. If T plays $x_2 = 1$ then in the smallest case F wins by $x_3 = 0$ and in the general case F wins by Claim 3 by a manriki created from x_1's left end to x_2's right end. If T plays $x_2 = 0$ then F wins. If T does not play x_2 then F wins by $x_2 = 0$.
- Case 2: *Mixed at* x_1 *but pure at* x_4 *(the next non-spare variable on the chain).* F can play $x_4 = 0$. If T plays $x_1 = 0$ or $x_2 = 0$ then F wins by $x_2 = 0$ or $x_1 = 0$. If T plays $x_1 = 1$ then F wins by $x_3 = 0$. If T plays $x_2 = 1$ then

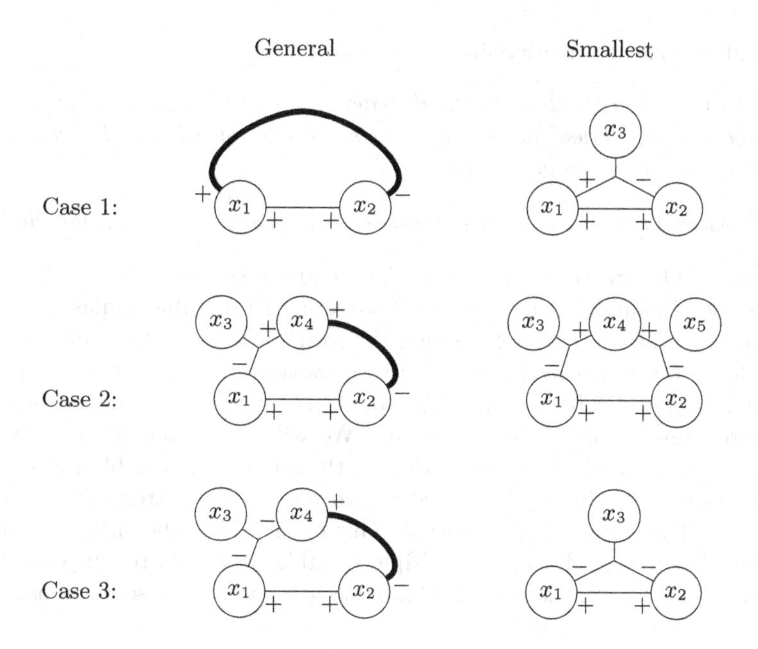

Fig. 5. Subformula *(4)* (Claim 4)

in the smallest case F wins by $x_5 = 0$ and in the general case F wins by a manriki created from x_4's right end to x_2's right end. If T plays x_3 then in the smallest case F wins by the manriki $(x_5 \vee \overline{x}_2) \wedge (x_2 \vee x_1)$ and in the general case F wins by a manriki created from x_4's right end to $(x_2 \vee x_1)$. If T plays any other variable then F wins by the manriki $(x_3 \vee \overline{x}_1) \wedge (x_1 \vee x_2)$.

- Case 3: *Mixed at both* x_1 *and* x_4. F can play $x_3 = 0$. In the smallest case, since F moves last, F can wait by playing other variables until T has to play x_1 or x_2, and then F can win by making $x_1 = x_2$. Now consider the general case. If T plays $x_1 = 0$ or $x_2 = 0$ then F wins by $x_2 = 0$ or $x_1 = 0$. If T plays $x_1 = 1$ then F wins by $x_4 = 1$. If T plays $x_2 = 1$ then F wins by a manriki created from x_2's right end to $(\overline{x}_4 \vee \overline{x}_1)$. If T plays $x_4 = 0$ then F wins by a manriki created from x_4's right end to $(x_2 \vee x_1)$. If T plays $x_4 = 1$ then F wins by $x_1 = 1$. If T plays any other variable then F wins by the manriki $(x_2 \vee x_1) \wedge (\overline{x}_1 \vee \overline{x}_4)$. □

Claim 5. *If subformula (5) exists,* F *has a winning strategy.*

The proof of Claim 5 is omitted due to space constraints.

Moreover, in all cases, there exists a subformula *(1)* or *(2)* or *(3)* within one round for Claim 4 and within two rounds for Claim 5.

3.2 Left-to-right Implication of Lemma 1

Definition 1. *A **cobweb** is a formula where none of the subformulas (1–5) exist (and each width-3 clause has at least one spare variable). Note that any subformula in a cobweb is also a cobweb.*

Observation 1. *A cobweb has a variable that occurs in at most one clause.*

The proof of Observation 1 is omitted due to space constraints.

Suppose F cannot ensure that at least one of the subformulas *(1–5)* exists within one round. So at the beginning the formula is a cobweb and in the first round, for every move by F there exists a move for T such that the residual formula is again a cobweb. In other words, T can ensure that the beginning cobweb remains a cobweb after a round. We will argue that T has a winning strategy. The proof will be by induction on the number of variables. In order for the induction to go through, we need to prove something stronger: "T can win even if F is allowed to use pass moves." This means F has the option of forgoing any turn, thus forcing T to play multiple variables in a row. In this case it does not make sense to consider which player has the last move, so we consider the game $\mathrm{G}^*_{3,\mathrm{F}\ldots}$ in this section.

First we consider a special case of cobweb that we call a jellyfish.

Definition 2. *A **jellyfish** is a connected cobweb with a width-2 clause. Its **eyes** are the variables in the width-2 clause.*

Lemma 2. *If the formula is a jellyfish then* T *has a winning strategy in* $\mathrm{G}^*_{3,\mathrm{F}\ldots}$ *even if* F *can use pass moves.*

The proof of Lemma 2 is omitted due to space constraints.

Definition 3. *A **winweb** is a cobweb such that* T *can ensure that it remains a cobweb after a round (where* F *is not allowed to use pass moves).*

Lemma 3. *Every subformula of a winweb is also a winweb.*

The proof of Lemma 3 is omitted due to space constraints.

The following lemma proves something stronger than the left-to-right implication of Lemma 1, because F can use pass moves.

Lemma 4. *If the formula is a winweb then* T *has a winning strategy in* $\mathrm{G}^*_{3,\mathrm{F}\ldots}$ *even if* F *can use pass moves.*

Proof. We prove this by induction on the number of variables.

Base case: The formula is a cobweb with one or two variables. In case of one variable the only possibility is an isolated variable with no clauses since subformula *(1)* does not exist. T has already won in this case. In case of two variables there exists either two isolated variables where T has already won or a width-2 clause which T can satisfy in one move.

Induction step: The formula (φ, X) is a winweb with at least three variables.

Suppose F played a pass move. There exists an isolated or spare variable since the formula is a cobweb (Observation 1). T can play that isolated/spare variable to remove the isolated variable or satisfy a clause. The residual formula is a subformula, which is a winweb by Lemma 3. Thus T can win the rest of the game by the induction hypothesis.

Now suppose F did not play a pass move. By the definition of winweb, T has a response such that the residual formula is a cobweb. Call this residual formula (φ', X') and let (φ_1, X_1), (φ_2, X_2), ..., (φ_k, X_k) be its connected components (so $\varphi' = \bigwedge_i \varphi_i$ and $X' = \bigcup_i X_i$). We claim that for each component individually, T has a winning strategy even if F can use pass moves:

- If (φ_i, X_i) has a width-2 clause then it is a jellyfish (since it is a connected cobweb) so by Lemma 2, T can win even if F can use pass moves.
- Suppose (φ_i, X_i) has no width-2 clause. Then it has only width-3 clauses since subformula *(1)* does not exist, and so it is a subformula of the winweb (φ, X) since no new width-3 clause can be created during the game. By Lemma 3, (φ_i, X_i) is also a winweb and hence by the induction hypothesis, T can win even if F can use pass moves.

We now explain how to combine T's winning strategies for the separate components to get a winning strategy for the rest of the game on (φ', X'). After F plays a variable in some X_i, T simply responds according to his winning strategy for component (φ_i, X_i), unless F played the last remaining variable in X_i. In the latter case, or if F played a pass move, T picks any other component (φ_j, X_j) with remaining variables and continues according to his winning strategy in that component, as if F had just played a pass move in that component. □

References

1. Ahlroth, L., Orponen, P.: Unordered constraint satisfaction games. In: Rovan, B., Sassone, V., Widmayer, P. (eds.) MFCS 2012. LNCS, vol. 7464, pp. 64–75. Springer, Heidelberg (2012). https://doi.org/10.1007/978-3-642-32589-2_9
2. Aspvall, B., Plass, M., Tarjan, R.: A linear-time algorithm for testing the truth of certain quantified Boolean formulas. Inf. Process. Lett. **8**(3), 121–123 (1979)
3. Byskov, J.: Maker-maker and maker-breaker games are PSPACE-complete. Technical report RS-04-14, BRICS, Department of Computer Science, Aarhus University (2004)
4. Calabro, C.: 2-TQBF is in P (2008). https://cseweb.ucsd.edu/~ccalabro/essays/complexity_of_2tqbf.pdf. Unpublished
5. Kutz, M.: The angel problem, positional games, and digraph roots. Ph.D. thesis, Freie Universität Berlin (2004). Chapter 2: Weak Positional Games
6. Kutz, M.: Weak positional games on hypergraphs of rank three. In: Proceedings of the 3rd European Conference on Combinatorics, Graph Theory, and Applications (EuroComb), pp. 31–36. Discrete Mathematics & Theoretical Computer Science (2005)
7. Rahman, M.L., Watson, T.: Complexity of unordered CNF games. In: Proceedings of the 29th International Symposium on Algorithms and Computation (ISAAC), pp. 9:1–9:12. Schloss Dagstuhl (2018)

8. Reingold, O.: Undirected connectivity in log-space. J. ACM **55**(4), 17:1–17:24 (2008)
9. Rozenman, E., Vadhan, S.: Derandomized squaring of graphs. In: Chekuri, C., Jansen, K., Rolim, J.D.P., Trevisan, L. (eds.) APPROX/RANDOM -2005. LNCS, vol. 3624, pp. 436–447. Springer, Heidelberg (2005). https://doi.org/10.1007/11538462_37
10. Schaefer, T.: Complexity of decision problems based on finite two-person perfect-information games. In: Proceedings of the 8th Symposium on Theory of Computing (STOC), pp. 41–49. ACM (1976)
11. Schaefer, T.: On the complexity of some two-person perfect-information games. J. Comput. Syst. Sci. **16**(2), 185–225 (1978)
12. Stockmeyer, L., Meyer, A.: Word problems requiring exponential time. In: Proceedings of the 5th Symposium on Theory of Computing (STOC), pp. 1–9. ACM (1973)

Quantum Computing

Lower Bounds for Testing Complete Positivity and Quantum Separability

Costin Bădescu(✉) and Ryan O'Donnell

Computer Science Department, Carnegie Mellon University, Pittsburgh, PA, USA
{cbadescu,odonnell}@cs.cmu.edu

Abstract. In this work we are interested in the problem of testing quantum entanglement. More specifically, we study the *separability* problem in quantum property testing, where one is given n copies of an unknown mixed quantum state ϱ on $\mathbb{C}^d \otimes \mathbb{C}^d$, and one wants to test whether ϱ is separable or ϵ-far from all separable states in trace distance. We prove that $n = \Omega(d^2/\epsilon^2)$ copies are necessary to test separability, assuming ϵ is not too small, viz. $\epsilon = \Omega(1/\sqrt{d})$.

We also study completely positive distributions on the grid $[d] \times [d]$, as a classical analogue of separable states. We analogously prove that $\Omega(d/\epsilon^2)$ samples from an unknown distribution p are necessary to decide whether p is completely positive or ϵ-far from all completely positive distributions in total variation distance.

1 Introduction

A bipartite quantum state ϱ on $\mathbb{C}^d \otimes \mathbb{C}^d$ is said to be *separable* if it can be written as a convex combination of product states, meaning states of the form $\rho_1 \otimes \rho_2$ where ρ_1 and ρ_2 are quantum states on $\mathbb{C}^d$. Separable quantum states are precisely those states which do not exhibit any form of quantum entanglement. These are the only states that can be prepared by separated parties who can only share classical information. Understanding the general structure and properties of the set of separable states in higher dimensions is a difficult problem and is the subject of much ongoing research. For instance, deciding whether a given $d^2 \times d^2$ matrix represents a separable state on $\mathbb{C}^d \otimes \mathbb{C}^d$ – also known as the *separability problem* in the quantum literature – is NP-hard [7]. In this work, we study the following property testing version of the separability problem:

> Given unrestricted measurement access to n copies of an unknown quantum state ϱ on $\mathbb{C}^d \otimes \mathbb{C}^d$, decide with high probability if ϱ is separable or ϵ-far from all separable states in trace distance.

Supported by NSF grant FET-1909310. This material is based upon work supported by the National Science Foundation under grant numbers listed above. Any opinions, findings and conclusions or recommendations expressed in this material are those of the author and do not necessarily reflect the views of the National Science Foundation (NSF).

Y. Kohayakawa and F. K. Miyazawa (Eds.): LATIN 2020, LNCS 12118, pp. 375–386, 2020.
https://doi.org/10.1007/978-3-030-61792-9_30

The ultimate goal is to determine the number of copies of ϱ that is necessary and sufficient to solve the problem, up to constant factors, as a function of d and ϵ.

By estimating (i.e., fully learning) ϱ using recent algorithms for quantum state tomography [9,14] and checking if the estimate is sufficiently close to a separable state, this problem can be solved using $O(d^4/\epsilon^2)$ copies of ϱ. In this paper, we prove a lower bound, showing that $\Omega(d^2/\epsilon^2)$ copies of ϱ are necessary when $\epsilon = \Omega(1/\sqrt{d})$; this reaches a lower bound of $\Omega(d^3)$ for $\epsilon = \Theta(1/\sqrt{d})$. Closing the gap between the known bounds seems like a difficult problem, and we have no particularly strong feeling about whether the tight bound is the upper bound, the lower bound, or something in between. (Indeed, at least one paper [2] contains some evidence that $\widetilde{\Theta}(d^3)$ might be the true complexity for constant ϵ).

Given the difficulty of closing the gap, we have sought a classical analogue of the separability testing problem to try as a first step. Analogies between quantum states and classical probability distributions have proven to be a helpful source of inspiration throughout quantum theory. Unfortunately, entanglement is understood to be a purely quantum phenomenon; every finitely-supported discrete distribution can be expressed as a convex combination of product point distributions, so there are no "entangled" distributions. But motivated by the characterization of separable quantum states using symmetric extensions and the quantum de Finetti theorem [4], we propose as a kind of analogue the study of mixtures of i.i.d. bivariate distributions, which arise in the classical de Finetti theorem. Doherty et al. [4] used the quantum de Finetti theorem to show that a quantum state ϱ on $\mathbb{C}^d \otimes \mathbb{C}^d$ is separable (i.e. a mixture of product states) if and only if ϱ has a symmetric extension to $\mathbb{C}^d \otimes (\mathbb{C}^d)^{\otimes k}$ for any positive integer k. Somewhat analogously, the classical de Finetti theorem states that a sequence of real random variables is a mixture of i.i.d. sequences of random variables if and only if it is exchangeable [3].

We call distributions which are mixtures of i.i.d. bivariate distributions *completely positive*, due to their connection with completely positive matrices. We show that, given sample access to an unknown distribution p over $[d] \times [d]$, at least $\Omega(d/\epsilon^2)$ samples are necessary to decide with high probability if p is completely positive or ϵ-far from all completely positive distributions in total variation distance. Our proof is a generalization of Paninski's lower bound for testing if a distribution is uniform [15].

Regarding upper bounds, one can again get a trivial upper bound of $O(d^2/\epsilon^2)$ samples for testing complete positivity, simply by fully estimating p to ϵ-accuracy in total variation distance. We again do not know how to close the gap between $\Omega(d/\epsilon^2)$ and $O(d^2/\epsilon^2)$, but we present evidence that the upper bound may be the true complexity. Specifically, a common strategy for trying to test a family D of distributions is to solve the problem of *learning* an unknown distribution promised to be in D. In the full version of the paper, we show that learning a completely positive distribution on $[d] \times [d]$ to accuracy ϵ requires $\Omega(d^2/\epsilon^2)$

samples. On the other hand, we are not able to show an analogous improved lower bound for learning separable quantum states.

1.1 Previous Work

The property testing version of the separability problem, as defined above, appears in [12], where a lower bound of $\Omega(d^2)$ is proven for constant ϵ. As in [12], our proof also reduces the problem of testing if a state is separable to the problem of testing if a state is the maximally mixed state. However, we do not pass through the notion of entanglement of formation, as [12] does, and instead rely on results about the convex structure of the set of separable states. This approach yields a more direct proof that certain random states are w.h.p. far from separable, which allows us to take advantage of a lower bound from [13] (see Theorem 1).

We believe that the separability testing problem has seen further study, but that there has been a lack of results due to its difficulty. There *is* a very extensive literature on the subject of entanglement detection (see e.g. [6,10]), which is concerned with establishing different criteria for detecting or verifying entanglement. However, it is not obvious how these results can be applied in the property testing setting. In particular, few of these criteria are specifically concerned with states that are far from separable in trace distance and many only apply to certain restricted classes of quantum states.

As regards our classical analogue – testing if a bipartite distribution is completely positive (mixture of i.i.d.) – we are not aware of previous work in the literature. The proof of our $\Omega(d/\epsilon^2)$ lower bound is inspired by, and generalizes, Paninski's lower bound for testing if a distribution is uniform [15]. The proof of our tight $\Omega(d^2/\epsilon^2)$ lower bound for learning completely positive distributions uses the Fano inequality method.

1.2 Outline

In Sect. 2 we cover background material on completely positive distributions, quantum states and separability, and the property testing framework that our results are concerned with. In Sect. 3, we prove that testing if a distribution p on $[d] \times [d]$ is completely positive or ϵ-far from all completely positive distributions in total variation distance requires $\Omega(d/\epsilon^2)$ samples from p; in the full version of the paper, we also show that *learning* completely positive distributions requires $\Omega(d^2/\epsilon^2)$ samples. Finally, in Sect. 4, we show that testing if a quantum state ϱ on $\mathbb{C}^d \otimes \mathbb{C}^d$ is separable or ϵ-far from all separable states in trace distance requires $\Omega(d^2/\epsilon^2)$ copies of ϱ when $\epsilon = \Omega(1/\sqrt{d})$.

2 Preliminaries

This section covers the mathematical background and notation used in the rest of the paper.

2.1 Completely Positive Distributions

There is a well-developed theory of completely positive and copositive matrices (see e.g. [5, Chapter 7]). In this section, we review some known material.

Let d be a positive integer. We consider distributions over the grid $[d]^2 = \{(1,1),(1,2),\ldots,(d,d)\}$ which we represent as matrices $A \in \mathbb{R}^{d\times d}$ with A_{ij} being the probability of sampling (i,j).

Example 1. If $p \in \mathbb{R}^d$ is a distribution on $[d] = \{1,\ldots,d\}$ represented as a column vector, then pp^T is the natural i.i.d. product probability distribution on $[d] \times [d]$ derived from p, with $p_i p_j$ being the probability of sampling (i,j).

Definition 1. *A matrix $A \in \mathbb{R}^{d\times d}$ is* completely positive *(CP) if there exist vectors $v_1,\ldots,v_k \in \mathbb{R}^d_{\geq 0}$ with nonnegative entries such that A can be expressed as a convex combination of their projections $v_1 v_1^T,\ldots,v_k v_k^T$, viz.*

$$A = \sum_{i=1}^{k} c_i v_i v_i^T \tag{1}$$

for some nonnegative real numbers $c_1,\ldots,c_k \in \mathbb{R}$ with $c_1 + \cdots + c_k = 1$.

A distribution on $[d]^2$ represented as a matrix A is completely positive *if A is a CP matrix.*

Remark 1. For a CP distribution A, the vectors v_i in Eq. (1) may be taken to be probability distributions, since one can replace v_i by $v_i/\|v_i\|_1$ and c_i by $c_i\|v_i\|_1^2$. Thus, CP distributions are precisely the mixtures of i.i.d. distributions.

It follows immediately from Definition 1 that a CP matrix A satisfies three basic properties:

(i) A is symmetric ($A^\mathsf{T} = A$),
(ii) $A_{ij} \geq 0$ for all $i,j \in [d]$, and
(iii) A is positive semidefinite (PSD), denoted $A \geq 0$.

A matrix satisfying these three properties is called *doubly nonnegative.* However, if $d \geq 5$, then there exist doubly nonnegative matrices which are not completely positive [11].

Example 2. Let J denote the $d \times d$ matrix with $J_{ij} = 1$ for all $i,j \in [d]$ and let $\mathrm{Unif}_{d^2} = J/d^2$ denote the uniform distribution on $[d]^2$. Since $\mathrm{Unif}_{d^2} = (\frac{1}{d},\ldots,\frac{1}{d})(\frac{1}{d},\ldots,\frac{1}{d})^\mathsf{T}$, the uniform distribution on $[d]^2$ is completely positive.

Let CP_d denote the set of completely positive $d \times d$ matrices and let CPD_d denote its subset of completely positive distributions on $[d]^2$. It is well known that CP_d is a cone and that its dual cone consists of *copositive* matrices, i.e. matrices M such that $x^\mathsf{T} M x \geq 0$ for all nonnegative vectors $x \in \mathbb{R}^d_{\geq 0}$. Thus, by cone duality, if $B \notin \mathrm{CP}_d$ is a non-CP matrix, then there exists a copositive matrix W such that $\mathrm{tr}(AW) \geq 0$ for all $A \in \mathrm{CP}_d$ and $\mathrm{tr}(BW) < 0$. This result yields witnesses certifying nonmembership in CPD_d. However, its usefulness is limited by the fact that it provides no quantitative information about how far a nonmember A is from the set CPD_d.

In what follows, we interpret distributions on $[d]^2$ as weighted directed graphs with self-loops and obtain a sufficient condition for a distribution to be ϵ-far in total variation distance from CPD_d in terms of the maximum value of a cut in the corresponding graph.

We interpret a distribution A on $[d]^2$ as a weighted directed graph G with vertices $V(G) = [d]$ and edges $E(G) = \{(i,j) \in [d]^2 \mid A_{ij} > 0\}$.

A *cut* $x \in \{\pm 1\}^d$ in G is a bipartition of the vertices $V(G) = E_1 \cup E_2$ with $E_1 = \{i \in [d] \mid x_i < 0\}$ and $E_2 = \{i \in [d] \mid x_i > 0\}$. The total weight of edges cut by this bipartition is

$$\sum_{(i,j)\in[d]^2} \frac{1 - x_i x_j}{2} A_{ij} = \mathop{\mathbf{E}}_{(i,j)\sim A} \frac{1 - x_i x_j}{2} = \frac{1}{2} - \frac{1}{2} \mathop{\mathbf{E}}_{(i,j)\sim A} x_i x_j = \frac{1}{2} - \frac{1}{2} x^{\mathsf{T}} A x.$$

In particular, if $A = pp^{\mathsf{T}}$ with $p \in \mathbb{R}^d$, then $x^{\mathsf{T}} A x = x^{\mathsf{T}} p p^{\mathsf{T}} x = (x^{\mathsf{T}} p)^2 \geq 0$.

By Remark 1, a CP distribution is a convex combination of matrices of the form pp^{T}. Thus, the following holds:

Proposition 1. *If A is a CP distribution, then the total weight of a cut in the graph represented by A is at most $\frac{1}{2}$.*

This fact allows us to prove the following result which gives a sufficient condition for a distribution to be ϵ-far from all CP distributions in ℓ^1 distance. (The matrix norms in the following are entrywise.)

Proposition 2. *Let A be a distribution on $[d]^2$. If there exists a cut $x \in \{\pm 1\}^d$ with $x^{\mathsf{T}} A x \leq -\epsilon$, then $\|B - A\|_1 \geq \epsilon$ for all $B \in \mathrm{CPD}_d$.*

Proof. Let $B \in \mathrm{CPD}_d$ be arbitrary. By Hölder's inequality, for all $U \in \mathbb{R}^{d\times d}$ with $\|U\|_\infty = 1$, $\|B - A\|_1 \geq \mathrm{tr}(U^{\mathsf{T}}(B - A)) = \mathrm{tr}(U^{\mathsf{T}} B) - \mathrm{tr}(U^{\mathsf{T}} A)$.

Let $U = xx^{\mathsf{T}}$. Since $x^{\mathsf{T}} B x \geq 0$ and $\mathrm{tr}(U^{\mathsf{T}} A) = x^{\mathsf{T}} A x \leq -\epsilon$, $\|B - A\|_1 \geq x^{\mathsf{T}} B x - x^{\mathsf{T}} A x \geq \epsilon$.

2.2 Quantum States and Separability

This section serves as a brief introduction to quantum states and separability. For a more comprehensive introduction, see e.g. [17].

We work over $\mathbb{C}$ and use bra–ket notation to denote vectors in $\mathbb{C}^d$, viz. for all vectors $x, y \in \mathbb{C}^d$ and matrices $A \in \mathbb{C}^{d\times d}$, $|x\rangle = x$, $\langle x| = x^\dagger = \bar{x}^{\mathsf{T}}$, $\langle x \otimes y| = \langle x| \otimes \langle y|$, $|x \otimes y\rangle = |x\rangle \otimes |y\rangle$, $\langle x|y\rangle = x^\dagger y$, $|x\rangle\langle y| = xy^\dagger$, and $\langle x|A|y\rangle = x^\dagger A y$.

Definition 2. *A* quantum state *ρ on $\mathbb{C}^d$ is a positive semidefinite matrix $\rho \in \mathbb{C}^{d\times d}$ with $\mathrm{tr}(\rho) = 1$. A* measurement *is a set $\{E_1, \ldots, E_k\}$ of positive semidefinite matrices on $\mathbb{C}^d$ with $E_1 + \cdots + E_k = \mathbb{1}$, where $\mathbb{1}$ denotes the identity matrix.*

Let ρ and $\{E_1, \ldots, E_k\}$ be as in the definition above and let $p_i = \mathrm{tr}(\rho E_i)$ for $i = 1, \ldots, k$. Since ρ and the E_i are PSD, $p_i \geq 0$ for all $i = 1, \ldots, k$, and

$$p_1 + \cdots + p_k = \mathrm{tr}(\rho E_1) + \cdots + \mathrm{tr}(\rho E_k) = \mathrm{tr}(\rho(E_1 + \cdots + E_k)) = \mathrm{tr}(\rho) = 1.$$

Hence, $(p_1, \ldots, p_k)$ is a distribution on $[k]$. Applying the measurement $\{E_i \mid i \in [k]\}$ to the quantum state ρ yields outcome $i \in [k]$ with probability $p_i = \mathrm{tr}(\rho E_i)$.

Example 3. $\frac{1}{d}$ is a quantum state on $\mathbb{C}^d$ called the *maximally mixed state*; it is analogous to the uniform distribution on $[d]$.

Definition 3. *A state of the form $\rho = |x\rangle\langle x|$ for some $x \in \mathbb{C}^d$ is called a* pure *state.*

Given quantum states ρ and σ on $\mathbb{C}^d$, the tensor product $\rho \otimes \sigma$ is a quantum state on $\mathbb{C}^d \otimes \mathbb{C}^d$. If ρ and σ represent the individual states of two isolated particles, then $\rho \otimes \sigma$ is the state of the physical system comprising both particles. Thus, the system composed of n identical copies of the state ρ is represented as the state $\rho^{\otimes n}$ on $(\mathbb{C}^d)^{\otimes n}$.

Definition 4. *A quantum state ϱ on $\mathbb{C}^d \otimes \mathbb{C}^d$ is* separable *if ϱ can be expressed as a convex combination of product states, viz.*

$$\varrho = \sum_{i=1}^{k} c_i \rho_i \otimes \sigma_i,$$

where ρ_i and σ_i are states on $\mathbb{C}^d$ for $i = 1, \ldots, k$ and $c_1, \ldots, c_k \in \mathbb{R}_{\geq 0}$ satisfy $c_1 + \ldots + c_k = 1$. Thus, the physical system represented by ϱ may be regarded as being in the state $\rho_i \otimes \sigma_i$ with probability c_i.

A state that is not separable is called entangled.

Example 4. Since $\frac{1}{d^2} = \frac{1}{d} \otimes \frac{1}{d}$, the maximally mixed state is separable.

Definition 5. *Let* Sep *denote the set of separable states on $\mathbb{C}^d \otimes \mathbb{C}^d$ and let* $\mathrm{Sep}_{\pm}$ *denote its* cylindrical symmetrization *(cf. [1, p. 81]), viz.* $\mathrm{Sep}_{\pm} = \mathrm{conv}(\mathrm{Sep} \cup (-\mathrm{Sep}))$*, where* $\mathrm{conv}(E)$ *denotes the convex hull of the set E.*

Similar to the duality between completely positive and copositive matrices, the set Sep generates a cone of separable operators whose dual is the cone of *block-positive* operators (see e.g. [1]). A block-positive operator acts as an entanglement witness certifying that a given quantum state is not separable. Thus, Proposition 4 in Sect. 4 is comparable to Proposition 2 in that it describes witnesses certifying that a quantum state is not just entangled but actually ϵ-far from all separable states in trace distance.

2.3 The Property Testing Framework

In the property testing model, we have a set $\mathcal{O}$ of objects and also a distance function dist : $\mathcal{O}\times\mathcal{O} \to \mathbb{R}$. A *property* $\mathcal{P}$ is a subset of $\mathcal{O}$ and the distance between an object $x \in \mathcal{O}$ and the property $\mathcal{P}$ is defined by $\operatorname{dist}(x, \mathcal{P}) = \inf_{y\in\mathcal{P}} \operatorname{dist}(x, y)$. An algorithm $\mathcal{T}$ is said to test $\mathcal{P}$ if, given some type of access to $x \in \mathcal{O}$ (e.g.. independent samples or identical copies), $\mathcal{T}$ accepts x w.h.p. when $x \in \mathcal{P}$ and $\mathcal{T}$ rejects x w.h.p. when $\operatorname{dist}(x, \mathcal{P}) \geq \epsilon$.

In Sect. 3, $\mathcal{O}$ is the set of distributions on $[d] \times [d]$, dist is the total variation distance, and $\mathcal{P} = \mathrm{CPD}_d \subseteq \mathcal{O}$ is the set of CP distributions. Given samples $\boldsymbol{x}_1, \ldots, \boldsymbol{x}_n$ from a distribution p on $[d]^2$, a testing algorithm $\mathcal{T}$ for CPD_d satisfies

$$\begin{aligned} p \in \mathrm{CPD}_d &\implies \mathbf{P}[\mathcal{T}(\boldsymbol{x}_1, \ldots, \boldsymbol{x}_n) \text{ accepts}] \geq \frac{2}{3}, \\ p\ \epsilon\text{-far from }\ \mathrm{CPD}_d &\implies \mathbf{P}[\mathcal{T}(\boldsymbol{x}_1, \ldots, \boldsymbol{x}_n) \text{ accepts}] \leq \frac{1}{3}. \end{aligned}$$

In Sect. 4, $\mathcal{O}$ is the set of quantum states on $\mathbb{C}^d \otimes \mathbb{C}^d$, $\operatorname{dist}(\varrho, \sigma) = \frac{1}{2}\|\varrho - \sigma\|_1$ is the trace distance between quantum states, and $\mathcal{P} = \mathrm{Sep}$ is the set of separable states on $\mathbb{C}^d \otimes \mathbb{C}^d$. Given measurement access to n copies $\varrho^{\otimes n}$ of a state $\varrho \in \mathbb{C}^d \otimes \mathbb{C}^d$, a testing algorithm for Sep is a two-outcome measurement $\{E_0, E_1\}$ on $(\mathbb{C}^d)^{\otimes n}$ satisfying:

$$\begin{aligned} \varrho \in \mathrm{Sep} &\implies \operatorname{tr}(E_1 \varrho^{\otimes n}) \geq \frac{2}{3}, \\ \varrho\ \epsilon\text{-far from Sep} &\implies \operatorname{tr}(E_1 \varrho^{\otimes n}) \leq \frac{1}{3}. \end{aligned}$$

3 Testing Complete Positivity

Let d be a positive integer. If d is odd, we can reduce to the case of $d - 1$ by using distributions that don't involve outcome $d \in [d]$, and the asymptotics of $\Omega(d/\epsilon^2)$ remain unchanged. Hence we may assume, without loss of generality, that d is even.

We begin by defining a family of distributions on $[d]^2$ which are ϵ-far from CPD_d. Let $S \subseteq [d]$ be a subset of size $|S| = \frac{d}{2}$. Thus, $|S^c| = \frac{d}{2}$ and $|S\times S^c \cup S^c \times S| = |S \times S^c| + |S^c \times S| = d^2/2$. Let $\phi_S : [d]^2 \to \mathbb{R}$ be the function defined by

$$\phi_S(x) = \begin{cases} 1 + \epsilon, & x \in S \times S^c \cup S^c \times S \\ 1 - \epsilon, & \text{otherwise.} \end{cases}$$

Hence, $\operatorname{avg}_{x\in[d]^2} \phi_S(x) = \frac{1}{d^2}\left(\frac{d^2}{2}(1 + \epsilon) + \frac{d^2}{2}(1 - \epsilon)\right) = 1$. So we may think of ϕ_S as a density function with respect to the uniform distribution on $[d]^2$.

Let $x \in \{\pm 1\}^d$ be defined as follows: for all $i \in [d]$, if $i \in S$, then $x_i = 1$, otherwise $x_i = -1$. Let A^S be the matrix defined by $A^S_{ij} = \phi_S((i, j))/d^2$. Thus,

A^S is a symmetric distribution on $[d]^2$ and x is a cut. The total weight of this cut is $1/2 + \epsilon/2$.

Therefore, for every subset $S \subseteq [d]$, the distribution A^S is *not* completely positive. Moreover, $x^\top A^S x = -\epsilon$, so, by Proposition 2, $\|A^S - B\|_1 \geq \epsilon$ for every CP distribution B (where the matrix norm is entry-wise). In other words, for every subset $S \subseteq [d]$ with $|S| = \frac{d}{2}$, A^S is a distribution on $[d]^2$ which is ϵ-far in ℓ^1 distance from every CP distribution on $[d]^2$.

Fix $\Omega = [d]^2$ and let $\phi : \Omega^n \to \mathbb{R}$ denote the function defined by

$$\phi(x) = \operatorname*{avg}_{\substack{S \subseteq [d] \\ |S| = d/2}} \phi_S(x_1) \cdots \phi_S(x_n).$$

Let $\mathcal{D}_n$ denote the distribution on Ω^n defined by the density ϕ and let $d_{\chi^2}(\_,\_)$ denote the χ^2-distance between probability distributions, i.e. for distributions $\mathcal{P}$ and $\mathcal{Q}$ on Ω,

$$d_{\chi^2}(\mathcal{P}, \mathcal{Q}) = \mathop{\mathbf{E}}_{\boldsymbol{x} \sim \mathcal{Q}} \left[\left(\frac{\mathcal{P}(\boldsymbol{x})}{\mathcal{Q}(\boldsymbol{x})} - 1 \right)^2 \right].$$

The following proposition, proved in the full version of the paper, will be shown to imply our lower bound:

Proposition 3. *If $d_{\chi^2}(\mathcal{D}_n, \mathrm{Unif}_{d^2}^{\otimes n}) \geq \frac{1}{3}$, then $n = \Omega(d/\epsilon^2)$.*

Let $d_{\mathrm{TV}}(\_,\_)$ denote the total variation distance between probability distributions. Let $p \in \mathrm{CPD}_d$ and let q be a distribution ϵ-far from CPD_d.

A testing algorithm $f : ([d]^2)^n \to \{0,1\}$ for complete positivity determines a probability event $E \subseteq ([d]^2)^n$ satisfying $p^{\otimes n}(E) \geq 2/3$ and $q^{\otimes n}(E) \leq 1/3$. Hence, $\mathrm{Unif}_{d^2}^{\otimes n}(E) \geq 2/3$ and, since $\mathcal{D}_n$ is supported on distributions ϵ-far from CPD_d, $\mathcal{D}_n(E) \leq 1/3$. Therefore, $d_{\mathrm{TV}}(\mathcal{D}_n, \mathrm{Unif}_{d^2}^{\otimes n}) \geq 1/3$ and the following corollary establishes the lower bound:

Corollary 1. *If $d_{\mathrm{TV}}(\mathcal{D}_n, \mathrm{Unif}_{d^2}^{\otimes n}) \geq 1/3$, then $n = \Omega(d/\epsilon^2)$.*

Proof. The inequality $(d/4n\epsilon^2 - 1)^{-1} \geq d_{\chi^2}(\mathcal{D}_n, \mathrm{Unif}_{d^2}^{\otimes n})$ is obtained in the proof of Proposition 3. Since $2d_{\mathrm{TV}}(\mu,\nu)^2 \leq d_{\chi^2}(\mu,\nu)$ holds for all distributions μ and ν, it follows that $(d/4n\epsilon^2 - 1)^{-1} \geq 2d_{\mathrm{TV}}(\mathcal{D}_n, \mathrm{Unif}_{d^2}^{\otimes n})^2 \geq 2/9$. Therefore, $n = \Omega(d/\epsilon^2)$.

4 Testing Separability

Let d be a positive integer. As in the previous section, we may assume, without loss of generality, that d is even.

Let $\mathcal{H} = \mathbb{C}^d \otimes \mathbb{C}^d$, let $\mathrm{U}(\mathcal{H})$ denote the set of unitary operators on $\mathcal{H}$, and recall that Sep denotes the set of separable states on $\mathcal{H}$. For all operators T on $\mathcal{H}$, let $\|T\|_p$ denote the Schatten p-norm of T, viz. $\|T\|_p = (\mathrm{tr}(|T|^p))^{\frac{1}{p}}$, where

$|T| = \sqrt{T^\dagger T}$ is the absolute value of the operator T. Let $d_{\mathrm{tr}}(\varrho, \sigma) = \frac{1}{2}\|\varrho - \sigma\|_1$ denote the trace distance between quantum states ϱ and σ.

We begin by defining a family of quantum states which are with high probability $O(\epsilon)$-far from Sep. For $0 \le \epsilon \le 1/2$, let D_ϵ be the diagonal matrix on $\mathcal{H}$ defined by

$$\mathsf{D}_\epsilon = \mathrm{diag}\left(\frac{1+2\epsilon}{d^2}, \ldots, \frac{1+2\epsilon}{d^2}, \frac{1-2\epsilon}{d^2}, \ldots, \frac{1-2\epsilon}{d^2}\right),$$

where $\mathrm{tr}(\mathsf{D}_\epsilon) = 1$, and let $\mathcal{D}$ denote the family of all quantum states on $\mathcal{H}$ with the same spectrum as D_ϵ, viz. $\mathcal{D} = \{U\mathsf{D}_\epsilon U^\dagger \mid U \in \mathrm{U}(\mathcal{H})\}$.

Our lower bound will rely on the next theorem which follows immediately from [13, Lemma 2.22 and Theorem 4.2[1]]:

Theorem 1. *$\Omega(d^2/\epsilon^2)$ copies are necessary to test whether a quantum state ϱ on $\mathcal{H}$ is the maximally mixed state or $\varrho \in \mathcal{D}$.*

If $\boldsymbol{U}$ is a random unitary on $\mathcal{H}$ distributed according to the Haar measure, then $\boldsymbol{\varrho} = \boldsymbol{U}\mathsf{D}_\epsilon\boldsymbol{U}^\dagger$ is a random element of $\mathcal{D}$. This induced probability measure is invariant under conjugation by a fixed unitary: for all $V \in \mathrm{U}(\mathcal{H})$, $V\boldsymbol{\varrho}V^\dagger$ has the same distribution as $\boldsymbol{\varrho}$. We want to show the following:

Lemma 1. *There is a universal constant C_0 such that for all $C_0/\sqrt{d} \le \epsilon \le 1/2$, the following holds when $\boldsymbol{\varrho} = \boldsymbol{U}\mathsf{D}_\epsilon\boldsymbol{U}^\dagger$ is a uniformly random state in $\mathcal{D}$ in the sense discussed above: $\mathbf{P}[\forall\sigma \in \mathrm{Sep},\ \|\boldsymbol{\varrho} - \sigma\|_1 \ge 2\epsilon] \ge \frac{2}{3}$.*

As ϵ tends to zero, the elements of $\mathcal{D}$ get closer to the maximally mixed state and eventually become separable, by the Gurvits–Barnum theorem [8]. Indeed, if $\epsilon \le 1/(2\sqrt{d^2-1})$, then $\mathcal{D} \subseteq \mathrm{Sep}$. Hence, some assumption on ϵ is necessary for Lemma 1 to hold.

Lemma 1 and Theorem 1 easily imply the desired lower bound:

Theorem 2. *Let ϱ be a quantum state on $\mathbb{C}^d \otimes \mathbb{C}^d$ and let $\epsilon = \Omega(1/\sqrt{d})$. Testing if ϱ is separable or ϵ-far from Sep in trace distance requires $\Omega(d^2/\epsilon^2)$ copies of ϱ.*

Proof. Let $\{E_0, E_1\}$ be a measurement corresponding to a separability testing algorithm using n copies of ϱ. To apply the lower bound in Theorem 1, we use $\{E_0, E_1\}$ to define an algorithm that decides w.h.p. if a state ϱ is equal to the maximally mixed state $\frac{\mathbb{1}}{d^2}$ or $\varrho \in \mathcal{D}$.

Let $\varrho^{\otimes n}$ be given with either $\varrho \in \mathcal{D}$ or $\varrho = \frac{\mathbb{1}}{d^2}$. Note that, for all $\varrho \in \mathcal{D}$, $d_{\mathrm{tr}}(\varrho, \frac{\mathbb{1}}{d^2}) \ge \epsilon$ holds. Let $\boldsymbol{U}$ be a random unitary. If ϱ is the maximally mixed state, then $V\varrho V^\dagger = \varrho$ for all $V \in \mathrm{U}(\mathcal{H})$, so $(\boldsymbol{U}\varrho\boldsymbol{U}^\dagger)^{\otimes n} = \varrho^{\otimes n}$. Otherwise, $\boldsymbol{U}\varrho\boldsymbol{U}^\dagger$ is a random state in $\mathcal{D}$.

Applying the separability test $\{E_0, E_1\}$ to $\boldsymbol{U}\varrho\boldsymbol{U}^\dagger$, we have that:

(i) if $\boldsymbol{U}\varrho\boldsymbol{U}^\dagger = \varrho = \frac{\mathbb{1}}{d^2}$, then $\boldsymbol{U}\varrho\boldsymbol{U}^\dagger$ is separable, so $\mathrm{tr}((\boldsymbol{U}\varrho\boldsymbol{U}^\dagger)^{\otimes n}E_1) = \mathrm{tr}(\varrho^{\otimes n}E_1) \ge \frac{2}{3}$.

[1] Note that d in [13] corresponds to d^2 in the present paper.

(ii) if $\varrho \in \mathcal{D}$, then the probability of error is $\mathbf{E}_{\boldsymbol{U}} \operatorname{tr}((\boldsymbol{U}\varrho\boldsymbol{U}^\dagger)^{\otimes n} E_1)$, which is at most $\mathbf{P}[\boldsymbol{U}\varrho\boldsymbol{U}^\dagger$ is ϵ-close to Sep$] + \mathbf{P}[$test fails $\mid \boldsymbol{U}\varrho\boldsymbol{U}^\dagger$ is ϵ-far from Sep$] \leq \frac{1}{3} + \frac{1}{3} \cdot \frac{2}{3} = \frac{5}{9}$, where the second inequality follows from Lemma 1.

Thus, using the separability test, we can distinguish w.h.p. between $\varrho = \frac{\mathbb{1}}{d^2}$ and $\varrho \in \mathcal{D}$ using n copies of ϱ. Therefore, by Theorem 1, $n = \Omega(d^2/\epsilon^2)$.

It remains to show that Lemma 1 holds. Its proof relies on two main facts: first, that Sep is approximated by a polytope with $\exp(O(d))$ vertices which are separable pure states; and, second, that a random element of $\mathcal{D}$ is ϵ-far from a fixed pure state except with probability $\exp(-O(d))$.

The first fact follows from the next lemma which is a rephrasing of [1, Lemma 9.4]:

Lemma 2. *There exists a constant $C > 0$ such that, for every dimension d, there is a family $\mathcal{N}$ of pure product states on $\mathcal{H}$ (i.e. states of the form $|x \otimes y\rangle\langle x \otimes y|$ with $x, y \in \mathbb{C}^d$) with $|\mathcal{N}| \leq C^d$ satisfying* $\operatorname{conv}(\mathcal{N} \cup -\mathcal{N}) \subseteq \mathrm{Sep}_\pm \subseteq 2\operatorname{conv}(\mathcal{N} \cup -\mathcal{N})$.

Now, we wish to upper bound the probability that a random element of $\mathcal{D}$ is ϵ-far from a fixed pure state. The following result provides a sufficient condition for a state σ on $\mathcal{H}$ to be ϵ-far from a state $\varrho \in \mathcal{D}$:

Proposition 4. *Let $\varrho \in \mathcal{D}$ be arbitrary and let $W = \dfrac{\mathbb{1}}{d^2} - \varrho$. For all quantum states σ on $\mathcal{H}$, if $\operatorname{tr}(\sigma W) \geq -\epsilon \|W\|_\infty$, then $\|\varrho - \sigma\|_1 \geq \epsilon$.*

Proof. Note that $\operatorname{tr}(\varrho W) = \frac{1}{d^2} - \operatorname{tr}(\varrho^2) = \frac{1}{d^2} - \frac{1+4\epsilon^2}{d^2} = -\frac{4\epsilon^2}{d^2}$ and $\|W\|_\infty = \left\|\frac{\mathbb{1}}{d^2} - \mathsf{D}_\epsilon\right\|_\infty = \frac{2\epsilon}{d^2}$. By Hölder's inequality for matrices, $\operatorname{tr}((\sigma - \varrho)W) \leq \|\sigma - \varrho\|_1 \cdot \|W\|_\infty$. Hence, $\|\sigma - \varrho\|_1 \geq \frac{\operatorname{tr}(\sigma W) - \operatorname{tr}(\varrho W)}{\|W\|_\infty} = 2\epsilon + \frac{\operatorname{tr}(\sigma W)}{\|W\|_\infty}$.

When $\sigma = |x\rangle\langle x|$ with $x \in \mathcal{H}$ and $\varrho = U\mathsf{D}_\epsilon U^\dagger$, we have

$$\operatorname{tr}(|x\rangle\langle x|W) = \langle x|W|x\rangle = \langle x|U\left(\tfrac{\mathbb{1}}{d^2} - \mathsf{D}_\epsilon\right)U^\dagger|x\rangle = \|W\|_\infty \cdot \langle x|U\mathsf{Z}U^\dagger|x\rangle, \quad (2)$$

where $\mathsf{Z} = \operatorname{diag}(-1, \ldots, -1, 1, \ldots, 1)$ is just $\mathbb{1}/d^2 - \mathsf{D}_\epsilon$ divided by $\|W\|_\infty$. Hence, $\|\varrho - |x\rangle\langle x|\|_1 \geq \epsilon$ holds if $\langle x|U\mathsf{Z}U^\dagger|x\rangle \geq -\epsilon$.

Since we are interested in the case when $\boldsymbol{\varrho} = \boldsymbol{U}\mathsf{D}_\epsilon\boldsymbol{U}^\dagger$ is random, it suffices to show that $\langle x|\boldsymbol{U}\mathsf{Z}\boldsymbol{U}^\dagger|x\rangle$ concentrates in the interval $[-\epsilon, \epsilon]$. This fact follows easily from the next lemma:

Lemma 3. *Let k be a positive even integer. If $\boldsymbol{u} \in \mathbb{C}^k$ is a uniformly random unit vector, then, for sufficiently large k, $\mathbf{P}\left[|\langle \boldsymbol{u}|Z|\boldsymbol{u}\rangle| \geq \frac{1}{2}ck^{-1/4}\right]$ is at most $4\exp(-\sqrt{k}c^2/8)$, where $Z = \operatorname{diag}(1, \ldots, 1, -1, \ldots, -1)$ is a $k \times k$ diagonal matrix with $\operatorname{tr}(Z) = 0$ and c may be any positive constant.*

Proof. Let $\boldsymbol{u} = (\boldsymbol{a}_1 + i\boldsymbol{b}_1, \ldots, \boldsymbol{a}_k + i\boldsymbol{b}_k) \in \mathbb{C}^k$ be a uniformly random unit vector with $\boldsymbol{a}_1, \ldots, \boldsymbol{a}_k, \boldsymbol{b}_1, \ldots, \boldsymbol{b}_k \in \mathbb{R}$ and let $\boldsymbol{v} \in \mathbb{R}^{2k}$ be defined by $\boldsymbol{v} = (\boldsymbol{a}_1, \ldots, \boldsymbol{a}_{\frac{k}{2}}, \boldsymbol{b}_1, \ldots, \boldsymbol{b}_{\frac{k}{2}}, \boldsymbol{a}_{\frac{k}{2}+1}, \ldots, \boldsymbol{a}_k, \boldsymbol{b}_{\frac{k}{2}+1}, \ldots, \boldsymbol{b}_k)$.

Let D be the $2k \times 2k$ diagonal matrix $D = \mathrm{diag}(1,\ldots,1,-1,\ldots,-1)$ with $\mathrm{tr}(D) = 0$. Thus, $\boldsymbol{v}$ is a uniformly random real unit vector such that $\langle \boldsymbol{v}|D|\boldsymbol{v}\rangle = \langle \boldsymbol{u}|Z|\boldsymbol{u}\rangle$.

Let $\boldsymbol{x}_1,\ldots,\boldsymbol{x}_k,\boldsymbol{y}_1,\ldots,\boldsymbol{y}_k \in \mathbb{R}$ be $2k$ standard Gaussian random variables. Let $\boldsymbol{X} = \boldsymbol{x}_1^2 + \cdots + \boldsymbol{x}_k^2$ and $\boldsymbol{Y} = \boldsymbol{y}_1^2 + \cdots + \boldsymbol{y}_k^2$. By the rotational symmetry of multivariate Gaussian random variables, $\boldsymbol{v}$ has the same distribution as $\frac{(\boldsymbol{x}_1,\ldots,\boldsymbol{x}_k,\boldsymbol{y}_1,\ldots,\boldsymbol{y}_k)}{\sqrt{\boldsymbol{X}+\boldsymbol{Y}}}$.

Hence, $\langle \boldsymbol{v}|D|\boldsymbol{v}\rangle$ and $\frac{\boldsymbol{X}-\boldsymbol{Y}}{\boldsymbol{X}+\boldsymbol{Y}}$ have the same distribution. Since $\boldsymbol{X}$ and $\boldsymbol{Y}$ are independent χ^2 random variables with k degrees of freedom each, it holds that (see e.g. [16, Example 2.11]) $\mathbf{P}\big[\big|\frac{X}{k} - 1\big| \geq t\big] \leq 2\exp(-kt^2/8)$, for all $t \in (0,1)$ and similarly for Y. Hence, for $t = ck^{-1/4}$, we have $\mathbf{P}\big[|\boldsymbol{X} - k| \geq ck^{3/4}\big] \leq 2\exp(-\sqrt{k}c^2/8)$.

If $|\boldsymbol{X} - k| < ck^{3/4}$ and $|\boldsymbol{Y} - k| < ck^{3/4}$, then, for k sufficiently large,

$$|\langle \boldsymbol{v}|D|\boldsymbol{v}\rangle| = \frac{|\boldsymbol{X}-\boldsymbol{Y}|}{\boldsymbol{X}+\boldsymbol{Y}} \leq \frac{2ck^{3/4}}{2k - 2ck^{3/4}} = \frac{c}{k^{1/4}-1} < \frac{1}{2}ck^{-1/4}.$$

Hence, $\mathbf{P}[|\langle \boldsymbol{v}|D|\boldsymbol{v}\rangle| < \frac{1}{2}ck^{-1/4}] \geq 1 - 4\exp(-\sqrt{k}c^2/8)$.

If $\boldsymbol{U}$ is a random unitary distributed according to the Haar measure on $\mathrm{U}(\mathcal{H})$ and $x \in \mathcal{H}$ is a fixed unit vector, then $\boldsymbol{u} = \boldsymbol{U}|x\rangle$ is a uniformly random unit vector in $\mathcal{H}$. Hence, we can apply Lemma 3 to $|\langle \boldsymbol{u}|\mathsf{Z}|\boldsymbol{u}\rangle|$ to get

$$\mathbf{P}[|\langle x|\boldsymbol{U}\mathsf{Z}\boldsymbol{U}^\dagger|x\rangle| \geq \epsilon] \leq 4\exp(-dc^2/8), \tag{3}$$

where c is an arbitrary positive constant and $\epsilon \geq \frac{1}{2}cd^{-1/2}$.

We now have all the elements needed to prove Lemma 1:

Proof (Proof of Lemma 1*).* Let $\boldsymbol{\varrho} = \boldsymbol{U}\mathsf{D}_\epsilon\boldsymbol{U}^\dagger$ be a uniformly random element of $\mathcal{D}$ and let $\boldsymbol{W} = \frac{\mathbb{1}}{d^2} - \boldsymbol{\varrho}$. Thus, assuming $\epsilon \geq cd^{-1/2}$,

$$\begin{aligned}
&\mathbf{P}[\forall\sigma \in \mathrm{Sep},\ d_{\mathrm{TV}}(\boldsymbol{\varrho},\sigma) \geq \epsilon] \\
&\quad= \mathbf{P}[\forall\sigma \in \mathrm{Sep},\ \|\boldsymbol{\varrho} - \sigma\|_1 \geq 2\epsilon] \\
&\quad\geq \mathbf{P}[\forall\sigma \in \mathrm{Sep},\ \mathrm{tr}(\sigma\boldsymbol{W}) \geq -2\epsilon\|\boldsymbol{W}\|_\infty] && \text{(by Proposition 4)} \\
&\quad\geq \mathbf{P}[\forall\sigma \in 2\,\mathrm{conv}(\mathcal{N} \cup -\mathcal{N}),\ \mathrm{tr}(\sigma\boldsymbol{W}) \geq -2\epsilon\|\boldsymbol{W}\|_\infty] && \text{(by Lemma 2)} \\
&\quad= \mathbf{P}[\forall|x\rangle\langle x| \in \mathcal{N} \cup -\mathcal{N},\ 2\,\mathrm{tr}(|x\rangle\langle x|\boldsymbol{W}) \geq -2\epsilon\|\boldsymbol{W}\|_\infty] && \text{(by convexity)} \\
&\quad= \mathbf{P}\Big[\forall|x\rangle\langle x| \in \mathcal{N},\ |\langle x|\boldsymbol{U}\mathsf{Z}\boldsymbol{U}^\dagger|x\rangle| \leq \epsilon\Big] && \text{(by Equation (2))} \\
&\quad\geq 1 - \sum_{|x\rangle\langle x| \in \mathcal{N}} \mathbf{P}\Big[|\langle x|\boldsymbol{U}\mathsf{Z}\boldsymbol{U}^\dagger|x\rangle| > \epsilon\Big] && \text{(by the union bound)} \\
&\quad\geq 1 - |\mathcal{N}| \cdot 4\exp(-dc^2/8) && \text{(by Equation (3))} \\
&\quad= 1 - 4\exp(d(\log C - c^2/8)) && \text{(since } |\mathcal{N}| = C^d\text{).}
\end{aligned}$$

Hence, if $c = \sqrt{8(\log C + 1)}$, then, for $d \geq \log 12$,

$$\mathbf{P}[\forall\sigma \in \mathrm{Sep},\ d_{\mathrm{TV}}(\boldsymbol{\varrho},\sigma) \geq \epsilon] \geq 1 - 4\exp(d(\log C - c^2/8)) = 1 - 4\exp(-d) \geq \frac{2}{3}.$$

References

1. Aubrun, G., Szarek, S.: Alice and Bob Meet Banach. American Mathematical Society, Providence (2017)
2. Aubrun, G., Szarek, S.: Dvoretzky's theorem and the complexity of entanglement detection. Discr. Anal. 1–20 (2017)
3. Diaconis, P.: Finite forms of de Finetti's theorem on exchangeability. Synthese **36**(2), 271–281 (1977). https://doi.org/10.1007/BF00486116
4. Doherty, A.C., Parrilo, P.A., Spedalieri, F.M.: Complete family of separability criteria. Phys. Rev. A **69**(2), 022308 (2004)
5. Gärtner, B., Matoušek, J.: Approximation Algorithms and Semidefinite Programming. Springer, Heidelberg (2012). https://doi.org/10.1007/978-3-642-22015-9
6. Gühne, O., Tóth, G.: Entanglement detection. Phys. Rep. **474**(1–6), 1–75 (2009)
7. Gurvits, L.: Classical complexity and quantum entanglement. J. Comput. Syst. Sci. **69**(3), 448–484 (2004)
8. Gurvits, L., Barnum, H.: Largest separable balls around the maximally mixed bipartite quantum state. Phys. Rev. A **66**(6), 062311 (2002)
9. Haah, J., Harrow, A.W., Ji, Z., Wu, X., Yu, N.: Sample-optimal tomography of quantum states. In: Proceedings of the 48th Annual ACM SIGACT Symposium on Theory of Computing, STOC 2016. ACM Press (2016)
10. Horodecki, R., Horodecki, P., Horodecki, M., Horodecki, K.: Quantum entanglement. Rev. Mod. Phys. **81**(2), 865–942 (2009)
11. Maxfield, J.E., Minc, H.: On the matrix equation $X'X = A$. In: Proceedings of the Edinburgh Mathematical Society, vol. 13, no. 02, p. 125 (1962)
12. Montanaro, A., de Wolf, R.: A survey of quantum property testing. Theory Comput. **1**(1), 1–81 (2016)
13. O'Donnell, R., Wright, J.: Quantum spectrum testing. In: Proceedings of the 47th Annual ACM SIGACT Symposium on Theory of Computing, STOC 2015. ACM Press (2015)
14. O'Donnell, R., Wright, J.: Efficient quantum tomography. In: Proceedings of the 48th Annual ACM SIGACT Symposium on Theory of Computing - STOC 2016. ACM Press (2016)
15. Paninski, L.: A coincidence-based test for uniformity given very sparsely sampled discrete data. IEEE Trans. Inf. Theory **54**(10), 4750–4755 (2008)
16. Wainwright, M.J.: High-Dimensional Statistics. Cambridge University Press, Cambridge (2019)
17. Watrous, J.: The Theory of Quantum Information. Cambridge University Press, Cambridge (2018)

Exponential-Time Quantum Algorithms for Graph Coloring Problems

Kazuya Shimizu[1] and Ryuhei Mori[1,2](✉)

[1] School of Computing, Tokyo Institute of Technology, Tokyo, Japan
shimizu.k.ap@m.titech.ac.jp, mori@c.titech.ac.jp
[2] Japan Science and Technology Agency, PRESTO, Tokyo, Japan

Abstract. The fastest known classical algorithm deciding the k-colorability of n-vertex graph requires running time $\Omega(2^n)$ for $k \geq 5$. In this work, we present an exponential-space quantum algorithm computing the chromatic number with running time $O(1.9140^n)$ using quantum random access memory (QRAM). Our approach is based on Ambainis et al.'s quantum dynamic programming with applications of Grover's search to branching algorithms. We also present a polynomial-space quantum algorithm not using QRAM for the graph 20-coloring problem with running time $O(1.9575^n)$. For the polynomial-space quantum algorithm, we essentially show $(4 - \epsilon)^n$-time *classical* algorithms that can be improved quadratically by Grover's search.

Keywords: Quantum algorithm · Graph coloring · Grover's search · Dynamic programming

1 Introduction

Exhaustive search is believed to be (almost) the fastest classical algorithm for many NP-complete problems including SAT, hitting set problem, etc. [8]. Grover's quantum search quadratically improves the running time of exhaustive search [15]. Hence, the best classical running time for many NP-complete problems can be quadratically improved by quantum algorithms. On the other hand, non-trivial faster classical algorithms are known for some NP-complete problems including the travelling salesman problem (TSP), the graph coloring problem, etc. For these problems, more complicated techniques, such as dynamic programming, arithmetic algorithm based on inclusion–exclusion principle, etc., are used in the fastest known classical algorithms. It is not obvious how to boost these classical algorithms by a quantum computer. Recently, Ambainis et al. showed a general idea of quantum dynamic programming using quantum random access memory (QRAM), and showed quantum speedup for many NP-complete and NP-hard problems including TSP, set cover, etc. [1]. Ambainis et al.'s work gives a new general method for exact exponential-time quantum algorithms.

In this work, we present exact exponential-time quantum algorithms for the graph coloring problem. The fastest known classical algorithm computes the

Y. Kohayakawa and F. K. Miyazawa (Eds.): LATIN 2020, LNCS 12118, pp. 387–398, 2020.
https://doi.org/10.1007/978-3-030-61792-9_31

Table 1. $O(2^{d_k^* n})$-time quantum algorithms not using QRAM for the graph k-coloring problem.

k	d_k^*	$2^{d_k^*}$	k	d_k^*	$2^{d_k^*}$	k	d_k^*	$2^{d_k^*}$
3	0.2051	1.1528	9	0.8041	1.7460	15	0.9488	1.9303
4	0.4039	1.3231	10	0.8298	1.7775	16	0.9488	1.9303
5	0.5553	1.4695	11	0.8298	1.7775	17	0.9488	1.9303
6	0.6099	1.5261	12	0.8676	1.8246	18	0.9536	1.9366
7	0.7234	1.6511	13	0.8874	1.8499	19	0.9690	1.9575
8	0.7299	1.6585	14	0.8938	1.8580	20	0.9691	1.9575

chromatic number of n-vertex graph with running time poly$(n)2^n$ on the random access memory (RAM) model. The main result of this work is the following theorem.

Theorem 1. *There is an exponential-space bounded-error quantum algorithm using QRAM for the chromatic number problem with running time*[1] $O^*((2^{37/35} 3^{3/7} 5^{-9/70} 7^{-5/28})^n) = O(1.9140^n)$.

The quantum algorithm in Theorem 1 is based on Ambainis et al.'s quantum dynamic programming for TSP with applications of Grover's search to Byskov's algorithm enumerating all maximal independent sets (MISs) of fixed size [7]. Byskov's algorithm is not naive exhaustive search, but is a branching algorithm (also referred as Branch & Reduce), for which Grover's search can be applied [12]. While RAM is widely accepted model of classical computation, QRAM is sometimes criticized due to the difficulty of implementation. In this paper, we also present quantum algorithms not using QRAM.

Theorem 2. *For $k \leq 20$, there exists $\epsilon > 0$ such that there are polynomial-space bounded-error quantum algorithms not using QRAM for the graph k-coloring problem with running time $2^{(1-\epsilon)n}$.*

Note that classical algorithms with running time $2^{(1-\epsilon)n}$ are known only for $k = 3, 4$ [2,7]. Running times of the quantum algorithms in Theorem 2 are shown in Table 1. For proving Theorem 2, we essentially show *classical* algorithms with running time $4^{(1-\epsilon)n}$ that can be improved quadratically by Grover's search. These classical algorithms are obtained by generalizing Byskov's techniques for reducing the graph $k(\geq 4)$-coloring problem to the graph 3-coloring problem [7]. The proofs of Theorem 1 and 2 in this paper require numerical calculations.

1.1 Related Work

Since a graph is k-colorable if and only if the set of vertices can be partitioned into k independent sets, many algorithms for the graph k-coloring problem use

[1] In this paper, $O^*(f(n))$ means $O(\text{poly}(n)f(n))$.

enumeration algorithms of independent sets. There is a simple branching algorithm enumerating all MISs in time $O^*(3^{n/3}) = O(1.4423^n)$ [11]. Lawler showed that 3-colorability can be decided in time $O^*(3^{n/3})$ by enumerating all MISs and checking the bipartiteness of the subgraph induced by the complement of each MIS [17]. Lawler also showed that the chromatic number can be computed in time $O(2.4423^n)$ by a simple dynamic programming.

Beigel and Eppstein showed an efficient algorithm for the graph 3-coloring problem with running time $O(1.3289^n)$ [2]. Byskov showed reduction algorithms from the graph $k(\geq 4)$-coloring problem to the graph 3-coloring problem [7]. By using Beigel and Eppstein's graph 3-coloring algorithm, Byskov showed classical algorithms for the graph 4-, 5- and 6-coloring problems with running time $O(1.7504^n)$, $O(2.1592^n)$ and $O(2.3289^n)$, respectively. Fomin et al. showed an algorithm for the graph 4-coloring problem with running time $O(1.7272^n)$ by using the path decomposition [10].

In 2006, Björklund and Husfeldt, and Koivisto showed an exponential-space $O^*(2^n)$-time algorithm for the chromatic number problem in the RAM model [3,16]. These algorithms are based on the inclusion–exclusion principle. They also showed that if there is a polynomial-space $O^*(\alpha^n)$-time algorithm counting the number of independent sets, then there is a polynomial-space $O^*((1+\alpha)^n)$-time algorithm computing the chromatic number [3,5]. Since the fastest known polynomial-space algorithm computes the number of independent sets with running time $O(1.2356^n)$ [13], there is a polynomial-space $O(2.2356^n)$-time algorithm computing the chromatic number.

There is almost no previous theoretical work on quantum algorithms for the graph coloring problems. Fürer mentioned that Grover's algorithm can be applied to branching algorithms, and hence, Beigel and Eppstein's algorithm for the graph 3-coloring problem can be improved to running time $O(\sqrt{1.3289}^n) = O(1.1528^n)$ [12]. The quantum algorithms for Theorem 2 are basically obtained by applying Grover's search to generalized Byskov's reduction algorithms on the basis of Fürer's observation.

For general NP-hard problems, Ambainis et al. showed exponential-space exponential-time quantum algorithms using QRAM for many NP-hard problems [1]. The quantum algorithm for Theorem 1 is based on Ambainis et al.'s algorithm for TSP with application of Grover's search to Byskov's algorithm enumerating MISs of fixed size on the basis of Fürer's observation.

1.2 Overview of Quantum Algorithms

Quantum Algorithm for the Chromatic Number Problem. Similarly to Ambainis et al.'s quantum algorithm for TSP, the quantum algorithm for Theorem 1 is a simple divide-and-conquer algorithm with dynamic programming approach. The basic classical algorithm was shown in [4, Proposition 3]. The chromatic number of a graph G is equal to a sum of the chromatic numbers of $G[S]$ and $G[V \setminus S]$ for some non-empty $S \subsetneq V$ unless G is one-colorable. If we can assume that S has size exactly $\lfloor n/2 \rfloor$ or $\lceil n/2 \rceil$, then we can consider a classical algorithm that recursively finds S of size $\lfloor n/2 \rfloor$ minimizing $\chi(G[S]) + \chi(G[V \setminus S])$. Let $T(n)$ be the running time of this algorithm. Then, it follows

$T(n) = \binom{n}{\lfloor n/2 \rfloor}(T(\lfloor n/2 \rfloor) + T(\lceil n/2 \rceil))$, so that we can apply Ambainis et al.'s quantum dynamic programming straightforwardly and obtain $O(1.7274^n)$-time quantum algorithm [1]. However, the balanced partition S satisfying $\chi(G) = \chi(G[S]) + \chi(G[V \setminus S])$ does not necessarily exist. Hence, we use the following useful fact.

Fact 1. *Let $a_1, \ldots, a_k$ be positive integers, and $n := \sum_{i=1}^{k} a_i$. Assume that $a_1 \geq a_i$ for all $i \in \{1, 2, \ldots, k\}$. Then, for any $m \in \{1, 2, \ldots, n-1\}$, there exists $S \subseteq \{2, 3, \ldots, k\}$ such that $\sum_{i \in S} a_i \leq m$ and $\sum_{i \in \{2,\ldots,k\} \setminus S} a_i \leq n - m - 1$.*

Proof. Let $t := \max\{j \in \{2, \ldots, k\} \mid \sum_{i=2}^{j} \leq m\}$. Let $S := \{2, 3, \ldots, t\}$. Then, $\sum_{i \in \{t+1, t+2, \ldots, k\}} a_i \leq \sum_{i \in \{1, t+2, t+3, \ldots, k\}} a_i = n - \sum_{i \in \{2, 3, \ldots, t+1\}} a_i \leq n - m - 1$. □

From Fact 1, we can consider the following quantum algorithm computing the chromatic number. First, the algorithm precomputes the chromatic number of all induced subgraphs with size at most $\lfloor n/4 \rfloor$. This precomputation is based on Lawler's formula

$$\chi(G) = 1 + \min_{I \in \mathrm{MIS}(G)} \chi(G[V \setminus I]) \tag{1}$$

where $\mathrm{MIS}(G)$ denotes the set of all MISs of G [17]. There is a classical algorithm enumerating all MISs with running time $O^*(3^{n/3})$. We will show in Sect. 3 that Grover's search can be applied to this algorithm, and hence, the quantum algorithm can search for all MISs with running time $O^*(3^{n/6})$. Here, computed chromatic numbers are stored to QRAM. Hence, we can apply Grover's search for computing the minimum in (1). The precomputation requires the running time $O^*\left(\sum_{i=1}^{\lfloor n/4 \rfloor} \binom{n}{i} 3^{i/6}\right) = O(1.8370^n)$. Then, the main part of the algorithm computes the chromatic number of G by using the formula

$$\chi(G) = 1 + \min_{I \in \mathrm{MIS}(G)} \min_{S \subseteq V \setminus I,\, |S| \leq \lfloor n/2 \rfloor,\, |V \setminus I \setminus S| \leq \lfloor n/2 \rfloor} \left\{ \chi(G[S]) + \chi(G[V \setminus I \setminus S]) \right\}$$

for $\chi(G) \geq 3$. This formula is justified by Fact 1 for $m = \lfloor n/2 \rfloor$. Grover's search is used for finding S. For computing $\chi(G[S])$ and $\chi(G \setminus I \setminus S])$, the above formula is used again. Then, we need the chromatic numbers of subgraphs of G of size at most $\lfloor n/4 \rfloor$, which were precomputed and stored to QRAM. The running time of the main part of this quantum algorithm is

$$O^*\left(3^{n/6} 2^{n/2} 3^{n/12} 2^{n/4}\right) = O(2.2134^n).$$

Quantum algorithm for Theorem 1 searches for all MISs of size t for each $t \in \{1, 2, \ldots, n\}$ separately. Therefore, the above estimate of the running time is larger than the actual running time since if MIS I of size t is chosen, the remaining graph $G[V \setminus I]$ has only $n - t$ vertices. Hence, the factor $2^{n/2}$ in the above estimate can be replaced by $2^{(n-t)/2}$. Then, precise analysis shows that the running time of the improved quantum algorithm is $O(1.9140^n)$.

Quantum Algorithms for the Graph k-Coloring Problem. We will derive classical algorithms that can be improved quadratically by Grover's search. In the classical algorithms, the graph $k(\geq 4)$-coloring problem is reduced to the graph k'-coloring problems for some $k' < k$. Since a graph G is $k(\geq 2)$-colorable if and only if there exists a subset S of vertices such that $G[S]$ is $\lfloor k/2 \rfloor$-colorable and $G[V \setminus S]$ is $\lceil k/2 \rceil$-colorable. Let us consider a classical algorithm that simply searches for $S \subseteq V$ satisfying the above condition. Let $T_k(n)$ be the running time of this algorithm for the graph k-coloring problem. Then, $T_k(n)$ satisfies

$$T_1(n) = T_2(n) = 1,$$

$$T_k(n) = \sum_{i=0}^{n} \binom{n}{i} (T_{\lfloor k/2 \rfloor}(i) + T_{\lceil k/2 \rceil}(n-i)), \qquad k \geq 3$$

where polynomial factors in n are ignored. Then, we obtain $T_4(n) = O^*(2^n)$, $T_8(n) = O^*(3^n)$ and $T_{16}(n) = O^*(4^n)$. Let us consider a quantum algorithm that uses Grover's search for finding S. Let $T_k^*(n)$ be the running time of the quantum algorithm. Then, it follows $T_k^*(n) = \sum_{i=0}^{n} \sqrt{\binom{n}{i}} (T^*_{\lfloor k/2 \rfloor}(i) + T^*_{\lceil k/2 \rceil}(n-i))$, which implies $T_k^*(n) = O^*(\sqrt{T_k(n)})$. Hence, we obtain $T_4^*(n) = O(1.4143^n)$, $T_8^*(n) = O(1.7321^n)$ and $T_{16}^*(n) = O^*(2^n)$. This yields a weaker version of Theorem 2 that is valid for $k \leq 8$ rather than $k \leq 20$. Better reduction algorithms are used for Theorem 2.

1.3 Organization

In Sect. 2, notations and known classical and quantum algorithms are introduced. In Sect. 3, we present details of quantum algorithm for branching algorithms. In Sect. 4, we prove Theorem 1. The proof of Theorem 2 is not presented in this paper due to the page limit, but included in the full version of this paper [18].

2 Preliminaries

2.1 Definitions and Notations

For a finite vertex set V, a set E of edges consists of subsets of V of size two. A pair (V, E) of finite vertex set V and a set E of edges is called an undirected simple graph. In this paper, we simply call a graph rather than an undirected simple graph. The number of vertices $|V|$ is denoted by n. A mapping $c\colon V \to \{1, 2, \ldots, k\}$ is called k-coloring if $c(v) \neq c(w)$ for all $\{v, w\} \in E$. For a graph G, the smallest k such that there exists a k-coloring is called the chromatic number of G, and denoted by $\chi(G)$. A subset $I \subseteq V$ of vertices is called an independent set if $\{v, w\} \notin E$ for all $v, w \in I$. An independent set I is said to be maximal if there is no strict superset of I that is an independent set. A maximal independent set of size t is called t-MIS. For $S \subseteq V$, $G[S]$ denotes a induced subgraph $(S, \{\{v, w\} \in E \mid v, w \in S\})$ of G. Let $h(\delta) := -\delta \log \delta - (1-\delta) \log(1-\delta)$

for $\delta \in [0,1]$ where $0 \log 0 = 0$. In this paper, the base of logarithm is 2. The notation $g(n) = O^*(f(n))$ means that $g(n) = O(n^c f(n))$ for some constant c. For $O^*(\lambda^n)$, we often round λ up to the fourth digit after the decimal point. In this case, we can use $O()$ rather than $O^*()$. For example, we often write $g(n) = O(1.4143^n)$ rather than $g(n) = O^*(2^{n/2})$. The notation $g(n) = \widetilde{O}(f(n))$ means that $g(n) = O((\log f(n))^c f(n))$ for some constant c.

2.2 Known Classical Algorithm for Enumerating All t-MISs

Byskov showed the following theorem.

Theorem 3 (Byskov [7]). *The maximum number of t-MISs of n-vertex graphs is*

$$I(n,t) := \lfloor n/t \rfloor^{(\lfloor n/t \rfloor + 1)t - n} (\lfloor n/t \rfloor + 1)^{n - \lfloor n/t \rfloor t}.$$

Furthermore, there is a classical algorithm enumerating all t-MISs of n-vertex graph in time $O^*(I(n,t))$.

We can straightforwardly obtain the following lemma and corollary.

Lemma 1. *For any constant $\delta \in (0,1)$, $I(n, \lfloor \delta n \rfloor) = O(2^{E(\delta)n})$ where*

$$E(\delta) := ((\lfloor \delta^{-1} \rfloor + 1)\delta - 1) \log \lfloor \delta^{-1} \rfloor + (1 - \lfloor \delta^{-1} \rfloor \delta) \log(\lfloor \delta^{-1} \rfloor + 1).$$

Here, $E(\delta)$ is concave (and hence, continuous) and piecewise linear for $\delta \in (0,1)$. The maximum of $E(\delta)$ is given at $\delta = 1/3$.

Proof. It is easy to see that $I(n, \lfloor \delta n \rfloor) = O(2^{E(\delta)n})$. If $\lfloor \delta^{-1} \rfloor$ in the definition of $E(\delta)$ is replaced by δ^{-1}, we obtain $-\delta \log \delta$, which is obviously concave for $\delta \in (0,1)$. Furthermore, $E(\delta) = -\delta \log \delta$ if δ is an inverse integer. For $\delta \in (1/(s+1), 1/s)$ where s is some positive integer, $E(\delta)$ is a linear function since $\lfloor \delta^{-1} \rfloor$ is constant in this domain. Therefore, it is sufficient to show that $E(\delta)$ is continuous at $\delta = 1/s$ for all $s \in \mathbb{Z}_{\geq 1}$ for showing that $E(\delta)$ is concave. Obviously, $E(\delta)$ is left-continuous. At $\delta = 1/s$ for some positive integer s, $E(\delta)$ is equal to $-\delta \log \delta = (\log s)/s$ even if $\lfloor \delta^{-1} \rfloor = s$ is replaced by $s - 1$. This means that $E(\delta)$ is right-continuous as well.

Finally, by comparing $E(1/2)$, $E(1/3)$ and $E(1/4)$, it is shown that the maximum of $E(\delta)$ is given at $\delta = 1/3$. □

Corollary 1. *For any $a \in \mathbb{R}_{\geq 0}$ and $t \in \mathbb{Z}_{\geq 3}$, the maximum of $E(\delta) - a\delta$ for $\delta \in [1/t, 1]$ is given at $\delta = 1/s$ for some $s \in \{3, 4, \dots, t\}$.*

2.3 Grover's Search

Here, Grover's search is briefly introduced without using quantum circuit, unitary oracle, etc.

Theorem 4 (Grover [15], Boyer et al. [6]). *Let* $A\colon \{1,2,\ldots,N\} \to \{0,1\}$ *be a bounded-error quantum algorithm with running time* T*. Then, there is a bounded-error quantum algorithm computing* $\bigvee_{x\in\{1,\ldots,N\}} A(x)$ *with running time* $\widetilde{O}(\sqrt{N}T)$*. If it is guaranteed that* $|A^{-1}(1)| \geq M$ *or* $|A^{-1}(1)| = 0$*, then there is a bounded-error quantum algorithm with running time* $\widetilde{O}(\sqrt{N/M}T)$*.*

Theorem 5 (Dürr and Høyer [9]). *Let* $A\colon \{1,2,\ldots,N\} \to \{1,2,\ldots,M\}$ *be a bounded-error quantum algorithm with running time* T*. Then, there is a bounded-error quantum algorithm computing* $\min_{x\in\{1,\ldots,N\}} A(x)$ *with running time* $\widetilde{O}(\sqrt{N}T)$*.*

2.4 QRAM

QRAM is the quantum analogue of RAM which can be accessed in a superposition [14]. QRAM has been used in many quantum algorithms [1]. RAM is the memory that can be accessed in constant or logarithmic time with respect to the memory size. For computing the minimum of $f(x,W)$ for all $x \in \{1,2,\ldots,N\}$ where W denotes a read-only RAM, we can replace RAM with QRAM and apply Grover's search for computing the minimum. Then, we obtain $\widetilde{O}(\sqrt{N}T)$-time quantum algorithm where T denotes the running time for computing f.

3 Grover's Search for Branching Algorithms

Fürer mentioned that Grover's search can be applied to branching algorithms [12]. Since the details of the quantum algorithm were not explicitly described in [12], we will show the details in this section. A branching algorithm is an algorithm which recursively reduce a problem into some problems of smaller parameters. We now consider decision problems with ℓ parameters $n_1, n_2, \ldots, n_\ell$ that are non-negative integers. If the parameters are sufficiently small, we do not apply any branching rule and solve this problem in some way. For a problem P with parameters $n_1, \ldots, n_\ell$ that are not sufficiently small, we choose a branching rule $b(P)$ such that P is reduced to $m_{b(P)}$ problems $P_1, P_2 \ldots, P_{m_{b(P)}}$ of the same class. Here, P_i has parameters $f_1^{b(P),i}(n_1), \ldots, f_\ell^{b(P),i}(n_\ell)$ for some function $f_j^{b(P),i}$ satisfying $f_j^{b(P),i}(n_j) \leq n_j$ for $i = 1,2,\ldots,m_{b(P)}$ and $j = 1,2,\ldots,\ell$. At least one of the parameters of P_i must be smaller than the same parameter of P for all $i \in 1,2,\ldots,m_{b(P)}$. The solution of P is true if and only if at least one of the solutions of $P_1, \ldots, P_{m_{b(P)}}$ is true. Hence, we will call this algorithm OR-branching algorithm. For a problem P of this class, we can consider a *computation tree* that represents the branchings of the reductions. The computation tree for P is a single node if P has sufficiently small parameters, so that no branching rule is performed, and is a rooted tree where children of the root node are the root nodes of the computation trees for $P_1, P_2, \ldots, P_{m_{b(P)}}$ if some branching rule $b(P)$ is applied to P. Let $L(n_1, \ldots, n_\ell)$ be the maximum number of leaves of the computation tree for P with parameters $n_1, \ldots, n_\ell$. Assume that the running

Algorithm 1. Algorithm computing s-th leaf of P

```
1: function LEAF(P, s)
2:     if P is a leaf then return P
3:     Compute the branching rule b ← b(P)
4:     for i ∈ {1, 2, ..., m_b − 1} do
5:         Compute P_i and its parameters n'_1, ..., n'_ℓ = f_1^{b,i}(n_1), ..., f_ℓ^{b,i}(n_ℓ)
6:         if s ≤ U(n'_1, ..., n'_ℓ) then return LEAF(P_i, s)
7:         else  s ← s − U(n'_1, ..., n'_ℓ)
8:     return LEAF(P_{m_b}, s)
```

time of the computation at a non-leaf node, including computations of $b(P)$, P_i, and $f_j^{b(P),i}$, is polynomial with respect to $n_1, \dots, n_\ell$. Then, the total running time of the OR-branching algorithm is at most $\mathrm{poly}(n_1, \dots, n_\ell)L(n_1, \dots, n_\ell)T$ where T is the running time for the computation at a leaf node. We can apply Grover's search to OR-branching algorithms if we have an upper bound of $L(n_1, \dots, n_\ell)$ with some properties.

Lemma 2. *Let $U(n_1, \dots, n_\ell)$ be an upper bound of $L(n_1, \dots, n_\ell)$ that can be computed in polynomial time with respect to the parameters, and satisfies*

$$U(n_1, \dots, n_\ell) \geq \sum_{i=1}^{m_b} U(f_1^{b,i}(n_1), \dots, f_\ell^{b,i}(n_\ell))$$

for any branching rule b. Then, there is a bounded-error quantum algorithm with running time $\mathrm{poly}(n_1, \dots, n_\ell)\sqrt{U(n_1, \dots, n_\ell)}T$.

Proof. If we can assign an integer $s \in \{1, 2, \dots, U(n_1, \dots, n_\ell)\}$ to every leaf of the computation tree, and can compute the corresponding leaf from given s in polynomial time with respect to the parameters, then, we can apply Grover's search for computing

$$a(P) = \bigvee_{Q \in W(P)} a(Q)$$

where $a(P)$ denotes the solution of a problem P and $W(P)$ denotes the set of all problems corresponding to leaves of the computation tree for P. Then, we obtain quantum algorithm with running time $\mathrm{poly}(n_1, \dots, n_\ell)\sqrt{U(n_1, \dots, n_\ell)}T$ [12]. The algorithm computing s-th leaf of a problem P is shown in Algorithm 1. We will show the validity of Algorithm 1.

Proposition 1. *For any problem Q that corresponds to a leaf node of the computation tree of a problem P with parameters $n_1, \dots, n_\ell$, there exists $s \in \{1, 2, \dots, U(n_1, \dots, n_\ell)\}$ such that* $\textsc{Leaf}(P, s) = Q$.

Proof. The proof is an induction on the depth of the computation tree for P. If the computation tree for P consists of a single node, then Algorithm 1 returns P for any s. Since $U(n_1, \dots, n_\ell)$ is an upper bound of $L(n_1, \dots, n_\ell)$, it follows

$U(n_1, \ldots, n_\ell) \geq 1$, and hence, $\{1, 2, \ldots, U(n_1, \ldots, n_\ell)\}$ is non-empty. Assume that the proposition holds for any P with the computation tree of depth at most d. We will consider a problem P with computation tree of depth $d+1$. Let i be the index of the branching at P that achieves Q. From the induction hypothesis, there exists $s' \in \{1, \ldots, U(f_1^{b,i}(n_1), \ldots, f_\ell^{b,i}(n_\ell))\}$ such that $\textsc{Leaf}(P_i, s') = Q$. Let $s := s' + \sum_{j=1}^{i-1} U(f_1^{b,j}(n_1), \ldots, f_\ell^{b,j}(n_\ell))$. Then, $\textsc{Leaf}(P, s) = Q$. Here, $s \leq \sum_{j=1}^{m_b} U(f_1^{b,j}(n_1), \ldots, f_\ell^{b,j}(n_\ell)) \leq U(n_1, \ldots, n_\ell)$. □

From Proposition 1 and a fact that $\textsc{Leaf}(P, s)$ always returns a problem corresponding to one of the leaf nodes for P, we obtain

$$a(P) = \bigvee_{s \in \{1, \ldots, U(n_1, \ldots, n_\ell)\}} a(\textsc{Leaf}(P, s)).$$

Since the depth of the computation tree for P is at most $\sum_{j=1}^{\ell} n_j$, the running time of $\textsc{Leaf}(P, s)$ is polynomial with respect to the parameters. Hence, there is a quantum algorithm computing $a(P)$ with running time $\text{poly}(n_1, \ldots, n_\ell)$ $\sqrt{U(n_1, \ldots, n_\ell)}T$. □

For a problem P whose solution is an integer, we can also consider a branching algorithm satisfying $a(P) = \min_{i=1}^{m_{b(P)}} a(P_i)$ for children $P_1, \ldots, P_{m_{b(p)}}$ of P. In this case, we will call this algorithm MIN-branching algorithm. Similarly to OR-branching algorithm, we can apply Grover's search to MIN-branching algorithm from Theorem 5.

In this paper, we apply Lemma 2 to Byskov's algorithm in Theorem 3. Byskov showed the upper bound $I(n, t)$ satisfying the conditions in Lemma 2 for the branching algorithm with two parameters n and t [7, Theorem 2]. Hence, we can apply Grover's search to Byskov's algorithm in Theorem 3. For example, since $\sum_{t=1}^{n} I(n, t) = O^*(3^{n/3})$, there is a bounded-error quantum algorithm searching for all MISs in time $O^*(3^{n/6})$.

4 Quantum Algorithms for the Chromatic Number Problem

The overview of the quantum algorithm was described in Sect. 1.2. The quantum algorithm for Theorem 1 is shown in Algorithm 2. For computing the chromatic number of $G[S]$, when MIS I of size t is chosen, we have to chose $T \subseteq S \setminus I$ satisfying $|T| \leq |S|/2$ and $|S \setminus I \setminus T| \leq |S|/2$ as mentioned in Sect. 1.2. This implies the condition $|S|/2 - t \leq |T| \leq |S|/2$. Hence, Algorithm 2 computes the chromatic number correctly. By analyzing the running time of Algorithm 2, we obtain the following theorem.

Theorem 6. *Algorithm 2 computes the chromatic number of n-vertex graph with running time* $O^*\left((2^{37/35}3^{3/7}5^{-9/70}7^{-5/28})^n\right) = O(1.9140^n)$ *with bounded error probability.*

Algorithm 2. Algorithm computing the chromatic number of G. Grover's search is used for mins.

1: **function** CHR(G)
2: **if** G is two colorable **then return** the chromatic number of G
3: $\chi[\emptyset] \leftarrow 0$
4: **for** $S \subseteq V$, $S \neq \emptyset$, $|S| \leq \lfloor n/4 \rfloor$ **do** (any order consistent with the inclusion relation)
5: $\chi[S] \leftarrow 1 + \min_{I \in \mathrm{MIS}(G[S])} \{\chi[S \setminus I]\}$
6: **return** CHR1(V)

7: **function** CHR1(S)
8: $c \leftarrow |S|$
9: **for** $t \in \{1, \ldots, |S|\}$, $s \in \{\max\{\lceil |S|/2 \rceil - t, 1\}, \ldots, \lfloor (|S| - t)/2 \rfloor\}$ **do**
10: $a \leftarrow \min_{I \in \mathrm{MIS}(G[S]), |I|=t} \min_{T \subseteq S \setminus I, |T|=s} (\mathrm{CHR2}(T) + \mathrm{CHR2}(S \setminus I \setminus T))$
11: $c \leftarrow \min\{c, a\}$
12: **return** $c + 1$

13: **function** CHR2(S)
14: **if** $G[S]$ is two colorable **then return** the chromatic number of $G[S]$
15: $c \leftarrow |S|$
16: **for** $t \in \{1, \ldots, |S|\}$, $s \in \{\max\{\lceil |S|/2 \rceil - t, 1\}, \ldots, \lfloor (|S| - t)/2 \rfloor\}$ **do**
17: $a \leftarrow \min_{I \in \mathrm{MIS}(G[S]), |I|=t} \min_{T \subseteq S \setminus I, |T|=s} (\chi[T] + \chi[S \setminus I \setminus T])$
18: $c \leftarrow \min\{c, a\}$
19: **return** $c + 1$

Proof. The running time of the precomputation is $O^*\left(\sum_{i=1}^{\lfloor n/4 \rfloor} \binom{n}{i} 3^{i/6}\right) = O^*(2^{h(1/4)n} 3^{n/24}) = O(1.8370^n)$. Let $T_1(n)$ be the running time of CHR1(V) and $T_2(m)$ be the running time of CHR2(S) for $S \subseteq V$ of size m. Then, we obtain

$$T_2(m) = \sum_{t=1}^{m} \sqrt{I(m,t)} \sum_{s=\max\{\lceil m/2 \rceil - t, 1\}}^{\lfloor (m-t)/2 \rfloor} \sqrt{\binom{m-t}{s}}, \tag{2}$$

$$\begin{aligned} T_1(n) &= \sum_{t=1}^{n} \sqrt{I(n,t)} \sum_{s=\max\{\lceil n/2 \rceil - t, 1\}}^{\lfloor (n-t)/2 \rfloor} \sqrt{\binom{n-t}{s}} (T_2(s) + T_2(n-t-s)) \\ &\leq \sum_{t=1}^{n} \sqrt{I(n,t)} \sum_{s=0}^{\min\{\lfloor n/2 \rfloor, n-t\}} \sqrt{\binom{n-t}{s}} T_2(s) \end{aligned} \tag{3}$$

by ignoring polynomial factors in n. Here, $T_2(m) \leq \sum_{t=1}^{m} \sqrt{I(m,t)} 2^{\frac{m-t}{2}}$ whose exponent is equal to $\max_{\delta \in [0,1]} \{(E(\delta) + (1-\delta))/2\}$. From Corollary 1, it is sufficient to take maximum among δ being an inverse integer. Numerical calculation shows that the maximum is given at $\delta = 1/5$ and hence $T_2(m) = O^*(80^{m/10}) = O(1.5500^m)$. Hence, the exponent of $T_1(n)$ is equal to

$$\max_{\delta \in [0,1/3],\, \lambda \in [0,1/2]} \left\{ \frac{1}{2} E(\delta) + \frac{1}{2} h\left(\frac{\lambda}{1-\delta}\right)(1-\delta) + \left(\frac{1}{10}\log 80\right)\lambda \right\}.$$

Here, we only consider maximum for $t \leq n/3$ since $I(n,t)$ is decreasing with respect to t for $t \geq n/3$, and the other part $\sum_s \sqrt{\binom{n-t}{s} T_2(s)}$ in (3) is also decreasing with respect to t. Numerical calculation shows that the maximum is given at $\delta = 1/7, \lambda = 1/2$. Hence, we obtain

$$\begin{aligned} T_1(n) &= O^*\left(\left(7^{1/14} 2^{h(7/12)3/7} 80^{1/20}\right)^n\right) \\ &= O^*\left((2^{37/35} 3^{3/7} 5^{-9/70} 7^{-5/28})^n\right) = O(1.9140^n). \end{aligned}$$ □

Careful readers may notice that the running times of the precomputation and the main computation are not balanced. If the quantum algorithm precomputes the chromatic numbers of induced subgraphs with size at most $(1/4+\epsilon)n$ for some $\epsilon > 0$, the precomputation and the main computation require more and less running time, respectively (we can use Fact 1 for unbalanced m). By optimizing ϵ such that the both running time are balanced, we may obtain improved running time. This idea improved the running time of the quantum algorithm for TSP [1], but does not improve the running time of Algorithm 2. Equation. (2) is dominated by $t = n/5$ and $s = (2/5)n$. Equation. (3) is dominated by $t = n/7$ and $s = n/2$. In order to exclude $s = (2/5)n$ in the summation in (2), the chromatic number of induced subgraph with size at most $(3/10)n$ must be precomputed. However, the running time of the precomputation in this case is $\sum_{i=1}^{(3/10)n} \binom{n}{i} 3^{i/6} = \Omega(1.9460^n)$. Hence, the running time of quantum algorithm is not improved.

Acknowledgment. This work was supported by JST PRESTO Grant Number JPMJPR1867 and JSPS KAKENHI Grant Numbers JP17K17711 and JP18H04090. The authors thank François Le Gall for the insightful comments.

References

1. Ambainis, A., Balodis, K., Iraids, J., Kokainis, M., Prūsis, K., Vihrovs, J.: Quantum speedups for exponential-time dynamic programming algorithms. In: Proceedings of the 30th Annual ACM-SIAM Symposium on Discrete Algorithms (SODA 2019). pp. 1783–1793. SIAM (2019)
2. Beigel, R., Eppstein, D.: 3-coloring in time $O(1.3289^n)$. J. Alg. **54**(2), 168–204 (2005)
3. Björklund, A., Husfeldt, T.: Inclusion-exclusion algorithms for counting set partitions. In: Proceedings of the 47th Annual IEEE Symposium on Foundations of Computer Science (FOCS 2006). pp. 575–582. IEEE (2006)
4. Björklund, A., Husfeldt, T.: Exact algorithms for exact satisfiability and number of perfect matchings. Algorithmica **52**(2), 226–249 (2008)
5. Björklund, A., Husfeldt, T., Koivisto, M.: Set partitioning via inclusion-exclusion. SIAM J. Comput. **39**(2), 546–563 (2009)

6. Boyer, M., Brassard, G., Høyer, P., Tapp, A.: Tight bounds on quantum searching. Fortschritte der Physik: Progress of Physics **46**(4–5), 493–505 (1998)
7. Byskov, J.M.: Enumerating maximal independent sets with applications to graph colouring. Oper. Res. Lett. **32**(6), 547–556 (2004)
8. Cygan, M., et al.: On problems as hard as CNF-SAT. ACM Trans. Alg. (TALG) **12**(3), 41 (2016)
9. Dürr, C., Høyer, P.: A quantum algorithm for finding the minimum. arXiv preprint quant-ph/9607014 (1996)
10. Fomin, F.V., Gaspers, S., Saurabh, S.: Improved exact algorithms for counting 3- and 4-colorings. In: Lin, G. (ed.) COCOON 2007. LNCS, vol. 4598, pp. 65–74. Springer, Heidelberg (2007). https://doi.org/10.1007/978-3-540-73545-8_9
11. Fomin, F.V., Kratsch, D.: Split and List. Exact Exponential Algorithms. TTCSAES, pp. 153–160. Springer, Heidelberg (2010). https://doi.org/10.1007/978-3-642-16533-7_9
12. Fürer, M.: Solving np-complete problems with quantum search. In: Laber, E.S., Bornstein, C., Nogueira, L.T., Faria, L. (eds.) LATIN 2008. LNCS, vol. 4957, pp. 784–792. Springer, Heidelberg (2008). https://doi.org/10.1007/978-3-540-78773-0_67
13. Gaspers, S., Lee, E.J.: Faster graph coloring in polynomial space. In: Cao, Y., Chen, J. (eds.) COCOON 2017. LNCS, vol. 10392, pp. 371–383. Springer, Cham (2017). https://doi.org/10.1007/978-3-319-62389-4_31
14. Giovannetti, V., Lloyd, S., Maccone, L.: Quantum random access memory. Phys. Rev. Lett. **100**(16), 160501 (2008)
15. Grover, L.K.: A fast quantum mechanical algorithm for database search. In: Proceedings of the 28th Annual ACM Symposium on Theory of Computing (STOC 1996). pp. 212–219. ACM (1996)
16. Koivisto, M.: An $O^*(2^n)$ algorithm for graph coloring and other partitioning problems via inclusion-exclusion. In: Proceedings of the 47th Annual IEEE Symposium on Foundations of Computer Science (FOCS 2006). pp. 583–590. IEEE (2006)
17. Lawler, E.L.: A note on the complexity of the chromatic number problem. Inf. Proc. Lett. **5**, 66–67 (1976)
18. Shimizu, K., Mori, R.: Exponential-time quantum algorithms for graph coloring problems. arXiv e-prints p. 1907.00529 (2019), https://arxiv.org/abs/1907.00529

Neural Networks and Biologically Inspired Computing

On Symmetry and Initialization for Neural Networks

Ido Nachum[1(✉)] and Amir Yehudayoff[2(✉)]

[1] EPFL - École polytechnique fédérale de Lausanne, Lausanne, Switzerland
ido.nachum@epfl.ch
[2] Technion - Israel Institute of Technology, Haifa, Israel
amir.yehudayoff@gmail.com

Abstract. This work provides an additional step in the theoretical understanding of neural networks. We consider neural networks with one hidden layer and show that when learning symmetric functions, one can choose initial conditions so that standard SGD training efficiently produces generalization guarantees. We empirically verify this and show that this does not hold when the initial conditions are chosen at random. The proof of convergence investigates the interaction between the two layers of the network. Our results highlight the importance of using symmetry in the design of neural networks.

Keywords: Neural networks · Symmetry

1 Introduction

Building a theory that can help to understand neural networks and guide their construction is one of the current challenges of machine learning. Here we wish to shed some light on the role symmetry plays in the construction of neural networks. It is well-known that symmetry can be used to enhance the performance of neural networks. For example, convolutional neural networks (CNNs) (see [26]) use the translational symmetry of images to classify images better than fully connected neural networks. Our focus is on the role of symmetry in the initialization stage. We show that symmetry-based initialization can be the difference between failure and success.

On a high-level, the study of neural networks can be partitioned to three different aspects.

Expressiveness Given an architecture, what are the functions it can approxi- mate well?

Training Given a network with a "proper" architecture, can the network fit the training data and in a reasonable time?

Generalization Given that the training seemed successful, will the true error be small as well?

Y. Kohayakawa and F. K. Miyazawa (Eds.): LATIN 2020, LNCS 12118, pp. 401–412, 2020.
https://doi.org/10.1007/978-3-030-61792-9_32

We study these aspects for the first "non trivial" case of neural networks, networks with one hidden layer. We are mostly interested in the initialization phase. If we take a network with the appropriate architecture, we can always initialize it to the desired function. A standard method (that induces a non trivial learning problem) is using random weights to initialize the network. A different reasonable choice is to require the initialization to be useful for an entire class of functions. We follow the latter option.

Our focus is on the role of symmetry. We consider the following class of symmetric functions

$$\mathbb{S} = \mathbb{S}_n = \Big\{ \sum_{i=0}^{n} a_i \cdot \mathbb{1}_{|x|=i} : a_1, \dots, a_n \in \{\pm 1\} \Big\},$$

where $x \in \{0,1\}^n$ and $|x| = \sum_i x_i$. The functions in this class are invariant under arbitrary permutations of the input's coordinates. The parity function $\pi(x) = (-1)^{|x|}$ and the majority function are well-known examples of symmetric functions.

Expressiveness for this class was explored by [30]. They showed that the parity function cannot be represented using a network with limited "connectivity". Contrastingly, if we use a fully connected network with one hidden layer and a common activation function (like *sign*, *sigmoid*, or *ReLU*) only $O(n)$ neurons are needed. We provide such explicit representations for all functions in $\mathbb{S}$; see Lemmas 1 and 2.

We also provide useful information on both the *training* phase and *generalization* capabilities of the neural network. We show that, with proper initialization, the training process (using standard SGD) efficiently converges to zero empirical error, and that consequently the network has small true error as well.

Theorem 1. *There exists a constant $c > 1$ so that the following holds. There exists a network with one hidden layer, cn neurons with sigmoid or ReLU activations, and an initialization such that for all distributions $\mathcal{D}$ over $X = \{0,1\}^n$ and all functions $f \in \mathbb{S}$ with sample size $m \geq c(n+\log(1/\delta))/\epsilon$, after performing poly(n) SGD updates with a fixed step size $h = 1/poly(n)$ it holds that*

$$\underset{x^m \sim \mathcal{D}^m}{P}\left(\left\{S : \underset{x \sim \mathcal{D}}{\Pr}(N_S(x) \neq f(x)) > \epsilon\right\}\right) < \delta$$

where $S = \{(x_1, f(x_1)), ..., (x_m, f(x_m))\}$ and $N_S(x)$ is the network after training over S.

The number of parameters in the network described in Theorem 1 is $\Omega(n^2)$. So in general one could expect overfitting when the sample size is as small as $O(n)$. Nevertheless, the theorem provides generalization guarantees, even for such a small sample size.

The initialization phase plays an important role in proving Theorem 1. To emphasize this, we report an empirical phenomenon (this is "folklore"). We show that a network cannot learn parity from a random initialization. On one hand, if the network size is big, we can bring the empirical error to zero (as suggested in [41]), but the true error is close to 1/2. On the other hand, if its size is too small, the network is not even able to achieve small empirical error. We observe a similar phenomenon also for a random symmetric function. An open question remains: why is it true that a sample of size polynomial in n does not suffice to learn parity (with random initialization)?

A similar phenomenon was theoretically explained by [37] and [40]. The parity function belongs to the class of all parities

$$\mathbb{P} = \mathbb{P}_n = \{\pi_s(x) = (-1)^{s \cdot x} : s \in X\}$$

where $\cdot$ is the standard inner product. This class is efficiently PAC-learnable with $O(n)$ samples using Gaussian elimination. A continuous version of $\mathbb{P}$ was studied by [37] and [40]. To study the training phase, they used a generalized notion of *statistical queries* (SQ); see [24]. In this framework, they show that most functions in the class $\mathbb{P}$ cannot be efficiently learned (roughly stated, learning the class requires an exponential amount of resources). This framework, however, does not seem to capture actual training of neural networks using SGD. For example, it is not clear if one SGD update corresponds to a single query in this model. In addition, typically one receives a dataset and performs the training by going over it many times, whereas the query model estimates the gradient using a fresh batch of samples in each iteration. The query model also assumes the noise to be adversarial, an assumption that does not necessarily hold in reality. Finally, the SQ-based lower bound holds for every initialization (in particular, for the initialization we use here), so it does not capture the efficient training process Theorem 1 describes.

Theorem 1 shows, however, that with symmetry-based initialization, parity can be efficiently learned. So, in a nutshell, parity can not be learned as part of $\mathbb{P}$, but it can be learned as part of $\mathbb{S}$. One could wonder why the hardness proof for $\mathbb{P}$ cannot be applied for $\mathbb{S}$ as both classes consist of many input sensitive functions. The answer lies in the fact that $\mathbb{P}$ has a far bigger statistical dimension than $\mathbb{S}$ (all functions in $\mathbb{P}$ are orthogonal to each other, unlike $\mathbb{S}$).

The proof of the theorem utilizes the different behavior of the two layers in the network. SGD is performed using a step size h that is polynomially small in n. The analysis shows that in a polynomial number of steps that is *independent* of the choice of h the following two properties hold: (i) the output neuron reaches a "good" state and (ii) the hidden layer does not change in a "meaningful" way. These two properties hold when h is small enough.

Here is a high level description of the proof. The ℓ neurons in the hidden layer define an "embedding" of the inputs space $X = \{0,1\}^n$ into $\mathbb{R}^\ell$ (a.k.a. the feature map). This embedding changes in time according to the training examples and process. The proof shows that if at any point in time this embedding has good enough margin, then training with standard SGD quickly converges. This is explained in more detail in Sect. 3. It remains an interesting open problem to understand this phenomenon in greater generality, using a cleaner and more abstract language.

1.1 Background

To better understand the context of our research, we survey previous related works.

The expressiveness and limitations of neural networks were studied in several works such as [6,17,33,43]. Constructions of small *ReLU* networks for the parity function appeared in several previous works, such as [8,9,22,44]. Constant depth circuits for the parity function were also studied in the context of computational complexity theory, see for example [2,19,21].

The training phase of neural networks was also studied in many works. Here we list several works that seem most related to ours. [15] analyzed SGD for general neural network architecture and showed that the training error can be nullified, e.g., for the class of bounded degree polynomials (see also [5]). [23] studied neural tangent kernels (NTK), an infinite width analogue of neural networks. [16] showed that randomly initialized shallow *ReLU* networks nullify the training error, as long as the number of samples is smaller than the number of neurons in the hidden layer. Their analysis only deals with optimization over the first layer (so that the weights of the output neuron are fixed). [12] provided another analysis of the latter two works. [4] showed that over-parametrized neural networks can achieve zero training error, as long as the data points are not too close to one another and the weights of the output neuron are fixed. [46] provided guarantees for zero training error, assuming the two classes are separated by a positive margin.

Convergence and generalization guarantees for neural networks were studied in the following works. [11] studied linearly separable data. [27] studied well separated distributions. [3] gave generalization guarantees in expectation for SGD. [7] gave data-dependent generalization bounds for GD. All these works optimized only over the hidden layer (the output layer is fixed after initialization).

Margins play an important role in learning, and we also use it in our proof. [10,38,39,42] gave generalization bounds for neural networks that are based on their margin when the training ends. From a practical perspective, [18,29,34], suggested different training algorithms that optimize the margin.

As discussed above, it seems difficult for neural networks to learn parities. [40] and [37] demonstrated this using the language statistical queries (SQ). This is a valuable language, but it misses some central aspects of training neural networks. SQ seems to be closely related to GD, but does not seem to capture SGD. SQ also shows that many of the parities functions $\otimes_{i \in S} x_i$ are difficult to learn, but it does not imply that *the* parity function $\otimes_{i \in [n]} x_i$ is difficult to learn. [1] demonstrated a similar phenomenon in a setting that is closer to the "real life" mechanics of neural networks.

We suggest that taking the symmetries of the learning problem into account can make the difference between failure and success. Several works suggested different neural architectures that take symmetries into account; see [13,20,45].

2 Representations

Here we describe efficient representations for symmetric functions by networks with one hidden layer. These representations are also useful later on, when we study the training process. We study two different activation functions, *sigmoid* and *ReLU* (similar statement can be proved for other activations, like arctan). Each activation function requires its own representation, as in the two lemmas below.

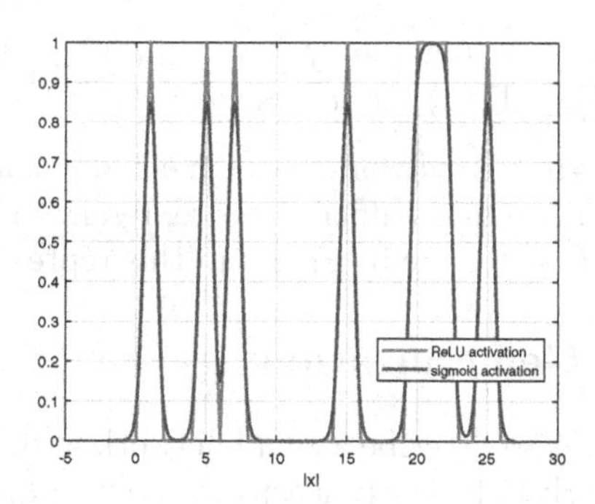

Fig. 1. Approximations of the symmetric function $f_A = \text{sign}(\sum_{i \in A} \mathbb{1}_{|x|=i} - 0.5)$ by *sigmoid* and *ReLU* activations for $A = \{1, 5, 7, 15, 20, 21, 22, 25\}$.

2.1 Sigmoid

We start with the activation $\sigma(\xi) = \frac{1}{1+\exp(-\xi)}$, since it helps to understand the construction for the *ReLU* activation. The building blocks of the symmetric functions are indicators of $|x| = i$ for $i \in \{0, 1, \ldots, n\}$. An indicator function is essentially the difference between two *sigmoid* functions:

$$\text{sign}(\mathbb{1}_{|x|=i} - 0.5) = \text{sign}(\Delta_i - 0.5),$$

where $\Delta_i(x) = \sigma(5(|x| - i + 0.5)) - \sigma(5(|x| - i - 0.5))$.

Lemma 1. *The symmetric function* $f_A = \text{sign}(\sum_{i \in A} \mathbb{1}_{|x|=i} - 0.5)$ *satisfies* $f_A(x) = \text{sign}(-0.5 + \sum_{i \in A} \Delta_i(x))$, *where* $A \subset [n]$.

A network with one hidden layer of $n + 2$ neurons with *sigmoid* activations and one bias neuron is sufficient to represent any function in $\mathbb{S}$. The coefficients of the *sigmoid* gates are $0, \pm 1$ in this representation. The proofs of this lemma and the subsequent lemmas are deferred to the full version of this paper.

2.2 ReLU

A *sigmoid* function can be represented using $ReLU(\xi) = \max\{0, \xi\}$ as the difference between two *ReLU*s

$$\sigma(5(x + 0.5)) \approx ReLU(x + 1) - ReLU(x)$$

Hence, an indicator function can be represented using $\text{sign}(\mathbb{1}_{|x|=i} - 0.5) = \text{sign}(\Gamma_i - 0.5)$ where

$$\Gamma_i(x) = ReLU(|x| - i + 1) - 2ReLU(|x| - i) + ReLU(|x| - i - 1).$$

Lemma 2. *The symmetric function* $f_A = \text{sign}(\sum_{i \in A} \mathbb{1}_{|x|=i} - 0.5)$ *satisfies* $f_A(x) = \text{sign}(-0.5 + \sum_{i \in A} \Gamma_i(x))$*, where* $A \subset [n]$.

The lemma shows that a network with one hidden layer of $n + 3$ *ReLU* neurons and one bias neuron is sufficient to represent any function in $\mathbb{S}$. The coefficients of the *ReLU* gates are $0, \pm 1, \pm 2$ in this representation.

3 Training and Generalization

The goal of this section is to describe a small network with one hidden layer that (when initialized properly) efficiently learns symmetric functions using a small number of examples (the training is done via SGD).

3.1 Specifications

Here we specify the architecture, initialization and loss function that is implicit in our main result (Theorem 1).

To guarantee convergence of SGD, we need to start with "good" initial conditions. The initialization we pick depends on the activation function it uses, and is chosen with resemblance to Lemma 2 for *ReLU*. On a high level, this indicates that understanding the class of functions we wish to study in term of "representation" can be helpful when choosing the architecture of a neural network in a learning context.

The network we consider has one hidden layer. We denote by w_{ij} the weight between coordinate j of the input and neuron i in the hidden layer. We denote W this matrix of weights. We denote by b_i the bias of neuron i of the hidden layer. We denote B this vector of weights. We denote by m_i is the weight from neuron i in the hidden layer to the output neuron. We denote M this vector of weights. We denote by b the bias of the output neuron.

Initialize the network as follows: The dimensions of W are $(n + 3) \times n$. For all $1 \le i \le (n + 3)$ and $1 \le j \le n$, we set $w_{ij} = 1$ and $b_i = -i + 2$. We set $M = 0$ and $b = 0$.

To run SGD, we need to choose a loss function. We use the *hinge loss*,

$$L(x, f) = \max\{0, -f(x)(v_x \cdot M + b) + \beta\},$$

where $v_x = ReLU(Wx + B)$ is the output of the hidden layer on input x and $\beta > 0$ is a parameter of confidence.

3.2 Margins

A key property in the analysis is the 'margin' of the hidden layer with respect to the function being learned.

A map $Y : V \to \{\pm 1\}$ over a finite set $V \subset \mathbb{R}^d$ is linearly[1] separable if there exists $w \in \mathbb{R}^d$ such that $\text{sign}(w \cdot v) = Y(v)$ for all $v \in V$. When the Euclidean norm of w is $\|w\| = 1$, the number $marg(w, Y) = \min_{v \in V} Y(v) w \cdot v$ is the margin of w with respect to Y. The number $marg(Y) = \sup_{w \in \mathbb{R}^d : \|w\|=1} marg(w, Y)$ is the margin of Y.

We are interested in the following set V in $\mathbb{R}^d$. Recall that W is the weight matrix between the input layer and the hidden layer, and that B is the relevant bias vector. Given W, B, we are interested in the set $V = \{v_x : x \in X\}$, where $v_x = ReLU(Wx + B)$. In words, we think of the neurons in the hidden layer as defining an "embedding" of X in Euclidean space. A similar construction works for other activation functions. We say that $Y : V \to \{\pm 1\}$ agrees with $f \in \mathbb{S}$ if for all $x \in X$ it holds that $Y(v_x) = f(x)$.

The following lemma bounds from below the margin of the initial V.

Lemma 3. *If Y is a partition that agrees with some function in $\mathbb{S}$ for the initialization described above then $marg(Y) \geq \Omega(1/\sqrt{n})$.*

Proof. By Lemmas 1 and 2, we see that any function in $\mathbb{S}$ can be represented with a vector of weights $M, b \in [-2, 2]^{\Theta(n)}$ of the output neuron together with a bias . These M, b induce a partition Y of V. Namely, $Y(v_x) M \cdot v_x + b > 0.25$ for all $x \in X$. Since $\|(M, b)\| = O(\sqrt{n})$ we have our desired result.

3.3 Freezing the Hidden Layer

Before analyzing the full behavior of SGD, we make an observation: if the weights of the hidden layer are fixed with the initialization described above, then Theorem 1 holds for SGD with batch size 1. This observation, unfortunately, does not suffice to prove Theorem 1. In the setting we consider, the training of the neural network uses SGD without fixing any weights. This more general case is handled in the next section. The rest of this subsection is devoted for explaining this observation.

[32] showed that the perceptron algorithm [35] makes a small number of mistakes for linearly separable data with large margin. For a comprehensive survey of the perceptron algorithm and its variants, see [31].

Running SGD with the hinge loss induces the same update rule as in a modified perceptron algorithm, Algorithm 1.

Novikoff's proof can be generalized to any $\beta > 0$ and batches of any size to yield the following theorem; see [14,25].

[1] A standard "lifting" that adds a coordinate with 1 to every vector allows to translate the affine case to the linear case.

Algorithm 1. The modified perceptron algorithm

Initialize: $w^{(0)} = \mathbf{0}$, $t = 0$, $\beta > 0$ and $h > 0$
while $\exists v \in V$ with $Y(v)w^{(t)} \cdot v \leq \beta$ **do**
  $w^{(t+1)} = w^{(t)} + Y(v)vh$
  $t = t + 1$
end while
return $w^{(t)}$

Theorem 2. *For* $Y : V \to \{\pm 1\}$ *with margin* $\gamma > 0$ *and step size* $h > 0$*, the modified perceptron algorithm performs at most* $\frac{2\beta h + (Rh)^2}{(\gamma h)^2}$ *updates and achieves a margin of at least* $\frac{\gamma \beta h}{2\beta h + (Rh)^2}$*, where* $R = \max_{v \in V} \|v\|$.

So, when the weights of the hidden layer are fixed, Lemma 3 implies that the number of SGD steps is at most polynomial in n.

3.4 Stability

When we run SGD on the entire network, the layers interact. For a *ReLU* network at time t, the update rule for W is as follows. If the network classifies the input x correctly with confidence more than β, no change is made. Otherwise, we change the weights in M by $\Delta M = yv_x h$, where y is the true label and h is the step size. If also neuron i of the hidden *fired* on x, we update its incoming weights by $\Delta W_{i,:} = ym_i xh$. These update rules define the following dynamical system: (a)

$$W^{(t+1)} = W^{(t)} + y\left(\left(M^{(t)}\right)\right) \tag{1}$$

$$W^{(t+1)} = W^{(t)} + y\left(\left(M^{(t)}\right)^T \circ H\left(W^{(t)}x + B^{(t)}\right)\right)x^T h \tag{2}$$

$$B^{(t+1)} = B^{(t)} + y\left(\left(M^{(t)}\right)^T \circ H\left(W^{(t)}x + B^{(t)}\right)\right)h \tag{3}$$

$$M^{(t+1)} = M^{(t)} + yReLU\left(W^{(t)}x + B^{(t)}\right)h \tag{4}$$

$$b^{(t+1)} = b^{(t)} + yh, \tag{5}$$

where H is the Heaviside step function and $\circ$ is the Hadamard pointwise product.

A key observation in the proof is that the weights of the last layer ((4) and (5)) are updated exactly as the modified perceptron algorithm. Another key statement in the proof is that if the network has reached a good representation of the input (i.e., the hidden layer has a large margin), then the interaction between the layers during the continued training does not impair this representation. This is summarized in the following lemma (we are not aware of a similar statement in the literature).

Lemma 4. *Let $M = 0$, $b = 0$, and $V = \{ReLU(Wx + B) : x \in X\}$ be a linearly separable embedding of X and with margin $\gamma > 0$ by the hidden layer of a neural network of depth two with ReLU activation and weights given by W, B. Let $R_X = \max_{x \in X} \|x\|$, let $R = \max_{v \in V} \|v\|$, and $0 < h \leq \frac{\gamma^{5/2}}{100 R^2 R_X}$ be the integration step. Assuming $R_X > 1$ and $\gamma \leq 1$, and using $\beta = R^2 h$ in the loss function, after t SGD iterations the following hold:*

- *Each $v \in V$ moves a distance of at most $O(R_X^2 h^2 R t^{3/2})$.*
- *The norm $\|M^{(t)}\|$ is at most $O(Rh\sqrt{t})$.*
- *The training ends in at most $O(R^2/\gamma^2)$ SGD updates.*

Intuitively, this type of lemma can be useful in many other contexts. The high level idea is to identify a "good geometric structure" that the network reaches and enables efficient learning.

4 Main Result

Proof (Proof of Theorem 1). There is an unknown distribution $\mathcal{D}$ over the space X. We pick i.i.d. examples $S = ((x_1, y_1), ..., (x_m, y_m))$ where $m \geq c(\frac{n + \log(1/\delta)}{\epsilon})$ according to $\mathcal{D}$, where $y_i = f(x_i)$ for some $f \in \mathbb{S}$. Run SGD for $O(n^4)$ steps, where the step size is $h = O(1/n^5)$ and the parameter of the loss function is $\beta = R^2 h$ with $R = n^{3/2}$.

We claim that it suffices to show that at the end of the training (i) the network correctly classifies all the sample points $x_1, \ldots, x_m$, and (ii) for every $x \in X$ such that there exists $1 \leq i \leq m$ with $|x| = |x_i|$, the network outputs y_i on x as well. Here is why. The initialization of the network embeds the space X into $n+4$ dimensional space (including the bias neuron of the hidden layer). Let $V^{(0)}$ be the initial embedding $V^{(0)} = \{ReLU(W^{(0)}x + B^{(0)}) : x \in X\}$. Although $|X| = 2^n$, the size of $V^{(0)}$ is $n+1$. The VC dimension of all the boolean functions over $V^{(0)}$ is $n+1$. Now, m samples suffice to yield ϵ true error for an ERM when the VC dimension is $n + 1$; see e.g. Theorem 6.7 in [36]. It remains to prove (i) and (ii) above.

By Lemma 3, at the beginning of the training, the partition of $V^{(0)}$ defined by the target $f \in \mathbb{S}$ has a margin of $\gamma = \Omega(1/\sqrt{n})$. We are interested in the eventual $V^* = \{ReLU(W^* x + B^*) : x \in X\}$ embedding of X as well. The modified perceptron algorithm together with Lemma 4 guarantees that after $K \leq 20R^2/\gamma^2 = O(n^4)$ updates, (M^*, b^*) separates the embedded sample $V_S^* = \{ReLU(W^* x_i + B^*) : 1 \leq i \leq m\}$ with a margin of at least $0.9\gamma/3$. This concludes the proof of (i).

It remains to prove (ii). Lemma 4 states that as long as less than $K = O(n^5)$ updates were made, the elements in V moved at most $O(1/n^2)$. At the end of the training, the embedded sample V_S is separated with a margin of at least $\gamma/3$ with respect to the hyperplane defined by M^* and B^*. Each v_x^* for $x \in X$ moved at most $O(1/n^2) < \gamma/4$. This means that if $|x| = |x_i|$ then the network has the same output on x and x_i. Since the network has zero empirical error, the output on this x is y_i as well.

A similar proof is available with *sigmoid* activation (with better convergence rate and larger allowed step size).

Remark 1. The generalization part of the above proof can be viewed as a consequence of sample compression [28]. Although the eventual network depends on *all* examples, the proof shows that its functionality depends on at most $n+1$ examples. Indeed, after the training, all examples with equal hamming weight have the same label.

Remark 2. The parameter $\beta = R^2 h$ we chose in the proof may seem odd and negligible. It is a construct in the proof that allows us to bound efficiently the distance that the elements in V have moved during the training. For all practical purposes $\beta = 0$ works as well.

5 Conclusion

This work demonstrates that symmetries can play a critical role when designing a neural network. We proved that any symmetric function can be learned by a shallow neural network, with proper initialization. We demonstrated by simulations that this neural network is stable under corruption of data, and that the small step size is the proof is necessary.

We also demonstrated that the parity function or a random symmetric function cannot be learned with random initialization. How to explain this empirical phenomenon is still an open question. The works [37] and [40] treated parities using the language of SQ. This language obscures the inner mechanism of the network training, so a more concrete explanation is currently missing.

We proved in a special case that the standard SGD training of a network efficiently produces low true error. The general problem that remains is proving similar results for general neural networks. A suggestion for future works is to try to identify favorable geometric states of the network that guarantee fast convergence and generalization.

References

1. Abbe, E., Sandon, C.: Provable limitations of deep learning. arXiv e-prints p. 1812.06369 (2018)
2. Ajtai, M.: $\sum_1^1$-formulae on finite structures. Ann. Pure Appl. Logic **24**, 1–48 (1983)
3. Allen-Zhu, Z., Li, Y., Liang, Y.: Learning and generalization in overparameterized neural networks, going beyond two layers. arXiv e-prints p. 1811.04918 (2018)
4. Allen-Zhu, Z., Li, Y., Song, Z.: A convergence theory for deep learning via over-parameterization. arXiv e-prints p. 1811.03962 (2018)
5. Andoni, A., Panigrahy, R., Valiant, G., Zhang, L.: Learning polynomials with neural networks. In: Xing, E.P., Jebara, T. (eds.) Proceedings of the 31st International Conference on Machine Learning. pp. 1908–1916 (2014)
6. Arora, R., Basu, A., Mianjy, P., Mukherjee, A.: Understanding deep neural networks with rectified linear units. arXiv e-prints p. 1611.01491 (2016)

7. Arora, S., Du, S.S., Hu, W., Li, Z., Wang, R.: Fine-grained analysis of optimization and generalization for overparameterized two-layer neural networks. arXiv e-prints p. 1901.08584 (2019)
8. Arslanov, M.Z., Ashigaliev, D.U., Ismail, E.: N-bit parity ordered neural networks. Neurocomput. **48**, 1053–1056 (2002)
9. Arslanov, M., Amirgalieva, Z.E., Kenshimov, C.A.: N-bit parity neural networks with minimum number of threshold neurons. Open Eng. **6**, 309–313 (2016)
10. Bartlett, P., Foster, D.J., Telgarsky, M.: Spectrally-normalized margin bounds for neural networks. arXiv e-prints p. 1706.08498 (2017)
11. Brutzkus, A., Globerson, A., Malach, E., Shalev-Shwartz, S.: SGD learns over-parameterized networks that provably generalize on linearly separable data. arXiv e-prints p. 1710.10174 (2018)
12. Chizat, L., Bach, F.: A note on lazy training in supervised differentiable programming. arXiv e-prints p. 1812.07956 (2018)
13. Cohen, T.S., Welling, M.: Group equivariant convolutional networks. arXiv e-prints p. 1602.07576 (2016)
14. Collobert, R., Bengio, S.: Links between perceptrons. In: Proceedings of the 21st International Conference on Machine Learning. p. 23 (2004)
15. Daniely, A.: SGD learns the conjugate kernel class of the network. arXiv e-prints p. 1702.08503 (2017)
16. Du, S.S., Zhai, X., Póczos, B., Singh, A.: Gradient descent provably optimizes over-parameterized neural networks. arXiv e-prints p. 1810.02054 (2018)
17. Eldan, R., Shamir, O.: The power of depth for feedforward neural networks. In: Feldman, V., Rakhlin, A., Shamir, O. (eds.) Proceedings of the 29th Annual Conference on Learning Theory. Proceedings of Machine Learning Research, vol. 49, pp. 907–940. PMLR, Columbia University, New York, USA (2016)
18. Elsayed, G.F., Krishnan, D., Mobahi, H., Regan, K., Bengio, S.: Large margin deep networks for classification. arXiv e-prints p. 1803.05598 (2018)
19. Furst, M., Saxe, J.B., Sipser, M.: Parity, circuits, and the polynomial-time hierarchy. In: Proceedings of the 22nd Symposium on the Foundations of Computer Science. pp. 260–270 (1981)
20. Gens, R., Domingos, P.M.: Deep symmetry networks. In: Ghahramani, Z., Welling, M., Cortes, C., Lawrence, N.D., Weinberger, K.Q. (eds.) Advances in Neural Information Processing Systems 27, NIPS 2014, pp. 2537–2545 (2014)
21. Håstad, J.: Computational Limitations of Small-depth Circuits. MIT Press, United States (1987)
22. Iyoda, E.M., Nobuhara, H., Hirota, K.: A solution for the n-bit parity problem using a single translated multiplicative neuron. Neural Process. Lett. **18**, 233–238 (2003)
23. Jacot, A., Gabriel, F., Hongler, C.: Neural tangent kernel: Convergence and generalization in neural networks. arXiv e-prints p. arXiv:1806.07572 (2018)
24. Kearns, M.: Efficient noise-tolerant learning from statistical queries. J. ACM **45**(6), 983–1006 (1998)
25. Krauth, W., Mezard, M.: Learning algorithms with optimal stability in neural networks. J. Phys. A: Math. General **20**, L745–L752 (1987)
26. Lecun, Y., Bottou, L., Bengio, Y., Haffner, P.: Gradient-based learning applied to document recognition. Proc. IEEE **86**, 2278–2324 (1998)
27. Li, Y., Liang, Y.: Learning overparameterized neural networks via stochastic gradient descent on structured data. arXiv e-prints p. 1808.01204 (2018)
28. Littlestone, N., Warmuth, M.K.: Relating data compression and learnability (1986), Unpublished manuscript, University of California Santa Cruz (1986)

29. Liu, W., Wen, Y., Yu, Z., Yang, M.M.: Large-margin softmax loss for convolutional neural networks. arXiv e-prints p. 1612.02295 (2016)
30. Minsky, M.L., Papert, S.A.: Perceptrons, Expanded edn. MIT Press, Cambridge, MA, USA (1988)
31. Moran, S., Nachum, I., Panasoff, I., Yehudayoff, A.: On the perceptron's compression. arXiv e-prints p. 1806.05403 (2018)
32. Novikoff, A.B.J.: On convergence proofs on perceptrons. Proceedings of the Symposium on the Mathematical Theory of Automata. **12**, 615–622 (1962)
33. Rahimi, A., Recht, B.: Random features for large-scale kernel machines. In: Platt, J.C., Koller, D., Singer, Y., Roweis, S.T. (eds.) Advances in Neural Information Processing Systems 20, pp. 1177–1184 (2008)
34. Romero, E., Alquezar, R.: Maximizing the margin with feedforward neural networks. In: Proceedings of the 2002 International Joint Conference on Neural Networks, IJCNN 2002. vol. 1, pp. 743–748 (2002)
35. Rosenblatt, F.: The perceptron: a probabilistic model for information storage and organization in the brain. Psychol. Rev. **65**, 386–408 (1958)
36. Shalev-Shwartz, S., Ben-David, S.: Understanding machine learning: From theory to algorithms. Cambridge University Press (2014)
37. Shamir, O.: Distribution-specific hardness of learning neural networks. arXiv e-prints p. 1609.01037 (2016)
38. Sokolic, J., Giryes, R., Sapiro, G., Rodrigues, M.R.D.: Margin preservation of deep neural networks. arXiv e-prints p. 1605.08254v1 (2016)
39. Sokolic, J., Giryes, R., Sapiro, G., Rodrigues, M.R.D.: Robust large margin deep neural networks. IEEE Trans. Signal Process. **65**, 4265–4280 (2017)
40. Song, L., Vempala, S., Wilmes, J., Xie, B.: On the complexity of learning neural networks. arXiv e-prints p. 1707.04615 (2017)
41. Soudry, D., Carmon, Y.: No bad local minima: Data independent training error guarantees for multilayer neural networks. arXiv e-prints p. 1605.08361 (2016)
42. Sun, S., Chen, W., Wang, L., Liu, T.Y.: Large margin deep neural networks: Theory and algorithms. arXiv e-prints p. 1506.05232 (2015)
43. Telgarsky, M.: Representation benefits of deep feedforward networks. arXiv e-prints p. 1509.08101 (2016)
44. Wilamowski, B., Hunter, D., Malinowski, A.: Solving parity-n problems with feedforward neural networks. In: Proceedings of the International Joint Conference on Neural Networks, IJCNN. vol. 4, pp. 2546–2551 (2003)
45. Zaheer, M., Kottur, S., Ravanbakhsh, S., Poczos, B., Salakhutdinov, R., Smola, A.: Deep sets. In: Guyon, I., et al., (eds.) Advances in Neural Information Processing Systems 30, pp. 3391–3401 (2017)
46. Zou, D., Cao, Y., Zhou, D., Gu., Q.: Stochastic gradient descent optimizes overparameterized deep relu networks. arXiv e-prints p. 1811.08888 (2018)

How to Color a French Flag

Biologically Inspired Algorithms for Scale-Invariant Patterning

Bertie Ancona[1(✉)], Ayesha Bajwa[1], Nancy Lynch[1], and Frederik Mallmann-Trenn[2]

[1] Massachusetts Institute of Technology, Cambridge, USA
{bancona,abajwa}@alum.mit.edu, lynch@csail.mit.edu
[2] King's College London, London, UK
frederik.mallmann-trenn@kcl.ac.uk

Abstract. In the *French flag problem*, initially uncolored cells on a grid must differentiate to become blue, white or red. The goal is for the cells to color the grid as a French flag, i.e., a three-colored triband, in a distributed manner. To solve a generalized version of the problem in a distributed computational setting, we consider two models: a biologically-inspired version that relies on morphogens (diffusing proteins acting as chemical signals) and a more abstract version based on reliable message passing between cellular agents.

Much of developmental biology research focuses on concentration-based approaches, since morphogen gradients are an underlying mechanism in tissue patterning. We show that both model types easily achieve a *French ribbon* - a French flag in the 1D case. However, extending the ribbon to the 2D flag in the concentration model is somewhat difficult unless each agent has additional positional information. Assuming that cells are identical, it is impossible to achieve a French flag or even a close approximation. In contrast, using a message-based approach in the 2D case only requires assuming that agents can be represented as logarithmic or constant size state machines.

We hope that our insights may lay some groundwork for what kind of message passing abstractions or guarantees, if any, may be useful in analogy to cells communicating at long and short distances to solve patterning problems. We also hope our models and findings may be of interest in the design of nano-robots.

Keywords: Distributed computing · French flag · Biologically inspired algorithms

1 Introduction

In the *French flag problem*, initially uncolored cells on a grid must differentiate to become blue, white or red, ultimately coloring the grid as a three-colored triband

The authors were supported in part by NSF Award Numbers CCF-1461559 and CCF-0939370.

Y. Kohayakawa and F. K. Miyazawa (Eds.): LATIN 2020, LNCS 12118, pp. 413–424, 2020.
https://doi.org/10.1007/978-3-030-61792-9_33

without centralized decision-making. Lewis Wolpert's original French flag problem formulation [19,20] has been applied and extended to understand how organisms determine cell fate, or final differentiated cell type, a question central to developmental biology. Wolpert's formulation of positional information models is both complementary to and contrasted with Turing's earlier formulation of reaction-diffusion instability [18], which relies on random asymmetries that arise from activator-inhibitor dynamics in a developmental system. Our methods make use of both positional information and initial asymmetry. However, we distinguish between absolute and relative positional information to probe whether full knowledge of the coordinates is needed to solve the problem, or if strictly less information suffices.

Broadly speaking, our work is inspired by the biological mechanisms leading to cell fate decisions in the original French flag problem. These long and short-distance mechanisms inform the design of algorithms and analyses of the problem in two distributed computing contexts. More precisely, we relate a reliable message passing model (Sect. 2.2) to local cell-cell communication, and a concentration-based model (Sect. 2.1) to morphogen gradients over long distances.

We analyze a generalized French flag problem for k colors in these two computational models. We aim to understand the resources and minimum set of assumptions required to solve the problem exactly or approximately. In particular, we study whether cells must know their exact positions and the grid dimensions in order to solve the k-flag problem. We hope that characterizing the resources and information required might have some translation back to the mechanisms enabling scale-invariant patterning.

We begin by studying the *French ribbon problem*, the 1D scenario in our models. Both exact and approximate solutions are possible, with a general tradeoff between precision and space complexity. While both models easily achieve a French ribbon, extending 1D decision-making to the 2D setting is provably difficult in the concentration model. We show that in a 2D grid with point sources at the corners, each agent knowing its absolute distance to every source is insufficient positional information to color the grid even approximately correctly. On the other hand, extending to the 2D setting is easy in the message passing model. We analyze numerous algorithms to demonstrate tradeoffs between time complexity, message size, memory size and precision of the obtained French flag.

We do not claim more accurate or thorough models than those proposed by the biology community. However, we hope this work may illuminate computational abstractions or guarantees that may be useful in analogy to cells communicating at long and short distances to solve patterning problems.

1.1 Biology Background and Related Work

A key principle of our models is that initial asymmetry and local communication eventually leads to long-distance transmission of the relative positional information of cellular agents, allowing for distributed decision-making. Morphogens, or molecules acting as chemical signals, underlie cell-cell communication over long distances. Two well-studied morphogens are *Bicoid (Bcd)* for anterior-posterior patterning in fruit flies [5,14], and *Sonic hedgehog (Shh)*, a morphogen for neural patterning in vertebrates, including humans [4,15]. Exactly how these morphogens produce scale-invariant patterns in organisms and tissues of varying size is an interesting biological question [8].

Mechanisms for local cell-cell communication include cell surface receptors and ligands, such as the Notch-Delta system previously studied in a distributed computing context [1]. There are also physical channels for signalling molecules, such as gap junctions in animal cells and plasmodesmata in plant cells [2]. We liken local signalling to message passing between neighboring agents.

Building on earlier work on gradients [12,16], Wolpert focused the French flag problem and model [19,20] on the concept of positional information and its generalization to other patterning mechanisms. Subsequent papers validated the importance of positional information through empirical studies in model species [5,14,17]. Turing had previously studied reaction-diffusion instability as a driver of morphogenesis [18], theorizing that periodic patterns could spontaneously arise from activator-inhibitor dynamics. Turing's paradigm is often contrasted with Wolpert's notion of positional information. The idea that cells may learn positional information via concentration has fundamentally altered the field of developmental biology [7,10]. The French flag problem has been studied using various models, including growth and repair simulation models [11] and reaction-diffusion experimental models [21].

1.2 Results

Here we summarize results in the two computational models. We first present our results for the concentration model, where we assume that each node on a line has access to just morphogens concentrations c_1 and c_2, each emitted from an endpoint of the line, and no other information. We define the model formally in Sect. 2.1.

On the positive side, it is possible to solve the French ribbon problem exactly.

Theorem 1. *Algorithm Exact Concentration Ribbon solves the concentration model k-ribbon for an n-agent line graph of arbitrary finite length a with constant time and communication complexity, given that agents have knowledge of morphogen concentrations c_1 and c_2, which have reached steady states, as well as the gradient function.*

On the negative side, we show that extending to the French flag (2D-case) with just four point-sources at the corners is infeasible. Here, symmetry prevents us from obtaining a ε-approximate algorithm in this model.

Theorem 2. *Consider the concentration model. Fix any $\varepsilon \in (0, 1/6)$. No algorithm can produce an ε-approximate French flag.*

The concentration model contrasts the message passing model, in which even exact solutions are possible. Results for the message passing model are summarized in Table 1 below, and the exact statements can be found in Sect. 4. Finally, we show in Sect. 4.1 how these algorithms can be extended to the 2D case.

Table 1. Comparison of k-ribbon algorithms in the message passing model. For brevity we ignore additive $O(k)$ terms in the round complexity. The time complexity of *Exact Count* is tight up to an additive $2k$ term, regardless of k and the starting agent. The memory and message complexity of *Bubble Sort* are independent of n and in fact constant assuming $k = O(1)$.

Algorithm	Rounds	Agent memory	Msgs	Msg bits	Exact	Reference
Exact count	$(2 - 1/k)n$	$3 \log n + O(1)$	$O(n)$	$O(\log n)$	✓	Thm. 3
Exact silent count	$3n$	$2 \log n + O(1)$	$O(n)$	$O(1)$	✓	Thm. 4
Bubble sort	$3n$	$O(\log k)$	$O(n^2)$	$O(\log k)$	✓	Thm. 7
Approx count	$2n$	$2 \log \log n + O(1)$	$O(n)$	$O(\log \log n)$	×	Thm. 6

2 Models and Notation

2.1 Concentration Model

For concentration-based solutions to the French flag problem, we assume that each agent receives concentration inputs from up to four source agents s_1, s_2, s_3, and s_4. The *measured concentration* a cell at 2D coordinate $C = (x, y)$ receives from source s_i, $i \in [4]$ is given by the following *gradient function*, which is assumed to be invertible and monotonically decreasing in $\text{dist}(C, s_i)$, the distance between cell C and the source s_i. For concreteness, consider the following power-law function

$$\lambda_i(C) = \frac{1}{\text{dist}(C, s_i)^\alpha} \tag{1}$$

where α is the power-law constant. This family of functions is also handy for the 1D case with coordinate $C = x$ and source s_i, $i \in [2]$ in Sect. 3, where we argue that coloring correctly can be reduced to comparing $\lambda_1(C)/\lambda_2(C)$ to 2^α and $2^{-\alpha}$.

Though we choose a power-law for convenience, our upper bounds and lower bounds hold for more general gradient functions satisfying the above constraints. Deriving precise thresholds for $\lambda_1(C)$ and $\lambda_2(C)$ is more difficult when the thresholds fall close together or when the gradient function is complicated. The more difficult these conditions, the less biologically practical it may be.

We do not assume any noise, so agents have arbitrarily good precision in measuring concentration. Additionally, we assume that the cells do not receive any other input apart from measured concentration. In particular, they do not have any other positional information such as knowledge of their coordinate or the total ribbon or flag size. We assume all agents behave identically, performing the same algorithms. No messages are passed between agents, so we consider only local computation for time complexity, assuming morphogen concentrations have reached steady state.

For the French ribbon, we assume that the two sources s_1 and s_2 are positioned at the ends of the line. We have two sources rather than one because a single source only gives an agent information about the distance of that agent to the source, without giving information about the agent's distance to the other side of the line.

For the French flag we assume the $s_i \in [4]$ are positioned at the four corners. We make this assumption in order to understand if the concentration model is 'strong' enough to solve the French flag problem without any additional communication. Assuming that additional sources are placed at convenient positions such as $(a/3, 0)$ for example, defies the idea of scale invariant systems. The corner points are already distinguished in that they only have two neighbors, and if one were to place a constant number of sources, these positions are somewhat natural.

2.2 Message Passing Model

We first consider a 1D version of the French flag problem which we call the *French ribbon problem*. We assume a line graph consisting of n nodes which we refer to as agents. We later consider the 2D version, the standard *French flag problem*, where the graph is a $a \times b$ grid on $n = a \cdot b$ agents.

Our message-passing model is similar to the standard LOCAL distributed model, with a few exceptions. Though agents have no knowledge of their global position, they do have a common sense of direction $dir \in \{up, down, left, right\}$. Additionally, agents know which of their neighbors exist, meaning they know whether they are endpoints of rows or columns (or both, if they are corners). Initially, all but one arbitrary agent called the *starting agent* s, representing the source of the communication signal, are *asleep* and thus perform no computation. Sleeping agents wake upon receiving a message.

The goal is to design algorithms that solve the French ribbon problem. Eventually, each agent must output a color so that the line is segmented into three colors: blue, white, and red from left to right. Formally, if b, w, and r denote the number of agents of each respective color, $\max\{|b - w|, |b - r|, |w - r|\} \leq 1$. In addition, each color should be in a single, contiguous sub-line of the graph—blue, white, red from left to right. We also define the more general 1D k-Ribbon problem in the same model, in which there are k distinct colors $\{1, ..., k\}$ which must form bands of approximately equal size, in increasing numerical order, along a line graph of n agents.

The 2D model is similar to the static, oriented 1D line graph model, but the system consists of an r by c grid of agents, oriented with *up* and *down* as well as $left$ and $right$. A solution to the French flag problem requires that every agent outputs a single color, such that the grid is divided into three vertical blocks. Every row must abide by the requirements of the French ribbon problem, such that the left side is blue and the right side is red. Furthermore, an agent should be the same color as the agent above and below it in its column. The 2D k-Flag problem generalizes in the same manner as above.

2.3 Approximation Definition

Intuitively speaking, the definition of approximation ensures two properties. First, agents that are clearly within one stripe should have the corresponding color. Second, agents that are close to a color border (c_1, c_2) should have either color c_1 or c_2.

We say a k-colored flag of dimensions $a \times b$ is an *ε-approximate* (French) flag if for every color $z \in \{1, ..., k\}$ the following hold. For each agent u with coordinates (x, y):

1. if $x \in \left[(\frac{z-1}{k} + \varepsilon) \cdot a, (\frac{z}{k} - \varepsilon) \cdot a\right]$, then the agent has color z.
2. if u has color z, then $x \in \left[(\frac{z-1}{k} - \varepsilon) \cdot a, (\frac{z}{k} + \varepsilon) \cdot a\right]$.

3 Concentration Model Results

3.1 1D Exact Concentration Ribbon

Algorithm Exact Concentration Ribbon. We consider an n-agent line of arbitrary finite length a in the concentration model. Assume morphogens m_1 and m_2 (with concentrations c_1 and c_2) are each secreted by one of the endpoint agents. We assume the underlying gradient function for concentration given position x is the inverse power law in α, which is assumed to be noiseless.

Assume that m_1 is secreted at $x = 0$ and m_2 is secreted at $x = a$, we have $c_1 = 1/x^\alpha$ and $c_2 = 1/(a - x)^\alpha$. The ratio of c_2 to c_1 is then $(a - x)^\alpha / x^\alpha$. Each agent computes this ratio independently from the measured values of c_1 and c_2. Let $ratio = c_2/c_1$. After calculating its measured ratio, each agent computes the smallest color z such that $ratio \geq ((z - 1)/(k - z))^\alpha$, decides color z, and halts.

The algorithm is size-invariant and works for a line graph of arbitrary finite length.

3.2 2D Concentration Lower Bound

In this section we sketch a proof of Theorem 2, showing that the concentration model, without absolute positional information, cannot produce a correct French flag (or even a good approximation) regardless of the gradient function.

Given an arbitrary flag G of dimensions $a \times b$, we show that we can construct a flag G' with dimensions $a' \times b'$ such that there are two agents in both flags that 1) have exactly the same distances from the respective sources and 2) must choose different colors. Since the two agents have the same respective distance to every source, they receive the same concentration input and cannot distinguish between settings, making it impossible to always color correctly. See Fig. 1 for an illustration. To show that such a flag G' exists, we frame the constraints as a system of equations and we show that there exists a valid solution.

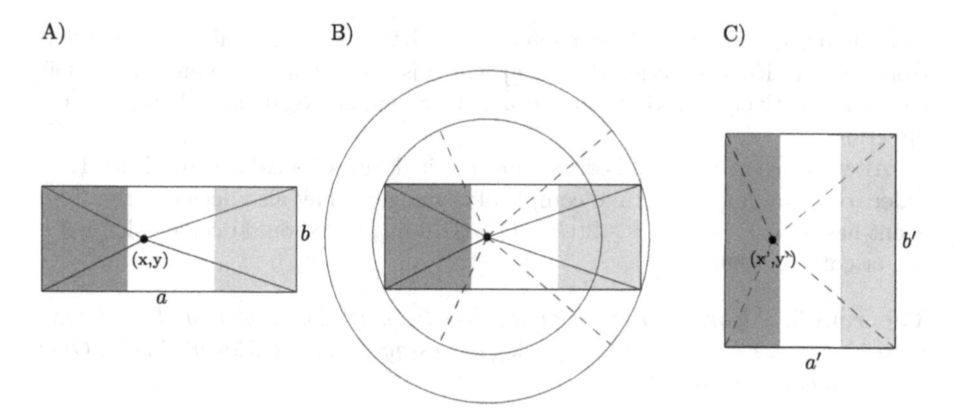

Fig. 1. A) depicts an arbitrary original flag. In the proof of Theorem 2 we argue how to construct a new flag as in C) such that there are two agents in both flags that have exactly the same distances from the respective sources and must also choose different colors. Since the two agents have the same respective distance to every source, they receive the same concentration input and cannot distinguish between the settings, making it impossible to always color correctly. We construct the new flag by changing the aspect ratio in a way that maintains the distances. B) depicts this transformation.

4 Message Passing Model

Before we present our algorithms, note there is a trivial algorithm that works as follows for $k = 3$. The starting agent sends a wakeup message to the leftmost and rightmost agents. Then start a counter from each of these agents. When an agent receives the counters n_ℓ and n_r, it can determine in which stripe it is by testing whether $n_\ell/n_r \geq 2$ or $n_\ell/n_r \leq 1/2$. This idea generalizes to arbitrary k.

The algorithms we present improve on the trivial algorithm in various ways. Table 1 summarizes the tradeoffs of our approaches in the message passing model. As a starting point, we observe that each agent can learn the number of agents to its left and right, from which information it can determine its own color [20]. This principle is central to some of our algorithms.

Note 1. An agent in the k-ribbon problem may determine its correct color knowing the number of agents on each side of it in line, and knowing which side should be color 1.

Algorithm Exact Count. The starting agent stores the value $n_{mid} \leftarrow 0$ and sends $n_{mid} + 1$ in both directions. Intuitively, the value measures the distance to the starting agent. All other agents upon waking store the received value as n_{mid} and forward the value $n_{mid} + 1$ to the next agent in the same direction. Each agent also stores $t \leftarrow n_{mid}$ and increments t every round after.

When the left endpoint receives a value for n_{mid}, it decides on color 1 and sends $n_\ell = 1$ to its right neighbor. When the right endpoint receives a value for

n_{mid}, it decides on color k and sends $n_r = 1$ to its left neighbor. Each agent stores n_d for either direction $d \in \{\ell, r\}$ which is the number of agents to the left (right, respectively). Upon receiving n_d, the agents forwards $n_d + 1$ in the same direction.

After an agent receives both n_ℓ and n_r, it decides its color using Note 1. In order to get an improved time complexity, an agent may also decide early: if an agent has a value n_d and $t \geq 2((k-1) \cdot n_d) - n_{mid}$, it should decide color 1 if d is ℓ or color k otherwise.

Theorem 3. *Algorithm Exact Count solves the k-ribbon problem and requires at most* $(2 - \frac{1}{k}) \cdot n + k$ *rounds,* $(4 - \frac{2}{k}) \cdot n \log n$ *message bits, and* $3 \log n + \log k + O(1)$ *bits of memory per agent.*

In reliable and synchronous models, it is well-known that silence conveys information. We improve the message bit complexity in Theorem 3 using the absence of a message as information, at a small cost to round complexity.

Algorithm Exact Silent Count. The starting agent sends the message 0 to the left and 1 to the right. If it is an endpoint, the starting agent sends a 0 and a 1 in the same, 2-bit message to its neighbor. Agents will forward any received messages in the same direction, except endpoints which will send the messages back.

The agents do additional processing. The endpoint on the d side sets $n_d \leftarrow 0$ upon waking and never modifies it. Otherwise, the first time an agent receives a message from direction d, it sets $n_{\bar{d}} \leftarrow 0$, and each round thereafter the agent increments $n_{\bar{d}}$, until it receives a message from the $\bar{d}$ direction, at which point it stops incrementing $n_{\bar{d}}$ and sets $n_{\bar{d}} \leftarrow n_{\bar{d}}/2$. When an agent has final values for n_ℓ and n_r, and has sent 0 to the left and 1 to the right, it decides its color based on its stored values of n_ℓ and n_r using Note 1 and halts. [1]

Theorem 4. *Algorithm Exact Silent Count solves the k-Ribbon problem and requires* $3n$ *rounds,* $6n$ *message bits, and* $2 \log n + \log k + O(1)$ *bits of memory per agent.*

Proof. We show correctness for the case when the starting agent is not an endpoint; we leave that end-case for the reader. W.l.o.g. consider an agent that first receives a 0 from the right. After $2n_\ell$ rounds, the 0 bit will return to the agent after having been forwarded to the left endpoint and back, so the stored value of n_ℓ at the end of the round will be correct. After $2n_r$ more rounds, the 0 bit is received again from the right and n_r is correctly set. Thus, as long as the agent receives the 0 bit 3 times, it will color itself correctly. The 0 bit must then travel from the starting agent to the left, back to the right endpoint, then back to the left endpoint; at that point, all agents to the left of the starting agent will correctly color themselves. As long as the agents to the right of the starting agent return the 0 bit leftward, this will occur. We thus have correctness, because all

[1] We note that a similar algorithm may use a single token rather than binary messages, at an additional constant-factor increase in round complexity.

agents only halt after forwarding the opposite bit back to the other side. The same argument applies to the 1 bit in the other direction.

A bit travels at most 3 times down the line, so all agents terminate after $3n$ rounds. Each round has 2 bits sent, so the message bit complexity is $6n$. Each agent stores k and two values in $\Theta(n)$, requiring only $2\log n + \log k + O(1)$ bits of memory each. □

Next, we use the approximation approach of Morris [13] and Flajolet [6] to reduce space complexity in exchange for a slight increase in error for the final k-ribbon. The randomized modification is made to our deterministic exact counting algorithm.

The following theorem gives the guarantees of each counter.

Theorem 5. [6] *Let* $\beta = 2^{2^{-\delta}}$. *Consider the counter procedure of [6], in which we maintain a counter c over n increments, and increase the counter by one only with probability* $(\frac{1}{\beta})^c$ *at each increment. Using* $\log\log n + \delta$ *bits for the counter, the expected value of the counter is* $\log_\beta((\beta - 1)\cdot n + \beta)$, *and the value of n we could recover from the counter has standard deviation at most* $n/2^{-\delta}$.

Algorithm Approximate Count. The starting agent sends a bit in either direction to wake all agents. When the endpoint in the d direction wakes up, it sets a counter c_d to 0, increments it as in [6], and sends the resulting value to its neighbor. Each agent upon receiving a message from direction d, stores the message as c_d, increments it in the same way and forwards the result to the next agent.

When an agent has received two values of c_d, it does the following: For each i in the sequence $1, \ldots, k$, if $c_\ell - c_r \le \log_\beta \frac{i}{k-i}$, then the agent decides on color i. If the agent has not decided on a color yet after all i, the agent decides on color k. After deciding on a color, the agent halts.

Theorem 6. *Fix any k. For n large enough, Algorithm Approximate Count solves the* ϵ*-approximate k-Ribbon problem for constant* $\epsilon < \frac{1}{2(k-1)}$ *with probability* $1 - \frac{1}{32k}$ *and requires* $2n$ *rounds,* $O(n\log\log n)$ *total message bits, and* $2\log\log n + O(1)$ *bits of memory per agent.*

We restrict $\epsilon < \frac{1}{2(k-1)}$ because otherwise the color thresholds would bleed into each other and we would have regions with more than two valid colors. The core idea of using an approximate counter as proposed in [6] is that when subtracting the counter from the left and from the right, we get for some β, ignoring small standard deviations,

$$\log_\beta((\beta-1)n_\ell + \beta) - \log_\beta((\beta-1)n_r + \beta) \approx \log(n_\ell/n_r).$$

Using thresholds for each color then gives the right color. Using monotonicity of the counters, we only need to consider $O(k)$ different counters which allows us to take a union bound over $O(k)$ of them, showing that all n counters are 'correct'. The proof can be found in the full version [3].

We next demonstrate how to use bubble sort to color the flag exactly. Assume blue, white and red are 1, 2, and 3 respectively.

Algorithm Bubble Sort. The algorithm is an application of the parallel sorting algorithm of [9]. The idea is to naively color agents, in alternating fashion with the colors of the flag, to ensure correct total counts of each color regardless of the ribbon length. The algorithm then performs swaps in parallel to ensure that blue elements ripple to the left, white elements to the middle, and red elements to the right. In an even round, any agent at an even position swaps the value (color) with its right neighbor if the right neighbor has a larger value. Odd rounds are analogous.

In order to avoid cases in which a agent would like to swap its color with both neighbors at same time, we also ensure through message passing that each agent knows whether it is at an odd or even position and whether the current round is odd or even.

Theorem 7. *Algorithm Bubble Sort solves the 1-D k-Ribbon problem and requires at most* $2n$ *rounds,* $n^2 \log k$ *message bits, and* $O(\log k)$ *bits of memory per agent.*

Proof. The algorithm of [9] requires at most n time-steps to sort an array using neighbor swaps in parallel. However, to assign each node a starting color, a message must be propagated from the leader to all nodes, requiring up to an additional n rounds.

Each round, up to half of the nodes send messages of size $\log k$ to broadcast their current values to one of their neighbors, for a total of at most $n^2 \log k$ message bits. Each node must store its own value using $\log k$ bits. □

4.1 Extending from Ribbon to Flag

We may solve the k-flag problem by extending any k-ribbon algorithm, with little loss in most parameters.

Algorithm Up & Down. The starting agent begins the k-ribbon algorithm for its row, and all agents in the row follow the algorithm to completion once awakened. However, after deciding on a color but before halting, each agent in the row tells its color to above and below neighbors. When an agent is awoken with a color, it decides that color and forwards the color either above or below before halting.

Theorem 8. *Given an algorithm for the k-ribbon problem which takes* $T(n, k)$ *rounds,* $M(n, k)$ *total message bits, and* $S(n, k)$ *bits of memory per agent, Algorithm Up & Down solves the k-flag problem on a* $a \times b$ *grid with at most* $a + T(b, k)$ *rounds,* $ab \log k + M(b, k)$ *total message bits, and* $S(b, k)$ *bits of memory per agent.*

Other reductions to the k-ribbon problem that optimize for round complexity rather than space and message bit complexity are left to the reader.

4.2 Message-Passing Lower Bounds

There are straightforward lower bounds for the 1D and 2D cases.

Theorem 9. *No algorithm exists that can solve the k-Ribbon problem on an oriented line graph if all agents are identical, even if endpoints know that they are endpoints, in less than* $(2 - \frac{1}{k}) \cdot n - 3$ *rounds.*

Theorem 10. *No algorithm exists to solve the k-flag problem on an* $a \times b$ *grid in less than* $\max\{(2 - \frac{1}{k}) \cdot b - k, a + b - 2\}$ *rounds.*

5 Conclusion

The 1D French ribbon problem can be solved exactly and approximately in both the concentration and the message passing models. However, the 2D French flag problem requires additional positional information in order to satisfy size invariance.

One direct extension of this work is a randomized version of the *Silent Count* algorithm (Theorem 4). An exciting new research direction is how other pattering problems can be solved in more general settings and under the influence of noise. Future work could develop models that better capture important biological constraints. For example, one could study models in which part of an organism (e.g., a finger or the beak of a bird) grows over time.

Acknowledgements. We thank Ama Koranteng, Adam Sealfon, and Vipul Vachharajani for valuable discussions and contributions.

References

1. Afek, Y., Alon, N., Barad, O., Hornstein, E., Barkai, N., Bar-Joseph, Z.: A biological solution to a fundamental distributed computing problem. Science **331**(6014), 183–185 (2011)
2. Alberts, B.: Molecular Biology of the Cell. CRC Press, Boca Raton (2017)
3. Ancona, A., Bajwa, A., Lynch, N., Mallmann-Trenn, F.: How to color a french flag-biologically inspired algorithms for scale-invariant patterning. arXiv e-prints p. 1905.00342 (2019)
4. Dessaud, E., McMahon, A.P., Briscoe, J.: Pattern formation in the vertebrate neural tube: a sonic hedgehog morphogen-regulated transcriptional network. Development **135**(15), 2489–2503 (2008)
5. Driever, W., Nüsslein-Volhard, C.: A gradient of bicoid protein in drosophila embryos. Cell **54**(1), 83–93 (1988)
6. Flajolet, P.: Approximate counting: a detailed analysis. BIT **25**(1), 113–134 (1985)
7. Green, J.B.A., Sharpe, J.: Positional information and reaction-diffusion: two big ideas in developmental biology combine. Development **142**(7), 1203–1211 (2015)
8. Gregor, T., Bialek, W., van Steveninck, R.R.D.R., Tank, D.W., Wieschaus, E.F.: Diffusion and scaling during early embryonic pattern formation. Proc. Natl. Acad. Sci. **102**(51), 18403–18407 (2005)

9. Habermann, N.: Parallel neighbor-sort (or the glory of the induction principle). Carnegie-Mellon University, Technical report (1972)
10. Jaeger, J., Martinez-Arias, A.: Getting the measure of positional information. PLoS Biol. **7**(3), e1000081 (2009)
11. Miller, J.F.: Evolving a self-repairing, self-regulating, french flag organism. In: Deb, K. (ed.) GECCO 2004. LNCS, vol. 3102, pp. 129–139. Springer, Heidelberg (2004). https://doi.org/10.1007/978-3-540-24854-5_12
12. Morgan, T.H.: "Polarity" considered as a phenomenon of gradation of materials. J. Exp. Zool. **2**, 495–506 (1905)
13. Morris, R.: Counting large numbers of events in small registers. Commun. ACM **21**(10), 840–842 (1978)
14. Nüsslein-Volhard, C., Wieschaus, E.: Mutations affecting segment number and polarity in drosophila. Nature **287**(5785), 795–801 (1980)
15. Patten, I., Placzek, M.: The role of sonic hedgehog in neural tube patterning. Cellular Molecular Life Sci. CMLS **57**(12), 1695–1708 (2000)
16. Stumpf, H.F.: Mechanism by which cells estimate their location within the body. Nature **212**(5060), 430–431 (1966)
17. Summerbell, D., Lewis, J.H., Wolpert, L.: Positional information in chick limb morphogenesis. Nature **244**(5417), 492–496 (1973)
18. Turing, A.M.: The chemical basis of morphogenesis. Philos. Trans. R. Soc. Biol. Sci. **237**(641), 37–72 (1952)
19. Wolpert, L.: The french flag problem: a contribution to the discussion on pattern development and regulation. Towards Theoret. Biol. **1**, 125–133 (1968)
20. Wolpert, L.: Positional information and the spatial pattern of cellular differentiation. J. Theor. Biol. **25**(1), 1–47 (1969)
21. Zadorin, A.S., et al.: Synthesis and materialization of a reaction–diffusion french flag pattern. Nat. Chem. **9**(10), 990–996 (2017)

Simple Intrinsic Simulation of Cellular Automata in Oritatami Molecular Folding Model

Daria Pchelina[1], Nicolas Schabanel[2](✉), Shinnosuke Seki[3], and Yuki Ubukata[4]

[1] École Normale Supérieure de Paris, Paris, France
[2] École Normale Supérieure de Lyon (LIP UMR5668, MC2, ENS de Lyon), Lyon, France
nicolas.schabanel@ens-lyon.fr
[3] University of Electro-Communications, 1-5-1 Chofugaoka, Chofu, Tokyo 1828585, Japan
[4] NTT DATA Corporation, Tokyo, Japan

Abstract. The Oritatami model was introduced by Geary et al. (2016) to study the computational potential of RNA cotranscriptional folding as first shown in wet-lab experiments by Geary et al. (Science 2014). In the Oritatami model, a molecule grows component by component (named beads) into the triangular grid and folds as it grows. More precisely, the δ last nascent beads are free to move and adopt the positions that maximize the number of bonds with the current folded structure. Geary et al. (2018) proved that the Oritatami model is capable of efficient Turing universal computation using a complicated construction that simulates Turing machines via tag systems. We propose here a simple Oritatami system which intrinsically simulates arbitrary 1D cellular automata. Being intrinsic, our simulation emulates the behavior of cellular automata in a readable way and in time linear in space and time of the simulated automaton. The Oritatami model has proven to be a fruitful framework to study molecular reconfigurability. Our construction relies on the development of new mechanisms which are simple enough that we believe that some simplification of them may be implemented in the wet lab. An implementation of our construction can be downloaded for testing.

Keywords: Molecular self-assembly · Co-transcriptional folding · Intrinsic universality · Cellular automata · Turing universality

1 Introduction

DNA computing encompasses the field which tries to implement computation at the molecular levels. A recent example is [17], which implements arbitrary 6-bit

N. Schabanel—His work is supported in part by the CNRS grants MOPREXPROG-MOL and AMARP from the Mission pour l'interdisciplinarité.
S. Seki—His work is supported in part by JSPS KAKENHI Grant-in-Aids for Challenging Research (Exploratory) No. 18K19779 and JST Program to Disseminate Tenure Tracking System, MEXT, Japan, No. 6F36.

Y. Kohayakawa and F. K. Miyazawa (Eds.): LATIN 2020, LNCS 12118, pp. 425–436, 2020.
https://doi.org/10.1007/978-3-030-61792-9_34

cellular automata onto DNA nanotubes, realising a first DNA-based universal computer (limited to 6 bits of memory). This success of the field was built by going back and forth between theory, models and experiments. The Oritatami model was introduced in 2016 by [4] to study the computational potential of RNA cotranscriptional folding as first shown in wet-lab experiments by [5].

In Oritatami systems, we consider a finite set of *bead types*, and a periodic sequence of *beads*, each of a specific bead type. Beads are attracted to each other according to a fixed symmetric relation. In any folding (*configuration*), a *bond* is formed between any pair of beads located at adjacent positions and attracting each other. At each step, the latest few beads in the sequence are allowed to explore all possible positions, and adopt only those positions that minimise the energy, or otherwise put, those positions that maximise the number of bonds in the folding. "Beads" are a metaphor for domains, i.e. subsequences, in RNA and DNA (and are thus not limited to 4 types only). The Oritatami model has proven to be a fruitful framework to study *molecular reconfigurability*, one of the most promising directions to reduce error in wetlab molecular implementation as error might be erased by reconfiguration later on. Indeed, programming Oritatami systems consists of designing molecules whose shape changes depending on their contexts, hence achieving some form of reconfiguration. Other models studying molecular reconfiguration include nubots [16] and signal passing tile assembly [10,11]. Previous work on Oritatami includes among others the implementation of a binary counter [4], the Heighway dragon fractal [7], folding of shapes at small scale [2], NP-hardness of the rule minimization [6,9], a study of its parameters [13], and polynomial-time Turing machine simulation [3].

Our Contribution. The universality result by Geary et al. in [3] relies on a complicated construction that simulates Turing machines via tag systems [1,18]. We propose here a simple Oritatami system which intrinsically simulates arbitrary 1D cellular automata. Being intrinsic [8,15], our simulation emulates the behavior of cellular automata in a readable way and in time which is linear in the space and time of the simulated automaton. Precisely, our main result is:

Theorem 1 (Main result). *There is a universal finite set of 183 bead types $\mathcal{B}$ such that for any 1D cellular automaton A with Q states and radius r, there is a delay-2 Oritatami system with bead types in $\mathcal{B}$ and periodic transcript with period precisely*

$$\frac{71}{3}\left((3+q)\cdot 2(Q_r)^2 + 8(2q \mod 3)\right) + 10q + 610 \sim \frac{142}{3}(Q_r)^2 \log_2 Q_r$$

that simulates A intrinsically with a supercell shaped as a lozenge with sides of size $O((Q_r)^2 \log Q_r)$, where $q = \lceil \log_2(2Q^{2r+1}) \rceil$ and $Q_r = 2^q \leqslant 4Q^{2r+1}$.

This improves the previous construction in [3] as the number of bead types is only 183 (instead of 542) and the delay is 2 (instead of 3). Furthermore, our construction relies on the development of new mechanisms which are now simple enough to believe that some simplification of them may be implemented in the wet lab.

An implementation of our construction can be downloaded for testing [14].

2 Model and Preliminary Results

2.1 Oritatami Model

Let B be a finite set of *bead types*. A *configuration* c of a bead type sequence $p \in B^* \cup B^{\mathbb{N}}$ is a directed self-avoiding path $c_0 c_1 c_2 \cdots$ in the triangular lattice $\mathbb{T}$,[1] where for all integer i, the vertex c_i of c is labeled by p_i and refers to the *position* in $\mathbb{T}$ of the $(i+1)$-th bead in the configuration. A *partial configuration* of p is a configuration of a prefix of p. The class of all the configurations obtained by applying an isometry of $\mathbb{T}$ to a given configuration is called a *conformation*.

For any partial configuration c of some sequence p, an *elongation* of c by k beads (or k-*elongation*) is a partial configuration of p of length $|c| + k$ extending by k positions the self-avoiding path of c. We denote by $\mathcal{C}_p$ the set of all partial configurations of p (the index p will be omitted whenever it is clear from the context). We denote by $c^{\triangleright k}$ the set of all k-elongations of a partial configuration c of sequence p.

Oritatami Systems. An *oritatami system* $\mathcal{O} = (p, \heartsuit, \delta)$ is composed of (1) a (possibly infinite) bead type sequence p, called the *transcript*, (2) an *attraction rule*, which is a symmetric relation $\heartsuit \subseteq B^2$, and (3) a parameter δ called the *delay*. $\mathcal{O}$ is said *periodic* if p is infinite and periodic. Periodicity ensures that the "program" p embedded in the oritatami system is finite (does not hardcode any specific behavior) and at the same time allows arbitrarily long computation.[2]

We say that two bead types a and b *attract* each other when $a \heartsuit b$. Furthermore, given a (partial) configuration c of a bead type sequence q, we say that there is a *bond* between two adjacent positions c_i and c_j of c in $\mathbb{T}$ if $q_i \heartsuit q_j$ and $|i - j| > 1$. The *number of bonds* of configuration c of q is denoted by $H(c) = |\{(i, j) : c_i \sim c_j,\ j > i + 1, \text{ and } q_i \heartsuit q_j\}|$.

Oritatami Dynamics. The folding of an oritatami system is controlled by the delay δ. Informally, the configuration grows from a *seed configuration* (the input), one bead at a time. This new bead adopts the position(s) that maximize(s) the potential number of bonds the configuration can make when elongated by δ beads in total. This dynamics is *oblivious* as it keeps no memory of the previously preferred positions [3].

Formally, given an Oritatami system $\mathcal{O} = (p, \heartsuit, \delta)$ and a *seed configuration* σ of a *seed bead type sequence* s, we denote by $\mathcal{C}_{\sigma,p}$ the set of all partial configurations of the sequence $s \cdot p$ elongating the seed configuration σ. The considered *dynamics* $\mathscr{D} : 2^{\mathcal{C}_{\sigma,p}} \to 2^{\mathcal{C}_{\sigma,p}}$ maps every subset S of partial configurations of length ℓ elongating σ of the sequence $s \cdot p$ to the subset $\mathscr{D}(S)$ of partial

[1] The triangular lattice is defined as $\mathbb{T} = (\mathbb{Z}^2, \sim)$, where $(x, y) \sim (u, v)$ if and only if $(u, v) \in \cup_{\epsilon=\pm 1}\{(x + \epsilon, y), (x, y + \epsilon), (x + \epsilon, y + \epsilon)\}$. Every position (x, y) in $\mathbb{T}$ is mapped in the euclidean plane to $x \cdot \vec{\mathsf{e}} + y \cdot \vec{\mathsf{sw}}$ using the vector basis $\vec{\mathsf{e}} = (1, 0)$ and $\vec{\mathsf{sw}} = \text{RotateClockwise}(\vec{\mathsf{e}}, 120°) = (-\frac{1}{2}, -\frac{\sqrt{3}}{2})$.

[2] Note that we do not impose here a maximal number of bonds per bead (called arity).

configurations of length $\ell + 1$ of $s \cdot p$ as follows:

$$\mathscr{D}(S) = \bigcup_{c \in S} \arg\max_{\gamma \in c^{\rhd 1}} \left(\max_{\eta \in \gamma^{\rhd(\delta-1)}} H(\eta) \right)$$

The possible configurations at time t of the oritatami system $\mathcal{O}$ are the elongations of the seed configuration σ by t beads in the set $\mathscr{D}^t(\{\sigma\})$.

We say that the Oritatami system is *deterministic* if at all time t, $\mathscr{D}^t(\{\sigma\})$ is either a singleton or the empty set. In this case, we denote by c^t the configuration at time t, such that: $c^0 = \sigma$ and $\mathscr{D}^t(\{\sigma\}) = \{c^t\}$ for all $t > 0$; we say that the partial configuration c^t *folds (co-transcriptionally) into* the partial configuration c^{t+1} deterministically. In this case, at time t, the $(t+1)$-th bead of p is placed in c^{t+1} at the position that maximises the number of bonds that can be made in a δ-elongation of c^t.

2.2 Sweeping 2-Fan-in 2-Fan-Out Cellular Automata

Our construction simulates intrinsically the space-time diagrams of a specific type of one-way cellular automata where each cell has fan-in and fan-out 2 as shown in Fig. 2, similar to the gates implemented in [17]. Formally, a *2-fan-in 2-fan-out automaton (2FA)* $\mathcal{A}$ is given by its set of states $[Q] = \{0, \ldots, Q-1\}$ and its transition function $f : [Q]^2 \to [Q]^2$. A finite configuration of $\mathcal{A}$ is an even-length word $c \in [Q]^*$, and its image by $\mathcal{A}$ is $c' = F(c)$ where $(c'_{2i}, c'_{2i+1}) = f(c_{2i-1}, c_{2i})$ for $i = 0..\frac{|c|}{2} - 1$, with the convention that $c_{-1} = c_{|c|} = 0$. Classically, any 1D cellular automaton with Q states and radius r can be simulated intrinsically by a 2FA with Q^{2r+1} states using a time rescaling by r.

Sweeping Simulation. Our construction simulates intrinsically any 2FA by sweeping down (even time step) and up (odd time step), see Fig. 2. As a consequence, every other step, the two inputs are read in reverse order and the transition function is applied with its arguments exchanged. Formally a configuration (c, d) of a *sweeping 2FA (S2FA)* $([Q], f)$ consists of an even-length word $c \in [Q]^*$ together with a direction $d \in \{\uparrow, \downarrow\}$, and has the following dynamics: $F(c, \downarrow) = (c', \uparrow)$ where $(c'_{2i}, c'_{2i+1}) = f(c_{2i-1}, c_{2i})$ for $i = 0..\frac{|c|}{2} - 1$; $F(c, \uparrow) = (c', \downarrow)$ where $(c'_{2i+1}, c'_{2i}) = f(c_{2i}, c_{2i-1})$ for $i = 0..|c|/2 - 1$. Clearly, any 2FA $([Q], f)$ can be simulated intrinsically in real time by the S2FA $([Q] \times \{\uparrow, \downarrow\}, g)$ where $g((x, \uparrow), (y, \uparrow)) = ((x', \downarrow), (y', \downarrow))$ with $(x', y') = f(x, y)$; and $g((x, \downarrow), (y, \downarrow)) = ((x', \uparrow), (y', \uparrow))$ with $(y', x') = f(y, x)$.

From now on, we consider a S2FA $\mathcal{A} = ([Q], f)$, where $Q = 2^q$ is a power of two with $q \geqslant 1$. We will denote by $(x'(x, y), y'(x, y))$ the value of $f(x, y)$.

3 Overview of the Construction

Due to space constraint, we will expose here the principle of the construction. The full description of the modules and of the attraction rule is given in the full version of the present article [12].

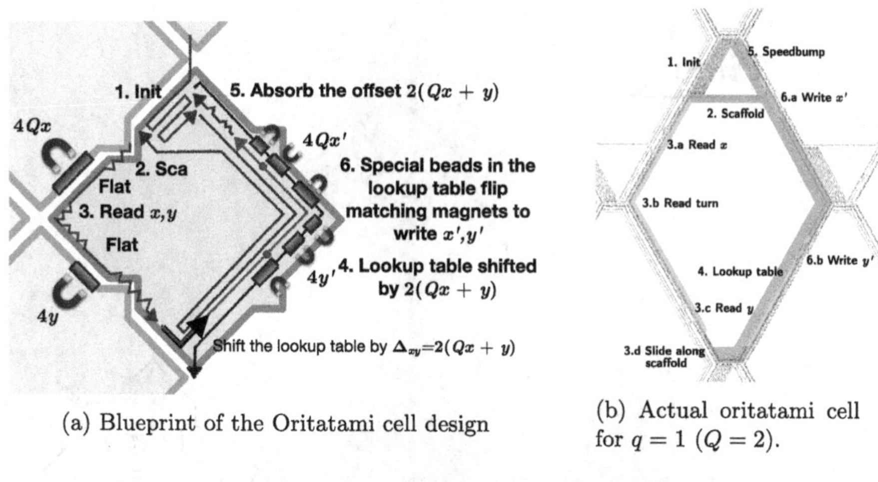

(a) Blueprint of the Oritatami cell design

(b) Actual oritatami cell for $q = 1$ $(Q = 2)$.

Fig. 1. The modules inside a cell: (Left) Schematic view; (Right) 1. Cell Init highlighted in yellow; 2. Scaffold in red; 3. Read in blue, green and purple; 4. Lookup Table in yellow and violet; 5. Speedbump in cyan; 6. Write in green. (Color figure online)

In our intrinsic simulation, each cell of the simulated S2FA is affinely mapped onto a supercell shaped as a hexagon with two short sides (N and S) of lengths 12 and 13, and four long sides (NE, NW, SE and SW) of lengths s and $s - 1$ where $s = O(Q^2 \log Q)$ (see Fig. 1b and 2). The states are encoded on the sides of the hexagons as described below. The simulation proceeds by building one after the other the supercells simulating each of the cells of the simulated S2FA according to the up-down order given in Fig. 2. Each supercell is the result of the folding of exactly one period of the transcript. The period of transcript consists of the sequence of 6 *modules*, each of them achieving one specific task:

$$\boxed{\mathsf{I}} \cdot \boxed{\mathsf{S}} \cdot \boxed{\mathsf{R}} \cdot \boxed{\mathsf{L}} \cdot \boxed{\mathsf{SB}} \cdot \boxed{\mathsf{W}}$$

The Modules. Their respective roles and positions inside the supercell are blueprinted in Fig. 1a. $\boxed{\mathsf{I}}$ is responsible for extending the configuration by one supercell and reversing the up-down order at the end of the current column of supercells (see [12]). $\boxed{\mathsf{S}}$ has two roles: providing a scaffold along which the next modules will fold, and ensuring that the molecule "resynchronizes" (will be defined later) before $\boxed{\mathsf{W}}$ writes the two outputs x' and y' on the output sides. $\boxed{\mathsf{R}}$ is responsible for reading the value of the two inputs x and y and translating accordingly the lookup table of the simulated S2FA, encoded in the next module $\boxed{\mathsf{L}}$. $\boxed{\mathsf{SB}}$ is responsible for "resynchronizing" the molecule along the scaffold, annihilating the translation of the lookup table induced by the reading of x and y by $\boxed{\mathsf{R}}$. Finally, $\boxed{\mathsf{W}}$ writes on the output sides of the supercell the values x' and y' dictated by the translated lookup table $\boxed{\mathsf{L}}$, and exits the supercell at the entrance of the next one.

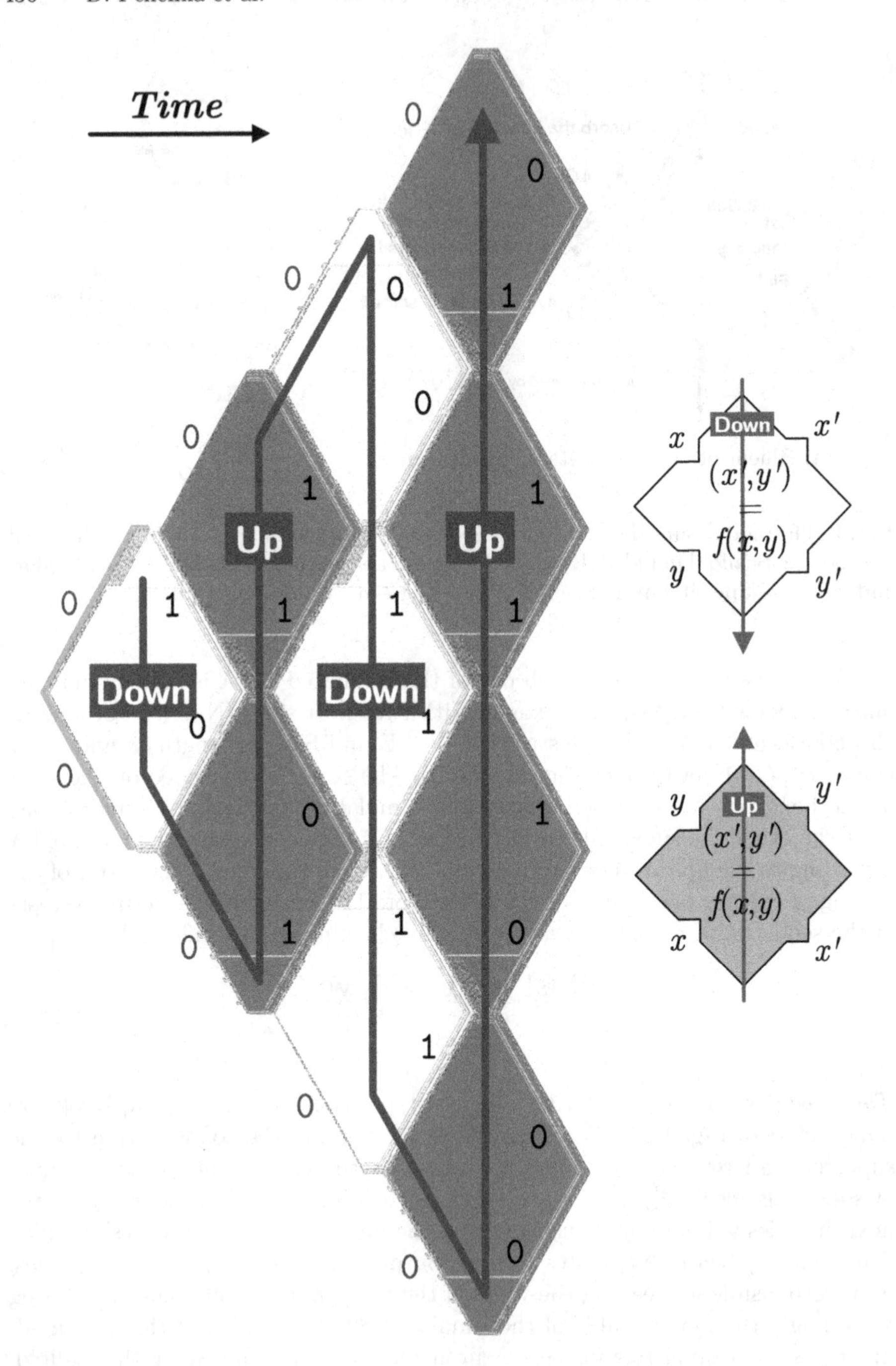

Fig. 2. The 10 first super-steps of the Oritatami simulation of the 2-state S2FA ($q = 1$) $f(x, y) = (y + 1 \bmod 2, x)$ from the seed configuration encoding input 00 (to the left in brown): $(00, \downarrow) \mapsto (0100, \uparrow) \mapsto (011010, \downarrow) \mapsto (00011110, \uparrow) \mapsto (0011110000, \downarrow)$; in white (resp. gray), the down- (resp. up-) hexagonal supercells.

Encoding x and y. The values of x and y are encoded along the sides of the supercells using "magnetic flipping flaps" of total lengths $4Qx$ and $4y$ respectively, as schematically shown on Fig. 1a. When the read module **R** folds along the side of the neighboring supercells, it gets flattened by these magnetic flaps; this shifts the progression of the molecule forward by exactly half the lengths of the flaps. It follows that the module **R** completes its folding $\Delta_{xy} = 2(Qx + y)$ beads further than it would in absence of the magnetic flaps. This, in turn, translates the position of the lookup table module **L** by Δ_{xy} along both output sides of the supercell, placing the entries corresponding to $x'(\Delta_{xy}) = x'(x, y)$ and $y'(\Delta_{xy}) = y'(x, y)$ in front of the flipping flaps of the module **W** to be folded next so that, when folded, the total magnetic length of the flipping flaps on each output side is $4Qx'$ and $4y'$ respectively (see Sect. 4 and Fig. 4 for details).

4 Description of the Key Mechanisms

Due to space constraints, we will focus on the new mechanisms involved in this construction. In particular, we will not discuss **I** because its behavior is just a direct translation of the Module G in [3] (see [12] for details). **S** is simply hardcoded and only its key part will be discussed next in Sect. 4.2.

4.1 Modules R, L, and W: The Read, Lookup, Write Mechanism

The previous section gave the principle of the interactions between these modules: the reading of x and y on the input sides by **R** results in shifting the lookup table **L** by $\Delta_{xy} = 2(Qx + y)$, which aligns the entries corresponding to $x'(x, y)$ and $y'(x, y)$ properly with the flaps of module **W** which, in turn, writes the corresponding $x'(x, y)$ and $y'(x, y)$ on the x'- and y'-output sides respectively using the magnetic flaps as illustrated in Fig. 4. Let us start with Module **L**. Refer to Fig. 3 for the alignment of the various parts involved.

Module **L**. Each output $x'(x, y)$ and $y'(x, y)$ is encoded in binary into q tables of Q^2 bits using bead types **Q0** and **Q1**. The entry indexed $Qx + y$ in the i-th table for x' (resp. y') contains the value of the i-th bit of $x'(x, y)$ (resp. $y'(x, y)$). More precisely, if we write $x'(x, y) = \sum_{i=0}^{q-1} b_i 2^i$ in binary, the table for x' consists of the sequence of bead types: $\mathsf{Lookup}_X = \left(\prod_{i=0}^{q-1}\prod_{x=0}^{Q-1}\prod_{y=0}^{Q-1}(\mathbf{Q}(b_i))^2\right)^R$, such that the bead types in $\mathsf{Lookup}_X{}^R$ at indices $0, 2Q^2, \ldots, (q-1)2Q^2$ shifted by $\Delta_{xy} = 2(Qx + y)$ are $\mathbf{Q}(b_0), \ldots, \mathbf{Q}(b_{q-1})$. Lookup_Y is defined similarly.

Module. **W** consists of a zigzag glider **T0..7** that runs along the two output sides of the supercell, together with q "magnetic flipping flaps," equally spaced by $2Q^2$ beads on each output side (see Fig. 3): q flaps of lengths $2^0Q, \ldots, 2^{q-1}Q$ on the x'-output side and of lengths $2^0, \ldots, 2^{q-1}$ on the y'-output side. We define a *magnetic flipping flap of length ℓ* as the bead type sequence:

$$\mathsf{SegFF}(\ell) = \mathbf{U0..5}\,(\mathbf{T4}\,\mathbf{P4}\,(\mathbf{T6}\,\mathbf{P0}\,\mathbf{T0}\,\mathbf{P3}\,\mathbf{T2}\,\mathbf{P2}\,\mathbf{T4}\,\mathbf{P1})^{\ell}\mathbf{T6}\,\mathbf{P0..4}\,\mathbf{U6..8}.$$

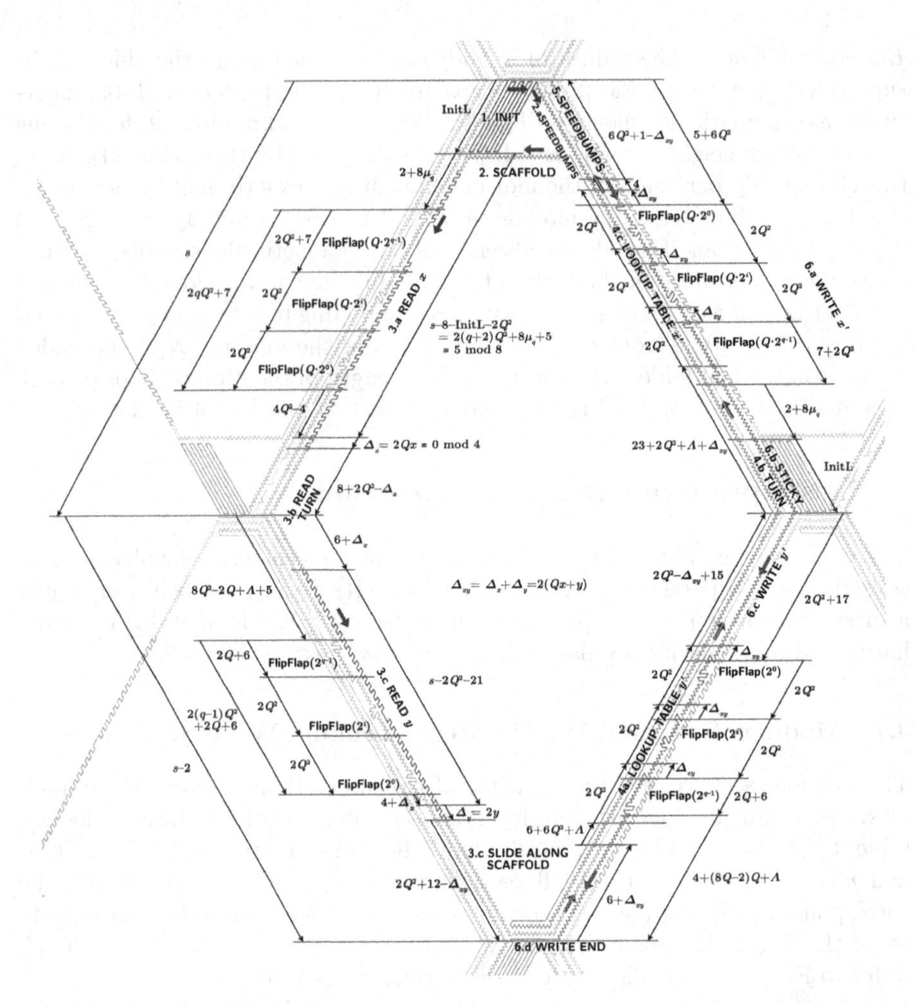

Fig. 3. Alignment of the various modules

Each flap is either activated (magnetic for R) or deactivated (neutral for R) depending on whether its "magnetic" beads **P0..3** point *outwards*, towards the upcoming neighboring supercell, or *inwards*, towards the inside of the supercell currently folding (see Fig. 4). Now, thanks to the alignment of the modules (see Fig. 3), the i-th flap of W starts folding in front of the entries $\Delta_{xy} + 2iQ^2$ of the lookup table on each output side, that is in front of the pair of beads $\mathbf{Q0}^2$ or $\mathbf{Q1}^2$ corresponding to the i-th bit of the value to write on this side. Now, a flap folds outwards (is activated) by default, unless its initial bead **U5** is attracted by a pair of beads **Q0** corresponding to a bit set to 0. It follows that the i-th flap of W on each side is activated if and only if the i-th of the output is 1; and as it is of length 2^iQ and 2^i for the x'- and y'-output side respectively, the total numbers of magnetic beads are $4Qx'(x,y)$ and $4y'(x,y)$ on each x'- and y'-output side, respectively.

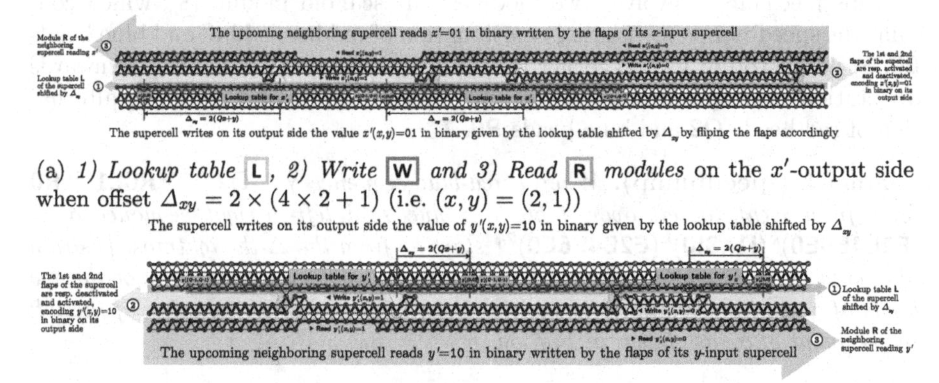

(a) *1) Lookup table* **L**, *2) Write* **W** *and 3) Read* **R** *modules* on the x'-output side when offset $\Delta_{xy} = 2 \times (4 \times 2 + 1)$ (i.e. $(x,y) = (2,1)$)

(b) *1) Lookup table* **L**, *2) Write* **W** *and 3) Read* **R** *modules* on the y'-output side when offset $\Delta_{xy} = 2 \times (4 \times 1 + 0)$ (i.e. $(x,y) = (1,0)$)

Fig. 4. *Illustration of the border between two neighboring supercells:* Interactions of 1) the lookup table and 2) the write modules within a supercell and 3) the read module of the upcoming neighboring supercell, when $Q = 4$ $(q = 2)$. The lookup tables for each bit follow each other and are $2Q^2$ beads long each (2 beads per bit). The write module folds into $q = 2$ flipping flaps on each output side of lengths 4×2^0 and 4×2^1 on the x'-output side and 2^0 and 2^1 on the y'-output side. We have highlighted in yellow the folding of bead **U5**, which decides the orientation of each flap (in- or out-wards if **Q0** or **Q1** is present resp.). We have highlighted in orange the folding of bead **U6**, which is attracted by all bead types **Q0..2** and restores the orientation of the write glider to defaults after each flap. (Color figure online)

Module **R**. We are now ready to conclude this mechanism by observing that the read module **R** folds along the write modules of the two neighboring input supercells, and that it gets flattened each time it folds along an activated flap (see Fig. 3 and 4), which extends its length by half the number of magnetic beads **P0..3** of the flap. It follows that the end of its folding is shifted forward by $(4Qx + 4y)/2 = \Delta_{xy}$, which in turn shifts forward the lookup table module **L** by Δ_{xy} as claimed. Refer to Fig. 6 for a complete view of the folding of **R**.

The full description of the modules **R**, **L** and **W** may be found in the full version [12] of the present article.

4.2 Modules SB and S: Resynchronization Using Speedbumps

In order for the period of the transcript to end precisely at the exit of the supercell, regardless of which inputs x and y were read by the read module **R**, we need to absorb the Δ_{xy} offset. Precisely, we need to absorb it before the write module W folds to ensure that it is properly aligned with the shifted lookup table. This is the role of the speedbump module SB. Its behavior is illustrated in Fig. 5.

This mechanism involves two modules: the scaffold module S, which contains the speedbumps (consisting of alternation of red beads **I0..3** and blue beads **E0..3**) at the top of its NE corner (assuming the supercell is in the downwards orientation); the speedbump module SB which consists of a matching alternation of red beads **Q2** and blue beads **R0..1**.

Lemma 1 (speedbump). *When a red-blue sequence* $\gamma = \mathbf{Q2}^{4k-1}(\mathbf{R0..1})^{4k}\mathbf{R0}$ *folds from right to left over a blue-red-blue seed left-to-right sequence* $\sigma = (\mathbf{E2E4E6E0})^k(\mathbf{I1..3I0})^k(\mathbf{E2E4E6E0})^{2k}$ *starting from the* Δ*-th rightmost position of* σ *with* $\Delta < 4k$*, the* Δ *leftmost blue beads of* γ *fold into a zigzag over the red beads of* σ*, and the folding of* γ *ends at the* $\lfloor \Delta/2 \rfloor$ *rightmost position of the left red segment of* σ*, as shown in Fig. 5b.*

Corollary 1. *When the folding of the speedbump module* SB *completes, the offset* Δ_{xy} *is totally absorbed.*

Proof (Sketch). Note that the maximum offset when the speedbump module SB starts to fold is $\Delta = 2(Q^2 - 1)$ corresponding to reading input $(x, y) = (Q - 1, Q - 1)$. The matching exponentially decreasing alternation of blue and red regions from 2^{2q} to 4 in S and SB (see [12]) ensures that the offset is divided by 2 until it reaches 0, absorbing the total offset as shown in Fig. 5.

Correctness of the Folding. Finally, the correctness of the folding is proved by induction using automated folding tree certificates (see [4,12]). The key is to choose carefully the size s of the supercell so that all modules are properly aligned regardless of the inputs x and y. This is ensured by enforcing the position of every pattern in every module modulo 8 in the supercell as explained in Fig. 3 and detailed in [12].

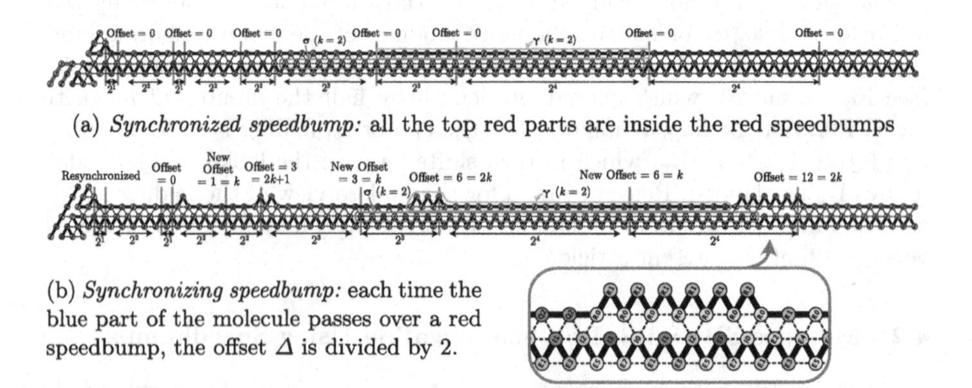

(a) *Synchronized speedbump:* all the top red parts are inside the red speedbumps

(b) *Synchronizing speedbump:* each time the blue part of the molecule passes over a red speedbump, the offset Δ is divided by 2.

Fig. 5. Speedbumps decrease exponentially the offset of the molecule folding on top (going from right to left) until it vanishes.

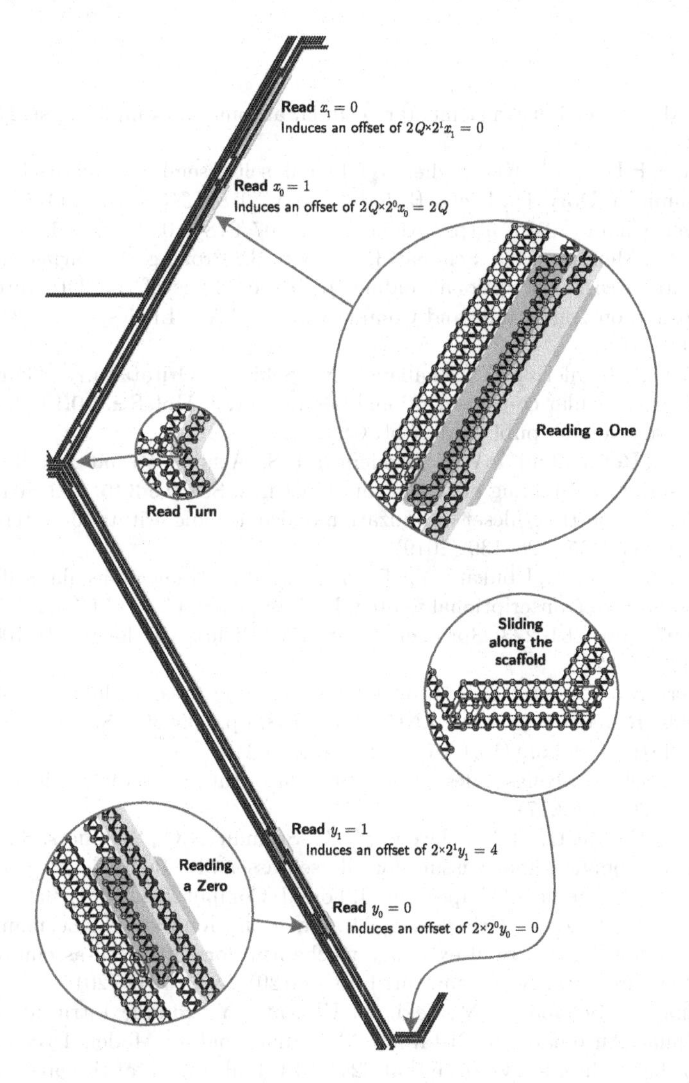

Fig. 6. *Read* module consists of 4 parts: ReadX, Turn, ReadY, and Slide along the scaffold. ReadX and ReadY get flattened each time it passes along an outward write magnetic flap (encoding a 1). This shifts the molecule by the length of the corresponding flap. The following part of the molecule is then shifted overall by $\Delta_{xy} = 2(Qx + y)$. This shift allows then to align the entry corresponding to (x, y) of the lookup table of each side with the upcoming write modules. The two glider-based parts "Turn" and "Slide along" are only there to ensure that after reading each input, the molecule turns at the expected position, regardless of the offset.

References

1. Cook, M.: Universality in elementary cellular automata. Complex Syst. **15**(1), 1–40 (2004)
2. Demaine, E.D., et al.: Know when to fold 'em: self-assembly of shapes by folding in Oritatami. In: Doty, D., Dietz, H. (eds.) DNA 2018. LNCS, vol. 11145, pp. 19–36. Springer, Cham (2018). https://doi.org/10.1007/978-3-030-00030-1_2
3. Geary, C., Meunier, P.E., Schabanel, N., Seki, S.: Proving the Turing universality of oritatami co-transcriptional folding. In: Proceedings of the 29th International Symposium on Algorithms and Computation, ISAAC. LIPIcs, vol. 123, pp. 23:1–23:13 (2018)
4. Geary, C., Meunier, P.E., Schabanel, N., Seki, S.: Oritatami: a computational model for molecular co-transcriptional folding. Int. J. Mol. Sci. **20**(9), 2259 (2019). Preliminary version published in MFCS 2016
5. Geary, C., Rothemund, P.W.K., Andersen, E.S.: A single-stranded architecture for co transcriptional folding of RNA nanostructures. Science **345**, 799–804 (2014)
6. Han, Y.S., Kim, H.: Ruleset optimization on isomorphic oritatami systems. Theor. Comput. Sci. **785**, 128–139 (2019)
7. Masuda, Y., Seki, S., Ubukata, Y.: Towards the algorithmic molecular self-assembly of fractals by co transcriptional folding. In: Câmpeanu, C. (ed.) CIAA 2018. LNCS, vol. 10977, pp. 261–273. Springer, Cham (2018). https://doi.org/10.1007/978-3-319-94812-6_22
8. Ollinger, N.: Two-states bilinear intrinsically universal cellular automata. In: Freivalds, R. (ed.) FCT 2001. LNCS, vol. 2138, pp. 396–399. Springer, Heidelberg (2001). https://doi.org/10.1007/3-540-44669-9_41
9. Ota, M., Seki, S.: Ruleset design problems for oritatami systems. Theor. Comput. Sci. **671**, 26–35 (2017)
10. Padilla, J.E., Patitz, M.J., Schweller, R.T., Seeman, N.C., Summers, S.M., Zhong, X.: Asynchronous signal passing for tile self-assembly: fuel efficient computation and efficient assembly of shapes. Int. J. Found. Comput. Sci. **25**(4), 459–488 (2014)
11. Padilla, J.E., Sha, R., Kristiansen, M., Chen, J., Jonoska, N., Seeman, N.C.: A signal-passing DNA strand exchange mechanism for active self-assembly of DNA nanostructures. Angew. Chem. Int. Edit. **54**(20), 5939–5942 (2015)
12. Pchelina, D., Schabanel, N., Seki, S., Ubukata, Y.: Simple Intrinsic Simulation of Cellular Automata in Oritatami Molecular Folding Model, December 2019. https://hal.archives-ouvertes.fr/hal-02410874. Full version of the present article
13. Rogers, T.A., Seki, S.: Oritatami system; a survey and the impossibility of simple simulation at small delays. Fund. Inform. **154**(1–4), 359–372 (2017)
14. Schabanel, N.: iOS CAOS simulator. hub.darcs.net/nikaoOoOoO/CAOSSimulator
15. Theyssier, G.: Automates Cellulaires: un Modèle de Complexités. Ph.D. thesis, École Normale Supérieure de Lyon (2005)
16. Woods, D., Chen, H., Goodfriend, S., Dabby, N., Winfree, E., Yin, P.: Active self-assembly of algorithmic shapes and patterns in polylogarithmic time. In: Proceedings of the 4th Conference on Innovations in Theoretical Computer Science, ITCS, pp. 353–354 (2013)
17. Woods, D., et al.: Diverse and robust molecular algorithms using reprogrammable DNA self-assembly. Nature **567**, 366–372 (2019)
18. Woods, D., Neary, T.: On the time complexity of 2-tag systems and small universal Turing machines. In: Proceedings of the 47th Annual IEEE Symposium on Foundations of Computer Science, FOCS, pp. 439–448 (2006)

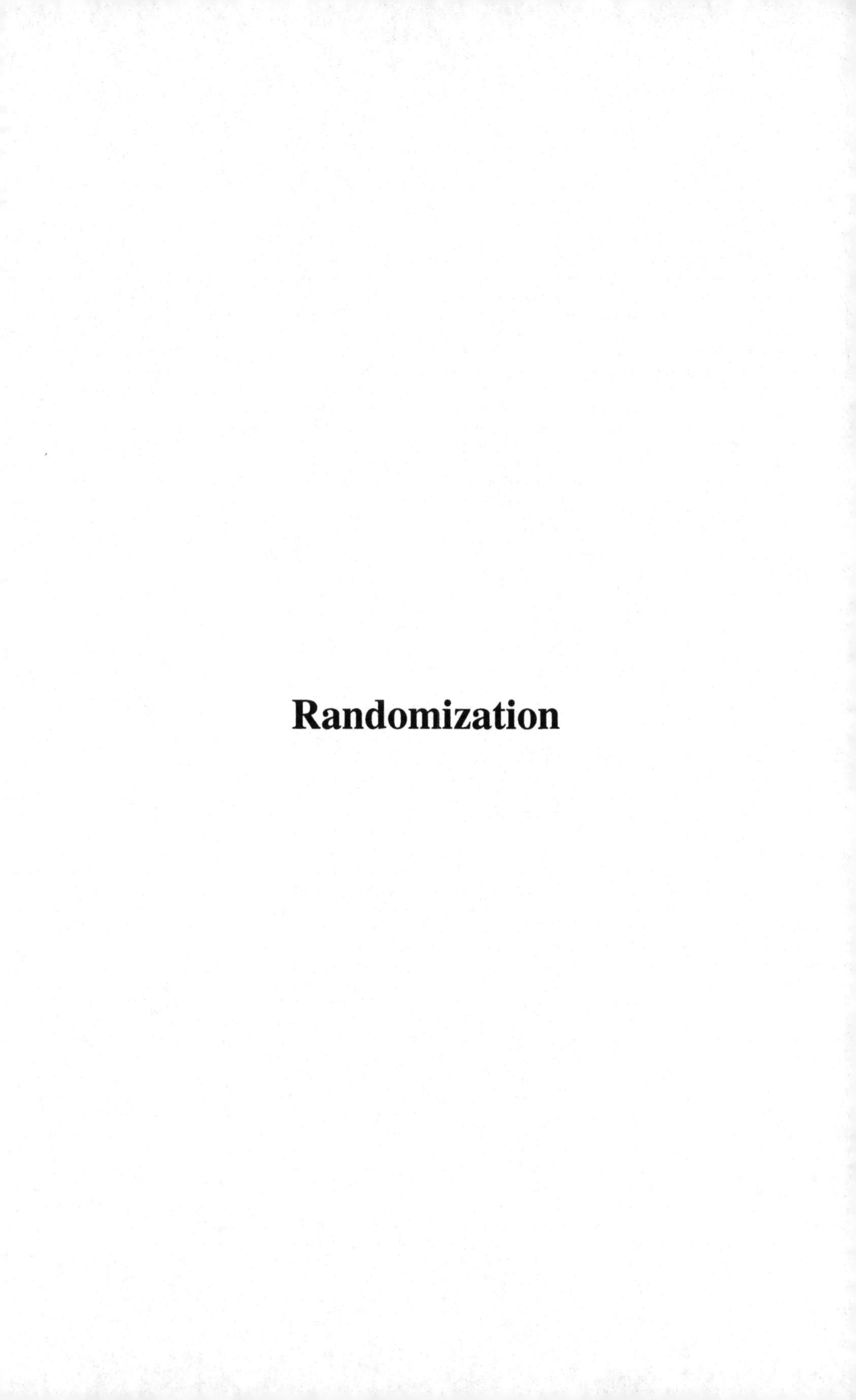

Randomization

Transmitting once to Elect a Leader on Wireless Networks

Ny Aina Andriambolamalala(✉) and Vlady Ravelomanana(✉)

IRIF UMR CNRS 8243, University of Paris, Paris, France
{ny-aina.andriambolamalala,vlad}@irif.fr

Abstract. Distributed wireless network's devices are battery-powered most of the time. Transmitting a message uses more energy than receiving one which spends more energy than internal computations. Therefore in this paper, we will focus on the energy complexity of leader election, a fundamental distributed computing problem. As the message's size impacts on the energy consumption, we highlight that our algorithms have almost optimal time complexities: each device is allowed to send only once $1-bit$ message and to listen to the network during at most 2 time slots. We will firstly work on Radio Networks on which the devices can detect when a node transmits alone: RNstrongCD where both senders and receivers have collision detection capability, RNsenderCD, RNreceiverCD and RNnoCD. If the nodes know their number n, our algorithm elects a leader in optimal $O(\log n)$ time slots with a probability of $1-1/poly(n)$. Then, if all nodes do not know n but know its upper bound u such that $\log u = \Theta(\log n)$, it has $O(\log^2 n)$ time complexity on RNnoCD and RNsenderCD. On RNreceiverCD and RNstrongCD, it has $O(\log^{(1+\alpha)} n)$ time complexity where $\alpha \in]0,1[$ is constant. For the Beeping Networks model on which the devices cannot detect single transmissions, it has $O(n^{\alpha})$ time complexity with probability $1-1/poly(n)$.

1 Introduction

Distributed leader election problem has been extensively studied over the years [11,12,17,19,20,26]. It consists in all the n devices of the distributed system, denoted $s_1, s_2, \ldots, s_n$, agreeing on one device to be the leader in a decentralized manner. The study of its energy complexity gained in importance with the design of Low-power Wireless Sensor Devices [14,23,25]. On such devices, transmitting uses more energy than listening to the network which spends more energy than internal computations [3,17,23,25]. For example, in [23], each device consumes respectively 1.8W, 0.6 W and 0.05 W when transmitting, receiving a message and having radio switched off. Such energy consumption also depends on the collision detection capability [6,17] and the message's size [2]. Energy complexity is the maximum over all devices of the time slot number during which any device is awake[1]. In this paper, each node exchanges only a single bit

[1] When it transmits or listens to the network.

Y. Kohayakawa and F. K. Miyazawa (Eds.): LATIN 2020, LNCS 12118, pp. 439–450, 2020.
https://doi.org/10.1007/978-3-030-61792-9_35

message, transmits at most once and listens to the network during at most two time slots. This is in contrast with the works in [4,11,21] where the messages have $O(\log n)$ or even larger size [6,10]. We found some similarities between our model and a distributed real-time communication model (Time Triggered Protocols) on which a unique sending slot is assigned to each node [18,22]. This latter model was widely used for designing energy-efficient algorithms on Wireless Sensor Networks [1,15]. We consider the single-hop[2] Networks defined below, the basic ingredient out of which larger multi-hop networks are built [21,26].

Single-Hop Radio Networks or RN. Introduced by Chlamtac and Kutten in the 80's [7], communications occur in synchronous time slots. At any time slot, a node independently decides whether to transmit, to listen to the network, or to remain idle or asleep. The network can have three status: SINGLE if exactly one node transmits, NULL if no node transmits and COLLISION if at least two nodes transmit. Only the SINGLE transmissions are received by the listening nodes. We consider four models: on RNstrongCD, both transmitters and listeners have collision detection capability and on RNsenderCD, only transmitters can detect collision. On RNreceiverCD or RNCD, only listeners can detect collision but transmitters can not and on RNnoCD, no device has collision detection.

Single-Hop Beeping Networks or BN. This was introduced in 2010 by Cornejo and Kuhn [8] and is strictly weaker than the RN model as far as the message length and the collision detection capability are concerned [2]. It makes little demands on the devices which need only be able to do carrier-sensing as well as differentiating between silence and the presence of a jamming signal on the network. The communications occur synchronously as in RN but the transmitting nodes cannot detect collisions and the listening nodes cannot distinguish between single and more beeps emitted by their neighbors.

To use randomness, we assume that the devices can generate discrete random variables (see for instance Devroye [9]). We also assume that these devices are initially anonymous and indistinguishable and can do any internal computations [6,20]. They cannot communicate on the network when in a sleeping state. However, they can choose to wake up or to sleep at any time slot. All presented algorithms in this paper succeed with high probability[3] or *w.h.p.* for short.

1.1 Related Works

Considering single-hop RNCD, in the 70's, Tsybakov [24], Capetanakis [5] designed deterministic leader election algorithms terminating in $O(\log n)$ time slots. Such algorithms are optimal since Greenberg and Winograd [13] have established a lower bound of $\Omega(\log n)$ on the time complexity for deterministic algorithms when all nodes known n. On the randomized side, Willard [26] designed protocols working in expected $O(\log \log n)$ and in $O(\log n)$ time slots

[2] The underlying graph of the network is complete.

[3] An event ε_n occurs *w.h.p.* if $\mathbb{P}[\varepsilon_n] \geq 1 - n^{-c}$ where c is a positive constant.

with high probability when n is unknown. Given an error rate ε, without the knowledge of n, Nakano and Olariu [21] provided a randomized algorithm terminating in $O(\log\log n) + o(\log\log n) + O(\log 1/\varepsilon)$ time slots with a probability exceeding $1 - \varepsilon$. They also provided a lower bound of $\Omega(\log n)$ for uniform protocols. Ghaffari, Lynch and Sastry [12] extended this lower bound to all protocols: given an upper bound u of n, they presented a lower bound of $\Omega(\min\{\log(u/n), \log(1/\varepsilon)\})$ for leader election algorithms succeeding with probability greater than $1 - \epsilon$. Amongst other results, Kardas, Klonowski and Pajak [17] designed a leader election algorithm for the RNstrongCD model where n is unknown, succeeding in $O(\log^{\epsilon} n)^4$ expected time slots with $O(\log\log\log n)$ energy complexity. When the nodes know $\Theta(n)$, Jurdziński, Kutyłowskiowski, and Zatopiański [16] designed an algorithm with $O(\log^* n)^5$ energy complexity and $O(\log n)$ time complexity on the RNCD model. Bender, Kopelowitz, Pettie and Young [4] then gave an $O(\log(\log^* n))$ upper bound for the energy complexity of leader election and approximate counting on RNCD when n is unknown. In [6], Chang, Kopelowitz, Pettie, Wang and Zhan presented a leader election protocol for RNreceiverCD (*resp.* RNnoCD) with $n^{o(1)}$ time complexity and $O(\log\log^* n)$ energy complexity (*resp.* $O(\log^* n)$) when n is unknown. Amongst several important results, they proved a $\Omega(\log\log^* n)$ (*resp.* $\Omega(\log^* n)$) lower bound for the energy complexity of leader election on RNCD (*resp.* RNnoCD) for such setting.

1.2 Our Results

The following Table 1 shows what differentiate our results from existing results.

Table 1. Difference between our results and existing results.

Existing results				
Assumptions	Model	Time	Energy	Probability
n known	RNCD [17]	$O(\log n)$	$O(\log\log\log n)$	$1 - O(1/n)$
$\Theta(n)$ known	RNCD [16]	$O(\log n)$	$O(\log^* n)$	$1 - O(1/n)$
n unknown	RNCD, RNnoCD [6]	$O(n^{o(1)})$	$O(\log^* n)$	$1 - O(1/n)$
n unknown	RNstrongCD, RNsenderCD	$O(n^{o(1)})$	$O(\log\log^* n)$	$1 - O(1/n)$
Our results				
n known	RNCD, RNnoCD	$O(\log n)$	3	$1 - O(1/n)$
n known	RNstrongCD, RNsenderCD, Sect. 2.1	$O(\log n)$	2	$1 - O(1/n)$
n unknown and $\Theta(\log n)$ known[a]	RNnoCD	$O(\log^2 n)$	3	$1 - O(1/n)$
	RNsenderCD, Sect. 2.2	$O(\log^2 n)$	2	$1 - O(1/n)$
	BN $\alpha \in]0, 1[$ Sect. 3	$O(n^{\alpha/(\alpha+1)} \times \log n)$	2	$1 - O(n^{-\alpha/(\alpha+1)})$

[a] An upper bound u of n is known and $\log u = \Theta(\log n)$.

[4] $\log^{\varepsilon} n = (\log n)^{\varepsilon}$ for any constant ε.

[5] $\log^* n$ represents the iterated logarithm of n.

Our algorithm's design is based on the nodes locally generating random values before communicating on the network in a deterministic way. Each node transmits at most once and listens to the network during at most 2 time slots. When n is known, the presented algorithm in Sect. 2.1 is optimal in view of both time [21] and energy complexities [17]. As the IDs of the nodes commonly fit in $O(\log n)$ bits, in Sect. 2.2, u is an upper bound of n such that $\log u = \Theta(\log n)$ *i.e.* $u \in]n, n^c]$. The result in [6] can be adapted to work on such setting with the same $O(\log^2 n)$ time complexity, $O(1)$ energy complexity and $O(\log n \log\log n)$ messages size. The best result on this scenario is the $O(\log n)$ time complexity with $O(\log^* n)$ energy presented in [16] when all nodes know $\Theta(n)$. Note that the assumption of knowing $\Theta(\log n)$ is slightly weaker than knowing $\Theta(n)$.

2 Radio Networks

2.1 The Nodes Initially Know the Exact Value of n

We start by assuming that all the nodes initially know the exact value of n and remembering that in the RNnoCD, only the listening nodes can differentiate from single, no transmitter or multiple transmitters. We take advantage of this ability to simulate loneliness detection [12] in such a model. Our goal is to cause the following events to occur during the execution of such an algorithm:
–(i) If $t_0 = 0$ is the initial time slot, there is a time slot $t_g = t_0 + g$ when exactly one node s_1 transmits alone while a set S_z of nodes listens to the network.
–(ii) Then, exactly one second node $s_2 \in S_z$ transmits alone at $t_g + 1$ (while s_1 listens to the network) to notify s_1 that it was elected. Thus, s_2 is the unique witness of the probable election of s_1. To fulfill this goal, our algorithm is based on each node locally generating random values and communicating on the network in a deterministic manner, to find out two consecutive unique[6] values. Therefore, we make each node generate one copy of a discrete random variable (r.v. for short) X such that if $X_1, X_2, \ldots, X_N$ are N independent copies of X: there are 2 unique consecutive values X_i, X_{i+1} with a constant probability. G(1/2)[7], the geometric distribution with parameter 1/2 respects such property.

We use $\lg a$ to denote the logarithm of a in base 2. We suppose that $\log a$, $\lg a$ and e^a are integers for any value a.

Remark 1. *We use some basic probability theories and the Chernoff bound formula to prove all the presented Lemmas in this Section. Due to space constraint, such proofs are not shown in this extended abstract.*

Lemma 1. *Let $X_1, X_2, \ldots, X_N$ be N independent copies of a r.v. distribution following* G(1/2)*. I is a discrete interval of integers and $|I|$ is the size of the interval I. We then have $I = \{I_0, I_1, \ldots I_{|I|-1}\}$ where $I_r = I_0 + r$ is an integer. Let p be the probability that $\exists (i, j) \in [1, N]^2$ such that $X_i = \lg N, X_j = \lg N - 1$ and $X_l \notin \{\lg N - 1, \lg N\} \forall l \notin \{i, j\}$. We have*

$$p > \frac{1}{5}\left(1 - O\left(\frac{1}{N}\right)\right)$$

[6] A random value is said to be unique if it is held by exactly one node.
[7] If X is a r.v. distributed as G(1/2), $q_k = \mathbb{P}[X = k] = 2^{-k-1}$ for all $k \geq 0$.

Overview of the Algorithm: It works on RNnoCD and RNCD when each node knows n. In what follows, each node s_i has a status denoted $\textsc{Status}(s_i)$, which can take one of the following values: NULL is the initial status, CANDIDATE if s_i is candidate to be the leader, ELIMINATED if s_i cannot be elected, MARKED if s_i is temporarily marked to do some computations and LEADER if s_i is the elected node. Any node s_i having $\textsc{Status}(s_i) = \textsc{Null}$ is designated as a NULL node and we do the same for all status. Each node is initially sleeping and is restricted to send a $1 - bit$ message only once. Our algorithm is designed to make each node s_i aware of its final $\textsc{Status}(s_i) \in \{\textsc{Leader}, \textsc{Eliminated}\}$.

Our main idea is to make each node s_i locally generate one random copy X_i of a r.v. X distributed as G(1/2). Then, all nodes *browse through*[8] the interval $I = [\lg n - 1, \lg n]$, in order to find out which two of them have consecutive unique random values. For instance, by Lemma 1, a sequence of $X_1, X_2, \ldots, X_n$ with unique node s_i (*resp.* s_j) holding $X_i = \lg n - 1$ (*resp.* $X_j = \lg n$) occurs with a constant probability. Thus such idea leads us to the election of s_i with a constant probability in $O(|I|)$ time slots. Then, to reach the high probability requirement, we have to execute such algorithm $O(\log n)$ times by keeping the energy complexity at a maximum of 3. To do so, our new algorithm is subdivided into $2 \log n + 1$ steps. All nodes are firstly distributed such that $O(n/\log n)$ nodes participate to each step and each node participates to only one step. During each such step, $O(n/\log n)$ nodes then do a leader election succeeding with a constant probability as described earlier.

Step 0: choice of step. Each node chooses uniformly at random in which step it will participate. Let S_z be the set of nodes participating to Step z, $\forall z > 0$.

Lemma 2. *$Card(S_z) \in [2n/5 \log n, 3n/5 \log n]$ with a probability greater than $1 - e^{-O(n/\log n)}$.*

Each step is subdivided into 3 Phases: candidacy, witnessing and browsing. For the sake of clarity, we describe the execution of Step 1 but this will be generalized for any Step z in the presentation of the algorithm. During the *candidacy* phase, each node in S_1 chooses to be CANDIDATE or ELIMINATED. Then, on the *witnessing*[9] phase, each ELIMINATED node in S_1 chooses at which time slot of the *browsing* phase it will witness for the election of a node. Finally, during the *browsing* phase, all nodes in S_1 *browse through* I to elect a leader. I is defined by Lemma 1 by replacing N with $O(n/\log n)$. For greater clarity, we present Phase 3 before Phase 2.

Step 1 Phase 1: candidacy. At t_0, each node $s_i \in S_1$ locally generates one independent copy X_i of a r.v. X distributed as G(1/2). Based on Lemma 1, all nodes in S_1 having $X_i \in I = [\lg(2n/5 \log n) - 1, \lg(3n/5 \log n)]$ then take the CANDIDATE status and the other nodes of S_1 become ELIMINATED.

Lemma 3. *There are $O(\log n)$ CANDIDATE nodes in each Step with a probability greater than $1 - O(\log n/n)$.*

[8] At each time slot $t_0, t_1, \ldots, t_g$, each node s_i checks if the corresponding value I_g in the interval I is equal to its X_i, then transmits or does some computations at t_g.

[9] Listening to verify an election at the time slot.

Step 1 Phase 3: browsing through I. This phase uses an odd/even time slots scheduling. Even time slots $\{t_0, t_2, \ldots, t_{2g}\}$ are dedicated for transmissions and odd time slots $\{t_1, t_3, \ldots, t_{2g+1}\}$ are used for feedback. At t_0, each CANDIDATE node $s_i \in S_1$ checks if $X_i = I_0$, then, transmits a $1 - bit$ message ($I_0 = \lg(2n/5\log n) - 1$). Each S_1's CANDIDATE node s_i having $X_i = I_1$ listens to the network. Then, at t_1, the nodes with $X_i = I_0$ listen in their turn and if the nodes that listened at t_0 received a message, they send a single bit feedback at t_1. A node that transmitted at t_0 and receives the feedback at t_1 becomes LEADER and the other nodes become ELIMINATED. Each CANDIDATE node $s_i \in S_1$ executes these computations at each time slot $t_0, t_1, \ldots, t_g, \ldots, t_{2|I|-1}$, checking if its $X_i = I_g$ at t_g. It is possible to have several consecutive unique random values in the interval I, involving the election of multiple leaders. In order to bypass such a problem, we add the following Phase 2 before Phase 3.

Step 1 Phase 2: witnessing an election at a time slot and flooding the next time slots. After Phase 1, the $O(n/\log n)$ ELIMINATED nodes in S_1 (Lemma 2 and Lemma 3) are distributed to witness the probable election of a leader at each time slot of Phase 3. Let T be the time complexity of Phase 3. We have $T = 2|I| \leq 6$. At the round t_0, after executing Phase 1, each ELIMINATED node in S_1 chooses uniformly at random or UAR one time slot t_w or time to witness from $\{t_0, t_2, \ldots, t_{T-2}\}$. So $t_w = UAR(\{t_0, t_2, t_4\})$[10]. These ELIMINATED nodes listen to the network at t_w and t_w+1 during Phase 3. They receive messages at both points if a leader is elected. So, to avoid another election, each node chooses a time to flood[11] $t_f = UAR(\{t_w + 2, \ldots, t_{(4|I|\log n)-1}\})$ and transmits at t_f. By flooding all the remaining time slots, no other CANDIDATE node can transmit alone. The following Fig. 1 illustrates the execution of one step of such algorithm whith 8 devices.

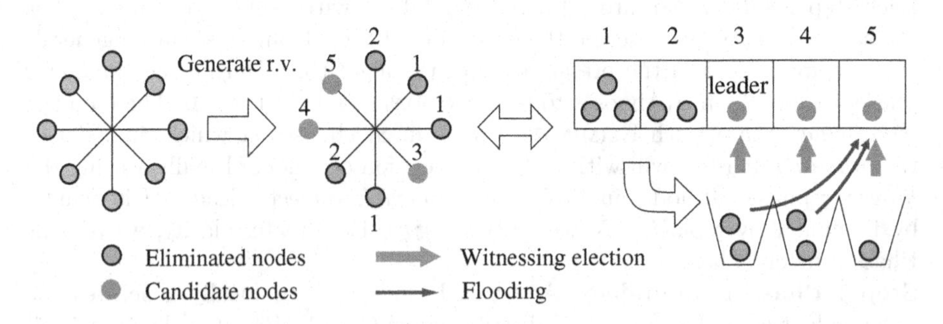

Fig. 1. Leader election for 8 devices and $I = [3, 5]$.

Lemma 4. *Each time slot of such leader election algorithm is witnessed and flooded by at least one node with a probability greater than* $1 - e^{-O(n/\log^2 n)}$.

[10] $UAR(B)$ returns one value picked uniformly at random from the set B.

[11] Sending a message at the time slot if a leader has already been elected.

Algorithm 1. LEADERELECTION(n).

Input : The exact value of n.
Output: Each node s_i with a $Status(s_i) \in \{Leader, Eliminated\}$.

1 **Step 0:** Each node enters a set $UAR(\{S_1, S_2, \ldots, S_{2\log n}\})$ where S_z is the set of nodes that will participate in Step z.
2 **Step 1 to Step 2 log n: for** z *from* 1 *to* $2\log n$ **do**
3   **Step z Phase 1:** Each node $s_i \in S_z$ locally generates a random value X_i distributed as G(1/2).
4   Each node $s_i \in S_z$ having $X_i \in I = [\lg(2n/5\log n) - 1, \lg(3n/5\log n)]$ sets $Status(s_i) \leftarrow Candidate$, $Status(s_i) \leftarrow Eliminated$ otherwise.
5   **Step z Phase 2:** Each ELIMINATED node in S_z sets $t_w \leftarrow UAR(\{t_{(z-1)2|I|}, t_{(z-1)2|I|+2}, \ldots, t_{2z|I|-2}\})$ and $t_f \leftarrow UAR(\{t_w + 2, \ldots, t_{(4\log n|I|)-1}\})$.
6   **Step z Phase 3:** Each node runs the BROWSE(I) procedure.
7   Each remaining CANDIDATE node s_i sets $status(s_i) \leftarrow Eliminated.$
8 **end**

Algorithm 2. BROWSE(I): called at Step z.

Input : Interval I.
Output: Each node s_i with $Status(s_i) \in \{Leader, Eliminated\}$.

1 **for** g *from* 0 *to* $|I| - 1$ **do**
2   Each node sets $t = t_{2(z-1)|I|+2g}$.
3   Each CANDIDATE node $s_i \in S_z$ with $X_i = I_g$ sends $1 - bit$ at time slot t and listens at $t + 1$.
4   Each CANDIDATE node $s_j \in S_z$ with $X_j = I_{g+1}$ listens at t.
5   **if** s_j *receives a message at* t **then**
6   s_j transmits $1 - bit$ message at $t + 1$.
7   **end**
8   **if** s_i *receives a message at* $t + 1$ **then**
9   s_i sets $Status(s_i) \leftarrow Leader.$
10   **end**
11   Each ELIMINATED node $s_e \in S_z$ having $t_w = t$ listens at t and $t + 1$.
12   **if** s_e *receives a message at both* t *and* $t + 1$ **then**
13   s_e sets $Status(s_e) \leftarrow Marked.$
14   **end**
15   Each MARKED node that has $t_f = t$ transmits at t and sets $Status(s_e) \leftarrow Eliminated.$
16 **end**

Lemma 5. *During the execution of* LEADERELECTION(n), *each node wakes up to transmit one bit at most once and listens to the network during at most two time slots.*

Remark 2. *On the RNsenderCD and RNstrongCD models, each node can know when it transmits alone. Then, the other nodes do not have to notify the leader that it was elected. Thus, we can cause the* CANDIDATE *nodes to never listen to the network and the* ELIMINATED *nodes to witness at only one time slot.*

Theorem 1. *In single-hop RNnoCD and RNCD (resp. RNstrongCD and RNsenderCD) networks of known large size n, there is a randomized Monte-Carlo leader election algorithm succeeding in $O(\log n)$ time slots with a probability of at least $1 - O\left(n^{-1/3}\right)$. Each node transmits $1-$bit message no more than once and listens to the network for a maximum of two (resp. one) time slots.*

Proof. By Lemma 1, a leader can be elected with a strictly positive constant probability by running the BROWSE(I) protocol once. Thus, the LEADERELECTION(n) algorithm elects a leader *w.h.p.* in $4|I|\log n = O(\log n)$ time slots as $|I| \leq 6$. According to Lemma 4, a maximum of one leader is elected with a probability greater than $1 - e^{-O\left(n/\log^2 n\right)} \geq 1 - O\left(n^{-1}\right)$ for sufficiently large n. During the execution of LEADERELECTION(n), each node wakes up during at most three time slots, transmitting once and listening to at most two time slots for the RNnoCD and RNCD (Lemma 5). Applying Remark 2 to LEADERELECTION(n), we have each node listening at exactly on time slot for the RNstrongCD and RNsenderCD models.

Remark 3. *To simplify Algorithm1, we made it run the* BROWSE(I) *protocol $2\log n$ times. Thus, it succeeds with probability $1 - O\left(n^{-1/3}\right)$. This can be improved to reach $1 - O\left(n^{-1}\right)$ by running* BROWSE(I) *$5\log n$ times.*

2.2 The Nodes Do Not Know n

When the nodes do not know n but know its upper bound u such that $\log u = \Theta(\log n)$ *i.e.* $u \in]n, n^c]$ where $c > 1$, we adapt the LEADERELECTION(n) algorithm as follows. On Step 0, each node chooses UAR to participate in one of the remaining $2\log u$ Steps. Let S_z be the set of nodes participating in Step z. By Lemma 2, $Card(S_z) \in [2n/5\log u, 3n/5\log u]$ *w.h.p.* Then on Phase 1 of each Step z, in order to have an interval containing the interval $I = [\lg(2n/5\log u) - 1, \lg(3n/5\log u)]$, each node sets a new interval $J = [\lg\left(2u^{1/c}/5\log u\right) - 1, \lg(3u/5\log u)]$. Then, on Phase 2, the nodes eliminated after Phase 1 set $t_w = UAR(\{t_{(z-1)2|J|}, t_{(z-1)2|J|+2}, \ldots, t_{2z|J|-2}\})$ and $t_f = UAR(\{t_w + 2, \ldots, t_{(4\log u|J|)-1}\})$. Finally, on Phase 3, all nodes run the BROWSE(J) protocol.

Theorem 2. *In single-hop RNnoCD and RNCD (resp. RNsenderCD and RNstrongCD) networks of unknown large size n, if an upper bound u of n is given in advance to all the nodes, there is a randomized Monte-Carlo leader election algorithm succeeding in $O(\log^2 n)$ time slots with a probability greater than $1 - O\left(n^{-1}\right)$. Each node transmits during no more than one time slot and listens to the network during at most two (resp. one) time slots.*

Proof. By applying $Card(J) = O(\log n)$ and Remark 3 to the proof of Theorem 1, we obtain *w.h.p.* a $O(\log^2 n)$ time complexity and an energy complexity of at most 3. □

3 Beeping Networks

In this section, we consider the BN model where neither a beeping node s_i nor listening nodes can detect if s_i beeps alone or not. The goal here is to make a node know that it beeped along *w.h.p.* without any feedback from the network. To do so, our main idea is based on the uniqueness of the maximum of n independent copies $Y_1, Y_2, \ldots, Y_n$ of the following new r.v.

Definition 1 *(Definition of the distribution of the r.v. Y). Throughout this paper, let $p_k = \mathbb{P}[Y = k]$ for all integers $k \geq 0$ defined for some $\alpha \in]0,1[$ as follows. Fix $\beta = 1/(1+\alpha)$,*

$$p_0 = e^{-1} \text{ and } p_k = \exp\left(-k^{\beta}\right) - \exp\left(-(k+1)^{\beta}\right) \text{ for all } k > 0\,. \tag{1}$$

If such a maximum is unique, we cause each node to generate a random copy of Y and our algorithm has to localize which node holds such a maximum. This latter node then becomes LEADER.

The following observation is crucial for our purpose:

Lemma 6. *Let $Y_1, Y_2, \cdots, Y_N$ be N independent copies of a r.v. distributed as described by (1) and $m = \max_{1 \leq i \leq N}\{Y_i\}$*
(a) $\mathbb{P}\left[Card\{l \text{ such that } Y_l = m\} = 1\right] \geq 1 - O\left(1/\log^{\alpha} N\right)$.

(b) Let $\mu = \mathbb{P}\left[(\log N - \log\log\log N)^{(1+\alpha)} \leq m \leq (\log N + \log\log N)^{(1+\alpha)}\right]$:

$$\mu \geq 1 - O\left(1/\log N\right).$$

(c) Set $L = [(\log N - \log\log\log N)^{(1+\alpha)}, (\log N + \log\log N)^{(1+\alpha)}]$, *and let q be the random variable such that $q = Card(\{l \text{ such that } Y_l \in L\})$.*

$$\mathbb{P}[q \geq 3\log\log N] \leq O\left(\frac{1}{\log N}\right).$$

Remark 4. *Due to space limitation, we do note give the full proof of this Lemma in this extended abstract. Such proof uses some basic probability properties, the standard Euler-Maclaurin formula and the Chernoff bounds formula.*

The time complexities of our algorithms on BN, when the nodes know and do not know n are quite similar. Thus, we immediately consider the case when the nodes do not know n.

3.1 The Nodes Do Not Know n

Each node knows an upper bound u of n such that $\log u = \Theta(\log n)$ *i.e.* $u \in]n, n^c]$ where $c > 1$ is known by the nodes. In order to reach the high probability requirement, we adapt LEADERELECTION(n) to work on BN model with the following 3 phases. For better clarity, we present Phase 3 before Phase 2.

- **Phase 1:** Let $V = \exp\left(u^{\alpha/(c(\alpha+1))}\right)$. Based on Lemma 6 (a), each node s_i generates V random copies $Y_{i,1}, Y_{i,2}, \ldots, Y_{i,V}$ of a r.v. Y distributed as described by (1) and saves $Y_i = \max_{h=1,2,\ldots V}\{Y_{i,h}\}$. Then, according to Lemma 6 (b), with $N = nV$, each node computes $L_0 = (\log V - \log\log\log V)^{(1+\alpha)}$ and $L_{Last} = (\log(uV) + \log\log V)^{(1+\alpha)}$. Each node s_i having Y_i in the interval of integers $L = [L_0, L_{Last}] = \{L_0, L_1, \ldots, L_{Last}\}$ such that $L_r = L_0 + r$, becomes CANDIDATE and the other nodes are ELIMINATED.
- **Phase 3:** Each CANDIDATE node *browses through* the interval L one value at a time as in the BROWSE(I) protocol but in reverse order from L_{Last} to L_0 in order to find out which holds the maximum. This latter node becomes LEADER. Firstly, if a CANDIDATE node s_i has $Y_i = L_{Last}$, it becomes LEADER and beeps at t_0. At each time slot $t_0, t_1, \ldots, t_g$, each CANDIDATE node checks if $Y_i = L_{Last} - (g+1)$. If a CANDIDATE node has $Y_i = L_{Last} - (g+1)$, it listens to the network at t_g. If it does not hear a beep at t_g, it beeps at t_{g+1} and becomes LEADER.

Some values in L may not be picked by any node *i.e.* there can be time slots during Phase 3 when node neither beeps nor listens to the network. The algorithm can then elect more than one leader. To circumvent that, we introduce a new witnessing procedure which consists of flooding all time slots after an election.

- **Phase 2:** All ELIMINATED nodes sets $t_w = UAR(\{t_0, \ldots, t_{|L|-1}\})$. They will listen to the network at t_w on Phase 3. Then, if a node beeps at a time slot t_g of Phase 3, all nodes hearing a beep: the CANDIDATE nodes with $Y_i = L_{|L|} - (g+1)$ and the ELIMINATED nodes with $t_w = t_g$, have to beep at t_{g+1} in order to notify the next CANDIDATE nodes (which become ELIMINATED) that a LEADER has already been elected.

After Phase 3, all remaining CANDIDATE nodes become ELIMINATED.

Theorem 3. *Fix $\alpha \in]0,1[$, in single-hop BN networks of large size n, if no node knows n, but an upper bound u of n is given in advance to all the nodes, there is a randomized Monte-Carlo leader election algorithm that elects a leader in $O(n^\alpha)$ time slots with a probability of $1 - O\left(n^{-\alpha^2/(\alpha+1)}\right)$. Each node transmits and listens during a maximum of one time slot.*

Proof. As for the proof of Theorem 1, the time complexity comes from the time spent to browse through the interval L which is $O(|L|) = O(n^\alpha)$. The success probability depends on two facts: the probability that the maximum of all generated random values in Phase 1 is unique and the probability that each time slot of Phase 3 is witnessed by at least one node. It is straightforward to see that these probabilities are greater than $1 - O\left(n^{-\alpha^2/(1+\alpha)}\right)$ and that each node transmits and listens to the network at most once. □

Conclusion

We designed leader election algorithms taking into account their energy consumption and their time complexities while each device can transmit $1 -$ *bit* message once and can listen to the network during a maximum of 2 time slots. Our algorithm design is based on each node locally generating random values with a probability distribution and communicating in a deterministic manner on the network to find out which node has a unique value. The latter node becomes the leader. Its time complexity only depends on the time slots spent to localize such a node. Assuming that the nodes are initially indistinguishable and know n, our randomized algorithm terminates in optimal $O(\log n)$ time slots *w.h.p.* in the Radio Networks with and without collision detection. If a common value $\alpha \in]0,1[$ is given to all nodes, it has $O(n^{\alpha/(\alpha+1)})$ time complexity for the Beeping Networks. For the realistic case when the nodes do not know n, if a common upper bound u such that $\log u = \Theta(\log n)$ is given in advance to all the nodes, our algorithms terminate in $O(\log^2 n)$ time slots for the RN models and $O(n^{\alpha})$ for BN. Some existing results can be adapted to reach $O(1)$ energy complexity on the models studied in this paper [4,6] but we present the first results with each node transmitting at most once and listening to the network during at most two time slots, exchanging $1 - bit$ messages. Optimal energy complexity has been reached in [6] for the Radio Networks models when the nodes have no information about the topology of the network, but designing a polynomial time leader election for the BN model matching such lower bounds is open.

References

1. Aby, A.T., Guitton, A., Lafourcade, P., Misson, M.: SLACK-MAC: adaptive MAC protocol for low duty-cycle wireless sensor networks. In: Mitton, N., Kantarci, M.E., Gallais, A., Papavassiliou, S. (eds.) ADHOCNETS 2015. LNICST, vol. 155, pp. 69–81. Springer, Cham (2015). https://doi.org/10.1007/978-3-319-25067-0_6
2. Afek, Y., Alon, N., Bar-Joseph, Z., Cornejo, A., Haeupler, B., Kuhn, F.: Beeping a maximal independent set. Distrib. Comput. **26**(4), 195–208 (2013)
3. Barnes, M., Conway, C., Mathews, J., Arvind, D.: ENS: an energy harvesting wireless sensor network platform. In: Proceedings of the 5th International Conference on Systems and Networks Communications, pp. 83–87. IEEE (2010)
4. Bender, M.A., Kopelowitz, T., Pettie, S., Young, M.: Contention resolution with log-logstar channel accesses. In: Proceedings of the 48th Annual ACM Symposium on Theory of Computing, pp. 499–508. ACM (2016)
5. Capetanakis, J.I.: Tree algorithms for packet broadcast channels. IEEE Trans. Inf. Theor. **25**(5), 505–515 (1979)
6. Chang, Y.J., Kopelowitz, T., Pettie, S., Wang, R., Zhan, W.: Exponential separations in the energy complexity of leader election. In: Proceedings of the 49th Annual ACM SIGACT Symposium on Theory of Computing, pp. 771–783. ACM (2017)
7. Chlamtac, I., Kutten, S.: On broadcasting in radio networks-problem analysis and protocol design. IEEE Trans. Commun. **33**(12), 1240–1246 (1985)
8. Cornejo, A., Kuhn, F.: Deploying wireless networks with beeps. In: Lynch, N.A., Shvartsman, A.A. (eds.) DISC 2010. LNCS, vol. 6343, pp. 148–162. Springer, Heidelberg (2010). https://doi.org/10.1007/978-3-642-15763-9_15
9. Devroye, L.: Non-Uniform Random Variate Generation. Devroye's web page (2003). http://www.nrbook.com/devroye/

10. Fraigniaud, P., Korman, A., Peleg, D.: Towards a complexity theory for local distributed computing. J. ACM (JACM) **60**(5), 35 (2013)
11. Ghaffari, M., Haeupler, B.: Near optimal leader election in multi-hop radio networks. In: Proceedings of the 24th Annual ACM-SIAM Symposium on Discrete Algorithms, pp. 748–766 (2013)
12. Ghaffari, M., Lynch, N., Sastry, S.: Leader election using loneliness detection. Distrib. Comput. **25**(6), 427–450 (2012)
13. Greenberg, A.G., Winograd, S.: A lower bound on the time needed in the worst case to resolve conflicts deterministically in multiple access channels. J. ACM **32**(3), 589–596 (1985)
14. Guo, C., Zhong, L.C., Rabaey, J.M.: Low power distributed mac for ad hoc sensor radio networks. In: Proceedings of the IEEE Global Telecommunications Conference, GLOBECOM 2001. vol. 5, pp. 2944–2948. IEEE (2001)
15. He, Y., Du, P., Li, K., Yong, S.: An optimization algorithm based on the Monte Carlo node localization of mobile sensor network. Int. J. Simul. Syst. Sci. Technol. **17**, 20 (2016)
16. Jurdziński, T., Kutyłowski, M., Zatopiański, J.: Weak communication in single-hop radio networks: adjusting algorithms to industrial standards. Concurr. Comput. Pract. Exper. **15**(11–12), 1117–1131 (2003)
17. Kardas, M., Klonowski, M., Pajak, D.: Energy-efficient leader election protocols for single-hop radio networks. In: Proceedings of the 42nd International Conference on Parallel Processing, ICPP, pp. 399–408. IEEE (2013)
18. Liu, F., Narayanan, A., Bai, Q.: Real-time systems (2000)
19. Metcalfe, R.M., Boggs, D.R.: Ethernet: distributed packet switching for local computer networks. Commun. ACM **19**(7), 395–404 (1976)
20. Nakano, K., Olariu, S.: Randomized leader election protocols in radio networks with no collision detection. In: Goos, G., Hartmanis, J., van Leeuwen, J., Lee, D.T., Teng, S.-H. (eds.) ISAAC 2000. LNCS, vol. 1969, pp. 362–373. Springer, Heidelberg (2000). https://doi.org/10.1007/3-540-40996-3_31
21. Nakano, K., Olariu, S.: Uniform leader election protocols for radio networks. IEEE Trans. Parallel Distrib. Syst. **13**(5), 516–526 (2002)
22. Oh, H., Han, T.D.: A demand-based slot assignment algorithm for energy-aware reliable data transmission in wireless sensor networks. Wire. Netw. **18**(5), 523–534 (2012)
23. Sivalingam, K.M., Srivastava, M.B., Agrawal, P.: Low power link and access protocols for wireless multimedia networks. In: Proceedings of the IEEE 47th Vehicular Technology Conference. Technology in Motion, vol. 3, pp. 1331–1335. IEEE (1997)
24. Tsybakov, B.S.: Free synchronous packet access in a broadcast channel with feedback. Problem. Inform. Trans. **14**(4), 259–280 (1978)
25. Vieira, M.A.M., Coelho, C.N., Da Silva, D., da Mata, J.M.: Survey on wireless sensor network devices. In: Proceedings of the 2003 IEEE Conference on Emerging Technologies and Factory Automation, vol. 1 (2003)
26. Willard, D.: Log-logarithmic selection resolution protocols in a multiple access channel. SIAM J. Comput. **15**(2), 468–477 (1986)

Asymptotics for Push on the Complete Graph

Rami Daknama, Konstantinos Panagiotou, and Simon Reisser(✉)

Ludwig-Maximilians Universität München, Munich, Germany
{kpanagio,reisser}@math.lmu.de

Abstract. We study the popular randomized rumour spreading protocol *push*. Initially, a node in a graph possesses some information, which is then spread in a round based manner. In each round, each informed node chooses uniformly at random one of its neighbours and passes the information to it. The quantity to investigate is the *runtime*, that is, the number of rounds until everybody has received the information.

In this work, we study the case where the underlying graph is complete with n nodes. Even in this most basic setting, specifying the limiting distribution of the runtime as well as determining asymptotically related quantities, like its expectation, have remained open problems since the protocol was introduced.

As our main result, we show that the limiting distribution of the runtime does not converge, and that it becomes, after the appropriate normalization, asymptotically periodic both on the $\log_2 n$ as well as on the $\ln n$ scale. In particular, the limiting distribution converges only if we restrict ourselves to suitable subsequences of $\mathbb{N}$, where simultaneously $\log_2 n - \lfloor \log_2 n \rfloor \to x$ and $\ln n - \lfloor \ln n \rfloor \to y$ for some fixed $x, y \in [0, 1)$. We are not aware of any other structure exhibiting such a behaviour. Apart from that, on such subsequences we show that the expected runtime is $\log_2 n + \ln n + h(x, y) + o(1)$, where h is explicitly given and numerically $|\sup h - \inf h| \approx 2 \cdot 10^{-4}$.

Keywords: Randomized rumour spreading · Asymptotics · Complete graph

1 Introduction

We consider the well-known and well-studied rumour spreading protocol *Push*. It has applications in replicated databases [6], multicast [1] and blockchain technology [20]. *Push* operates on graphs and proceeds in rounds as follows. In the beginning, one node has a piece of information. In subsequent rounds each informed node chooses a neighbour independently and uniformly at random and informs it. For a graph $G = (V, E)$ with $|V| = n$ and a node $v \in V$ we denote by $X(G, v)$ the (random) number of rounds needed to inform all nodes, where at the beginning of the first round only v knows the information. We call $X(G, v)$

Y. Kohayakawa and F. K. Miyazawa (Eds.): LATIN 2020, LNCS 12118, pp. 451–463, 2020.
https://doi.org/10.1007/978-3-030-61792-9_36

the *runtime* (on G with start node v). The most basic case, and the one that we study here, is when G is the complete graph K_n. Since in that case the initially informed node makes no difference, we will abbreviate $X(K_n, v) = X_n$ for any starting node v.

Related Work. There are several works studying the runtime of *push* on the complete graph. The first paper considering this protocol is by Frieze and Grimmett [12], who showed that with high probability (whp), that is, with probability $1 - o(1)$ as $n \to \infty$, that $X_n = \log_2 n + \ln n + o(\ln n)$. Moreover, they obtained bounds for (very) large deviations of X_n from its expectation. In [21], Pittel improved upon the results in [12], in particular, he showed that for any $f : \mathbb{N} \to \mathbb{R}^+$ with $f = \omega(1)$, whp, $|X_n - \log_2 n - \ln n| \leq f(n)$. The currently most precise result in this context was obtained by Doerr and Künnemann [7], who considered in great detail the distribution of X_n. They showed that X_n can be stochastically bounded (from both sides) by coupon collector type problems. This gives a lot of control regarding the distribution of X_n, and it allowed them to derive, for example, very sharp bounds for tail probabilities. Apart from that, it enabled them to consider related quantities, as for example the expectation of X_n. Among other results, their bounds on the distribution of X_n imply that

$$\lfloor \log_2 n \rfloor + \ln n - 1.116 \leq \mathbb{E}[X_n] \leq \lceil \log_2 n \rceil + \ln n + 2.765, \tag{1}$$

which pins down the expectation up to a constant additive term. Besides on complete graphs, *push* has been extensively studied on several other graph classes. For example, Erdös-Rényi random graphs [9,10], random regular graphs and expander graphs [5,11,19]. More general bounds that only depend on some graph parameter have also been derived, e.g. the diameter [9], graph conductance [3,4,13,18] and node expansion [4,14,15,22].

Results. In order to state our main result we need some definitions first. Set

$$g = g^{(1)} : [0,1] \to [0,1], \quad x \mapsto xe^{x-1}$$

and $g^{(i)} : [0,1] \to [0,1]$, $g^{(i)} = g \circ g^{(i-1)}$, $i \geq 2$. As we will see later, the function g describes, for a wide range of the parameters, the evolution of the number of uninformed nodes; in particular, if at the beginning of some round there are xn uninformed nodes, then at the end of the same round there will be (roughly) $g(x)n$ uninformed nodes, and after i rounds there will be (roughly) $g^{(i)}(x)n$ uninformed nodes. This fact is not new – at least for bounded i – and has been observed long ago, see for example [21, Lem. 2]. For $x \in \mathbb{R}$ define the function

$$c(x) = -x + \lim_{a \to \infty, a \in \mathbb{N}} \lim_{b \to \infty, b \in \mathbb{N}} -a + b + \ln\left(g^{(b)}(1 - 2^{-a-x})\right), \tag{2}$$

whose actual meaning will become clear later. We will show that the double limit exists, so that this indeed defines a function $c : \mathbb{R} \to \mathbb{R}$. Moreover, we will show

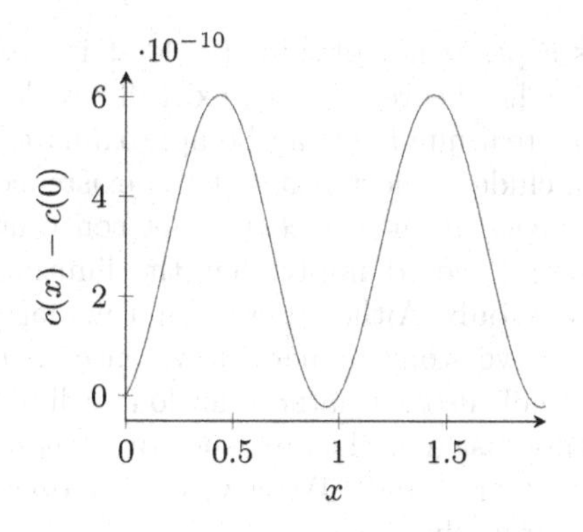

Fig. 1. This figure shows the function $c(x) - c(0)$, $c(0) \approx 0.105$, plotted for values of x between 0 and 2. The periodic nature of the function and the small amplitude is evident.

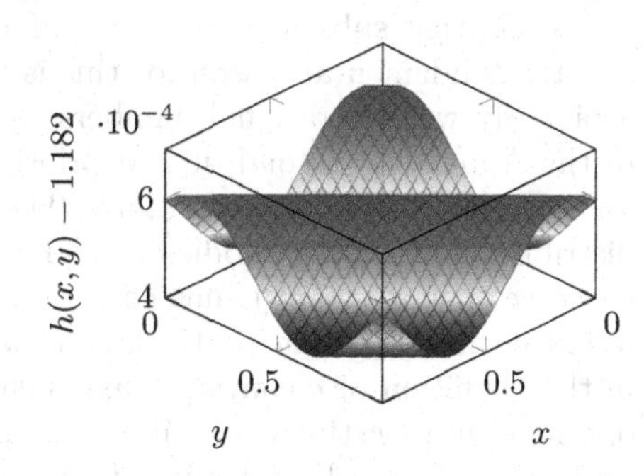

Fig. 2. The figure shows the function $h(x, y) - 1.182$ plotted for values of x and y between 0 and 1. This function is not periodic.

that c is continuous and periodic with period 1, that is, if we write $\{x\} = x - \lfloor x \rfloor$ then $c(x) = c(\{x\})$, and that (numerically) $|\sup c - \inf c| \approx 10^{-9}$, cf. Fig. 1.

The Gumbel distribution will play a prominent role in our considerations. We say that a real valued random variable G follows a Gum(α) distribution with parameter $\alpha \in \mathbb{R}$, $G \sim \text{Gum}(\alpha)$, if for all $x \in \mathbb{R}$

$$P[G \leq x] = e^{-e^{-x-\alpha}}, \quad x \in \mathbb{R}.$$

Finally, let γ denote the Euler-Mascheroni constant. With all these ingredients we can now state our main result, which specifies – see also below – the distribution of the runtime of *push* on the complete graph.

Lemma 1. *Let* $G \sim \text{Gum}(\gamma)$. *Then, as* $n \to \infty$

$$\sup_{k \in \mathbb{N}} \Big| P[X_n \geq k] - P\Big[\lceil G + \log_2 n + \ln n + \gamma + c(\{\log_2 n\}) \rceil \geq k\Big] \Big| = o(1).$$

This lemma does not look completely innocent, and it actually has striking consequences. It readily implies the following result, which establishes that the limiting distribution X_n is periodic both on the $\log_2 n$ and on the $\ln n$ scale. In order to formulate it, we need a version of the Gumbel distribution where we restrict ourselves to integers only. More specifically, we say that a random variable G follows a *discrete Gumbel* distribution, $G \sim \text{dGum}(\alpha)$, if the domain of G is $\mathbb{Z}$ and

$$P[G \leq k] = e^{-e^{-k-\alpha}}, \quad k \in \mathbb{Z}.$$

Theorem 1. *Let* $x, y \in [0, 1)$ *and* $(n_i)_{i \in \mathbb{N}}$ *be a strictly increasing sequence of natural numbers, such that* $\log_2 n_i - \lfloor \log_2 n_i \rfloor \to x$ *and* $\ln n_i - \lfloor \ln n_i \rfloor \to y$ *as* $i \to \infty$. *Then in distribution, as* $i \to \infty$

$$X_{n_i} - \big(\lfloor \log_2 n_i \rfloor + \lfloor \ln n_i \rfloor\big) \to \text{dGum}(-x - y - c(x)).$$

Some remarks are in place. First, it is a priori not obvious (at least it was not to us) that subsequences as required in the theorem indeed exist. They do, and the fundamental reason for this is that real numbers can be approximated arbitrarily well by rational numbers; we include a short proof of the existence in the Appendix. Second, it is a priori not clear that $x + c(x)$ is not constant for $x \in [0, 1)$. If it was constant, Theorem 1 would imply that the limiting distribution of X_n is periodic on the $\ln n$ scale only. Although we didn't manage to *prove* that $x + c(x)$ is not constant, we have stong numerical evidence that it indeed is not so. In particular, as we shall also see later, the double limit in the definition of c converges exponentially fast and thus it is not difficult to obtain accurate estimates for it and explicit error bounds. We leave it as an open problem to study the behavior of c more accurately.

Our next result addresses the expectation of X_n. Bounds given in previous works, for example in [7], guarantee that $X_n - \log_2 n - \ln n$ is uniformly integrable. This allows us to conclude that the expectation also converges.

Theorem 2. *Let $x, y \in [0, 1)$ and $(n_i)_{i \in \mathbb{N}}$ be a strictly increasing sequence of natural numbers, such that $\log_2 n_i - \lfloor \log_2 n_i \rfloor \to x$ and $\ln n_i - \lfloor \ln n_i \rfloor \to y$ as $i \to \infty$. Then, as $i \to \infty$*

$$\mathbb{E}\big[X_{n_i}\big] - \big(\lfloor \log_2 n_i \rfloor + \lfloor \ln n_i \rfloor\big) \to \mathbb{E}\big[\mathrm{dGum}(-x - y - c(x))\big].$$

For any random variable X with support in $\mathbb{Z}$ and finite first moment we have that $\mathbb{E}[X] = \sum_{k \geq 1}(P[X \geq k] - P[-X \geq k])$. Thus, for $\alpha \in \mathbb{R}$

$$\mathbb{E}[\mathrm{dGum}(\alpha)] = \sum_{k \geq 1} \left(1 - e^{-e^{-k-\alpha+1}}\right) - \sum_{k \geq 1} e^{-e^{k-\alpha}}.$$

This converges also exponentially fast and allows for an effective numeric treatment. In particular, for $x, y \in [0, 1)$ and a strictly increasing sequence of natural numbers $(n_i)_{i \in \mathbb{N}}$ such that $\{\log_2 n_i\} \to x$ and $\{\ln n_i\} \to y$ we obtain that

$$\mathbb{E}\big[X_{n_i}\big] = \log_2 n_i + \ln n_i + h(x, y) + o(1)$$

with $h(x, y) = \mathbb{E}\big[\mathrm{dGum}(-x - y - c(x))\big] - x - y$ and numerically $|\sup h - \inf h| \approx 2 \cdot 10^{-4}$, cf. Fig. 2. In summary, improving (1), we get for all $n \in \mathbb{N}$

$$\log_2 n + \ln n + 1.18242 \leq \mathbb{E}[X_n] \leq \log_2 n + \ln n + 1.18263,$$

and these bounds are best possible as $\inf_{0 \leq x, y \leq 1} h(x, y) = 1.18242\ldots$ and $\sup_{0 \leq x, y \leq 1} h(x, y) = 1.18262\ldots$.

Outline. In the next section we give an outline of the proof of our main results, in which we highlight the intuitive behaviour of *push*. At the beginning of the rumour spreding process *push* is dominated by an exponential growth of the informed nodes (Lemma 3). For the main part, where most nodes get informed, it closely follows a deterministic recursion (Lemma 2) and at the end it is described by a coupon collector type problem (Lemma 4). Based on these lemmas we give the rigorous proof of our claims in Sect. 3. Some ideas to the proof of these three important lemmas can be also be found there, in Subsects. 3.3 and 3.4. Due to restrictions on the number of pages we had to heavily abbreviate there.

Further Notation. Unless stated otherwise, all asymptotic behaviour in this paper is for $n \to \infty$. Consider a graph $G = (V, E)$. For $t \in \mathbb{N}_0$ $(= \mathbb{N} \cup \{0\})$ we denote by $I_t \subseteq V$ the set of informed nodes at the end of round t; in particular $|I_0| = 1$. Analogously we write $U_t = V \backslash I_t$ for the set of uninformed nodes. For an event A, we sometimes write $P_A[\cdot]$ instead of $P[\cdot \mid A]$ to denote the conditional probability and we write $\mathbb{E}_A[\cdot] = \mathbb{E}[\cdot \mid A]$. If we condition on I_t, then we also abbreviate $P[\cdot \mid I_t] = P_t[\cdot]$ and $\mathbb{E}[\cdot \mid I_t] = \mathbb{E}_t[\cdot]$.

2 Proof Overview

Let us start the proof of Lemma 1 about the distribution of the runtime of *push* on K_n with a simple observation, that is more or less explicit also in previous works. Note that as long as the *total number* of pushes performed is $o(\sqrt{n})$, then whp no node will be informed twice – this is a simple consequence of the famous birthday paradox. That is, whp as long as $|I_t| = o(\sqrt{n})$, every node in I_t will inform a currently uninformed node and thus $|I_{t+1}| = 2|I_t|$. In particular, whp

$$|I_{t_0}| = 2^{t_0}, \quad \text{where} \quad t_0 := \lfloor 0.49 \cdot \log_2 n \rfloor. \tag{3}$$

Soon after round t_0 things get more complicated. We continue with a definition. Apart from the functions $g^{(i)}$ defined in the previous section, we will also need the following functions. Set

$$f = f^{(1)} : [0, 1] \to [0, 1], \quad x \mapsto 1 - e^{-x}(1 - x)$$

and $f^{(i)} : [0, 1] \to [0, 1], f^{(i)} = f \circ f^{(i-1)}, i \geq 2$. Some elementary properties of f are: f is strictly increasing and concave, and $f^{(b)}(x) \to 1$ as $b \to \infty$ for all $x \in (0, 1]$. Moreover, $f^{(i)}(x) = 1 - g^{(i)}(1 - x)$ for all $x \in [0, 1]$ and $i \in \mathbb{N}$. It is also not hard to establish, see also [21], that f captures the behavior of the expected number of informed nodes after one round of the protocol. Moreover, $|I_{t+1}|$ is typically close to $f(|I_t|/n)n$. Here we will need a more explicit qualitative control of how $|I_t|$ behaves, since our aim is to specify the limiting distribution. We show the following statement, which implies that if we start in round t_0 (set $T = t_0$ in that lemma) then whp for *all* succeeding rounds $t_0 + t$ the number of informed nodes is close to $f^{(t)}(|I_{t_0}/n|)n$.

Lemma 2. *Let* $0 < c < 0.49$ *and* $T \geq c \log_2 n$. *Then*

$$P_T \left[\bigcap_{t \in \mathbb{N}_0} \left\{ \left| |I_{T+t}| - f^{(t)}\left(|I_T|/n\right) n \right| \leq n^{1-c/4} \right\} \right] = 1 - O(n^{-c^2/10}).$$

Thus, the key to understanding $|I_t|$ is to understand how f behaves when iterated very many times. Note that when the number of informed nodes is xn for some very small x, then the e^{-x} term in the definition of f can be approximated by $1 - x$ and therefore $f(x) \approx 1 - (1 - x)^2 \approx 2x$. This crude estimate suggests that the number of informed nodes doubles every round as long as there are only few informed nodes, and we know already that the doubling is perfect if $xn = o(\sqrt{n})$. Our next lemma actually shows that the doubling continues to be *almost* perfect, as long as the total number of nodes is not close to n.

Lemma 3. *Let $a, T \in \mathbb{N}$ be such that $2^{-a} < 0.1$ and $T \leq \lfloor 0.49 \cdot \log_2 n \rfloor$. Set $t_1 := \lfloor \log_2 n \rfloor - a$. Then*

$$\left| 2^{t_1} - f^{(t_1 - T)}\left(2^T/n\right) n \right| \leq 2^{-2a+1} n.$$

Combining the previous lemmas we have thus established that for any $a \in \mathbb{N}$ with $2^{-a} < 0.1$ whp

$$(1 - 2^{-a+2}) \cdot 2^{t_1} \leq |I_{t_1}| \leq 2^{t_1}, \quad t_1 := \lfloor \log_2 n \rfloor - a. \tag{4}$$

Here we can think of a being very large (but fixed) and then the two bounds are very close to each other; in particular, $|I_{t_1}| \approx 2^{\lfloor \log_2 n \rfloor - a}$ and thus I_t contains a linear number of nodes. Up to that point we have studied the behaviour of the process up to time t_1. Next we perform another b steps, where again b is fixed. Applying Lemma 2 once more and using that $f^{(b)}(x)$ is increasing and is less than 1 for $x < 1$ yields with room to spare that, setting $t_2 := t_1 + b$, whp

$$\left(1 - n^{-1/6}\right) f^{(b)}\left((1 - 2^{-a+2}) 2^{t_1}/n\right) \leq |I_{t_2}|/n \leq \left(1 + n^{-1/6}\right) f^{(b)}\left(2^{t_1}/n\right). \tag{5}$$

In essence, this says that if we write $x = \log_2 n - \lfloor \log_2 n \rfloor = \{\log_2 n\}$, then (we begin getting informal and obtain that)

$$|I_{t_2}| \approx f^{(b)}\left(2^{t_1}/n\right) n = f^{(b)}\left(2^{-a-x}\right) n, \quad \text{where} \quad t_2 = \lfloor \log_2 n \rfloor - a + b.$$

In particular, choosing a priori b large enough makes the fraction $|I_{t_2}|/n$ arbitrarily close to 1, that is, almost all nodes except for a tiny fraction are informed. All in all, up to time t_2 we have very fine control of the number of informed nodes, and we also see how the quantity $\{\log_2 n\}$ slowly sneaks in.

After time t_2 the behavior changes once more. There is an interesting connection to the well-known Coupon Collector Problem (CCP), which was also exploited in [7]. In order to formulate the connection, note that the number of pushes that are needed to inform a node, having N informed nodes, is (in distribution) equal to the number of coupons needed to draw the $(N + 1)$st distinct coupon. It is a classic result that, appropriately normalized, the total number of coupons tends to a Gumbel distribution. However, translating the number of required pushes to the number of rounds – the quantity we are interested in – is not straightforward. In particular, the number of pushes in one round depends on the current number of informed nodes. On the other hand, after round t_2 there are $n - o(n)$ informed nodes, so that we may hope to approximate the remaining number of rounds with n^{-1} times the number of coupons in the CCP. The next lemma establishes the precise bridge between the two problems. For two sequences of random variables $(X_n)_{n\in\mathbb{N}}$ and $(Y_n)_{n\in\mathbb{N}}$ we write $X_n \precsim Y_n$ if there is a function $h : \mathbb{N} \to \mathbb{R}^+$ with $h = o(1)$ such that $P[X_n \geq x] \leq P[Y_n \geq x] + h(n)$ for all $n \in \mathbb{N}, x \in \mathbb{R}$; $X_n \succsim Y_n$ is defined with "$\geq$" instead of "$\leq$".

Lemma 4. *Let* $G \sim \mathrm{Gum}(\gamma), b > 2a \in \mathbb{N}$ *and assume that* $\ell \cdot n \leq |I_{\lfloor \log_2 n \rfloor - a + b}| \leq u \cdot n$ *for some* $\ell, u \in [0, 1)$. *Then*

$$X_n - \lfloor \log_2 n \rfloor + a - b \succsim \lceil \ln n + \ln(1/u - 1) + \gamma + G \rceil$$

and

$$X_n - \lfloor \log_2 n \rfloor + a - b \precsim \lceil \ln n + \ln(1/\ell - 1) + \ln(\ell/(e\ell - e + 1)) + \gamma + G \rceil .$$

Note that the previous discussion guarantees that ℓ, u in Lemma 4 are very close to 1 and very close to each other. So, the term $\ln(\ell/(e\ell - e + 1))$ is very close to 0. We obtain that in distribution

$$X_n - \lfloor \log_2 n \rfloor + a - b \approx \lceil \ln n + \ln(1/u - 1) + \gamma + G \rceil, \quad \text{where} \quad u = f^{(b)}(2^{-a-x}),$$

and equivalently with $x = \log_2 n - \lfloor \log_2 n \rfloor$

$$X_n \approx \left\lceil \log_2 n + \ln n - a + b + \ln\left(g^{(b)}(2^{-a-x})\right) - x + \gamma + G \right\rceil . \qquad (6)$$

Here we now encounter the mysterious function c from (2). The next lemma collects some important properties of it that will turn out to be very helpful.

Lemma 5. *The function*

$$c(x) = \lim_{a\to\infty, a\in\mathbb{N}} \lim_{b\to\infty, b\in\mathbb{N}} -a + b + \ln\left(g^{(b)}(1 - 2^{-a-x})\right) - x$$

is well-defined, continuous and periodic with period 1.

With all these facts at hand, the proof of Lemma 1 is completed by considering the random variable on the right-hand side of (6); in particular, the dependence on $y = \ln n - \lfloor \ln n \rfloor$ arises naturally. The complete details of the proof, which is based on Lemmas 2–4 and follows the strategy outlined here can be found in Sect. 3 (together with the proofs of the lemmas).

As described in the introduction, apart from the limiting distribution we are interested in the asymptotic expectation of the runtime. A key ingredient towards the proof of Theorem 2 is uniform integrability, which can be shown by using the distributional bounds from [7]. Uniform integrability is a sufficient condition that convergence in distribution implies convergence of the means.

Lemma 6. *The random variable* $Y_n := X_n - \lfloor \log_2 n \rfloor - \lfloor \ln n \rfloor$ *is uniformly integrable, that is*

$$\lim_{N\to\infty} \sup_{n\in\mathbb{N}} \mathbb{E}\left[|Y_n| \;\middle|\; \mathbb{1}[|Y_n| > N]\right] = 0.$$

3 Proof of the Main Result

In this section we complete the proof of Lemma 1 outlined in Sect. 2. Afterwards we give the (short) proofs for Theorems 1 and 2.

3.1 Proof of Lemma 1

As the outline was indeed rigorous until (5) we take the proof up from there. Choose the quantities $a, b \in \mathbb{N}$ such that $2a < b$ and recall that $t_1 = \lfloor \log_2 n \rfloor - a$. Set furthermore for brevity

$$\ell = \left(1 - n^{-1/6}\right) f^{(b)}((1 - 2^{-a+2})2^{t_1}/n) \quad \text{and} \quad u = \left(1 + n^{-1/6}\right) f^{(b)}(2^{t_1}/n).$$

Then (5) states that, for $t_2 = \lfloor \log_2 n \rfloor - a + b$, we have that $\ell \le n^{-1} |I_{t_2}| \le u$, and Lemma 4 yields, for $Y_n = X_n - \lfloor \log_2 n \rfloor + a - b$, that

$$Y_n \precsim \lceil \ln n + \ln(1/\ell - 1) + \ln(\ell/(e\ell - e + 1)) + \gamma + G \rceil$$

and

$$Y_n \succsim \lceil \ln n + \ln(1/u - 1) + \gamma + G \rceil.$$

The next lemma establishes that both ℓ, u tend to 1 as a gets large, and moreover that the difference $\ln(1/\ell - 1) - \ln(1/u - 1)$ can be made arbitrarily small. Its proof is omitted due to space limitations.

Lemma 7. *For ℓ, u defined as above (where $b > 2a$)*

$$\lim_{a \to \infty} \sup_{n \in \mathbb{N}} |\ln \ell| = \lim_{a \to \infty} \sup_{n \in \mathbb{N}} |\ln u| = \lim_{a \to \infty} \sup_{n \in \mathbb{N}} |\ln(\ell/(e\ell - e + 1))| = 0.$$

Furthermore,

$$\lim_{a \to \infty} \sup_{n \in \mathbb{N}} |\ln(1 - \ell) - \ln(1 - u)| = 0.$$

Thus, as $n \to \infty$,

$$\ln(1 - u) = \ln\left(1 - f^{(b)}\left(2^{t_1}/n\right)\right) + o(1) = \ln\left(g^{(b)}\left(1 - 2^{-a - \{\log_2 n\}}\right)\right) + o(1).$$

Let $\varepsilon > 0$. Lemma 7 readily implies that there are $a_0, n_0 \in \mathbb{N}$ such that for all $a > a_0$ and $n > n_0$,

$$Y_n \succsim \left\lceil \ln n + \ln\left(g^{(b)}\left(1 - 2^{-a - \{\log_2 n\}}\right)\right) + \gamma + G - \varepsilon \right\rceil$$

and similarly also

$$Y_n \precsim \left\lceil \ln n + \ln\left(g^{(b)}\left(1 - 2^{-a - \{\log_2 n\}}\right)\right) + \gamma + G + \varepsilon \right\rceil.$$

Lemma 5 guarantees that there is an $a_1 \geq a_0$ such that for all $a \geq a_1$

$$\left|\ln\left(g^{(b)}\left(1-2^{-a-\{\log_2 n\}}\right)\right) - a + b - (c(\{\log_2 n\}) + \{\log_2 n\})\right| \leq \varepsilon.$$

Thus for all $a > a_1$ and $n > n_0$

$$X_n \gtrsim \lceil \log_2 n + \ln n + c(\{\log_2 n\}) + \gamma + G - 2\varepsilon \rceil,$$

as well as $X_n \lesssim \lceil \log_2 n + \ln n + c(\{\log_2 n\}) + \gamma + G + 2\varepsilon \rceil$. Thus we are left with getting rid of the ε terms in the previous equations. The following lemma accomplishes exactly that and therefore implies the claim of Lemma 1. Its proof is omitted due to space limitations.

Lemma 8. *Let* $h : \mathbb{N} \to \mathbb{R}^+$ *and* $G \sim \mathrm{Gum}(\gamma)$. *Then*

$$\forall \varepsilon > 0 : X_n \lesssim \lceil h(n) + G + \varepsilon \rceil \implies X_n \lesssim \lceil h(n) + G \rceil.$$

The respective statement also holds for "$\gtrsim$".

3.2 Proof of Theorems 1 and 2

Proof (Theorem 1). Let $(n_i)_{i\in\mathbb{N}}$ be a strictly increasing subsequence of $\mathbb{N}$ such that $\log_2 n_i - \lfloor \log_2 n_i \rfloor \to x$ and $\ln n_i - \lfloor \ln n_i \rfloor \to y$. Substituting $k = \lfloor \log_2 n_i \rfloor + \lfloor \ln n_i \rfloor + 1 + t$ for any $t \in \mathbb{Z}$ in Lemma 1 and using the continuity of c and Lemma 8 we get that

$$\left| P\big[X_{n_i} \geq \lfloor \log_2 n_i \rfloor + \lfloor \ln n_i \rfloor + 1 + t\big] - P\big[G + x + y + \gamma + c(x) > t\big] \right| = o(1).$$

Using the distribution function of $G \sim \mathrm{Gum}(\gamma)$ we get

$$P\big[X_{n_i} \geq \lfloor \log_2 n_i \rfloor + \lfloor \ln n_i \rfloor + 1 + t\big] \stackrel{i\to\infty}{\longrightarrow} 1 - \exp\big(-\exp\left(-t + x + y + c(x)\right)\big).$$

Proof (Theorem 2). Lemma 6 states that $X_n - \lfloor \log_2 n \rfloor - \lfloor \ln n \rfloor$ is uniformly integrable and Theorem 1 established its convergence in distribution to $\mathrm{dGum}(-x - y - c(x))$. Together this implies

$$\mathbb{E}\big[X_n - \lfloor \log_2 n \rfloor - \lfloor \ln n \rfloor\big] \to \mathbb{E}\big[\mathrm{dGum}(-x - y - c(x))\big].$$

3.3 Proof of Lemma 2

We will use Chernoff-type bounds to control the distribution of $|I_{t+1}|$. We will exploit an approach initiated in [5] that is based on so-called self-bounding functions. For $x = (x_1, \ldots, x_n)$ we write $x^{(i)} = (x_1, \ldots, x_{i-1}, x_{i+1}, \ldots, x_n)$.

Definition 1 (Self-bounding function, [2,17]). *A non-negative function* $h : \mathbb{N}^n \to \mathbb{R}$ *is self-bounding if there are functions* $h_i : \mathbb{N}^{n-1} \to \mathbb{R}$ *such that for all* $x \in \mathbb{N}^n$ *and all* $i = 1, \ldots, n$,

$$0 \leq h(x) - h_i(x^{(i)}) \leq 1 \quad \textit{and} \quad \sum_{1\leq i\leq n} (h(x) - h_i(x^{(i)})) \leq h(x).$$

Lemma 9 (Exponential inequalities for self-bounding functions, [2]). *Let $n \in \mathbb{N}$ and $h : \mathbb{N}^n \to \mathbb{R}$ be a self-bounding function. Let $Y_1, \ldots, Y_n$ be independent random variables that take values in $\mathbb{N}$. Let $Y = h(Y_1, \ldots, Y_n)$. Then for $s \geq 0$ and $0 < t < \mathbb{E}[Y]$*

$$P[Y \geq \mathbb{E}[Y] + s] \leq \exp\left(\frac{-s^2}{2\mathbb{E}[Y] + 2s/3}\right), \quad P[Y \leq \mathbb{E}[Y] - t] \leq \exp\left(\frac{-t^2}{2\mathbb{E}[Y]}\right).$$

Our next lemma from [5] asserts that we can apply the framework of self-bounding functions to study I_t.

Lemma 10. *There is $m \in \mathbb{N}$, independent random variables $Y_1, ..., Y_m$ in $\mathbb{N}$ and a self-bounding function $h : \mathbb{N}^m \to \mathbb{R}$ such that conditioned on I_t,*

$$|I_{t+1}| = h(Y_1, \ldots, Y_m).$$

Lemma 11 is a concentration result for the number of informed nodes, and it follows by applying the Chernoff-type bounds from Lemma 9. The proof of Lemma 2 depends heavily on it.

Lemma 11. *Let $0 < c \leq 1$, let $t_0 \in \mathbb{N}$ and assume that $|I_{t_0}| \geq n^c$. For $t \in \mathbb{N}$ and $\varepsilon > 0$ let C_t denote the event that*

$$\big||I_{t+1}| - \mathbb{E}_t[|I_{t+1}|]\big| \leq (\mathbb{E}_t[|I_{t+1}|])^{1/2+\varepsilon}.$$

Let $C = \cap_{t \geq t_0} C_t$. Then $P_{t_0}[C] = 1 - \mathcal{O}\left(n^{-c\varepsilon}\right)$.

3.4 Proof of Lemma 4

The next well known theorem, that is linked to the Coupon Collector Problem, states that a sum of n independent geometrically distributed random variables that are centered around their expectation and rescaled with a factor $1/n$ converge to a Gumbel distribution.

Theorem 3 ([8])**.** *Let $T_1, \ldots, T_{n-1}$ be independent random variables such that $T_i \sim \text{Geo}((n-i)/(n-1))$ for $1 \leq i < n$. Then, in distribution*

$$n^{-1} \sum_{i=1}^{n-1} (T_i - \mathbb{E}[T_i]) \to \text{Gum}(\gamma).$$

Unfortunately we can not directly apply Theorem 3 to our setting, as we will have to deal with a sum of independent geometric random variables that are not normalized with the 'correct' factor n^{-1}. However, the next lemma is a more general version of Theorem 3 that is applicable to our setting.

Lemma 12. *Let $T_1, \ldots, T_{n-1}$ be independent random variables such that $T_i \sim \text{Geo}((n-i)/(n-1))$ for $1 \leq i < n$. Let furthermore $\varepsilon > 0$ and $s : \mathbb{N} \to [1, n]$ be a function such that $s(n-i) \geq \big(1 - o(1)\big)(n - c \cdot i)$ for any positive integer $i < \varepsilon n$. Then, in distribution*

$$\sum_{(1-\varepsilon)n \leq i < n} \frac{T_i - \mathbb{E}[T_i]}{s(i)} \to \text{Gum}(\gamma).$$

Let us briefly outline the proof of Lemma 4. We have already shown bounds for the number of informed nodes after $\lfloor \log_2 n \rfloor - a + b$ rounds in (5). Starting from these bounds we will use the Coupon Collector Problem to compute the number of *pushes* that are needed to inform all remaining uninformed nodes. This will yield sums of independent geometric random variables (one summand for each uninformed node). Then we will translate these numbers of pushes into numbers of rounds, which results in almost normalised sums of geometric random variables that Lemma 12 assures to converge to a Gumbel distribution. We will end up with upper and lower bounds to the distribution function of *push*.

A Existence of Subsequence

Let $x, y \in [0,1]$. In this section we show that there is an unbounded sequence of natural numbers $(n_i)_{i\in\mathbb{N}}$ such that $\log_2 n_i - \lfloor \log_2 n_i \rfloor \to x$ and $\ln n_i - \lfloor \ln n_i \rfloor \to y$ as $i \to \infty$. To this end, set $z = y - x \ln 2$. According to a Theorem of Kronecker, see e.g. [16, Thm. 440], for all $i \in \mathbb{N}$, there are $p_i, q_i \in \mathbb{N}$ such that

$$\left| q_i \ln 2 - p_i - z \right| \le i^{-1}. \tag{7}$$

Actually even more is true: there are infinitely many $p_i, q_i \in \mathbb{N}$ that solve (7). To see this, assume that there are only finitely many, then there is $k, \ell \in \mathbb{N}$ such that $k \ln 2 = \ell + z$, otherwise there would be some $i \in \mathbb{N}$ where (7) has no solution. However, according to a Theorem of Hurwitz, see e.g. [16, Thm. 193], there are infinitely many $r_j, s_j \in \mathbb{N}$ such that $\left| r_j \ln 2 - s_j \right| \le r_j^{-2}$. But then

$$\left| r_j \ln 2 - s_j \right| = \left| (r_j + k) \ln 2 - (s_j + \ell) - z \right| \le r_j^{-2},$$

a contradiction, thus there are infinitely many solutions to (7). We continue with that equation, which we can restate, as $i \to \infty$,

$$q_i \ln 2 + x \ln 2 = p_i + y + O\left(i^{-1}\right).$$

Taking the exponential on both sides thus yields $2^{q_i + x} = e^{p_i + y + O(i^{-1})}$ as $i \to \infty$. Set $n_i = \lfloor 2^{q_i + x} \rfloor$ for all $i \in \mathbb{N}$, where we choose q_i such that $q_i \ge i$ from the infinitely many solutions to (7). Then $n_i \in \mathbb{N}$ for all $i \in \mathbb{N}$ and

$$\log_2 n_i - \lfloor \log_2 n_i \rfloor = x + O(2^{-i}) \quad \text{as well as} \quad \ln n_i - \lfloor \ln n_i \rfloor = y + O\left(i^{-1}\right).$$

Thus the subsequence of natural numbers that is induced by $\log_2 n_i - \lfloor \log_2 n_i \rfloor \to x$ and $\ln n_i - \lfloor \ln n_i \rfloor \to y$ is non-empty and unbounded.

References

1. Birman, K.P., Hayden, M., Ozkasap, O., Xiao, Z., Budiu, M., Minsky, Y.: Bimodal multicast. ACM Trans. Comput. Syst. **17**(2), 41–88 (1999)
2. Boucheron, S., Lugosi, G., Massart, P.: A sharp concentration inequality with applications. Random Struct. Algorithms **16**(3), 277–292 (2000)

3. Chierichetti, F., Lattanzi, S., Panconesi, A.: Almost tight bounds for rumour spreading with conductance. In: Proceedings of the 42nd ACM Symposium on Theory of Computing, pp. 399–408. ACM (2010)
4. Chierichetti, F., Lattanzi, S., Panconesi, A.: Rumour spreading and graph conductance. In: Proceedings of the 21st Annual ACM-SIAM Symposium on Discrete Algorithms, pp. 1657–1663. SIAM (2010)
5. Daknama, R., Panagiotou, K., Reisser, S.: Robustness of randomized rumour spreading. In: Proceedings of the 27th Annual European Symposium on Algorithms, ESA. Leibniz International Proceedings in Informatics (LIPIcs), vol. 144, pp. 36:1–36:15. Schloss Dagstuhl-Leibniz-Zentrum fuer Informatik, Dagstuhl, Germany (2019)
6. Demers, A., et al.: Epidemic algorithms for replicated database maintenance. In: Proceedings of the 6th Annual ACM Symposium on Principles of Distributed Computing, PODC, pp. 1–12. ACM (1987)
7. Doerr, B., Künnemann, M.: Tight analysis of randomized rumor spreading in complete graphs. In: Proceedings of the 11th Workshop on Analytic Algorithmics and Combinatorics, ANALCO, pp. 82–91. SIAM (2014)
8. Erdős, P., Rényi, A.: On a classical problem of probability theory. Magyar Tud. Akad. Mat. Kutató Int. Közl. **6**, 215–220 (1961)
9. Feige, U., Peleg, D., Raghavan, P., Upfal, E.: Randomized broadcast in networks. Random Struct. Algorithms **1**(4), 447–460 (1990)
10. Fountoulakis, N., Huber, A., Panagiotou, K.: Reliable broadcasting in random networks and the effect of density. In: Proceedings of the 29th IEEE International Conference on Computer Communications, IEEE Annual Joint Conference, INFOCOM, pp. 2552–2560 (2010)
11. Fountoulakis, N., Panagiotou, K.: Rumor spreading on random regular graphs and expanders. In: Serna, M., Shaltiel, R., Jansen, K., Rolim, J. (eds.) APPROX/RANDOM - 2010. LNCS, vol. 6302, pp. 560–573. Springer, Heidelberg (2010). https://doi.org/10.1007/978-3-642-15369-3_42
12. Frieze, A.M., Grimmett, G.R.: The shortest-path problem for graphs with random arc-lengths. Discrete Appl. Math. **10**(1), 57–77 (1985)
13. Giakkoupis, G.: Tight bounds for rumor spreading in graphs of a given conductance. In: Proceedings of the 28th International Symposium on Theoretical Aspects of Computer Science, STACS, Dortmund, Germany, pp. 57–68 (2011)
14. Giakkoupis, G.: Tight bounds for rumor spreading with vertex expansion. In: Proceedings of the 25th Annual ACM-SIAM Symposium on Discrete Algorithms, SODA, Portland, Oregon, USA, pp. 801–815 (2014)
15. Giakkoupis, G., Sauerwald, T.: Rumor spreading and vertex expansion. In: Proceedings of the 23rd Annual ACM-SIAM Symposium on Discrete Algorithms, SODA, Kyoto, Japan, 17–19 January 2012, pp. 1623–1641 (2012)
16. Hardy, G.H.: Wright: An Introduction to the Theory of Numbers. Oxford University Press, Oxford (1979)
17. Lugosi, G.: Concentration-of-Measure Inequalities. Lecture notes (2009)
18. Mosk-Aoyama, D., Shah, D.: Fast distributed algorithms for computing separable functions. IEEE Trans. Inf. Theory **54**(7), 2997–3007 (2008)
19. Panagiotou, K., Pérez-Giménez, X., Sauerwald, T., Sun, H.: Randomized rumour spreading: the effect of the network topology. Comb. Probab. Comput. **24**(2), 457–479 (2015)
20. Patsonakis, C., Roussopoulos, M.: Revisiting asynchronous rumor spreading in the blockchain era. In: Proceedings of the IEEE 25th International Conference on Parallel and Distributed Systems, ICPADS, pp. 284–293 (2019)

21. Pittel, B.: On spreading a rumor. SIAM J. Appl. Math. **47**(1), 213–223 (1987)
22. Sauerwald, T., Stauffer, A.: Rumor spreading and vertex expansion on regular graphs. In: Proceedings of the 22nd Annual ACM-SIAM Symposium on Discrete Algorithms, SODA 2011, San Francisco, California, USA, 23–25 January 2011, pp. 462–475 (2011)

The Hardness of Sampling Connected Subgraphs

Andrew Read-McFarland(✉) and Daniel Štefankovič

University of Rochester, Rochester, NY, USA
areadmcf@ur.rochester.edu, stefanko@cs.rochester.edu

Abstract. We consider the problem of sampling connected induced subgraphs of a given input graph G. Our first result is that an efficient algorithm to approximately sample connected induced subgraphs of a given size (the size is specified in the input) does not exist unless **RP** = **NP**. We then focus on the problem of approximately sampling connected induced subgraphs with a bias, more precisely we consider a distribution where the probability of a connected subgraph induced by $S \subseteq V(G)$ is proportional to $\lambda^{|S|}$. When the input graph G has maximum degree d we identify a threshold $\lambda_d = \frac{(d-1)^{(d-1)}}{d^d}$. For $0 < \lambda < \lambda_d$ there exists a trivial efficient sampler for the problem, and for $\lambda_d < \lambda < 1$ an efficient approximate sampler does not exist unless **RP** = **NP**. Finally, we show local Markov chains are unlikely to be effective at approximately sampling connected subgraphs.

1 Introduction

Sampling a subgraph allows us to examine small sections of a graph without having to look at the potentially massive graph as a whole [4,11,13]. When we can approximately sample connected subgraphs, we gain information about the occurrence of configurations in the graph [2,11]. There are several variants of what sampling a connected subgraph means, the most common of which is to examine spanning subgraphs as done in [5], where we sample edges such that the graph is connected and every vertex is reachable. Another variant is counting the number of induced subgraphs of a graph G that are isomorphic to another graph H [17]. Exactly counting these connected induced copies is essential for polynomial time execution for Barvinok's algorithm, as done in [16].

In this paper we are concerned with fully polynomial approximate samplers (FPAS), rather than the more common fully polynomial randomized approximation scheme (FPRAS) as we wish to sample connected subgraphs induced by vertices rather than count them. Within this paradigm we consider two models of sampling: fixed size and with bias $\lambda > 0$ (where each graph of size k is sampled with probability proportional to λ^k). In Sect. 2 we look at the fixed size case, where the connected subgraph we sample always has k vertices (for a given k). This has been studied in an applied setting by [13] with various algorithms given. The related problem of exactly counting connected induced subgraphs

Y. Kohayakawa and F. K. Miyazawa (Eds.): LATIN 2020, LNCS 12118, pp. 464–475, 2020.
https://doi.org/10.1007/978-3-030-61792-9_37

on k vertices is $\#W[1]$-hard [7] and is also considered in combinatorial settings [9,23]. We show in Theorem 1 that if there is an FPAS for the uniform distribution of fixed sized connected subgraphs, then $\mathbf{RP} = \mathbf{NP}$. We then prove the even stronger result that an FPAS on graphs of maximum degree three implies $\mathbf{RP} = \mathbf{NP}$ in Theorem 2.

Next we consider sampling with bias λ. Specifically, Theorem 3 shows sampling with bias λ is efficient on a graph with maximum degree d for any $\lambda < \lambda_d = \frac{(d-1)^{d-1}}{d^d}$ and Theorem 4 proves an FPAS for connected induced subgraphs with bias $\lambda \in (\lambda_d, 1)$ implies $\mathbf{RP} = \mathbf{NP}$.

Finally, in Sect. 5 we give a tree such that no local Markov chain can efficiently sample connected subgraphs of fixed size, and similarly with bias $1 > \lambda > 0$ with Theorems 5 and 6 respectively. This hints that local Markov chains likely are not effective, as they do not perform well even on trees, where we know the problem to be easy (using dynamic programming).

The following examples are mentioned to motivate sampling with bias λ and the study of computational thresholds in this setting. The variant with sampling biased by size is considered, for example, in the hardcore model in statistical physics [6,12]. Weitz shows that on a graph of maximum degree d for all $\lambda < \lambda_c = \frac{(d-1)^{d-1}}{(d-2)^d}$ we can efficiently approximately count independent sets [24]. Sly then showed for all $\lambda > \lambda_c$ we cannot efficiently approximately count independent sets unless $\mathbf{RP} = \mathbf{NP}$ [21]. Closer to our setting, Savoie et al. sample simply connected subgraphs (that is, connected subgraphs with no "holes") on a grid, with bias λ on the perimeter [19].

We generally take a subgraph of G to be induced by a subset of the vertices of G. However, in the proofs of Theorems 1, 2, and 4 we also induce subgraphs of G induced by edges of G. We formally define both below.

Definition 1. *For a graph G and $S \subseteq V(G)$, let $G[S]$ denote the subgraph of G induced by S. Formally, $V(G[S]) = S$ and $E(G[S]) = \{\{u, v\} \mid \{u, v\} \in E(G)$ and $u, v \in S\}$.*

Similarly, let $G[R]$ for $R \subseteq E(G)$ be defined as $V(G[R]) = \bigcup_{\{u,v\} \in R} \{u, v\}$ and $E(G[R]) = R$.

We will use the following formal definition of FPAS (see, e.g. [3]).

Definition 2. *An algorithm $\mathcal{A}$ is a Fully Polynomial Approximate Sampler (FPAS) for a problem $\mathcal{B}$ if for any $\delta > 0$ and input to $\mathcal{B}$ the distribution of the output of $\mathcal{A}$ is within δ of the distribution of $\mathcal{B}$ (on the given input) and $\mathcal{A}$ runs in time polynomial with respect to its input and $\log \delta^{-1}$. By distance we mean the total variation distance, $d_{TV}(\mu, \nu) = \frac{1}{2}||\mu - \nu||_1$.*

2 Sampling Fixed Size Connected Subgraphs

In this section we show an FPAS for connected subgraphs of a given size is possible only if $\mathbf{RP} = \mathbf{NP}$. Now we give the formal definition of our sampling problem, which asks for a uniformly random sample from the set of all connected subgraphs of a given size.

Definition 3. *Let **Connected Induced Subgraphs of Given Size** or **CISGS** be the problem that on input (G, K) (for a graph G and non-negative integer K) outputs uniformly random $L \subseteq V(G)$ such that $|L| = K$ and $G[L]$ is connected.*

We show a FPAS for CISGS solves the Steiner Tree problem (see below).

Definition 4. *(see [10]) Let **Steiner Tree** or **ST** be the decision problem of whether there is a connected subgraph of G such that all vertices of a set S (the vertices in S are called terminals) are included and the total weight of all the edges used is no more than ℓ. Formally,* $\mathrm{ST} = \{(G, S, \phi, \ell) \mid \exists R \subseteq E(G)$ *such that $G[R]$ is connected, $S \subseteq V(G[R])$, and $\sum_{r \in R} \phi(r) \leq \ell\}$.*

We will use the **NP**-Hardness of ST many times throughout the paper, which Karp shows [10]. However we use the stronger result that when $\phi(e) = 1$ for each edge (often called the Cardinality Steiner Problem) ST is hard [10,25].

Theorem 1. *If an* FPAS *exists for* CISGS, *then* **RP** = **NP**.

Proof. Let (G, S, ϕ, ℓ) be an instance of ST, where G has n vertices and m edges, and $\phi(e) = 1$ for all $e \in E(G)$. Now let us give a brief outline of the proof. Given an instance of ST we construct a graph G' such that an FPAS for CISGS on G' allows us to obtain a solution to ST with high probability. We have 4 main sections of this proof to accomplish this.

1. First we construct G' and pick K based upon G, S, and ℓ.
2. Then we create a function f that maps connected subgraphs of G' to connected subgraphs of G.
3. Next we give a combinatorial argument to show at least 2/3 of the connected subgraphs of G' of size K map to solutions of ST on G.
4. Finally we make the complexity argument to show the Theorem's claim.

Now let us construct G' such that if we can sample connected subgraphs of G' in polynomial time we will solve ST in polynomial time on G (using a randomized algorithm). Let $k = n^3$, $c = k^2$, and $K = |S| \cdot k/2 + \ell \cdot c$.

Intuitively, G' replaces vertices in S with complete graphs of size k whose connected subgraphs provide high entropy (so that typical subgraphs will include these vertices). On the other hand, it also replaces edges with paths of length c so that including long paths consumes many vertices and therefore lowers the entropy (hence typical solutions will avoid long paths). Let A be the set of nodes in the "S-gadgets" (each a complete graph on k vertices), B be the vertices forming the elongated edges, and let C be the unchanged vertices of $V(G)$. Formally, let G' be such that $V(G') = A \cup B \cup C$ where

1. $A = \bigcup_{s \in S, i \in [1,k]} \{v_{s,i}\}$ (where $[1, k] = \{1, ..., k\}$),
2. $B = \bigcup_{e \in E(G), i \in [1,c]} \{v_{e,i}\}$, and
3. $C = (V(G) - S)$.

Note we use $v_{s,i}$ and $v_{e,i}$ to give names to vertices we are creating. Now let A' be the edges between the nodes in A, B' be the edges between the nodes in B, and C' be the edges between A, B, and C. Thus $E(G') = A' \cup B' \cup C'$, where

1. $A' = \bigcup_{s \in S} \{\{v_{s,i}, v_{s,j}\} \mid i \neq j,\ i, j \in [1, k]\}$ and
2. $B' = \bigcup_{e \in E(G), i \in [1, c-1]} \{\{v_{e,i}, v_{e,i+1}\}\}$ (making paths for each $e \in E(G)$).
3. Finally, we need to connect the paths to the original vertices and the complete graphs. Fix an arbitrary ordering of $V(G')$ and let

$$C' = \bigcup_{\{u,v\}=e \in E(G)} \{\{v_{e,1}, \min(u', v')\}, \{v_{e,c}, \max(u', v')\}\}$$

where $u' = v_{u,1}$ if $u \in S$ and otherwise is simply u, and the same holds for v'. Note the min and max are with respect to the ordering we picked.

Let us now define a function f that maps a connected subset L of $V(G')$ (that is, $G[L]$ is connected) to $R \subseteq E(G)$ such that $G[R]$ is connected (note $G[R]$ is a subgraph induced by edges rather than vertices). Informally, if a path of vertices corresponding to an edge in G is fully included in L, then we will include that edge, and otherwise we will not. Formally, let L be a subset of $V(G')$ such that $G[L]$ is connected. Then $f(L) = R$ where $R \subseteq E(G)$ such that $e \in R$ if and only if $v_{e,1}, v_{e,2}, ..., v_{e,c} \in L$.

A subset of vertices L falls into one of two cases:

(1) $\sum_{e \in G[f(L)]} \phi(e) \leq \ell$ and $G[f(L)]$ includes all points in S
(2) $G[f(L)]$ excludes some point in S.

Note that for $\sum_{e \in G[f(L)]} \phi(e) > \ell$ we must use $c(\ell + 1)$ vertices on edges in G'. This is impossible, as we sample $K = |S| \cdot k/2 + \ell \cdot c < \ell \cdot c + c$ vertices.

Note a subset L from case (1) yields a solution to ST, whereas a subset L from case (2) does not. For convenience, let C_1 be the set of all L such that $G[f(L)]$ falls into case (1), and C_2 be likewise for case (2). Note that if $(G, S, \phi, \ell) \notin$ ST, then $C_1 = \emptyset$. With that in mind, let $(G, S, \phi, \ell) \in$ ST and let us bound the size of C_1. Since $(G, S, \phi, \ell) \in$ ST, there is a subset of edges with weight less than or equal to ℓ such that all nodes in S are included and the graph is connected. Thus in G' we can include the paths that correspond to those edges, which requires $\ell \cdot c$ vertices, and then use the remaining $|S| \cdot k/2$ nodes in the clusters for each $s \in S$. We can use $k/2$ in each of the complete graphs created for vertices in S, and so $|C_1| \geq \binom{k}{k/2}^{|S|}$.

Now let us show $|C_2| \leq 2^{k(|S|-1)}|S| \cdot c^{2m}$. This follows since there are $|S|$ ways to pick an $s \in S$ to omit, and then $2^{k(|S|-1)}$ ways to include or exclude the $k(|S| - 1)$ points in the remaining $|S| - 1$ gadgets. Note that the number of ways to allocate any amount of vertices to edges is at most c^{2m} as for each of the m edges we can choose the length of the partial paths on either side, which can both be at most length c. Now let us show $|C_1| \geq 2|C_2|$.

Since for $n \geq 75$ we have $n^3 \geq 1 + \log_2(n) + n \log_2(n^3 + 1) + 4n^2 \log_2(n^3)$. Thus, we can substitute in $k = n^3$, $|S| \leq n$, and $m \leq n^2$ to get

$k \geq 1 + \log_2(|S|)|S| \log_2(k+1) + 4m \log_2(k)$.

Then, by exponentiating both sides and multiplying by $\frac{2^{k(|S|-1)}}{(k+1)^{|S|}}$ we obtain $\left(\frac{2^k}{k+1}\right)^{|S|} \geq 2 \cdot 2^{k(|S|-1)}|S| \cdot c^{2m}$. Since $\binom{k}{k/2} \geq \frac{2^k}{k+1}$, this gives $|C_1| \geq 2|C_2|$.

Thus a random sample of a connected subgraph of size K from G' falls into case (1) with probability $\geq 2/3$ (recall we assume $(G, S, \phi, \ell) \in$ ST). However, we are using an FPAS, and so our distribution is within δ of the uniform distribution. Thus, we obtain a sample from case (1) with probability $\geq 2/3 - \delta$, and so any $\delta < 1/6$ is sufficient. Our reduction at this point is quite simple; we sample L from G', and then accept if and only if $G[f(L)]$ has weight $\leq \ell$ and includes all terminals. Thus, if a solution with weight $\leq \ell$ does not exist, we will never accept, and if one does we accept with probability $> 1/2$. Finally, to show that the size of G' is polynomial with respect to (G, S, ϕ, ℓ), note G' has $O(mn^6)$ vertices. Thus, if an FPAS exists for CISGS, we have an **RP** algorithm that solves ST, and so **RP** = **NP**. □

Since ST is hard for planar graphs [18] (with maximum degree 4), so the above proof shows hardness for planar graphs as well by simply modifying the complete graphs on k vertices to be a single vertex with k adjacent vertices. Now let us extend this further to show if an FPAS exists for CISGS on graphs of maximum degree three, then **RP** = **NP**.

First, note Steiner Tree is hard for graphs of maximum degree three by a simple reduction of splitting vertices and connecting them with a 0 weight edge. In the proof of Theorem 2 we require all terminals to have maximum degree two (as we will attach a tree to each), so note any vertex with degree three can be split into two, one of which has degree two (and we consider that one to be a terminal and the other not to be).

The basic idea of the proof is the same as Theorem 1 but instead of complete graph gadgets, we will have binary trees of size k. Let us give some definitions for use in analyzing the number of connected subgraphs of a tree.

Definition 5. *For a graph G and $v \in V(G)$, the connected* ***rooted subgraphs*** *of G at v are the subgraphs of G that include v, together with the empty subgraph.*

Definition 6. *Let GT_d be an infinite d-ary tree with root vertex v_{GT_d}.*

Definition 7. *Let G be a tree with arbitrary root v. Then for all $w \in V(G)$ let the* ***height*** *of w be the length of the path between w and v.*

Definition 8. *Let $T_{h,k}$ denote the number of connected subtrees of GT_2 rooted at v_{GT_2} of size k with maximum height h.*

Suppose that $h = \lfloor \log_2(n) \rfloor$, then we can compute $T_{h,k}$ for any k in polynomial time, as we can recursively compute $T_{h,k} = \sum_{i=0}^{k} T_{h-1,i}T_{h-1,k-i-1}$ (note that the numbers have polynomially many (in n) bits). Thus for a given h, we can compute the k such that $T_{h,k}$ is maximal.

Definition 9. *Let k_h denote the index such that $T_{h,k_h} \geq T_{h,k}$ for all $0 \leq k \leq 2^{h+1} - 1$.*

Definition 10. *Let T_h denote the number of connected subtrees of GT_2 rooted at v_{GT_2} with maximum height h. That is, $T_h = \sum_{k=0}^{2^{h+1}-1} T_{h,k}$.*

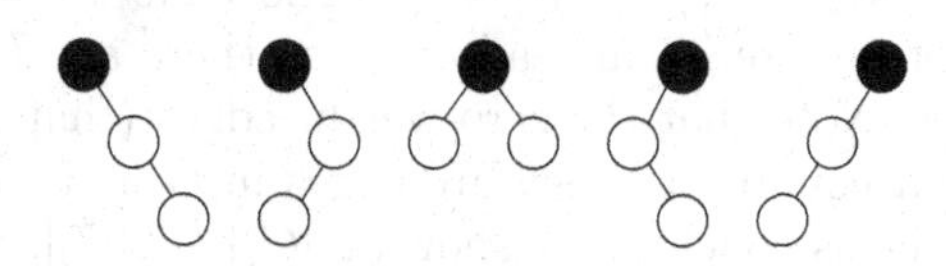

Fig. 1. The 5 configurations of $T_{2,3}$, with v_{GT_2} being the black vertex.

Figure 1 shows the 5 configurations of $T_{2,3}$. Also note we can iteratively calculate $T_h = 1 + T_{h-1}^2$ as we can either include no vertices (by omitting the root), or include the root and have any height $h-1$ subtree on either side. Now let us move on to the proof.

Theorem 2. *If an* FPAS *to* CISGS *exists on graphs of maximum degree three, then* **RP** = **NP**.

Proof. The proof follows similarly to that of Theorem 1 except let $h = \lfloor \log_2(n^3) \rfloor$, $k = 2^h$, and $K = T_{h,k_h} \cdot |S| + \ell \cdot c + n^2 + |S|$.

We assume that $\phi(e)$ is 0 or 1 for each edge e and G has maximum degree 3. As mentioned above, we assume every terminal has degree ≤ 2. Now, we construct G' with the same idea as that of Theorem 1, but using trees instead of complete graph gadgets. If $\phi(e) = 1$, then as before we replace it with a path of length c, but if $\phi(e) = 0$, e is a single node in G' rather than a path of length c (note there are at most n^2 edges with weight 0). The sets B, C, B', and C' are essentially the same as in the proof of Theorem 1 (except for the additional zero weight edges). However, we use trees for gadgets rather than complete graphs so let

$$A = S \cup \bigcup_{s\in S, i\in[1,k-1]} \{v_{s,i}\} \text{ and}$$
$$A' = \bigcup_{s\in S} \left(\{\{v_{s,1}, s\}\} \cup \{\{v_{s,i}, v_{s,\lfloor i/2 \rfloor}\} \mid i \in [2, k-1]\}\right).$$

Let f be as in Theorem 1 and note that a sample L falls into one of 2 cases as before (where again, $G[f(L)]$ is a subgraph induced by edges):

1. $\phi(G[f(L)]) \leq \ell$ and $G[f(L)]$ includes all points in S
2. $G[f(L)]$ excludes some point in S.

Let C_1 be the set of such L that fall into case (1) and C_2 be likewise for case (2). Let us first bound $|C_1| \geq \left(\frac{T_h}{2^{h+1}}\right)^{|S|}$. This is since $\sum_{k=0}^{2^{h+1}-1} T_{h,k} = T_h$ and since T_{h,k_h} has maximum value, it must be at least the average value. Thus we can allocate $c \cdot \ell$ vertices to the edges of G' and $k_h + 1$ to each of the "S-trees" (k_h for the tree, 1 for s) and so there are at least $\left(\frac{T_h}{2^{h+1}}\right)^{|S|}$ distinct connected subgraphs created in this manner. Note that we might need to use weight 0 edges in this construction which equates to using a single vertex for each weight 0 edge. However, we have n^2 "extra" vertices in K to be used specifically for this (as there are no more than n^2 weight 0 edges).

Now let us show $|C_2| \leq T_h^{|S|-1}|S| \cdot c^{2m}$. The logic for this is the same as in Theorem 1 except we use T_h instead of 2^k as there are T_h ways to allocate the vertices to a tree. Note that the zero weight edges (that are represented by a single vertex) are accounted as they are edges in G and so contribute to the value of m. Finally let us conclude by showing $|C_1| \geq 2|C_2|$. Since for $n \geq 73$

$$n^3 \geq 1 + n + n\lfloor \log_2(n^3) \rfloor + \log_2(n) + 4n^2 \lfloor \log_2(n^3) \rfloor,$$

by substituting in terms and exponentiating both sides we get

$$2^{2^h} \geq 2 \cdot 2^{|S|(h+1)} \cdot |S| \cdot c^{2m}.$$

Now note $T_h \geq 2^{2^h}$ as $T_0 = 2$ and $T_h = 1 + T_{h-1}^2 \geq T_{h-1}^2$. Thus by making another substitution and multiplying by $\frac{T_h^{|S|-1}}{2^{|S|(h+1)}}$ we have

$$\left(\frac{T_h}{2^{h+1}}\right)^{|S|} \geq 2 \cdot T_h^{|S|-1}|S| \cdot c^{2m}$$

Therefore $|C_1| \geq 2|C_2|$ and so if $(G, S, \phi, \ell) \in \text{ST}$ then the probability that the sampler gives a solution with weight $\leq \ell$ is at least $2/3 - \delta$ as in Theorem 1. Additionally if no such solution exists, this algorithm will never give one. Since the whole process runs in polynomial time, we have an **RP** algorithm that solves ST, and so **RP** = **NP** if an FPAS exists for CISGS on graphs with maximum degree three. □

3 Trees and Efficient Sampling with Bias

We showed earlier that it is hard to sample connected subgraphs of general graphs, planar graphs, and even for bounded degree graphs. In this section we will show that for bounded degree graphs as long as the bias parameter λ is small enough, we can sample connected subgraphs with bias λ allowing arbitrarily small error ε in time polynomial in n and $1/\varepsilon$ (for a fixed λ).

We analyzed earlier $T_{h,k}$, but let us now extend this definition to letting h be unbounded for a fixed k.

Definition 11. *Let $\tilde{T}_{k,d}$ be the number of connected subtrees of GT_d of size k rooted at v_{GT_d}.*

We will need the following result of Stanley [22] reformulated in our setting.

Lemma 1. $\tilde{T}_{k,d} = \binom{dk}{k}\frac{1}{(d-1)k+1}$.

Proof. Letting a full d-ary tree mean every node either has d children or is a leaf, in his Proposition 6.2.2 [22, p. 172] Stanley shows the number of full d-ary trees with n vertices and m leaves is equal to $\frac{1}{n}\binom{n}{j}$ if $n = dj+1$ and $m = (d-1)j+1$ for some j and 0 otherwise. Note that if a full d-ary tree has a nodes each with d children, then there are $(d-1)a+1$ leaves. Thus, we wish $n = dk+1$ and $m = (d-1)k+1$, so it is clear to see that if we remove all leaves from such trees we can obtain every d-ary tree on k vertices, and likewise every d-ary tree can have every node without d children add leaves until it has d children to obtain all such full trees. Thus the number of d-ary trees on k vertices is $\binom{dk+1}{k}/(dk+1) = \binom{dk}{k}\frac{1}{(d-1)k+1}$. □

Now we move on to show that a d-ary tree has more rooted connected subgraphs of a fixed size than any maximum degree d graph.

Definition 12. *Let $C_{k,d,G,v}$ be the number of connected subgraphs on k vertices with maximum degree d of a graph G rooted at vertex $v \in V(G)$. Let $C_{k,d,G}$ be as above, but for unrooted subgraphs.*

Lemma 2. *For all k, d, G, v, $C_{k,d,G,v} \leq \tilde{T}_{k,d}$.*

Proof. Let G be a graph with n nodes, and fix k, d, v. Let T be the SAW (self-avoiding walk) tree of G rooted at v (see [8,24]). That is, each node in T corresponds to a path in G starting at v. Thus T has maximum height n, has maximum degree $\leq d$, and since it has no cycles it must be a subtree of GT_d.

Now let $S \subseteq V(G)$ such that $G[S]$ is connected and $v \in S$, and let T' be a spanning tree of $G[S]$. Note T' is a subtree of T as every node in T' is a node in T, and T' is unique as any other S' cannot generate T' because it must necessarily omit some vertex in S. Thus $C_{k,d,G,v} \leq C_{k,d,T,v} \leq \tilde{T}_{k,d}$. □

Now we will shall show that for small enough λ, the total weight, $\sum_{k=0}^{\infty} \tilde{T}_{k,d}\lambda^k$ converges.

Lemma 3. *Fix d, and let $\lambda = c\frac{(d-1)^{d-1}}{d^d}$ where $c < 1$. Then for any $s \geq 0$ $\sum_{k=s}^{\infty} \tilde{T}_{k,d}\lambda^k \leq \frac{c^s}{1-c}$.*

Proof. By Lemma 1 we have that $\sum_{k=0}^{\infty} \tilde{T}_{k,d}\lambda^k = \sum_{k=0}^{\infty}\binom{dk}{k}\frac{\lambda^k}{(d-1)k+1}$. Then $\binom{dk}{k}\cdot\lambda^k = c^k\binom{dk}{k}(d-1)^{(d-1)k}/((d-1)+1)^{dk}$ which by the binomial theorem is

$$c^k\frac{\binom{dk}{k}(d-1)^{(d-1)k}}{\sum_{i=0}^{dk}\binom{dk}{i}(d-1)^i}.$$

Since the numerator occurs in the sum in the denominator, the fraction is less than 1. Thus, $\sum_{k=0}^{\infty} \tilde{T}_{k,d}\lambda^k \leq \sum_{k=0}^{\infty}\frac{c^k}{(d-1)k+1} \leq \sum_{k=0}^{\infty} c^k = \frac{1}{1-c}$.

Additionally, for $s \geq 0$, $\sum_{k=s}^{\infty} \tilde{T}_{k,d}\lambda^k \leq \sum_{k=s}^{\infty}\frac{c^k}{(d-1)k+1} \leq \frac{c^s}{1-c}$, by the same logic as above, giving us our result. □

We now show there is a sampler for any $\lambda < \frac{(d-1)^{d-1}}{d^d}$ and $\epsilon > 0$ that runs in time polynomial with respect to n and $1/\epsilon$. Note that this is not a FPAS.

Theorem 3. *For any $c < 1$ and any constant d the following is true. There exists an algorithm that for any $\lambda < c\frac{(d-1)^{d-1}}{d^d}$ and graph G with maximum degree d samples connected subgraphs with size bias λ in polynomial (in n and $1/\varepsilon$) time with error at most $\varepsilon > 0$.*

Proof. We will pick some s such that the probability that we would obtain a graph with size $\geq s$ is less than ε, and so we can only consider graphs of size $< s$. Thus, we want

$$\frac{\sum_{k=s}^{\infty} C_{k,d,G,v}\lambda^k}{\sum_{k=0}^{\infty} C_{k,d,G,v}\lambda^k} < \varepsilon.$$

This term is less than $\sum_{k=s}^{\infty} C_{k,d,G,v}\lambda^k$ as the denominator is at least 1 (because of the empty set). By Lemma 2 we have that this term is again less than $\sum_{k=s}^{\infty} \tilde{T}_{k,d}\lambda^k$. By Lemma 3 we have that this sum is no more than $\frac{c^s}{1-c}$. Now we simply need to pick s such that $\frac{c^s}{1-c}$ is less than ε. Therefore, as long as $s > \log_c(\varepsilon(1-c))$ we have that the chance of randomly sampling a subgraph of size greater than s is less than ε. So, we can have a sampling algorithm that only samples up to size s and since $C_{k,d,G,v} \leq \tilde{T}_{k,d} \leq \binom{dk}{k} \leq (\frac{e \cdot d \cdot k}{k})^k = (e \cdot d)^k$, there are $O((e \cdot d)^{\log_c(\varepsilon(1-c))}) = O((\varepsilon(1-c))^{\log_c(ed)})$ graphs we need to sample allowing error ε. Since this is a polynomial number of graphs, we can inductively enumerate them (up to size s) to calculate their weights and approximate Ω.

Additionally, if we wish to remove the rooted aspect of the subgraphs, note that $C_{k,d,G} \leq \sum_{v \in V(G)} C_{k,d,G,v} \leq n \max_{v \in V(G)} C_{k,d,G,v}$, and since $\tilde{T}_{k,d} \geq C_{k,d,G,v}$ for any v, we have $C_{k,d,G} \leq n\tilde{T}_{k,d} \leq n(e \cdot d)^k$. Then we only need to consider sampling from $O(n(\varepsilon(1-c))^{\log_c(ed)})$ subgraphs. Thus, sampling unrooted connected subgraphs still only requires examining a polynomial (in n and $1/\varepsilon$) number of subgraphs. □

4 Hardness of Sampling with Bias

We will now show that even when we sample connected subgraphs with bias λ rather than having fixed size, the problem is hard for $1 > \lambda > \frac{(d-1)^{d-1}}{d^d}$. However we need to give the analogous definition for CISGS.

Definition 13. *Let* ***Connected Induced Subgraphs With Bias*** *or* **CISWB** *be the problem that on input (G, λ) (for a graph G and $\lambda \in \mathbb{R}_{\geq 0}$) outputs L such that $L \subseteq V(G)$, $G[L]$ is connected, and L occurs with probability $\lambda^{|L|}/Z$ where $Z = \sum_{L' \subseteq V(G), G[L'] \text{is connected}} \lambda^{|L'|}$.*

Now let us show that an efficient algorithm for CISWB would give an effective solution to ST.

Theorem 4. *If there is an* FPAS *to* CISWB *for (G, λ) where G has maximum degree d and $1 > \lambda > (d-1)^{d-1}/d^d$, then* **RP** = **NP**.

The proof is extremely similar to those of Theorems 1 and 2, so for brevity we have removed the proof of Theorem 4.

5 Markov Chains and Sampling

Now we will show that Markov chains are not likely to be useful in sampling connected subgraphs of either fixed size or with bias λ. To do this we will show that local Markov chains cannot be rapidly mixing while also sampling connected subgraphs from the desired distribution. Here we use local to mean neighboring states share $k-1$ vertices, that is for two neighboring states X and Y, $X - \{x\} = Y - \{y\}$ for some $x \in X$ and $y \in Y$ (see, e.g., [15]). The notion of local can be extended to mean neighboring states must share at least one vertex and our proofs would follow accordingly, but we use sharing $k-1$ vertices for simplicity. Our proof uses conductance (see, e.g., [20]) to show slow mixing, the standard definition is given below. We use the standard notions of $P(i,j)$ to mean the probability we move from state i to state j and $\pi(i)$ to be the probability of being in state i according to the stationary distribution.

Definition 14. *The* **conductance** *of a Markov chain M on state space Ω is $\Phi_M = \min_{U \subseteq \Omega, C_U \leq 1/2} \Phi(U)$ where $\Phi(U) = F_U/C_U$ and $F_U = \sum_{i \in U, j \in \overline{U}} P(i,j)\pi(i)$, $C_U = \sum_{i \in U} \pi(i)$.*

We use conductance to bound τ, the mixing time of M. Formally, $\tau = \min\{t : \sum_{j \in \Omega} |P^{t'}(i,j) - \pi(j)| \leq 1/e$ for all $t' \geq t$ and $i \in \Omega\}$. It is a well known result that $1/\tau \leq 8\Phi(M)$ for an ergodic chain M (see, e.g., [1], we use this result so that M can be non-reversible, see [14] for a similar argument). Therefore, in the following proofs we give a tree G such that for any local ergodic chain M, $\Phi(M)$ is tiny.

Theorem 5. *There is a tree G with maximum degree 3 such that the following is true. Let M be a local ergodic Markov chain whose states are $S \subseteq V(G)$ such that $|S| = k$ and $G[S]$ is connected and the stationary distribution is uniform. Then the mixing time of M is exponential in k.*

Proof. Let G be a graph on $4n$ vertices consisting of 2 binary trees on n nodes with a path of length $2n$ in between them. Nodes 1 through n are in one tree, nodes $n+1$ to $3n$ are a path from the first tree to the second tree, and nodes $3n+1$ through $4n$ are the second tree. We will use a conductance argument to show slow mixing, and so we will give some U such that $\Phi(U) \leq \frac{2}{2^{k/2}}$.

Let $U = \{U' \mid U' \in \Omega, \forall v \in U', v \leq 2n, |U'| = k\}$. Clearly $|U|$ is no more than 1/2 of the total number of connected subsets of size k as we can see $|\bar{U}| \geq |U|$, thus $C_U \leq 1/2$. Note that the only set in U that can move out of U in 1 move is $\{2n, 2n-1, ..., 2n-(k-1)\}$ as we require the vertex $2n$ to be included to add the vertex $2n+1$. Thus, $F_U \leq \frac{1}{|\Omega|}$.

Now let us give a lower bound on $|\Omega|$ by counting the number of configurations in the trees alone. Consider taking a connected subset of size $k/2$ rooted at the root of the tree such that no leaves are in the subset. Thus there are

k vertices left to choose from (as each vertex has degree 3 and $k/2 - 1$ edges are used internally), and so there are at least $\binom{k}{k/2}$ connected subsets with this specific configuration. Thus $|\Omega| \geq \binom{k}{k/2} \geq 2^{k/2}$ and so $\Phi(U) \leq \frac{2}{2^{k/2}}$. Therefore the mixing time is exponential in k. $\square$

Note that Theorem 5 implies that the chain is not rapidly mixing for $k = \omega(\log n)$. Now let us give an analogous proof for sampling with bias λ.

Theorem 6. *Fix d and $1 > \lambda > \frac{(d-1)^{(d-1)}}{d^d}$, then there is a graph G with maximum degree $d+1$ such that the following is true. Let M be a local ergodic Markov chain whose states are $S \subseteq V(G)$ and $G[S]$ is connected and the stationary distribution is such that S occurs with probability $\lambda^{|S|}/Z$. Then the mixing time of M is exponential in n.*

The proof is very similar to that of Theorem 5 and so in consideration of space we have removed it.

Further Questions

- We showed hardness for CISWB on a general graph for $1 > \lambda > \frac{(d-1)^{d-1}}{d^d}$. Is there a polynomial solution for CISWB on an infinite grid (rooted at an arbitrary vertex) for some $1 > \lambda > \frac{(d-1)^{d-1}}{d^d}$?
- We have hardness results for CISWB with $\lambda < 1$. Is there a similar threshold for $\lambda > 1$?
- Similarly, we can sample connected subgraphs of a bounded degree graph with bias $\lambda < \frac{(d-1)^{d-1}}{d^d}$ for any error ε. Is there some threshold for $\lambda > 1$ where this is also true?
- In Sect. 5 we showed Markov chains are likely not useful in randomly sampling trees. What sets of graphs can they randomly sample and rapidly mix?

References

1. Aldous, D., Fill, J.A.: Reversible Markov chains and random walks on graphs (2002). https://www.stat.berkeley.edu/users/aldous/RWG/book.pdf. Unfinished monograph, recompiled 2014
2. Baskerville, K., Grassberger, P., Paczuski, M.: Graph animals, subgraph sampling, and motif search in large networks. Phys. Rev. E **76**(3), 036107, 13 (2007)
3. Frieze, A.: Notes on Counting and rapidly mixing Markov chains. http://www.math.cmu.edu/~af1p/Mixing.html
4. Grochow, J.A., Kellis, M.: Network motif discovery using subgraph enumeration and symmetry-breaking. In: Speed, T., Huang, H. (eds.) RECOMB 2007. LNCS, vol. 4453, pp. 92–106. Springer, Heidelberg (2007). https://doi.org/10.1007/978-3-540-71681-5_7
5. Guo, H., Jerrum, M.: A polynomial-time approximation algorithm for all-terminal network reliability. In: Proceedings of the 45th International Colloquium on Automata, Languages, and Programming, ICALP 2018, Prague, Czech Republic, 9–13 July 2018, pp. 68:1–68:12 (2018)

6. Ising, E.: Contribution to the theory of ferromagnetism. Z. Phys. **31**, 253–258 (1925)
7. Jerrum, M., Meeks, K.: The parameterised complexity of counting connected subgraphs and graph motifs. J. Comput. Syst. Sci. **81**(4), 702–716 (2015)
8. Jung, K., Shah, D.: Inference in binary pair-wise Markov random fields through self-avoiding walks. arXiv e-prints p. cs/0610111 (2006)
9. Kangas, K., Kaski, P., Koivisto, M., Korhonen, J.H.: On the number of connected sets in bounded degree graphs. In: Kratsch, D., Todinca, I. (eds.) WG 2014. LNCS, vol. 8747, pp. 336–347. Springer, Cham (2014). https://doi.org/10.1007/978-3-319-12340-0_28
10. Karp, R.M.: Reducibility among combinatorial problems. In: Miller, R.E., Thatcher, J.W. (eds.) Complexity of Computer Computations, pp. 85–103. Plenum Press, Boston (1972)
11. Kashtan, N., Milo, R., Itzkovitz, S., Alon, U.: Efficient sampling algorithm for estimating subgraph concentrations and detecting network motifs. Bioinformatics **20**(11), 1746–1758 (2004)
12. Lenz, W.: Beitrag zum Verständnis der magnetischen Erscheinungen in festen Körpern. Z. Phys. **21**, 613–615 (1920)
13. Lu, X., Bressan, S.: Sampling connected induced subgraphs uniformly at random. In: Ailamaki, A., Bowers, S. (eds.) SSDBM 2012. LNCS, vol. 7338, pp. 195–212. Springer, Heidelberg (2012). https://doi.org/10.1007/978-3-642-31235-9_13
14. Łuczak, T., Vigoda, E.: Torpid mixing of the Wang-Swendsen-Kotecký algorithm for sampling colorings. J. Discret. Algorithms **3**(1), 92–100 (2005)
15. Mossel, E., Weitz, D., Wormald, N.: On the hardness of sampling independent sets beyond the tree threshold. Probab. Theory Relat. Fields **143**(3), 401–439 (2009)
16. Patel, V., Regts, G.: Deterministic polynomial-time approximation algorithms for partition functions and graph polynomials. SIAM J. Comput. **46**(6), 1893–1919 (2017)
17. Patel, V., Regts, G.: Computing the number of induced copies of a fixed graph in a bounded degree graph. Algorithmica **81**(5), 1844–1858 (2018)
18. Garey, M.R., Johnson, D.: The rectilinear steiner tree problem is NP-complete. SIAM J. Appl. Math. **32**, 826–834 (1977)
19. Savoie, W., et al.: Phototactic supersmarticles. Artif. Life Robot. **23**(4), 459–468 (2018). https://doi.org/10.1007/s10015-018-0473-7
20. Sinclair, A.: Algorithms for Random Generation and Counting: A Markov Chain Approach. Birkhauser Verlag, Basel (1993)
21. Sly, A.: Computational transition at the uniqueness threshold. In: Proceedings of the 51st IEEE Annual Symposium on Foundations of Computer Science, FOCS, pp. 287–296 (2010)
22. Stanley, R.P.: Enumerative Combinatorics: vol. 2, 1st edn. Cambridge University Press, New York (1999)
23. Vince, A.: Counting connected sets and connected partitions of a graph. Australas. J. Comb. **67**(2), 281–293 (2017)
24. Weitz, D.: Counting independent sets up to the tree threshold. In: Proceedings of the 38th Annual ACM Symposium on Theory of Computing, STOC, pp. 140–149. ACM, New York (2006)
25. White, K., Farber, M., Pulleyblank, W.: Steiner trees, connected domination and strongly chordal graphs. Networks **15**(1), 109–124 (1985)

17. [illegible]
20. [illegible]

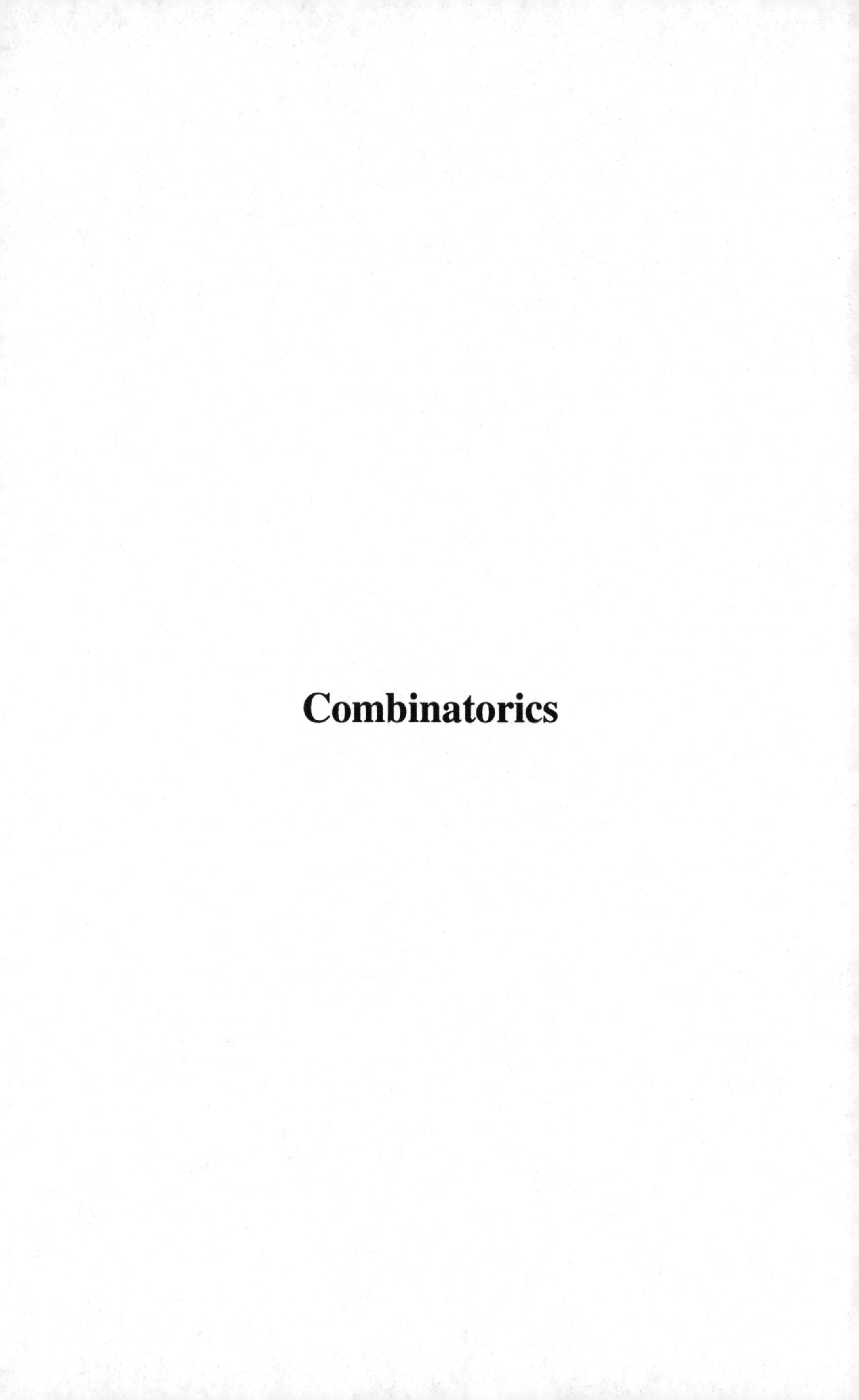

Combinatorics

Lower Bounds for Max-Cut via Semidefinite Programming

Charles Carlson[1](✉), Alexandra Kolla[1], Ray Li[2], Nitya Mani[3], Benny Sudakov[4], and Luca Trevisan[5]

[1] Department of Computer Science, University of Colorado Boulder, Boulder, CO 80302, USA
{charles.carlson,alexandra.kolla}@colorado.edu
[2] Department of Computer Science, Stanford University, Stanford, CA 94305, USA
rayyli@cs.stanford.edu
[3] Department of Mathematics and Computer Science, Stanford University, Stanford, CA 94305, USA
nmani@cs.stanford.edu
[4] Department of Mathematics, ETH, 8092 Zurich, Switzerland
benjamin.sudakov@math.ethz.ch
[5] Department of Decision Sciences, Bocconi University, Milan, Italy
l.trevisan@unibocconi.it

Abstract. For a graph G, let $f(G)$ denote the size of the maximum cut in G. The problem of estimating $f(G)$ as a function of the number of vertices and edges of G has a long history and was extensively studied in the last fifty years. In this paper we propose an approach, based on semidefinite programming (SDP), to prove lower bounds on $f(G)$. We use this approach to find large cuts in graphs with few triangles and in K_r-free graphs.

Keywords: Max-Cut · Semidefinite programming · K_r-free graphs

1 Introduction

The celebrated Max-Cut problem asks for the largest bipartite subgraph of a graph G, i.e., for a partition of the vertex set of G into disjoint sets V_1 and V_2

Alexandra Kolla was supported by NSF CAREER grant 1452923 as well as NSF AF grant 1814385.
Ray Li was supported by an NSF GRF grant DGE-1656518 and by NSF grant CCF-1814629.
Nitya Mani was supported in part by a Stanford Undergraduate Advising and Research Major Grant.
Luca Trevisan was supported by the NSF under grant CCF 181543 and his work on this project has received funding from the European Research Council (ERC) under the European Union's Horizon 2020 research and innovation programme (grant agreement No. 834861).
Benny Sudakov was supported in part by SNSF grant 200021_196965.

Y. Kohayakawa and F. K. Miyazawa (Eds.): LATIN 2020, LNCS 12118, pp. 479–490, 2020.
https://doi.org/10.1007/978-3-030-61792-9_38

so that the number of edges of G crossing V_1 and V_2 is maximal. This problem has been the subject of extensive research, both from a largely algorithmic perspective in computer science and from an extremal perspective in combinatorics. Throughout, let G denote a graph with n vertices and m edges with maximal cut of size $f(G)$. The extremal version of Max-Cut problem asks to give bounds on $f(G)$ solely as a function of m and n. This question was first raised more than fifty years ago by Erdős [9] and has attracted a lot of attention since then (see, e.g., [1,3,5–7,10,11,16,17] and their references).

It is well known that every graph G with m edges has a cut of size at least $m/2$. To see this, consider a random partition of vertices of the vertices G into two parts V_1, V_2 and estimate the expected number of edges between V_1 and V_2. On the other hand, already in 1960's Erdős [9] observed that the constant $1/2$ cannot be improved even if we consider very restricted families of graphs, e.g., graphs that contain no short cycles. Therefore the main question, which has been studied by many researchers, is to estimate the error term $f(G) - m/2$, which we call *surplus*, for various families of graphs G.

The elementary bound $f(G) \geq m/2$ was improved by Edwards [7,8] who showed that every graph with m edges has a cut of size at least $\frac{m}{2} + \frac{\sqrt{8m+1}-1}{8}$. This result is easily seen to be tight in case G is a complete graph on an odd number of vertices, that is, whenever $m = \binom{k}{2}$ for some odd integer k. Estimates on the second error term for other values of m can be found in [4] and [5].

Although the $\sqrt{m}$ error term is tight in general, it was observed by Erdős and Lovász [10] that for triangle-free graph it can be improved to at least $m^{2/3+o(1)}$. This naturally yiels a motivating question: what is the best surplus which can always be achieved if we assume that our family of graphs is *H-free*, i.e., no graph contains a fixed graph H as a subgraph. It is not difficult to show (see, e.g. [2]) that for every fixed graph H there is some $\epsilon = \epsilon(H) > 0$ such that $f(G) \geq \frac{m}{2} + \Omega(m^{1/2+\epsilon})$ for all H-free graphs with m edges. However, the problem of estimating the error term more precisely is not easy, even for relatively simple graphs H. It is plausible to conjecture (see [3]) that for every fixed graph H there is a constant c_H such that every H-free graph G with m edges has a cut with surplus at least $\Theta(m^{c_H})$, i.e., there is both a lower bound and an infinite sequence of example showing that exponent c_H can not be improved. This conjecture is very difficult. Even in the case $H = K_3$ determining the correct error term took almost twenty years. Following the works of [10,15,16], Alon [1] proved that every m-edge triangle free graph has a cut with surplus of order $m^{4/5}$ and that this is tight up to constant factors. There are several other forbidden graphs H for which we know quite accurately the error term for the extremal Max-Cut problem in H-free graphs. For example, it was proved in [3], that if $H = C_r$ for $r = 4, 6, 10$ then $c_H = \frac{r+1}{r+2}$. The answer is also known in the case when H is a complete bipartite graph $K_{2,s}$ or $K_{3,s}$ (see [3] for details).

Due to space constraints several proofs will appear in the full version of this paper.

1.1 New Approach to Max-Cut Using Semidefinite Programming

Many extremal results for the Max-Cut problem rely on quite elaborate probabilistic arguments. A well known example of such an argument is a proof by Shearer [16] that if G is a triangle-free graph with n vertices and m edges, and if $d_1, d_2, \ldots, d_n$ are the degrees of its vertices, then $f(G) \geq \frac{m}{2} + O(\sum_{i=1}^{n} \sqrt{d_i})$. The proof is quite intricate and is based on first choosing a random cut and then randomly redistributing some of the vertices, depending on how many their neighbors are on the same side as the chosen vertex in the initial cut. Shearer's arguments were further extended, with more technically involved proofs, in [3] to show that the same lower bound remains valid for graphs G with relatively sparse neighborhoods (i.e., graphs which locally have few triangles).

In this article we propose a different approach to give lower bounds on the Max-Cut of sparse H-free graphs using approximation by semidefinite programming (SDP). This approach is intuitive and computationally simple. The main idea was inspired by the celebrated approximation algorithm of Goemans and Williamson [13] of the Max-Cut: given a graph G with m edges, we first construct an explicit solution for the standard Max-Cut SDP relexation of G which has value at least $(\frac{1}{2}+W)m$ for some positive surplus W. We then apply a Goemans-Williamson randomized rounding, based on the sign of the scalar product with random unit vector, to extract a cut in G whose surplus is within constant factor of W. Using this approach we prove the following result.

Theorem 1. *Let $G = (V, E)$ be a graph with n vertices and m edges. For every $i \in [n]$, let V_i be some subset of neighbors of vertex i and $\varepsilon_i \leq \frac{1}{\sqrt{|V_i|}}$. Then,*

$$f(G) \geq \frac{m}{2} + \sum_{i=1}^{n} \frac{\varepsilon_i |V_i|}{4\pi} - \sum_{(i,j)\in E} \frac{\varepsilon_i \varepsilon_j |V_i \cap V_j|}{2}.$$

This results implies the Shearer's bound [16]. To see this, set V_i to the neighbors of i and $\varepsilon_i = \frac{1}{\sqrt{d_i}}$ for all i. Then, if G is triangle-free graph, then $|V_i \cap V_j| = 0$ for every pair of adjacent vertices i, j.

The fact that we apply Goemans-Williamson SDP rounding in this setting is perhaps surprising for a few reasons. In general, our result obtains a surplus of $\Omega(W)$ from an SDP solution with surplus W, which is not possible in general. The best cut that can be guaranteed from any kind of rounding of a Max-Cut SDP solution with value $(\frac{1}{2} + W)m$ is $(\frac{1}{2} + \Omega(\frac{W}{\log W}))m$ (see [14]). Furthermore, this is achieved using the RPR^2 rounding, not the Geomans-Williamson rounding. Nevertheless, we show that our explicit Max-Cut solution has additional properties that circumvents these issues and permits a better analysis.

1.2 New Lower Bound for Max-Cut of Triangle Sparse Graphs

Using Theorem 1, we give a new result on the Max-Cut of triangle sparse graphs that is more convenient to use than previous similar results. A graph G is d-*degenerate* if there exists an ordering of the vertices $1, \ldots, n$ such that vertex i

has at most d neighbors $j < i$. Degeneracy is a broader notion of graph sparseness than maximum degree: all maximum degree d graphs are d-degenerate, but the star graph is 1-degenerate while having maximum degree $n-1$. Theorem 1 gives the following useful corollary on the Max-Cut of d-degenerate graphs.

Corollary 1. *Let $\varepsilon \leq \frac{1}{\sqrt{d}}$. Let G be a d-degenerate graph with m edges and t triangles. Then*

$$f(G) \geq \frac{m}{2} + \frac{\varepsilon m}{4\pi} - \frac{\varepsilon^2 t}{2}.$$

Indeed, let $1, \ldots, n$ be an ordering of the vertices such that any i has at most d neighbors $j < i$, and let V_i be this set of neighbors. Let $\varepsilon_i = \varepsilon$ for all i. In this way, $\sum_i |V_i|$ counts every edge exactly once and $\sum_{(i,j)\in E} |V_i \cap V_j|$ counts every triangle exactly once, and the result follows. This shows that graphs with few triangles have cuts with surplus similar to triangle-free graphs.

This result is new and more convenient to use than existing results in this vein, because it relies only on the global count of the number of triangles, rather than a local triangle sparseness property assumed by prior results. For example, it was shown that (using Lemma 3.3 of [3]) a d-degenerate graph with a local triangle-sparseness property, namely that every large induced subgraph with a common neighbor is sparse, has Max-Cut at least $\frac{m}{2} + \Omega(\frac{m}{\sqrt{d}})$. However, we can achieve the same result with only the guarantee that the global number of triangles is small. In particular, when there are at most $O(m\sqrt{d})$ triangles, which is always the case with the local triangle-sparseness assumption above, setting $\varepsilon = \Theta(\frac{1}{\sqrt{d}})$ in Corollary 1 gives that the Max-Cut is again at least $\frac{m}{2} + \Omega(\frac{m}{\sqrt{d}})$.

1.3 Corollary: Lower Bounds for Max-Cut of H-free Degenerate Graphs

We illustrate usefulness of the above results by giving the following lower bound on the Max-Cut of K_r-free graphs.

Theorem 2. *Let $r \geq 3$. There exists a constant $c = c(r) > 0$ such that, for all K_r-free d-degenerate graphs G with m edges,*

$$f(G) \geq \left(\frac{1}{2} + \frac{c}{d^{1-1/(2r-4)}}\right) m.$$

Lower bounds such as Theorem 2 giving a surplus of the form $c \cdot \frac{m}{d^\alpha}$ are more fine-grained than those that depend only on the number of edges. Accordingly, they are useful for obtaining lower bounds of the Max-Cut independent of the degeneracy: many tight Max-Cut lower bounds in H free graphs of the form $\frac{m}{2} + cm^\alpha$ first establish that $f(G) \geq \frac{m}{2} + c \cdot \frac{m}{\sqrt{d}}$ for all H-free graphs, and then case-work on the degeneracy [3].

In the case of $r = 4$ one can use our arguments together with Alon's result on Max-Cut in triangle-free graphs to improve Theorem 2 further to $m/2 + cm/d^{2/3}$. While Theorem 2 gives nontrivial bounds for K_r-free graphs, we believe that a stronger statement is true and propose the following conjecture.

Conjecture 1. For any graph H, there exists a constant $c = c(H) > 0$ such that, for all H-free d-degenerate graphs with $m \geq 1$ edges,

$$f(G) \geq \left(\frac{1}{2} + \frac{c}{\sqrt{d}}\right) m. \tag{1}$$

Our Theorem 1 implies this conjecture for various graphs H, e.g., $K_{2,s}, K_{3,s}, C_r$ and for any graph H which contains a vertex whose deletion makes it acyclic. This was already observed in [3] using the weaker, locally triangle-sparse form of Corollary 1 described earlier. Conjecture 1 provides a natural route to proving a closely related conjecture of Alon, Bollobás, Krivelevich, and Sudakov [2].

Conjecture 2 (See concluding remarks of [2]*).* For any graph H, there exists constants $\varepsilon = \varepsilon(H) > 0$ and $c = c(H) > 0$ such that, for all H-free graphs with $m \geq 1$ edges,

$$f(G) \geq \frac{m}{2} + cm^{3/4+\varepsilon}.$$

Since every graph with m edges is obviously $\sqrt{2m}$-degenerate, the Conjecture 1 implies immediately a weaker form of Conjecture 2 with surplus of order $m^{3/4}$. With some extra technical work we can show that it actually implies the full conjecture, achieving a surplus of $m^{3/4+\varepsilon}$ for any graph H. For many graphs H for which Conjecture 2 is known, (1) was implicitly established for H-free graphs [3], making Conjecture 1 a plausible stepping stone to Conjecture 2. As further evidence of the plausibility of Conjecture 1, we show that Conjecture 2 implies a weaker form of Conjecture 1, namely that any H-free graph has Max-Cut $\frac{m}{2} + cm \cdot d^{-5/7}$. Using similar techniques, we can obtain nontrivial, unconditional results on the Max-Cut of d-degenerate H-free graphs for particular graphs H.

Conjecture 1, if true, gives a surplus of $\Omega(\frac{m}{\sqrt{d}})$ that is optimal up to a multiplicative constant factor for every fixed graph H containing a cycle. To see this, consider an Erdős-Rényi random graph $G(n,p)$ with $p = n^{-1+\delta}$. Using standard Chernoff-type estimates, one can easily show that with high probability that this graph is $O(np)$-degenerate and its Max-Cut has size at most $\frac{1}{4}\binom{n}{2}p + O(n\sqrt{np})$. Moreover, if $\delta = \delta(H) > 0$ is small enough, then with high probability $G(n,p)$ contains very few copies of H, which can be destroyed by deleting few vertices, without changing the degeneracy and surplus of the Max-Cut.

2 Lower Bounds for Max-Cut Using SDP

In this section we give a lower bound for $f(G)$ in graphs with few triangles, showing Theorem 1. To prove this result, we make heavy use of the SDP relaxation of the Max-Cut problem, formulated below for a graph $G = (V, E)$:

$$\begin{aligned} \text{maximize} \quad & \sum_{(i,j)\in E} \frac{1}{2}(1 - \langle v^{(i)}, v^{(j)} \rangle) \\ \text{subject to} \quad & \|v^{(i)}\|^2 = 1 \, \forall i \in V. \end{aligned} \tag{2}$$

We leverage the classical Goemans-Williamson [13] rounding algorithm which that gives an integral solution from a vector solution to the Max-Cut SDP.

Proof of Theorem 1. For $i \in [n]$, define $\tilde{v}^{(i)} \in \mathbb{R}^n$ by

$$\tilde{v}_j^{(i)} = \begin{cases} 1 & i = j \\ -\varepsilon_i & j \in V_i \\ 0 & otherwise. \end{cases}.$$

For $i \in [n]$, let $v^{(i)} \stackrel{\text{def}}{=} \frac{\tilde{v}^{(i)}}{\|\tilde{v}^{(i)}\|} \in \mathbb{R}^n$. Then $1 \leq \|\tilde{v}^{(i)}\| \leq 1+\varepsilon_i^2|V_i| \leq 2$ for all i. For each edge (i,j) with $i \in V_j$, we have

$$v_i^{(i)} v_i^{(j)} = \frac{1}{\|\tilde{v}^{(i)}\|} \cdot \frac{-\varepsilon_j}{\|\tilde{v}^{(j)}\|} \leq \frac{-\varepsilon_j}{4}.$$

For $k \in V_i \cap V_j$, we have $v_k^{(i)} v_k^{(j)} \leq \varepsilon_i \varepsilon_j$. For $k \notin \{i,j\} \cup (V_i \cap V_j)$, we have $v_k^{(i)} v_k^{(j)} = 0$ as $v_k^{(i)} = 0$ or $v_k^{(j)} = 0$. Thus, for all edges (i,j),

$$\langle v^{(i)}, v^{(j)} \rangle \leq -\frac{\varepsilon_i}{4} \mathbb{1}_{V_j}(i) - \frac{\varepsilon_j}{4} \mathbb{1}_{V_i}(j) + |V_i \cap V_j| \varepsilon_i \varepsilon_j.$$

Here, $\mathbb{1}_S(i)$ is 1 if $i \in S$ and 0 otherwise. Vectors $v^{(1)}, \ldots, v^{(n)}$ form a vector solution to the SDP (2). We now round this solution using the Goemans-Williamson [13] rounding. Let w denote a uniformly random unit vector, $A = \{i \in [n] : \langle v^{(i)}, w \rangle \geq 0\}$, and $B = [n] \setminus A$. The angle between vectors $v^{(i)}, v^{(j)}$ is equal to $\cos^{-1}(\langle v^{(i)}, v^{(j)} \rangle)$, so the probability an edge (i,j) is cut is

$$\begin{aligned} \mathbf{Pr}[(i,j) \text{ cut}] &= \frac{\cos^{-1}(\langle v^{(i)}, v^{(j)} \rangle)}{\pi} = \frac{1}{2} - \frac{\sin^{-1}(\langle v^{(i)}, v^{(j)} \rangle)}{\pi} \\ &\geq \frac{1}{2} - \frac{1}{\pi} \sin^{-1}\left(|V_i \cap V_j| \varepsilon_i \varepsilon_j - \frac{\varepsilon_i}{4} \mathbb{1}_{V_j}(i) - \frac{\varepsilon_j}{4} \mathbb{1}_{V_i}(j) \right) \\ &\geq \frac{1}{2} + \frac{\varepsilon_i}{4\pi} \mathbb{1}_{V_j}(i) + \frac{\varepsilon_j}{4\pi} \mathbb{1}_{V_i}(j) - \frac{|V_i \cap V_j| \varepsilon_i \varepsilon_j}{2}. \end{aligned}$$

In the last inequality, we used that $\sin^{-1}(a-b) \leq \frac{\pi}{2} a - b$ for $a, b \in [0,1]$. This is true as $\sin^{-1}(x) \leq \frac{\pi}{2} x$ when x is positive and $\sin^{-1}(x) \leq x$ when x is negative. Thus, the expected size of the cut given by $A \sqcup B$ is, by linearity of expectation,

$$\begin{aligned} \sum_{(i,j) \in E} \mathbf{Pr}[(i,j) \text{ cut}] &\geq \sum_{\substack{(i,j) \in E \\ i<j}} \left(\frac{1}{2} + \frac{\varepsilon_i}{4} \mathbb{1}_{V_j}(i) + \frac{\varepsilon_j}{4} \mathbb{1}_{V_i}(j) - \frac{|V_i \cap V_j| \varepsilon_i \varepsilon_j}{2} \right) \\ &= \frac{m}{2} + \sum_{i=1}^{n} \frac{|V_i| \varepsilon_i}{4\pi} - \sum_{(i,j) \in E} \frac{|V_i \cap V_j| \varepsilon_i \varepsilon_j}{2}. \end{aligned}$$

□

In the proof of Theorem 2 we use the following version of Corollary 1.

Corollary 2. *There exists an absolute constant $c > 0$ such that the following holds. For all $d \geq 1$ and $\varepsilon \leq \frac{1}{\sqrt{d}}$, if a d-degenerate graph $G = (V, E)$ has m edges and at most $\frac{m}{8\varepsilon}$ triangles then*

$$f(G) \geq \left(\frac{1}{2} + c\varepsilon\right) \cdot m.$$

3 Decomposition of Degenerate Graphs

In a graph $G = (V, E)$, let $n(G)$ and $m(G)$ denote the number of vertices and edges, respectively. For a vertex subset $V' \subset V$, let $G[V']$ denote the subgraph induced by V'. We show that d-degenerate graphs with many triangles have small subsets of neighborhoods with many edges.

Lemma 1. *Let $d \geq 1$ and $\varepsilon > 0$, and let $G = (V, E)$ be a d-degenerate graph with at least $\frac{m(G)}{\varepsilon}$ triangles. There exists a subset V' of at most d vertices with a common neighbor in G such that the induced subgraph $G[V']$ has at least $\frac{|V'|}{\varepsilon}$ edges.*

This lemma helps us partition the vertices of any d-degenerate graph in a useful way. Repeatedly applying this lemma, we can peel off small subsets of neighborhoods with many edges until we are left with an vertex subset containing many triangles.

Lemma 2. *Let $\varepsilon > 0$. Let $G = (V, E)$ be a d-degenerate graph on n vertices with m edges. Then there exists a partition $V_1, \ldots, V_{k+1}$ of the vertex set V with the following properties.*

1. *For $i = 1, \ldots, k$, the vertex subset V_i has at most d vertices and has a common neighbor, and the induced subgraph $G[V_i]$ has at least $\frac{|V_i|}{\varepsilon}$ edges.*
2. *The induced subgraph $G[V_{k+1}]$ has at most $\frac{m(G[V_{k+1}])}{\varepsilon}$ triangles.*

3.1 Large Max-Cut from Decompositions

For a d-degenerate graph $G = (V, E)$, in a partition $V_1, \ldots, V_{k+1}$ of V given by Lemma 2, the induced subgraph $G[V_{k+1}]$ has few triangles, and thus, by Corollary 1, has a cut with good surplus. This allows us to obtain the following technical result regarding the Max-Cut of H-free d-degenerate graphs.

Lemma 3. *There exists an absolute constant $c > 0$ such that the following holds. Let H be a graph and H' be obtained by deleting any vertex of H. Let $0 < \varepsilon < \frac{1}{\sqrt{d}}$. For any H-free d-degenerate graph $G = (V, E)$, one of the following holds:*

– *We have*

$$f(G) \geq \left(\frac{1}{2} + c\varepsilon\right) m. \tag{3}$$

– *There exist graphs $G_1, \ldots, G_k$ such that five conditions hold: (i) graphs G_i are H'-free for all i, (ii) $n(G_i) \leq d$ for all i, (iii) $m(G_i) \geq \frac{n(G_i)}{8\varepsilon}$ for all i, (iv) $n(G_1) + \cdots + n(G_k) \geq \frac{m}{6d}$, and (v)*

$$f(G) \geq \frac{m(G)}{2} + \sum_{i=1}^{k} \left(f(G_i) - \frac{m(G_i)}{2}\right). \tag{4}$$

Proof. Let $c_1 < 1$ be the parameter given by Corollary 2. Let $c = \frac{c_1}{6}$. Let $G = (V, E)$ be a d-degenerate H-free graph. Applying Lemma 2 with parameter 8ε, we can find a partition $V_1, \ldots, V_{k+1}$ of the vertex set V with the following properties.

1. For $i = 1, \ldots, k$, the vertex subset V_i has at most d vertices and has a common neighbor, and the induced subgraph $G[V_i]$ at least $\frac{|V_i|}{8\varepsilon}$ edges.
2. The subgraph $G[V_{k+1}]$ has at most $\frac{m(G[V_{k+1}])}{8\varepsilon}$ triangles.

For $i = 1, \ldots, k+1$, let $G_i \stackrel{\text{def}}{=} G[V_i]$ and let $m_i \stackrel{\text{def}}{=} m(G_i)$. For $i = 1, \ldots, k$, since G is H-free and each V_i is a subset of some vertex neighborhood in G, the graphs G_i are H'-free. For $i = 1, \ldots, k$, fix a maximal cut of G_i with associated vertex partition $V_i = A_i \sqcup B_i$. By the second property above, the graph G_{k+1} has at most $\frac{m_{k+1}}{8\varepsilon}$ triangles. Applying Corollary 2 with parameter ε, we can find a cut of G_{k+1} of size at least $(\frac{1}{2} + c_1\varepsilon)m_{k+1}$ with associated vertex partition $V_{k+1} = A_{k+1} \sqcup B_{k+1}$.

We now construct a cut of G by randomly combining the cuts obtained above for each G_i. Independently, for each $i = 1, \ldots, k+1$, we add either A_i or B_i to vertex set A, each with probability $\frac{1}{2}$. Setting $B = V \setminus A$, gives a cut of G. As $V_1, \ldots, V_{k+1}$ partition V, each of the $m - (m_1 + \cdots + m_{k+1})$ edges that is not in one of the induced graphs $G_1, \ldots, G_{k+1}$ has exactly one endpoint in each of A, B with probability 1/2. This allows us to compute the expected size of the cut (a lower bound on $f(G)$ as there is some instantiation of this random process that achieves this expected size).

$$\begin{aligned} f(G) &\geq \frac{1}{2}(m - (m_1 + \cdots + m_{k+1})) + \left(\frac{1}{2} + c_1\varepsilon\right) \cdot m_{k+1} + \sum_{i=1}^{k} f(G_i) \\ &= \frac{m}{2} + c_1\varepsilon m_{k+1} + \sum_{i=1}^{k} \left(f(G_i) - \frac{m_i}{2}\right). \end{aligned} \tag{5}$$

We bound (5) based on the distribution of edges in G in 3 cases:

– $m_{k+1} \geq \frac{m}{6}$. Since $f(G_i) \geq \frac{m_i}{2}$ for all $i = 1, \ldots, k$, (3) holds:

$$f(G) \geq \frac{m}{2} + c_1\varepsilon m_{k+1} \geq \left(\frac{1}{2} + c\varepsilon\right) \cdot m.$$

- The number of edges between $V_1 \cup \cdots \cup V_k$ and V_{k+1} is at least $\frac{2m}{3}$. Then, the cut given by vertex partition $V = A' \sqcup B'$ with $A' = V_1 \cup \cdots \cup V_k$ and $B' = V_{k+1}$ has at least $\frac{2m}{3}$ edges, in which case $f(G) \geq \frac{2m}{3} > (\frac{1}{2} + \frac{c_1 \varepsilon}{6}) \cdot m$, so (3) holds.
- $G' = G[V_1 \cup \cdots \cup V_k]$ has at least $\frac{m}{6}$ edges. We show (4) holds. By construction, for $i = 1, \ldots, k$, the graph G_i is H' free, has at most d vertices, and has at least $\frac{m_i}{8\varepsilon}$ edges. Since G is d-degenerate, G' is as well, so

$$\frac{m}{6} \leq m(G') \leq d \cdot n(G') = d \cdot \sum_{i=1}^{k} n(G_i),$$

Hence $n(G_1) + \cdots + n(G_k) \geq \frac{m}{6d}$. Lastly, by (5), we have

$$f(G) \geq \frac{m}{2} + \sum_{i=1}^{k} \left(f(G_i) - \frac{m_i}{2} \right).$$

This covers all cases, and in each case we showed either (3) or (4) holds. □

Lemma 3 allows us to convert Max-Cut lower bounds on H-free graphs to Max-Cut lower bounds on H-free d-degenerate graphs.

Lemma 4. *Let H be a graph and H' be obtained by deleting any vertex of H. Suppose that there exists constants $a = a(H') \in [\frac{1}{2}, 1]$ and $c' = c'(H') > 0$ such that for all H'-free graphs G with $m' \geq 1$ edges, $f(G) \geq \frac{m'}{2} + c' \cdot (m')^a$. Then there exists a constant $c = c(H) > 0$ such that for all H-free d-degenerate graphs G with $m \geq 1$ edges,*

$$f(G) \geq \left(\frac{1}{2} + c d^{-\frac{2-a}{1+a}} \right) \cdot m.$$

4 Max-Cut in K_r-free Graphs

In this section we specialize Lemma 3 to the case $H = K_r$ to prove Theorem 2. Let $\chi(G)$ denote the chromatic number of a graph G, the minimum number of colors needed to properly color the vertices of the graph so that no two adjacent vertices receive the same color. We first obtain a nontrivial upper bound on the chromatic number of a K_r-free graph G, giving an lower bound (Lemma 7) on the Max-Cut of K_r-free graphs. The lower bound on the Max-Cut of general K_r-free graphs enables us to apply Lemma 3 to give a lower bound on the Max-Cut of d-degenerate K_r-free graphs per Theorem 2. The following well known lemma gives a lower bound on the Max-Cut using the chromatic number.

Lemma 5. *(see e.g. Lemma 2.1 of [2]) Given a graph $G = (V, E)$ with m edges and chromatic number $\chi(G) \leq t$, we have $f(G) \geq (\frac{1}{2} + \frac{1}{2t})m$.*

We can bound the chromatic number of K_r-free graphs by repeatedly applying a standard bound [12] on the off-diagonal Ramsey number $R(r, \cdot)$.

Lemma 6. *Let $r \geq 3$ and $G = (V, E)$ be a K_r-free graph on n vertices. Then,*

$$\chi(G) \leq 4n^{(r-2)/(r-1)}.$$

Lemma 5 and Lemma 6 give the following immediate corollary.

Lemma 7. *If G is a K_r-free graph with at most n vertices and m edges, then*

$$f(G) \geq \left(\frac{1}{2} + \frac{1}{8n^{(r-2)/(r-1)}}\right) m.$$

The above bounds allow us to prove Theorem 2.

Proof of Theorem 2. Let G be a d-degenerate K_r-free graph and $\varepsilon = d^{-1+\frac{1}{2r-4}}$. Let c_2 be the parameter from Lemma 3. Let $c = \min(c_2, \frac{1}{388})$. Applying Lemma 3 with parameter ε, one of two properties hold. If (3) holds, we are done by our choice of ε. If (4) holds, there exist K_{r-1}-free graphs $G_1, \ldots, G_k$ such that G_i has at most d vertices and at least $\frac{n(G_i)}{8\varepsilon}$ edges, $n(G_1) + \cdots + n(G_k) \geq \frac{m}{6d}$, and

$$f(G) \geq \frac{m}{2} + \sum_{i=1}^{k} \left(f(G_i) - \frac{m(G_i)}{2}\right).$$

For all i, we have

$$f(G_i) - \frac{m(G_i)}{2} \geq \frac{m(G_i)}{8n(G_i)^{(r-3)/(r-2)}} \geq \frac{n(G_i)}{64\varepsilon n(G_i)^{(r-3)/(r-2)}} = \frac{\varepsilon d n(G_i)}{64}.$$

In the first inequality, we used Lemma 7. In the second inequality, we used $m(G_i) \geq \frac{n(G_i)}{8\varepsilon}$. In the third inequality, we used $n(G_i) \leq d$ and the definition of ε. Hence, as $d(n(G_1) + \cdots + n(G_k)) \geq \frac{m}{6}$, we have as desired that

$$f(G) \geq \frac{m}{2} + \sum_{i=1}^{k} \frac{\varepsilon d n(G_i)}{64} \geq \frac{m}{2} + \frac{\varepsilon m}{388} \geq \left(\frac{1}{2} + cd^{-1+\frac{1}{2r-4}}\right) \cdot m.$$

□

Remark 1. As we already mentioned in the introduction, we can improve the result of Theorem 2 in the case that $r = 4$ using Lemma 4. By a result of [1], we may apply Lemma 4 with $H = K_4$ and $H' = K_3$ and $a = 4/5$, so that for some absolute $c > 0$, any K_4-free d-degenerate graph G with $m \geq 1$ edges satisfies

$$f(G) \geq \left(\frac{1}{2} + cd^{-\frac{2-(4/5)}{1+(4/5)}}\right) \cdot m = \left(\frac{1}{2} + cd^{-2/3}\right) \cdot m.$$

5 Concluding Remarks

In this paper we presented an approach, based on semidefinite programming (SDP), to prove lower bounds on Max-Cut and used it to find large cuts in graphs with few triangles and in K_r-free graphs. A closely related problem of interest is bounding the Max-t-Cut of a graph, i.e. the largest t-colorable (t-partite) subgraph of a given graph. Our results imply good lower bounds for this problem as well. Indeed, by taking a cut for a graph G with m edges and surplus W, one can produce a t-cut for G of size $\frac{t-1}{t}m + \Omega(W)$ as follows. Let A, B be the two parts of the original cut. If $t = 2s$ is even, simply split randomly both A, B into s parts. If $t = 2s+1$ is odd, then put every vertex of A randomly in the parts $1, \ldots, s$ with probability $2/(2s+1)$ and in the part $2s+1$ with probability $1/(2s+1)$. Similarly, put every vertex of B randomly in the parts $s+1, \ldots, 2s$ with probability $2/(2s+1)$ and in the part $2s+1$ with probability $1/(2s+1)$. An easy computation (which we omit here) shows that the expected size of the resulting t-cut is $\frac{t-1}{t}m + \Omega(W)$.

The main open question left by our work is Conjecture 1. Proving this conjecture will require some major new ideas. Even showing that any d-degenerate H-free graph with m edges has a cut with surplus at least $m/d^{1-\delta}$ for some fixed δ (independent of H) is out of reach of current techniques.

Acknowledgements. The authors thank Jacob Fox and Matthew Kwan for helpful discussions and feedback. The authors thank Joshua Brakensiek for finding an error in an earlier draft of this paper. The authors thank Joshua Brakensiek and Yuval Wigderson for helpful feedback on an earlier draft of the paper.

References

1. Alon, N.: Bipartite subgraphs. Combinatorica **16**, 301–311 (1996)
2. Alon, N., Bollobás, B., Krivelevich, M., Sudakov, B.: Maximum cuts and judicious partitions in graphs without short cycles. J. Combin. Theory Ser. B **88**, 329–346 (2003)
3. Alon, N., Krivelevich, M., Sudakov, B.: Max cut in H-free graphs. Combin. Prob. Comput. **14**, 629–647 (2005)
4. Alon, N., Halperin, E.: Bipartite subgraphs of integer weighted graphs. Discret. Math. **181**, 19–29 (1998)
5. Bollobás, B., Scott, A.D.: Better bounds for max cut. In: Bollobás, B. (ed.) Contemporary Combinatorics. Bolyai Society Mathematical Studies, pp. 185–246. Springer, Heidelberg (2002)
6. Conlon, D., Fox, J., Kwan, M., Sudakov, B.: Hypergraph cuts above the average. Isr. J. Math. **233**(1), 67–111 (2019). https://doi.org/10.1007/s11856-019-1897-z
7. Edwards, C.S.: Some extremal properties of bipartite subgraphs. Canad. J. Math. **3**, 475–485 (1973)
8. Edwards, C.S.: An improved lower bound for the number of edges in a largest bipartite subgraph. In: Proceedings of the 2nd Czechoslovak Symposium on Graph Theory, Prague, pp. 167–181 (1975)
9. Erdős, P.: On even subgraphs of graphs. Mat. Lapok **18**, 283–288 (1967)

10. Erdős, P.: Problems and results in graph theory and combinatorial analysis. In: Proceedings of the 5th British Combinatorial Conference, pp. 169–192 (1975)
11. Erdős, P., Faudree, R., Pach, J., Spencer, J.: How to make a graph bipartite. J. Combin. Theory Ser. B **45**, 86–98 (1988)
12. Erdős, P., Szekeres, G.: A combinatorial problem in geometry. Compos. Math. **2**, 463–470 (1935)
13. Goemans, M.X., Williamson, D.P.: Improved approximation algorithms for maximum cut and satisfiability problems using semidefinite programming. J. ACM **42**, 1115–1145 (1995)
14. O'Donnell, R., Wu, Y.: An optimal SDP algorithm for Max-Cut, and equally optimal long code tests. In: Proceedings of the 40th ACM Symposium on Theory of Computing, pp. 335–344 (2008)
15. Poljak, S. Tuza, Zs.: Bipartite subgraphs of triangle-free graphs. SIAM J. Discret. Math. **7**, 307–313 (1994)
16. Shearer, J.: A note on bipartite subgraphs of triangle-free graphs. Rand. Struct. Alg. **3**, 223–226 (1992)
17. Sudakov, B.: Making a K_4-free graph bipartite. Combinatorica **27**, 509–518 (2007)

Quasi-Random Words and Limits of Word Sequences

Hiệp Hàn[1], Marcos Kiwi[2,3](✉), and Matías Pavez-Signé[3]

[1] Departamento de Matemática y Ciencia de la Computación, Universidad de Santiago de Chile, Santiago, Chile
hiep.han@usach.cl

[2] Centro de Modelamiento Matemático (UMI CNRS 2807), Santiago, Chile

[3] Departamento de Ingeniería Matemática, Universidad de Chile, Santiago, Chile
{mk,mpavez}@dim.uchile.cl

Abstract. Words are sequences of letters over a finite alphabet. We study two intimately related topics for this object: quasi-randomness and limit theory. With respect to the first topic we investigate the notion of uniform distribution of letters over intervals, and in the spirit of the famous Chung-Graham-Wilson theorem for graphs we provide a list of word properties which are equivalent to uniformity. In particular, we show that uniformity is equivalent to counting 3-letter subsequences.

Inspired by graph limit theory we then investigate limits of convergent word sequences, those in which all subsequence densities converge. We show that convergent word sequences have a natural limit, namely Lebesgue measurable functions of the form $f : [0,1] \to [0,1]$. Via this theory we show that every hereditary word property is testable, address the problem of finite forcibility for word limits and establish as a byproduct a new model of random word sequences.

Along the lines of the proof of the existence of word limits, we can also establish the existence of limits for higher dimensional structures. In particular, we obtain an alternative proof of the result by Hoppen, Kohayakawa, Moreira and Rath (2011) proving the existence of permutons.

1 Introduction

Roughly speaking, quasi-random structures are deterministic objects which share many characteristic properties of their random counterparts. Formalizing this concept has turned out to be tremendously fruitful in several areas, among others, number theory, graph theory, extremal combinatorics, the design of algorithms and complexity theory. This often follows from the fact that if an object is quasi-random, then it immediately enjoys many other properties satisfied by its random counterpart.

The first author was supported by the FONDECYT Regular grant 1191838. The second author was supported by CONICYT via PIA Concurso Apoyo a Centros Científicos y Tecnológicos de Excelencia con financiamiento basal AFB170001. The third author was partially supported by CONICYT Doctoral Fellowship 21171132.

Y. Kohayakawa and F. K. Miyazawa (Eds.): LATIN 2020, LNCS 12118, pp. 491–503, 2020.
https://doi.org/10.1007/978-3-030-61792-9_39

Seminal work on quasi-randomness concerned graphs [9,28,31]. Afterward, other combinatorial objects were considered, which include subsets of $\mathbb{Z}_n$ [11,16], hypergraphs [1,10,17,32], finite groups [18], and permutations [13]. Curiously, in the rich history of quasi-randomness, *words*, i.e., sequences of letters from a finite alphabet, one of the most basic combinatorial object with many applications, do not seem to have been explicitly investigated. We overcome this apparent neglect, put forth a notion of quasi-random words and show it is equivalent to several other properties. We then show how our notion of quasi-random words relates to other topics as explained next.

For the primal example of graphs, the notion of quasi-randomness was first studied by Thomason [31] who investigated a quantitative version of the following. A sequence $(G_n)_{n\to\infty}$ of graphs G_n on n-vertices has *uniform edge distribution*, if for some $p \in [0,1]$ we have

$$e(U) = p\binom{|U|}{2} + o(n^2) \qquad \text{for all } U \subseteq V(G_n). \tag{1}$$

The former is not only a.a.s. satisfied by the random graph $G(n,p)$, but is moreover considered one of its emblematic properties. In a cornerstone result of the area, Chung, Graham and Wilson [9] relate several properties of $G(n,p)$ to (1). For example, (1) implies the *counting* property, meaning that in a sequence $(G_n)_{n\to\infty}$ with property (1) the number $N_F(G_n)$ of labeled copies of a fixed graph F in G_n is asymptotically close to what is expected from $G(n,p)$, i.e.,

$$N_F(G_n) = p^{e_F} n^{v_F} + o(n^{v_F}) \qquad \text{for all graphs } F, \tag{2}$$

where $v_F = |V(F)|$ and $e_F = |E(F)|$. The converse of this implication also holds; and in a rather surprisingly strong form. Indeed, to imply the full strength of uniform edge distribution it is sufficient to require the counting property (2) to hold for $F = C_4$, the cycle of length four, and for $F = K_2$, i.e., to know the global edge density. In other words, $(G_n)_{n\to\infty}$ satisfies (1) if for some $p \in [0,1]$ we have

$$e(G_n) = p\binom{n}{2} + o(n^2) \qquad \text{and} \qquad N_{C_4}(G_n) = p^4 n^4 + o(n^4).$$

Hence, graph sequences $(G_n)_{n\to\infty}$ which satisfy one of the three properties must satisfy them all and such sequences are called quasi-random. We refer to [9,31] for further characterizations of graph quasi-randomness and to [22] for a more recent survey on the subject.

In contrast to the classical topic of quasi-randomness the research of limits for discrete structures was launched rather recently by Chayes, Lovász, Sós, Szegedy and Vesztergombi [8,25], and has become a very active topic of research since. Central to the area is the notion of convergent graph sequences $(G_n)_{n\to\infty}$, i.e., sequences of graphs which, roughly speaking, become more and more "similar" as $n = |V(G_n)|$ grows. The measure of similarity used in [8,25] is given in terms of homomorphism density

$$t(F,G) = \frac{1}{(v_G)^{v_F}} \hom(F,G),$$

where $\hom(F,G)$ denotes the number of homomorphisms of F into G, i.e., edge preserving maps from $V(F)$ to $V(G)$. A sequence of graphs $(G_n)_{n\to\infty}$ is then called convergent if $\big(t(F,G_n)\big)_{n\to\infty}$ converges for all F. Note that quasi-randomness deals with the particular case when $\lim_{n\to\infty} t(F,G_n) = p^{e_F}$ for every F, see (2). For convergent graph sequences, Lovász and Szegedy [25] show the existence of natural limit objects, called *graphons*, which are symmetric Lebesgue measurable functions of the form $W : [0,1]^2 \to [0,1]$. A graphon W is the limit of $(G_n)_{n\to\infty}$ if

$$\lim_{n\to\infty} t(F,G_n) = t(F,W) \qquad \text{for every } F,$$

where for a k-vertex graph F

$$t(F,W) = \int_0^1 \cdots \int_0^1 \prod_{ij\in E(F)} W(x_i,x_j)\,\mathrm{d}x_1 \ldots \mathrm{d}x_k.$$

For example, it follows from (2) that quasi-random graph sequences with edge density $p + o(1)$ converge to the constant p function.

Graphons can be used to define new models of random graphs, which is an interesting and important consequence of the theory. For a graphon W and an $n \in \mathbb{N}$, define the W-random graph $G(n,W)$ on the vertex set $[n] = \{1,2,\ldots,n\}$ created by first choosing $X_1,\ldots,X_n$ uniformly from $[0,1]$ and then connecting the pair i and j with probability $W(X_i,X_j)$, independently of other pairs. If W is constant and equal to p this coincides with $G(n,p)$, and for general W it was shown [25] that the sequence $G(n,W)$ converges to W a.s. as $n \to \infty$.

Note: Due to space limitations all proofs have been omitted from this extended abstract. For a version of this article that includes them, see [19].

2 Main Contributions

We continue previously mentioned investigations and study quasi-randomness for words and limits of convergent word sequences. Surprisingly, in the rich literature of quasi-randomness and in the one concerning limits of discrete structures, explicit investigation of this fundamental object has been overlooked so far.

A word $\boldsymbol{w}$ of length n is an ordered sequence $\boldsymbol{w} = (w_1, w_2, \ldots, w_n)$ of letters $w_i \in \Sigma$ from a fixed size alphabet Σ. For the sake of presentation we restrict our consideration to the two letter alphabet $\Sigma = \{0,1\}$, but most of our results and their proofs have straightforward generalizations to finite size alphabets.

2.1 Quasi-Random Words

Concerning quasi-randomness for words, our central notion is that of uniform distribution of letters over intervals. Specifically, a word $\boldsymbol{w} = (w_1 \dots w_n) \in \{0,1\}^n$ is called *(d,ε)-uniform* if for every interval $I \subseteq [n]$ we have[1]

$$\sum_{i \in I} w_i = |\{i \in I \colon w_i = 1\}| = d|I| \pm \varepsilon n. \tag{3}$$

We say that $\boldsymbol{w}$ is *ε-uniform* if $\boldsymbol{w}$ is (d,ε)-uniform for some d. Thus, uniformity states that up to an error term of εn the number of 1-entries of $\boldsymbol{w}$ in each interval I is roughly $d|I|$, a property which binomial random words with parameter d satisfy with high probability. In a different context, the notion of uniformity has been studied previously by Cooper [13] who gave a list of equivalent properties. A word $(w_1, \dots, w_n) \in \{0,1\}^n$ can also be seen as the set $W = \{i \colon w_i = 1\} \subseteq \mathbb{Z}_n$ and from this point of view our notion should be compared to the classical notion of quasi-randomness of subsets of $\mathbb{Z}_n$ studied by Chung and Graham in [11] and extended to the notion of U_k-uniformity by Gowers in [16]. With respect to this line of research we note that our notion of uniformity is weaker than all of the ones studied in [11,16]. Indeed, the weakest of them concerns U_2-uniformity and may be rephrased as follows: $W \subseteq \mathbb{Z}_n$ has U_2-norm at most $\varepsilon > 0$ if for all $A \subseteq \mathbb{Z}_n$ and all but εn elements $x \in \mathbb{Z}_n$ we have $|W \cap (A+x)| = |W|\frac{|A|}{n} \pm \varepsilon n$ where $A+x = \{a+x \colon a \in A\}$. Thus, e.g., the word $0101\dots01$ is uniform in our sense but its corresponding set does not have small U_2-norm.

As in the graph case there is a counting property related to uniformity. Given a word $\boldsymbol{w} = (w_1 \dots w_n)$ and $I = \{i_1, \dots, i_\ell\} \subseteq [n]$, where $i_1 < i_2 < \dots < i_\ell$, let $\mathrm{sub}(I, \boldsymbol{w})$ be the length ℓ subsequence $\boldsymbol{u} = (u_1 \dots u_\ell)$ of $\boldsymbol{w}$ such that $u_j = w_{i_j}$. We show that uniformity implies adequate subsequence count, i.e., for any fixed $\boldsymbol{u}$ the number of subsequences equal to $\boldsymbol{u}$ in a large uniform word $\boldsymbol{w}$, denoted by $\binom{\boldsymbol{w}}{\boldsymbol{u}}$, is roughly as expected from a random word with the same density of 1-entries as $\boldsymbol{w}$. It is then natural to ask whether the converse also holds. Our main result concerning quasi-random words states that uniformity is indeed already enforced by counting of length three subsequences. Let $\|\boldsymbol{w}\|_1 = \sum_{i \in [n]} w_i$ denote the number of 1-entries in $\boldsymbol{w}$, then our result reads as follows.

Theorem 1. *For every $\varepsilon > 0$, $d \in [0,1]$, and $\ell \in \mathbb{N}$, there is an n_0 such that for all $n > n_0$ the following holds.*

- *If $\boldsymbol{w} \in \{0,1\}^n$ is (d,ε)-uniform, then for each $\boldsymbol{u} \in \{0,1\}^\ell$,*

$$\binom{\boldsymbol{w}}{\boldsymbol{u}} = d^{\|\boldsymbol{u}\|_1}(1-d)^{\ell - \|\boldsymbol{u}\|_1}\binom{n}{\ell} \pm 5\varepsilon n^\ell.$$

- *Conversely, if $\boldsymbol{w} \in \{0,1\}^n$ is such that for all $\boldsymbol{u} \in \{0,1\}^3$ we have*

$$\binom{\boldsymbol{w}}{\boldsymbol{u}} = d^{\|\boldsymbol{u}\|_1}(1-d)^{3 - \|\boldsymbol{u}\|_1}\binom{n}{3} \pm \varepsilon n^3,$$

then $\boldsymbol{w}$ is $(d, 18\varepsilon^{1/3})$-uniform.

[1] We write $a \pm x$ to denote a number contained in the interval $[a-x, a+x]$.

Note that in the second part of the theorem the density of 1-entries is implicitly given. This is because $\binom{\boldsymbol{w}}{(111)} = \binom{\|\boldsymbol{w}\|_1}{3}$ and therefore the condition $\binom{\boldsymbol{w}}{(111)} \approx d^3\binom{n}{3}$ implies that $\|\boldsymbol{w}\|_1 \approx dn$. We also note that length three subsequences in the theorem cannot be replaced by length two subsequences and in this sense the result is best possible. Indeed, the word $(0\ldots01\ldots10\ldots0)$ consisting of $(1-d)\frac{n}{2}$ zeroes followed by dn ones followed by $(1-d)\frac{n}{2}$ zeroes contains the "right" number of every length two subsequences without being uniform.

From Theorem 1 and a result from Cooper [13, Theorem 2.3] we obtain a list of properties equivalent to uniformity (see Theorem 2 below). To state the result let $\boldsymbol{w}[j]$ denote the j-th letter of the word $\boldsymbol{w}$. Furthermore, by the Cayley digraph $\Gamma = \Gamma(\boldsymbol{w})$ of a word $\boldsymbol{w} = (w_1, \ldots, w_n)$ we mean the graph on the vertex set $\mathbb{Z}_n$ in which i and j form an edge if and only if $w_{i-j \pmod n} = 1$. Given a word $\boldsymbol{u} \in \{0,1\}^{\ell+1}$, a sequence of vertices $(v_1, \ldots, v_{\ell+1})$ is an increasing $\boldsymbol{u}$-path in $\Gamma = \Gamma(\boldsymbol{w})$ if the numbers $i_1, \ldots, i_\ell \in [n]$ defined by $v_{k+1} = v_k + i_k \pmod n$ satisfy $i_1 < \cdots < i_\ell$ and for each $k \in [\ell]$ the pair $v_k v_{k+1}$ is an edge in Γ if $u_k = w_{i_k} = 1$ and a non-edge if $u_k = w_{i_k} = 0$.

Theorem 2. *For a sequence $(\boldsymbol{w}_n)_{n\to\infty}$ of words such that $\boldsymbol{w}_n \in \{0,1\}^n$ and $\|\boldsymbol{w}_n\|_1 = dn + o(n)$ for some $d \in [0,1]$, the following are equivalent:*

- *(Uniformity) $(\boldsymbol{w}_n)_{n\to\infty}$ is $(d, o(1))$-uniform.*
- *(Counting) For all $\ell \in \mathbb{N}$ and all $\boldsymbol{u} \in \{0,1\}^\ell$ we have*
$$\tbinom{\boldsymbol{w}_n}{\boldsymbol{u}} = d^{\|\boldsymbol{u}\|_1}(1-d)^{\ell-\|\boldsymbol{u}\|_1}\tbinom{n}{\ell} + o(n^\ell).$$
- *(Minimizer) For all $\boldsymbol{u} \in \{0,1\}^3$ we have*
$$\tbinom{\boldsymbol{w}_n}{\boldsymbol{u}} = d^{\|\boldsymbol{u}\|_1}(1-d)^{3-\|\boldsymbol{u}\|_1}\tbinom{n}{3} + o(n^3).$$
- *(Exponential sums) For any fixed $\alpha > 0$ and for all non-zero $k \in \mathbb{Z}_n$ we have*
$$\tfrac{1}{n}\textstyle\sum_{j\in[n]} \boldsymbol{w}_n[j] \cdot \exp\left(\tfrac{2\pi i}{n}kj\right) = o(1)|k|^\alpha.$$
- *(Equidistribution) For every Lipschitz function $f : \mathbb{R}/\mathbb{Z} \to \mathbb{C}$*
$$\tfrac{1}{n}\textstyle\sum_{j=1}^n \boldsymbol{w}_n[j] \cdot f(\tfrac{j}{n}) = d\int_{\mathbb{R}/\mathbb{Z}} f + o(1)\|f\|_{\mathrm{Lip}}.$$
- *(Cayley graph) For all $\boldsymbol{u} \in \{0,1\}^3$ the number of increasing $\boldsymbol{u}$-paths in $\Gamma(\boldsymbol{w}_n)$ is*
$$d^{\|\boldsymbol{u}\|_1}(1-d)^{3-\|\boldsymbol{u}\|_1} n\tbinom{n}{3} + o(n^4).$$

We will say that a word sequence is *quasi-random* if it satisfies one of (hence all) the properties of Theorem 2.

2.2 Convergent Word Sequences and Word Limits

Over the last two decades it has been recognized that quasi-randomness and limits of discrete structures are intimately related subjects. Being interesting on their own right, limit theories have also unveiled many connections between various branches of mathematics and theoretical computer science. Thus, as a natural continuation of the investigation on quasi-randomness, we study convergent word sequences and their limits, a topic which to the best of our knowledge, has only been briefly mentioned by Szegedy [29].

The notion of convergence we consider is specified in terms of convergence of subsequence densities. Given $\boldsymbol{w} \in \{0,1\}^n$ and $\boldsymbol{u} \in \{0,1\}^\ell$, let $t(\boldsymbol{u}, \boldsymbol{w}) = \binom{\boldsymbol{w}}{\boldsymbol{u}}\binom{n}{\ell}^{-1}$ be the density of occurrences in $\boldsymbol{w}$ of the subsequence $\boldsymbol{u}$. Alternatively, if we let $\mathrm{sub}(\ell, \boldsymbol{w})$, with $\ell \leq n$, denote the length ℓ subsequence of $\boldsymbol{w}$ corresponding to $\mathrm{sub}(I, \boldsymbol{w})$, for I uniformly chosen among all subsets of $[n]$ of size ℓ, then $t(\boldsymbol{u}, \boldsymbol{w}) = \mathbb{P}(\mathrm{sub}(\ell, \boldsymbol{w})) = \boldsymbol{u})$. A sequence of words $(\boldsymbol{w}_n)_{n\to\infty}$ is called *convergent* if for every finite word $\boldsymbol{u}$ the sequence $\big(t(\boldsymbol{u}, \boldsymbol{w}_n)\big)_{n\to\infty}$ converges. In what follows, we will only consider sequences of words such that the length of the words tend to infinity. This, however, is not much of a restriction since convergent word sequences with bounded lengths must be constant eventually and limits considerations for these sequences are simple.

We show that convergent word sequences have natural limit objects, which turn out to be Lebesgue measurable functions of the form $f : [0,1] \to [0,1]$. Formally, write $f^1 = f$ and $f^0 = 1 - f$ for a function $f : [0,1] \to [0,1]$ and for a word $\boldsymbol{u} \in \{0,1\}^\ell$ define

$$t(\boldsymbol{u}, f) = \ell! \int_{0 \leq x_1 < \cdots < x_\ell \leq 1} \prod_{i \in [\ell]} f^{u_i}(x_i)\, \mathrm{d}x_1 \ldots \mathrm{d}x_\ell. \tag{4}$$

We say that $(\boldsymbol{w}_n)_{n\to\infty}$ *converges to* f and that f is the *limit* of $(\boldsymbol{w}_n)_{n\to\infty}$, if for every word $\boldsymbol{u}$ we have

$$\lim_{n\to\infty} t(\boldsymbol{u}, \boldsymbol{w}_n) = t(\boldsymbol{u}, f).$$

In particular, $(\boldsymbol{w}_n)_{n\to\infty}$ is convergent in this case. Furthermore, let $\mathcal{W}$ be the set of all Lebesgue measurable functions of the form $f : [0,1] \to [0,1]$ in which, moreover, functions are identified when they are equal almost everywhere. We show that each convergent word sequence converges to a unique $f \in \mathcal{W}$ and that, conversely, for each $f \in \mathcal{W}$ there is a word sequence which converges to f.

Theorem 3 (Limits of convergent word sequences).

- *For each convergent word sequence $(\boldsymbol{w}_n)_{n\to\infty}$ there is an $f \in \mathcal{W}$ such that $(\boldsymbol{w}_n)_{n\to\infty}$ converges to f. Moreover, if $(\boldsymbol{w}_n)_{n\to\infty}$ converges to g then f and g are equal almost everywhere.*
- *Conversely, for every $f \in \mathcal{W}$ there is a word sequence $(\boldsymbol{w}_n)_{n\to\infty}$ which converges to f.*

Theorem 3 can be phrased in topological terms as follows. Given a word $\boldsymbol{u}$, one can think of $t(\boldsymbol{u}, \cdot)$ as a function from $\mathcal{W}$ to $[0,1]$. Then, endow $\mathcal{W}$ with the initial topology with respect to the family of maps $t(\boldsymbol{u}, \cdot)$, with $\boldsymbol{u} \in \{0,1\}^\ell$ and $\ell \in \mathbb{N}$, that is, the smallest topology that makes all these maps continuous.

We show that this initial topology is metrizable and, moreover, compact (thereby proving Theorem 3). Specifically, given $h : [0,1] \to [-1,1]$ define the *interval-norm*

$$\|h\|_\square = \sup_{I \subseteq [0,1]} \left| \int_I h(x)\,\mathrm{d}x \right|,$$

where the supremum is taken over all intervals $I \subseteq [0,1]$. The *interval-metric* is then defined by $d_\square(f,g) = \|f-g\|_\square$ for every $f, g : [0,1] \to [0,1]$, and we write

$$f_n \overset{\square}{\to} f \qquad \text{if} \qquad \lim_{n\to\infty} d_\square(f_n, f) = 0.$$

The following result states that the interval-norm controls subsequence counts, in particular. As a byproduct of the lemma, we obtain the first part of Theorem 1. concerning counting subsequences in uniform words.

Lemma 1. *For $f, g \in \mathcal{W}$ and $\boldsymbol{u} \in \{0,1\}^\ell$ we have*

$$\big|t(\boldsymbol{u}, f) - t(\boldsymbol{u}, g)\big| \le \ell^2 \cdot d_\square(f,g).$$

Moreover, if $\boldsymbol{w} \in \{0,1\}^n$ is ε-uniform and $n = n(\varepsilon, \ell)$ is sufficiently large, then for some $d \in [0,1]$ and all $\boldsymbol{u} \in \{0,1\}^\ell$ we have $\binom{\boldsymbol{w}}{\boldsymbol{u}} = d^{\|\boldsymbol{u}\|_1}(1-d)^{\ell - \|\boldsymbol{u}\|_1}\binom{n}{\ell} \pm 5\varepsilon n^\ell$.

It follows immediately that if a sequence $(f_n)_{n\to\infty}$ in $\mathcal{W}$ is Cauchy with respect to $d_\square$, then it is also t-convergent. A technical novelty of the proof strategy we follow is that it yields the converse without relying on compactness arguments.

Proposition 1. *If $(f_n)_{n\to\infty}$ is a sequence in $\mathcal{W}$ which is t-convergent, then it is a Cauchy sequence with respect to $d_\square$. Moreover, if $f_n \overset{t}{\to} f$ for some $f \in \mathcal{W}$, then $f_n \overset{\square}{\to} f$.*

To prove the result we use that for any polynomial $P(x) \in \mathbb{R}[x]$ we can write $\int_0^1 (f_n(x) - f_m(x))P(x)\,\mathrm{d}x$ as a linear combination of subsequence densities. By approximating $\mathbf{1}_{[a,b]}(x)$ by a polynomial $P_{a,b}(x) \in \mathbb{R}[x]$, with error term uniform in $0 \le a < b \le 1$, we may show that $\int_0^1 (f_n(x) - f_m(x))\mathbf{1}_{[a,b]}(x)\,\mathrm{d}x$ can be approximated by $\int_0^1 (f_n(x) - f_m(x))P_{a,b}(x)\,\mathrm{d}x$, thence by a linear combination of subsequence densities, implying our claim. In order to prove this approximation result, we rely on Bernstein polynomials whose first use was precisely to give a constructive proof for the Stone-Weierstrass approximation theorem.

The compactness of the metric space $(\mathcal{W}, d_\square)$ can be easily established via the Banach–Alaoglu theorem in $L^\infty([0,1])$. One can also constructively establish the compactness of $(\mathcal{W}, d_\square)$ by using the regularity lemma for words [5]. Instead, we follow a different strategy. We introduce a probabilistic point of view for the convergence in $d_\square$ that is based on a new model of random words that naturally arises from this theory, which is interesting on its own.

Theorem 4. *The metric space* $(\mathcal{W}, d_\square)$ *is compact.*

The last theorem thus establishes the existence of the limit object claimed in the first part of Theorem 3.

Our overall approach is in line with what has been done for graphons [25] and permutons [21]. However, there are important technical differences, specially concerning the (in our case, more direct) proofs of the equivalence between distinct notions of convergence. In contrast with other technically more involved limit theories, the simplicity of the underlying combinatorial objects we consider (words) yields concise arguments, elegant proofs, simple limit objects, and is based on far fewer concepts. Yet despite the technically comparatively simpler theory, as illustrated in the remaining sections, many interesting aspects common to other structures and some specific to words appear in our investigation.

2.3 Testing Hereditary Word Properties

The concept of self-testing/correcting programs was introduced by Blum et al. [6, 7] and greatly expanded by the concept of graph property testing proposed by Goldreich, Goldwasser and Ron [14] (for an in depth coverage of property testing see the book by Goldreich [15]). An insightful connection between testable graph properties and regularity was established by Alon and Shapira [3] and further refined in [2,4]. It was then observed that similar and related results can be obtained via limit theories (for the case of testing graph properties, the reader is referred to [24], and for the case of (weakly) testing permutation properties, to [20]). Thus, it is not surprising that analogue results can be established for word properties. On the other hand, it is noteworthy that such consequences can be obtained very concisely.

We next state our main result concerning testing word properties. Formally, for $\boldsymbol{u}, \boldsymbol{w} \in \{0,1\}^n$ let $d_1(\boldsymbol{w}, \boldsymbol{u}) = \frac{1}{n}\sum_{i\in[n]} |w_i - u_i|$. A *word property* is simply a collection of words. A word property $\mathcal{P}$ is said to be *testable* if there is another word property $\mathcal{P}'$ (called *test property for* $\mathcal{P}$) satisfying the following conditions:

(Completeness) For every $\boldsymbol{w} \in \mathcal{P}$ of length n and every $\ell \in [n]$, $\mathbb{P}(\mathrm{sub}(\ell, \boldsymbol{w}) \in \mathcal{P}') \geq \frac{2}{3}$.
(Soundness) For every $\varepsilon > 0$ there is an $\ell(\varepsilon) \geq 1$ such that if $\boldsymbol{w} \in \{0,1\}^n$ with $d_1(\boldsymbol{w}, \mathcal{P}) = \min_{\boldsymbol{u}\in\mathcal{P}\cap\{0,1\}^n} d_1(\boldsymbol{w}, \boldsymbol{u}) \geq \varepsilon$, then $\mathbb{P}(\mathrm{sub}(\ell, \boldsymbol{w}) \in \mathcal{P}') \leq \frac{1}{3}$ for all $\ell(\varepsilon) \leq \ell \leq n$.

Variants of the notion of testability can be considered. However, the one stated is sort of the most restrictive. On the other hand, the notion can be strengthened by replacing the 2/3 in the completeness part by $1 - \varepsilon$ and 1/3 in the soundness part by ε. The notion can be weakened letting the test property $\mathcal{P}'$ depend on ε. These variants do not change the concept of testability. A word property $\mathcal{P}$ is called *hereditary* if for each $\boldsymbol{w} \in \mathcal{P}$, every subsequence $\boldsymbol{u}$ of $\boldsymbol{w}$ also belongs to $\mathcal{P}$.

Theorem 5. *Every hereditary word property is testable.*

Since our notion of testability is very restrictive (it consists in sampling uniformly a constant number of characters from the word being tested) it straightforwardly yields efficient (polynomial time) testing procedures.

Examples of hereditary properties are: (1) the collection $\mathcal{P}_{\mathcal{F}}$ of words that do not contain as subsequence any word in $\mathcal{F}$ where $\mathcal{F}$ is a family of words ($\mathcal{F}$ might even be infinite), and (2) for given $\mathcal{P}_1, ..., \mathcal{P}_k$ hereditary word properties, the collection $\mathcal{P}_{\text{col}}$ of words that can be k-colored (i.e., each of its letters assigned a color in $[k]$) so that for all $c \in [k]$ the induced c colored sub-word is in $\mathcal{P}_c$.

2.4 Finite Forcibility

Finite forcibility was introduced by Lovász and Sós [26] (see also Lovász and Szegedy [23]) while studying a generalization of quasi-random graphs. We say that $f \in \mathcal{W}$ is *finitely forcible* if there is a finite list of words $\boldsymbol{u}_1, \ldots \boldsymbol{u}_m$ such that any function $h : [0,1] \to [0,1]$ which satisfies $t(\boldsymbol{u}_i, h) = t(\boldsymbol{u}_i, f)$ for all $i \in [m]$ must agree with f almost everywhere. A direct consequence of Theorem 1 concerning quasi-random words is that the constant functions are finitely forcible (by words of length three). We can generalize this result as follows:

Theorem 6. *Piecewise polynomial functions are finitely forcible. Specifically, if there is an interval partition $\{I_1, ..., I_k\}$ of $[0,1]$, polynomials $P_1(x), ..., P_k(x)$ of degrees $d_1, ..., d_k$, respectively, and $f \in \mathcal{W}$ is such that $f(x) = P_i(x)$ for all $i \in [k]$ and $x \in I_i$, then there is a list of words $\boldsymbol{u}_1, \ldots, \boldsymbol{u}_m$, with $m \leq 2^{1+2k+2\sum_i d_i} + 2^{\binom{k}{2}(1+\max_i d_i)}$ such that any function $h : [0,1] \to [0,1]$ which satisfies $t(\boldsymbol{u}_i, h) = t(\boldsymbol{u}_i, f)$ for all $i \in [m]$ must agree with f almost everywhere.*

2.5 Extensions

We have studied quasi-randomness for binary words and limits of convergent binary word sequences. However, our results (except for the ones concerning testing word properties) can be easily extended to any alphabet of finite size. Indeed, for a word $\boldsymbol{w} \in \Sigma^n$ and an interval $I \subseteq [n]$ let $N_a(\boldsymbol{w}, I)$ denote the number of occurrences of $a \in \Sigma$ in $\text{sub}(I, \boldsymbol{w})$ and let $N_a(\boldsymbol{w}) = N_a(\boldsymbol{w}, [n])$. As for the binary alphabet case, denote by $\binom{\boldsymbol{w}}{\boldsymbol{u}}$ the number of subsequences of $\boldsymbol{w}$ which coincide with $\boldsymbol{u}$.

A sequence $(\boldsymbol{w}_n)_{n\to\infty}$ of words $\boldsymbol{w}_n \in \Sigma^n$ is called $o(1)$*-uniform* if for each $a \in \Sigma$ there is a density d_a such that $N_a(\boldsymbol{w}_n, I) = d_a|I| + o(1)n$ holds for each interval $I \subseteq [n]$. By fixing any $a \in \Sigma$ and replacing every other letter in $\boldsymbol{w}_n$, say, by b, we obtain a sequence of words over the alphabet $\{a, b\}$ and from Theorem 1 we deduce that a is uniformly distributed. This observation coupled with a large alphabet analogue of Lemma 1 yields the following result.

Corollary 1. *Let $(\boldsymbol{w}_n)_{n\to\infty}$ be a sequence of words $\boldsymbol{w}_n \in \Sigma^n$ over the finite size alphabet Σ. If $(\boldsymbol{w}_n)_{n\to\infty}$ is $o(1)$-uniform, then there are $d_1, \ldots, d_{|\Sigma|}$ such that* *for every $\ell \in \mathbb{N}$ and every word $\boldsymbol{u} \in \Sigma^\ell$ we have $\binom{\boldsymbol{w}_n}{\boldsymbol{u}} = \prod_{i\in\Sigma} d_i^{N_i(\boldsymbol{u})}\binom{n}{\ell} + o(n^\ell)$. Conversely, if for some $d_1, \ldots, d_{|\Sigma|}$ we have $\binom{\boldsymbol{w}_n}{\boldsymbol{u}} = \prod_{i\in\Sigma} d_i^{N_i(\boldsymbol{u})}\binom{n}{3} + o(n^3)$ for all words $\boldsymbol{u} \in \Sigma^3$, then $(\boldsymbol{w}_n)_{n\to\infty}$ is $o(1)$-uniform.*

Similarly, one can obtain an analog of Theorem 3 concerning limits of convergent word sequences for larger alphabets. A sequence $(\boldsymbol{w}_n)_{n\to\infty}$ of words $\boldsymbol{w}_n \in \Sigma^n$ over the alphabet $\Sigma = \{a_1, \ldots, a_k\}$ is convergent if for all $\ell \in \mathbb{N}$ and $\boldsymbol{u} \in \Sigma^\ell$ the subsequence density $\left(\binom{\boldsymbol{w}_n}{\boldsymbol{u}}/\binom{n}{\ell}\right)_{n\to\infty}$ converges and we say that $(\boldsymbol{w}_n)_{n\to\infty}$ converges to $\boldsymbol{f} = (f^{a_1}, \ldots, f^{a_k})$ if $\left(\binom{\boldsymbol{w}_n}{\boldsymbol{u}}/\binom{n}{\ell}\right)_{n\to\infty}$ converges to

$$t(\boldsymbol{u}, \boldsymbol{f}) = \ell! \int_{0\le x_1<\cdots<x_\ell\le 1} \prod_{i\in[\ell]} f^{u_i}(x_i)\,\mathrm{d}x_1 \ldots \mathrm{d}x_\ell.$$

Corollary 2 (Limits of convergent k-letter word sequences). *Let $\Sigma = \{a_1, \ldots, a_k\}$.*

- *Each convergent sequence $(\boldsymbol{w}_n)_{n\to\infty}$ of words over Σ converges to some vector $\boldsymbol{f} = (f^{a_1}, \ldots, f^{a_k})$ with $f^{a_i} \in \mathcal{W}$ and $f^{a_1}(x) + \cdots + f^{a_k}(x) = 1$ for almost all $x \in [0, 1]$. Moreover, if $(\boldsymbol{w}_n)_{n\to\infty}$ converges to $\boldsymbol{g} = (g^{a_1}, \ldots, g^{a_k})$ then f^{a_i} and g^{a_i}, $i \in [k]$, are equal almost everywhere.*
- *Conversely, for every vector $\boldsymbol{f} = (f^{a_1}, \ldots, f^{a_k})$ of functions $f^{a_i} \in \mathcal{W}$ which satisfies $f^{a_1}(x) + \cdots + f^{a_k}(x) = 1$ for almost all $x \in [0, 1]$ there is a sequence $(\boldsymbol{w}_n)_{n\to\infty}$ of words over Σ which converges to $\boldsymbol{f}$.*

2.6 Permutons from Words Limits

Given $n \in \mathbb{N}$, we denote by $\mathfrak{S}_n$ the set of permutations of order n and $\mathfrak{S} = \bigcup_{n\ge 1} \mathfrak{S}_n$ the set of all finite permutations. Also, for $\sigma \in \mathfrak{S}_n$ and $\tau \in \mathfrak{S}_k$ we let $\Lambda(\tau, \sigma)$ be the number of copies of τ in σ, that is, the number of k-tuples $1 \le x_1 < \cdots < x_k \le n$ such that for every $i, j \in [k]$

$$\sigma(x_i) \le \sigma(x_j) \quad \text{iff} \quad \tau(i) \le \tau(j).$$

The density of copies of τ in σ, denoted by $t(\tau, \sigma)$, is the probability that σ restricted to a randomly chosen k-tuple of $[n]$ yields a copy of τ. A sequence $(\sigma_n)_{n\to\infty}$ of permutations, with $\sigma_n \in \mathfrak{S}_n$ for each $n \in \mathbb{N}$, is said to be convergent if $\lim_{n\to\infty} t(\tau, \sigma_n)$ exists for every permutation $\tau \in \mathfrak{S}$. Hoppen et al. [21] proved that every convergent sequence of permutations converges to a suitable analytic object called *permuton*, which are probability measures on the Borel σ-algebra on $[0, 1] \times [0, 1]$ with uniform marginals, the collection of which we henceforth denote $\mathcal{Z}$, and also extend the map $t(\tau, \cdot)$ to the whole of $\mathcal{Z}$. Then, Hoppen et al. define a metric $d_\square$ on $\mathcal{Z}$ so that for all $\tau \in \mathfrak{S}$ the maps $t(\tau, \cdot)$ are Lipschitz continuous with respect to $d_\square$. They also show that $(\mathcal{Z}, d_\square)$ is compact and, as a consequence, establish that t-convergence and convergence in $d_\square$ are equivalent. In particular, they prove that for every convergent sequence of permutations

$(\sigma_n)_{n\to\infty}$ there is a permuton $\mu \in \mathcal{Z}$ such that $t(\tau, \sigma_n) \to t(\tau, \mu)$ for all $\tau \in \mathfrak{S}$. We give new proofs of these results by using a more direct approach based on Theorem 3 and the Stone–Weirestrass theorem. In particular, without relying on compactness argument (in contrast to [21]) we establish the following.

Proposition 2. *If $(\mu_n)_{n\to\infty}$ is a sequence in $\mathcal{Z}$ which is t-convergent, then it is a Cauchy sequence with respect to $d_\square$.*

The result's proof argument is an extension, relying in multi-variate Bernstein polynomials, of the one used to establish Proposition 1.

3 Concluding Remarks

A variety of applications use data structures and algorithms on strings/words. In many settings, it is reasonable to assume that strings are generated by a random source of known characteristics. Several basic (generic) probabilistic models have been proposed and are often encountered in the analysis of problems on words, among others, memoryless Markov, mixing and ergodic sources (for detailed discussion see [30]). Our investigations suggest that a new probabilistic model for generating strings under which to analyze the behavior of algorithms on words is, for $f \in \mathcal{W}$, the sequence of distribution on words $(\mathrm{sub}(n, f))_{n\in\mathbb{N}}$. For instance, one may consider variants of classical long-standing open problems on words such as the Longest Common Subsequence (LCS) problem, for which (in the mid 70's) it was shown [12] that two random words uniformly chosen in $\{0,1\}^n$ have a LCS of size proportional to n plus low order terms. The exact value of the proportionality constant remains unknown, although good upper and lower bounds have been established [27]. Generalizing this model, one may consider two random strings $\mathrm{sub}(n, f_1)$ and $\mathrm{sub}(n, f_2)$ and ask for conditions on $f_1, f_2 \in \mathcal{W}$ so that the expected length of the LCS is of size $o(n)$.

Acknowledgments. We would like to thank Svante Janson, Yoshiharu Kohayakawa and Jaime San Martín for valuable discussions and suggestions. We are also greatful to the anonymous reviewers whose comments have greatly improved this manuscript.

References

1. Aigner-Horev, E., Conlon, D., Hàn, H., Person, Y., Schacht, M.: Quasirandomness in hypergraphs. Electron. J. Comb. **25**(3), 3–34 (2018)
2. Alon, N., Fischer, E., Newman, I., Shapira, A.: A combinatorial characterization of the testable graph properties: it's all about regularity. SIAM J. Comput. **39**(1), 143–167 (2009)
3. Alon, N., Shapira, A.: Every monotone graph property is testable. In: Proceedings of the 37th Annual ACM Symposium on Theory of Computing, STOC 2005, pp. 128–137. ACM (2005)
4. Alon, N., Shapira, A.: A characterization of the (natural) graph properties testable with one-sided error. SIAM J. Comput. **37**(6), 1703–1727 (2008)

5. Axenovich, M., Person, Y., Puzynina, S.: A regularity lemma and twins in words. J. Combin. Theory Ser. A **120**(4), 733–743 (2013)
6. Blum, M., Kannan, S.: Designing programs that check their work. J. ACM **42**, 269–291 (1995)
7. Blum, M., Luby, M., Rubinfeld, R.: Self-testing/correcting with applications to numerical problems. J. Comput. Syst. Sci. **47**, 549–595 (1993)
8. Borgs, C., Chayes, J., Lovász, L., Sós, V., Vesztergombi, K.: Convergent sequences of dense graphs I: subgraph frequencies, metric properties and testing. Adv. Math. **219**(6), 1801–1851 (2008)
9. Chung, F.R.K., Graham, R.L., Wilson, R.M.: Quasi-random graphs. Combinatorica **9**(4), 345–362 (1989)
10. Chung, F., Graham, R.: Quasi-random hypergraphs. Proc. Natl. Acad. Sci. **86**(21), 8175–8177 (1989)
11. Chung, F., Graham, R.: Quasi-random subsets of $\mathbb{Z}_n$. J. Comb. Theory Ser. A **61**(1), 64–86 (1992)
12. Chvátal, V., Sankoff, D.: Longest common subsequences of two random sequences. J. Appl. Probab. **12**(2), 306–315 (1975)
13. Cooper, J.N.: Quasirandom permutations. J. Combin. Theory Ser. A **106**(1), 123–143 (2004)
14. Goldreich, O., Goldwasser, S., Ron, D.: Property testing and its connection to learning and approximation. J. ACM **45**(4), 653–750 (1998)
15. Goldreich, O.: Introduction to Property Testing. Cambridge University Press, Cambridge (2017)
16. Gowers, W.: A new proof of Szemerédi's theorem. Geom. Funct. Anal. **11**(3), 465–588 (2001)
17. Gowers, W.: Quasirandomness, counting and regularity for 3-uniform hypergraphs. Comb. Probab. Comput. **15**(1–2), 143–184 (2006)
18. Gowers, W.: Quasirandom groups. Comb. Probab. Comput. **17**(3), 363–387 (2008)
19. Hàn, H., Kiwi, M., Pavez-Signé, M.: Quasi-random words and limits of word sequences. arXiv e-prints, arXiv:2003.03664, March 2020
20. Hoppen, C., Kohayakawa, Y., Moreira, C.G., Sampaio, R.M.: Testing permutation properties through subpermutations. Theoret. Comput. Sci. **412**(29), 3555–3567 (2011)
21. Hoppen, C., Kohayakawa, Y., Moreira, C.G., Ráth, B., Sampaio, R.M.: Limits of permutation sequences. J. Combin. Theory Ser. B **103**(1), 93–113 (2013)
22. Krivelevich, M., Sudakov, B.: Pseudo-random graphs. In: Győri, E., Katona, G.O.H., Lovász, L., Fleiner, T. (eds.) More Sets, Graphs and Numbers: A Salute to Vera Sós and András Hajnal, pp. 199–262. Springer, Heidelberg (2006). https://doi.org/10.1007/978-3-540-32439-3_10
23. Lovász, L., Szegedy, B.: Finitely forcible graphons. J. Comb. Theory Ser. B **101**(5), 269–301 (2011)
24. Lovász, L., Szegedy, B.: Testing properties of graphs and functions. Israel J. Math. **178**(1), 113–156 (2010)
25. Lovász, L., Szegedy, B.: Limits of dense graph sequences. J. Comb. Theory Ser. B **96**(6), 933–957 (2006)
26. Lovász, L., Sós, V.T.: Generalized quasirandom graphs. J. Comb. Theory Ser. B **98**(1), 146–163 (2008)
27. Lueker, G.S.: Improved bounds on the average length of longest common subsequences. J. ACM **56**(3), 17:1–17:38 (2009)
28. Rödl, V.: On universality of graphs with uniformly distributed edges. Discret. Math. **59**(1–2), 125–134 (1986)

29. Szegedy, B.: From graph limits to higher order Fourier analysis. In: Proceedings of the International Congress of Mathematicians, vol. 3, pp. 3197–3218. World Scientific (2018)
30. Szpankowski, W.: Average Case Analysis of Algorithms on Sequences. Series in Discrete Mathematics and Optimization. Wiley-Interscience (2001)
31. Thomason, A.: Pseudo-random graphs. In: Barlotti, A., Biliotti, M., Cossu, A., Korchmaros, G., Tallini, G. (eds.) Annals of Discrete Mathematics (33). North-Holland Mathematics Studies, vol. 144, pp. 307–331. North-Holland (1987)
32. Towsner, H.: σ-algebras for quasirandom hypergraphs. Random Struct. Algorithms **50**(1), 114–139 (2017)

Thresholds in the Lattice of Subspaces of $\mathbb{F}_q^n$

Benjamin Rossman(✉)

Duke University, Durham, NC 27708, USA
benjamin.rossman@duke.edu

Abstract. Let Q be an ideal (downward-closed set) in the lattice of linear subspaces of $\mathbb{F}_q^n$, ordered by inclusion. For $0 \leqslant k \leqslant n$, let $\mu_k(Q)$ denote the fraction of k-dimensional subspaces that belong to Q. We show that these densities satisfy

$$\mu_k(Q) = \frac{1}{1+z} \quad \Longrightarrow \quad \mu_{k+1}(Q) \leqslant \frac{1}{1+qz}.$$

This implies a sharp threshold theorem: if $\mu_k(Q) \leqslant 1-\varepsilon$, then $\mu_\ell(Q) \leqslant \varepsilon$ for $\ell = k + O(\log_q(1/\varepsilon))$.

Keywords: Subspace lattice · Sharp threshold · q-analog · Kruskal-Katona

1 Introduction

Let $\mathcal{L}_q(n)$ be the lattice of linear subspaces of $\mathbb{F}_q^n$, ordered by inclusion. Let Q be a nontrivial ideal in $\mathcal{L}_q(n)$ (that is, a nonempty proper subset of $\mathcal{L}_q(n)$ such that $A \in Q$ implies $B \in Q$ for all $B \subset A$). For $0 \leqslant k \leqslant n$, let $\mu_k(Q)$ denote the fraction of k-dimensional subspaces that belong to Q. Densities $\mu_k(Q)$ are known to be non-increasing: thus,

$$1 = \mu_0(Q) \geqslant \cdots \geqslant \mu_{t-1}(Q) \geqslant 1/2 > \mu_t(Q) \geqslant \cdots \geqslant \mu_n(Q) = 0$$

for a unique t. This paper addresses the question: How quickly must $\mu_k(Q)$ transition from $1 - o(1)$ to $o(1)$?

It follows from known results (described in Sect. 2) that

$$\mu_{\lfloor (t-1)/c \rfloor}(Q) \geqslant 2^{-1/c} \quad \text{and} \quad \mu_{\lceil ct \rceil}(Q) \leqslant 2^{-c}$$

for all $c \geqslant 1$. This is the q-analog of the Bollobás-Thomason Theorem [3], which speaks of ideals in the boolean lattice $\mathcal{P}(n)$ of subsets of $\{1, \ldots, n\}$.

On the one hand, $\mathcal{L}_q(n)$ is the q-analog of $\mathcal{P}(n)$; on the other hand, it is a sub-lattice of $\mathcal{P}(q^n)$. This raises the question: Do k-subspace densities of ideals

Y. Kohayakawa and F. K. Miyazawa (Eds.): LATIN 2020, LNCS 12118, pp. 504–515, 2020.
https://doi.org/10.1007/978-3-030-61792-9_40

in $\mathcal{L}_q(n)$ scale like k-subset densities in $\mathcal{P}(n)$ or like q^k-subset densities in $\mathcal{P}(q^n)$? Quantitatively, the latter suggests we should expect that

$$\mu_{t-1-c}(Q) \geqslant 1 - q^{-c} \quad \text{and} \quad \mu_{t+c}(Q) \leqslant q^{-c}.$$

for all integers $c \geqslant 1$. This is precisely what we show.

Our main result actually concerns shadows in the subspace lattice. Let $\mathcal{L}_q(n,k)$ denote the set of k-dimensional subspaces of $\mathbb{F}_q^n$. For $1 \leqslant k \leqslant n$ and $S \subseteq \mathcal{L}_q(n,k)$, the *shadow* of S is the set $\triangle S \subseteq \mathcal{L}_q(n,k-1)$ defined by $\triangle S := \{B \in L_{n,k-1} : \exists A \in S,\ A \subset B\}$. We show:

Theorem 1. *For all $1 \leqslant k \leqslant n$ and $S \subseteq \mathcal{L}_q(n,k)$, if $\mu_k(S) = (1+z)^{-1}$ where $z \in \mathbb{R}_{\geqslant 0}$, then*

$$\mu_{k-1}(\triangle S) \geqslant \left(1 + \frac{q(q^{k-1}-1)(q^{n-k}-1)}{(q^k-1)(q^{n-k+1}-1)} \cdot z\right)^{-1} \geqslant \left(1 + \frac{z}{q}\right)^{-1}.$$

The first inequality in Theorem 1 is tight in two cases:

- when S is the set of k-dimensional subspaces of a fixed $n-1$-dimensional space ($z = \frac{q^n - q^{n-k}}{q^{n-k}-1}$), as well as
- when S is the set of k-dimensional subspaces not containing a fixed 1-dimensional space ($z = \frac{q^k-1}{q^n-q^k}$).

For values of z between $\frac{q^k-1}{q^n-q^k}$ and $\frac{q^n-q^{n-k}}{q^{n-k}-1}$, Theorem 2 improves the lower bound on $\mu_{k-1}(\triangle S)$ given by a q-analog of the Kruskal-Katona Theorem due to Chowdhury and Patkós [5].

A sharp threshold theorem for $\mathcal{L}_q(n)$ follows immediately from Theorem 1 and the observation that $\triangle(Q \cap \mathcal{L}_q(n,k)) \subseteq Q \cap \mathcal{L}_q(n,k-1)$ for ideals Q.

Theorem 2. *For every ideal Q in $\mathcal{L}_q(n)$ and $1 \leqslant k \leqslant n-1$, if $\mu_k(Q) = (1+z)^{-1}$, then $\mu_{k-1}(Q) \geqslant (1 + (z/q))^{-1}$ and $\mu_{k+1}(Q) \leqslant (1+qz)^{-1}$. As a consequence, if $\mu_k(Q) \leqslant 1 - \varepsilon$, then $\mu_\ell(Q) \leqslant \varepsilon$ for $\ell = k + O(\log_q(1/\varepsilon))$.*

The rest of the paper is organized as follows. In Sect. 2 we describe the previous q-analogs of the Kruskal-Katona and Bollobás-Thomason Theorems and their dual versions. In Sect. 3 we prove Theorem 1 using well-known tools (the Expander Mixing Lemma and bounds on the eigenvalues of Grassmann graphs). In Sect. 4 we discuss the tightness of the results. Finally, in Sect. 5 we give an application of Theorem 2 to a problem in query complexity.

2 q-Analogs of Kruskal-Katona and Bollobás-Thomason

For $x \in \mathbb{R}_{\geqslant 0}$, let $[x]_q := \frac{q^x-1}{q-1}$. The (Gaussian) q-binomial coefficient $\begin{bmatrix} x \\ k \end{bmatrix}_q$ is defined by

$$\begin{bmatrix} x \\ k \end{bmatrix}_q := \prod_{i=0}^{k-1} \frac{[x-i]_q}{[k-i]_q}.$$

Note that $[0]_q = 0$ and $[1]_q = 1$ and $|\mathcal{L}_q(n,k)| = \left[{n \atop k}\right]_q = \left[{n \atop n-k}\right]_q$ for integers $n \geqslant k$.

Chowdhury and Patkós [5] proved a q-analog the Kruskal-Katona Theorem [8,11], specifically a version due Keevash [10]. (See [15] for an alternative proof.)

Theorem 3 (**q-Kruskal-Katona**)**.** *For all $1 \leqslant k \leqslant n$ and $S \subseteq \mathcal{L}_q(n,k)$, if $|S| = \left[{x \atop k}\right]_q$, then $|\Delta S| \geqslant \left[{x \atop k-1}\right]_q$. Moreover, this bound is tight when S is the set of k-dimensional subspaces of a fixed ℓ-dimensional space where $k \leqslant \ell \leqslant n$.*

Note that the parameter n (the dimension of the ambient vector space) plays no role in this bound, in contrast to Theorem 1. It turns out Theorem 3 is slack when $n - 1 < x < n$; this is precisely where Theorem 1 gives an improvement (as we discuss in Sect. 4).

Combining Theorem 3 with the inequality $(\left[{x \atop k-1}\right]_q / \left[{n \atop k-1}\right]_q)^k \geqslant (\left[{x \atop k}\right]_q / \left[{n \atop k}\right]_q)^{k-1}$ for all $k \leqslant x \leqslant n$, we have the following q-analog of the Bollobás-Thomason Theorem [3] for the boolean lattice $\mathcal{P}(n)$.

Theorem 4 (**q-Bollobás-Thomason**)**.** *For every ideal Q in $\mathcal{L}_q(n)$,*

$$\mu_1(Q) \geqslant \mu_2(Q)^{1/2} \geqslant \mu_3(Q)^{1/3} \geqslant \cdots \geqslant \mu_n(Q)^{1/n}.$$

In particular, if $\mu_{t-1}(Q) \geqslant 1/2 > \mu_t(Q)$, then $\mu_{\lfloor (t-1)/c \rfloor}(Q) \geqslant 2^{-1/c}$ and $\mu_{\lceil ct \rceil}(Q) \leqslant 2^{-c}$ for all $c \geqslant 1$.

If we regard Q as a sequence of ideals in $\mathcal{L}_q(n)$, one for each n, then Theorem 4 implies that every nontrivial Q has a *threshold function* $t(n)$, meaning that $\mu_{k(n)}(Q) = 1 - o(1)$ for all $k(n) = o(t(n))$ and $\mu_{\ell(n)}(Q) = o(1)$ for all $\ell(n) = \omega(t(n))$. In the boolean lattice $\mathcal{P}(n)$, nothing more can be said in general, although certain classes of ideals in $\mathcal{P}(n)$, such as monotone graph properties when $n = \binom{m}{2}$, are known to have *sharp thresholds* such that $\mu_{k(n)}(Q) = 1 - o(1)$ and $\mu_{\ell(n)}(Q) = o(1)$ for some $k(n) = t(n) - o(t(n))$ and $\ell(n) = t(n) + o(t(n))$ (see [6]). In the same sense, Theorem 2 shows that every sequence of nontrivial ideals in $\mathcal{L}_q(n)$ has a sharp threshold.

2.1 Dual Versions of Theorems 3 and 4

For a subspace A of $\mathbb{F}_q^n$, the orthogonal complement is defined by

$$A^\perp := \{b \in \mathbb{F}_q^n : \sum_{i=1}^{n} a_i b_i = 0 \text{ for all } a \in A\}.$$

Note that $\dim(A^\perp) = n - \dim(A)$ and $(A^\perp)^\perp = A$ and $B \subseteq A \Longrightarrow A^\perp \subseteq B^\perp$.

For every ideal Q in $\mathcal{L}_q(n)$, there is a dual ideal $Q^* := \{A \in \mathcal{L}_q(n) : A^\perp \notin Q\}$ satisfying $\mu_k(Q^*) = 1 - \mu_{n-k}(Q)$. Applying Theorem 4 to Q^* yields:

Theorem 5 (Dual q-Bollobás-Thomason). *For every ideal Q in $\mathcal{L}_q(n)$,*

$$1 - \mu_{n-1}(Q) \geqslant (1 - \mu_{n-2}(Q))^{1/2} \geqslant (1 - \mu_{n-3}(Q))^{1/3} \geqslant \cdots \geqslant (1 - \mu_0(Q))^{1/n}.$$

In particular, $\mu_{\lfloor c(t-1)+(1-c)n \rfloor}(Q) \geqslant 1 - 2^{-c}$ and $\mu_{\lceil t/c+(1-1/c)n \rceil}(Q) \leqslant 1 - 2^{-1/c}$ for all $c \geqslant 1$. (This improves Theorem 4 when $t \geqslant n/2$.)

Similarly, there is a dual version of Theorem 3. It may be helpful to include the proof, since we will use a similar argument in Sect. 3.

Theorem 6 (Dual q-Kruskal-Katona). *For all $1 \leqslant k \leqslant n$ and $n - k + 1 \leqslant y \leqslant n$ and $S \subseteq \mathcal{L}_q(n, k)$, if $|S| = \left[{n \atop k}\right]_q - \left[{y \atop n-k}\right]_q$, then $|\Delta S| \geqslant \left[{n \atop k-1}\right]_q - \left[{y \atop n-k+1}\right]_q$.*

Proof. We will assume $|\Delta S| < \left[{n \atop k-1}\right]_q - \left[{y \atop n-k+1}\right]_q$ and prove that $|S| < \left[{n \atop k}\right]_q - \left[{y \atop n-k}\right]_q$. Define $T \subseteq \mathcal{L}_q(n, n-k+1)$ by

$$T := \{B^\perp : B \in \mathcal{L}_q(n, k-1) \setminus \Delta S\}.$$

Note that $|T| = \left[{n \atop k-1}\right]_q - |\Delta S| = \left[{y \atop n-k+1}\right]_q$. Therefore, Theorem 3 implies $|\Delta T| > \left[{y \atop n-k}\right]_q$.

For all $A \in \mathcal{L}_q(n, k)$, observe that

$$\begin{aligned} A^\perp \in \Delta T &\Longleftrightarrow \exists B \in \mathcal{L}_q(n, k-1) \setminus \Delta S,\ A^\perp \subset B^\perp \\ &\Longleftrightarrow \exists B \in \mathcal{L}_q(n, k-1) \setminus \Delta S,\ B \subset A \\ &\Longrightarrow A \notin S. \end{aligned}$$

Therefore, $S \subseteq \{A \in \mathcal{L}_q(n, k) : A^\perp \notin \Delta T\}$. We conclude that $|S| = \left[{n \atop k}\right]_q - |\Delta T| < \left[{n \atop k}\right]_q - \left[{y \atop n-k}\right]_q$, as required.

3 Proof of Theorem 1

The proof of Theorem 1 involves bounding the edge-expansion of sets in the Grassmann graph $J_q(n, k)$. We state the required definitions and lemmas below. (See [12] for a much deeper study of expansion of Grassman graphs.)

Definition 7. For a d-regular graph $G = (V, E)$ and $S \subseteq V$, the *edge-expansion* of S is defined by

$$\Phi_G(S) := \frac{|E(S, \overline{S})|}{d|S|}$$

where $E(S, \overline{S})$ is the set of edges between S and $\overline{S} = V \setminus S$.

Lemma 8 (Expander Mixing Lemma [1]). *Let $G = (V, E)$ be a d-regular graph and suppose the second largest eigenvalue (in absolute value) of the adjacency matrix of G is at most λ. Then for all $S \subseteq V$,*

$$\left(1 - \frac{\lambda}{d}\right)\left(1 - \frac{|S|}{|V|}\right) \leqslant \Phi_G(S) \leqslant \left(1 + \frac{\lambda}{d}\right)\left(1 - \frac{|S|}{|V|}\right).$$

Definition 9. For $1 \leqslant k \leqslant n$, the *Grassmann graph* $J_q(n,k)$ is the $q[k]_q[n-k]_q$-regular graph with vertex set $\mathcal{L}_q(n,k)$ and edge set

$$E_{J_q(n,k)} := \{(A_1, A_2) \in \mathcal{L}_q(n,k) \times \mathcal{L}_q(n,k) : \dim(A_1 \cap A_2) = k-1\}.$$

Lemma 10 (Spectrum of $J_q(n,k)$ [4]). *The adjacency matrix of $J_q(n,k)$ has eigenvalue $q^{i+1}[k-i]_q[n-k-i]_q - [i]_q$ with multiplicity $\left[{n \atop i}\right]_q - \left[{n \atop i-1}\right]_q$ for each $0 \leqslant i \leqslant \min(k, n-k)$. In particular, the second largest eigenvalue (in absolute value) equals 1 if $k \in \{1, n-1\}$ and equals $q^2[k-1]_q[n-k-1]_q - 1$ if $2 \leqslant k \leqslant n-2$.*

Lemmas 8 and 10 give the following lower bound on $\Phi_{J_q(n,k)}(S)$.

Lemma 11. *For all $2 \leqslant k \leqslant n-2$ and $S \subseteq \mathcal{L}_q(n,k)$,*

$$\Phi_{J_q(n,k)}(S) \geqslant \frac{[n]_q}{q[k]_q[n-k]_q}(1 - \mu_k(S)).$$

Proof. Lemma 8 implies the lower bound

$$\Phi_{J_q(n,k)}(S) \geqslant \left(1 - \frac{q^2[k-1]_q[n-k-1]_q - 1}{[k]_q[n-k]_q}\right)(1 - \mu_k(S)).$$

By a straightforward calculation,

$$1 - \frac{q^2[k-1]_q[n-k-1]_q - 1}{q[k]_q[n-k]_q} = \frac{q^{n+1} - q^n - q + 1}{q^{n+1} - q^{k+1} - q^{n-k+1} + q} = \frac{[n]_q}{q[k]_q[n-k]_q}.$$

We next show an upper bound on $\Phi_{J_q(n,k)}(S)$ in terms of the ratio $\mu_k(S)/\mu_{k-1}(\triangle S)$.

Lemma 12. *For all $1 \leqslant k \leqslant n$ and $\emptyset \subset S \subseteq \mathcal{L}_q(n,k)$,*

$$\Phi_{J_q(n,k)}(S) \leqslant \frac{[n-k+1]_q}{q[n-k]_q}\left(1 - \frac{\mu_k(S)}{\mu_{k-1}(\triangle S)}\right).$$

Proof. For $B \in \triangle S$, let $S_B := \{A \in S : B \subset A\}$. We have $\sum_{B \in \triangle S} |S_B| = [k]_q|S|$ and, by the Cauchy-Schwarz inequality,

$$\sum_{B \in \triangle S} |S_B|^2 \geqslant \frac{(\sum_{B \in \triangle S} |S_B|)^2}{|\triangle S|} = \frac{([k]_q|S|)^2}{|\triangle S|}.$$

Therefore,

$$\begin{aligned} |E_{J_q(n,k)}(S, \overline{S})| &= \sum_{B \in \triangle S} |S_B \times \overline{S}_B| = \sum_{B \in \triangle S} |S_B|([n-k+1]_q - |S_B|) \\ &\leqslant [k]_q|S|\left([n-k+1]_q - \frac{[k]_q|S|}{|\triangle S|}\right). \end{aligned}$$

We now have

$$\Phi_{J_q(n,k)}(S) = \frac{|E_{J_q(n,k)}(S,\overline{S})|}{q[k]_q[n-k]_q|S|} \leqslant \frac{[n-k+1]_q}{q[n-k]_q} - \frac{[k]_q}{q[n-k]_q} \cdot \frac{|S|}{|\triangle S|}.$$

The lemma now follows from the equality

$$\frac{[k]_q}{q[n-k]_q} \cdot \frac{|S|}{|\triangle S|} = \frac{[k]_q \begin{bmatrix} n \\ k \end{bmatrix}_q}{q[n-k]_q \begin{bmatrix} n \\ k-1 \end{bmatrix}_q} \cdot \frac{\mu_k(S)}{\mu_{k-1}(\triangle S)} = \frac{[n-k+1]_q}{q[n-k]_q} \cdot \frac{\mu_k(S)}{\mu_{k-1}(\triangle S)}.$$

We are ready to prove:

Theorem 1 (restated). *For all $1 \leqslant k \leqslant n$ and $S \subseteq \mathcal{L}_q(n,k)$, if $\mu_k(S) = (1+z)^{-1}$ where $z \in \mathbb{R}_{\geqslant 0}$, then*

$$\mu_{k-1}(\triangle S) \geqslant \left(1 + \frac{q(q^{k-1}-1)(q^{n-k}-1)}{(q^k-1)(q^{n-k+1}-1)} \cdot z\right)^{-1} \geqslant \left(1 + \frac{z}{q}\right)^{-1}.$$

Proof. The second inequality is by a straightforward calculation:

$$\frac{q(q^{k-1}-1)(q^{n-k}-1)}{(q^k-1)(q^{n-k+1}-1)} = \frac{1}{q}\left(1 - \frac{(q-1)(q^{n-k+1}+q^k-q-1)}{(q^k-1)(q^{n-k+1}-1)}\right) \leqslant \frac{1}{q}.$$

For the first inequality, consider the case that $k \in \{1, n\}$. In both cases, we have $\mu_{k-1}(\triangle S) = 1$ for every nonempty $S \subseteq \mathcal{L}_q(n,k)$. Therefore, the inequality holds (moreover, with equality since $[k-1]_q[n-k]_q = 0$).

Next, consider the case that $2 \leqslant k \leqslant n-2$. In this case, Lemmas 11 and 12 imply

$$\frac{[n]_q}{q[k]_q[n-k]_q}(1-\mu_k(S)) \leqslant \Phi_{J_q(n,k)}(S) \leqslant \frac{[n-k+1]_q}{q[n-k]_q}\left(1 - \frac{\mu_k(S)}{\mu_{k-1}(\triangle S)}\right).$$

Therefore,

$$\frac{[n]_q}{[k]_q[n-k+1]_q}(1-\mu_k(S)) \leqslant 1 - \frac{\mu_k(S)}{\mu_{k-1}(\triangle S)}.$$

Substituting $(1+z)^{-1}$ for $\mu_k(S)$, this rearranges to

$$\begin{aligned}\mu_{k-1}(\triangle S) &\geqslant \left(1 + \left(1 - \frac{[n]_q}{[k]_q[n-k+1]_q}\right) z\right)^{-1} \\ &= \left(1 + \frac{q(q^{k-1}-1)(q^{n-k}-1)}{(q^k-1)(q^{n-k+1}-1)} \cdot z\right)^{-1}.\end{aligned}$$

We derive the remaining case $k = n-1$ from the case $k = 2$ via duality. Letting $S \subseteq \mathcal{L}_q(n, n-1)$, we will assume that

$$\mu_{n-2}(S) < \left(1 + \frac{q(q-1)(q^{n-2}-1)}{(q^2-1)(q^{n-1}-1)} \cdot z\right)^{-1}$$

and show that $\mu_{n-1}(S) < (1+z)^{-1}$. Let $T := \{B^\perp : B \in \mathcal{L}_q(n, n-2) \setminus \Delta S\}$ and note that

$$\mu_2(T) = 1 - \mu_{n-2}(\Delta S) > 1 - \left(1 + \frac{q(q-1)(q^{n-2}-1)}{(q^2-1)(q^{n-1}-1)} \cdot z\right)^{-1}$$
$$= \left(1 + \frac{(q^2-1)(q^{n-1}-1)}{z \cdot q(q-1)(q^{n-2}-1)}\right)^{-1}.$$

From the case $k = 2$, we have

$$\mu_1(\Delta T) \geqslant \left(1 + \frac{q(q-1)(q^{n-2}-1)}{(q^2-1)(q^{n-1}-1)} \cdot (\mu_2(T)^{-1} - 1)\right)^{-1} > \left(1 + \frac{1}{z}\right)^{-1}.$$

Since $S \subseteq \{A \in \mathcal{L}_q(n, n-1) : A^\perp \notin \Delta T\}$ (as in the proof of Theorem 6), it follows that

$$\mu_{n-1}(S) \leqslant 1 - \mu_1(\Delta T) < 1 - \left(1 + \frac{1}{z}\right)^{-1} = (1+z)^{-1},$$

as required.

We remark that Theorem 1 is self-dual: for any $1 \leqslant k \leqslant n$ and $S \subseteq \mathcal{L}_q(n, k)$, we get the same inequality between $\mu_k(S)$ and $\mu_{k-1}(\Delta S)$ as between $1 - \mu_{n-k}(\Delta T)$ and $1 - \mu_{n-k+1}(T)$ where $T := \{B^\perp : B \in \mathcal{L}_q(n, k-1) \setminus \Delta S\}$.

4 Tightness of the Result

Fix a flag $V_0 \subset V_1 \subset \cdots \subset V_n = \mathbb{F}_q^n$. (Without loss of generality, we may take $V_k = \{u \in \mathbb{F}_q^n : u_{k+1} = \cdots = u_n = 0\}$.) For $1 \leqslant j \leqslant n$, let $Q_{\hat{j}}$ be the ideal

$$Q_{\hat{j}} := \{A \in \mathcal{L}_q(n) : A \cap (V_j - V_{j-1}) = \emptyset\}.$$

In particular, $Q_{\hat{1}}$ is the set of subspaces of $\mathbb{F}_q^n$ that do not contain V_1, while $Q_{\hat{n}}$ is the set of subspaces contained in V_{n-1}.

Densities $\mu_k(Q_{\hat{j}})$ satisfy the following (in)equalities:

(i) $\mu_{n-j}(Q_{\hat{j}}) > 1/2 > \mu_{n-j+1}(Q_{\hat{j}})$,

(ii) if $2 \leqslant k \leqslant n-1$ and $\mu_k(Q_{\hat{j}}) = (1+z)^{-1}$, then

$$\left(1 + \frac{z}{q^2}\right)^{-1} \geqslant \mu_{k-1}(Q_{\hat{j}}) \geqslant \left(1 + \frac{z}{q}\right)^{-1},$$

(iii) $\mu_k(Q_{\hat{n}}) = \dfrac{\begin{bmatrix} n-1 \\ k \end{bmatrix}_q}{\begin{bmatrix} n \\ k \end{bmatrix}_q} = \dfrac{[n-k]_q}{[n]_q} = \left(1 + \dfrac{q^{n-k}(q^k-1)}{(q^{n-k}-1)}\right)^{-1}$,

(iv) $\mu_k(Q_{\hat{1}}) = 1 - \dfrac{\begin{bmatrix} n-1 \\ k-1 \end{bmatrix}_q}{\begin{bmatrix} n \\ k \end{bmatrix}_q} = 1 - \dfrac{[k]_q}{[n]_q} = \left(1 + \dfrac{(q^k-1)}{q^k(q^{n-k}-1)}\right)^{-1}$.

Inequalities (i) and (ii) show that Theorem 2 is essentially tight, no matter where in $\{1,\ldots,n\}$ the threshold for Q occurs. Equations (iii) and (iv) show that the first inequality of Theorem 1 is tight both:

- when S is the set of k-dimensional subspaces of a fixed $n-1$-dimensional space, as well as
- when S is the set of k-dimensional subspaces not containing a fixed 1-dimensional space.

The first example is also tight for q-Kruskal-Katona (Theorem 3), while the second example is tight for the Dual q-Kruskal-Katona (Theorem 6). Taking the maximum of the bounds given by Theorem 1, 3 and 6, we get:

Corollary 13. *For all* $1 \leqslant k \leqslant n$ *and* $\emptyset \subset S \subset \mathcal{L}_q(n,k)$,

$$|\Delta S| \geqslant \begin{cases} \begin{bmatrix} x \\ k-1 \end{bmatrix}_q & \text{if } |S| = \begin{bmatrix} x \\ k \end{bmatrix}_q,\ k \leqslant x \leqslant n-1, \\ \begin{bmatrix} n \\ k-1 \end{bmatrix}_q \left(1 + \frac{z \cdot (q^{k-1}-1)}{q^{k-1}(q^{n-k+1}-1)}\right)^{-1} & \text{if } |S| = \begin{bmatrix} n \\ k \end{bmatrix}_q \left(1 + \frac{z \cdot (q^{k}-1)}{q^{k}(q^{n-k}-1)}\right)^{-1},\ 1 \leqslant z \leqslant q^n, \\ \begin{bmatrix} n \\ k-1 \end{bmatrix}_q - \begin{bmatrix} y \\ n-k+1 \end{bmatrix}_q & \text{if } |S| = \begin{bmatrix} n \\ k \end{bmatrix}_q - \begin{bmatrix} y \\ n-k \end{bmatrix}_q,\ n-k+1 \leqslant y \leqslant n-1. \end{cases}$$

Corollary 13 is known to be tight when x or y are integers (or $z \in \{1, q^n\}$, coinciding with cases $y = n-1$ and $x = n-1$). In other cases, determining the optimal lower bound for $|\Delta S|$ in terms of $|S|$ remains an open problem. In contrast, note that the original Kruskal-Katona Theorem [8,11] completely solves the shadow minimization problem in the boolean lattice: if S is a family of k-element sets and $|S| = \binom{n_k}{k} + \binom{n_{k-1}}{k-1} + \cdots + \binom{n_j}{j}$ where $n_k > n_{k-1} > \cdots > n_j = j \geqslant 1$, then $|\Delta S| \geqslant \binom{n_k}{k-1} + \binom{n_{k-1}}{k-2} + \cdots + \binom{n_j}{j-1}$ and this bound is tight. Moreover, a family of nested solutions is given by the subsets of $\{1,\ldots,n\}$ in co-lexicographic order. The situation in $\mathcal{L}_q(n)$ appears more complicated, as nested solutions to the shadow minimization problem in $\mathcal{L}_q(n)$ are known not to exist [2,7,13].

5 Application to a Query Problem

In this section, we present an application of Theorem 2 to a problem in query complexity. In this problem, A is a *hidden* nontrivial subspace of $\mathbb{F}_2^n$ and the goal is to learn a nonzero element of A with probability $\geqslant 1/2$ by making m simultaneous (non-adaptive) monotone queries. What is the minimum m for which this is possible? An upper bound of $O(n^2)$ follows from the Valiant-Vazirani isolation technique [14] (see [9]). The following theorem gives a matching lower bound of $\Omega(n^2)$. (We adopt the convention of writing random variables in boldface.)

Theorem 14. *Let* $(\mathbf{Q}_1, \ldots, \mathbf{Q}_m)$ *be a joint distribution over ideals in the subspace lattice of* $\mathbb{F}_2^n$ *and let* f *be a function* $\{0,1\}^m \to \mathbb{F}_2^n \setminus \{\overline{0}\}$*. Suppose that for every nontrivial subspace* A *of* $\mathbb{F}_2^n$*, it holds that*

$$\mathbb{P}[\, f(1_{\{A \in \mathbf{Q}_1\}}, \ldots, 1_{\{A \in \mathbf{Q}_m\}}) \in A\,] \geqslant 1/2$$

where $1_{\{A \in Q_i\}}$ *is the indicator function for the event that* $A \in \mathbf{Q}_i$*. Then* $m = \Omega(n^2)$.

This result answers a question of Kawachi, Watanabe and the author [9], who proved the special case of Theorem 14 where ideals $\mathbf{Q}_i$ are restricted to be of the form $\mathbf{Q}_i = \{A \in \mathcal{L}_2(n) : A \cap \mathbf{U}_i = \emptyset\}$ for an arbitrary joint distribution $(\mathbf{U}_1, \ldots, \mathbf{U}_m)$ of subsets $\mathbf{U}_i \subseteq \mathbb{F}_2^n$. In the remainder of this section, we prove Theorem 14 by combining our threshold theorem for $\mathcal{L}_2(n)$ with a few lemmas from the paper [9].

By Yao's principle [16], it suffices to exhibit a *random* nontrivial subspace $\mathbf{A}$ of $\mathbb{F}_2^n$ such that, for all *fixed* ideals $Q_1, \ldots, Q_m$ and every function $f : \{0,1\}^m \to \mathbb{F}_2^n \setminus \{\overline{0}\}$, if

$$\mathbb{P}[\, f(1_{\{\mathbf{A} \in Q_1\}}, \ldots, 1_{\{\mathbf{A} \in Q_m\}}) \in \mathbf{A}\,] \geqslant 1/2,$$

then $m = \Omega(n^2)$. We define $\mathbf{A}$ as follows: first, choose $\mathbf{k} \in \{1, \ldots, \lfloor n/2 \rfloor\}$ uniformly at random; then let $\mathbf{A}$ be a uniform random $\mathbf{k}$-dimensional subspace of $\mathbb{F}_2^n$ (i.e., a uniform random element of $\mathcal{L}_2(n, \mathbf{k})$).

We next state three lemmas (adapted from [9]) concerning the entropy of random variables that depend on $\mathbf{A}$. A reminder of the definition of the (conditional) entropy function: for discrete random variables $\mathbf{X}$ and $\mathbf{Y}$, let

$$\mathbb{H}[\,\mathbf{X}\,] := \sum_{x \in \mathrm{Supp}(\mathbf{X})} \mathbb{P}[\,\mathbf{X} = x\,] \cdot \log(1/\mathbb{P}[\,\mathbf{X} = x\,]),$$

$$\mathbb{H}[\,\mathbf{X} \mid \mathbf{Y}\,] := \sum_{y \in \mathrm{Supp}(\mathbf{Y})} \mathbb{P}[\,\mathbf{Y} = y\,] \cdot \mathbb{H}[\,\mathbf{X} \mid \mathbf{Y} = y\,],$$

where $\log(\cdot)$ is the base-2 logarithm.

Lemma 15. *For every ideal* Q *in* $\mathcal{L}_2(n)$*, we have*

$$\mathbb{H}[\,1_{\{\mathbf{A} \in Q\}} \mid \mathbf{k}\,] = O(1/n).$$

Proof. If Q is trivial, then this conditional entropy is 0. So we assume Q in nontrivial and let $t \in \{1, \ldots, n\}$ be the unique threshold such that $\mu_{t-1}(Q) \geqslant 1/2 > \mu_t(Q)$. Let $0 \leqslant y \leqslant 1 < z$ be the unique real numbers such that $\mu_{t-1}(Q) =$

$(1+y)^{-1}$ and $\mu_t(Q) = (1+z)^{-1}$. By our threshold theorem for ideals in $\mathcal{L}_2(n)$ (Theorem 2), for all $i \in \{1, \dots, t-1\}$ and $j \in \{1, \dots, n-t\}$,

$$\mu_{t-1-i}(Q) \geqslant (1+y2^{-i})^{-1} \geqslant (1+2^{-i})^{-1} \geqslant 1-2^{-i},$$
$$\mu_{t+j}(Q) \leqslant (1+z2^{j})^{-1} < (1+2^{j})^{-1} \leqslant 2^{-j}.$$

For $k \in \{0, \dots, n\}$, let $\mathbf{S}_k$ is a uniform random k-dimensional subspace of $\mathbb{F}_2^n$. It follows that

$$\begin{aligned}
\mathbb{H}[\,1_{\{\mathbf{S}_k \in Q\}}\,] &= \mu_k(Q)\log(\tfrac{1}{\mu_k(Q)}) + (1-\mu_k(Q))\log(\tfrac{1}{1-\mu_k(Q)}) \\
&\leqslant \begin{cases} 1 & \text{if } k \in \{t-1, t\}, \\ i2^{-i} + (1-2^{-i})\log(\tfrac{1}{1-2^{-i}}) & \text{if } k = t-1-i, \\ j2^{-j} + (1-2^{-j})\log(\tfrac{1}{1-2^{-j}}) & \text{if } k = t+j, \end{cases} \\
&\leqslant \begin{cases} 1 & \text{if } k \in \{t-1, t\}, \\ i2^{1-i} & \text{if } k = t-1-i, \\ j2^{1-j} & \text{if } k = t+j. \end{cases}
\end{aligned}$$

We now obtain the desired bound as follows:

$$\begin{aligned}
\mathbb{H}[\,1_{\{\mathbf{A} \in Q\}} \,|\, \mathbf{k}\,] &= \sum_{k=1}^{\lfloor n/2 \rfloor} \mathbb{P}[\,\mathbf{k} = k\,] \cdot \mathbb{H}[\,1_{\{\mathbf{S}_k \in Q\}}\,] \\
&\leqslant \frac{1}{\lfloor n/2 \rfloor}\Bigg(2 + \sum_{k=1}^{\min(t-2, \lfloor n/2 \rfloor)} \mathbb{H}[\,1_{\{\mathbf{S}_k \in Q\}}\,] + \sum_{k=t+1}^{\lfloor n/2 \rfloor} \mathbb{H}[\,1_{\{\mathbf{S}_k \in Q\}}\,]\Bigg) \\
&\leqslant \frac{2}{n+1}\Bigg(2 + 2\sum_{i=1}^{\infty} i2^{1-i}\Bigg) \\
&= \frac{20}{n+1}.
\end{aligned}$$

Lemma 16. *Let* $\mathbf{v}$ *be a random vector in* $\mathbb{F}_2^n \setminus \{\overline{0}\}$, *not necessarily independent of* $\mathbf{A}$. *Then*

$$\mathbb{P}[\,\mathbf{v} \in \mathbf{A}\,] \leqslant \frac{4}{n}\,\mathbb{H}[\,\mathbf{v}\,] + \frac{1}{2^{n/4}}.$$

Proof. Let

$$U := \big\{x \in \mathbb{F}_2^n \setminus \{\overline{0}\} : \mathbb{P}[\,\mathbf{v} = x\,] \geqslant 2^{-n/4}\big\}.$$

Note that

$$\mathbb{P}[\,\mathbf{v} \in \mathbf{A}\,] \leqslant \mathbb{P}[\,\mathbf{v} \notin U\,] + \mathbb{P}[\,\mathbf{A} \cap U \neq \emptyset\,].$$

(If $\mathbf{v} \in \mathbf{A}$, then either $\mathbf{v} \notin U$ or $\mathbf{v} \in \mathbf{A} \cap U$.) We bound these two terms separately.

First, we have

$$\begin{aligned}\mathbb{P}[\mathbf{v} \notin U] = \sum_{x \in (\mathbb{F}_2^n \setminus \{\bar{0}\}) \setminus U} \mathbb{P}[\mathbf{v} = x] &\leqslant \sum_{x \in (\mathbb{F}_2^n \setminus \{\bar{0}\}) \setminus U} \mathbb{P}[\mathbf{v} = x] \frac{\log(1/\mathbb{P}[\mathbf{v} = x])}{n/4} \\ &\leqslant \sum_{x \in \mathbb{F}_2^n \setminus \{\bar{0}\}} \mathbb{P}[\mathbf{v} = x] \frac{\log(1/\mathbb{P}[\mathbf{v} = x])}{n/4} \\ &= \frac{4}{n} \mathbb{H}[\mathbf{v}].\end{aligned}$$

For the second term, observing that $|U| \leqslant 2^{n/4}$ and $\mathbb{P}[x \in \mathbf{A}] \leqslant \frac{2^{\lfloor n/2 \rfloor}-1}{2^n-1} \leqslant \frac{1}{2^{n/2}}$ for every $x \in \mathbb{F}_2^n \setminus \{\bar{0}\}$, we have

$$\mathbb{P}[\mathbf{A} \cap U \neq \emptyset] \leqslant \sum_{x \in U} \mathbb{P}[x \in \mathbf{A}] \leqslant |U| \cdot \frac{1}{2^{n/2}} \leqslant \frac{1}{2^{n/4}}.$$

This complete the proof.

Lemma 17. *For every function $f : \{0,1\}^m \to \mathbb{F}_2^n \setminus \{\bar{0}\}$ and ideals $Q_1, \dots, Q_m \subseteq \mathcal{L}_2(n)$,*

$$\mathbb{P}[f(1_{\{\mathbf{A} \in Q_1\}}, \dots, 1_{\{\mathbf{A} \in Q_m\}}) \in \mathbf{A}] \leqslant O\left(\frac{m + n \log n}{n^2}\right).$$

Proof. By standard entropy inequalities and Lemma 15,

$$\begin{aligned}\mathbb{H}[f(1_{\{\mathbf{A} \in Q_1\}}, \dots, 1_{\{\mathbf{A} \in Q_m\}})] &\leqslant \mathbb{H}[1_{\{\mathbf{A} \in Q_1\}}, \dots, 1_{\{\mathbf{A} \in Q_m\}}] \\ &\leqslant \mathbb{H}[1_{\{\mathbf{A} \in Q_1\}}, \dots, 1_{\{\mathbf{A} \in Q_m\}}, \mathbf{k}] \\ &= \mathbb{H}[\mathbf{k}] + \mathbb{H}[1_{\{\mathbf{A} \in Q_1\}}, \dots, 1_{\{\mathbf{A} \in Q_m\}} \mid \mathbf{k}] \\ &\leqslant \log(\lfloor n/2 \rfloor) + \sum_{i=1}^{m} \mathbb{H}[1_{\{\mathbf{A} \in Q_i\}} \mid \mathbf{k}] \\ &\leqslant O(\log n) + O(m/n).\end{aligned}$$

Combining the above with Lemma 16, we get the stated bound

$$\begin{aligned}\mathbb{P}[f(1_{\{\mathbf{A} \in Q_1\}}, \dots, 1_{\{\mathbf{A} \in Q_m\}}) \in \mathbf{A}] &\leqslant \frac{4}{n} \mathbb{H}[f(1_{\{\mathbf{A} \in Q_1\}}, \dots, 1_{\{\mathbf{A} \in Q_m\}})] + \frac{1}{2^{n/4}} \\ &\leqslant \frac{4}{n}\Big(O(\log n) + O(m/n)\Big) + \frac{1}{2^{n/4}} \\ &= O\left(\frac{m + n \log n}{n^2}\right).\end{aligned}$$

Theorem 14 follows directly from Lemma 17, as

$$\mathbb{P}[f(1_{\{\mathbf{A} \in Q_1\}}, \dots, 1_{\{\mathbf{A} \in Q_m\}}) \in \mathbf{A}] \geqslant 1/2 \implies m = \Omega(n^2).$$

References

1. Alon, N., Chung, F.R.: Explicit construction of linear sized tolerant networks. Discret. Math. **72**(1–3), 15–19 (1988)
2. Bezrukov, S., Blokhuis, A.: A Kruskal-Katona type theorem for the linear lattice. Eur. J. Comb. **20**(2), 123–130 (1999)
3. Bollobás, B., Thomason, A.G.: Threshold functions. Combinatorica **7**(1), 35–38 (1987)
4. Brouwer, A.E., Haemers, W.H.: Distance-regular graphs. In: Brouwer, A.E., Haemers, W.H. (eds.) Spectra of Graphs, pp. 177–185. Springer, New York (2012). https://doi.org/10.1007/978-1-4614-1939-6_12
5. Chowdhury, A., Patkós, B.: Shadows and intersections in vector spaces. J. Comb. Theory Ser. A **117**(8), 1095–1106 (2010)
6. Friedgut, E., Kalai, G.: Every monotone graph property has a sharp threshold. Proc. Am. Math. Soc. **124**(10), 2993–3002 (1996)
7. Harper, L., Hergert, F.: The isoperimetric problem in finite projective planes. Congressus Numerantium **103**, 225–232 (1994)
8. Katona, G.: A theorem of finite sets. In: Gessel, I., Rota, G.C. (eds.) Classic Papers in Combinatorics, pp. 381–401. Springer, Boston (2009). https://doi.org/10.1007/978-0-8176-4842-8_27
9. Kawachi, A., Rossman, B., Watanabe, O.: The query complexity of witness finding. Theory Comput. Syst. **61**(2), 305–321 (2017)
10. Keevash, P.: Shadows and intersections: stability and new proofs. Adv. Math. **218**(5), 1685–1703 (2008)
11. Kruskal, J.B.: The number of simplices in a complex. In: Bellmann, R.E. (ed.) Mathematical Optimization Techniques, vol. 10, pp. 251–278. University of California Press (1963)
12. Subhash, K., Minzer, D., Safra, M.: Pseudorandom sets in Grassmann graph have near-perfect expansion. In: Proceedings of the 59th Annual IEEE Symposium on Foundations of Computer Science, FOCS, pp. 592–601 (2018)
13. Ure, P.K.: A study of $(0, n, n+1)$-sets and other solutions of the isoperimetric problem in finite projective planes. Ph.D. thesis, California Institute of Technology (1996)
14. Valiant, L.G., Vazirani, V.V.: NP is as easy as detecting unique solutions. In: Proceedings of the 17th Annual ACM Symposium on Theory of Computing, STOC, pp. 458–463 (1985)
15. Wang, J.: Intersecting antichains and shadows in linear lattices. J. Comb. Theory Ser. A **118**(7), 2092–2101 (2011)
16. Yao, A.C.C.: Probabilistic computations: toward a unified measure of complexity. In: Proceedings of the 18th Annual Symposium on Foundations of Computer Science, FOCS, pp. 222–227. IEEE (1977)

Analytic and Enumerative Combinatorics

On Minimal-Perimeter Lattice Animals

Gill Barequet(✉) and Gil Ben-Shachar

Department of Computer Science, The Technion—Israel Institute of Technology, 3200003 Haifa, Israel
{barequet,gilbe}@cs.technion.ac.il

Abstract. A *lattice animal* is a connected set of cells on a lattice. The *perimeter* of a lattice animal A consists of all the cells that do not belong to A, but that have a least one neighboring cell of A. We consider *minimal-perimeter* lattice animals, that is, animals whose periemeter is minimal for all animals of the same area, and provide a set of conditions that are sufficient for a lattice to have the property that inflating all minimal-perimeter animals of a certain size yields (without repetitions) all minimal-perimeter animals of a new, larger size. We demonstrate this result for polyhexes (animals on the two-dimensional hexagonal lattice).

There is still more to be done in the studies of benzenoid isomers and the constant-isomer series in particular. The Dias paradigm is interesting, as well as the accompanying topological characteristics of the pertinent benzenoids. It is stressed that these patterns have not been proved rigorously.

Cyvin S.J., Cyvin B.N., Brunvoll J. (1993) Enumeration of benzenoid chemical isomers with a study of constant-isomer series. In: *Computer Chemistry*, part of *Topics in Current Chemistry* book series, vol. 166. Springer, Berlin, Heidelberg (p. 117).

1 Introduction

An *animal* on a d-dimensional lattice is a connected set of lattice cells, where connectivity is through $(d-1)$-dimensional faces of the cells. Specifically, in two dimensions, connectivity is through lattice edges. Two animals are considered identical if one can be obtained from the other by *translation* only, without rotations or flipping. (Such animals are called "fixed" animals in the literature.)

Lattice animals attracted interest as combinatorial objects [10] and as a model in statistical physics and chemistry [17]. In this paper, we consider lattices in two dimensions, specifically, the hexagonal, triangular, and square lattices, where animals are called polyhexes, polyiamonds, and polyominoes, respectively. We focus on the application of our results to the hexagonal lattice, and explain how to make them applicable also to the triangular lattice.

Let $A^{\mathcal{L}}(n)$ denote the number of lattice animals of size n, that is, animals composed of n cells, on a lattice $\mathcal{L}$. A major research problem in the study of lattices is understanding the nature of $A^{\mathcal{L}}(n)$, either by finding a formula for it as a function of n, or by evaluating it for specific values of n. This problem is

Y. Kohayakawa and F. K. Miyazawa (Eds.): LATIN 2020, LNCS 12118, pp. 519–531, 2020.
https://doi.org/10.1007/978-3-030-61792-9_41

to this date still open for any nontrivial lattice. Redelmeier [15] introduced the first algorithm for counting all polyominoes of a given size, with no polyomino being generated more than once. Later, Mertens [14] showed that Redelmeier's algorithm can be utilized for any lattice. The first algorithm for counting lattice animals without generating all of them was introduced by Jensen [13]. Using his method, the number of animals on the 2-dimensional square, hexagonal, and triangular lattices were computed up to size 56, 46, and 75, respectively.

An important measure of lattice animals is the size of their *perimeter* (sometimes called "site perimeter"). The perimeter of a lattice animal is defined as the set of empty cells adjacent to the animal cells. This definition is motivated by models in statistical physics. In such discrete models, the plane or space is made of small cells (squares or cubes, respectively), and quanta of material or energy "jump" from a cell to a neighboring cell with some probability. Thus, the perimeter of a cluster determines where units of material or energy can move to, and guide the statistical model of the flow.

Asinowski et al. [2,3] provided formulae for polyominoes and polycubes with perimeter size close to the maximum possible. On the other extreme reside animals with the *minimum* possible perimeter size for their area. The study of polyominoes of a minimal perimeter dates back to Wang and Wang [19], who gave an infinite sequence of cells on the square lattice, the first n of which (for any n) form a minimal-perimeter polyomino. Later, Altshuler et al. [1], and independently Sieben [16], studied the closely-related problem of the *maximum* area of a polyomino with p perimeter cells, and provided a closed formula for the minimum perimeter of an n-cell polyomino.

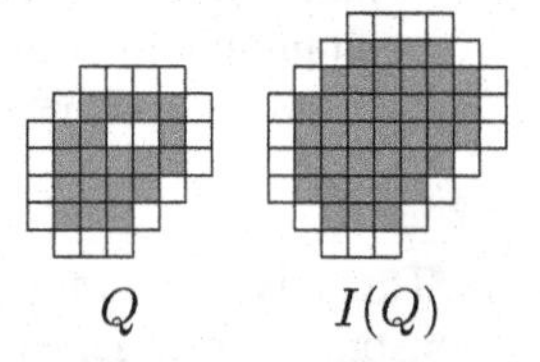

Fig. 1. A polyomino Q and its inflated polyomino $I(Q)$. Polyomino cells are colored gray, perimeter cells are colored white.

Recently, Barequet and Ben-Shachar [4,5] studied properties of minimal-perimeter polyominoes. A key notion in their findings is the *inflation* operation. Simply put, inflating a polyomino is creating the union of a polyomino and the set of its perimeter cells (see Fig. 1). Barequet and Ben-Shachar showed that inflating all the minimal-perimeter polyominoes of some size yields all the minimal-perimeter polyominoes of some larger size in a bijective manner. In this paper, we generalize this result to other lattices and find a sufficient set of conditions for such a bijection to exist.

In the literature, minimal-perimeter animals were studied also on other lattices. For animals on the triangular lattice (polyiamonds), the main result is due to Fülep and Sieben [11], who characterized all the polyiamonds with maximum area for their perimeter, and provided a formula for the minimum perimeter of a polyiamond of size n. However, there has been much more intensive research of minimal-perimeter animals on the hexagonal lattice (polyhexes), mainly in the literature on organic chemistry. There has been a vast amount of work on molecules called *benzenoid hydrocarbons*. It is a known natural fact that molecules made of carbon atoms are structured as shapes on the hexagonal lattice, that is, exactly as polyhexes. Benzenoids hydrocarbons are made of only carbon and hydrogen atoms. In such a molecule, the carbon atoms are arranged as a polyhex and the hydrogen atoms are arranged around the carbons, at the perimeter of the polyhex. The number of hydrogen atoms is exactly the size of the perimeter of the imaginary polyhex. Figure 2 shows a schematic drawing of Naphthalene (molecular formula $C_{10}H_8$), a simple benzenoid hydrocarbon made of 10 carbon atoms and 8 hydrogen atoms. Note that different configurations of atoms exist for the same molecular formula—these are called *isomers*. In the field of organic chemistry, a major goal is to enumerate all the different isomers of a given formula. In a series of papers (culminated in Reference [9]), Dias provided the basic theory of the enumeration of benzenoids hydrocarbons.

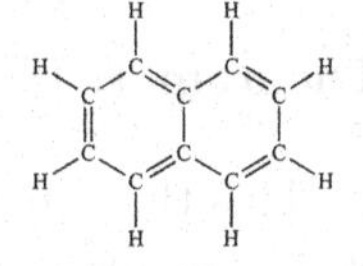

Fig. 2. The Naphthalene molecule ($C_{10}H_8$).

A comprehensive review of the subject is given by Brubvoll and Cyvin [6]. Several other works [7,8,12] also dealt with the properties and enumeration of such animals. Inflating is called by chemists *circumscribing*. For example, circumscribing the Naphthalene molecule yields a molecule known as Circumnaphthalene. In the chemistry literature, it is well known that inflating all isomers of some molecular formulae creates all isomers that correspond to another molecular formula. (The sequences of molecular formulae that have the same number of isomers created by circumscribing are known as *constant-isomer series.*) Although this fact is well known, to the best of our knowledge, no rigorous proof of it was ever given. This is exactly the analogue of a theorem proven by the authors of this paper for polyominoes [4].

In this paper, we generalize the fact that inflation induces a bijection between sets of minimal-perimeter animals from the square lattice to other lattices, specifically, to the hexagonal lattice. By this, we prove the long-observed (but never proven) phenomenon of "constant-isomer chains," that is, that inflating isomers of benzenoid hydrocarbon molecules (in our terminology, inflating minimum-perimeter polyhexes) yields all the isomers of a larger molecule.

2 Preliminaries

Let $\mathcal{L}$ be a lattice, and let Q be an animal on $\mathcal{L}$. The *perimeter* of Q, denoted by $\mathcal{P}(Q)$, is the set of all empty lattice cells that are neighbors of at least one cell of Q. Similarly, the *border* of Q, denoted by $\mathcal{B}(Q)$, is the set of cells of Q that are neighbors of at least one empty cell. The *inflated* version of Q is defined as $I(Q) := Q \cup \mathcal{P}(Q)$. Similarly, the *deflated* version of Q is defined as $D(Q) := Q \backslash \mathcal{B}(Q)$. These operations are demonstrated in Fig. 3.

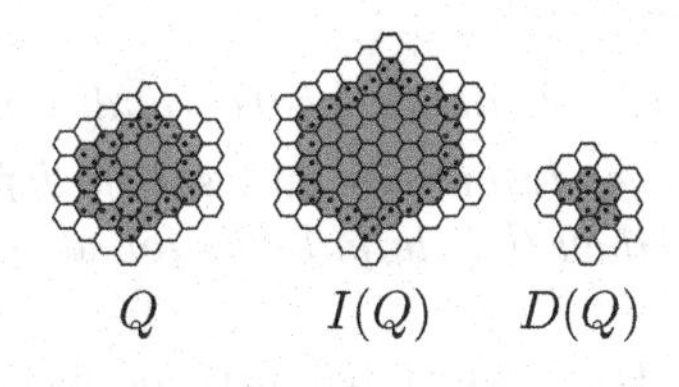

Fig. 3. A polyhex Q, its inflated polyhex $I(Q)$, and its deflated polyhex $D(Q)$.

Denote by $\epsilon^{\mathcal{L}}(n)$ the minimum size (number of cells) of the perimeter of n-cell animals on $\mathcal{L}$, and by $M_n^{\mathcal{L}}$ the set of all minimal-perimeter n-cell animals on $\mathcal{L}$.

Let $\mathcal{S}$ be the two-dimensional square lattice. Animals on $\mathcal{S}$ are usually called *polyominoes*. For this lattice, we know the following.

Theorem 1. [4, Thm. 4] $|M_n^{\mathcal{S}}| = \left|M_{n+\epsilon^{\mathcal{S}}(n)}^{\mathcal{S}}\right|$ *(for $n \geq 3$).*

This theorem is a corollary of another theorem that states that the inflation operation induces bijections between sets of minimal-perimeter polyominoes. This is demonstrated in Fig. 4.

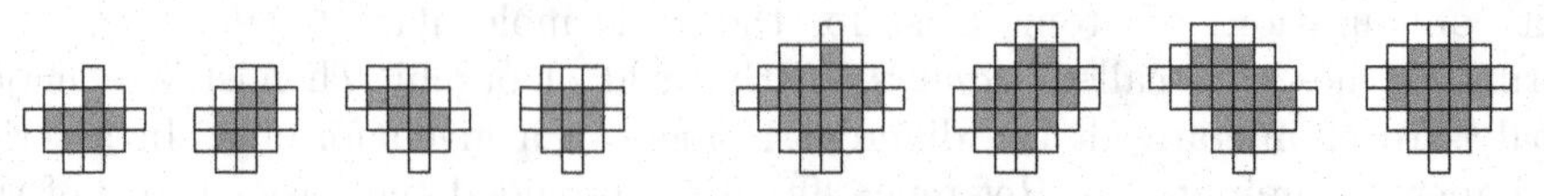

(a) All four minimal-perimeter polyominoes of size 7 (up to rotations) (b) All four minimal-perimeter polyominoes of size 17 (up to rotations)

Fig. 4. A demonstration of Theorem 1.

3 Minimal-Perimeter Animals

Our main result consists of a certain set of conditions, which is sufficient for minimal-perimeter animals to satisfy a claim similar to the one stated in Theorem 1. Throughout this section, we consider animals on some specific lattice $\mathcal{L}$.

3.1 A Bijection

Theorem 2. *Consider the following set of conditions.*

(1) The function $\epsilon^{\mathcal{L}}(n)$ is weakly monotone increasing.
(2) There exists some constant $c \geq 0$, for which, for any minimal-perimeter animal Q, we have that $|\mathcal{P}(Q)| = |\mathcal{B}(Q)| + c$ and $|\mathcal{P}(I(Q))| \leq |\mathcal{P}(Q)| + c$.
(3) If Q is a minimal-perimeter animal, then $D(Q)$ is a valid (connected) animal.

If all the above conditions hold for $\mathcal{L}$, then $|M_n^{\mathcal{L}}| = \left|M_{n+\epsilon^{\mathcal{L}}(n)}^{\mathcal{L}}\right|$. If these conditions are not satisfied for only a finite amount of sizes of animals on $\mathcal{L}$, then the claim holds for all sizes greater than some nominal size n_0. □

Remark. Obviously, no lattice fulfills condition (2) with $c < 0$, and only trivial lattices (e.g., the 1-dimensional lattice) fulfill it with $c = 0$.

The remainder of this section is devoted to proving the theorem above. We begin with proving that inflation preserves perimeter minimality.

Lemma 1. *If Q is a minimal-perimeter animal, then $I(Q)$ is a minimal-perimeter animal as well.*

Proof. Let Q be a minimal-perimeter animal. Assume to the contrary that $I(Q)$ is not a minimal-perimeter animal, thus, there exists an animal Q', such that $|Q'| = |I(Q)|$ and $|\mathcal{P}(Q')| < |\mathcal{P}(I(Q))|$. By Condition (2) of Theorem 2, we know that $|\mathcal{P}(I(Q))| \leq |\mathcal{P}(Q)| + c$, thus, $|\mathcal{P}(Q')| < |\mathcal{P}(Q)| + c$, and since Q' is a minimal-perimeter animal, we also know by the same condition that $|\mathcal{P}(Q')| = |\mathcal{B}(Q')| + c$, and, thus, that $|\mathcal{B}(Q')| < |\mathcal{P}(Q)|$. Consider now the animal $D(Q')$. Recall that $|Q'| = |I(Q)| = |Q| + |\mathcal{P}(Q)|$, hence, the size of $D(Q')$ is at least $|Q| + 1$, and $|\mathcal{P}(D(Q'))| < |\mathcal{P}(Q)| = \epsilon^{\mathcal{L}}(|Q|)$ (since the perimeter of $D(Q')$ is a subset of the border of Q'). This is a contradiction to Condition (1), which states that the sequence $\epsilon^{\mathcal{L}}(n)$ is monotone increasing. Therefore, the animal Q' cannot exist, and $I(Q)$ is a minimal-perimeter animal. □

We now proceed to demonstrate the effect of repeated inflation on the size of minimal-perimeter animals.

Lemma 2. *The minimum size of the perimeter of animals of area $n + k\epsilon^{\mathcal{L}}(n) + ck(k-1)/2$ (for $n > 1$ and any $k \in \mathbb{N}$) is $\epsilon(n) + ck$.*

Proof. We repeatedly inflate a minimal-perimeter animal Q, whose initial size is n. The size of the perimeter of Q is $\epsilon^{\mathcal{L}}(n)$, thus, inflating it creates a new animal of size $n + \epsilon^{\mathcal{L}}(n)$, and the size of the border of $I(Q)$ is $\epsilon^{\mathcal{L}}(n)$, thus, by Condition (2), the size of the perimeter of $I(Q)$ is $\epsilon^{\mathcal{L}}(n) + c$. By repeating this operation, the kth inflation step will increase the size of the animal by $\epsilon^{\mathcal{L}}(n) + (k-1)c$ and will increase the size of the perimeter by c. Summing up these amounts yields the claim. □

Next, we prove that inflation preserves difference, that is, inflating two different minimal-perimeter animals (of equal or different sizes) always produces two different new animals. (This is not true for non-minimal-perimeter animals.)

Lemma 3. *Let Q_1, Q_2 be two different minimal-perimeter animals. Then, regardless of whether or not Q_1, Q_2 have the same area, the animals $I(Q_1)$ and $I(Q_2)$ are different as well.*

Proof. Assume to the contrary that $Q = I(Q_1) = I(Q_2)$, i.e., that $Q = Q_1 \cup \mathcal{P}(Q_1) = Q_2 \cup \mathcal{P}(Q_2)$. In addition, since $Q_1 \neq Q_2$, and since a cell cannot belong simultaneously to both an animal and to its perimeter, this means that $\mathcal{P}(Q_1) \neq \mathcal{P}(Q_2)$. The border of Q is a subset of both $\mathcal{P}(Q_1)$ and $\mathcal{P}(Q_2)$, that is, $\mathcal{B}(Q) \subset \mathcal{P}(Q_1) \cap \mathcal{P}(Q_2)$. Since $\mathcal{P}(Q_1) \neq \mathcal{P}(Q_2)$, we have that either $|\mathcal{B}(Q)| < |\mathcal{P}(Q_1)|$ or $|\mathcal{B}(Q)| < |\mathcal{P}(Q_2)|$; assume without loss of generality the former case. Now, consider the animal $D(Q)$. Its size is $|Q| - |\mathcal{B}(Q)|$. The size of Q is $|Q_1| + |\mathcal{P}(Q_1)|$, thus, $|D(Q)| > |Q_1|$, and since the perimeter of $D(Q)$ is a subset of the border of Q, we have that $|\mathcal{P}(D(Q))| < |\mathcal{P}(Q_1)|$. However, Q_1 is a minimal-perimeter animal, which is a contradiction to Condition (1) of Theorem 2, which states that $\epsilon^{\mathcal{L}}(n)$ is monotone increasing. □

To complete the cycle, we also prove that for any minimal-perimeter animal $Q \in M^{\mathcal{L}}_{n+\epsilon^{\mathcal{L}}(n)}$, there is a minimal-perimeter source in $M^{\mathcal{L}}_n$, i.e., an animal Q' whose inflation yields Q. Specifically, this animal is $D(Q)$.

Lemma 4. *For any $Q \in M^{\mathcal{L}}_{n+\epsilon^{\mathcal{L}}(n)}$, we also have that $I(D(Q)) = Q$.*

Proof. Since $Q \in M^{\mathcal{L}}_{n+\epsilon(n)}$, we have by Lemma 2 that $|\mathcal{P}(Q)| = \epsilon(n) + c$. Combining this with the equality $|\mathcal{P}(Q)| = |\mathcal{B}(Q)| + c$, we obtain that $|\mathcal{B}(Q)| = \epsilon(n)$, thus, $|D(Q)| = n$ and $|\mathcal{P}(D(Q))| \geq \epsilon(n)$. Since the perimeter of $D(Q)$ is a subset of the border of Q, and $|\mathcal{B}(Q)| = \epsilon(n)$, we conclude that the perimeter of $D(Q)$ and the border of Q are the same set of cells, and, thus, $I(D(Q)) = Q$. □

Let us now wrap up the proof of Theorem 2. In Lemma 1 we have shown that for any minimal-perimeter animal $Q \in M_n$, we have that $I(Q) \in M^{\mathcal{L}}_{n+\epsilon^{\mathcal{L}}(n)}$. In addition, Lemma 3 states that the inflation of two different minimal-perimeter animals results in two other different minimal-perimeter animals. Combining the two lemmata, we obtain that $\left|M^{\mathcal{L}}_n\right| \leq \left|M^{\mathcal{L}}_{n+\epsilon^{\mathcal{L}}(n)}\right|$. On the other hand, in Lemma 4 we have shown that if $Q \in M^{\mathcal{L}}_{n+\epsilon^{\mathcal{L}}(n)}$, then $I(D(Q)) = Q$, and, thus, for any animal in $M^{\mathcal{L}}_{n+\epsilon^{\mathcal{L}}(n)}$, there is a unique source in $M^{\mathcal{L}}_n$ (specifically, $D(Q)$), whose inflation yields Q. Hence, $\left|M^{\mathcal{L}}_n\right| \geq \left|M^{\mathcal{L}}_{n+\epsilon^{\mathcal{L}}(n)}\right|$. Combining the two relations, we conclude that $\left|M^{\mathcal{L}}_n\right| = \left|M^{\mathcal{L}}_{n+\epsilon^{\mathcal{L}}(n)}\right|$.

3.2 Inflation Chains

Theorem 2 implies that there exist infinitely-many chains of sets of minimal-perimeter animals, each one obtained from the previous one by inflation, while the cardinalities of all sets in a single chain are identical. Obviously, there are sets of minimal-perimeter animals that are not created by inflating any other set. We call the size of animals in such sets an *inflation-chain root.* Using the definitions and proofs in the previous section, we are able to characterize which sizes are the inflation-chain roots. The result is stated in the following theorem, and its full proof is given in the full version of the paper.

Theorem 3. *Let $\mathcal{L}$ be a lattice for which the three premises of Theorem 2 are satisfied, and, in addition, the following condition holds.*

(4) The inflation operation preserves (for an animal) the property of having a maximum size for a given perimeter.

Then, if n is the minimum animal area for a minimal-perimeter size p, or equivalently, if there exists a perimeter size p, such that $n = \min\left\{n \in \mathbb{N} \mid \epsilon^{\mathcal{L}}(n) = p\right\}$, then n is an inflation-chain root. □

4 Application to Polyhexes

Denote the two-dimensional hexagonal lattice by $\mathcal{H}$. In this section, we show that the conditions of Theorem 2 hold for the lattice $\mathcal{H}$.

4.1 Condition 1: Monotonicity

Condition (1) was proven independently, first by Vainsencher and Bruckstien [18], and later by Fülep and Sieben [11]. We will use the latter, stronger proof which also provides a formula for $\epsilon^{\mathcal{H}}(n)$.

Theorem 4. [11, Thm. 5.12] $\epsilon^{\mathcal{H}}(n) = \lceil \sqrt{12n-3} \rceil + 3$. □

Clearly, the function $\epsilon^{\mathcal{H}}(n)$ is weakly monotone increasing.

4.2 Condition 2: Constant Inflation

To show that Condition (2) holds, we will analyze the different patterns that may appear in the border and perimeter of minimal-perimeter polyhexes. We can classify every border or perimeter cell by one of exactly 24 patterns, distinguished by the number and positions of their adjacent occupied cells. The 24 existing patterns are shown in Fig. 5.

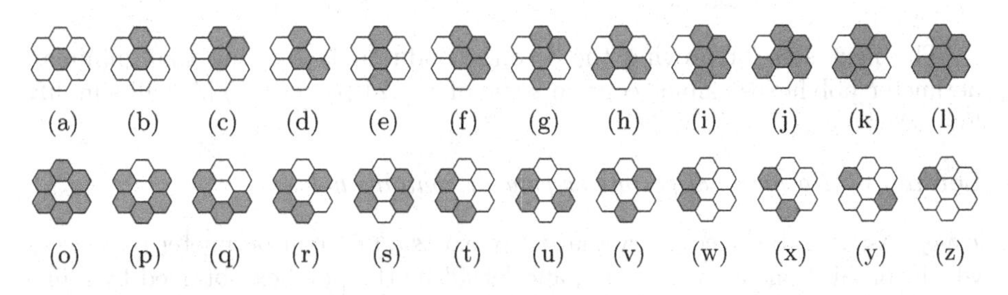

Fig. 5. All possible patterns (up to symmetric cases) of border (first row) and perimeter (second row) cells. The gray cells are polyhex cells, while the white cells are perimeter cells. Each pattern consists of a cell in the middle, and the possible distribution of cells surrounding it.

Asinowski et al. [2] defined the *excess* of a perimeter cell to be the number of adjacent occupied cell minus one. We extend this definition to border cells, and, in a similar manner, we define the excess of a border cell as the number of adjacent empty cells minus one. Following these definitions, we define the *perimeter excess* of a polyhex Q, $e_P(Q)$, to be the sum of excesses over all perimeter cells of Q, and similarly, the border excess of Q, $e_B(Q)$, is defined to be the sum of excesses over all border cells of Q.

The following formula is universal for all polyhexes.

Lemma 5. *For every polyhex Q, we have that*

$$|\mathcal{P}(Q)| + e_P(Q) = |\mathcal{B}(Q)| + e_B(Q) \tag{1}$$

Proof. Consider the (one or more) polygons bounding the polyhex Q. The two sides of the equation are equal to the total length of the polygon(s) in terms of polyhex edges. Indeed, this length can be computed by iterating over either the border or the perimeter cells of Q. In both cases, each cell contributes one edge plus its excess to the total length. The claim follows. □

Our next goal is to express the excess of a polyhex Q as a function of the numbers of cells of Q of each pattern. We denote the number of cells of a specific pattern in Q by $\#\square$, where '$\square$' is one of the 24 patterns listed in Fig. 5. The excess (either border or perimeter excess) of Pattern $\square$ is denoted by $e(\square)$. (For simplicity, we omit the dependency on Q in the notations of $\#\square$ and $e(\square)$. This should be understood from the context.) The border excess can be expressed as $e_{\mathcal{B}}(Q) = \sum_{\square \in \{a,\dots,l\}} e(\square)\#\square$, and, similarly, the perimeter excess can be expressed as $e_{\mathcal{P}}(Q) = \sum_{\square \in \{o,\dots,z\}} e(\square)\#\square$. By plugging these equations into Eq. (1), we obtain that

$$|\mathcal{P}(Q)| + \sum_{\square \in \{o,\dots,z\}} e(\square)\#\square = |\mathcal{B}(Q)| + \sum_{\square \in \{a,\dots,l\}} e(\square)\#\square\ . \qquad (2)$$

The next step of proving the second condition is showing that minimal-perimeter polyhexes cannot contain some of the 24 patterns. This will simplify Eq. (2).

Lemma 6. *No minimal-perimeter polyhex contains holes.*

Proof. Assume to the contrary that there exists a minimal-perimeter polyhex Q which contains one or more holes, and let Q' be the polyhex obtained by filling one of the holes in Q. Clearly, $|Q'| > |Q|$, and by filling the hole we eliminated some perimeter cells and did not create new perimeter cells. Hence, $|\mathcal{P}(Q')| < |\mathcal{P}(Q)|$. This contradicts the fact that $\epsilon^{\mathcal{H}}(n)$ is monotone increasing, as implied by Theorem 4. □

Another important observation is that minimal-perimeter polyhexes tend to be "compact." We formalize this observation in the following lemma.

A *bridge* is a cell whose removal unites two holes or renders the polyhex disconnected (specifically, Patterns (b), (d), (e), (g), (h), (j), and (k)). Similarly, a *perimeter bridge* is an empty cell whose addition to the polyhex creates a hole in the latter (specifically, Patterns (p), (r), (s), (u), (v), (x), and (y)).

Lemma 7. *Minimal-perimeter polyhexes contain neither bridges nor perimeter bridges.* □

The proof is given in the full version of the paper.

As a consequence of Lemma 6, Pattern (o) cannot appear in any minimal-perimeter polyhex. In addition, Lemma 7 tells us that the Border Patterns (b), (d), (e), (g), (h), (j), and (k), as well as the Perimeter Patterns (p), (r), (s), (u), (v), (x), and (y) cannot appear in any minimal-perimeter polyhex. (Note that the central cells in Patterns (b) and (p) are not bridges by themselves, however, the adjacent cells are bridges.) Finally, Pattern (a) appears only in the singleton cell (the unique polyhex of size 1), which can be disregarded. Ignoring all the patterns mentioned above, we conclude that

$$|\mathcal{P}(Q)| + 3\#q + 2\#t + \#w = |\mathcal{B}(Q)| + 3\#c + 2\#f + \#i. \tag{3}$$

Note that Patterns (l) and (z) have excess 0, and, thus, although they may appear in minimal-perimeter polyhexes, they do not appear in the equation.

Consider a polyhex having only the six feasible patterns (those that appear in Eq. (3)). Let us examine the single polygon bounding the polyhex, specifically, let us count the number of vertices and the sum of internal angles which appear in this polygon as a function of the numbers of appearances of the different patterns. We are able to show that the total number of vertices is

$$3\#c + 2\#f + \#i + 3\#q + 2\#t + \#w,$$

and that the sum of internal angles is

$$(3\#c + 2\#f + \#i)120^\circ + (3\#q + 2\#t + \#w)240^\circ. \tag{4}$$

The full details of these calculations are given in the full version of the paper. On the other hand, it is known that the sum of internal angles is equal to

$$(3\#c + 2\#f + \#i + 3\#q + 2\#t + \#w - 2)180^\circ. \tag{5}$$

Equating the terms in Formulae (4) and (5), we obtain that

$$3\#c + 2\#f + \#i = 3\#q + 2\#t + \#w + 6.$$

Plugging this into Eq. (3), we conclude that $|\mathcal{P}(Q)| = |\mathcal{B}(Q)| + 6$, as required.

We also need to show the second part of Condition (2), that is, that if Q is a minimal-perimeter polyhex, then $|\mathcal{P}(I(Q))| \leq |\mathcal{P}(Q)| + 6$. To this aim, note that $\mathcal{B}(I(Q)) \subset \mathcal{P}(Q)$, thus, it is sufficient to show that $|\mathcal{P}(I(Q))| \leq |\mathcal{B}(I(Q))| + 6$. Obviously, Eq. (2) holds for the polyhex $I(Q)$, thus, in order to prove the relation, we only need to show that there are no bridges in $I(Q)$. The proof is given in the full version of the paper. We wrap up this discussion with the following lemma.

Lemma 8. *If Q is a minimal-perimeter polyhex, then $I(Q)$ does not contain any polyhex bridge.* □

4.3 Condition 3: Deflation Resistance

The last condition which we need to show states that deflating a minimal-perimeter polyhex results in another (smaller) valid polyhex. The intuition behind this condition is that a minimal-perimeter polyhex is "compact," having a shape which does not become disconnected by deflation. The next lemma formalizes this notion of compactness. The proof is provided in the full version of the paper.

Lemma 9. *For any minimal-perimeter polyhex Q, the shape D(Q) is a valid polyhex.* □

To conclude, we have shown that all the premises of Theorem 2 are satisfied for the hexagonal lattice, and, thus, inflating a set of all the minimal-perimeter polyhexes of a certain size yields another set of minimal-perimeter polyhexes of another, larger size. This result is demonstrated in Fig. 6.

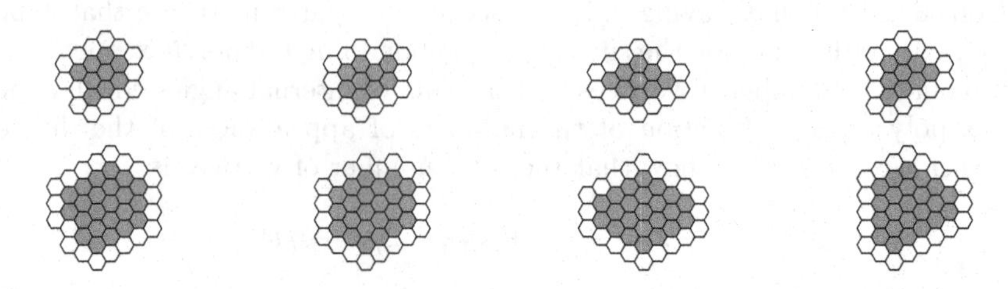

Fig. 6. A demonstration of Theorem 2 for polyhexes. The top row contains all polyhexes (up to rotations and reflections) in $M_9^{\mathcal{H}}$ (minimal-perimeter polyhexes of area 9), while the bottom row contains their inflated versions, all members of $M_{23}^{\mathcal{H}}$.

We also characterized inflation-chain roots of polyhexes. As is mentioned above, the premises of Theorem 3 are satisfied for polyhexes [16,18], and, thus, the inflation-chain roots are those which have the minimum size for a given minimal-perimeter size. An easy consequence of Theorem 4 is that the formula $\left\lfloor \frac{(p-4)^2}{12} + \frac{5}{4} \right\rfloor$ generates all these inflation-chain roots. This result is demonstrated in Fig. 7.

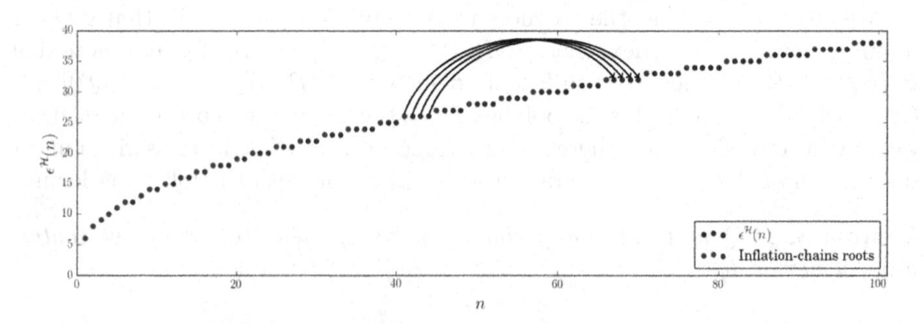

Fig. 7. The relation between the minimum perimeter of polyhexes, $\epsilon^{\mathcal{H}}(n)$, and the inflation-chain roots. The points represent the minimum perimeter of a polyhex of size n, and sizes which are inflation-chain roots are colored in red. The arrows show the mapping between sizes of minimal-perimeter polyhexes (induced by the inflation operation).

5 Polyiamonds

Polyiamonds are sets of edge-connected triangles on the regular triangular lattice, which is made of *two* types of cells. Due to this complication, inflating a minimal-perimeter polyiamond does not necessarily result in a minimal-perimeter polyiamond. Indeed, the second condition of Theorem 2 does not hold for polyiamonds. This fact is not surprising, since inflating minimal-perimeter polyiamonds creates "jagged" polyiamonds, which do not have a minimal perimeter (see Fig. 8(b)).

However, we can fix this situation by modifying the definition of the perimeter of a polyiamond so that the perimeter will include all cells that share a *vertex* (instead of an edge) of the (boundary of the) polyiamond. Theorem 2 holds under the new definition. The reason for this is surprisingly simple: The modified definition merely mimics the inflation of animals on the graph dual to that of the triangular lattice. (Recall that graph duality maps vertices to faces (cells), and vice versa, and edges to edges.) However, the dual of the triangular lattice is the hexagonal lattice, for which we have already shown in Sect. 4 that all the premises of Theorem 2 hold. Thus, applying the modified inflation operator ($I^*(\cdot)$) to the triangular lattice induces a bijection between sets of minimal-perimeter polyiamonds. This operation is demonstrated in Fig. 8.

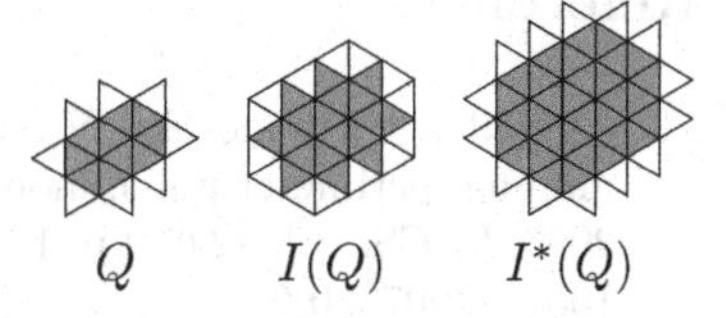

Fig. 8. Inflating polyiamonds. The polyiamond Q is of minimum perimeter, but $I(Q)$ is not. However, the polyiamond $I^*(Q)$, obtained by adding to Q all the cells sharing a *vertex* with Q, is a minimal-perimeter polyiamond.

6 Conclusion

In this paper, we have generalized a result which states that inflation induces a bijection between sets of minimal-perimeter polyominoes, to any lattice satisfying three conditions. We have shown that this generalization holds for the hexagonal lattice, and in some sense (with a modified definition of perimeter) also for the triangular lattice. The most important contribution of this paper is providing a proof for this phenomenon for polyhexes, which was observed in the chemistry literature more than 30 years ago but was never proven.

However, we do not believe that this set of conditions is necessary. Empirically, it seems that by inflating all the minimal-perimeter polycubes (animals on the 3-dimensional cubical lattice) of a given size, we obtain all the minimal-perimeter polycubes of some larger size. However, Condition (2) does not hold for this lattice. Moreover, we believe that as stated, Theorem 2 applies only to

2-dimensional lattices! A simple conclusion from Lemma 2 is that if the premises of Theorem 2 hold for animals on a lattice $\mathcal{L}$, then $\epsilon^{\mathcal{L}}(n) = \Theta(\sqrt{n})$. We find it is reasonable to assume that for a d-dimensional lattice $\mathcal{L}_d$, the relation between the size of a minimal-perimeter animal and its perimeter is roughly equal to the relation between a d-dimensional sphere and its surface area. Hence, we conjecture that $\epsilon^{\mathcal{L}_d}(n) = \Theta(n^{\frac{d-1}{d}})$, and, thus, Theorem 2 does not hold in higher dimensions.

References

1. Altshuler, Y., Yanovsky, V., Vainsencher, D., Wagner, I.A., Bruckstein, A.M.: On minimal perimeter polyminoes. In: Kuba, A., Nyúl, L.G., Palágyi, K. (eds.) DGCI 2006. LNCS, vol. 4245, pp. 17–28. Springer, Heidelberg (2006). https://doi.org/10.1007/11907350_2
2. Asinowski, A., Barequet, G., Zheng, Y.: Enumerating polyominoes with fixed perimeter defect. In: Proceedings of the 9th European Conference on Combinatorics, Graph Theory, and Applications, Vienna, Austria, vol. 61, pp. 61–67. Elsevier (2017)
3. Asinowski, A., Barequet, G., Zheng, Y.: Polycubes with small perimeter defect. In: Proceedings of the 29th Annual ACM-SIAM Symposium on Discrete Algorithms, New Orleans, LA, pp. 93–100, January 2018
4. Barequet, G., Ben-Shachar, G.: Properties of minimal-perimeter polyominoes. In: Wang, L., Zhu, D. (eds.) COCOON 2018. LNCS, vol. 10976, pp. 120–129. Springer, Cham (2018). https://doi.org/10.1007/978-3-319-94776-1_11
5. Barequet, G., Ben-Shachar, G.: Minimal-perimeter polyominoes: chains, roots, and algorithms. In: Pal, S.P., Vijayakumar, A. (eds.) CALDAM 2019. LNCS, vol. 11394, pp. 109–123. Springer, Cham (2019). https://doi.org/10.1007/978-3-030-11509-8_10
6. Brunvoll, J., Cyvin, S.: What do we know about the numbers of benzenoid isomers? Zeitschrift für Naturforschung A **45**(1), 69–80 (1990)
7. Cyvin, S.J., Brunvoll, J.: Series of benzenoid hydrocarbons with a constant number of isomers. Chem. Phys. Lett. **176**(5), 413–416 (1991)
8. Dias, J.: New general formulations for constant-isomer series of polycyclic benzenoids. Polycyclic Aromat. Compd. **30**, 1–8 (2010)
9. Dias, J.: Handbook of Polycyclic Dydrocarbons. Part A: Benzenoid Hydrocarbons. Elsevier, New York (1987)
10. Eden, M.: A two-dimensional growth process. In: Neyman, J. (ed.) Proceedings of the 4th Berkeley Symposium on Mathematical Statistics and Probability, vol. 4, pp. 223–239 (1961)
11. Fülep, G., Sieben, N.: Polyiamonds and polyhexes with minimum site-perimeter and achievement games. Electron. J. Comb. **17**(1), 65 (2010)
12. Harary, F., Harborth, H.: Extremal animals. J. Comb. Inf. Syst. Sci. **1**(1), 1–8 (1976)
13. Jensen, I., Guttmann, A.: Statistics of lattice animals (polyominoes) and polygons. J. Phys. A: Math. General **33**(29), L257 (2000)

14. Mertens, S.: Lattice animals: a fast enumeration algorithm and new perimeter polynomials. J. Stat. Phys. **58**(5), 1095–1108 (1990)
15. Redelmeier, D.H.: Counting polyominoes: yet another attack. Discret. Math. **36**(2), 191–203 (1981)
16. Sieben, N.: Polyominoes with minimum site-perimeter and full set achievement games. Eur. J. Comb. **29**(1), 108–117 (2008)
17. Temperley, H.: Combinatorial problems suggested by the statistical mechanics of domains and of rubber-like molecules. Phys. Rev. **103**(1), 1 (1956)
18. Vainsencher, D., Bruckstein, A.M.: On isoperimetrically optimal polyforms. Theor. Comput. Sci. **406**(1–2), 146–159 (2008)
19. Wang, D.L., Wang, P.: Discrete isoperimetric problems. SIAM J. Appl. Math. **32**(4), 860–870 (1977)

Improved Upper Bounds on the Growth Constants of Polyominoes and Polycubes

Gill Barequet[1(✉)] and Mira Shalah[2]

[1] Department of Computer Science, The Technion—Israel Institute of Technology, 3200003 Haifa, Israel
barequet@cs.technion.ac.il
[2] Department of Computer Science, Stanford University, Stanford, CA, USA
mira@cs.stanford.edu

Abstract. A d-dimensional polycube is a face-connected set of cells on $\mathbb{Z}^d$. Let $A_d(n)$ denote the number of d-dimensional polycubes (distinct up to translations) with n cubes, and λ_d denote their growth constant $\lim_{n\to\infty} \frac{A_d(n+1)}{A_d(n)}$. We revisit and extend the method for the best known upper bound on $A_2(n)$. Our contributions: We (1) prove that $\lambda_2 \le 4.5252$; (2) prove that $\lambda_d \le (2d-2)e + o(1)$ for $d \ge 2$ (already improving significantly the upper bound on λ_3 to 9.8073); and (3) implement an iterative process in 3D, improving further the upper bound on λ_3 to 9.3835.

Keywords: Klarner's constant · Square lattice · Cubical lattice

1 Introduction

Polyominoes are edge-connected sets of squares on $\mathbb{Z}^d$. The size of a polyomino is the number of squares it contains. Figure 1 shows all polyominoes of size up to 4. Likewise, *polycubes* are facet-connected sets of d-D unit cubes, where connectivity is through $(d-1)$-D faces. Two *fixed* polycubes are considered identical if one can be obtained by a translation of the other. In this work, we consider only fixed polycubes.

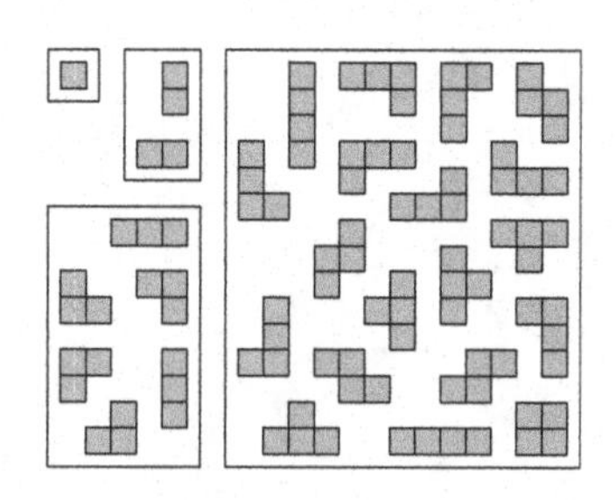

Fig. 1. Polyominoes of sizes $1 \le n \le 4$

The fundamental combinatorial problem concerning polycubes is "How many polycubes with n cubes are there?" This problem originated in parallel in the theory of *percolation* [9,22], the analysis of chemical graphs [14,17], and the graph-theoretic treatment of cell-growth problems [12] more than half a century ago. However, despite much research in those areas, most of what is known relies primarily on heuristics and empirical studies, and very little is known rigorously, even for the low-dimensional lattices.

Work on this paper by the first author has been supported in part by ISF Grant 575/15 and by BSF Grant 2017684.

Y. Kohayakawa and F. K. Miyazawa (Eds.): LATIN 2020, LNCS 12118, pp. 532–545, 2020.
https://doi.org/10.1007/978-3-030-61792-9_42

Let $A_d(n)$ (seq. A001168 [1]) denote the number of polycubes of size n. Since no analytic formula for $A_d(n)$ is known for any dimension $d > 1$, many researchers have focused on efficient algorithms for *counting* polycubes by size, primarily on the square lattice. These methods are based on either explicitly enumerating all polycubes [18,21], or on implicit enumeration [7,11]. The sequence $A_2(n)$ has been determined so far up to $n = 56$ [11]. Enumerating polycubes in higher dimensions is an even more elusive problem. Most notably, Aleksandrowicz and Barequet [2,3] extended polycube counting by efficiently generalizing Redelmeier's algorithm [21] to higher dimensions. The most comprehensive listing of $A_3(n)$ (up to $n = 19$) is by Luther and Mertens [15].

A key fact was discovered by Klarner [12], namely, that the limit $\lambda_2 := \lim_{n\to\infty} \sqrt[n]{A_2(n)}$ exists. This is a straightforward consequence of the fact that the sequence $(\log A_2(n))$ is supper-additive, i.e., $A_2(n)A_2(m) \le A_2(n+m)$ for all $n, m \in \mathbb{N}$. Hence, λ_2 was coined as "Klarner's constant." Later, Madras [16] proved the existence of the asymptotic growth rate, namely, $\lim_{n\to\infty} A_2(n+1)/A_2(n)$, which thus equals λ_2. These results hold, in fact, in any dimension.

A great deal of attention has been given to estimating the values of λ_d, especially for $d = 2, 3$. Their exact values are not known and have remained elusive for many years. Based on interpolation methods, applied to the known values of the sequences $(A_2(n))$ and $(A_3(n))$, it is estimated (without a rigorous proof), that $\lambda_2 \approx 4.06$ [11] and $\lambda_3 \approx 8.34$ [10]. There have been several attempts to bound λ_2 from below, with significant progress over the years [4,5,8,12,19,20], but almost nothing is known for higher dimensions. For $d = 2$, it has been proven that $\lambda_2 \ge 4.0025$ [5]. For $d > 2$, the only known way to set a lower bound on λ_d is by using the fact [12] that $\lambda_d = \lim_{n\to\infty} \sqrt[n]{A_d(n)} = \sup_{n\ge1} \sqrt[n]{dA_d(n)}$. In particular, for d=3, the value $A_3(19)$ [15] yields the lower bound $\lambda_3 \ge \sqrt[19]{3A_3(19)} \approx 6.3795$, which is quite far from the best estimate of λ_3 mentioned above.

On the other hand, only one procedure (Eden [8]) is known for bounding λ_d from above. This procedure (explained in detail in the next section) shows that $\lambda_2 \le 6.7500$, $\lambda_3 \le 12.2071$ and that $\lambda_d \le (2d-1)e$. It was shown [6] that $\lambda_d \sim 2ed - o(d)$ as d tends to infinity, and conjectured (based on an unproven assumption) that λ_d is asymptotically equal to $(2d-3)e + O(1/d)$.

As we detail Sect. 2, Klarner and Rivest [13] enhanced Eden's method, proving that $\lambda_2 \le 4.6496$. We extend this enhancement to higher dimensions and show that it results in the two-variable rational generating function $g^{(d)}(x, y) = \sum_{n,m=0}^{\infty} l_d(n, m)x^n y^m = \sum_{n=0}^{\infty} xy^n \left((1 + x)^{2(d-1)} + x^2\right)^n = \frac{x}{1-y\left((1+x)^{2(d-1)}+x^2\right)}$, and its diagonal function $\sum_{n=0}^{\infty} l_d(n, n)x^n y^n$ generates a sequence which dominates $(A_d(n))$. It was shown [13] that $l_2(n, n) \le 4.8285^n$. Similarly, we prove that $l_3(n, n) \le 9.8073^n$, giving the first nontrivial upper bound on λ_3. We also prove that $l_d(n, n) \le ((2d-2)e + 1/(2d-2))^n$, implying that $\lambda_d \le (2d-2)e + 1/(2d-2)$. This is the first generalization of this method to higher dimensions.

We also revisit the approach used by Klarner and Rivest [13] to further improve the upper bound on λ_2 to 4.6495. We are not aware of any published attempt to reproduce their result. With the computing resources available to us, we improve the upper bound on λ_2 to 4.5252. We also extend the approach to $d = 3$, and prove that $\lambda_3 \le 9.3835$.

2 Previous Work

For two d-dimensional cubes with centers $c_1 = (x_1, \ldots, x_d)$ and $c_2 = (y_1, \ldots, y_d)$, we say that c_1 is *lexicographically smaller* than c_2 if $x_i < y_i$ for the first index i where they differ. Let P be an n-cell polycube in d dimensions. P can be uniquely encoded with a binary string W_P of length $(2d{-}1)n{-}1$ [6,8], as follows. W_P is initialized by the empty string. Perform a breadth-first search on G, the adjacency graph of cells of P and its neighboring empty cells, starting at cell 1 (the smallest cell of P). In the course of this procedure, every cell $c \in P$ is reached through some incoming edge e since G is connected. Clearly, c is connected to at most $2d{-}1$ neighboring cells. The procedure traverses all these outgoing edges according to a fixed order determined by their orientations relative to e. Then, if such an edge leads to a cell of P which has not been labeled yet, this cell is assigned the next unused number, and we update $W_P := W_P{\cdot}1$ ("$\cdot$" is the concatenation operator). Otherwise, if the cell does not belong to P, or it is already assigned a number, we set $W_P := W_P{\cdot}0$. Since each cell is assigned a number only once, this procedure maps polycubes in a one-to-one manner into binary sequences with $n{-}1$ ones and $(2d{-}2)n$ zeros. Hence, using Stirling's formula, $A_d(n) \leq \binom{(2d-1)(n-1)}{n-1} \leq \left(\frac{(2d-1)^{2d-1}}{(2d-2)^{2d-2}}\right)^n$. For polyominoes, this procedure is equivalent to assigning an element of $\mathsf{E} = \{e_1, \ldots, e_8\}$ (Fig. 2) to each cell of P (in the same order). In three dimensions, one obtains that $\lambda_3 \leq 5^5/4^4 \leq 12.2071$. In general, since $\frac{(2d-1)^{2d-1}}{(2d-2)^{2d-2}} = (2d-1)\left(1 + \frac{1}{2d-2}\right)^{2d-2} < (2d{-}1)e$, we have that $\lambda_d \leq (2d{-}1)e$ [6]. (In the full version of the paper, we show that this bound can easily be improved to $(2d{-}1.5)e$ with a more thorough analysis.)

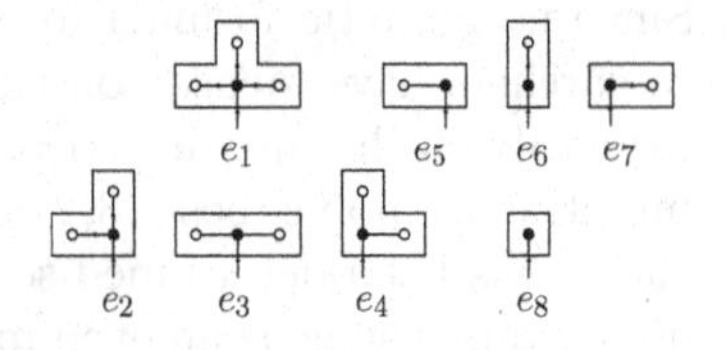

Fig. 2. Eden's twigs [13, Fig. 3]

For Klarner and Rivest's improvement [13], refer to Fig. 3. Around any square u on the square lattice, there are eight L-shaped 4-sets of squares, called the "L-contexts" of u. The *status* of a cell refers to whether or not the cell belongs to the polyomino. Klarner and Rivest designed a set of "*twigs*" L (Fig. 4), which is more compact than E, and showed that every n-cell polyomino P corresponds to a unique n-term sequence of elements of L, while not every such sequence represents a polyomino, implying that $\lambda_2 < |L| = 5$, a substantial improvement over 6.75. The key idea behind the design of L was that one could perform the same search on G, assigning each cell one of the eight L-contexts, s.t. the statuses of *all* cells in the L-contexts would already be encoded by the algorithm. Therefore, while E encodes all $2^3{=}8$ possible status configurations of *three* neighbors for every cell, L encodes only the statuses of *two* neighbors. The sequence L_p, encoding a polyomino P, can be constructed algorithmically as follows. Maintain a queue (initially empty) of *white* (*open*) cells, and a list $\mathcal{D}$ of *black* (*dead*) cells.

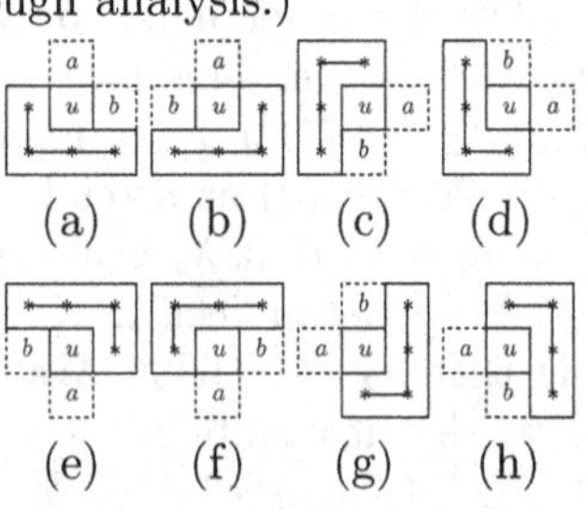

Fig. 3. L-contexts [13, Fig. 6]

Black (resp., white) cells are cells that have (resp., have not) been visited by the algorithm. Start from the lexicographically-smallest cell of P, putting it in the queue. (Assuming, w.l.o.g., that the cells are ordered first by their y-coordinate.) The L-context assigned to this cell is the one shown in Fig. 3(a) since the cells in this neighborhood of the cell do not to belong to P. The addition of twigs to the configuration T constructed so far proceeds as follows until the queue is empty. Dequeue u, the oldest cell in the queue. Let a, b denote the cells connected to u (which are not in its assigned L-context, see Fig. 3), c ($\neq u$) denote the cell connected to both a, b, and ℓ denote the last label assigned to a cell of P (initially $\ell = 0$).

Refer to Fig. 4. The twig L assigned to u is L_1 if $a, b, c \notin P$ (or $a, b, c \in \mathcal{D}$); L_2 if $b, c \notin P$ (or $b, c \in \mathcal{D}$) and $a \in P$; L_3 if $b \notin P$ (or $b \in \mathcal{D}$) and $a, c \in P$; L_4 if $a \notin P$ (or $a \in \mathcal{D}$) and $b \in P$; or L_5 if $a, b \in P$. A new configuration $T * L$ is then constructed:

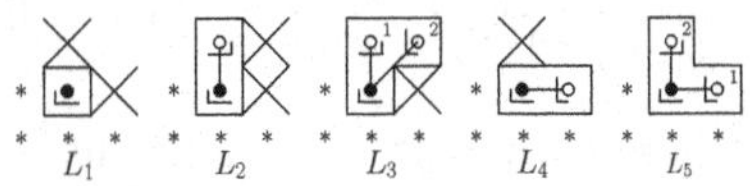

Fig. 4. Twig set L [13, Fig. 7]

1. The root cell of L (black in Fig. 4) is placed over u, s.t. the orientation (L-context) of L and u coincide (possibly with a reflection and/or rotation of L).
2. The white cell, where L was added, turns black (*dead*). This step is legal only if no other cells of L overlap cells of T and no cell of L occupies a forbidden cell.
3. The (*forbidden*) cells of L marked with an X become forbidden in T.
4. The white cells of L are added (in their indicated order) to the queue.

Note that when $a \in P$ or $b \notin P$, we have to encode whether or not $c \in P$ (twigs L_3 and L_2, resp.), so that when a is inserted to the queue, the statuses of all cells in its indicated L-context are encoded. Note also that the order of the white cells in L_3 and L_5 is necessary for ensuring the uniqueness of the construction. For the second white cell in either L_3 or L_5, the statuses of all cells in its indicated L-context are known only after the algorithm visits the first open cell. An example of this process is provided in the full version of the paper.

Having a set of five twigs implies that $\lambda_2 \leq 5$. This bound can be improved by a more delicate analysis, which assigns different weights to different elements of L, as follows. Each twig $L \in \mathsf{L}$ is assigned a weight $w(L) = x^a y^b$, where a (resp., b) denotes the number of cells minus 1 (resp., black cells) in P. (Obviously, for twigs, $b = 1$.) Thus, $w(L_1) = y$, $w(L_2) = xy$, $w(L_3) = x^2 y$, $w(L_4) = xy$, and $w(L_5) = x^2 y$. The weight of the empty sequence is defined as x, and the weight of a sequence $S = (\ell_1, \ldots, \ell_k) \in \mathsf{L}^k$ is defined as $W(S) = x \cdot w(\ell_1) \cdot \ldots \cdot w(\ell_k)$.

Let P be a polyomino of size n, and let $L_p = \{\ell_1, \ldots, \ell_n\} \in \mathsf{L}^n$ denote the sequence encoding P. For each $\ell_i \in \mathsf{L}$, we have that $w(\ell_i) = x^{a_i} y$, such that $a_i \in \{0, 1, 2\}$ equals the number of open cells in ℓ_i. Thus, $w(L_p) = x \cdot x^{\sum_{i=1}^{n} a_i} y^n$. Moreover, $\sum_{i=1}^{n} a_i = n - 1$ because each cell of P (other than the smallest cell) becomes open only once, and is thus accounted for by some a_j in the sum. The smallest cell is accounted for by the term x in $w(L_p)$. Therefore, $w(L_p) = x^n y^n$.

Now, let L^k denote the set of all sequences of $k \geq 0$ elements of L. The sum of weights of all finite sequences of elements of L is

$$\sum_{k=0}^{\infty} \sum_{S \in \mathsf{L}^k} W(S) = \sum_{k=0}^{\infty} x \left(\sum_{\ell \in \mathsf{L}} w(\ell) \right)^k = x \left(1 - \sum_{\ell \in \mathsf{L}} w(\ell) \right)^{-1} . \quad (1)$$

Since $\sum_{\ell \in \mathsf{L}} w(\ell) = y(2x^2 + 2x + 1)$, the generating function given in (1) is

$$\sum_{m,n=0}^{\infty} l(m,n) x^m y^n = \frac{x}{1 - y(2x^2 + 2x + 1)} = \sum_{n=0}^{\infty} x y^n (2x^2 + 2x + 1)^n , \quad (2)$$

where $l(m,n)$ is the coefficient of $x^m y^n$. Due to the injection from polyominoes of size n into sequences of elements of L of weight $x^n y^n$, the coefficient $l(n,n)$ in Eq. (2) is an upper bound on $A_2(n)$, hence, its nth root bounds λ_2 from above. In the next two sections, we generalize this method to higher dimensions.

3 Twigs in Higher Dimensions

***d* = 3.** Let us generalize the twigs idea to 3-space. Refer to Fig. 5. Let $o = (0,0,0)$ be the lexicographically-smallest cell of the polycube. By definition, all cubes that lie in the planes $x_1{=}-1$ and $x_2{=}1$ do not belong to the polycube. We define the "+L-context" of o to be the six cells around o shown in asterisks in Fig. 5. Note that the set of 2-dimensional twigs L (Fig. 4) captures all possible occupancy configurations of the neighbors of o that lie in the $x_1 x_2$ plane. For the remaining neighbors of o (cells $(0,0,-1)$ and $(0,0,1)$), there are four possible encodings of whether or not they belong to the polycube. This yields the set $\mathsf{L}^{(3)}$ of 17 three-dimensional twigs shown in Fig. 6.

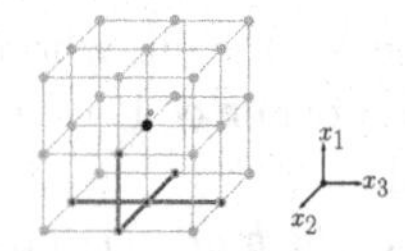

Fig. 5. The +L context (bold black lines) of a cell o on $\mathbb{N}^3$

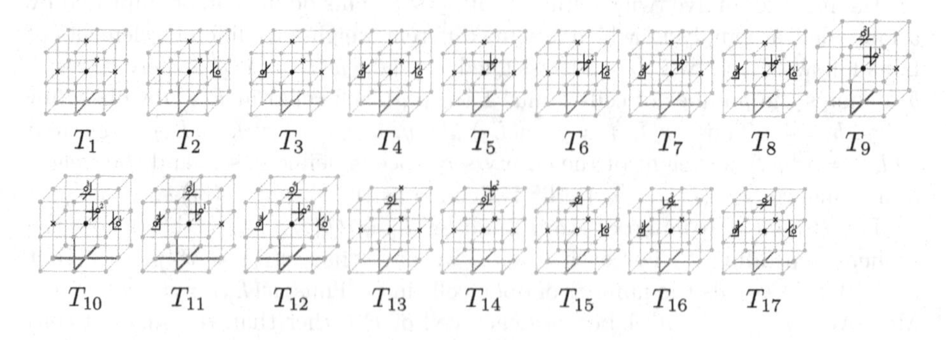

Fig. 6. 3-dimensional twigs

Similarly to the plane, the cells of a twig are either black or white, and the $+\mathsf{L}$ context and linear order of the open cells is indicated. Similarly to $L_1, \ldots, L_5$ (in Fig. 2), the twigs $T_1, \ldots, T_{17}$ (in Fig. 4) are a complete set of building blocks for polycubes since they cover all possible situations (a formal proof is given in the full paper). Every n-cell polycube P corresponds to a unique n-term sequence of elements of $\mathsf{L}^{(3)}$. The sequence corresponding to a polycube can be constructed as in two dimensions. Every twig is assigned a weight in the same manner, and we get that $\sum_{\ell \in \mathsf{L}^{(3)}} w(\ell) = y(1+4x+7x^2+4x^3+x^4) = y\left((x+1)^4 + x^2\right)$. Thus, the generating function given by Eq. (1) is $\frac{x}{1-y(1+4x+7x^2+4x^3+x^4)} = \sum_{n=0}^{\infty} xy^n(1 + 4x + 7x^2 + 4x^3 + x^4)^n$. We provide the analysis of this function in Sect. 4.

$\boldsymbol{d > 3}$. Our construction proceeds inductively for $d > 3$. The base is d=2, where we fix a square in the x_1x_2 plane (Fig. 5) together with its L-context. In general, when we go from $d-1$ ($d \geq 3$) to d dimensions, a cube gains two neighbors in the new dimension x_d. Let $o = (0, 0, \ldots, 0)$ be a d-D cube ($d > 2$). We define the $+_d\mathsf{L}$-context of o recursively. The base is $+_2\mathsf{L} := \mathsf{L}$ and $+_3\mathsf{L} := +\mathsf{L}$, and the recursion is $+_d\mathsf{L} := +_{d-1}\mathsf{L} \cup \{c_1, c_2\}$, where $c_{1,2} = (-1, 0, \ldots, 0, \pm1)$. The geometric interpretation of the $+_d\mathsf{L}$-context of o is an L-shape around o in the x_1x_2 plane, which intersects $d-2$ lines in the $x_1 = -1$ plane at $(-1, 0, \ldots, 0)$.

The set $\mathsf{L}^{(d)}$ (with $\mathsf{L}^{(2)} = \mathsf{L}$) consists of all 2^{2d-2} occupancy options for the neighbors of o (that are not in its $+_d\mathsf{L}$-context): In dimensions $x_3, \ldots, x_d$, the construction covers all $2^{2(d-2)}$ options for the two neighbors of o. In the x_1x_2 plane, the occupancies of the neighbors of o are captured by L (Fig. 4), and the only problematic case is when cell c=$(1, 0, \ldots, 0)$ is white and all other neighbors of o are not (twigs L_2, L_3 in Fig. 4, and T_{13}, T_{14} in Fig. 6). It is, thus, necessary to encode the status of $(1, -1, 0, \ldots, 0)$ since it is contained in the $+_d\mathsf{L}$-context of c. This yields $2^{2(d-2)} \cdot 2^2 + 1 = 2^{2(d-1)} + 1$ twigs, which compares favorably with Eden's generalized construction, having about twice the number of twigs (2^{2d-1}).

To prove that our construction is more efficient, we will show that for any white cell u in every twig in $\mathsf{L}^{(d)}$, there are $4+2(d-2)=2d$ cells around u which can be ignored when visiting u. Those cells will form its $+_d\mathsf{L}$-context. Except the second white cell in the problematic twig, all white cells are neighbors of o. If a new neighbor of o, $(0, 0, \ldots, 0, \pm1)$, is open, the "L shape" in its $+_d\mathsf{L}$-context is formed by c_1 (or c_2), $(-1, 0, \ldots, 0)$, o, and $(1, 0, \ldots, 0)$; the other cells in its $+_d\mathsf{L}$-context are $(0, \pm1, 0, \ldots, 0)$, $\ldots$, $(0, \ldots, 0, \pm1, 0)$. The statuses of these cells are known by construction. For the two other possible white neighbors of o, namely, $\nu_{1,2}$=$(0, \ldots, 0, \pm1, 0, \ldots, 0)$, we have that $+_d\mathsf{L} = +_{d-1}\mathsf{L} \cup (0, \ldots, 0, \pm1)$ since o is exactly where the 'L' and the '+' in the $+_{d-1}\mathsf{L}$-context of $\nu_{1,2}$ intersect. Thus, $+_{d-1}\mathsf{L} \cup (0, \ldots, 0, \pm1)$ is the $+_d\mathsf{L}$-context of o: the statuses of the cells in its $+_{d-1}\mathsf{L}$-context and of $(0, \ldots, 0, \pm1)$ are known by induction and construction. We also address the second white cell p=$(1, -1, 0, \ldots, 0)$ in the problematic twig. It will be visited after the first open cell of the twig, q=$(0, 1, 0, \ldots, 0)$, is visited and assigned a twig. When this happens, the statuses of all its neighbors are known. Then, $(1, 0, \ldots, 0)$, o, $(0, 1, 0, \ldots, 0)$, and $(0, 2, 0, \ldots, 0)$ form the L-shape near p, and together with the remaining $2(d-1)$ neighbors of q, form the $+_d\mathsf{L}$-context of p. The statuses of these cells are also already known by construction.

Finally, we compute the weight function $W^{(d)}(x,y) = \sum_{t \in \mathsf{L}^{(d)}} w(t)$. Since o has $2(d-1)$ neighbors not in its $+_d\mathsf{L}$-context, there are $\binom{2(d-1)}{i}$ twigs in $\mathsf{L}^{(d)}$ with i white cells and one black cell (o), and the weight of each such twig is $x^i y$. Recall the problematic case, resulting in an additional twig with 3 cells (1 black and 2 white), whose weight is, thus, $x^2 y$. Hence, $W^{(d)}(x,y) = \sum_{t \in \mathsf{L}^{(d)}} w(t) = \sum_{i=0}^{2(d-1)} \left[\binom{2(d-1)}{i} x^i y\right] + x^2 y = y((x+1)^{2(d-1)} + x^2)$. Substituting $W^{(d)}(x,y)$ in the generating function of Eq. (1), we obtain $g^{(d)}(x,y) = l_d(n,m)x^m y^n = \sum_{n=0}^{\infty} xy^n \left((1+x)^{2(d-1)} + x^2\right)^n = \frac{x}{1-y\left((1+x)^{2(d-1)}+x^2\right)}$. As in 2D, polycubes of size n are mapped uniquely to sequences of elements of $\mathsf{L}^{(d)}$ having weight $x^n y^n$.

4 Analysis of the Generating Functions

It can be easily observed that $l_d(n,n)$, the coefficient of $x^n y^n$ in $g^{(d)}(x,y)$, is the coefficient of x^{n-1} in $\left((1+x)^{2(d-1)}+x^2\right)^n$. We now show how to compute $l_d(n,n)$. Let $h^{(d)}(x) = \left((1+x)^{2(d-1)}+x^2\right)^n$.

$\boldsymbol{d = 2}$ [13]. In the plane, $h^{(2)}(x) = \left((1+x)^2 + x^2\right)^n = (1 + 2x + 2x^2)^n$. By the Multinomial Theorem, we have $(1 + 2x + 2x^2)^n = \sum_{i_1,i_2} \left[\binom{n}{n-i_1-i_2, i_1, i_2} (2x)^{i_1} (2x^2)^{i_2}\right]$.

In order to compute the coefficient of x^{n-1}, we require that $i_1+2i_2 = n-1$. Thus,

$l_2(n,n) = \sum_{i_2} \left[\binom{n}{i_2+1, n-2i_2-1, i_2} 2^{n-i_2-1}\right] = \frac{2^n}{2} \sum_{i_2} \left[\binom{n}{i_2+1, n-2i_2-1, i_2} \left(\frac{1}{2}\right)^{i_2}\right] = \frac{2^n}{\sqrt{2}} \sum_{i_2} \left[\binom{n}{i_2+1, n-2i_2-1, i_2} \left(\frac{1}{\sqrt{2}}\right)^{i_2} \left(\frac{1}{\sqrt{2}}\right)^{i_2+1}\right] <_* \frac{2^n}{\sqrt{2}} \left(\frac{1}{\sqrt{2}} + \frac{1}{\sqrt{2}} + 1\right)^n = \frac{\left(2(\sqrt{2}+1)\right)^n}{\sqrt{2}}$. (The relation "$<_*$" is because the summation in its left-hand side contains only a subset of the terms whose sum is equal to the exponential term on the right-hand side.) Hence, $\lambda_2 \le 2(\sqrt{2}+1) \approx 4.82843$.

$\boldsymbol{d = 3}$

Theorem 1. $\lambda_3 \le 9.8073$

Proof. Similarly to two dimensions, let $h^{(3)}(x) = \left((1+x)^4 + x^2\right)^n = (1 + 4x + 7x^2 + 4x^3 + x^4)^n = \sum_{i_1,i_2,i_3,i_4} \left[\binom{n}{(n-\sum_{j=1}^4 i_j), i_1, i_2, i_3, i_4} 4^{i_1} 7^{i_2} 4^{i_3} x^{i_1+2i_2+3i_3+4i_4}\right]$. We now require that $i_1+2i_2+3i_3+4i_4 = n-1$. By substituting i_1 in the right-hand side of the equality above, and using the Multinomial Theorem, we obtain $l_3(n,n) = \sum \left[\binom{n}{i_2+2i_3+3i_4+1, n-1-2i_2-3i_3-4i_4, i_2, i_3, i_4} 4^{n-1-2i_2-3i_3-4i_4} 7^{i_2} 4^{i_3}\right] = \frac{4^n}{4} \sum_{i_2,i_3,i_4} \left[\binom{n}{i_2+2i_3+3i_4+1, n-1-2i_2-3i_3-4i_4, i_2, i_3, i_4} \left(\frac{7}{4^2}\right)^{i_2} \left(\frac{4}{4^3}\right)^{i_3} \left(\frac{1}{4^4}\right)^{i_4}\right] < \frac{4^n}{4} \left(\frac{7}{4^2} + \frac{1}{4^2} + \frac{1}{4^4} + 1 + 1\right)^n = \frac{1}{4} \left(\frac{641}{64}\right)^n$. Thus, $\lambda_3 \le \frac{641}{64} \approx 10.016$, already improving significantly on the known upper bound of $\lambda_3 \le 12.2071$ (see Sect. 2).

However, we can do better than that. Let $b > 0$ be some constant, whose value will be specified later, and rewrite the multinomial expression above as $l_3(n,n) =$

$$\frac{4^n}{4}\sum_{i_2,i_3,i_4}\left[\binom{n}{i_2+2i_3+3i_4+1, n-1-2i_2-3i_3-4i_4, i_2, i_3, i_4}\underbrace{\left(\frac{7}{(b\frac{4}{b})^2}\right)^{i_2}\left(\frac{4}{(b\frac{4}{b})^3}\right)^{i_3}\left(\frac{1}{(b\frac{4}{b})^4}\right)^{i_4}}_{c(b)}\right].$$

Let us now re-arrange the three terms in $c(b)$ as follows.

$$\begin{aligned}
c(b) &= \left(\frac{1}{b^2}\right)^{i_2}\left(\frac{7}{(\frac{4}{b})^2}\right)^{i_2}\left(\frac{1}{b^3}\right)^{i_3}\left(\frac{4}{(\frac{4}{b})^3}\right)^{i_3}\left(\frac{1}{b^4}\right)^{i_4}\left(\frac{1}{(\frac{4}{b})^4}\right)^{i_4}\\
&= \left(\frac{1}{b}\right)^{i_2}\left(\frac{1}{b}\right)^{i_2}\left(\frac{7}{(\frac{4}{b})^2}\right)^{i_2}\left(\frac{1}{b^2}\right)^{i_3}\left(\frac{1}{b}\right)^{i_3}\left(\frac{4}{(\frac{4}{b})^3}\right)^{i_3}\left(\frac{1}{b^3}\right)^{i_4}\left(\frac{1}{b}\right)^{i_4}\left(\frac{1}{(\frac{4}{b})^4}\right)^{i_4}\\
&= \left(\frac{1}{b}\right)^{i_2}\left(\frac{7}{\frac{16}{b}}\right)^{i_2}\left(\frac{1}{b}\right)^{2i_3}\left(\frac{4}{\frac{4^3}{b^2}}\right)^{i_3}\left(\frac{1}{b}\right)^{3i_4}\left(\frac{1}{\frac{4^4}{b^3}}\right)^{i_4}\\
&= \left(\frac{1}{b}\right)^{i_2+2i_3+3i_4}\left(\frac{7}{\frac{16}{b}}\right)^{i_2}\left(\frac{4}{\frac{4^3}{b^2}}\right)^{i_3}\left(\frac{1}{\frac{4^4}{b^3}}\right)^{i_4}.
\end{aligned}$$

Thus, $l_3(n,n) = \frac{4^n}{4}\sum_{i_2,i_3,i_4}\left[\binom{n}{i_2+2i_3+3i_4+1, n-1-2i_2-3i_3-4i_4, i_2, i_3, i_4}\left(\frac{1}{b}\right)^{i_2+2i_3+3i_4}\right.$ $\left.\left(\frac{7}{\frac{16}{b}}\right)^{i_2}\left(\frac{4}{\frac{4^3}{b^2}}\right)^{i_3}\left(\frac{1}{\frac{4^4}{b^3}}\right)^{i_4}\right] < 4^n\left(\frac{1}{b}+1+\frac{7b}{16}+\frac{b^2}{4^2}+\frac{b^3}{4^4}\right)^n$,where the last relation is again due to the Multinomial Theorem and due to the partial summation.

Our trick is to choose the value of b (by assigning appropriate weights to the five components) that minimizes the sum of the summands of in the partial summation. Define $f(b) = \frac{1}{b}+1+\frac{7b}{16}+\frac{b^2}{4^2}+\frac{b^3}{4^4}$. Our goal, then, is to minimize $f(b)$. Elementary calculus shows that $f(b)$ assumes its minimum at $b_0 = 1.274306378$ and that $f(b_0) = 2.451823893$. Recall that $l_3(n,n) < 4^n f^n(b)$ for *any* b, in particular, for $b = b_0$. Hence, finally, $l_3(n,n) < 4^n \cdot 2.451823893^n = 9.807295572^n$. □

Higher values of d

Theorem 2. $\lambda_d \leq (2d-2)e + 1/(2d-2)$

Proof. The proof for a general value of $d > 3$ is similar to that for $d = 2, 3$. For simplicity of exposition, let us fix $a = 2(d-1)$. We have that $h^{(d)}(x) = ((1+x)^a + x^2)^n = \left(1 + ax + \left(\binom{a}{2}+1\right)x^2 + \sum_{j=3}^{a}\binom{a}{j}x^j\right)^n =$ $\sum_{i_1,\ldots,i_a}\left[\binom{n}{(n-\sum_{j=1}^{a} i_j), i_1, \ldots, i_a} a^{i_1}\left(\binom{a}{2}+1\right)^{i_2}\left(\prod_{j=3}^{a}\binom{a}{j}^{i_j}\right)x^{i_1+2i_2+\cdots+ai_a}\right]$. Again, we require that $i_1 + 2i_2 + \cdots + ai_a = n-1$, that is, $i_1 = n - 1 - \sum_{j=2}^{a}(j \cdot i_j)$. Thus, $l_d(n,n) = \sum_{i_2,\ldots,i_a}\left[\binom{n}{(\sum_{j=2}^{a}(j-1)i_j+1), (n-1-\sum_{j=2}^{a}(j \cdot i_j)), i_2, \ldots, i_a}\right.$

$a^{n-1-\sum_{j=2}^{a}(j\cdot i_j)}\left(\binom{a}{2}+1\right)^{i_2}\left(\prod_{j=3}^{a}\binom{a}{j}^{i_j}\right)\Big]$. Therefore, $l_d(n,n) = \frac{a^n}{a}\sum_{i_2,\ldots,i_a}$ $\left[\binom{n}{(\sum_{j=2}^{a}(j-1)i_j+1),(n-1-\sum_{j=2}^{a}(j\cdot i_j)),i_2,\ldots,i_a}\frac{\left(\binom{a}{2}+1\right)^{i_2}}{a^{2i_2}}\prod_{j=3}^{a}\frac{\binom{a}{j}^{i_j}}{a^{ji_j}}\right] = \frac{a^n}{a}\sum_{i_2,\ldots,i_a}$ $\left[\binom{n}{(\sum_{j=2}^{a}(j-1)i_j+1),(n-1-\sum_{j=2}^{a}(j\cdot i_j)),i_2,\ldots,i_a}\left(\frac{\binom{a}{2}+1}{a^2}\right)^{i_2}\prod_{j=3}^{a}\left(\frac{\binom{a}{j}}{a^j}\right)^{i_j}\right]$. It is well known that for all values of m and k, such that $1 \leq k \leq m$, we have that $\binom{m}{k} \leq \frac{m^k}{k!}$. Hence, for $j = 3,\ldots,a$, we have that $\frac{\binom{a}{j}}{a^j} \leq \frac{1}{j!}$. It is also known that $e = \sum_{j=0}^{\infty}\frac{1}{j!}$. Therefore, $l_d(n,n) \leq$ $\frac{a^n}{a}\sum_{i_2,\ldots,i_a}\left[\binom{n}{(\sum_{j=2}^{a}(j-1)i_j+1),(n-1-\sum_{j=2}^{a}(j\cdot i_j)),i_2,\ldots,i_a}\left(\frac{1}{2}+\frac{1}{a^2}\right)^{i_2}\prod_{j=3}^{a}\left(\frac{1}{j!}\right)^{i_j}\right] <$ $a^n\left(1+1+\left(\frac{1}{2}+\frac{1}{a^2}\right)+\sum_{j=3}^{a}\frac{1}{j!}\right)^n = a^n\left(\frac{1}{a^2}+\sum_{j=0}^{a}\frac{1}{j!}\right)^n < (ae+1/a)^n$.

(The relation "$<$" above is again because the summation in its left-hand side contains only a subset of the terms whose sum is equal to the exponential term on the right-hand side, and the factor $1/a$ in its left-hand side.) Consequently, $\lambda_d \leq (2d-2)e + \frac{1}{2d-2}$. □

5 Improving Further the Upper Bounds on λ_2 and λ_3

5.1 General

Klarner and Rivest [13] developed their idea further, noting that it is possible to start with a configuration containing a single open cell (as shown in Fig. 7), and keep adding twigs and updating the configuration, to construct from L increasingly larger sets $\mathsf{C}_1 = \mathsf{L}, \mathsf{C}_2, \mathsf{C}_3, \ldots$, where set C_i contains all possible twigs with i black cells (and possibly some white cells) or less than i black cells (and no white cells).

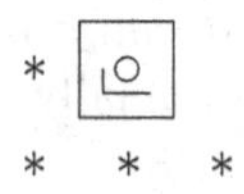

Fig. 7. A twig with one open cell

The process for building all twigs with i black cells is as follows:

1. Set $\mathsf{C}_i := \emptyset$, $B := \{\bar{s}\}$ (the twig in Fig. 7), and $W_i(x,y) := 0$;
2. `If` $B = \emptyset$, `then` output C_i and halt;
3. Remove some twig T from B;
4. `If` T contains no open cells or exactly i dead cells, `then` add T to C_i, set $W := W + w(T)$, and `goto` Step 2;

5. For $j = 1, \ldots, 5$ do
   Set $T_j := T * L_j$;
   If T_j meets condition $(*)$ below, then add T_j to B;
   od
6. Goto Step 2.

Condition $(*)$: *None of the cells of L_i (except the black cell) overlaps with neither any cell (black or white) of T nor with any cell of T marked with an* X.

Condition $(*)$ guarantees that adding a new twig to the configuration will not cause any overlap of cells.

Observation 3. $A_2(i) \le |C_i|$. □

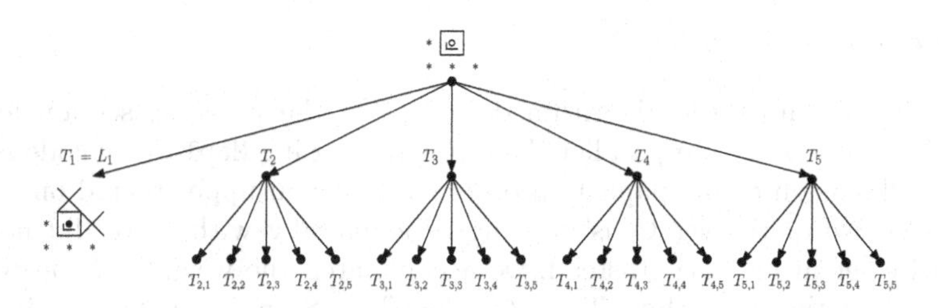

Fig. 8. The tree modeling the algorithm that generates C_i. The root r is a twig with one open cell; its L-context is shown in Fig. 3(a). For $i, j = 1, \ldots, 5$, set $T_i = L_i = r*L_i$, and $T_{i,j} = T_i * L_j$. The twig T_1 is a leaf because it has no open cells.

This relation is justified by the facts that every polyomino of size i can be built with *some* sequence of i twigs, and that the algorithm constructs *all* valid sequences of i twigs. The algorithm can be viewed as a breadth-first-search traversal of an infinite tree (Fig. 8) rooted at twig $\bar{s}$ (Fig. 7). All other vertices of the tree are twigs that can be "grown" from the root by repeatedly applying operation '$*$' (Sect. 2). The tree contains an edge directed from a twig T_1 to twig T_2 if $T_2 = T_1 * L_i$ (for some $L_i \in \mathsf{L}$). Hence, each vertex of the tree has at most five outgoing edges, and the leaves are all twigs which have no open cells.

The key idea is that, given a polyomino P, it is possible to encode P with a sequence of elements of C_i, for any $i \ge 1$, and any such sequence can be converted into a sequence of elements of L.

Observation 4. *The set of converted sequences of elements of C_{i+1} is a proper subset of the set of converted sequences of elements of C_i, since the former contains less invalid sequences (not representing polyominoes) than the latter.* □

Similarly to twigs in L, every twig $T \in C_i$ is assigned a weight $w(T) := x^a y^b$ (where a denotes the number of cells in T minus 1, and b denotes the number of black cells in T), and, thus, it can be shown that every polyomino of size n gives rise to a unique sequence of elements of C_i of weight $x^n y^n$. Letting

$W_i(x,y) = \sum_{T \in \mathsf{C}_i} w(T)$, we can plug $W_i(x,y)$ into the generating function in Eq. (1) and obtain $\sum_{m,n} c_i(m,n)x^m y^n = x/(1-W_i(x,y))$. Again, we are interested in the diagonal term $c_i(n,n)$ of the series expansion $\sum_{m,n} c_i(m,n)x^m y^n$. According to Observation 4, the sets $\mathsf{C}_1, \mathsf{C}_2, \ldots$ yield a sequence of improving (decreasing) upper bounds on λ_2. Thus, as i increases, the upper bound decreases. Therefore, the goal is to compute an upper bound on $c_i(n,n)$. The main computational challenge in this approach is to construct algorithmically the sets C_i (in order to compute $W_i(x,y)$), as $|\mathsf{C}_i|$ is increasing exponentially with i, like $A_2(i)$ does. Klarner and Rivest carried their approach to the limit of the resources they had available at the time, and computed C_i up to $i = 10$. Their computations are summarized in Table 1.

5.2 Two Dimensions

Theorem 5. $\lambda_2 \leq 4.5252$ □

We implemented the algorithm described in the previous section for constructing the sets C_i in a parallel C++ program, using `Maple` (the code is given in the full version of the paper), in order to derive an upper bound on $c_i(n,n)$. Since the size of the set C_i is growing exponentially with i, we did not keep the entire set in memory. Instead, we accumulated the weights of the twigs as in Step 4 in the algorithm. The "`for` loop" in Step 5 can be run in parallel since there are no dependencies between the twigs $T_1, \ldots, T_5$, as illustrated in Fig. 8. We used `OpenMP` and `OpenMPI` to run the program in parallel on a high-performance computer cluster. We used 33 computing nodes, each having 12 cores, for a total of 396 cores. The time for computing C_{10} was negligible even without parallelizing the program. Results were systematically improved by increasing i, the number of dead cells of the twigs. However, as the size of C_i increases roughly by a factor of 4 as i is incremented by 1, constructing C_{i+1} requires more than four times the computing power needed to construct C_i. The improved upper bound $\lambda_2 \leq 4.5252$ was obtained by using twigs with 21 dead cells. Computing C_{21} took roughly seven hours. Our results, alongside Klarner and Rivest's results, are summarized in Table 1. The two sets of results differ for $i = 6, \ldots, 10$. We address these differences in the full version of the paper. The weight functions $W_1(x,y), \ldots, W_{21}(x,y)$ are also provided in the full version.

For $i \geq 6$, the numbers of twigs ($|\mathsf{C}_i|$) we found are slightly (but consistently) larger than the numbers reported by Klarner and Rivest [13] (see Table 1). As a result, the upper bound we computed for C_{10} is slightly larger than the bound they reported. Since they provided neither the program that generated the sets C_i, nor the functions $W_6(x,y), \ldots, W_{10}(x,y)$ which they obtained, we had no means for comparing our results to theirs, but only to speculate the following.

Klarner and Rivest claimed that they used the following version of (∗):

1. None of the cells of L_i (except its root) overlaps with any of the cells or forbidden cells of T; and
2. None of the forbidden cells of L_i overlaps with any cells of T.

Table 1. Left: Results obtained by Klarner and Rivest [13]; Right: Our results.

i	$\|C_i\|$		$1/\sigma_i$		Time (Hours) Ours
	Ref. [13]	Ours	Ref. [13]	Ours	
1	5	5	4.828428	4.828427124	
2	21	21	4.828428	4.828427124	
3	93	93	4.828428	4.828427124	
4	409	409	4.796156	4.796155640	
5	1,803	1,803	4.765534	4.765532996	
6	7,929	**7,937**	4.738062	**4.738743624**	
7	34,928	**35,084**	4.714292	**4.716641912**	
8	151,897	**153,458**	4.690920	**4.695386599**	
9	656,363	**668,128**	4.669409	**4.676042980**	
10	2,821,227	**2,899,941**	4.649551	**4.658412767**	
11		**12,557,503**		**4.642235017**	
12		**54,137,703**		**4.627069746**	
13		**232,203,877**		**4.612780890**	
14		**991,607,177**		**4.599355259**	
15		**4,218,349,778**		**4.586741250**	
16		**17,881,987,659**		**4.574877902**	
17		**75,568,307,191**		**4.563716381**	
18		**318,489,941,731**		**4.553209881**	0:04
19		**1,339,093,701,964**		**4.543308340**	0:20
20		**5,617,897,764,831**		**4.533962650**	1:30
21		**23,521,568,438,976**		**4.525128839**	7:00

Table 2. Our results, 3D

i	$\|C_i^3\|$	$1/\sigma_i$
1	17	9.807295572
2	273	9.807295567
3	3,745	9.701430690
4	51113	9.631827042
5	693,725	9.573610717
6	9,047,959	9.517471577
7	114,736,608	9.467046484
8	1,428,690,351	9.422618063
9	17,538,443,750	**9.383460515**

However, although they did not state the following explicitly, they probably did not use the second part of condition (∗) in their program. We motivate this claim in the full version of the paper, explaining and emphasizing that using the second part of this condition *as-is* is incorrect. As mentioned above, our results agree with those of Klarner and Rivest's only up to $i = 5$, and we were unable to trace further the causes for the differences for $i \geq 6$.

5.3 Three Dimensions

We applied the process described above to construct sets $\mathsf{C}_1^3, \mathsf{C}_2^3, \ldots$ of larger 3-dimensional twigs. Again, we began with a single open cell on the cubical lattice, and constructed all twigs with i dead cells or fewer dead cells and no open cells. We were able to reach twigs with $i = 9$ dead cells, obtaining a set of about $17.5 \cdot 10^9$ twigs, by which we proved that $\lambda_3 \leq 9.3835$. Computing C_9^3 took 3 hours on the same cluster mentioned in Sect. 5.2.

Our results are summarized in Table 2.

References

1. The On-line Encyclopedia of Integer Sequences. http://oeis.org
2. Aleksandrowicz, G., Barequet, G.: Counting d-dimensional polycubes and nonrectangular planar polyominoes. Int. J. Comput. Geom. Appl. **19**, 215–229 (2009)
3. Aleksandrowicz, G., Barequet, G.: Counting polycubes without the dimensionality curse. Discret. Math. **309**, 576–583 (2009)
4. Barequet, G., Moffie, M., Ribó, A., Rote, G.: Counting polyominoes on twisted cylinders. INTEGERS: Electron. J. Comb. Number Theory **6**, 37 (2006)
5. Barequet, G., Rote, G., Shalah, M.: λ>4: an improved lower bound on the growth constant of polyominoes. Commun. ACM **59**, 88–95 (2016)
6. Barequet, R., Barequet, G., Rote, G.: Formulae and growth rates of high-dimensional polycubes. Combinatorica **30**, 257–275 (2010)
7. Conway, A.: Enumerating 2D percolation series by the finite-lattice method: theory. J. Phys. A: Math. General **28**, 335–349 (1995)
8. Eden, M.: A two-dimensional growth process. In: Neyman, J. (ed.) Proceedings of the 4th Berkeley Symposium on Mathematical Statistics and Probability, vol. 4, pp. 223–239 (1961)
9. Gaunt, D., Sykes, M., Ruskin, H.: Percolation processes in d-dimensions. J. Phys. A: Math. General **9**, 1899–1911 (1976)
10. Guttmann, A. (ed.): Polygons, Polyominoes, and Polycubes, vol. 775. Springer, Dordrecht (2009). https://doi.org/10.1007/978-1-4020-9927-4
11. Jensen, I.: Counting polyominoes: a parallel implementation for cluster computing. In: Sloot, P.M.A., Abramson, D., Bogdanov, A.V., Gorbachev, Y.E., Dongarra, J.J., Zomaya, A.Y. (eds.) ICCS 2003. LNCS, vol. 2659, pp. 203–212. Springer, Heidelberg (2003). https://doi.org/10.1007/3-540-44863-2_21
12. Klarner, D.: Cell growth problems. Can. J. Math. **19**, 851–863 (1967)
13. Klarner, D., Rivest, R.: A procedure for improving the upper bound for the number of n-ominoes. Can. J. Math. **25**, 585–602 (1973)
14. Lubensky, T., Isaacson, J.: Statistics of lattice animals and dilute branched polymers. Phys. Rev. A **20**, 2130–2146 (1979)
15. Luther, S., Mertens, S.: Counting lattice animals in high dimensions. J. Stat. Mech.: Theory Exp. **9**, 546–565 (2011)
16. Madras, N.: A pattern theorem for lattice clusters. Ann. Comb. **3**, 357–384 (1999)

17. Madras, N., et al.: The free energy of a collapsing branched polymer. J. Phys. A: Math. General **23**, 5327–5350 (1990)
18. Mertens, S., Lautenbacher, M.: Counting lattice animals: a parallel attack. J. Stat. Phys. **66**, 669–678 (1992)
19. Rands, B., Welsh, D.: Animals, trees and renewal sequences. IMA J. Appl. Math. **27**, 1–17 (1981)
20. Read, R.: Contributions to the cell growth problem. Can. J. Math. **14**, 1–20 (1962)
21. Redelmeier, D.: Counting polyominoes: yet another attack. Discret. Math. **36**, 191–203 (1981)
22. Sykes, M., Glen, M.: Percolation processes in two dimensions: I. low-density series expansions. J. Phys. A: Math. Gen. **9**, 87–95 (1976)

On the Collection of Fringe Subtrees in Random Binary Trees

Louisa Seelbach Benkner[1(✉)] and Stephan Wagner[2,3]

[1] Department für Elektrotechnik und Informatik, Universität Siegen, Hölderlinstrasse 3, 57076 Siegen, Germany
seelbach@eti.uni-siegen.de

[2] Department of Mathematical Sciences, Stellenbosch University, Private Bag X1, Matieland 7602, South Africa
swagner@sun.ac.za

[3] Department of Mathematics, Uppsala Universitet, Box 480, 751 06 Uppsala, Sweden
stephan.wagner@math.uu.se

Abstract. A fringe subtree of a rooted tree is a subtree consisting of one of the nodes and all its descendants. In this paper, we are specifically interested in the number of non-isomorphic trees that appear in the collection of all fringe subtrees of a binary tree. This number is analysed under two different random models: uniformly random binary trees and random binary search trees.

In the case of uniformly random binary trees, we show that the number of non-isomorphic fringe subtrees lies between $c_1 n/\sqrt{\ln n}(1+o(1))$ and $c_2 n/\sqrt{\ln n}(1+o(1))$ for two constants $c_1 \approx 1.0591261434$ and $c_2 \approx 1.0761505454$, both in expectation and with high probability, where n denotes the size (number of leaves) of the uniformly random binary tree. A similar result is proven for random binary search trees, but the order of magnitude is $n/\ln n$ in this case.

Our proof technique can also be used to strengthen known results on the number of distinct fringe subtrees (distinct in the sense of ordered trees). This quantity is of the same order of magnitude in both cases, but with slightly different constants in the upper and lower bounds.

Keywords: Uniformly random binary trees · Random binary search trees · Fringe subtrees · Tree compression

1 Introduction

A subtree of a rooted tree that consists of a node and all its descendants is called a *fringe subtree*. Fringe subtrees are a natural object of study in the context of

This project has received funding from the European Union's Horizon 2020 research and innovation programme under the Marie Skłodowska-Curie grant agreement No 731143 and the DFG research project LO 748/10-1 (QUANT-KOMP).

Y. Kohayakawa and F. K. Miyazawa (Eds.): LATIN 2020, LNCS 12118, pp. 546–558, 2020.
https://doi.org/10.1007/978-3-030-61792-9_43

random trees, and there are numerous results for various random tree models, see e.g. [3,9,11,13].

Fringe subtrees are of particular interest in computer science: One of the most important and widely used lossless compression methods for rooted trees is to represent a tree as a directed acyclic graph, which is obtained by merging nodes that are roots of identical fringe subtrees. This compressed representation of the tree is often shortly referred to as *minimal DAG* and its size (number of nodes) is the number of distinct fringe subtrees occurring in the tree. Compression by minimal DAGs has found numerous applications in various areas of computer science, as for example in compiler construction [2, Chapter 6.1 and 8.5], unification [24], symbolic model checking (binary decision diagrams) [7], information theory [20,28] and XML compression and querying [8,19].

In this work, we investigate the number of fringe subtrees in random binary trees, i.e. random trees such that each node has either exactly two or no children. So far, this problem has mainly been studied with respect to ordered fringe subtrees in random ordered binary trees: A *uniformly random ordered binary tree* of size n (with n leaves) is a random tree whose probability distribution is the uniform probability distribution on the set of ordered binary trees of size n. In [18], Flajolet, Sipala and Steyaert proved that the expected number of distinct ordered fringe subtrees in a uniformly random ordered binary tree of size n is asymptotically equal to $c \cdot n/\sqrt{\ln n}$, where c is the constant $2\sqrt{\ln 4/\pi}$. This result of Flajolet et al. was extended to unranked labelled trees in [6] (for a different constant c). Moreover, an alternative proof to the result of Flajolet et al. was presented in [25] in the context of simply-generated families of trees.

Another important type of random trees are so-called *random binary search trees*: A random binary search tree of size n is a binary search tree built by inserting the keys $\{1, \ldots, n\}$ according to a uniformly chosen random permutation on $\{1, \ldots, n\}$. Random binary search trees naturally arise in theoretical computer science, see e.g. [12]. In [16], Flajolet, Gourdon and Martinez proved that the expected number of distinct ordered fringe subtrees in a random binary search tree of size n is $O(n/\ln n)$. This result was improved in [10] by Devroye, who showed that the asymptotics $\Theta(n/\ln n)$ holds. Moreover, the result of Devroye was generalized from random binary search trees to a broader class of random ordered binary trees in [26], where the problem of estimating the expected number of distinct ordered fringe subtrees in random binary trees was considered in the context of so-called leaf-centric binary tree sources, which were introduced in [22,28] as a general framework for modeling probability distributions on the set of ordered binary trees of size n.

In this work, we focus on estimating the number of *non-isomorphic* fringe subtrees in random ordered binary trees, where we call two binary trees non-isomorphic if they are distinct as unordered binary trees. This question arises quite naturally for example in the context of XML compression: Here, one distinguishes between so-called document-centric XML, for which the corresponding XML document trees are ordered, and data-centric XML, for which the corresponding XML document trees are unordered. Understanding the interplay between ordered and unordered structures has thus received considerable attention in the context of XML (see, for example, [1,5,29]). In particular, in [23], it was investigated whether tree compression can benefit from unorderedness. For this reason, so-called *unordered minimal DAGs* were considered. An unordered minimal DAG of a binary tree is a directed acyclic graph obtained by merging nodes that are roots of isomorphic fringe subtrees, i.e. of fringe subtrees which are identical as unordered trees. From such an unordered minimal DAG, an unordered representation of the original tree can be uniquely retrieved. The size of this compressed representation is the number of non-isomorphic fringe subtrees occurring in the tree. So far, only some worst-case estimates comparing the size of a minimal DAG to the size of its corresponding unordered minimal DAG are known: Among other things, it was shown in [23] that the size of an unordered minimal DAG of a binary tree can be exponentially smaller than the size of the corresponding (ordered) minimal DAG.

However, no average-case estimates comparing the size of the minimal DAG of a binary tree to the size of the corresponding unordered minimal DAG are known so far. In particular, in [23] it is stated as an open problem to estimate the expected number of non-isomorphic fringe subtrees in a uniformly random ordered binary tree of size n and conjectured that this number asymptotically grows as $\Theta(n/\sqrt{\ln n})$.

In this work, as one of our main theorems, we settle this open conjecture by proving upper and lower bounds of order $n/\sqrt{\ln n}$ for the number of non-isomorphic fringe subtrees which hold both in expectation and with high probability (i.e., with probability tending to 1 as $n \to \infty$). Our approach can also be used to obtain an analogous result for random binary search trees, though the order of magnitude changes to $\Theta(n/\ln n)$. Again, we have upper and lower bounds in expectation and with high probability. Our two main theorems read as follows.

Theorem 1. *Let F_n be the total number of non-isomorphic fringe subtrees in a uniformly random ordered binary tree with n leaves. For two constants $c_1 \approx 1.0591261434$ and $c_2 \approx 1.0761505454$, the following holds:*

(i) $c_1 \frac{n}{\sqrt{\ln n}}(1+o(1)) \leq \mathbb{E}(F_n) \leq c_2 \frac{n}{\sqrt{\ln n}}(1+o(1))$,

(ii) $c_1 \frac{n}{\sqrt{\ln n}}(1+o(1)) \leq F_n \leq c_2 \frac{n}{\sqrt{\ln n}}(1+o(1))$ *with high probability.*

Theorem 2. *Let G_n be the total number of non-isomorphic fringe subtrees in a random binary search tree with n leaves. For two constants $c_3 \approx 1.5470025923$ and $c_4 \approx 1.8191392203$, the following holds:*

(i) $c_3 \frac{n}{\ln n}(1+o(1)) \leq \mathbb{E}(G_n) \leq c_4 \frac{n}{\ln n}(1+o(1))$,

(ii) $c_3 \frac{n}{\ln n}(1+o(1)) \leq G_n \leq c_4 \frac{n}{\ln n}(1+o(1))$ *with high probability.*

To prove the above Theorems 1 and 2, we refine techniques from [25]. Our proof technique also applies to the problem of estimating the number of distinct ordered fringe subtrees in uniformly random binary trees or in random binary search trees. In this case, upper and lower bounds for the expected value have already been proven by other authors. Our new contribution is to show that they also hold with high probability.

Theorem 3. *Let H_n denote the total number of distinct fringe subtrees in a uniformly random ordered binary tree with n leaves. Then, for the constant* $c = 2\sqrt{\ln 4/\pi} \approx 1.3285649405$, *the following holds:*

(i) $\mathbb{E}(H_n) = c\frac{n}{\sqrt{\ln n}}(1 + o(1))$,
(ii) $H_n = c\frac{n}{\sqrt{\ln n}}(1 + o(1))$ *with high probability.*

Here, the first part (i) was already shown in [18] and [25], part (ii) is new. Similarly, we are able to strengthen the results of [10] and [26]:

Theorem 4. *Let J_n be the total number of distinct fringe subtrees in a random binary search tree with n leaves. For two constants* $c_5 \approx 2.4071298335$ *and* $c_6 \approx 2.7725887222$, *the following holds:*

(i) $c_5\frac{n}{\ln n}(1 + o(1)) \leq \mathbb{E}(J_n) \leq c_6\frac{n}{\ln n}(1 + o(1))$,
(ii) $c_5\frac{n}{\ln n}(1 + o(1)) \leq J_n \leq c_6\frac{n}{\ln n}(1 + o(1))$ *with high probability.*

The upper bound in part (i) can already be found in [16] and [10]. Moreover, a lower bound of the form $\mathbb{E}(J_n) \geq \frac{\alpha n}{\ln n}(1 + o(1))$ was already shown in [10] for the constant $\alpha = (\ln 3)/2 \approx 0.5493061443$ and in [26] for the constant $\alpha \approx 0.6017824584$. So our new contributions are part (ii) and the improvement of the lower bound on $\mathbb{E}(J_n)$.

2 Preliminaries

Let $\mathcal{T}$ denote the set of ordered binary trees, i.e. of ordered rooted trees such that each node has either exactly two or no children. We define the *size* $|t|$ of a binary tree $t \in \mathcal{T}$ as the number of leaves of t and by $\mathcal{T}_k$ we denote the set of binary trees of size k for every integer $k \geq 1$. It is well known that $|\mathcal{T}_k| = C_{k-1}$, where C_k denotes the k-th *Catalan number* [17]: We have

$$C_k = \frac{1}{k+1}\binom{2k}{k} \sim \frac{4^k}{\sqrt{\pi}k^{3/2}}(1 + O(1/k)), \tag{1}$$

where the asymptotic growth of the Catalan numbers follows from Stirling's Formula [17]. Analogously, let $\mathcal{U}$ denote the set of unordered binary trees, i.e. of unordered rooted trees such that each node has either exactly two or no children. The *size* $|u|$ of an unordered tree $u \in \mathcal{U}$ is again the number of leaves of u and by $\mathcal{U}_k$ we denote the set of unordered binary trees of size k. We have $|\mathcal{U}_k| = W_k$, where W_k denotes the k-th *Wedderburn-Etherington number*. Their asymptotic growth is

$$W_k \sim A \cdot k^{-3/2} \cdot b^k, \tag{2}$$

for certain positive constants A, b [4,15]. In particular, we have $b \approx 2.4832535362$.

A *fringe subtree* of a binary tree is a subtree consisting of a node and all its descendants. For a binary tree t and a given node $v \in t$, let $t(v)$ denote the fringe subtree of t rooted at v. Two fringe subtrees are called *distinct* if they are distinct as ordered binary trees.

Every tree $t \in \mathcal{T}$ can be considered as an element of $\mathcal{U}$ by simply forgetting the ordering on t's nodes. If two binary trees t_1, t_2 correspond to the same unordered tree $u \in \mathcal{U}$, we call them *isomorphic*: Thus, we obtain a partition of $\mathcal{T}$ into isomorphism classes. If two binary trees $t_1, t_2 \in \mathcal{T}$ belong to the same isomorphism class, we can obtain t_1 from t_2 and vice versa by reordering the children of some of t_1's (respectively, t_2's) inner nodes. An inner node v of an ordered or unordered binary tree t is called a *symmetrical node* if the fringe subtrees rooted at v's children are isomorphic. Let $\mathrm{sym}(t)$ denote the number of symmetrical nodes of t. The cardinality of the automorphism group of t is given by $|\mathrm{Aut}(t)| = 2^{\mathrm{sym}(t)}$. Thus, by the orbit-stabilizer theorem, there are $2^{k-1-\mathrm{sym}(t)}$ many ordered binary trees in the isomorphism class of $t \in \mathcal{T}_k$, and likewise $2^{k-1-\mathrm{sym}(t)}$ many ordered representations of $t \in \mathcal{U}_k$.

We consider two types of probability distributions on the set of ordered binary trees of size n:

(i) The *uniform probability distribution* on $\mathcal{T}_n$, that is, every binary tree of size n is assigned the same probability $\frac{1}{C_{n-1}}$. A random variable taking values in $\mathcal{T}_n$ according to the uniform probability distribution is called a *uniformly random (ordered) binary tree* of size n.

(ii) The probability distribution induced by the so-called *Binary Search Tree Model* (see e.g. [12,16]): The corresponding probability mass function $P_{\mathrm{bst}} : \mathcal{T}_n \to [0,1]$ is given by

$$P_{\mathrm{bst}}(t) = \prod_{\substack{v \in t \\ |t(v)|>1}} \frac{1}{|t(v)|-1}, \tag{3}$$

for every $n \geq 1$. A random variable taking values in $\mathcal{T}_n$ according to this probability mass function is called a *random binary search tree* of size n.

Before we prove our main results, we need two preliminary lemmas:

Lemma 1. *Let a, ε be positive real numbers with $\varepsilon < \frac{1}{3}$. For every positive integer k with $a \ln n \leq k \leq n^{\varepsilon}$, let $\mathcal{S}_k \subset \mathcal{T}_k$ be a set of ordered binary trees with k leaves. We denote the cardinality of $\mathcal{S}_k$ by s_k. Let $X_{n,k}$ denote the (random) number of fringe subtrees with k leaves in a uniformly random ordered binary tree with n leaves that belong to $\mathcal{S}_k$. Moreover, let $Y_{n,\varepsilon}$ denote the (random) number of arbitrary fringe subtrees with more than n^{ε} leaves in a uniformly random ordered binary tree with n leaves. We have*

(1) $\mathbb{E}(X_{n,k}) = s_k 4^{1-k} n(1+O(k/n))$ *for all* k *with* $a \ln n \le k \le n^{\varepsilon}$, *the* O*-constant being independent of* k,
(2) $\mathbb{V}(X_{n,k}) = s_k 4^{1-k} n(1 + O(k^{-1/2}))$ *for all* k *with* $a \ln n \le k \le n^{\varepsilon}$, *again with an* O*-constant that is independent of* k,
(3) $\mathbb{E}(Y_{n,\varepsilon}) = O(n^{1-\varepsilon/2})$ *and*
(4) with high probability, the following statements hold:
   (i) $|\sum_k X_{n,k} - \mathbb{E}(X_{n,k})| \le \sum_k s_k^{1/2} 2^{-k} n^{1/2+\varepsilon}$, *where the sums are taken over all* k *with* $a \ln n \le k \le n^{\varepsilon}$,
   (ii) $Y_{n,\varepsilon} \le n^{1-\varepsilon/3}$.

Lemma 2. *Let* a, ε *be positive real numbers with* $\varepsilon < \frac{1}{3}$ *and let* n *and* k *denote positive integers. Moreover, for every* k, *let* $\mathcal{S}_k \subset \mathcal{T}_k$ *be a set of ordered binary trees with* k *leaves and let* p_k *denote the probability that a random binary search tree is contained in* $\mathcal{S}_k$, *that is,* $p_k = \sum P_{bst}(t)$, *where the sum is taken over all binary trees in* $\mathcal{S}_k$. *Let* $X_{n,k}$ *denote the (random) number of fringe subtrees with* k *leaves in a random binary search tree with* n *leaves that belong to* $\mathcal{S}_k$. *Moreover, let* $Y_{n,\varepsilon}$ *denote the (random) number of arbitrary fringe subtrees with more than* n^{ε} *leaves in a random binary search tree with* n *leaves. We have*

(1) $\mathbb{E}(X_{n,k}) = \frac{2p_k n}{k(k+1)}$ *for* $1 \le k < n$,
(2) $\mathbb{V}(X_{n,k}) = O(p_k n/k^2)$ *for all* k *with* $a \ln n \le k \le n^{\varepsilon}$, *where the* O*-constant is independent of* k,
(3) $\mathbb{E}(Y_{n,\varepsilon}) = 2n/\lceil n^{\varepsilon}\rceil - 1 = O(n^{1-\varepsilon})$ *and*
(4) with high probability, the following statements hold:
   (i) $|\sum_k X_{n,k} - \mathbb{E}(X_{n,k})| \le \sum_k p_k^{1/2} k^{-1} n^{1/2+\varepsilon}$, *where the sums are taken over all* k *with* $a \ln n \le k \le n^{\varepsilon}$,
   (ii) $Y_{n,\varepsilon} \le n^{1-\varepsilon/2}$.

For the proofs of Lemma 1 and Lemma 2, see the long version of the paper [27].

3 Fringe Subtrees in Uniformly Random Binary Trees

3.1 Ordered Fringe Subtrees

We provide the proof of Theorem 3 first, since it is simplest and provides us with a template for the other proofs. Basically, it is a refinement of the proof for the corresponding special case of Theorem 3.1 in [25]. In the following sections, we refine the argument further to prove Theorems 1, 2 and 4. For further details, see the long version of the paper [27].

Proof (Proof of Theorem 3). We prove the statement in two steps: In the first step, we show that the upper bound $H_n \le cn/\sqrt{\ln n}(1 + o(1))$ holds for $c = 2\sqrt{\ln 4/\pi}$ both in expectation and with high probability. In the second step, we prove the corresponding lower bound.

The Upper Bound: Let $k_0 = \log_4 n$. The number H_n of distinct fringe subtrees in a uniformly random ordered binary tree with n leaves equals (i) the number of such distinct fringe subtrees of size at most k_0 plus (ii) the number of such distinct fringe subtrees of size greater than k_0. We upper-bound (i) by the number of all ordered binary trees of size at most k_0 (irrespective of their occurrence as fringe subtrees) and (ii) by the total number of such fringe subtrees occurring in the tree to obtain, using the notation of Lemma 1,

$$H_n \leq \sum_{k \leq k_0} C_{k-1} + \Big(\sum_{k_0 < k \leq n^\varepsilon} X_{n,k} \Big) + Y_{n,\varepsilon}.$$

Here, $\mathcal{S}_k$ is the full set $\mathcal{T}_k$, so that $s_k = C_{k-1}$. The first sum is $O(n/(\ln n)^{3/2})$ by (1). This upper bound holds deterministically. In order to estimate the other two terms, we apply Lemma 1 with $a = \frac{1}{\ln 4}$ and $\varepsilon = \frac{1}{6}$. We thus find that the two terms are bounded from above by $\frac{2\sqrt{\ln 4}}{\sqrt{\pi}} \cdot \frac{n}{\sqrt{\ln n}} + O(n/(\ln n)^{3/2})$, both in expectation and with high probability.

The Lower Bound: Again, let $k_0 = \log_4 n$ and $\varepsilon = \frac{1}{6}$. In order to lower-bound the number H_n of distinct fringe subtrees in a uniformly random ordered tree with n leaves, we only count distinct fringe subtrees of sizes k with $k_0 < k \leq n^\varepsilon$. To this end, let $X_{n,k}^{(2)}$ denote the number of pairs of identical fringe subtrees of size k in a uniformly random ordered binary tree of size n. Each such pair can be obtained by taking an ordered tree with $n - 2k + 2$ leaves, picking two leaves, and replacing them by the same ordered binary tree of size k. The total number of such pairs of identical fringe subtrees of size k is thus

$$C_{n-2k+1} \cdot \binom{n-2k+2}{2} \cdot C_{k-1} = \frac{4^{n-k}}{2\pi k^{3/2}} (n - 2k + 1)^{1/2} (1 + O(1/k)).$$

By dividing by C_{n-1}, i.e. the total number of binary trees of size n, we thus obtain the expected value: $\mathbb{E}(X_{n,k}^{(2)}) = O(4^{-k} n^2 k^{-3/2})$ and consequently $\sum \mathbb{E}(X_{n,k}^{(2)}) = O(n/(\ln n)^{3/2})$, where the sum is taken over all k with $k_0 < k \leq n^\varepsilon$. If a binary tree of size k occurs m times as a fringe subtree in a uniformly random binary tree of size n, it contributes $m - \binom{m}{2}$ to the random variable $X_{n,k} - X_{n,k}^{(2)}$. Since $m - \binom{m}{2} \leq 1$ for all non-negative integers m, we find that $X_{n,k} - X_{n,k}^{(2)}$ is a lower bound on the number of distinct fringe subtrees with k leaves. Hence, we have

$$H_n \geq \sum_{k_0 < k \leq n^\varepsilon} X_{n,k} - \sum_{k_0 < k \leq n^\varepsilon} X_{n,k}^{(2)}.$$

The second sum is $O(n/(\ln n)^{3/2})$ in expectation and thus with high probability as well by the Markov inequality. As the first sum is $\frac{2\sqrt{\ln 4}}{\sqrt{\pi}} \cdot \frac{n}{\sqrt{\ln n}}(1 + o(1))$, both in expectation and with high probability by our estimate from the first part of the proof, the statement of Theorem 3 follows. ∎

As the main idea of the proof is to split the number of distinct fringe subtrees into the number of distinct fringe subtrees of size at most k_0 plus the number of distinct fringe subtrees of size greater than k_0 for some suitably chosen integer

k_0, this type of argument is called a *cut-point argument* and the integer k_0 is called the *cut-point* (see [16]). This basic technique is applied in several previous papers to similar problems (see for instance [10,16,25,26]). Moreover, we remark that the statement of Theorem 3 can be easily generalized to simply generated families of trees.

3.2 Unordered Fringe Subtrees

In this subsection, we prove Theorem 1. For this, we refine the cut-point argument we applied in the proof of Theorem 3: In particular, for the lower bound on F_n, we need a result due to Bóna and Flajolet [4] on the number of automorphisms of a uniformly random ordered binary tree. It is stated for random phylogenetic trees in [4], but the two probabilistic models are equivalent.

Theorem 5 ([4], Theorem 2). *Consider a uniformly random ordered binary tree T_k with k leaves, and let $A_k = |\mathrm{Aut}(T_k)|$ be the cardinality of its automorphism group. The logarithm of this random variable satisfies a central limit theorem: For certain positive constants γ and σ_1, we have*

$$\mathbb{P}(A_k \leq 2^{\gamma k + \sigma_1 \sqrt{k} x}) \overset{k\to\infty}{\to} \frac{1}{\sqrt{2\pi}} \int_{-\infty}^{x} e^{-t^2/2}\, dt$$

for every real number x. The numerical value of the constant γ is 0.2710416936.

With Theorem 5, we are able to upper-bound the probability that two fringe subtrees of the same size are isomorphic in our proof of Theorem 1:

Proof (Proof of Theorem 1*).* We prove the statement in two steps: First, we show that the upper bound on F_n stated in Theorem 1 holds both in expectation and with high probability, then we prove the respective lower bound.

The Upper Bound: The proof for the upper bound in Theorem 1 exactly matches the first part of the proof of Theorem 3, except that we choose a different cut-point: Let $k_0 = \log_b n$, where $b \approx 2.4832535362$ is the constant in the asymptotic formula (2) for the Wedderburn-Etherington numbers. We then find

$$F_n \leq \sum_{k<k_0} W_k + \Big(\sum_{k_0 \leq k \leq n^\epsilon} X_{n,k} \Big) + Y_{n,\epsilon} = \frac{2\sqrt{\ln b}}{\sqrt{\pi}} \cdot \frac{n}{\sqrt{\ln n}} + O(n(\ln n)^{-3/2}),$$

both in expectation and with high probability, where the estimates for $X_{n,k}$ and $Y_{n,\varepsilon}$ follow again from Lemma 1. We have $2\sqrt{\ln b}/\sqrt{\pi} \approx 1.0761505454$.

The Lower Bound: As a consequence of Theorem 5, the probability that the cardinality of the automorphism group of a uniformly random binary tree T_k of size k satisfies $|\mathrm{Aut}(T_k)| \leq 2^{\gamma k - k^{3/4}}$ tends to 0 as $k \to \infty$. We define $\mathcal{S}_k$ as the set of ordered trees with k leaves that do not satisfy this inequality, so that $s_k = |\mathcal{S}_k| = C_{k-1}(1 + o(1))$. Our lower bound is based on counting only fringe subtrees in $\mathcal{S}_k$ for suitable k. The reason for this choice is that we have an upper bound on the number of ordered binary trees in the same isomorphism class for every tree in $\mathcal{S}_k$. Recall that the number of possible ordered representations of an unordered binary tree t with k leaves is given by $2^{k-1}/|\mathrm{Aut}(t)|$ by the orbit-stabiliser theorem. Hence, the number of ordered binary trees in the same isomorphism class as a tree $t \in \mathcal{S}_k$ is bounded above by $2^{k-1-\gamma k+k^{3/4}}$.

Now set $k_1 = \frac{1+\delta}{1+\gamma}\log_2 n$ for some positive constant $\delta < \frac{2}{3}$, and consider only fringe subtrees that belong to $\mathcal{S}_k$, where $k_1 \le k \le n^{\delta/2}$. By Lemma 1, the number of such fringe subtrees in a random ordered binary tree with n leaves is $s_k 4^{1-k} n(1 + O(k/n + s_k^{-1/2} 2^k n^{(\delta-1)/2}))$ both in expectation and with high probability. Since $s_k = C_{k-1}(1+o(1))$, the number of fringe subtrees that belong to $\mathcal{S}_k$ in a random ordered binary tree of size n becomes $\frac{n}{\sqrt{\pi k^3}}(1+o(1))$. We show that most of these trees are the only representatives of their isomorphism classes as fringe subtrees. To this end, we consider all fringe subtrees in $\mathcal{S}_k$ for some k that satisfies $k_1 \le k \le n^{\delta/2}$. Let the sizes of the isomorphism classes of trees in $\mathcal{S}_k$ be $r_1, r_2, \ldots, r_\ell$, so that $r_1 + r_2 + \cdots + r_\ell = s_k$. By definition of $\mathcal{S}_k$, we have $r_i \le 2^{k-1-\gamma k+k^{3/4}}$ for every i. Let us condition on the event that their number $X_{n,k}$ is equal to N for some $N \le n$. Each of these N fringe subtrees $S_1, S_2, \ldots, S_N$ follows a uniform distribution among the elements of $\mathcal{S}_k$, so the probability of being in an isomorphism class with r_i elements is r_i/s_k. Moreover, the N fringe subtrees are also all independent. Let $X_{n,k}^{(2)}$ be the number of pairs of isomorphic trees among the fringe subtrees with k leaves. We have

$$\mathbb{E}(X_{n,k}^{(2)}|X_{n,k} = N) = \binom{N}{2}\sum_i \left(\frac{r_i}{s_k}\right)^2 \le \frac{n^2}{2s_k^2}\sum_i r_i^2 \le \frac{n^2}{s_k}2^{k-2-\gamma k+k^{3/4}}.$$

Since this holds for all N, the law of total expectation yields

$$\mathbb{E}(X_{n,k}^{(2)}) \le \frac{n^2}{s_k}2^{k-2-\gamma k+k^{3/4}} = \sqrt{\pi}n^2k^{3/2}2^{-k-\gamma k+k^{3/4}}(1+o(1)).$$

Since $k \ge k_1 = \frac{1+\delta}{1+\gamma}\log_2 n$, we find that

$$\mathbb{E}(X_{n,k}^{(2)}) \le n^2 2^{-(1+\gamma)k+O(k^{3/4})} \le n^{1-\delta}\exp\left(O((\ln n)^{3/4})\right).$$

Thus

$$\sum_{k_1 \le k \le n^{\delta/2}} \mathbb{E}(X_{n,k}^{(2)}) \le n^{1-\delta/2}\exp\left(O((\ln n)^{3/4})\right) = o(n/\sqrt{\ln n}).$$

As in the previous proof, we see that $X_{n,k} - X_{n,k}^{(2)}$ is a lower bound on the number of non-isomorphic fringe subtrees with k leaves. This gives us

$$F_n \ge \sum_{k_1 \le k \le n^{\delta/2}} X_{n,k} - \sum_{k_1 \le k \le n^{\delta/2}} X_{n,k}^{(2)}.$$

The second sum is negligible since it is $o(n/\sqrt{\ln n})$ in expectation and thus also with high probability by the Markov inequality. For the first sum, a calculation similar to that for the upper bound shows that it is

$$\frac{2\sqrt{(1+\gamma)\ln 2}}{\sqrt{\pi(1+\delta)}} \cdot \frac{n}{\sqrt{\ln n}}(1+o(1)),$$

both in expectation and with high probability. Since δ is arbitrary, we can choose any constant smaller than $\frac{2\sqrt{(1+\gamma)\ln 2}}{\sqrt{\pi}} \approx 1.0591261434$ for c_1. ■

4 Fringe Subtrees in Random Binary Search Trees

In order to show the respective lower bounds of Theorem 2 and Theorem 4, we need two theorems similar to Theorem 5: The first one shows that the logarithm of the random variable $B_k = P_{\text{bst}}(T_k)^{-1}$, where T_k denotes a random binary search tree of size k, satisfies a central limit theorem and is needed to estimate the probability that two fringe subtrees in a random binary search tree are identical. The second one transfers the statement of Theorem 5 from uniformly random binary trees to random binary search trees and is needed in order to estimate the probability that two fringe subtrees in a random binary search tree are isomorphic. The first of these two central limit theorems is shown in [14]:

Theorem 6 ([14], Theorem 4.1). *Consider a random binary search tree T_k with k leaves, and let $B_k = P_{bst}(T_k)^{-1}$. The logarithm of this random variable satisfies a central limit theorem: For certain positive constants μ and σ_2, we have*

$$\mathbb{P}\left(B_k \leq 2^{\mu k+\sigma_2\sqrt{k}x}\right) \overset{k\to\infty}{\to} \frac{1}{\sqrt{2\pi}}\int_{-\infty}^{x} e^{-t^2/2}\,dt$$

for every real number x. The numerical value of the constant μ is

$$\mu = \sum_{k=1}^{\infty} \frac{2\log_2 k}{(k+1)(k+2)} \approx 1.7363771368.$$

The second of these two central limit theorems follows from a general theorem devised by Holmgren and Janson [21]: The proof of Theorem 7 can be found in the long version of the paper [27].

Theorem 7. *Consider a random binary search tree T_k with k leaves, and let $A_k = |\mathrm{Aut}(T_k)|$ be the cardinality of its automorphism group. The logarithm of this random variable satisfies a central limit theorem: for certain positive constants ν and σ_3, we have*

$$\mathbb{P}(A_k \leq 2^{\nu k+\sigma_3\sqrt{k}x}) \overset{k\to\infty}{\to} \frac{1}{\sqrt{2\pi}}\int_{-\infty}^{x} e^{-t^2/2}\,dt$$

for every real number x. The numerical value of ν is $\nu \approx 0.3795493473$.

For the proofs of Theorems 2 and 4, we refer to the long version of the paper [27]: The techniques used in the proofs are mostly the same as in the proof of Theorem 1. In order to show the corresponding upper bounds, we make use of the cut-point technique presented in the proofs of Theorems 3 and 1, combined with Lemma 2. For the lower bounds, we suitably define, as in the proof of Theorem 1, respective sets $\mathcal{S}_k$ using Theorems 6 and 7. We then lower-bound the number of distinct (non-isomorphic, respectively) fringe subtrees by the number of such fringe subtrees of size k that belong to the respective set $\mathcal{S}_k$. The sets $\mathcal{S}_k$ and the range of k are again chosen in a way that allows us to bound the probability that two fringe subtrees from the set $\mathcal{S}_k$ are identical (isomorphic, respectively).

5 Open Problems

The following natural question arises from our results: Is it possible to determine constants $\alpha_1, \alpha_2, \alpha_3$ with $c_1 \leq \alpha_1 \leq c_2$, $c_3 \leq \alpha_2 \leq c_4$ and $c_5 \leq \alpha_3 \leq c_6$, such that

$$\mathbb{E}(F_n) = \frac{\alpha_1 n}{\sqrt{\log n}}(1+o(1)),\ \mathbb{E}(G_n) = \frac{\alpha_2 n}{\log n}(1+o(1)),\ \mathbb{E}(J_n) = \frac{\alpha_3 n}{\log n}(1+o(1)),$$

respectively, and

$$\frac{F_n}{n/\sqrt{\log n}} \xrightarrow{P} \alpha_1,\quad \frac{G_n}{n/\log n} \xrightarrow{P} \alpha_2,\ \text{and}\ \frac{J_n}{n/\log n} \xrightarrow{P} \alpha_3\ ?$$

In order to prove such estimates, it seems essential to gain a better understanding of the random variables $P_{\mathrm{bst}}(T_k)^{-1}$ and $|\mathrm{Aut}(T_k)|$, in particular their distributions further away from the mean values, for random binary search trees or uniformly random ordered binary trees T_k of size k.

References

1. Abiteboul, S., Bourhis, P., Vianu, V.: Highly expressive query languages for unordered data trees. Theory Comput. Syst. **57**(4), 927–966 (2015)
2. Aho, A.V., Sethi, R., Ullman, J.D.: Compilers: Principles, Techniques, and Tools. Addison-Wesley Series in Computer Science/World Student Series Edition. Addison-Wesley (1986)
3. Aldous, D.: Asymptotic fringe distributions for general families of random trees. Ann. Appl. Probab. **1**(2), 228–266 (1991)
4. Bóna, M., Flajolet, P.: Isomorphism and symmetries in random phylogenetic trees. J. Appl. Probab. **46**(4), 1005–1019 (2009)
5. Boneva, I., Ciucanu, R., Staworko, S.: Schemas for unordered XML on a DIME. Theory Comput. Syst. **57**(2), 337–376 (2015)
6. Bousquet-Mélou, M., Lohrey, M., Maneth, S., Noeth, E.: XML compression via DAGs. Theory Comput. Syst. **57**(4), 1322–1371 (2015)
7. Bryant, R.E.: Symbolic boolean manipulation with ordered binary-decision diagrams. ACM Comput. Surv. **24**(3), 293–318 (1992)

8. Buneman, P., Grohe, M., Koch, C.: Path queries on compressed XML. In: Freytag, J.C., et al. (eds.) Proceedings of the 29th Conference on Very Large Data Bases, VLDB 2003, pp. 141–152. Morgan Kaufmann (2003)
9. Dennert, F., Grübel, R.: On the subtree size profile of binary search trees. Comb. Probab. Comput. **19**(4), 561–578 (2010)
10. Devroye, L.: On the richness of the collection of subtrees in random binary search trees. Inf. Process. Lett. **65**(4), 195–199 (1998)
11. Devroye, L., Janson, S.: Protected nodes and fringe subtrees in some random trees. Electron. Commun. Probab. **19**, 1–10 (2014)
12. Drmota, M.: Random Trees: An Interplay Between Combinatorics and Probability, 1st edn. Springer, Heidelberg (2009). https://doi.org/10.1007/978-3-211-75357-6
13. Feng, Q., Mahmoud, H.M.: On the variety of shapes on the fringe of a random recursive tree. J. Appl. Probab. **47**(1), 191–200 (2010)
14. Fill, J.A.: On the distribution of binary search trees under the random permutation model. Random Struct. Algorithms **8**(1), 1–25 (1996)
15. Finch, S.R., Rota, G.C.: Mathematical Constants. Encyclopedia of Mathematics and Its Applications. Cambridge University Press, Cambridge (2003)
16. Flajolet, P., Gourdon, X., Martínez, C.: Patterns in random binary search trees. Random Struct. Algorithms **11**(3), 223–244 (1997)
17. Flajolet, P., Sedgewick, R.: Analytic Combinatorics. Cambridge University Press, Cambridge (2009)
18. Flajolet, P., Sipala, P., Steyaert, J.-M.: Analytic variations on the common subexpression problem. In: Paterson, M.S. (ed.) ICALP 1990. LNCS, vol. 443, pp. 220–234. Springer, Heidelberg (1990). https://doi.org/10.1007/BFb0032034
19. Frick, M., Grohe, M., Koch, C.: Query evaluation on compressed trees (extended abstract). In: Proceedings of the 18th Annual IEEE Symposium on Logic in Computer Science, LICS 2003, pp. 188–197. IEEE Computer Society Press (2003)
20. Ganardi, M., Hucke, D., Lohrey, M., Benkner, L.S.: Universal tree source coding using grammar-based compression. IEEE Trans. Inf. Theory **65**(10), 6399–6413 (2019)
21. Holmgren, C., Janson, S.: Limit laws for functions of fringe trees for binary search trees and random recursive trees. Electron. J. Probab. **20**, 1–51 (2015)
22. Kieffer, J.C., Yang, E.H., Szpankowski, W.: Structural complexity of random binary trees. In: Proceedings of the 2009 IEEE International Symposium on Information Theory, ISIT 2009, pp. 635–639. IEEE (2009)
23. Lohrey, M., Maneth, S., Reh, C.P.: Compression of unordered XML trees. In: Proceedings of the 20th International Conference on Database Theory, ICDT 2017, Venice, Italy, 21–24 March 2017, pp. 18:1–18:17 (2017)
24. Paterson, M., Wegman, M.N.: Linear unification. J. Comput. Syst. Sci. **16**(2), 158–167 (1978)
25. Ralaivaosaona, D., Wagner, S.G.: Repeated fringe subtrees in random rooted trees. In: Proceedings of the 12th Workshop on Analytic Algorithmics and Combinatorics, ANALCO 2015, pp. 78–88. SIAM (2015)
26. Seelbach Benkner, L., Lohrey, M.: Average case analysis of leaf-centric binary tree sources. In: Proceedings of the 43rd International Symposium on Mathematical Foundations of Computer Science, MFCS 2018, Liverpool, UK, 27–31 August 2018, pp. 16:1–16:15 (2018)
27. Seelbach Benkner, L., Wagner, S.: On the collection of fringe subtrees in random binary trees. arXiv e-prints arXiv:2003.03323 (2020). https://arxiv.org/abs/2003.03323

28. Zhang, J., Yang, E.H., Kieffer, J.C.: A universal grammar-based code for lossless compression of binary trees. IEEE Trans. Inf. Theory **60**(3), 1373–1386 (2014)
29. Zhang, S., Du, Z., Wang, J.T.: New techniques for mining frequent patterns in unordered trees. IEEE Trans. Cybern. **45**(6), 1113–1125 (2015)

A Method to Prove the Nonrationality of Some Combinatorial Generating Functions

Miklós Bóna(✉)

University of Florida, Gainesville, FL 32611, USA
bona@ufl.edu

Abstract. We are presenting a new method to prove that certain combinatorial generating functions are not rational. We show several applications of our method, such as permutation patterns, t-stack sortable permutations, and classic examples of algebraic generating functions, such as lattice paths.

Keywords: Generating functions · Rational functions · Permutations · Trees · Lattice paths

1 Introduction

When solving an enumeration problem, we often attempt to determine the number $f(n)$ of some structures of size n, or the generating function $F(z) = \sum_{n\geq 0} f(n)z^n$ of the corresponding sequence. When our efforts fail, we may be interested in why the problem at hand is so difficult. In this paper, we will present a method that can in some cases show that the generating function $F(z)$ is not a *rational function* (ratio of two polynomials) of z. This is equivalent to the statement that there does not exist a recurrence relation

$$f(n) = a_1 f(n-1) + a_2 f(n-2) + \cdots + a_r f(n-r),$$

for all n, where r is a *fixed* positive integer, and the a_i are *constants*. So, when our method works, it will provide some justification as to why the enumeration problem is difficult.

The *exponential order* or *exponential growth rate* of the sequence of the numbers $f(n)$ is defined to be $\limsup_n (f(n)^{1/n})$. The *Fundamental Theorem of Analytic Combinatorics* [4] states that if $F(z) = \sum_{n\geq 0} f(n)z^n$ is analytic at $z = 0$, then the exponential order of the sequence $f(n)$ is $1/r$, where r is the distance between zero and the singularity of $F(z)$ that is closest to 0. We call that singularity the *dominant* singularity of F. The type of that singularity can be used to determine the subexponential terms in the asymptotics of $f(n)$.

Partially supported by a Simons Collaboration Grant.

Y. Kohayakawa and F. K. Miyazawa (Eds.): LATIN 2020, LNCS 12118, pp. 559–570, 2020.
https://doi.org/10.1007/978-3-030-61792-9_44

All power series in this paper are assumed to have nonnegative real coefficients as they are combinatorial generating functions. For such a power series f, if $R > 0$ is the radius of convergence, then Pringsheim's theorem (Theorem IV.6 in [4]) shows that the positive real number R itself is a singularity of f. We will use this fact without explicitly mentioning it in what follows.

As we will discuss it in Sect. 5, when a generating function f is not explicitly known, it is usually quite difficult to prove that f does *not* belong to a certain class of power series. This is not surprising, in view of Rice's theorem [10], that states that for any non-trivial property of functions, no general and effective method can decide whether an algorithm computes a function with that property. This paper provides a tool to prove that certain power series are *not* rational functions, even though we do not know their explicit form, or even, their radius of convergence.

2 Supercriticality

Definition 1. *Let F and G be two generating functions with nonnegative real coefficients that are analytic at 0, and let us assume that $G(0) = 0$. Then the relation*

$$F(z) = \frac{1}{1 - G(z)} \tag{1}$$

is called supercritical *if $G(R_G) > 1$, where R_G is the radius of convergence of G.*

See [4] for a detailed discussion of supercritical relations. For our purposes, it is the following property of such relations that is crucial.

Proposition 1. *Let F and G be two generating functions with nonnegative real coefficients that are analytic at 0, and let us assume that the relation $F(z) = 1/(1 - G(z))$ is supercritical. Then the exponential growth rate of the coefficients of F is strictly larger than the exponential growth rate of the coefficients of G.*

Proof. As the coefficients of $G(z)$ are nonnegative, $G(R_G) > 1$ implies that $G(\alpha) = 1$ for some $\alpha \in (0, R_G)$. So, if the relation between F and G described above is supercritical, then the radius R_F of convergence of F is less than the radius of convergence R_G of G, and so the exponential growth rate $1/R_F$ of the coefficients of F is larger than the exponential growth rate $1/R_G$ of the coefficients of G. □

Theorem 1. *Let $G(z)$ be a rational power series with nonnegative real coefficients that satisfies $G(0) = 0$. Then the relation*

$$F(z) = \frac{1}{1 - G(z)} \tag{2}$$

is supercritical.

Proof. If G is rational, then its dominant singularity R_G is a pole, so $G(R_G) = \infty > 1$. □

Therefore, in order to prove that some generating function $F(z)$ is *not rational*, it suffices to prove that (2) is **not** supercritical. Note that (2) is equivalent to $F(z) = \sum_{n\geq 0} G(z)^n$. So the combinatorial meaning of (2) is the following. Let $G(z)$ be the generating function for the number of ways to carry out a task on an n-element set. Then $F(z)$ is the generating function for the number of ways to split the set $\{1, 2, \cdots, n\}$ into an unspecified number of non-empty intervals, and then to carry out the first task on each of those intervals. In other words, $G(z)$ counts structures that are the *irreducible building blocks* of the structures counted by $F(z)$. In analytic combinatorics, the symbolic notation for this relation is $\mathcal{F} = SEQ(\mathcal{G})$. If the relation between the builiding blocks (the $\mathcal{G}$-structures) and their sequences (the $\mathcal{F}$-structures) is supercritical, then the sequence enumerating the $\mathcal{F}$-structures has a higher exponential order. This leads to the following theorem, which we will apply several times in this paper.

Theorem 2. *Let g_n be the number of objects of size $n > 0$ of a certain kind, and let f_n be the number of* sequences *built up from various objects of that kind so that the total size of all objects in the sequence is n. If there exists a positive integer k so that for all nonnegative integers n, the inequality $f_n \leq g_{n+k}$ holds, then the generating function $F(z) = \sum_{n\geq 0} f_n z^n$ is not rational.*

Proof. If $F(z)$ were rational, then by Theorem 1, the exponential order of the sequence f_n would be higher than that of the sequence g_n, contradicting the assumption that $f_n \leq g_{n+k}$ holds for all n. □

Note that if an injection can be found from the set of all $\mathcal{F}$-structures of size n to the set of all $\mathcal{G}$-structures of size $n+1$, that proves that $f_n \leq g_{n+1}$, and so Theorem 2 can be applied with $k = 1$ to prove nonrationality of $F(z)$.

3 Permutation Patterns

We say that a permutation p *contains* the pattern $q = q_1 q_2 \cdots q_k$ if there is a k-element set of indices $i_1 < i_2 < \cdots < i_k$ so that $p_{i_r} < p_{i_s}$ if and only if $q_r < q_s$. If p does not contain q, then we say that p *avoids* q. For example, $p = 3752416$ contains $q = 2413$, as the first, second, fourth, and seventh entries of p form the subsequence 3726, which is order-isomorphic to $q = 2413$. A recent survey on permutation patterns can be found in [12] and a book on the subject is [2]. Let $\text{Av}_n(q)$ be the number of permutations of length n that avoid the pattern q. In general, it is very difficult to compute, or even describe, the numbers $\text{Av}_n(q)$, or their sequence as n goes to infinity. Accordingly, the explicit form of the generating function $A_q(z) = \sum_{n\geq 0} \text{Av}_n(q) z^n$ is only known for very few patterns. Still, there are known examples when $A_q(z)$ is algebraic, (when q is of length three or when $q = 1342$), and there are known examples when $A_q(z)$ is not algebraic (when q is the monotone pattern $12 \cdots k$, where $k \geq 4$ is an even integer).

We say that a permutation p is *skew indecomposable* if it is not possible to cut p into two parts so that each entry before the cut is larger than each entry after the cut. For instance, $p = 3142$ is skew indecomposable, but $r = 346512$ is not as we can cut it into two parts by cutting between entries 5 and 1, to obtain $3465|12$.

If p is not skew indecomposable, then there is a unique way to cut p into nonempty skew indecomposable strings $s_1, s_2, \cdots, s_\ell$ of consecutive entries so that each entry of s_i is larger than each entry of s_j if $i < j$. We call these strings s_i the *skew blocks* of p. For instance, $p = 67|435|2|1$ has four skew blocks, while skew indecomposable permutations have one skew block.

Theorem 3. *Let q be a skew indecomposable pattern that does not end in its largest entry. Then $A_q(z)$ is not rational.*

Proof. It is clear that p avoids q if and only if each of the skew blocks of p avoids q. This means that

$$A_q(z) = \frac{1}{1 - A_{q,1}(z)} \tag{3}$$

holds, where $A_{q,1}(z)$ is the generating function of the skew indecomposable q-avoiders. Let $\mathrm{Av}_{n,1}(q)$ denote the number of skew indecomposable q-avoiders of length n.

Let p be of length n, and let p avoid q. Now affix a new entry $n+1$ at the end of p. The new permutation p' still avoids q, but is also skew indecomposable.

This proves the inequality

$$\mathrm{Av}_n(q) \leq \mathrm{Av}_{n+1,1}(q)$$

for all n. This, in turn immediately implies the nonrationality of $A_q(z)$ by Theorem 2 if we select $k = 1$, $f_n = \mathrm{Av}_n(q)$, and $g_n = \mathrm{Av}_{n,1}(q)$ □

Note that we know that both sequences have a finite exponential order, since it is proved in [7] that $\mathrm{Av}_{n,1}(q) \leq \mathrm{Av}_n(q) \leq c_q^n$ for some constant c_q.

Using some straightforward symmetries, Theorem 3 can be strengthened as follows.

Theorem 4. *Let $q = q_1 q_2 \cdots q_k$ be any skew indecomposable permutation pattern so that $q_1 \neq 1$ or $q_k \neq q$. Then $A_q(z)$ is not rational.*

We will return to this result in Sect. 5, when we discuss how it fits into the "big picture".

4 Stack Sorting

Stack sorting of permutations has been defined in [6]. It is concerned with the operation of sorting permutations by passing them through a *stack*. This operation has many variations, and we have surveyed them in Chap. 8 of [2]. However, in this paper, we will restrict our attention to the most vigorously studied version, that is sometimes called *West* stack sorting, or *right-greedy* stack sorting.

There are at least two reasons for which this version of stack sorting is the subject of more work than other versions. First, there are three equivalent and natural ways of defining this stack sorting operation, which enables us to use at least three different sets of methods when proving results about stack sorting. Second, there are numerous conjectures about the operation that are very easy to state, yet very difficult to prove.

4.1 Three Equivalent Definitions

The Original Definition. In order to stack sort $p = p_1p_2\cdots p_n$, we consider the entries of the input permutation p one by one. First take p_1, and put it in the stack. Second, we take p_2. If $p_2 < p_1$, then it is allowed for p_2 to go in the stack on top of p_1, so we put p_2 there. If $p_2 > p_1$, however, then first we take p_1 out of the stack, and put it to the first position of the output permutation, and *then* we put p_2 into the stack. We continue this way: at step i, we compare p_i with the element $r = p_{a_{i-1}}$ currently on the top of the stack. If $p_i < r$, then p_i goes on the top of the stack; if not, then r goes to the next (that is, the leftmost) empty position of the output permutation, and p_i gets compared to the new element that is currently on the top of the stack. The algorithm ends when all n entries passed through the stack and are in the output permutation $s(p)$. See Fig. 1 for an illustration.

Definition 2. *If the output permutation* $s(p)$ *defined by the above algorithm is the identity permutation* $123\cdots n$*, then we say that* p *is* stack sortable.

output	stack	input
		3142
	3	142
	1 3	42
1	3	42
13	4	2
13	2 4	
132	4	
1324		

Fig. 1. Stack sorting 3142

The Recursive Definition. It follows from Definition 2 that the maximal entry n cannot enter the stack unless the stack is empty, that, is, all the entries that precede n in p are already in the output. Once n enters the stack, it will stay there until all other entries pass through the stack, at which point n will enter the output as its last entry. This proves the following.

Proposition 2. *Let $p = LMR$ be a permutation, where L denotes the string of entries on the left of the maximum entry M of p, and R denotes the string of entries on the right of M. Then the equality $s(p) = s(L)s(R)M$ holds.*

Note that if we define an operation S on all finite permutations by the rules

1. $S(\emptyset) = \emptyset$ and $S(1) = 1$, and
2. $S(p) = S(L)S(R)M$,

then these rules uniquely define $S(p)$ for every permutation p of any length. On the other hand, $s(p)$ satisfies both rules above, so by induction on the length of p, we have that $s(p) = S(p)$ for all p. So the above two rules *define* the stack sorting operation.

The Definition Using Trees. Let $p = p_1p_2 \cdots p_n$ be a permutation. The *decreasing binary tree* of p, which we denote by $T(p)$, is defined as follows. The root of $T(p)$ is a vertex labeled n, the largest entry of p. If a is the largest entry of p on the left of n, and b is the largest entry of p on the right of n, then the root will have two children, the left one will be labeled a, and the right one labeled b. If n is the first (resp. last) entry of p, then the root will have only one child, and that is a left (resp. right) child, and it will necessarily be labeled $n-1$ as $n-1$ must be the largest of all remaining elements. Define the rest of $T(p)$ recursively, by taking $T(L)$ and $T(R)$, where, as before, L and R are the substrings of p on the two sides of n, and affixing them to a and b.

Note that $T(p)$ is indeed a binary tree, that is, each vertex has 0, 1, or 2 children. Also note that each child is a left child or a right child of its parent, even if that child is an only child. Given $T(p)$, we can easily recover p by reading T according to the tree traversal method called *in-order*. In other words, first we read the left subtree of $T(p)$, then the root, and then the right subtree of $T(p)$. We read the subtrees according to this very same rule. See Fig. 2 for an illustration.

On the other hand, we can recover $s(p)$ by reading the vertices of $T(p)$ in *postorder*, that is, we first read the left subtree of the root, then the right subtree of the root, and then the root itself. The subtrees of the root are by the this same rule. It is a direct consequence of Proposition 2 that we indeed obtain $s(p)$ in this way.

Example 1. If $p = 328794615$, then reading the vertices of $T(p)$ shown in Fig. 2 in postorder, we obtain that $s(p) = 237841569$.

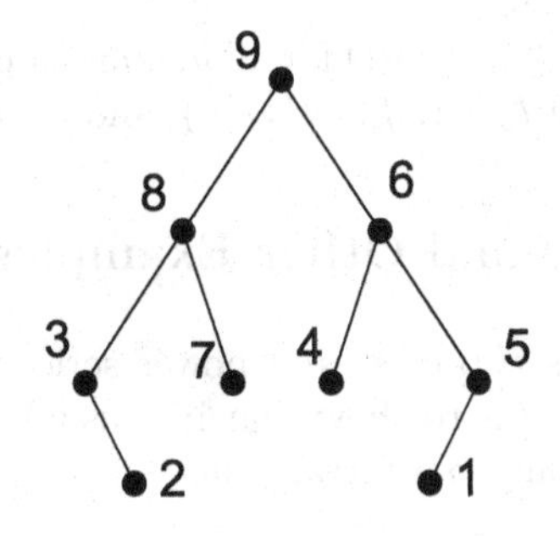

Fig. 2. The tree $T(p)$ for $p = 328794615$.

***t*-stack Sortable Permutations.** A permutation is called *t-stack sortable* if sending it through the stack t times results in the identity permutation. In other words, p is t-stack sortable if $s^t(p)$ is the identity permutation. Enumerating t-stack sortable permutations of length n for general t is extremely difficult. For $t = 1$, their number is the Catalan number $C_n = \binom{2n}{n}/(n+1)$, and for $t = 2$, their number is $W_2(n) = \frac{2}{(n+1)(2n+1)}\binom{3n}{n}$. (The latter is very difficult to prove.) For $t > 2$, not only we lack exact formulas, but also, we do not even know the exponential growth rate of the counting sequences.

Let $W_t(n)$ be the number of t-stack sortable permutations of length n, and let $W_t(z) = \sum_{n \geq 0} W_t(n) z^n$.

Theorem 5. *Let t be any positive integer. Then the generating function $W_t(z)$ is not rational.*

Proof. Let us say that a permutation p is *indecomposable* if it is not possible to cut p into two parts so that each entry before the cut is smaller than each entry after the cut. For instance, $p = 2413$ is indecomposable, but $r = 431265$ is not as we can cut it into two parts by cutting between entries 2 and 6, to obtain $4312|65$. Let $W_{t,1}(n)$ be the number of t-stack sortable permutations of length n that consist of one block, in other words, which are indecomposable. Let $W_{t,1}(z)$ be their generating function. It is then easy to see that

$$W_t(z) = \frac{1}{1 - W_{t,1}(z)}, \tag{4}$$

since the stack sorting operation will remove each block of smaller entries from the stack before entries from the next block can move in. In other words, a permutation is t-stack sortable if and only if all its blocks are.

Now let p be a t-stack sortable permutation of length n. Affix a new entry $n+1$ to its front, to get a new permutation that is indecomposable, and obviously t-stack sortable, since the first pass through the stack results in the permutation $s(p)(n+1)$, which is $(t-1)$-stack sortable since $s(p)$ is, and the entry $n+1$ will not destroy that property, always entering the stack after all other entries cleared it.

Therefore, $W_{t,1}(n+1) \leq W_t(n+1)$ for all n, and our claim follows immediately from Theorem 2, selecting $k = 1$, $f_n = W_t(n)$, and $g_n = W_{t,1}(n)$. □

5 The Big Picture and Other Examples

In this section, we discuss two classes of power series that contain the class of rational functions. Our goal is to show that it is usually difficult to prove that a power series does *not* belong to a certain class.

5.1 Algebraic Power Series

Definition 3. *The formal power series* $f \in \mathbf{C}[[z]]$ *is called* algebraic *if there exist polynomials* $P_0(z), P_1(z), \cdots, P_d(z) \in \mathbf{C}[z]$ *that are not all equal to zero so that*

$$P_0(z) + P_1(z)f(z) + \cdots + P_d(z)f^d(z) = 0. \tag{5}$$

The smallest $d > 0$ *for which such polynomials exist is called the* degree *of* f.

For instance, $1 - \sqrt{1-2z}$ and $\sqrt[3]{1+z}$ are algebraic power series. Trivially, all rational power series are algebraic, of degree one.

If we do not have an explicit form of a power series, it is usually difficult to prove that the power series is *not* algebraic. The one tool we are aware of is the following theorem of Jungen. If $f(z) \in \mathbf{C}[[z]]$ is algebraic and $f_n \sim cn^r\alpha^n$ for some constants $c \neq 0$ and $0 > r \in \mathbf{R}$, then $r = s + \frac{1}{2}$, for some integer s.

In his seminal book [11], Richard Stanley lists six general families of combinatorial objects, proves that they are counted by the same sequences, and shows that the generating function of those sequences is, with some basic assumptions, algebraic. Among these objects, we find lattice paths with certain steps, plane trees with prescribed down-degrees, legal sequence of parentheses, dissections of polygons with noncrossing diagonals, and sequences of integers with certain conditions. Of course, the fact that these power series are algebraic does not automatically mean that they are not rational, but we will present another application of our method to show that those sequences do not have rational generating functions.

Let S be a set of positive integers (finite or infinite). Let $f_S(n)$ be the number of lattice paths from $(0,0)$ to $(n,0)$ using steps $(1,k)$ where $k+1 \in S$, or $(1,-1)$ that never go below the horizontal axis. Let us call such lattice paths S-paths, and let

$$f_S(z) = \sum_{n \geq 0} f_S(n)z^n.$$

Stanley [11] mentions that $f_S(z)$ is algebraic if and only if S differs by a finite set from an infinite union of arithmetic progressions of positive integers. In particular, if S is finite, then $f_S(z)$ is algebraic.

Theorem 6. *If $S \neq \{1\}$, then $f_S(z)$ is not rational.*

Proof. Let $g_S(n)$ be number of S-paths from $(0,0)$ to $(n,0)$ that do not touch the horizontal axis, except in their starting and ending point, and let $g_S(z) = \sum_{n \geq 0} g_S(n) z^n$. Then clearly,

$$f_S(z) = \frac{1}{1 - g_S(z)}. \quad (6)$$

As in the preceding sections, we will show that $f_S(z)$ is not rational by proving that (6) is not a supercritical relation. Again, we will achieve that by showing that the sequence $g_S(n)$ has the same exponential order as the sequence $f_S(n)$.

In order to do that, let s be the minimal element of S, and let us take an S-path p from $(0,0)$ to $(n-s-1,0)$. Affix a $(1,s)$-step to the front of p, and affix s steps of type $(1,-1)$ to the end of p. Finally, translate the obtained path so that it starts at $(0,0)$. Then the new path p' will end in $(n,0)$, and will not touch the horizontal axis other than in its endpoints. As the map $p \rightarrow p'$ is obviously an injection from the set of S-paths from $(0,0)$ to $(n-s-1,0)$ to the set of S-paths from $(0,0)$ to $(n,0)$ that do not touch the horizontal axis, we have just proved the inequality $f_S(n-s-1) \leq g_S(n)$. Therefore, our claim follows immediately from Theorem 2, with $k = s+1$. □

5.2 *d*-finite Power Series

Definition 4. *We say that the power series $u(z) \in \mathbf{C}[[z]]$ is d-finite if there exists a positive integer d and polynomials $p_0(n), p_1(n), \cdots, p_d(n)$ so that $p_d \neq 0$ and*

$$p_d(z)u^{(d)}(z) + p_{d-1}(z)u^{(d-1)}(z) + \cdots + p_1(z)u'(z) + p_0(z)u(z) = 0, \quad (7)$$

Here $u^{(j)} = \frac{d^j u}{dz^j}$.

In other words, the derivatives of $u(z)$ span a *finite dimensional* vector space over the field of rational functions. The combinatorial importance of this class of power series is the following.

Definition 5. *A sequence $f : \mathbf{N} \rightarrow \mathbf{C}$ is called P-recursive if there exist polynomials $P_0, P_1, \cdots, P_k \in \mathbf{C}[n]$, with $P_k \neq 0$ so that*

$$P_k(n+k)f(n+k) + P_{k-1}(n+k-1)f(n+k-1) + \cdots + P_0(n)f(n) = 0 \quad (8)$$

for all natural numbers n.

Theorem 7 *[11]. The sequence $f(n)$ is P-recursive if and only if its generating function $\sum_{n \geq 0} f(n) z^n$ is d-finite.*

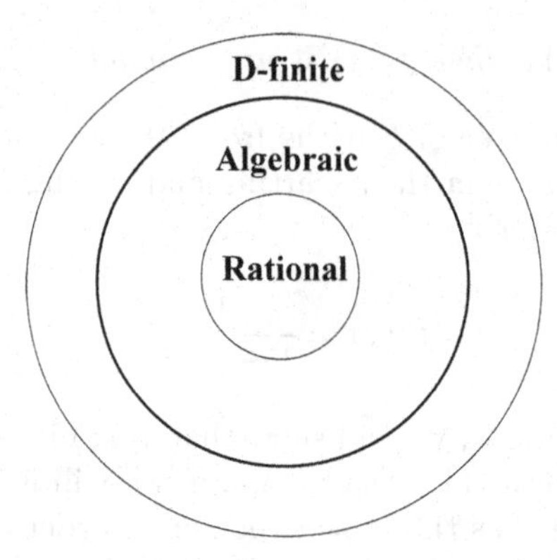

Fig. 3. Types of power series.

Note that rational power series are obviously d-finite as their coefficient sequences satisfy a fixed-term recurrence relation with *constant* coefficients, while the coefficient sequence of a d-finite power series is P-recursive, so it satisfies a fixed-term recurrence relation with *polynomial* coefficients. It is not quite this obvious, but it is not difficult to prove that algebraic power series are d-finite as well. See Fig. 3 for an illustration.

If we do not know the explicit form of a power series $f(z)$, then it is usually very difficult to prove that it is not d-finite. One way to prove that is to show that the coefficients of f grow too fast, but that is not the case if f is the generating function of some set of permutations of length n, since then $f_n \leq n!$, and the sequence $n!$ is P-recursive. Another way to prove that f is not d-finite is to show that it has infinitely many singularities.

In their important 1997 paper [9], John Noonan and Doron Zeilberger asked if $A_q(z)$ was a d-finite generating function for all q (see Sect. 3 for the definition of $A_q(z)$). While we cannot answer that question, our Theorem 4 answers a weaker version of that question for most patterns.

6 Further Directions

Proposition 3. *Let $B(z)$ be a power series with nonnegative real coefficients and convergence radius $r > 0$. If $B(r) < \infty$, then $B(z)$ is not a rational power series.*

Proof. As we mentioned in the introduction, r is a singular point of $B(z)$. If $B(r) < \infty$, then the singular point r cannot be a pole, and therefore, $B(z)$ is not a rational power series. □

The problem with using the simple Proposition 3 is that often we do not know the exact value of r, or we do not know the coefficients of $B(z)$ well enough to decide if $B(r)$ is finite or not. In this section, we will show examples on how these difficulties could potentially be overcome.

Let us keep the notation of Sect. 3, and let $\mathrm{Av}_{n,i}(q)$ denote the number of permutations of length n that avoid q and have i skew blocks. The following important result is proved in [3].

Lemma 1. *Let q be a skew indecomposable pattern that does not end in its largest entry. Then for all n, the inequality*

$$\mathrm{Av}_{n,2}(q) \leq \mathrm{Av}_{n,1}(q) \tag{9}$$

holds.

Let $A_{2,q}(z) = \sum_{n\geq 0} \mathrm{Av}_{n,2}(q)z^n$. Note that $A_{2,q}(z) = (A_{1,q}(z))^2$.

Corollary 1. *Let q be as in Lemma 1. Then $A_{1,q}(r) < 1$, and $A_q(r)$ is finite.*

Proof. Let us assume that $r > 1$. Let $z \in (0, r)$ so that $A_{1,r}(z) > 1$. Then

$$\sum_{n\geq 1} \mathrm{Av}_{n,1}(q)z_0^n = A_{1,q}(z_0) < A_{1,q}(z_0)^2 = A_{2,q}(z_0) = \sum_{n\geq 2} \mathrm{Av}_{n,2}(q)z_0^n,$$

contradicting (9), since the coefficients of both power series are all nonnegative.
So $A_{1,q}(r) < 1$, and therefore, by (3), it indeed holds that $A_q(z)$ is finite. □

Here is a potential application of Proposition 3. Note that Theorem 3 applies to the pattern 12453, but not to the pattern 13425. On the other hand, numerical evidence seems to suggest that $\mathrm{Av}_n(13425) \leq \mathrm{Av}_n(12453)$. If this inequality could be proved for all n, then, using the results in [1], it could be shown that for both patterns, the power series $A_q(z)$ has convergence radius $1/(9 + 4\sqrt{2})$. That would prove that for $r = 1/(9 + 4\sqrt{2})$, the chain of inequalities

$$A_{13425}(z) = \sum_{n\geq 0} \mathrm{Av}_n(13425)r^n \leq \sum_{n\geq 0} \mathrm{Av}_n(12453)r^n < \infty$$

holds, which would in turn prove that the generating function $A_{13425}(q)$ is not rational. That would be the first example of such a result for a pattern not covered by Theorem 4.

Here is another potential application of Proposition 3.

Theorem 8. *Let $K_{n,j}(q)$ be the number of permutations of length n that contain at most j copies of the pattern q. If there exists an absolute constant C so that*

$$K_{n,j}(q) \leq C\mathrm{Av}_n(q)$$

for all n, then the generating function

$$K_{j,q}(z) = \sum_{n\geq 0} K_{n,j}(q)z^n$$

is not rational.

Proof. This follows from the straightforward fact that the sequences $\mathrm{Av}_n(q)$ and $K_{n,j}(q)$ have the same exponential order. □

As far as the existence of the appropriate constant C in Theorem 8 goes, we know that such C exists if $q = 123$ and $j = 1$ [8], or $j = 2$ [5]. We conjecture that such C exists for all monotone q, and for any positive j.

Finally, here is another, related method to prove non-rationality of certain power series. It is proved in [4] that if the relation (1) between F and G is supercritical, then the expected number of irreducible components in a structure of size n is asymptotically equal to cn, for a specific constant c. So, if we can prove that in a given problem, the average number of irreducible components in a structure of size n is not asymptotically equal to n, or, equivalently, the average size of an irreducible component is not asymptotically equal to a constant, then the relation (1) between $F(z)$ and $G(z)$ is not supercritical, and therefore, the power series $F(z)$ and $G(z)$ are not rational.

References

1. Bóna, M.: The limit of a Stanley-Wilf sequence is not always rational, and layered patterns beat monotone patterns. J. Combin. Theory Ser. A **110**(2), 223–235 (2005)
2. Bóna, M.: Combinatorics of Permutations, 2nd edn. CRC Press, Boca Raton (2012)
3. Bóna, M.: Supercritical sequences, and the nonrationality of most principal permutation classes. Eur. J. Combin. **83**, 103020 (2020)
4. Flajolet, P., Sedgewick, R.: Analytic Combinatorics. Cambridge University Press, Cambridge (2009)
5. Fulmek, M.: Enumeration of permutations containing a prescribed number of occurrences of a pattern of length three. Adv. Appl. Math. **30**(4), 607–632 (2003)
6. Knuth, D.E.: The Art of Computer Programming: Volume 3: Sorting and Searching. Addison-Wesley, Reading (1973)
7. Marcus, A., Tardos, G.: Excluded permutation matrices and the Stanley-Wilf conjecture. J. Combin. Theory Ser. A **107**(1), 153–160 (2004)
8. Noonan, J.: The number of permutations containing exactly one increasing subsequence of length three. Discrete Math. **152**(1–3), 307–313 (1996)
9. Noonan, J., Zeilberger, D.: The enumeration of permutations with a prescribed number of "forbidden" patterns. Adv. Appl. Math. **17**(4), 381–407 (1997)
10. Rice, H.G.: Classes of recursively enumerable sets and their decision problems. Trans. Am. Math. Soc. **74**, 358–366 (1953)
11. Stanley, R.: Enumerative Combinatorics, vol. 2. Cambridge University Press, Cambridge (1997)
12. Vatter, V.: Permutation classes. In: Handbook of Enumerative Combinatorics, Miklós Bóna, editor. CRC Press, Boca Raton (2015)

Binary Decision Diagrams: From Tree Compaction to Sampling

Julien Clément[1(✉)] and Antoine Genitrini[2]

[1] Normandie Univ, UNICAEN, ENSICAEN, CNRS, GREYC, 14000 Caen, France
Julien.Clement@unicaen.fr
[2] Sorbonne Université, CNRS, LIP6, UMR 7606, 75005 Paris, France
Antoine.Genitrini@lip6.fr

Abstract. Any Boolean function corresponds with a complete full binary decision tree. This tree can in turn be represented in a maximally compact form as a direct acyclic graph where common subtrees are factored and shared, keeping only one copy of each unique subtree. This yields the celebrated and widely used structure called reduced ordered binary decision diagram (ROBDD). We propose to revisit the classical compaction process to give a new way of enumerating ROBDDs of a given size without considering fully expanded trees and the compaction step. Our method also provides an unranking procedure for the set of ROBDDs. As a by-product we get a random uniform and exhaustive sampler for ROBDDs for a given number of variables and size.

1 Introduction

The representation of a Boolean function as a binary decision tree has been used for decades. Its main benefit, compared to other representations like a truth table or a Boolean circuit, comes from the underlying *divide-and-conquer* paradigm. Thirty years ago a new data structure emerged, based on the compaction of binary decision tree, and hereafter denoted as Binary Decision Diagrams (or BDDs) [1]. Its take-off has been so spectacular that many variants of compacted structures have been developed, and called through many acronyms as presented in [14]. One way to represent the different diagrams consists in their embedding as

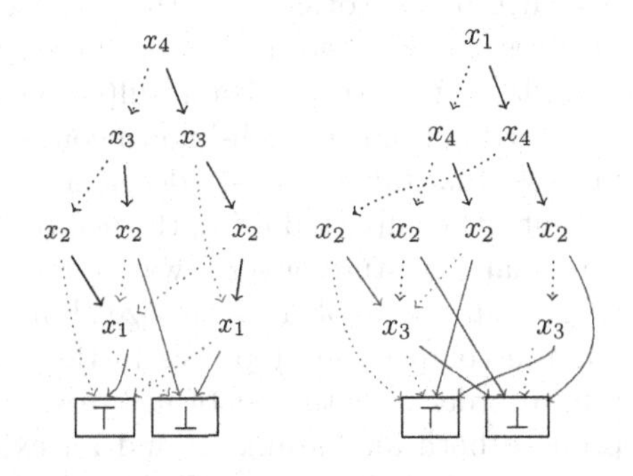

Fig. 1. Two Reduced Ordered Binary Decision Diagrams associated to the same Boolean function. Nodes are labeled with Boolean variables; left dotted edges (resp. right solid edges) are 0 links (resp. 1 links).

This work was partially supported by the ANR projects METACONC ANR-15-CE40-0014 and PING/ACK ANR-18-CE40-0011. An implementation of the results is provided at https://github.com/agenitrini/BDDgen.

Y. Kohayakawa and F. K. Miyazawa (Eds.): LATIN 2020, LNCS 12118, pp. 571–583, 2020.
https://doi.org/10.1007/978-3-030-61792-9_45

directed acyclic graphs (or DAGs). One reason for the existence of all these variants of diagrams is due to the fact that each DAG correspondence has its own internal agency of the nodes and thus each representation is oriented towards a specific constraint. For example, the case of Reduced Ordered Binary Decision Diagrams (ROBDDs) is such that the variables do appear at most once and in the same order along any path from the source to a sink of the DAG, and furthermore, no two occurrences of the same subgraph do appear in the structure. For such structures and others, like QOBDDs or ZBDDs for example, there is a canonical representation of each Boolean function.

In his book [9] Knuth proves or recalls combinatorial results, like properties for the profile of a BDD, or the way to combine two structures to represent a more complex function. However, one notes an unseemly fact. There are no results about the distribution of the Boolean functions according to their ROBDD size. In fact in contrast to (e.g.) binary trees where there is a recursive characterization that allows to well specify the trees, we have no local-constraint here for ROBDDs ans thus a similar recurrence is unexpected. Very recently, there is a first study exploring experimentally, numerically, and theoretically the typical and worst-case ROBDD sizes in [12]. We aim at obtaining the same kind of combinatorial results but here we design a partition of the decision diagrams that allows us to go much further in terms of size. In particular we obtain an exhaustive enumeration of the diagrams according to their size up to 9 variables. This was unreachable through the exhaustive approach proposed in [12] due to the double exponential complexity of the problem: there are 2^{2^k} Boolean functions with k variables. Our C++ implementation fully manages the case of 9 variables (see Fig. 2) that corresponds to $2^{512} \approx 10^{154}$ functions. In particular for 9 Boolean variables, our implementation shows one seventh of all ROBDDs are of size 132 (the possible sizes range from 3 to 143). Furthermore, ROBDDs of size between 125 and 143 represents more than 99.8% of all ROBDDs, in accordance with theoretical results from [8,13].

Starting from the well-known compaction process (that takes a binary decision tree and outputs its compacted form, the ROBDD), our combinatorial study gives a way of construction for ROBDDs of a given size, but without the compaction step. We further define a total order over the set of ROBDDs and we propose both an unranking and an exhaustive generation algorithm. The first one gives as a by-product a uniform random sampler for ROBDDs of a given number of variables and size. One strength of our approach is that it allows to sample *uniformly* ROBDDs of "small" size, for instance of linear size w.r.t the number k of variables, very efficiently in contrast to a naive rejection algorithm. The usual uniform distribution on Boolean functions [13] yields with high probability ROBDDs of near maximal size of order $2^k/k$, although ROBDDs encountered in applications, when tractable,

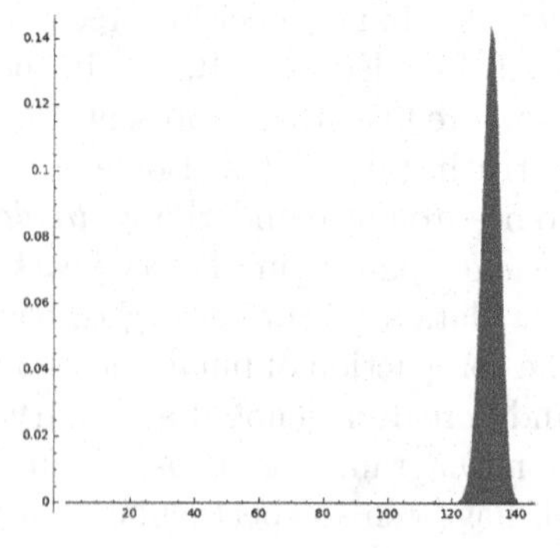

Fig. 2. Proportion of BDDs over 9 variables according to their size

are smaller. As a perspective, once the unranking method is well understood, and in particular the poset underlying the ROBDDs, then we might be able to bias the distribution to sample only in a specific subclass, e.g. ROBDDs corresponding to a particular class of formulas (e.g. read-once formulas).

Our results have practical applications in several contexts, in particular for testing structures and algorithms. The study given in [3] executes tests for an algorithm whose parameter is a binary decision diagram. It is based on QuickCheck [2], the famous software, taking as an entry a random generator and generating test cases for test suites. Using our uniform generator, we aim at obtaining statistical testing, in the sense that the underlying distribution of the samples is uniform, thus allowing to extract statistics thanks to the tests. Another application of our approach allows to derive exhaustive testing for small structures, like the study in [10], that we also can conduct inside QuickCheck.

In this paper, we focus exclusively on ROBDDs which is one of the first and simplest variants. Section 2 introduces the combinatorics underlying the decision tree compaction, leading in Sect. 3 to a way to unambiguously specify the structure of reduced ordered binary decision diagrams. We apply this strategy in Sect. 4 and obtain an unranking algorithm for ROBDDs.

2 Decision Diagrams as Compacted Trees

This section defines precisely our combinatorial context. Many definitions are detailed in the monograph of Wegener [14] and in the dedicated volume [9] of Knuth.

In this section we first recall a one-to-one correspondence between the representation of a Boolean function as a binary decision tree (built on a specific variable ordering and seen as a *plane tree*, i.e., the children of an internal nodes are ordered) and a reduced ordered binary decision diagram (ROBDDs) also seen as a plane structure. This approach is *non-classical* in the context of BDDs, but it allows the formalization of an equivalence relationship on ROBDDs that is the key of our enumeration: in fact our approach foundation relies on breaking down the symmetry in ROBDDs. We consider Boolean function on k variables. We recall there are 2^{2^k} such Boolean functions. The compaction process is now formalized.

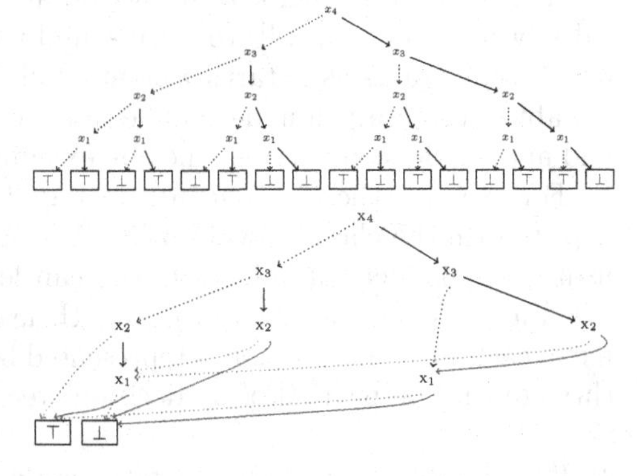

Fig. 3. A decision tree and its postorder compaction

Compaction and Plane Decision Diagrams. A Boolean function can be represented thanks to a binary decision diagram, which is a rooted, directed, acyclic graph, which consists of decision nodes and terminal nodes. There are two types of terminal nodes $\top$ and $\bot$ corresponding to truth values (resp. 1 and 0). Each decision node ν is labeled by a Boolean variable x_ν and has two child nodes (called *low child* and *high child*). The edge from node ν to a low (or high) child represents an assignment of x_ν to 0 and is represented as a dotted line (respectively 1, represented as a solid line). In the following we represent ROBDDs and decision trees (or BDDs in general) as *plane structures*, i.e., for a node we consider its low child to be its left child and the high child to be its right child.

In a (plane) full binary decision tree, no subtree is shared. By contrast we may decrease the number of decision nodes by factoring and sharing common substructures. Representing a function with its full decision tree is not space efficient. In Fig. 3 we depict, on top, a decision tree of a Boolean function on 4 variables. In the bottom of the figure we represent the compaction of the latter decision tree by using the classical common subexpression recognition notion (cf. e.g. [4,7]) based on a postorder traversal of the tree.

Definition 1 (Compaction). *Let T be the (plane) binary decision tree of a function f. The* DAG *T is modified through a postorder traversal. When the node ν is under visit, ν being a child of a node ρ. If an identical subtree than T_ν, the one rooted in ν, has already be seen during the traversal, rooted in a node μ, then T_ν is removed from T and the node ρ gets a pointer to μ (replacing the edge to ν). Once T has been traversed, the resulting* DAG *is the plane* ROBDD *of f.*

In our figures of ROBDDs we draw the pointers in red (there is an exception for the edges to the terminal nodes as we remark in Fig. 3 also drawn in red).

In a classical setting, ROBDDs are obtained by applying repetitively reduction rules (well detailed in [14]) to OBDDs, and the process is confluent. Our approach conceptually takes as a starting point a full decision tree with a given ordering on variables (meaning all nodes at the same level are labeled by the same variable) and applies the compaction rules by examining nodes of the tree in postorder.

For example the plane ROBDD in Fig. 3 corresponds to the leftmost ROBDD depicted (in the classical way) in Fig.1. Note that for a given Boolean function, using two distinct variable orderings can lead to two ROBDDs of different sizes (see Fig. 1 for such a situation). Nonetheless, an ordering of the variables being fixed, each Boolean function is represented by exactly one single ROBDD obtained through the compaction of its decision tree for this order.

In the rest of the paper, we consider only plane ROBDDs. From now we thus call them BDDs. We also assume the set of variables $X = \{\ldots > x_k > \ldots > x_1\}$ *is totally ordered.*

Our first goal aims at giving an effective method to enumerate BDDs with a chosen number k of variables and size n. A first naive approach is: (1) enumerate all the 2^{2^k} Boolean functions by construction of the decision trees; (2) apply the compaction procedure; and (3) finally filter the BDDs of size equal to the target size n. This algorithm ceases to be practical for k larger than 4 (see [12]).

In this paper, we propose a new combinatorial description of BDDs providing the basis for an enumeration algorithm avoiding the enumeration of all Boolean functions on k variables.

3 Recursive Decomposition

This section introduces a canonical and unambiguous decomposition of the BDDs yielding a recursive algorithm for their enumeration.

Automaton Point of View. Let us introduce an equivalent representation for a BDD. A BDD can indeed be described as a deterministic finite automaton with additional constraints and properties. This point of view gives a convenient formal characterization of the decomposition of BDDs used in our algorithms.

Definition 2 (BDD as an automaton). *A* BDD *B of index k is a tuple (Q, I, r, δ) where*

- *Q is the set of nodes of the* BDD*. Q contains two special sink nodes $\bot$ and $\top$.*
- *$I : Q \to \{0, \dots, k\}$ is the index function which associates with every node its index. By convention the index of both sink nodes is 0.*
- *$r \in Q$ is the root and has index $I(r) = k$.*
- *$\delta : Q \setminus \{\bot, \top\} \times \{0, 1\} \to Q$ is the full transition function.*

There are constraints on δ translating the classical ones of the BDDs*:*

- *for any node $\nu \in Q \setminus \{\bot, \top\}$, $\delta(\nu, 0) \neq \delta(\nu, 1)$.*
- *for any distinct nodes μ and ν with the same index, we have $\delta(\mu, 0) \neq \delta(\nu, 0)$ or $\delta(\mu, 1) \neq \delta(\nu, 1)$.*
- *the graph underlying δ forms a* DAG *with a unique node of in-degree 0, the root r.*
- *if $\tau = \delta(\nu, \alpha)$ for some $\alpha \in \{0, 1\}$ then $I(\tau) < I(\nu)$.*

We say τ is the low child of ν (respectively high child of ν) if $\delta(\nu, 0) = \tau$ (resp. $\delta(\nu, 1) = \tau$).

Definition 3 (Spine of a BDD, tree and non-tree edges). *Let a* BDD *$B = (Q, I, r, \delta)$ of root-index k. The* spine *of B is the spanning tree obtained by a* depth-first search *of the (plane)* BDD *(where low child is accessed before the high one), and omitting the sinks $\bot$ and $\top$. For a* BDD *B, the edges of the spine forms the set of* tree edges *(drawn in black). The other edges form the set of* non-tree edges *(drawn in red). We describe the spine T as a tuple $T = (Q', I, r, \delta')$ with set of nodes $Q' = Q \setminus \{\bot, \top\}$ (with the same index function I as for B). The edges of the spine are described using a partial transition function $\delta' : Q' \times \{0, 1\} \to Q' \cup \{\text{NIL}\}$ where* NIL *is a special symbol designating an undefined transition.*

Using standard terminology for depth-first search, *non-tree* edges are either *forward or cross* edges. We remark that, by definition, a DAG admits no cycles and still in the standard notation, it has no *backward* edges.

Undefined values of the transition function δ' can conveniently be seen as half edges. Since in a BDD every non-sink node has two children, the spine of a BDD of size n has $(n - 2)$ nodes and $(n - 1)$ half edges (drawn in red in Fig. 1). The four possible types of a node are depicted, as the roots in Fig. 4.

Definition 4 (Valid tree). *A binary tree is said to be* valid *if it is the spine of some* BDD*. The set of spines of size n is denoted as $\mathcal{T}_n$.*

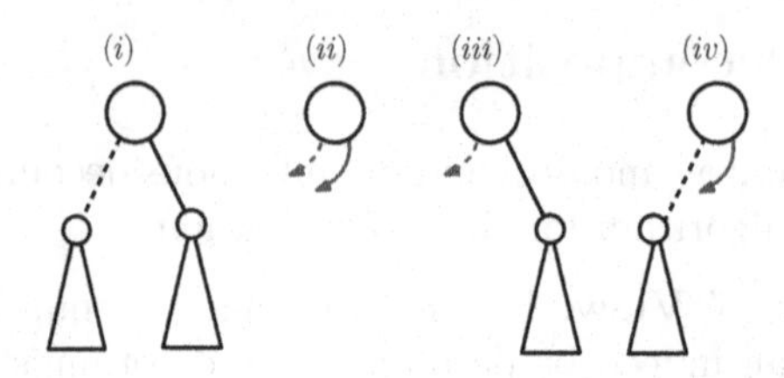

Fig. 4. The four cases for a node of a spine. From left to right: an internal node with both transitions defined, two half edges, one low (left) half edge, one high (right) half edge (cf. Proposition 1)

See Fig. 5 for examples of valid and invalid trees. To the best of our knowledge, there is no way to characterize valid trees, apart from exhibiting a ROBDD admitting this tree as a spine. We will discuss this point later.

For enumerating BDDs it will prove convenient to introduce the profile list of a set of nodes and some other useful notation for lists manipulation.

Definition 5. *The* profile *of* $\mathcal{N}$, *denoted by* $\text{profile}(\mathcal{N})$, *is a list with* $(k+1)$ *components* $\boldsymbol{p} = (p_0, \ldots, p_k)$ *where* $k = \max_{\nu \in \mathcal{N}} I(\nu)$ *is the maximal index and* p_i *is the number of nodes of index* i *in* $\mathcal{N}$.

This definition extends naturally to trees, graphs, etc. We also equip the set of lists with a '+' operation: let two lists $\boldsymbol{v} = (v_0, \ldots, v_m)$ and $\boldsymbol{v}' = (v'_0, \ldots, v'_n)$ with $n \geq m$ (w.l.o.g.), the sum $\boldsymbol{v} + \boldsymbol{v}'$ is equal to $\boldsymbol{w} = (w_0, \ldots, w_n)$ where for all $0 \leq i \leq m$, $w_i = v_i + v'_i$ and otherwise, when $m < i \leq n$, $w_i = v'_i$.

In the following, we will use two orderings on the nodes of a plane BDD induced by depth-first search, and called postordering and preordering. Since the structure is plane these orderings correspond exactly with the classical postorder traversal and the preorder traversal of its spanning tree. In a tree, for a node ν with low child ν_0 and high child ν_1, the postorder traversal visits the subtree rooted at ν_0 then, the one rooted at ν_1 and finally ν. The preorder traversal first visits the node ν, then the subtree rooted at ν_0 and finally the subtree rooted at ν_1. We use the notation $\mu \prec_{\text{post}} \nu$ (resp. $\mu \prec_{\text{pre}} \nu$) if the node μ is visited before ν using the postorder (resp. preorder) traversal.

We characterize now how the partial transition function of the spine is related to the full transition function of the BDD. Introducing the pool and level set of a node, we describe the valid choices for non-tree edges to yield a BDD.

Definition 6 (Pool and level set). *Let* T *be the spine of a* BDD. *The pool of a node* $\nu \in T$ *is*

$$\mathcal{P}_T(\nu) = \{\tau \in T \mid \tau \prec_{pre} \nu \text{ and } I(\tau) < I(\nu)\} \cup \{\bot, \top\}.$$

The pool profile $p_T(\nu)$ *of a node* ν *in a spine* T *is* $p_T(\nu) = \text{profile}(\mathcal{P}_T(\nu))$.
The level set *of* ν *is* $\mathcal{S}_T(\nu) = \{\tau \in T \mid \tau \prec_{pre} \nu \text{ and } I(\tau) = I(\nu)\}$, *and the* level rank $s_T(\nu) = |\mathcal{S}_T(\nu)|$ *of a node* ν *is the rank of* ν *among the set of nodes with the same index.*

Informally the pool of a node ν of a tree T is the set of nodes we could choose as a low child for ν without invalidating the spine. The first component of a pool profile is always 2 since both sinks $\bot$ or $\top$ are present in the pool of any node of the spine (providing the underlying BDD is not reduced to 1 or 0).

Proposition 1. *Let $T = (Q', I, r, \delta')$ be a valid spine with set of nodes Q', root r and partial transition function $\delta' : Q' \times \{0, 1\} \to Q' \cup \{\text{NIL}\}$. The full transition function $\delta : Q' \times \{0, 1\} \to Q' \cup \{\bot, \top\}$ is the transition function of a BDD with spine T if and only for any node $\nu \in Q'$, noting $\nu_0 = \delta(\nu, 0)$ and $\nu_1 = \delta(\nu, 1)$, the pair (ν_0, ν_1) satisfies*

(i) if $\delta'(\nu, 0) \neq \text{NIL}$ and $\delta'(\nu, 1) \neq \text{NIL}$ then $\nu_\alpha = \delta'(\nu, \alpha)$ for $\alpha \in \{0, 1\}$.
(ii) if $\delta'(\nu, 0) = \delta'(\nu, 1) = \text{NIL}$, then

$$\nu_\alpha \prec_{pre} \nu \text{ and } I(\nu_\alpha) < I(\nu) \text{ for } \alpha \in \{0, 1\} \text{ and } \nu_0 \neq \nu_1,$$

and there is no node $\tau \neq \nu$ with the same index as ν such that $\delta(\tau, \cdot) = \delta(\nu, \cdot)$.
(iii) if $\delta'(\nu, 0) = \text{NIL}$ and $\delta'(\nu, 1) \neq \text{NIL}$, then

$$\nu_0 \prec_{pre} \nu,\ I(\nu_0) < I(\nu) \text{ and } \nu_1 = \delta'(\nu, 1).$$

(iv) if $\delta'(\nu, 0) \neq \text{NIL}$ and $\delta'(\nu, 1) = \text{NIL}$, then $\nu_0 = \delta'(\nu, 0)$ and

$$\nu_1 \prec_{post} \nu,\ \nu_1 \neq \nu_0 \text{ and } I(\nu_1) < I(\nu).$$

Proof. Since $\delta(\cdot, \cdot)$ must extend $\delta'(\cdot, \cdot)$, case (i) is trivial since we must only extend the transition function where $\delta(\nu, \alpha) = \text{NIL}$. In case (ii), we have to choose for (ν_0, ν_1) two nodes in the pool of ν (ν is an external node of the spine). We use the preorder traversal (but since ν is an external node, the postorder would also be fine). Moreover $\nu_0 \neq \nu_1$ and no node with the same index as ν can have the exact same descendants (ν_0, ν_1) in accordance with Definition 2. In case (iii), the low child must be chosen in the pool of ν since we preserve the spine. In case (iv), the high child of ν is also chosen in the pool ν or in the pool of ν_0 (and must still be different from ν_0 by Definition 2). □

4 Counting and Generating BDDs

In this section, we sketch algorithms in order to count and sample BDDs of a given size n and given number k of variables.

Counting BDDs. Given a spine T, we can compute the number of BDDs corresponding with this spine. Thus counting BDDs of a certain size n will consists in building all valid spines of size $(n-2)$ and completing the transition function of the spine in all possible ways according to Proposition 1.

Definition 7 (Weight). *Let $T = (Q', I, \delta', r)$ be a spine, the* weight $w_T(\nu)$ *of a node $\nu \in Q'$ is the number of possibilities for completing the transition function $\delta'(\mu, \cdot)$ and yielding a BDD with spine T. The* cumulated weight *of a subtree T_ν rooted at $\nu \in T$ is $W_T(\nu) = \prod_{\tau \in T_\nu} w_T(\tau)$. We write $W(T) = W_T(r)$ to denote the cumulated weight of the whole spine T rooted at r.*

Note that the number of choices for the missing transitions out of a node ν are the ones remaining after previous choices have been made for other nodes of the spine.

Proposition 2 (Weight of a node). *Let T be a spine T, the* weight *of a node $\nu \in T$ is*

$$w_T(\nu) = \begin{cases} 1 & \text{if } \delta'(\nu,0) \neq \text{NIL and } \delta'(\nu,1) \neq \text{NIL} \\ \|p_T(\nu)\|(\|p_T(\nu)\| - 1) - s_T(\nu) & \text{if } \delta'(\nu,0) = \delta'(\nu,1) = \text{NIL} \\ \|p_T(\nu) + \text{profile}(T')\| & \text{if } \delta'(\nu,0) \neq \text{NIL and } \delta'(\nu,1) = \text{NIL} \\ \|p_T(\nu)\| & \text{if } \delta'(\nu,0) = \text{NIL and } \delta'(\nu,1) \neq \text{NIL} \end{cases}$$

where $p_T(\nu)$ is the pool profile of node ν, $T' = T_{\nu_0}$ is the subtree (when defined) rooted at $\nu_0 = \delta'(\nu, 0)$, and, for a list $\mathbf{p} = (p_0, \dots, p_k)$, we denote $\|\mathbf{p}\| = \sum_{i=0}^{k} p_i$.

In the third case, by $p_T(\nu) + \text{profile}(T')$, we mean the profile of the set of nodes visited before ν with the postorder traversal of T and of index strictly smaller than $I(\nu)$.

Proof. This is a direct application of Proposition 1. □

This formula allows to detect if a tree is a valid spine. Indeed as soon as the weight of a node is zero or negative, there is no way to define a total transition function δ for a BDD. Note that this situation can only happen for external nodes having two half edges, since for any node $\nu \in Q'$ and any spine T, $\|p_T(\nu)\| \geq 2$.

In Fig. 5, the binary tree on the left is invalid and cannot be the spine of any BDD. The two other trees on the right have weights 4 and 24, i.e., are resp. the spines of exactly 4 and 24 BDDs. It is an open problem to characterize the set of valid trees (apart from exhibiting corresponding BDDs).

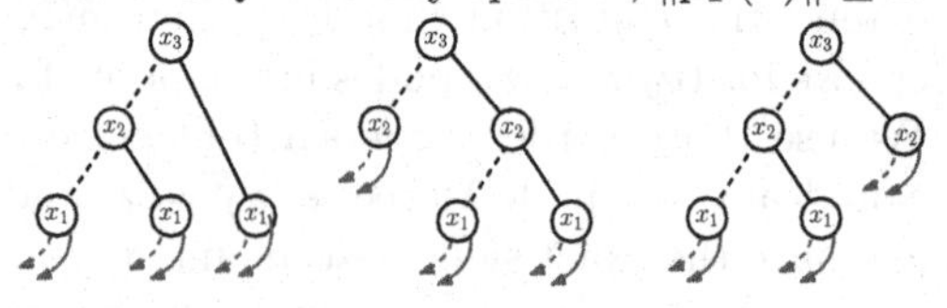

Fig. 5. Three examples of binary trees (first one is invalid, the two other have respective weights 4 and 24).

Proposition 2 gives access to the total weight of the spine $W(T)$ using a recursive procedure. A natural way to proceed algorithmically is to use a recursive postorder traversal of the tree maintaining at each node the weight in a multiplicative manner. To do so we need to keep track in the traversal of the pool profile and level rank of the current node.

Initially the pool of the root is reduced to the set $\{\bot, \top\}$. Thus the initial pool profile of the root of index k is initialized to $(2, 0, \dots, 0)$ of length k. The level rank of the root of the spine is 0.

Proposition 3. *Let $N(n, k)$ be the number of* BDDs *of index k and size n*

$$N(n, k) = \sum_{T \in \mathcal{T}_{n-2,k}} W(T),$$

where $\mathcal{T}_{m,k}$ is the set of valid spines with m nodes for BDDs *of index k.*

Proof. The weight of a spine is the number of ways of extending the transition function of T (Proposition 1), hence the number of BDDs for this given spine. □

Combinatorial Description of Spines. The set of spines is not straightforward to characterize in a combinatorial way. Indeed we need context to decide if the weight of a particular node in a tree is 0 or less, which in turn yields that the tree is not valid . To enumerate spines, we build recursively binary trees, and, while computing weights for its nodes, as soon we can decide the (partially built) tree is not valid, the tree is discarded.

To decompose (or count) spines of any size or index, $\mathcal{T} = \bigcup_{n\geq 1}\bigcup_{k\geq 1}\mathcal{T}_{n,k}$, we introduce a partition over subtrees which can occur in a spine $T \in \mathcal{T}$. The goal is to identify identical subtrees occurring within different spines and with the same weight to avoid redundant computations.

The combinatorial description we are about to present originates from the following observation: let us fix a spine T and a node $\nu \in T$. From Proposition 2, to compute the cumulated weight of the subtree T_ν rooted at ν, the sole knowledge of the pool profile $p_T(\nu)$ and the level rank $s_T(\nu)$ is sufficient.

Let S and S' be two subtrees with respective roots ν and ν' in some spines T and T', we denote $S \equiv S'$ if the following three conditions are satisfied:

- both trees have the same size: $|S| = |S'|$;
- the roots of both trees have the same pool profile: $p_T(\nu) = p_{T'}(\nu')$;
- the roots of both trees have the same level rank: $s_T(\nu) = s_{T'}(\nu')$.

The set $\mathcal{T}_{m,\boldsymbol{p},s}$ is the class equivalence for the relation '$\equiv$' and gathers trees (as a set, without multiplicities) which are possible subtrees of size m in any spine, knowing only the pool profile $\boldsymbol{p}$ and level rank s of the root of the subtree. More formally:

$$\mathcal{T}_{m,\boldsymbol{p},s} = \{T_\nu \mid (\exists T \in \mathcal{T})\ (\exists \nu \in T)\ p_T(\nu) = \boldsymbol{p} \text{ and } s_T(\nu) = s\}.$$

Note that we have $\mathcal{T}_{n,k} = \mathcal{T}_{n,(2,0,\ldots,0),0}$, where $(2, 0, \ldots, 0)$ has k components.

Proposition 4. *The set $\mathcal{T}_{m,\boldsymbol{p},s}$ of subtrees of size m rooted at a node having pool profile $\boldsymbol{p} = (p_0, \ldots, p_{k-1})$ and level rank s occurring in the set of spines $\mathcal{T}$ is decomposed without any ambiguity. We decompose a subtree $T \in \mathcal{T}_{m,\boldsymbol{p},s}$ as a tuple (ν, T', T'') where the root ν has index k and T' and T'' are its left and right (possibly empty) subtrees of respective sizes i and $m-1-i$, with $0 \leq i \leq m-1$, and verifying (when non empty)*

(i) $T' \in \displaystyle\bigcup_{k_0\in\{1,\ldots,k-1\}} \mathcal{T}_{i,(p_0,\ldots,p_{k_0-1}),p_{k_0}}$

(ii) $T'' \in \displaystyle\bigcup_{k_1\in\{1,\ldots,k-1\}} \mathcal{T}_{m-i-1,(p'_0,\ldots,p'_{k_1-1}),p'_{k_1}}$, *with* $\boldsymbol{p}' = \boldsymbol{p} + \text{profile}(T')$

(iii) if $m = 1$ *then* $\left(\sum_{i=0}^{k} p_i\right) \cdot \left(-1 + \sum_{i=0}^{k} p_i\right) - s > 0.$

This proposition ensures that we can decompose unambiguously subtrees occurring in spines in accordance with the equivalence relation '$\equiv$'. Practically this means that instead of considering all possible subtrees for all possible spines, we can compute cumulative weights for each representative of the equivalence relation (which are fewer although still of exponential cardinality).

Algorithm COUNT$(n, \boldsymbol{p}, s)$ in Algorithm 1 enumerates spines of BDDs and, at the same time, computes their cumulated weights. It takes as arguments a size n for considering all subtrees of size n, assuming an initial pool profile $\boldsymbol{p} = (p_0, \ldots, p_{k-1})$, level rank s and index k for the root of these trees. It returns in an associative array a list of pairs $(\boldsymbol{t}, w)$ where

- $\boldsymbol{t} = (t_0, t_1, \ldots, t_{k-1}, t_k)$ ranges over the set of profiles of trees in $\mathcal{T}_{m,\boldsymbol{p},s}$, i.e., $\boldsymbol{t} \in \{\text{profile(T)} \mid \text{T} \in \mathcal{T}_{\text{m},\boldsymbol{p}=(\text{p}_0,\ldots,\text{p}_{\text{k}-1}),\text{s}}\}$.
- w is the sum over all equivalent trees of size m with profile $\boldsymbol{t}$ of their cumulated weights when the root has pool profile $\boldsymbol{p}$ and level rank s (which gives enough information to compute the cumulated weight for each tree using Proposition 2).

Note that any subtree T with a root of index i has a profile $\boldsymbol{t} = (t_0, \ldots, t_i)$ with $t_0 = 0$ and $t_i = 1$.

Proposition 5. *The number $N(n, k)$ of* BDDs *of size n and of index k is computed thanks to Algorithm* COUNT() *and is equal to*

$$N(n, k) = \sum_{(\boldsymbol{t},w)\in\text{COUNT}(n-2,(2,0,\ldots,0),0)} w,$$

where $(2, 0, \ldots, 0)$ has k components and corresponds with a pool reduced to the two sink nodes $\bot$ and $\top$ of index 0.

Proof. Indeed $\boldsymbol{t}$ ranges over all possible profiles for spines of size $(n - 2)$ and we sum the weights of all spines for these profiles. Hence we compute exactly the number of BDDs of size n. □

An important refinement for this algorithm is to remark when summing over all spines, we consider subtrees of the same size whose root shares the same pool profile and same level rank, hence the same context. In order to avoid performing the same exact computations twice (or more) we can use *memoization* technique (that is storing intermediary results). It is an important trick to reduce the time complexity, although at the cost of some memory consumption.

Complexity of the Counting Algorithm. First, we remark the numbers involved in the computations are (very) big numbers (as seen before, of order 2^{2^k}).

Proposition 6. *The complexity (in the number of arithmetic operations) of the computations of the Algorithm 1 to evaluate $N(n, k)$ is $O\left(\frac{1}{k}2^{3k^2/2+k}\right)$.*

Algorithm 1. Algorithm COUNT(). THE INITIAL POOL PROFILE OF THE ROOT (OF INDEX k) IS $(2, 0, \ldots, 0)$ OF LENGTH k.

```
function COUNT(n, p = (p_0, ..., p_{k-1}), s)
    d ← {}                                                   ▷ Empty dictionary
    if n = 0 then
        S ← (Σ_{j=0}^{k-1} p_j) · (Σ_{j=0}^{k-1} p_j − 1)
        if S > 0 then d ← {e^(k) : S}                        ▷ See *
    else
        for i ← 0 to n − 1 do                ▷ Left/right subtrees of size i/n − i − 1
            d_0 ← {}
            if i = 0 then d_0 ← {ε : Σ_{i=0}^{k-1} p_i}     ▷ left subtree is empty, see *
            else
                for k_0 ← 1 to k − 1 do                      ▷ left node has index k_0
                    d_0 ← d_0 ∪ COUNT(i, (p_0, ..., p_{k_0−2}), p_{k_0−1})
            for (ℓ, w_0) ← d_0 do
                d_1 ← {}
                p' ← p + ℓ
                if n − 1 − i = 0 then d_1 ← d_1 ∪ {ε : −1 + Σ_{i=0}^{k-1} p'_i}  ▷ right subtree is
empty
                else
                    for k_1 ← 1 to k − 1 do                  ▷ right node has index k_1
                        d_1 ← d_1 ∪ COUNT(n − 1 − i, (p'_0, ..., p'_{k_1−2}), p'_{k_1−1})
                for (r, w_1) ← d_1 do
                    w ← w_0 · w_1
                    t ← ℓ + r + e^(k)                        ▷ index profile of the subtree
                    if t ∈ d then d[t] ← d[t] + w            ▷ update if t is already a key in d
                    else d ← d ∪ {t : w}                     ▷ t is a new key in d
    return d
```

* For an integer $k \geq 0$, the list $e^{(k)} = (0, \ldots, 0, 1)$ is the list with $(k+1)$ components where the last entry is 1 and all others are 0. The empty list of size 0 is denoted ϵ.

For Boolean functions in k variables, although the time complexity of our algorithm is of exponential growth $2^{3k^2/2}$. However the state space of Boolean functions is 2^{2^k} thus our computation is still much better than the exhaustive construction.

Unranking BDDs. Using the classical recursive method for the generation of structures [15] we base our generation approach on the combinatorial counting approach. Since the class of objects under study seems not admissible in the sense given in Analytic Combinatorics [6], we cannot directly apply the advanced techniques presented in [5] nor the approaches by Martínez and Molinero [11]. Thus we devise an unranking algorithm for BDDs and get as by-products algorithms for uniform random sampling and exhaustive generation.

The ranking/unranking techniques for objects of a combinatorial class $\mathcal{C}$ of size N consists in building a bijection between any $c \in \mathcal{C}$ and an integer (its *rank*) in the interval $[0..N-1]$ (if we starts from 0). This leads trivially to a uniform sampling algorithm by drawing uniformly first an integer and then building the corresponding object.

Proposition 7. *Once the pre-computations are done, the unranking (or uniform random sampling) algorithm needs $O(n \cdot |\mathcal{T}_{n,k}|)$ arithmetic operations to build a* BDD *of index k and size n.*

First, remark that the worst case happens when n is of order the largest possible size of a BDD over k variables $O(\frac{2^{k/2}}{k})$ (cf. [9, p. 102]) which corresponds to the generic case according to Fig. 2. Furthermore, the number of profiles is of order $2^{\frac{k^2}{2}}$. To generate a BDD given its rank, we first identify the correct profile of its spine (by enumeration). Then according to this target profile, recursively, for each node, we traverse at most all spines with this profile, in order to decompose the substructures in its left and right part, yielding the upper bound.

As a conclusion, note the process of enumerating, counting and sampling we introduced can be adapted to subclasses of functions (for instance those for which all variables are essential), but also to other strategies of compaction, like those used for Quasi-Reduced BDD*s and Zero-suppressed* BDD*s. A natural question is also to provide an algorithm enumerating valid spines and not all invalid ones as well to get more efficient enumeration and unranking algorithms for* ROBDD*s. These questions will be addressed in future work.*

Acknowledgement. We thank the anonymous reviewers whose comments and suggestions helped improve and clarify this manuscript.

References

1. Bryant, R.E.: Graph-based algorithms for boolean function manipulation. IEEE Trans. Comput. **35**(8), 677–691 (1986)
2. Claessen, K., Hughes, J.: Quickcheck: a lightweight tool for random testing of haskell programs. ACM SIGPLAN Notices **35**(9), 268–279 (2000)
3. Dybjer, P., Haiyan, Q., Takeyama, M.: Verifying haskell programs by combining testing and proving. In: Proceedings of the 3rd International Conference on Quality Software, QSIC, pp. 272–279 (2003)
4. Flajolet, P., Sipala, P., Steyaert, J.-M.: Analytic variations on the common subexpression problem. In: Paterson, M.S. (ed.) ICALP 1990. LNCS, vol. 443, pp. 220–234. Springer, Heidelberg (1990). https://doi.org/10.1007/BFb0032034
5. Flajolet, P., Zimmermann, P., Cutsem, B.V.: A calculus for the random generation of labelled combinatorial structures. Theor. Comput. Sci. **132**(2), 1–35 (1994)
6. Flajolet, P., Sedgewick, R.: Analytic Combinatorics, 1st edn. Cambridge University Press, New York, NY, USA (2009)
7. Genitrini, A., Gittenberger, B., Kauers, M., Wallner, M.: Asymptotic enumeration of compacted binary trees of bounded right height. J. Comb. Theory, Ser. A **172**, 105177 (2020)
8. Gröpl, C., Prömel, H.J., Srivastav, A.: Ordered binary decision diagrams and the shannon effect. Discrete Appl. Math. **142**(1), 67–85 (2004)

9. Knuth, D.E.: The Art of Computer Programming, Volume 4A, Combinatorial Algorithms. Addison-Wesley Professional, Boston (2011)
10. Marinov, D., Andoni, A., Daniliuc, D., Khurshid, S., Rinard, M.: An evaluation of exhaustive testing for data structures. Technical Report, MIT-LCS-TR-921 (2003)
11. Martínez, C., Molinero, X.: A generic approach for the unranking of labeled combinatorial classes. Random Struct. Algorithms **19**(3–4), 472–497 (2001)
12. Newton, J., Verna, D.: A theoretical and numerical analysis of the worst-case size of reduced ordered binary decision diagrams. ACM Trans. Comput. Logic **20**(1), 6:1–6:36 (2019)
13. Vuillemin, J., Béal, F.: On the BDD of a random boolean function. In: Proceedings of the 9th Asian Computing Science on Advances in Computer Science, ASIAN'2004, pp. 483–493 (2004)
14. Wegener, I.: Branching Programs and Binary Decision Diagrams. SIAM (2000)
15. Wilf, H.S., Nijenhuis, A.: Combinatorial algorithms: An update. SIAM (1989)

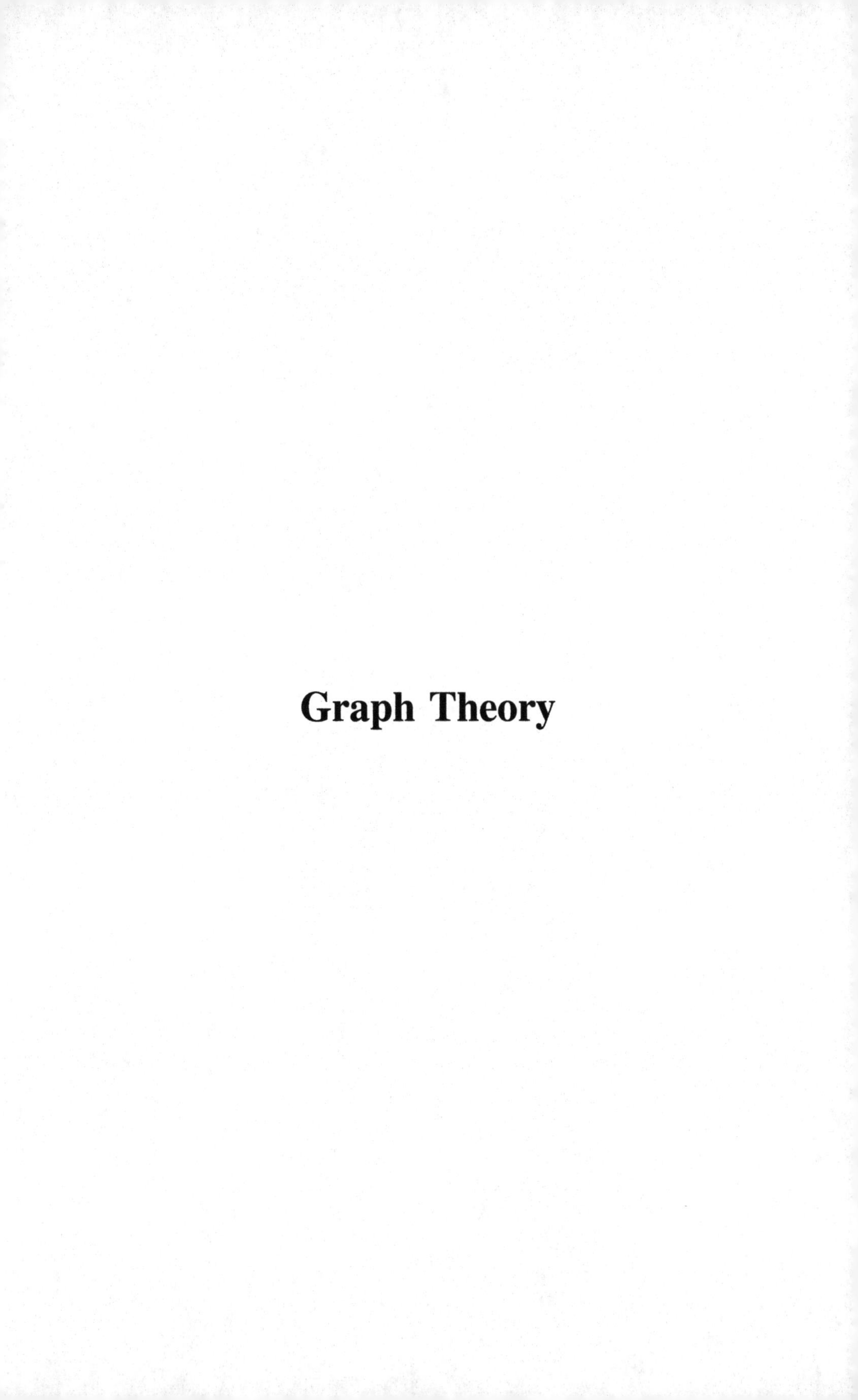

Graph Theory

Graph Sandwich Problem for the Property of Being Well-Covered and Partitionable into k Independent Sets and ℓ Cliques

Sancrey Rodrigues Alves[1], Fernanda Couto[2], Luerbio Faria[3], Sylvain Gravier[6], Sulamita Klein[4], and Uéverton S. Souza[5](✉)

[1] Fundação de Apoio à Escola Técnica do Estado do Rio de Janeiro, Rio de Janeiro, Brazil
sancrey@cos.ufrj.br

[2] Universidade Federal Rural do Rio de Janeiro, Rio de Janeiro, Brazil
nandavdc@gmail.com

[3] Universidade do Estado do Rio de Janeiro, Rio de Janeiro, Brazil
luerbio@cos.ufrj.br

[4] CNRS, Université Grenoble Alpes, Grenoble, France
sula@cos.ufrj.br

[5] Universidade Federal do Rio de Janeiro, Rio de Janeiro, Brazil
ueverton@ic.uff.br

[6] Universidade Federal Fluminense, Rio de Janeiro, Brazil
sylvain.gravier@univ-grenoble-alpes.fr

Abstract. A (k,ℓ)-*partition* of a graph G is a partition of its vertex set into k independent sets and ℓ cliques. A graph is (k,ℓ) if it admits a (k,ℓ)-partition. A graph is *well-covered* if every maximal independent set is also maximum. A graph is (k,ℓ)-*well-covered* if it is both (k,ℓ) and well-covered. In 2018, Alves et al. provided a complete mapping of the complexity of the (k,ℓ)-Well-Covered Graph problem, in which given a graph G, it is asked whether G is a (k,ℓ)-well-covered graph. Such a problem is polynomial-time solvable for the subclasses $(0,1)$, $(0,2)$, $(1,0)$, $(1,1)$, $(1,2)$, and $(2,0)$, and NP-hard or coNP-hard, otherwise. In the Graph Sandwich Problem for Property Π we are given a pair of graphs $G^1=(V,E^1)$ and $G^2=(V,E^2)$ with $E^1 \subseteq E^2$, and asked whether there is a graph $G=(V,E)$ with $E^1 \subseteq E \subseteq E^2$, such that G satisfies the property Π. It is well-known that recognizing whether a graph G satisfies a property Π is equivalent to the particular graph sandwich problem where $E^1=E^2$. Therefore, in this paper we extend previous studies on the recognition of (k,ℓ)-well-covered graphs by presenting a complexity analysis of Graph Sandwich Problem for the property of being (k,ℓ)-well-covered. Focusing on the classes that are tractable for the problem of recognizing (k,ℓ)-well-covered graphs, we prove that Graph Sandwich for (k,ℓ)-well-covered is polynomial-time solvable when $(k,\ell)=(0,1),(1,0),(1,1)$ or $(0,2)$, and NP-complete if we consider the property of being $(1,2)$-well-covered.

This work was supported by FAPERJ, CNPq and CAPES Brazilian Research Agencies.

Y. Kohayakawa and F. K. Miyazawa (Eds.): LATIN 2020, LNCS 12118, pp. 587–599, 2020.
https://doi.org/10.1007/978-3-030-61792-9_46

Keywords: Well-covered · (k, ℓ)-graph · Sandwich problem · Recognition

1 Introduction

A (k, ℓ)-*partition* of a graph $G = (V, E)$ is a partition of V into k independent sets $S^1, \ldots, S^k$ and ℓ cliques $K^1, \ldots, K^\ell$. By definition, some of these sets might be empty. A graph is (k, ℓ) if it admits a (k, ℓ)-partition. The P vs NP-complete dichotomy of recognizing (k, ℓ)-graphs is well known [3]: the problem is in P if $\max\{k, \ell\} \leq 2$, and NP-complete otherwise. (k, ℓ)-graphs and its subclasses have been extensively studied in the literature. For instance, list partitions of (k, ℓ)-graphs were studied by Feder et al. [6]. In another paper, Feder et al. [7] proved that recognizing graphs that are both chordal and (k, ℓ) is in P, and, in 2005, Demange et al. [5] presented efficient algorithms to recognize cographs that are partitionable into k independent sets and ℓ cliques.

Well-covered graphs were first introduced by Plummer [10] in 1970 as the class of graphs in which every maximal independent set has the same cardinality. In other words, every maximal independent set is maximum.

The problem of recognizing a well-covered graph, which we denote by Well-Covered Graph, was proved to be coNP-complete by Chvátal and Slater [4] and, independently, by Sankaranarayana and Stewart [12], but, when restricted to some graph classes, for instance, claw-free graphs, it is polynomial-time solvable [9,13]. Dealing with well-covered graphs is interesting because the polynomial greedy algorithm for maximal independent sets always returns a maximum independent set. Parameterized complexity analysis of recognizing well-covered graphs can be found in [1,2].

Let $k, \ell \geq 0$ be two fixed integers not simultaneously zero. A graph is (k, ℓ)-*well-covered* if it is both, (k, ℓ) and well-covered.

Golumbic, Kaplan and Shamir [8] stated the Graph Sandwich Problem for Property Π. The input is a pair of graphs $G^1 = (V, E^1)$ and $G^2 = (V, E^2)$ with $E^1 \subseteq E^2$, and the question is whether there is a graph $G = (V, E)$ with $E^1 \subseteq E \subseteq E^2$, such that G satisfies the property Π. For the sake of understanding we use to name the set E^1 as the *forced* edges, the set $E^0 = E^2 \backslash E^1$ as the *optional* edges, and the set $E^3 = \{e : e \notin E^2\}$ as the *forbidden* set of edges. Thus, the *optional* graph $G^0 = (V, E^0)$, and the *forbidden* graph $G^3 = (V, E^3)$ are defined.

Motivated by the relevance of well-covered and (k, ℓ)-graphs, in this paper we are interested in exploring the time complexity of Graph Sandwich for the property of being (k, ℓ)-well-covered. More precisely, in this paper we focus on the following two decision problems restricted to (k, ℓ)-well-covered-graphs.

(k, ℓ)-Well-Covered Graph
Input: A graph G.
Question: Is G a (k, ℓ)-well-covered graph?

GRAPH SANDWICH FOR (k,ℓ)-WELL-COVERED
Input: Graphs $G^1 = (V, E^1)$ and $G^2 = (V, E^2)$ with $E^1 \subseteq E^2$.
Question: Is there a graph $G = (V, E)$ with $E^1 \subseteq E \subseteq E^2$, such that G is (k,ℓ)-well-covered?

When a recognition problem for a property Π is NP-hard (resp. coNP-hard), we can consider the sets $E^1 = E^2$ to obtain that the graph sandwich problem for the property Π is also NP-hard (resp. coNP-hard). In 2018, Alves et al. [1] proved that the recognition of (k,ℓ)-well-covered graphs can be done in polynomial time for the cases $(0,1)$, $(0,2)$, $(1,0)$, $(1,1)$, $(1,2)$, and $(2,0)$ and NP-hard or coNP-hard otherwise. Therefore, the only cases where GRAPH SANDWICH FOR (k,ℓ)-WELL-COVERED can be no longer hard are in these six polynomial cases.

In this paper we prove that GRAPH SANDWICH FOR (k,ℓ)-WELL-COVERED is polynomial-time solvable when $(k,\ell) = (0,1)$, $(1,0)$, $(1,1)$ or $(0,2)$ but it is NP-complete when $(k,\ell) = (1,2)$ (see Table 1). Our polynomial-time algorithms generalize previous studies on (k,ℓ)-well-covered graphs' recognition, and our NP-completeness proof points out a contrast between the complexity of RECOGNITION and GRAPH SANDWICH problems for the property of being a $(1,2)$-well-covered graph. We left the problem for the property of being $(2,0)$-well-covered open. Due to space constraints, proofs of statements marked with '♣' are omitted.

Table 1. Complexity of GRAPH SANDWICH FOR (k,ℓ)-WELL-COVERED. coNPc stands for coNP-complete, NPh stands for NP-hard, NPc stands for NP-complete, and (co)NPh stands for both NP-hard and coNP-hard.

k	ℓ			
	0	1	2	≥ 3
0	-	P	P	NPc
1	P	P	NPc	NPc
2	?	coNPc	coNPc	(co)NPh
≥ 3	NPh	(co)NPh	(co)NPh	(co)NPh

Characterizations of some (k,ℓ)-well-covered graphs

Let $G = (V, E)$ be a graph, $v \in V$, and $S \subseteq V$, we define the *neighborhood* $N(v) = \{u \in V : uv \in E\}$ of v in G, the *neighborhood* $N_S(v) = \{u \in S : uv \in E\}$ of v in S, the *degree* $d(v) = |N(v)|$ of v in G, and the *degree* $d_S(v) = |N_S(v)|$ of v in S.

Next, we present polynomial-time characterizations for (k,ℓ)-WELL-COVERED GRAPHS. Notice that every $(0,1)$-graph, as well as $(1,0)$-graph, is well-covered. In addition, for the case $(0,2)$, i.e., co-bipartite graphs, it is easy to see that the following proposition holds.

Proposition 1. *A graph $G = (V, E)$ is $(0,2)$-well-covered if and only if G is $(0,2)$ and either G is a complete graph, or G has no universal vertex.*

Ravindra [11] presented the following characterization of $(2,0)$-well-covered graphs.

Proposition 2. *[11] A graph $G = (V, E)$ is $(2,0)$-well-covered if and only if G is $(2,0)$ and there is a perfect matching M of G such that for each $e = uv \in M$ the induced graph $G[N(u) \cup N(v)]$, by the union of the open neighbors of u and v, is a complete bipartite graph.*

Alves et al. [1] provided the following characterization for $(1,1)$-well-covered graphs.

Proposition 3. *[1] A graph $G = (V, E)$ is $(1,1)$-well-covered if and only if there is a partition $V = (K, S)$ for V where K is a clique, S is a independent set, and either $d_S(v) = 1$ for each vertex $v \in K$ or $d_S(v) = 0$ for each vertex $v \in K$.*

In order to complete the structural characterizations for polynomial-time recognizable (k,ℓ)-well-covered graphs, next we give a structural characterization of $(1,2)$-well-covered graphs.

Proposition 4. (♣) *Let $G = (V, E)$ be a graph with partition $V = (S, K^1, K^2)$ where S is maximal. Then, $G = (V, E)$ is a $(1,2)$-well covered graph if and only if G satisfies the following conditions:*

1. *If $v \in K^1 \cup K^2$, then $1 \leq |N_S(v)| \leq 2$;*
2. *Given $v \in K^i$ with $|N_S(v)| = 2$. Then $\exists\ u \in K^j, i \neq j$, with $uv \notin E$. In addition, $\forall u \in K^j$ with $uv \notin E$, then $N_S(u) \subseteq N_S(v)$; $(i \neq j$ and $i, j \in \{1,2\})$*
3. *Given $v \in K^i$ with $|N_S(v)| = 1$. If $u \in K^j$, with $uv \notin E$, then $|N_S(u) \cup N_S(v)| = 2$. $(i \neq j$ and $i, j \in \{1,2\})$*

2 Sandwich for (k,ℓ)-well-covered-Polynomial Cases

In order to check whether (G^1, G^2) is a YES-instance of the GRAPH SANDWICH FOR (k,ℓ)-WELL-COVERED when $(k,\ell) = (0,1)$ or $(1,0)$, it is enough to check whether, respectively, either G^2 is a complete graph, or $E^1 = \emptyset$.

Next, we will deal with cases $(0,2)$ and $(1,1)$.

2.1 GRAPH SANDWICH FOR $(0,2)$-WELL-COVERED

First, consider the following algorithm for GRAPH SANDWICH FOR $(0,2)$-WELL-COVERED.

```
Algorithm 1.
Input: graphs G¹ = (V,E¹) and G² = (V,E²) with E¹ ⊆ E²;
Begin
1   If (G² = (V,E²) is not (0,2)) then
2      Return no;
3   Else
```

```
4       If (G² = (V,E²) is a complete graph) then
5          Return yes;
6       Else
7          If (G¹ = (V,E¹) has a universal vertex) then
8             Return no
9         Else
10            Return yes;
End.
```

Theorem 1. *Algorithm 1 correctly asserts whether there is a graph $G = (V, E)$ with $E^1 \subseteq E \subseteq E^2$ such that G is $(0, 2)$-well-covered.*

Proof. First, note that the property of being co-bipartite is closed under edge addition, thus if $G^2 = (V, E^2)$ is not $(0, 2)$ then no spanning subgraph of G^2 will be a co-bipartite graph (line 1–2). Since every $(0, 1)$-graph (complete graph) is well-covered, we can assume that G^2 is not complete, otherwise the answer of the problem is positive (line 3–5). If G^2 is $(0, 2)$ but it is not $(0, 1)$, and G^1 has a universal vertex, then every graph $G = (V, E)$ with $E^1 \subseteq E \subseteq E^2$ has a universal vertex, thus, by Proposition 1, the answer is negative (line 6–8). Finally, assume that G^2 is $(0, 2)$ but it is not $(0, 1)$, and G^1 has no universal vertex. Take a $(0, 2)$-partition, (K^1, K^2), of G^2. Notice that every edge of E^3 is crossing from K^1 to K^2. If $v \in K^i$ dominates all vertices of K^j $(i \neq j)$ in G^1, then there is a vertex $w \in K^i$ such that $vw \in E^2$, otherwise v is a universal vertex in G^1. Therefore, we can update the $(0, 2)$-partition for $(K^i \setminus \{v\}, K^j \cup \{v\})$. This procedure can be applied successively until obtain a $(0, 2)$-partition in which for every vertex $v \in K^i$ there is at least one vertex $w \in K^j$ $(i \neq j)$ such that $vw \in E^0 \cup E^3$, which certifies that there is a co-bipartite graph $G = (V, E)$ with $E^1 \subseteq E \subseteq E^2$ having no universal vertex. Thus, the answer is positive (line 9–10). □

2.2 Graph Sandwich for $(1, 1)$-well-covered

Lemma 1. *There is a polynomial-time algorithm that either correctly solves* Graph Sandwich for $(1, 1)$-well-covered, *or outputs a partition $(S', K' \cup T')$ of V such that:*

1. *S' is an independent set;*
2. *$K' \cup T'$ induces a clique of G^2;*
3. *$K', T' \neq \emptyset$;*
4. *there are no edge of G^2 between the vertices of T' and of S';*
5. *each vertex $v \in K'$ is incident to at most one edge $vu \in E^1$ such that $u \in S'$, and at least one edge $vw \in E^2$ such that $w \in S'$.*

Proof. Recall that $(1, 1)$-graphs is exactly the class of split graphs. In 1995 [8], Golumbic, Kaplan and Shamir presented a polynomial-time algorithm for Graph Sandwich for split graphs. This algorithm is based on reducing the problem to an instance $I = (U, C)$ of 2SAT. The proposed construction consists of creating the set of variables

$$U = \{v_K, v_S : v \in V\}$$

and the set of clauses

$$C = \{(v_K \vee v_S), (\overline{v_K} \vee \overline{v_S}) : v \in V\} \cup \{(u_K \vee v_K) : uv \in E^1\} \cup \{(u_S \vee v_S) : uv \in E^3\}.$$

It is easy to see that $I = (U, C)$ is satisfiable if and only if V can be partitioned into K, S such that S induces an independent set of G^1 and K induces a clique of G^2, where v_K (resp. v_S) represents that v should be add to K (resp. S).

Considering the GRAPH SANDWICH FOR $(1,1)$-WELL-COVERED, every YES-instance of such a problem is also a YES-instance of GRAPH SANDWICH FOR SPLIT GRAPHS. However, by the characterization provided in Proposition 3, we know that each vertex of the clique K must be a neighbor of at most one vertex in S, in the solution graph G. Thus, in order to obey this restriction we add the following set of clauses to $I = (U, C)$:

$$\{(\overline{v_S} \vee \overline{w_S}) : u, v, w \in V \text{ and } uv, uw \in E^1\}.$$

Observe that if $I = (U, C)$ is not satisfiable, then (G^1, G^2) is a NO-instance of GRAPH SANDWICH FOR $(1,1)$-WELL-COVERED. However, if $I = (U, C)$ is satisfiable, then (G^1, G^2) can be partitioned into K, S such that S induces an independent set of G^1, K induces a clique of G^2, and every vertex $v \in K$ has at most one neighbor in S, in the graph G^1. Therefore, we can set $S' = S$, $T' = \{v : v \in K \text{and has no neighbor in } S, \text{ in the graph} G^2\}$, and $K' = \{K \setminus T'\}$. Now, if $T' = \emptyset$ then (G^1, G^2) is a YES-instance, since we can use the optional edges conveniently to satisfy the condition that all vertices have exactly one neighbor in S, and if $K' = \emptyset$ then (G^1, G^2) is also a YES-instance, since no vertex in K will be adjacent to a vertex in S. Therefore, we can either solve GRAPH SANDWICH FOR $(1,1)$-WELL-COVERED or output a partition $(S', K' \cup T')$ as required. □

Lemma 2. *Let (G^1, G^2) be an instance of* GRAPH SANDWICH FOR $(1,1)$-WELL-COVERED, *and $V = (S', K' \cup T')$ be a partition of V as described in Lemma 1. It can be checked in polynomial time whether there is a graph $G = (V, E)$ with $E^1 \subseteq E \subseteq E^2$, such that the set of vertices of G can be partitioned into (K, S) with $S' \subseteq S$, where K is a clique, S is an independent set, and either $d_S(v) = 1$ for each vertex $v \in K$, or $d_S(v) = 0$ for each vertex $v \in K$.*

Proof. Initially, label each vertex $v \in K' \cup T'$ with $label(v) = 0$. Now, label each vertex $v \in K'$ having a forced edge to S' with label equal to one ($label(v) = 1$). After that, for each vertex $v \in K' \cup T'$ with $label(v)$=0, such that $\exists u \in N(v)$ with $uv \in E^1$ and $label(u) = 1$, do $label(v) = 2$.

Since $S' \subseteq S$, no vertex v with $label(v) = 1$ can be in S, otherwise S is not an independent set. Consequently, no vertex v with $label(v) = 2$ can be in S, otherwise some vertices v with $label(v) = 1$ have two vertices in S. Therefore, if every vertex v of $K' \cup T'$ has $label(v) \neq 0$, then we can safely return NO. Moreover, if there is a vertex $v \in K' \cup T'$ such that $label(v) = 0$, we output YES, because $(S, K) = (S' \cup \{v\}, K' \cup T' \setminus \{v\})$, and $G = (V, E)$ with $E = E^1 \cup \{uw : u, w \in K' \cup T' \setminus \{v\}\} \cup \{vu : u \in K' \cup T'$, and $label(u) \neq 1\}$ is a solution. □

Lemma 3. *Let* (G^1, G^2) *be an instance of* GRAPH SANDWICH FOR (1, 1)-WELL-COVERED, *and let* $(S', K' \cup T')$ *be a partition of* V *as described in Lemma 1. If there are three vertices* a, b, c *such that* $ab, bc \in E^1$, $b \in K'$, *and* $a, c \in T'$, *then we can solve* (G^1, G^2) *in polynomial time.*

Proof. Let (S, K) be a partition as described in Proposition 3. If a and c are both in S then b has two neighbors in S, a contradiction. If, either, a or c is in K then $S' \subseteq S$. Therefore, by Lemma 2, we conclude that we can solve the instance (G^1, G^2) in polynomial time. □

Theorem 2. GRAPH SANDWICH FOR (1, 1)-WELL-COVERED *can be solved in polynomial time.*

Proof. Let (G^1, G^2) be an instance of GRAPH SANDWICH FOR (1, 1)-WELL-COVERED. Without loss of generality, assume that the algorithm presented in Lemma 1 outputs a partition $(S', K' \cup T')$ of V. By Lemma 2, we can also consider that there is no sandwich graph $G = (V, E)$ such that $V(G)$ can be partitioned into (K, S) according to Proposition 3, with $S' \subseteq S$. Hence, there is no vertex b having two neighbors a and c in G^1 such that $a, c \in T'$ (see Lemma 3).

Suppose that (G^1, G^2) has a sandwich graph $G = (V, E)$. Let (K, S) be a partition of $V(G)$ according to Proposition 3. Note that if $T' \cap K \neq \emptyset$, then $S' \subseteq S$, which is a contradiction. Therefore, $T' \subseteq S$. In addition, for every vertex v at a distance at most two, considering the graph G^1, of some vertex that must be in S, it holds that v must be contained in K, otherwise either S is not an independent set or some vertex of K will have two neighbors in S. Analogously, if a vertex v has a forbidden edge to a vertex that must be in K, then v must be in S.

Let A be the set of vertices of S' that must be in K, and let B be the set of vertices of $K' \cup T'$ that must be in S by the successive application of the rules described above. Note that $T' \subseteq B$, and such sets A, B can be easily found in polynomial time using search algorithms.

If $A \cap B \neq \emptyset$ then (G^1, G^2) is a *no*-instance. Otherwise, we can set the following paritition of V:

$$S'' = \{B \cup S' \setminus A\},$$

$$T'' = \{v : v \in A \text{ and has no neighbor in } S'' \text{ in the graph } G^2\},$$

and

$$K'' = \{(K' \setminus B) \cup (A \setminus T'')\}.$$

If (G^1, G^2) is a YES-instance then no vertex $v \in K''$ has two neighbors in S'' in the graph G^1, otherwise we have a contradiction (see Proposition 3).

If $T'' = \emptyset$ then every vertex in K'' has at least one edge of G^2 to some vertex in S''. Thus, when $T'' = \emptyset$ we can easily construct the solution graph G for (G^1, G^2) in polynomial time. Now, suppose that $T'' \neq \emptyset$. Since $T'' \subseteq A$ it follows that S'' must be in S for a (K, S) partition of $V(G)$ according to Proposition 3, if any. Thus by Lemma 2 it holds that (G^1, G^2) can be solved in polynomial time. □

3 Sandwich for (k, ℓ)-well-covered-NP-complete Case

Next, we will deal with $(1, 2)$-well-covered graphs. The NP-completeness of such a case points out a contrast between the complexity of RECOGNITION and GRAPH SANDWICH problems for (k, ℓ)-well-covered graphs.

GRAPH SANDWICH FOR $(1, 2)$-WELL-COVERED
In order to prove the NP-completeness of GRAPH SANDWICH FOR $(1, 2)$-WELL-COVERED, we present a reduction from POSITIVE 1-IN-3 SAT, a well-known NP-complete problem. First, we present some auxiliary definitions and preliminary results.

Let $G = (V, E)$ be a graph and $R, T \subseteq V$. We say that T *2–dominates* R if each vertex of R has at least 2 neighbors in T. Let $I = (U, C)$ be an instance of POSITIVE 1-IN-3 SAT where U is the set of variables and C is the set of clauses. Let $T \subseteq U$, we say that T *2–dominates* C if each clause of C has at least two literals in T, in this case T is called a *2-dominating* set of variables.

Lemma 4. *Let $I = (U, C)$ be an instance of* POSITIVE 1-IN-3 SAT, *where $U = \{u_1, u_2, u_3, \ldots, u_n\}$ and $C = \{c_1, c_2, c_3, \ldots, c_m\}$, and let $T \subseteq U$ such that T 2-dominates C. Then, I can be solved in time $O(2^{|T|}mn)$.*

Proof. We can check each of the $2^{|T|}$ truth assignments of T. For a given truth assignment of T, each clause contains at most one variable with undefined value. Thus, it is easy to see that, in linear time, one can check whether such an assignment can be extended into a satisfiable assignment for I. □

From Lemma 4, we may assume that the hard instances of POSITIVE 1-IN-3 SAT have no *2-dominating* set of variables of bounded size.

Theorem 3. GRAPH SANDWICH FOR $(1, 2)$-WELL-COVERED *is* NP-*complete.*

Proof. GRAPH SANDWICH FOR $(1, 2)$-WELL-COVERED is in NP, since given a graph G and a partition $V = (S, K^1, K^2)$, we can check in polynomial time the conditions (1), (2), (3) of the characterization presented in Proposition 4.

Let $I = (U, C)$ be an instance of POSITIVE 1-IN-3 SAT. By Lemma 4, we may assume that I does not have a 2-dominating set of variables of size smaller than 8, and that every truth assignment of I, if any, requires at least 2 literals set as true. In addition, we consider that every variable occurs in at least one clause.

We construct an instance (G^1, G^2) of GRAPH SANDWICH FOR $(1, 2)$-WELL-COVERED, such that I is satisfiable if and only if there is a $(1, 2)$-well-covered sandwich graph G for $G^1 = (V, E^1), G^2 = (V, E^2)$, as follows.

1. First, initialize $V = \{a, b, c\}$, and $E^1, E^2, E^3 = \emptyset$;
2. For each $u_i \in U$ add u_i to V.
3. For each $c_j \in C$ add c_j to V.
4. For each $c_j = (u_x \vee u_y \vee u_z) \in C$, add $u_i c_j, \forall\, i \in \{x, y, z\}$, to E^1.
5. For each pair $c_j, c_\ell \in C, j \neq \ell$, add $c_j c_\ell$ to E^1.
6. For each $c_j \in C$, add $c_j a, c_j b, c_j c$ to E^1.
7. For each $u_i \in U$ add $u_i a, u_i b, u_i c$ to E^0.
8. Add every possible edge $u_i u_j$ to E^0. (Remark $E^0 = E^2 \setminus E^1$)
9. Set $E^3 = (\{uv : u \neq v \text{ and } u, v \in V\} \setminus E^2)$

This completes the construction of $\big(G^1 = (V, E^1), G^2 = (V, E^2)\big)$.

For the sake of reader's convenience, we offer in Figure 1 a framework for the construction.

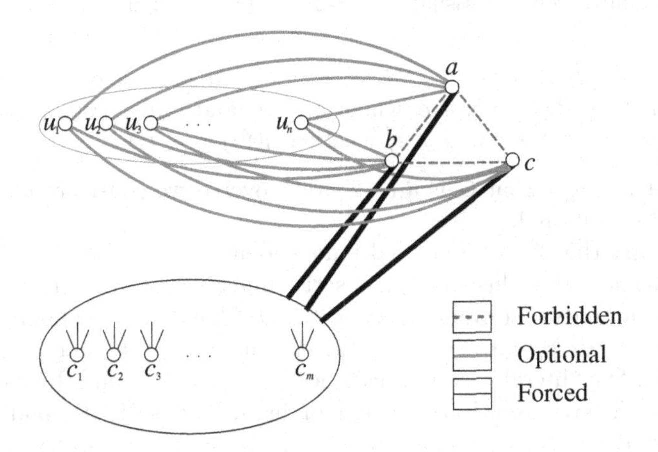

Fig. 1. Graph Sandwich for (1, 2)-well-covered instance, (G^1, G^2), obtained from an instance $I = (U, C)$ of Positive 1-in-3 Sat. Here are depicted all optional edges in thick green lines, and all forced edges in thin black straight lines of $\big(G^1 = (V, E^1), G^2 = (V, E^2)\big)$. We depict only the forbidden edges ab, ac, and bc in dashed red straight lines, omitting the forbidden edges between a variable vertex x and a clause vertex c_j such that x does not occur in c_j.

Now, suppose that $I = (U, C)$ is satisfiable. Let $\eta : U \to \{T, F\}$ be a satisfiable truth assignment for I. We build a $(1, 2)$-well-covered sandwich graph $G = (V, E)$ for $\big(G^1 = (V, E^1), G^2 = (V, E^2)\big)$ from η as follows (see Fig. 2):

- first, set $E = E^1$;
- set $u \in S$ if and only if $\eta(u) = T$;
- set $u \in K^2$ if and only if $\eta(u) = F$;
- set vertices c_j in K^1 $\forall c_j$;
- set vertices a in S; b in K^1; and c in K^2;
- if $\eta(u_i) = F$ then add to E the edges $u_i a, u_i b, u_i c$;

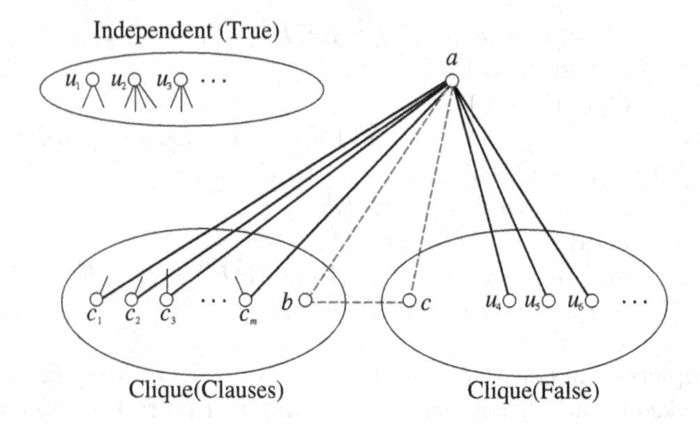

Fig. 2. Diagram showing a $(1, 2)$-well-covered-partition (S, K^1, K^2) of a sandwich graph G obtained from a satisfiable 1-IN-3 SAT truth assignment $\eta : U \to \{T, F\}$.

- add to E two edges, $u_i b$, $u_j c$, where $u_i \neq u_j$ and $\eta(u_i) = \eta(u_j) = T$;
- finally, add to E the edges $u_i u_j$ whether $\eta(u_i) = \eta(u_j) = F$.

In order to prove that G is a $(1, 2)$-well-covered we consider the characterization of Proposition 4.

Notice first that S is a maximal independent set and that K^1, and K^2 are cliques. Also note that, because η is a satisfiable truth assignment for POSITIVE 1-IN-3 SAT, it holds that every vertex of $K^1 \cup K^2$ has 1 or 2 neighbors in S. Since each clause vertex c_j has $N_S(c_j) = \{u_i, a\}$ where u_i is the only true literal of c_j in η, each false literal vertex u_i satisfies $N_S(u_i) = \{a\}$, and, by construction, both, b and c, have exactly one neighbor in S. This satisfies condition (1) of Proposition 4.

Now, observe that edges of E^3 between cliques K^1 and K^2 are: the edge bc, and the edges between c_j and u_ℓ, where $\eta(u_\ell) = F$ and it does not occur in c_j. If $\eta(u_\ell) = F$ and u_ℓ does not occur in c_j, then $|N_S(c_j) \cup N_S(u_\ell)| = |\{a, u_i\} \cup \{a\}| = 2$, where u_i is the true variable in clause c_j. Since, by construction, b and c have no common neighbor, then $|N_S(b) \cup N_S(c)| = |\{u_i\} \cup \{u_j\}| = 2$. Thus, conditions (2), (3) of Proposition 4 are satisfied, and G is $(1, 2)$-well-covered graph.

Conversely, suppose that there is a $(1, 2)$-well-covered sandwich graph $G = (V, E)$ for (G^1, G^2). Since $G = (V, E)$ is a $(1, 2)$-well-covered graph, there is a partition of V into a maximal independent set S and cliques K^1, K^2, as described in Proposition 4 . We proceed by defining a satisfiable 1-IN-3SAT assignment $\eta : U \to \{T, F\}$ for $I = (V, E)$, where the boolean variable $u_i = T$ if and only if vertex $u_i \in S$.

It remains to show that η is 1-IN-3SAT-satisfiable.

Claim 1. *Every vertex clause* $c_j \in K^1 \cup K^2$.

Proof. Since every clause vertex c_j satisfies that ac_j, bc_j and $cc_j \in E^1$, if $c_j \in S$, then vertices a, b, c belong together to $K^1 \cup K^2$, what is a contradiction, since $ab, ac, bc \in E^3$. Therefore, $c_j \in K^1 \cup K^2$.

Claim 2. *There is only one vertex x in $\{a, b, c\}$, such that $x \in S$.*

Proof. First, notice that a, b, c are false twins. Since $ab, ac, bc \in E^3$, then a, b, c cannot all be in $K^1 \cup K^2$. By Claim 1, we have that $c_j \in K^1 \cup K^2$. Hence, a, b, c cannot all be in S, otherwise $|N_S(c_j)| \geq 3$, contradicting Proposition 4. Suppose there are exactly two vertices of a, b, c in S, say $a, b \in S$. Then, all the literal vertices belong to $K^1 \cup K^2$, otherwise $|N_S(c_j)| \geq 3$ for some c_j, because every clause vertex belong to $K^1 \cup K^2$, and c_j is adjacent to a, and b. Thus $S = \{a, b\}$, what contradicts the maximality of S. Therefore $|S \cap \{a, b, c\}| = 1$.

Now, we may assume that: $a \in S$; $c_j \in K^1 \cup K^2 (\forall\ c_j)$; $b \in K^1$; $c \in K^2$.

Claim 3. *Each clause vertex c_j lies at a same clique, say K^1.*

Proof. Suppose that there are two clause vertices lying at distinct cliques, say $c_j \in K^1$ and $c_\ell \in K^2$. Since $a \in S$, each clause vertex belongs to $K^1 \cup K^2$, each clause vertex c_j is adjacent to a, then, by Proposition 4, it follows that each vertex clause c_j is adjacent to at most one variable vertex in S, and the other two variable neighbors are in $K^1 \cup K^2$. Notice now that the set of literal vertices in $K^1 \cup K^2$ forms a 2-dominating set of variables for I. Since each clause vertex has forbidden edges for all but three variables, we have at most three variable vertices in each clique, which implies that $I = (U, C)$ has a 2-dominating set of variables of size at most six, a contradiction. Thus, all the clause vertices belong to just one clique, that we will assume to be K^1.

Claim 4. *If a literal vertex $u_i \in K^1 \cup K^2$, then $u_i \in K^2$.*

Proof. Since each clause vertex belongs to K^1, if $u_i \in K^1$ then every clause has u_i as literal. Thus, by setting u_i as true and the other literals as false, we obtain a satisfiable assignment for $I = (U, C)$, a contradiction.

To conclude the proof of Theorem 3, it remains to prove the following claim.

Claim 5. *For any clause vertex c_j of G, it holds that there is exactly one variable vertex u_r adjacent to c_j such that $u_r \in S$.*

Proof. Since each clause vertex c_j is adjacent to a, by Proposition 4(1), there are at most an additional neighbor of c_j in S. Hence, in order to prove the claim, it is enough to prove that each clause vertex c_j has one additional neighbor in S. From previous claims we know that $b \in K^1$ and that $c \in K^2$. By Proposition 4, it follows that $|N_S(b) \cup N_S(c)| = 2$. Hence, there are two variable vertices, say u_1 and u_2 in S, such that, in G, it holds that $\{u_1, u_2\} = (N_S(b) \cup N_S(c))$. Since there are clauses c_i and c_j containing, respectively, variables u_1 and u_2, clause vertices c_i and c_j have two neighbors in S, satisfying that $N_S(c_i) = \{a, u_1\}$ and $N_S(c_j) = \{a, u_2\}$.

Suppose there is a clause vertex c_ℓ with $N_S(c_\ell) = \{a\}$. Recall that $K^2 \cap U$ is a 2-dominating set of variables. Therefore, by assumption, $|K^2 \cap U| \geq 8$. Hence, there is a variable $u \in K^2 \cap U$ that does not occur in clauses c_i, c_j and c_ℓ (notice that u_1, u_2 are in S). Therefore, $c_i u$, $c_j u$, $c_\ell u \in E^3$. (see Fig. 3)

Since $N_S(c_i) = \{u_1, a\}$, from Proposition 4(2) $N_S(u) \subseteq \{u_1, a\}$. Moreover, since $N_S(c_j) = \{u_2, a\}$, then, from Proposition 4(2), $N_S(u) \subseteq \{u_2, a\}$. Hence, $N_S(u) = \{a\}$. But, since $N_S(c_\ell) = \{a\}$ and $c_\ell u \in E^3$, from Proposition 4(3) $|N_S(c_\ell) \cup N_S(u)|$ should be equal to two, a contradiction. Thus, every clause vertex has a variable neighbor in S, and the claim holds.

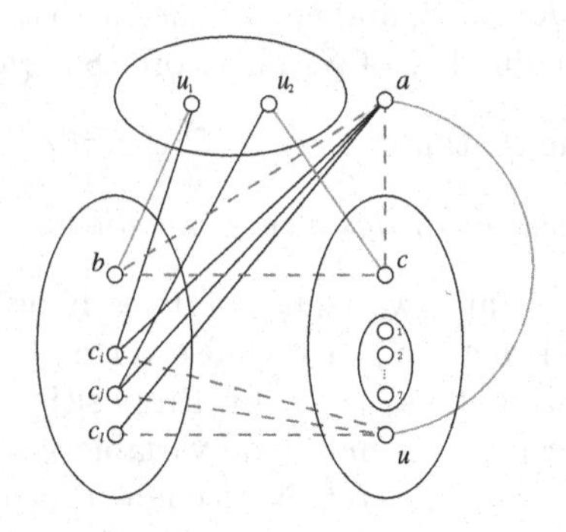

Fig. 3. A sandwich graph $G = (V, E)$ with well-covered-partition (S, K^1, K^2).

Therefore, the defined assignment η is a satisfiable assignment for POSITIVE 1-IN-3 SAT, and this concludes the proof of Theorem 3. □

References

1. Alves, S.R., Dabrowski, K.K., Faria, L., Klein, S., Sau, I., Souza, U.S.: On the (parameterized) complexity of recognizing well-covered (r, ℓ)-graph. Theor. Comput. Sci. **746**, 36–48 (2018)
2. Araújo, R.T., Costa, E.R., Klein, S., Sampaio, R.M., Souza, U.S.: FPT algorithms to recognize well covered graphs. Discrete Math. Theoretical Comput. Sci., **21**(1) (2019)
3. Brandstädt, A.: Partitions of graphs into one or two independent sets and cliques. Discrete Math. **152**(1–3), 47–54 (1996)
4. Chvátal, V., Slater, P.J.: A note on well-covered graphs. Ann. Discrete Math. **55**, 179–181 (1993)
5. Demange, M., Ekim, T., De Werra, D.: Partitioning cographs into cliques and stable sets. Discrete Optim. **2**(2), 145–153 (2005)
6. Feder, T., Hell, P., Klein, S., Motwani, R.: List partitions. SIAM J. Discrete Math. **16**(3), 449–478 (2003)
7. Feder, T., Hell, P., Klein, S., Nogueira, L.T., Protti, F.: List matrix partitions of chordal graphs. Theoretical Comput. Sci. **349**(1), 52–66 (2005)
8. Golumbic, M.C., Kaplan, H., Shamir, R.: Graph sandwich problems. J. Algorithms **19**(3), 449-473 (1995)
9. Lesk, M., Plummer, M.D., Pulleyblank, W.R.: Equi-matchable graphs. In: Graph Theory and Combinatorics. Academic Press, Cambridge, pp. 239–254 (1984)

10. Plummer, M.D.: Some covering concepts in graphs. J. Comb. Theory **8**(1), 91–98 (1970)
11. Ravindra, G.: Well-covered graphs. J. Comb. Inf. Syst. Sci. **2**(1), 20–21 (1977)
12. Sankaranarayana, R.S., Stewart, L.K.: Complexity results for well-covered graphs. Networks **22**(3), 247–262 (1992)
13. Tankus, D., Tarsi, M.: Well-covered claw-free graphs. J. Comb. Theory, Ser. B **66**(2), 293–302 (1996)

On the Maximum Number of Edges in Chordal Graphs of Bounded Degree and Matching Number

Jean R. S. Blair[1], Pinar Heggernes[2], Paloma T. Lima[2(✉)], and Daniel Lokshtanov[3]

[1] Department of Electrical Engineering and Computer Science, United States Military Academy, West Point, New York, USA
jean.blair@westpoint.edu
[2] Department of Informatics, University of Bergen, Bergen, Norway
{pinar.heggernes,paloma.lima}@uib.no
[3] Department of Computer Science, University of California, Santa Barbara, USA
daniello@ucsb.edu

Abstract. We determine the maximum number of edges that a chordal graph G can have if its degree, $\Delta(G)$, and its matching number, $\nu(G)$, are bounded. To do so, we show that for every $d, \nu \in \mathbb{N}$, there exists a chordal graph G with $\Delta(G) < d$ and $\nu(G) < \nu$ whose number of edges matches the upper bound, while having a simple structure: it is a disjoint union of cliques and stars.

1 Introduction

A problem that dates back to 1960 is to determine the maximum number of edges that a graph can have if its maximum degree and matching number are each bounded. It is important to note that this problem does not impose any constraint on the number of vertices of the graph. Because of that, in general, if one of the two parameters is not bounded, there is no upper bound on the number of edges that a graph can have. One can simply construct graphs formed by stars (trees that have only a single vertex of degree greater than one) or single edges. A star with unbounded number of leaves has matching number one but unbounded degree, while a graph that is a disjoint union of an unbounded number of edges has bounded degree but unbounded matching number. By Vizing's Theorem, every graph can have its edge set partitioned into a family of at most $\Delta(G) + 1$ matchings, where $\Delta(G)$ denotes the degree of the graph G. Thus, bounding both the maximum degree and the matching number is actually enough to bound the number of edges that a graph can have. Chvátal and Hanson [4] gave a tight upper bound on this value, in the case where no further restrictions are imposed to the graphs considered. Later on, Balachandran and Khare [1] gave a constructive proof of the same result, which made it possible to identify the structure of the graphs achieving the given bound on the number of edges. Such graphs are called edge-extremal graphs. They contain collections of stars and, in some cases, induced cycles of length four.

Y. Kohayakawa and F. K. Miyazawa (Eds.): LATIN 2020, LNCS 12118, pp. 600–612, 2020.
https://doi.org/10.1007/978-3-030-61792-9_47

An interesting problem that arises from these results is to investigate how the number of edges in the edge-extremal graphs is affected if we impose some additional structural property on the graphs considered. More specifically, what happens if we restrict the question to graph classes in which cycles of length four or stars are forbidden induced subgraphs? Natural candidates for such graph classes are chordal graphs and claw-free graphs. In the past few years, bounds for this problem have indeed been established for claw-free graphs in the work of Dibek et al. [5]. Furthermore, the problem is resolved on other graph classes, such as bipartite graphs, split graphs, disjoint unions of split graphs and unit interval graphs in the work of Måland [9]. However, on chordal graphs, the problem had so far remained unresolved. Chordal graphs form an extremely well-studied graph class, both from a structural and from an algorithmic point of view, with many and various applications. Hence, a large number of computer science papers are published every year on chordal graphs and their subclasses.

In this work, we determine the maximum number of edges that a chordal graph can have, given the constraints on its maximum degree and matching number. Given $d, \nu \in \mathbb{N}$, we denote by $\mathcal{M}_{chordal}(d, \nu)$ the set of chordal graphs such that $\Delta(G) < d$ and $\nu(G) < \nu$. A graph in $\mathcal{M}_{chordal}(d, \nu)$ achieving this maximum number of edges is called an edge-extremal graph. In order to establish the upper bound on the number of edges of an edge-extremal graph in $\mathcal{M}_{chordal}(d, \nu)$ we show that, among them, there is one that has a very simple structure: it is a disjoint union of cliques and stars of a given size.

Theorem 1. *There exists an edge-extremal graph in $\mathcal{M}_{chordal}(d, \nu)$ that is a disjoint union of cliques and stars.*

Section 3 is entirely devoted to the proof of Theorem 1[1]. Once the structure of this special edge-extremal graph is known, we are able to establish the following upper bound on the number of edges of a graph is $\mathcal{M}_{chordal}(d, \nu)$.

Theorem 2. *Given $d, \nu \in \mathbb{N}$, the maximum number of edges of a graph in $\mathcal{M}_{chordal}(d, \nu)$ is given by:*

$$\begin{cases} (d-1)(\nu-1), & \text{if } d \text{ is even} \\ (d-1)(\nu-1) + \lfloor \frac{d-1}{2} \rfloor \lfloor \frac{\nu-1}{\lceil \frac{d-1}{2} \rceil} \rfloor, & \text{if } d \text{ is odd} \end{cases}$$

Moreover, a graph achieving this number of edges is

$$\begin{cases} (\nu-1)K_{1,d-1}, & \text{if } d \text{ is even} \\ rK_{1,d-1} + qK_d, & \text{if } d \text{ is odd,} \end{cases}$$

where $\nu - 1 = q\lceil \frac{d-1}{2} \rceil + r$, with $r \geq 0$.

We also show that this result is tight in the sense that the same bound does not hold for any superclass of chordal graphs that is defined by a finite collection

[1] Statements marked with ♠ had their proofs omitted due to space constraints.

of forbidden induced cycles. It is also worth mentioning that this problem is related to the famous problem of computing Ramsey numbers, the general case being equivalent to determining Ramsey numbers for line graphs [2].

2 Preliminaries

The graphs considered are simple and undirected. We denote by V_G and E_G the vertex set and edge set of G, respectively. Given $x \in V_G$, we denote by $N_G(x)$ the neighborhood of x, that is, the set of vertices that are adjacent to x. Two vertices $x, y \in V_G$ are *true twins* if $N_G(x) \cup \{x\} = N_G(y) \cup \{y\}$. Given $x \in V_G$ and $X \subseteq V_G \setminus \{x\}$, we say x is *universal* to X if $X \subseteq N_G(x)$. For a set $X \subset V_G$, $N_G(X)$ denotes the set of vertices in $V_G \setminus X$ that have at least one neighbor in X. The *degree* of x is denoted by $\deg_G(x)$ and is defined as $|N_G(x)|$. The *degree of a graph* G is the maximum degree of a vertex in G and it is denoted by $\Delta(G)$. A vertex x is a *leaf* of G if $\deg_G(x) = 1$.

Given $S \subseteq V_G$, the *subgraph induced by* S is denoted by $G[S]$, and has S as its vertex set and $\{uv \mid u, v \in S \text{ and } uv \in E_G\}$ as its edge set. A *clique* is a set $K \subseteq V_G$ such that $G[K]$ is a complete graph. A clique is *maximal* if it is not properly contained in another clique. An *independent set* is a set S such that $G[S]$ has no edges. A vertex $v \in V_G$ is a *simplicial vertex* if $N_G(v)$ is a clique. Given a set $S \subseteq V_G$, we denote the graph $G[V_G \setminus S]$ by $G \setminus S$. If $S = \{v\}$, we denote the graph $G[V_G \setminus \{v\}]$ simply by $G \setminus v$. The set S is a *separator* if $G \setminus S$ has a larger number of connected components than G.

A set $M \subseteq E_G$ is a *matching* if no two edges in M share a common vertex and M is a *perfect matching* if every vertex of V_G is the endpoint of an edge in M. The *matching number of* G, denoted by $\nu(G)$, is the largest size of a matching in G. A graph G is a *factor-critical graph* if for every $v \in V_G$, $G \setminus v$ has a perfect matching.

Given a family $\mathcal{H}$ of graphs, we say that G is an *$\mathcal{H}$-free graph* if G does not contain an induced subgraph that is isomorphic to a graph in $\mathcal{H}$. If $\mathcal{H} = \{H\}$, we say G is an H-free graph. A *tree* is a connected acyclic graph. A *star* is a tree with at most one vertex that is not a leaf, and for $k \in \mathbb{N}$, a *k-star*, denoted by $K_{1,k}$, is a star with k leaves. A graph is a *complete graph* on n vertices, denoted by K_n, if there is an edge between every pair of its vertices. Given two graphs G and H, the *disjoint union of G and H*, denoted by $G + H$ is the graph with vertex set $V_G \cup V_H$ and edge set $E_G \cup E_H$. We denote by rH the graph that is the disjoint union of r copies of a graph H. A graph G is a *bipartite graph* if V_G can be partitioned into two independent sets. A bipartite graph with bipartition (A, B) is a *chain graph* if there exists an ordering $v_1 v_2 \ldots v_r$ of the vertices of A such that $N_G(v_r) \subseteq \ldots \subseteq N_G(v_1)$. This property of the vertices of A is called the *nested neighborhood* property. Bipartite chain graphs are also known to be the bipartite $2K_2$-free graphs.

A graph is a *chordal graph* if it has no induced cycle of length at least four. Chordal graphs constitute a widely studied graph class, with many different characterisations. Given a graph G, let T be a tree such that every vertex of T is a maximal clique of G. The vertices of $\mathcal{T}$ are referred to as *bags* and denoted with capital letters. For simplicity, we denote the set of vertices of G associated with a vertex of $\mathcal{T}$ with the same capital letter. Let $T_v = \{A \in V_{\mathcal{T}} \mid v \in A\}$. The tree $\mathcal{T}$ is a *clique tree* of G if for every $v \in V_G$, T_v is a subtree of $\mathcal{T}$. A characterisation of chordal graphs due to Gavril [7] states that a graph is chordal if and only if it has a clique tree. If $\mathcal{T}$ is a clique tree of a chordal graph G and $AB \in E_{\mathcal{T}}$, then $A \cap B$ is a separator for the graph G. Another important characterisation of chordal graphs is in terms of vertex orderings and simplicial vertices. An ordering $v_1 v_2 \dots v_n$ of the vertices of G is a *perfect elimination ordering* for G if for every i, the vertex v_i is simplicial in the graph $G[\{v_{i+1}, \dots, v_n\}]$. A characterisation of chordal graphs due to Fulkerson and Gross [6] states that a graph is chordal if and only if it has a perfect elimination ordering. See [3] for an overview of the properties of chordal graphs and clique trees.

Given two integers d and ν and a graph class $\mathcal{C}$, we denote by $\mathcal{M}_{\mathcal{C}}(d, \nu)$ the set of all graphs G in $\mathcal{C}$ such that $\Delta(G) < d$ and $\nu(G) < \nu$. A graph in $\mathcal{M}_{\mathcal{C}}(d, \nu)$ that has the maximum number of edges is called an *edge-extremal graph*. When the graph class considered is the class of all graphs, we write simply $\mathcal{M}(d, \nu)$. The following lemma establishes a connection between edge-extremal graphs and factor-critical graphs in some graph classes. Even though the statement we present here is different from the one stated in [1], the proof in [1] suffices to prove the result as stated below.

Lemma 1 ([1]). Let $\mathcal{C}$ be a graph class that is closed under vertex deletion and closed under taking disjoint union with stars. Let G be an edge-extremal graph in $\mathcal{M}_{\mathcal{C}}(d, \nu)$ with maximum number of connected components that are $(d-1)$-stars. Then every connected component of G that is not a $(d-1)$-star is factor-critical.

The following statement gives a summary of the results obtained by Balachandran and Khare [1].

Theorem 3 ([1]). Given $d, \nu \in \mathbb{N}$, the maximum number of edges of a graph in $\mathcal{M}(d, \nu)$ is given by $(d-1)(\nu-1) + \lfloor \frac{d-1}{2} \rfloor \lfloor \frac{\nu-1}{\lceil \frac{d-1}{2} \rceil} \rfloor$. Moreover, a graph achieving this number of edges is

$$\begin{cases} rK_{1,d-1} + qK'_d, & \text{if } d \text{ is even} \\ rK_{1,d-1} + qK_d, & \text{if } d \text{ is odd,} \end{cases}$$

where $\nu - 1 = q\lceil \frac{d-1}{2} \rceil + r$, with $r \geq 0$, and K'_d is the graph obtained from K_d by the removal of the edges of a perfect matching and addition of a new vertex adjacent to $d-1$ vertices.

In Sect. 3, we show the corresponding bounds for $\mathcal{M}_{chordal}(d, \nu)$ and obtain graphs that achieve these bounds. We remark that, in Theorem 3, the graph $rK_{1,d-1} + qK_d$, obtained when d is odd, is already a chordal graph. Thus, for odd d, the edge-extremal chordal graphs have the same number of edges as the edge-extremal general graphs. Our proof, however, does not rely on this fact and has a unified approach, that works regardless of the parity of d.

3 Chordal Graphs

In this section we present our main result. The strategy to determine the maximum number of edges that a graph in $\mathcal{M}_{chordal}(d,\nu)$ can have is to show that among the edge-extremal graphs in $\mathcal{M}_{chordal}(d,\nu)$, there is one that has a very simple structure: it is a disjoint union of cliques and stars of a given size.

Theorem 4 (restated). *There exists an edge-extremal graph in $\mathcal{M}_{chordal}(d,\nu)$ that is a disjoint union of cliques and stars.*

Overview of the proof. The proof is by contradiction. We start with an edge-extremal graph of $\mathcal{M}_{chordal}(d,\nu)$ that is, in some sense, closest to being a disjoint union of cliques and stars. From that, we will perform a series of modifications in the graph in order to obtain another graph of $\mathcal{M}_{chordal}(d,\nu)$ that has at least as many edges as the one we started with, but that is closer to being a disjoint union of cliques and stars, which will be a contradiction with our initial choice. To perform the modifications, we will consider a specific clique tree of our edge-extremal graph and exploit the structure of this graph around one of its cliques, given by a carefully chosen node of the tree. A crucial part of the proof is to ensure that, after each modification, the obtained graph still belongs to $\mathcal{M}_{chordal}(d,\nu)$. In this vein, Lemmas 3 and 4 will precisely show that the two modifications we describe can indeed be performed without disrupting membership in $\mathcal{M}_{chordal}(d,\nu)$. In this way, we obtain a new edge-extremal graph that, as a result, has several structural properties that will be exploited to conclude the proof.

Proof of Theorem 1. Assume for a contradiction that there is no edge-extremal graph in $\mathcal{M}_{chordal}(d,\nu)$ that is a disjoint union of cliques and stars. Let H be an edge-extremal graph in $\mathcal{M}_{chordal}(d,\nu)$ with maximum number of $(d-1)$-stars and subject to that, with maximum number of connected components. Let W be a connected component of H that is not a clique nor a star and let $\nu_1 = \nu(W)+1$. By Lemma 1, W is a factor-critical graph and therefore $|V_W| = 2\nu_1 - 1$. Note that $W \in \mathcal{M}_{chordal}(d,\nu_1)$ and, in fact, W is edge-extremal in $\mathcal{M}_{chordal}(d,\nu_1)$. Among all the edge-extremal graphs in $\mathcal{M}_{chordal}(d,\nu_1)$ with $2\nu_1 - 1$ vertices, let G be the one that has a clique tree with minimum number of leaves. Note that, in particular, G is connected, by the maximality of the number of connected components of the graph H.

Let $\mathcal{T}$ be a clique tree of G achieving the minimum number of leaves. We consider $\mathcal{T}$ rooted in an arbitrary bag R. Let X be a node of $\mathcal{T}$. We denote by T_X the subtree of $\mathcal{T}$ rooted at the node X. We define a subgraph G_X associated with each node X of $\mathcal{T}$ in the following way. If $X = R$, then $G_X = G$. Otherwise, let S be the separator of G given by the intersection between X and its parent in $\mathcal{T}$ and let V_{T_X} be the set of vertices appearing in the bags of T_X. The subgraph G_X associated with the node X is given by $G[V_{T_X} \setminus S]$. Observe that if X is a leaf of $\mathcal{T}$, then G_X is a complete graph. Let B be a bottommost bag in $\mathcal{T}$ such that G_B is not a complete graph. Note that such a node indeed exists since G is not a complete graph itself. Let $B_1, \ldots, B_k$ be the children of B in $\mathcal{T}$ and let

$S_i = B \cap B_i$. For simplicity, from now on we denote $C_i = V_{T_{B_i}} \setminus S_i$. Note that $G[C_i] = G_{B_i}$ is a complete graph for every i.

Observation 1(♠). For every i, the subgraph of G induced by the edges $E_i = \{xy \mid x \in S_i \text{ and } y \in C_i\}$ is a chain graph.

Observation 2(♠). For every i, the subtree T_{B_i} is a path.

In what follows, we want to modify the graph G in such a way to obtain a graph that is still chordal, has the same number of vertices as G and belongs to $\mathcal{M}_{chordal}(d, \nu_1)$, but either has more edges than G, or is disconnected or has a clique tree with smaller number of leaves. Either one of these outcomes will contradict the choice of G. Note that since G is a factor-critical graph, the addition of edges to G does not increase its matching number. Moreover, for any $k \in \mathbb{N}$, the removal of k vertices from G and addition of k new vertices does not increase its matching number either, since the total number of vertices remains unchanged. Therefore, all the modifications that are to be performed in what follows will not lead to a graph with larger matching number than G.

For every $v \in B$ let $f_G(v, i)$ denote the number of neighbors that vertex v has in the clique C_i, that is, $f_G(v, i) = |N_G(v) \cap C_i|$ and let $u_{i,1}, \ldots, u_{i,|C_i|}$ be an ordering of the vertices of C_i such that $\deg_G(u_{i,1}) \geq \deg_G(u_{i,2}) \geq \ldots \geq \deg_G(u_{i,|C_i|})$.

We first state and prove the following lemma that can be understood as the converse of Observation 3 and that will be useful throughout the paper to show that a graph is chordal.

Lemma 2 (♠). Let H be any graph and $B, C_1, \ldots, C_k$ be cliques of H such that

- $N_H(C_i) \subseteq B$, for every $1 \leq i \leq k$;
- $H[V_H \setminus (\cup_{i=1}^{k} C_i)]$ is a chordal graph.

If the subgraph of H induced by the edges $E_i = \{xy \mid x \in B \text{ and } y \in C_i\}$ is a chain graph for every $1 \leq i \leq k$, then H is a chordal graph.

We are now ready to state the two modifications that will be used repeatedly throughout the proof of Theorem 1.

Modification 1. Let $B, C_1, \ldots, C_k$ be subsets of the vertex set of the chordal graph G as previously described and let $v \in B$. For $1 \leq i \leq k$, if $0 < f_G(v, i) < |C_i|$ and v has a neighbor that does not belong to $G[V_{T_B}]$, we do the following (see Fig. 1a):

(i) Add an edge between v and the vertex $u_{i,f_G(v,i)+1}$;
(ii) Delete the edge from v to one of its neighbors outside $G[V_{T_B}]$. This neighbor is chosen in the following way: consider the subtree T_v of $\mathcal{T}$ formed by the bags that contain the vertex v. Let L be a leaf of T_v that is not in the subtree rooted in B. Such a leaf exists since v has a neighbor outside $G[V_{T_B}]$. Let L' be the bag that is adjacent to L in T_v. Since $L \nsubseteq L'$, there exists $u \in L \setminus L'$. Let u be the chosen neighbor of v and delete the edge uv.

Lemma 3. *Modification 3 preserves both membership in $\mathcal{M}_{chordal}(d, \nu)$ and number of edges.*

Proof. Let G' be the graph obtained with the application of Modification 3. First, note that the addition of the edge $vu_{i,f_G(v,i)+1}$ preserves the nested neighborhood property in the bipartite graph induced by the edges between B and C_i. Thus, since G is chordal and by Lemma 2, the addition of this edge does not disrupt membership in the class of chordal graphs. Therefore, to show that G' is chordal it suffices to show that the removal of the edge uv preserves chordality. We do so by providing a clique tree to $G - uv$. This clique tree is obtained from $\mathcal{T}$ as follows. Let $L'' = L \setminus \{u\}$. If $L'' \neq L'$, add L'' between L and L' in the tree $\mathcal{T}$ and delete v from L. If $L'' = L'$, just delete v from L in $\mathcal{T}$. Also, note that this operation does not change the number of leaves in $\mathcal{T}$. Hence, we obtain that the graph G' is chordal. Note that the degree of v does not change with this modification. The only vertex whose degree was increased by Modification 3 is $u_{i,f_G(v,i)+1}$. However, note that since $vu_{i,f_G(v,i)} \in E_G$ and $vu_{i,f_G(v,i)+1} \notin E_G$, we have that $\deg_G(u_{i,f_G(v,i)+1}) < \deg_G(u_{i,f_G(v,i)})$. Thus, $\deg_{G'}(u_{i,f_G(v,i)+1}) \leq \deg_{G'}(u_{i,f_G(v,i)}) = \deg_G(u_{i,f_G(v,i)}) < d$. We conclude the proof by observing that $|E_{G'}| = |E_G|$, since exactly one edge was deleted and exactly one edge was added by this modification. $\Diamond$

Modification 2. Let $B, C_1, \ldots, C_k$ be subsets of the vertex set of the chordal graph G as previously described and let $v \in B$. For $1 \leq i \leq k$, if $0 < f_G(v,i) < |C_i|$ and $f_G(v,j) > 0$ with $j > i$, we do the following (see Fig. 1b):

(i) Delete the edge $vu_{j,f_G(v,j)}$;
(ii) Add the edge $vu_{i,f_G(v,i)+1}$.

Lemma 4 (♠). Modification 3 preserves both membership in $\mathcal{M}_{chordal}(d, \nu)$ and number of edges.

Recall that our graph G is an edge-extremal graph in $\mathcal{M}_{chordal}(d, \nu_1)$, since it is a connected component of an edge-extremal graph $H \in \mathcal{M}_{chordal}(d, \nu)$, where H has maximum number of connected components among the edge-extremal graphs of $\mathcal{M}_{chordal}(d, \nu)$. Let G^* be the graph obtained from G by exhaustive applications of Modification 3 followed by exhaustive applications of Modification 3. It follows immediately from Lemmas 3 and 4 that $G^* \in \mathcal{M}_{chordal}(d, \nu_1)$ and that G^* is edge-extremal in this set. Moreover, if the graph obtained after the application of any modification is disconnected, we reach a contradiction with the maximality of the number of components of H. Therefore, we can assume G^* is connected. The following lemma describes the major structural property of G^* that will be exploited in the remainder of the proof.

Lemma 5. *Let G^* be the graph obtained from G by exhaustive applications of Modification 3 followed by exhaustive applications of Modification 3. Then, for every $v \in V_{G^*} \cap B$ and every i, if v has at least one neighbor in C_i, one of the following conditions hold:*

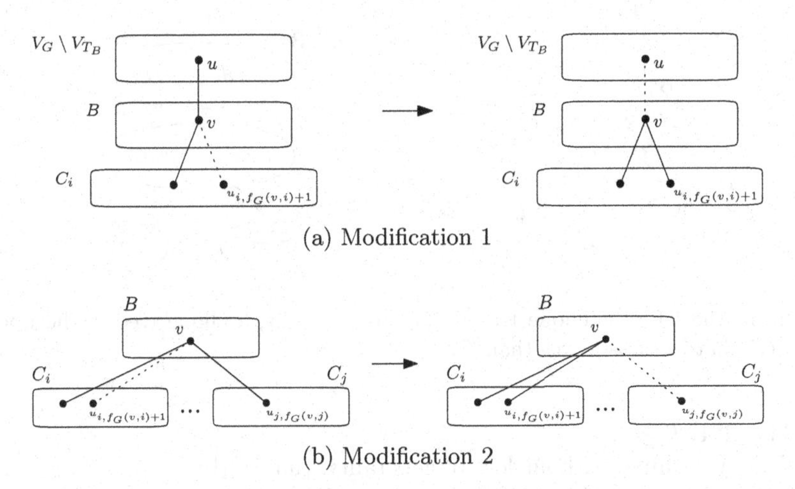

(a) Modification 1

(b) Modification 2

Fig. 1. The dotted lines between two vertices indicate non-edges.

(a) $C_i \subseteq N_{G^*}(v)$;
(b) $\deg_{G^*}(v) = \Delta(G^*)$ *and* $N_{G^*}(v) \subseteq B \cup C_1 \cup \ldots \cup C_i$.

Proof. First, let G' be the graph obtained from G by exhaustive applications of Modification 3. Since this modification can no longer be applied, then for every $v \in B$ and every i such that $f(v,i) > 0$, we have that either $f(v,i) = |C_i|$ or $f(v,j) = 0$ for every $j > i$. Thus, for every $v \in B$, there exists at most one index ℓ such that $0 < f(v,\ell) < |C_\ell|$. Now we apply Modification 3 exhaustively to G' and obtain the graph G^*. Recall that $f_{G'}(v,i) = |N_{G'}(v) \cap C_i|$. Observe that for every $v \in B$, if $f_{G'}(v,i) = 0$, then $f_{G^*}(v,i) = 0$ and if $f_{G'}(v,i) = |C_i|$, then $f_{G^*}(v,i) = |C_i|$. Furthermore, since Modification 3 can no longer be applied, if a vertex v is such that $0 < f_{G^*}(v,i) < |C_i|$, then v has no neighbors outside $B \cup C_1 \cup \ldots \cup C_i$. That is, if condition (a) does not hold, then $N_{G^*}(v) \subseteq B \cup C_1 \cup \ldots \cup C_i$. It remains to show that, in this case, $\deg_{G^*}(v) = \Delta(G^*)$. Indeed, if $\deg_{G^*}(v) < \Delta(G^*)$, we can add to G^* the edge $vu_{i,f_{G^*}(v,i)+1}$. The addition of this edge does not change the maximum degree of G^* by assumption. Moreover, by Lemma 2, it also preserves chordality. Since by Lemmas 3 and 4, we have that $|E_{G^*}| = |E_G|$ and that $G^* \in \mathcal{M}_{chordal}(d, \nu_1)$, this leads to a contradiction with the fact that G is edge-extremal in $\mathcal{M}_{chordal}(d, \nu_1)$. $\Diamond$

Since the graph G^* is such that $\Delta(G^*) < d$ and $|E_{G^*}| = |E_G|$, we can replace the connected component G in our edge-extremal graph H by G^*. This replacement will be convenient since Lemma 5 provides useful information on the structure of G^*. More concretely, *in the rest of the proof we shall assume that* $B, C_1, \ldots, C_k$ *satisfy the conclusion of Lemma* 5.

Let b be the size of the clique B, let $\Delta = \Delta(G^*)$ and recall that S_i is the separator between the bag B_i and B. We are now going to conclude the proof of Theorem 1 with a case analysis.

Case 1: There exists i such that $|C_i| + b \leq \Delta + 1$.

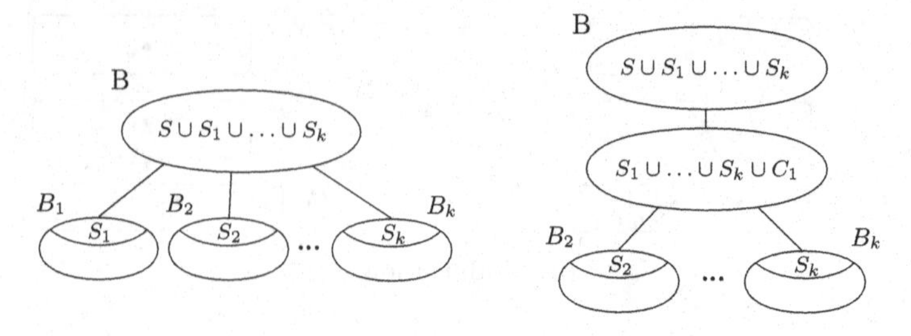

Fig. 2. To the left, the clique tree T and to the right, a clique tree of the updated graph G^* that has less leaves than T.

Case 1.1: $k \geq 2$.

We may assume, without loss of generality, that $|C_1| \leq |C_2| \leq \ldots \leq |C_k|$. In particular, this implies that $|C_1| + b \leq \Delta + 1$. We will show that, in this case, all the vertices of C_1 are adjacent to all the vertices of $S_1 \cup \ldots \cup S_k$. This will lead to a contradiction with the number of leaves of the clique tree of G. Suppose for a contradiction that there exists $v \in S_1 \cup \ldots \cup S_k$ that is not universal to C_1. This implies that $f_{G^*}(v, 1) < |C_1|$.

We will show that the graph G^* can be modified in order to obtain another edge-extremal graph, also in $\mathcal{M}_{chordal}(d, \nu_1)$, in which v is adjacent to every vertex of $S_1 \cup \ldots \cup S_k$.

First, note that it cannot be the case that $f_{G^*}(v, 1) > 0$, since by Lemma 5, if $0 < f_{G^*}(v, 1) < |C_1|$, then v has maximum degree and has no neighbors outside $B \cup C_1$. However, this is a contradiction, since $|C_1| + b \leq \Delta + 1$. Thus, it holds that $f_{G^*}(v, 1) = 0$.

In what follows, we will modify the graph G^* and the deletion of some edges might disrupt the membership in the class of chordal graphs. In these cases, we will use the following modification in order to restore it.

Modification 3. Let H be any graph satisfying the conditions of Lemma 2. We do the folowing:

(i) Delete from H all the edges xy such that $x \in B$ and $y \in C_i$ for some i;
(ii) For each $v \in B$ and each $1 \leq i \leq k$, if $f_H(v, i) > 0$, add the edges between v and the vertices $u_{i,1}, \ldots, u_{i,f_H(v,i)}$.

Lemma 6 (♠). Modification 3 preserves membership in the class of chordal graphs and number of edges.

We now modify G^* as follows. Let j be the largest index for which $f_{G^*}(v, j) > 0$. If $f_{G^*}(v, j) = |C_j|$, since $|C_1| \leq |C_j|$, we can delete $|C_1|$ edges between v and C_j and add all the edges between v and C_1. We then apply Modification 3 to the obtained graph in order to obtain a graph that, by Lemma 6, is chordal. Note that the only vertices whose degree has increased are the ones in C_1. However, since $|C_1| + b \leq \Delta + 1$, we conclude that the maximum degree of G^* did not increase.

If $f_{G^*}(v, j) < |C_j|$, then, by Lemma 5, v has maximum degree and has no neighbors outside $B \cup C_1 \cup \ldots \cup C_j$. This implies that $\sum_{\ell=2}^{k} f_{G^*}(v, \ell) = \Delta - b + 1$ (recall that $f_{G^*}(v, 1) = 0$). Since $|C_1| \leq \Delta - b + 1$ by assumption, we can delete $|C_1|$ edges between v and vertices of $C_2 \cup \ldots \cup C_j$ and add all the edges between v and C_1. We then apply Modification 3 to the obtained graph in order to obtain a graph that, by Lemma 6, is chordal. Again, the only vertices whose degree has increased in this process are the ones from C_1, thus we conclude the obtained graph still has degree at most Δ.

Finally note that in both cases, the modifications do not change the number of edges of G^*, since $\sum_{\ell=1}^{k} f_{G^*}(v, \ell)$ remains the same. We perform this change for every $v \in S_1 \cup \ldots \cup S_k$ such that $f_{G^*}(v, 1) > 0$ and obtain a new edge-extremal graph in $\mathcal{M}_{chordal}(d, \nu_1)$ such that all the vertices of C_1 are adjacent to all the vertices of $S_1 \cup \ldots \cup S_k$. Recall that among all the edge-extremal graphs in $\mathcal{M}_{chordal}(d, \nu_1)$ with $2\nu_1 - 1$ vertices, G was the one that had a clique tree with minimum number of leaves. This new graph, however, has a clique tree that has less leaves than the clique tree $\mathcal{T}$ of G. This is because the clique $C_1 \cup S_1 \cup \ldots \cup S_k$ is contained in B and contains the intersection between B and each child of B (see Fig. 2). This contradicts the minimality of the number of leaves of $\mathcal{T}$.

Case 1.2: $k = 1$.

Since $B \cup C_1$ is not a clique by assumption, there exists $v \in S_1$ that is not universal to C_1 in G^*. By Lemma 5, v has maximum degree and no neighbors outside $B \cup C_1$. Hence, $\deg_{G^*}(v) \leq b - 1 + |C_1| - 1$, which implies that $\Delta \leq b + |C_1| - 2$. This is a contradiction with the assumption of Case 1 that $|C_1| + b \leq \Delta + 1$.

Case 2: For every i, $|C_i| + b > \Delta + 1$.

Let $v \in S_1 \cup \ldots \cup S_k$. Let a_v be the smallest index such that $f_{G^*}(v, a_v) > 0$. Note that v cannot be universal to C_{a_v} in G^*, since by assumption $|C_{a_v}| + b > \Delta + 1$. By Lemma 5, $\deg_{G^*}(v) = \Delta$ and $N_{G^*}(v) \subseteq B \cup C_{a_v}$. This implies that for every $v \in S_1 \cup \ldots \cup S_k$, there exists a unique index a_v such that $f_{G^*}(v, a_v) > 0$. That is, for any $j \neq a_v$, $f_{G^*}(v, j) = 0$, and thus $S_i \cap S_j = \emptyset$ if $i \neq j$. Also, since $N_{G^*}(v) \subseteq B \cup C_{a_v}$ and v has degree Δ, we have that $f_{G^*}(v, a_v) = \Delta - b + 1$. That is, if $a_v = a_u$, then u and v are true twins in G^*. Moreover, for any $1 \leq i < j \leq k$, $|N_{G^*}(S_i) \cap C_i| = |N_{G^*}(S_j) \cap C_j|$. Let S be the separator between B and its parent in the clique tree $\mathcal{T}$. Since for every $v \in S_1 \cup \ldots \cup S_k$, $N_{G^*}(v) \subseteq B \cup C_{a_v}$, we know that $S \cap S_i = \emptyset$, for every i. Also, since the graph G^* is connected, $S \neq \emptyset$. See Fig. 3.

Let $u \in N_{G^*}(S_i) \cap C_i$. Suppose for a contradiction that $\deg_{G^*}(u) < \Delta$. Let G_1 be the graph obtained from G^* by the deletion of one vertex of S and addition of a new vertex w in S_i, such that $N_{G_1}[w] = B \cup (N(S_i) \cap C_i)$.

Claim 1(♠). $|E_{G_1}| \geq |E_{G^*}|$ and $\Delta(G_1) = \Delta(G^*)$.

If G_1 is disconnected or has more edges than G^*, we have a contradiction. We repeat the above modification until either the graph obtained is disconnected, that is, until $S = \emptyset$, or until for every i, the degree of the vertices in $N_{G_1}(S_i) \cap C_i$

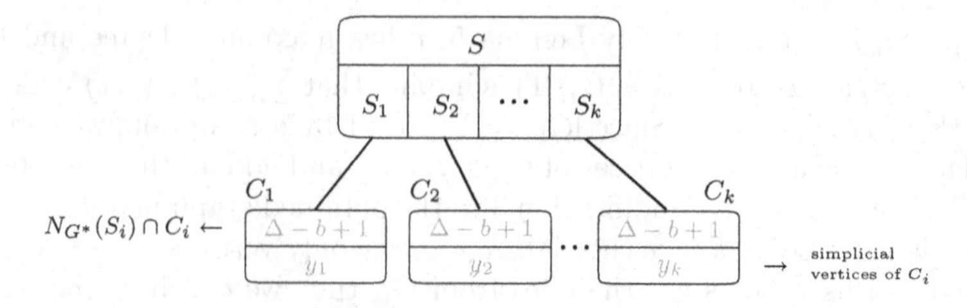

Fig. 3. Graph G^* in case 2. Thick lines indicate all possible edges between the sets. Gray text indicates the cardinality of the vertex set.

is Δ. Let G_2 be the graph obtained after exhaustive application of the above modification. If G_2 is disconnected, we have a contradiction with the maximality of the number of connected components of our initial edge-extremal graph. Otherwise, by Claim 3, $|E_{G_2}| \geq |E_{G^*}|$ and $\Delta(G_2) = \Delta(G^*)$. Therefore, we can now replace G^* by G_2 in the edge-extremal graph H. Note that G_2 is such that:

1. For every $1 \leq i < j \leq k$, $S_i \cap S_j = \emptyset$;
2. For every $1 \leq i \leq k$, the vertices of S_i and of $N_{G_2}(S_i) \cap C_i$ have degree Δ and $|N_{G_2}(S_i) \cap C_i| = \Delta - b + 1$.

Case 2.1: $k \geq 2$.

Let y_i be the number of simplicial vertices in the clique C_i. Assume without loss of generality that $y_1 \geq y_2$. We perform the following modifications in the graph G_2: deletion of one simplicial vertex from C_2 and one vertex from S_1 and addition of one vertex to S_2 and one simplicial vertex to C_1. Note that, after this modification, the only vertices that had their degree changed are the simplicial vertices from C_1 and C_2. Since these simplicial vertices did not have maximum degree before, the degree of the obtained graph does not exceed the degree of G_2. Note that $y_2 - 1 + \Delta - b + 1 + \Delta$ edges were removed by the deletion of the two vertices and $y_1 + \Delta - b + 1 + \Delta$ were added by the addition of the other two vertices. However, since $y_1 \geq y_2$, we have that the obtained graph has strictly more edges than G_2, which is a contradiction.

Case 2.2: $k = 1$.

Since all vertices in S_1 and in $N_{G_2}(S_1) \cap C_1$ have maximum degree, we can perform the following modification in G_2: delete all vertices of S_1 and add $|S_1|$ vertices to $N_{G_2}(S_1) \cap C_1$. The graph obtained after this modification has the same number of edges as G_2, since $|S_1|$ vertices of degree Δ were removed and the same amount of vertices with the same degree was added. However, the obtained graph is disconnected, which is a contradiction with the maximality of the number of connected components of the edge-extremal graph H.

This concludes the proof of Theorem 1. □

By Theorem 1, we know that there is an edge-extremal graph in $\mathcal{M}_{chordal}(d, \nu)$ that is a disjoint union of cliques and stars. The next lemma gives a tight upper bound on the number of edges of such an edge-extremal graph when d is even.

Lemma 7 (♠). Let G be a graph in $\mathcal{M}_{chordal}(d, \nu)$ that is a disjoint union of cliques and stars. If d is even, then $|E_G| \leq (d-1)(\nu-1)$.

By Theorem 3, we already know the maximum number of edges that a graph that is a disjoint union of cliques and stars can have when d is odd. From Theorem 1 and Lemma 7, we obtain our main result, Theorem 2 (see page 601), which establishes the upper bound on the number of edges that a chordal graph of $\mathcal{M}_{chordal}(d, \nu)$ can have and shows that the obtained bound is tight.

4 Final Remarks and Open Problems

In this work, we determined the maximum number of edges that a chordal graph can have if its maximum degree and matching number are bounded. We also exhibit examples of graphs achieving this bound.

An interesting question that remains open comes from the fact that the graph K_i' used in Theorem 3 has an induced C_4. *For each* d and ν, what is the maximum number of edges of a graph in $\mathcal{M}_{C_4-free}(d, \nu)$? We point out that the bound on the number of edges for chordal graphs does not hold for C_4-free graphs, as can be seen by the graph P, obtained from the famous Petersen graph by the subdivision of one edge. We have that $\Delta(P) = 3$, $\nu(P) = 5$ and $|E_P| = 16$. In this case, the bound given by Theorem 1 when $d = 4$ and $\nu = 6$ is 15. This idea can be further generalized to create examples in the class of $\mathcal{H}$-free graphs, where $\mathcal{H}$ is any finite collection of cycles. Indeed, let r be the size of a largest cycle of $\mathcal{H}$. A result due to Kochol [8] about snarks implies that for any $r \geq 5$ there exists an infinite family of 3-regular graphs of girth r that have a perfect matching. Let G be one such graph and let H be the graph obtained from G by the subdivision of one edge. The graph H is clearly $\mathcal{H}$-free and is such that $\Delta(H) = 3$, $\nu(H) = \nu(G)$ and $|E_H| = 3\nu(H) + 1$, while the bound given by Theorem 1 when $d = 4$ and $\nu = \nu(H) + 1$ is $3\nu(H)$.

References

1. Balachandran, N., Khare, N.: Graphs with restricted valency and matching number. Discrete Math. **309**, 4176–4180 (2009)
2. Belmonte, R., Heggernes, P., van 't Hof, P., Saei, R.: Ramsey numbers for line graphs and perfect graphs. In: Proceedings of the 18th Annual International Conference on Computing and Combinatorics, COCOON. pp. 204–215 (2012)
3. Blair, J.R.S., Peyton, B.: An introduction to chordal graphs and clique trees. In: George, A., Gilbert, J.R., Liu, J.W.H. (eds.) Graph Theory and Sparse Matrix Computation. The IMA Volumes in Mathematics and its Applications, vol 56. Springer, New York (1993) https://doi.org/10.1007/978-1-4613-8369-7_1
4. Chvátal, V., Hanson, D.: Degrees and matchings. J. Combin. Theory Ser. B **20**, 128–138 (1976)

5. Dibek, C., Ekim, T., Heggernes, P.: Maximum number of edges in claw-free graphs whose maximum degree and matching number are bounded. Discrete Math. **340**, 927–934 (2017)
6. Fulkerson, D.R., Gross, O.A.: Incidence matrices and interval graphs. Pacific J. Math. **15**, 835–855 (1965)
7. Gavril, F.: The intersection graphs of subtrees in trees are exactly the chordal graphs. J. Combin. Theory Ser. B **16**, 47–56 (1974)
8. Kochol, M.: Snarks without small cycles. J. Combin. Theory Ser. B **67**, 34–47 (1996)
9. Måland, E.: Maximum Number of Edges in Graph Classes under Degree and Matching Constraints. Master's thesis, University of Bergen, Norway (2015)

Steiner Trees for Hereditary Graph Classes

Hans L. Bodlaender[1], Nick Brettell[2(✉)], Matthew Johnson[2], Giacomo Paesani[2], Daniël Paulusma[2], and Erik Jan van Leeuwen[1]

[1] Department of Information and Computing Sciences, Utrecht University, Utrecht, The Netherlands
{h.l.bodlaender,e.j.vanleeuwen}@uu.nl
[2] Department of Computer Science, Durham University, Durham, UK
{nicholas.j.brettell,matthew.johnson2,giacomo.paesani, daniel.paulusma}@durham.ac.uk

Abstract. We consider the classical problems (Edge) Steiner Tree and Vertex Steiner Tree after restricting the input to some class of graphs characterized by a small set of forbidden induced subgraphs. We show a dichotomy for the former problem restricted to (H_1, H_2)-free graphs and a dichotomy for the latter problem restricted to H-free graphs. We find that there exists an infinite family of graphs H such that Vertex Steiner Tree is polynomial-time solvable for H-free graphs, whereas there exist only two graphs H for which this holds for Edge Steiner Tree. We also find that Edge Steiner Tree is polynomial-time solvable for (H_1, H_2)-free graphs if and only if the treewidth of the class of (H_1, H_2)-free graphs is bounded (subject to $\mathsf{P} \neq \mathsf{NP}$). To obtain the latter result, we determine all pairs (H_1, H_2) for which the class of (H_1, H_2)-free graphs has bounded treewidth.

1 Introduction

Let $G = (V, E)$ be a connected graph and $U \subseteq V$ be a set of *terminal* vertices. A *Steiner tree* for U (of G) is a tree in G that contains all vertices of U. An *edge weighting* of G is a function $w_E : E \rightarrow \mathbb{R}^+$. For a tree T in G, the *edge weight* $w_E(T)$ of T is the sum $\sum_{e \in E(T)} w_E(e)$. We consider the classical problem:

> Edge Steiner Tree
> *Instance:* a connected graph $G = (V, E)$ with an weighting w_E, a subset $U \subseteq V$ of terminals and a positive integer k.
> *Question:* does G have a Steiner tree T_U for U with $w_E(T_U) \leq k$?

This is often known simply as Steiner Tree, but we wish to distinguish it from a closely related problem. A *vertex weighting* of G is a function $w_V : V \rightarrow \mathbb{R}^+$. For a tree T in G, the *vertex weight* $w_V(T)$ of T is the sum $\sum_{v \in V(T)} w(v)$. The following problem is sometimes known as Node-Weighted Steiner Tree.

Supported by the Leverhulme Trust (RPG-2016-258) and the Royal Society (IES\R1\191223).

Y. Kohayakawa and F. K. Miyazawa (Eds.): LATIN 2020, LNCS 12118, pp. 613–624, 2020.
https://doi.org/10.1007/978-3-030-61792-9_48

VERTEX STEINER TREE
Instance: a connected graph $G = (V, E)$ with a vertex weighting w_V, a subset $U \subseteq V$ and a positive integer k.
Question: does G have a Steiner tree T_U for U with $w_V(T_U) \leq k$?

Note that EDGE STEINER TREE is a generalization of the SPANNING TREE problem (set $U = V(G)$). We refer to the textbooks of Du and Hu [7] and Prömel and Steger [14] for further background information on Steiner trees.

We consider the problems EDGE STEINER TREE and VERTEX STEINER TREE separately so that, for any graph under consideration, we have either an edge or vertex weighting but not both, so we will generally denote weightings by w without any subscript. Moreover, when we use the following terminology there is no ambiguity. We say that a Steiner tree of least possible weight is *minimum*, and that an instance of a problem is *unweighted* if the weighting is constant. It is well known that the unweighted versions of EDGE STEINER TREE and VERTEX STEINER TREE are NP-complete [8,12], and we note that these unweighted problems are polynomially equivalent. We denote instances of the weighted problems by (G, w, U, k) and of the unweighted problems by (G, U, k).

Our Focus. We focus on the complexity of EDGE STEINER TREE and VERTEX STEINER TREE for *hereditary* graph classes, i.e., graph classes closed under vertex deletion. We do this from a *systematic* point of view. It is well known, and readily seen, that a graph class $\mathcal{G}$ is hereditary if and only if it can be characterized by a set $\mathcal{H}$ of forbidden induced subgraphs. That is, a graph G belongs to $\mathcal{G}$ if and only if G has no induced subgraph isomorphic to some graph in $\mathcal{H}$. We normally require $\mathcal{H}$ to be minimal, in which case it is unique and we denote it by $\mathcal{H}_\mathcal{G}$. We note that $\mathcal{H}_\mathcal{G}$ may have infinite size; for example, if $\mathcal{G}$ is the class of bipartite graphs, then $\mathcal{H}_\mathcal{G} = \{C_3, C_5, \ldots,\}$, where C_r denotes the cycle on r vertices. For a systematic complexity study of a graph problem, we may first consider *monogenic graph classes* or *bigenic* graph classes, which are classes $\mathcal{G}$ with $|\mathcal{H}_\mathcal{G}| = 1$ or $|\mathcal{H}_\mathcal{G}| = 2$, respectively. This is the approach we follow here.

Our Results. We prove a dichotomy for EDGE STEINER TREE for bigenic graph classes in Sect. 2 and a dichotomy for VERTEX STEINER TREE for monogenic graph classes in Sect. 3. We denote the *disjoint union* of two vertex-disjoint graphs G and H by $G + H = (V(G) \cup V(H), E(G) \cup E(H))$, and the disjoint union of s copies of G by sG. A *linear forest* is a disjoint union of paths. For a graph H, a graph is *H-free* if it has no induced subgraph isomorphic to H. For a set of graphs $\{H_1, \ldots, H_p\}$, a graph is *$(H_1, \ldots, H_p)$-free* if it is H_i-free for every $i \in \{1, \ldots, p\}$. We let K_r and P_r denote the complete graph and path on r vertices. The *complete bipartite* graph $K_{s,t}$ is the graph whose vertex set can be partitioned into two sets S and T of size s and t, such that for any two distinct vertices u, v, we have $uv \in E$ if and only if $u \in S$ and $v \in T$. We call $K_{1,3}$ the *claw*. In the first dichotomy, the roles of H_1 and H_2 are interchangeable.

Theorem 1. *Let H_1 and H_2 be two graphs.* EDGE STEINER TREE *is polynomial-time solvable for (H_1, H_2)-free graphs if*

1. $H_1 = K_r$ *for some* $r \in \{1, 2\}$.
2. $H_1 = K_3$ *and* $H_2 = K_{1,3}$.
3. $H_1 = K_r$ *for some* $r \geq 3$ *and* $H_2 = P_3$.
4. $H_1 = K_r$ *for some* $r \geq 3$ *and* $H_2 = sP_1$ *for some* $s \geq 1$,

and otherwise it is NP*-complete.*

Theorem 2. *Let* H *be a graph. For every* $s \geq 0$, VERTEX STEINER TREE *is polynomial-time solvable for* H*-free graphs if* H *is an induced subgraph of* $sP_1 + P_4$*; otherwise even unweighted* VERTEX STEINER TREE *is* NP*-complete.*

We make the following observations about these two results:
1. We prove Theorem 1 by pinpointing a strong correspondence to the notion of treewidth. We show, in fact, that EDGE STEINER TREE can be solved in polynomial time for (H_1, H_2)-free graphs if and only if the treewidth of the class of (H_1, H_2)-free graphs is bounded. Although VERTEX STEINER TREE is polynomial-time solvable for graph classes of bounded mim-width [1] and thus also for graph classes of bounded treewidth, such a 1-to-1 correspondence does not hold for VERTEX STEINER TREE, for treewidth or mim-width. To see this, observe that complete graphs, and hence P_4-free graphs, have unbounded treewidth, whereas cobipartite graphs, and hence $3P_1$-free graphs, have unbounded mim-width. In Sect. 4 we discuss this connection between EDGE STEINER TREE and treewidth further.
2. The restriction of Theorem 1 to monogenic graph classes yields only two (trivial) graphs H, namely $H = P_1$ or $H = P_2$, for which the restriction of EDGE STEINER TREE to H-free graphs can be solved in polynomial time. In contrast, by Theorem 2, VERTEX STEINER TREE can, when restricted to H-free graphs, be solved in polynomial time for an infinite family of linear forests H, namely $H = sP_1 + P_4$ $(s \geq 0)$.
3. Theorem 2 is also a dichotomy for the unweighted VERTEX STEINER TREE problem. Moreover, as the unweighted versions of EDGE STEINER TREE and VERTEX STEINER TREE are polynomially equivalent, Theorem 2 is also a classification of the unweighted version of EDGE STEINER TREE.

2 The Proof of Theorem 1

In this section we give a proof for our first dichotomy, which is for EDGE STEINER TREE for (H_1, H_2)-free graphs. We note that this is not the first systematic study of EDGE STEINER TREE. For example, Renjitha and Sadagopan [15] proved that unweighted EDGE STEINER TREE is NP-complete for $K_{1,5}$-free split graphs, but can be solved in polynomial time for $K_{1,4}$-free split graphs. We present a number of other results from the literature, which we collect in Sect. 2.1, together with some lemmas that follow from these results. Then in Sect. 2.2 we discuss the notion of treewidth; as we shall see, this notion will play an important role. We then use these results to prove Theorem 1.

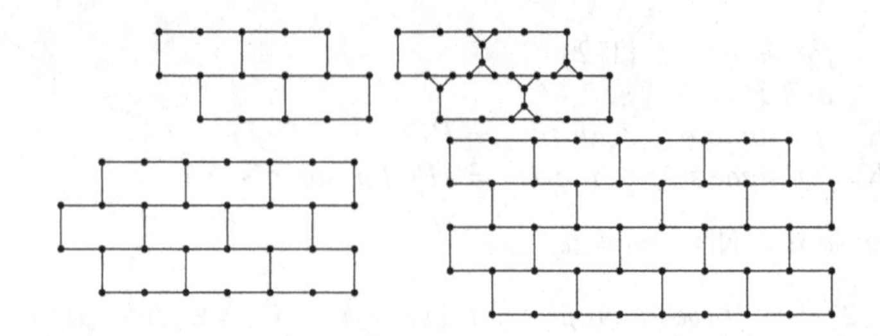

Fig. 1. A wall of height 2, its wye-net transformation, walls of height 3 and 4.

2.1 Preliminaries

The NP-completeness of EDGE STEINER TREE on complete graphs follows from the result [12] that the general problem is NP-complete: to obtain a reduction add any missing edges and give them sufficiently large weight such that they will never be used in any solution. Bern and Plasman proved the following stronger result.

Lemma 1. *[2]* EDGE STEINER TREE *is* NP-*complete for complete graphs where every edge has weight* 1 *or* 2.

To subdivide an edge $e = uv$ means to delete e and add a vertex w and edges uw and vw. Let r be a positive integer. To say that e is subdivided r times means that e is replaced by a path $P_e = uw_1 \cdots w_r v$ of $r + 1$ edges. The *r-subdivision* of a graph H is the graph obtained from H after subdividing each edge exactly r times. If we say that a graph is a *subdivision* of H, then we mean it can be obtained from H using subdivisions (the number of subdivisions can be different for each edge and some edges might not be subdivided at all). A graph G contains a graph H as a *subdivision* if G contains a subdivision of H as a subgraph.

Proposition 1. *If* EDGE STEINER TREE *is* NP-*complete on a class* $\mathcal{C}$ *of graphs, then, for every* $r \geq 0$*, it is so on the class of* r*-subdivisions of graphs in* $\mathcal{C}$*.*

We make the following observation (proof omitted).

Lemma 2. EDGE STEINER TREE *is* NP-*complete for complete bipartite graphs.*

The following follows by inspection of the reduction of Garey and Johnson for RECTILINEAR STEINER TREE [10]. Let n and m be positive integers. An $n \times m$ grid graph has vertex set $\{v^{i,j} \mid 1 \leq i \leq n, 1 \leq j \leq m\}$ and $v^{i,j}$ has neighbours $v^{i-1,j}$ (if $i > 1$), $v^{i+1,j}$ (if $i < n$), $v^{i,j-1}$ (if $j > 1$), and $v^{i,j+1}$ (if $j < m$).

Theorem 3. *[10] Unweighted* EDGE STEINER TREE *is* NP-*complete for grid graphs.*

A *wall* is a graph which can be thought of as a hexagonal grid. See Fig. 1 for three examples of walls of different *heights.* We refer to [6] for a formal definition. Note that walls of height at least 2 have maximum degree 3. From a wall of height h we obtain a *net-wall* by doing the following for each wall vertex u with three neighbours v_1, v_2, v_3: replace u and its incident edges with three new vertices u_1, u_2, u_3 and edges u_1v_1, u_2v_2, u_3v_3, u_1u_2, u_1u_3, u_2u_3. We call this a *wye-net transformation*, reminiscent of the well-known wye-delta transformation. Note that a net-wall is $K_{1,3}$-free but contains an induced *net*, which is the graph obtained from a triangle on vertices a_1, a_2, a_3 and three new vertices b_1, b_2, b_3 after adding the edge a_ib_i for $i = 1, 2, 3$.

We have two results on these classes.

Lemma 3. *For every* $r \geq 0$, EDGE STEINER TREE *is* NP*-complete for* r*-subdivisions of walls.*

Proof. We reduce from unweighted EDGE STEINER TREE on grid graphs, which is NP-hard by Theorem 3. Let (G, U, k) be an instance of unweighted EDGE STEINER TREE where G is an $n \times m$ grid graph. Think of $v^{1,1}$ as the top-left corner of the grid, and in $v^{i,j}$, i indicates the row of the grid containing the vertex, while j indicates the column.

From G, we obtain a graph W as follows. Two vertices of G are exceptional: $v^{n,1}$ is always exceptional, $v^{1,m}$ is exceptional if n is even, and $v^{1,1}$ is exceptional if n is odd. For every vertex $v^{i,j}$ of G that is not exceptional, W contains vertices $v_{\uparrow}^{i,j}$ and $v_{\downarrow}^{i,j}$ that are joined by an edge. We call these edges *new*. We also add to W vertices $v_{\uparrow}^{n,1}$, and $v_{\downarrow}^{1,m}$ (if $v^{1,m}$ is exceptional) or $v_{\downarrow}^{1,1}$ (otherwise). We add an edge from $v_{\downarrow}^{i,j}$ to $v_{\uparrow}^{i+1,j}$, for $1 \leq i \leq n-1$, $1 \leq i \leq m$. For $1 \leq i \leq n$, $1 \leq j \leq m-1$, if i is odd and n is even or if i is even and n is odd, we add an edge from $v_{\downarrow}^{i,j}$ to $v_{\uparrow}^{i,j+1}$, and otherwise, we add an edge from $v_{\uparrow}^{i,j}$ to $v_{\downarrow}^{i,j+1}$. The edges that are not new are *original.*

We note that W is a wall obtained from G by splitting each vertex in two (except the exceptional vertices that lie in a corner of the grid), and that there is a bijection between the original edges of W and the edges of G. We define an edge weighting w' for W by letting the weight of each original edge be 1 and the weight of each new edge be ε, where $\varepsilon > 0$ is chosen so that the sum of the weights of all new edges is less than 1. We define a set of terminals U' for W: if $v^{i,j}$ is in U, then U' contains each of $v_{\downarrow}^{i,j}$ and $v_{\uparrow}^{i,j}$ that exists (one or other will not exist if $v^{i,j}$ is exceptional).

We claim that there is a Steiner tree of k edges in G for terminal set U if and only if there is a Steiner tree of weight $k + \delta$ in (W, w') for terminal set U', where $0 \leq \delta < 1$. Indeed, any Steiner tree T in G for terminal set U of k edges corresponds naturally to a Steiner tree T' for U' in (W, w') of weight less than $k+1$ by adding all new edges to T and letting T' be a spanning tree of the component of the resulting subgraph of W that contains U'. Conversely, any Steiner tree T' for U' in (W, w') of weight $k+\delta$, $0 \leq \delta < 1$, corresponds naturally to a Steiner tree T for U in G of k edges by removing all new edges from T' and letting T be a spanning tree of the resulting subgraph of G. Effectively, this mimics the splitting and contraction operations which can be seen as the way in which we obtain W from G and vice versa.

The lemma now follows immediately from Proposition 1. □

The next lemma has a similar proof (omitted due to space restrictions).

Lemma 4. *For every $r \geq 0$,* Edge Steiner Tree *is* NP*-complete for r-subdivisions of net-walls.*

2.2 Treewidth and Implications

A *tree decomposition* of a graph $G = (V, E)$ is a tree T whose vertices, which are called *nodes*, are subsets of V and has the following properties: for each $v \in V$, the nodes of T that contain v induce a non-empty connected subgraph, and, for each edge $vw \in E$, there is at least one node of T that contains v and w.

The sets of vertices of G that form the nodes of T are called *bags*. The *width* of T is one less than the size of its largest bag. The *treewidth* of G is the minimum width of its tree decompositions. A graph class $\mathcal{G}$ has *bounded treewidth* if there exists a constant c such that each graph in $\mathcal{G}$ has treewidth at most c; otherwise $\mathcal{G}$ has *unbounded treewidth*. As trees with at least one edge form exactly the class of graphs with treewidth 1, the treewidth of a graph can be seen as a measure that indicates how close a graph is to being a tree. Many discrete optimization problems can be solved in polynomial time on every graph class of bounded treewidth. The Edge Steiner Tree problem is an example of such a problem (see, for instance, [5] or, for a faster algorithm [3]).

Lemma 5. *[3,5]* Edge Steiner Tree *can be solved in polynomial time on every graph class of bounded treewidth.*

We also need the well-known Robertson-Seymour Grid-Minor Theorem (also called the Excluded Grid Theorem), which can be formulated for walls.

Theorem 4. *[16] For every integer h, there exists a constant c_h such that a graph has treewidth at least c_h if and only if it contains a wall of height h as a subdivision.*

We will use two lemmas, both of which follow immediately from Theorem 4.

Lemma 6. *For every $r \geq 0$, the class of r-subdivided walls has unbounded treewidth.*

Lemma 7. *For every $r \geq 0$, the class of r-subdivided net-walls has unbounded treewidth.*

We need the following classification of the boundedness of treewidth for (H_1, H_2)-free graphs (in which we may exchange the roles of H_1 and H_2). Note that this classification coincides with the classification of Theorem 1.

Theorem 5. *Let H_1 and H_2 be two graphs. Then the class of (H_1, H_2)-free graphs has bounded treewidth if and only if*

1. $H_1 = K_r$ *for some* $r \in \{1, 2\}$.
2. $H_1 = K_3$ *and* $H_2 = K_{1,3}$.
3. $H_1 = K_r$ *for some* $r \geq 3$ *and* $H_2 = P_3$.
4. $H_1 = K_r$ *for some* $r \geq 3$ *and* $H_2 = sP_1$ *for some* $s \geq 1$.

Proof. We first prove that in each of the Cases 1–4, the class of (H_1, H_2)-free graphs has bounded treewidth. Let G be an (H_1, H_2)-free graph. First suppose that $H_1 = K_r$ for some $r \in \{1, 2\}$. Then G has no edges and so has treewidth 0. If $H_1 = K_3$ and $H_2 = K_{1,3}$, then G has maximum degree at most 2, that is, G is the disjoint union of paths and cycles. Hence G has treewidth at most 2. If $H_1 = K_r$ for some $r \geq 3$, and $H_2 = P_3$, then G is the disjoint union of complete graphs, each of size at most $r - 1$. Hence G has treewidth at most $r - 1$. Finally if $H_1 = K_r$, for some $r \geq 3$, and $H_2 = sP_1$, for some $s \geq 1$, then, by Ramsey's Theorem, the number of vertices of G is bounded by some constant $R(r, s)$. Hence G has treewidth at most $R(r, s)$.

We will now show that the class of (H_1, H_2)-free graphs has unbounded treewidth if Cases 1–4 do not apply. First suppose that neither H_1 nor H_2 is a complete graph. Then the class of (H_1, H_2)-free graphs contains the class of all complete graphs. As the treewidth of a complete graph K_r is readily seen to be equal to $r - 1$, the class of complete graphs, and thus the class of (H_1, H_2)-free graphs, has unbounded treewidth. From now on, assume that $H_1 = K_r$ for some $r \geq 1$. As Case 1 does not apply, we find that $r \geq 3$.

Suppose that H_2 contains a cycle C_s as an induced subgraph for some $s \geq 1$. As $H_1 = K_r$ for some $r \geq 3$, the class of (H_1, H_2)-free graphs contains the class of (C_3, C_s)-free subgraphs. As the latter graph class contains the class of $(s+1)$-subdivided walls, which have unbounded treewidth due to Lemma 6, the class of (H_1, H_2)-free graphs has unbounded treewidth.

Note that if H_2 contains a cycle as a subgraph, then it also contains a cycle as an induced subgraph. So now suppose that H_2 contains no cycle, that is, H_2 is a forest. First assume that H_2 contains an induced $P_1 + P_2$. Recall that $H_1 = K_r$ for some $r \geq 3$. Then the class of (H_1, H_2)-free graphs contains the class of complete bipartite graphs. As this class has unbounded treewidth, the class of (H_1, H_2)-free graphs has unbounded treewidth. From hereon we assume that H_2 is a $(P_1 + P_2)$-free forest.

Suppose that H_2 has a vertex of degree at least 3. In other words, as H_2 is a forest, the claw $K_{1,3}$ is an induced subgraph of H_2. Recall that $H_1 = K_r$ for some $r \geq 3$. First assume that $r = 3$. As Case 2 does not apply, H_2 properly contains an induced $K_{1,3}$. As H_2 is a forest, this means that H_2 contains an induced $P_1 + P_2$, which is not possible. We conclude that $r \geq 4$. Then the class of (H_1, H_2)-free graphs contains the class of net-walls. As the latter graph class has unbounded treewidth due to Lemma 7, the class of (H_1, H_2)-free graphs has unbounded treewidth.

From the above we may assume that H_2 does not contain any vertex of degree 3. This means that H_2 is a linear forest, that is, a disjoint union of paths.

As Case 4 does not apply, H_2 has an edge. Every (P_1+P_2)-free linear forest with an edge is either a P_2 or a P_3. However, this is not possible, as Case 1 (with the roles of H_1 and H_2 reversed) and Case 3 do not apply. We conclude that this case cannot happen. □

We are now ready to prove Theorem 1.

Theorem 1 (restated). *Let H_1 and H_2 be two graphs.* Edge Steiner Tree *is polynomial-time solvable for (H_1, H_2)-free graphs if and only if*

1. *$H_1 = K_r$ for some $r \in \{1, 2\}$.*
2. *$H_1 = K_3$ and $H_2 = K_{1,3}$.*
3. *$H_1 = K_r$ for some $r \geq 3$ and $H_2 = P_3$.*
4. *$H_1 = K_r$ for some $r \geq 3$ and $H_2 = sP_1$ for some $s \geq 1$,*

and otherwise it is NP*-complete.*

Proof (Sketch). Let $\mathcal{G}$ denote the class of (H_1, H_2)-free graphs under consideration. If one of Cases 1–4 applies, then $\mathcal{G}$ has bounded treewidth by Theorem 5; we apply Lemma 5. We now show NP-completeness in all remaining cases.

Suppose neither H_1 nor H_2 is a complete graph. Then $\mathcal{G}$ contains all complete graphs, and we apply Lemma 1. From now on, assume that $H_1 = K_r$ for some $r \geq 1$. As Case 1 does not apply, we find that $r \geq 3$. Suppose that H_2 contains a cycle C_s as an induced subgraph for some $s \geq 1$. Then $\mathcal{G}$ contains all (C_3, C_s)-free graphs. The latter class includes all $(s+1)$-subdivided walls, so we use Lemma 3.

Suppose that H_2 contains no cycle, that is, H_2 is a forest. If H_2 contains an induced $P_1 + P_2$, then $\mathcal{G}$ contains all complete bipartite graphs, and we apply Lemma 2. Now suppose that H_2 has a vertex of degree at least 3. As H is a forest, the claw $K_{1,3}$ is an induced subgraph of H_2. If $r = 3$, then as Case 2 does not apply, H_2 properly contains an induced $K_{1,3}$, which means that H_2 contains an induced $P_1 + P_2$, a contradiction. If $r \geq 4$, then $\mathcal{G}$ contains all net-walls, and we can apply Lemma 4. Now suppose that H_2 does not contain any vertex of degree 3; then H_2 is a linear forest. As Case 4 does not apply, H_2 has an edge. Every $(P_1 + P_2)$-free linear forest with an edge is a P_2 or a P_3. However, this is not possible, as Case 1 and Case 3 do not apply. □

3 The Proof of Theorem 2

In this section we give a proof of our second dichotomy. We state useful past results in Sect. 3.1 followed by some new results for P_4-free graphs in Sect. 3.2 and we show how to combine these results to obtain the proof of Theorem 2.

3.1 Known Results

The first result we need is due to Brandstädt and Müller. A graph is *chordal bipartite* if it has no induced cycles of length 3 or of length at least 5; that is, a graph is chordal bipartite if it is $(C_3, C_5, C_6, \ldots)$-free.

Theorem 6. *[4] The unweighted* VERTEX STEINER TREE *problem is* NP-*complete for chordal bipartite graphs.*

The second result that we need is due to Farber, Pulleyblank and White. A graph is *split* if its vertex set can be partitioned into a clique and an independent set. It is well known that the class of split graphs coincides with the class of $(2P_2, C_4, C_5)$-free graphs [9].

Theorem 7. *[8] The unweighted* VERTEX STEINER TREE *problem is* NP-*complete for split graphs.*

3.2 New Results

We start with the following lemma (proof omitted).

Lemma 8. *The unweighted* VERTEX STEINER TREE *problem is* NP-*complete for line graphs.*

Recall that a subgraph G' of a graph G is spanning if $V(G') = V(G)$. Let G_1 and G_2 be two graphs. The *join* operation adds an edge between every vertex of G_1 and every vertex of G_2. The *disjoint union* operation takes the disjoint union of G_1 and G_2. A graph G is a *cograph* if G can be generated from K_1 by a sequence of join and disjoint union operations. A graph is a cograph if and only if it is P_4-free. This implies the following well-known lemma.

Lemma 9. *Every connected P_4-free graph on at least two vertices has a spanning complete bipartite subgraph.*

Let G be a graph. For a set S, the graph $G[S] = (S, \{uv \in E(G)\ u, v \in S\})$ denotes the subgraph of G *induced by* S. Note that $G[S]$ can be obtained from G by deleting every vertex of $V(G) \setminus S$. If G has a vertex weighting w, then $w(S) = \sum_{u \in S} w(u)$ denotes the *weight* of S.

Lemma 10. *For every $s \geq 0$,* VERTEX STEINER TREE *can be solved in time $O(n^{2s^2-s+5})$ for connected $(sP_1 + P_4)$-free graphs on n vertices.*

Proof. Let $s \geq 0$ be an integer. Let $G = (V, E)$ be a connected $(sP_1 + P_4)$-free graph with a vertex weighting $w : V \to \mathbb{R}^+$ and set of terminals U. We show how to solve the optimization version of VERTEX STEINER TREE on G. Let $R \subseteq V \setminus U$ be such that $G[U \cup R]$ is connected and, subject to this condition, $U \cup R$ has minimum weight. Thus any spanning tree of $G[U \cup R]$ is an optimal solution. Let us consider the possible size of R.

First suppose that $G[U \cup R]$ is P_4-free. Then, by Lemma 9, $G[U \cup R]$ has a spanning complete bipartite subgraph. That is, there is a bipartition (A, B) of $U \cup R$ such that every vertex in A is joined to every vertex in B (and neither A nor B is the empty set). If U intersects both A and B, then $G[U]$ is connected and $|R| = 0$. So let us assume that $U \subseteq A$, and so $R \supseteq B$. Then $R \cap A = \emptyset$ since

$G[U \cup B]$ is connected. As we know that every vertex in $A = U$ is joined to every vertex in $B = R$, we find that $|R| = 1$.

Suppose instead that $G[U \cup R]$ contains an induced path P on four vertices. We call the connected components of $G[U]$ *bad* if they do not intersect P or the neighbours of P in G. There are at most $s-1$ bad components; else, G contains an $sP_1 + P_4$. Let U^* be a subset of U that includes one vertex from each of these bad components. Then each vertex of $G[U \cup R]$ belongs either to U or P or is an internal vertex of a shortest path in $G[U \cup R]$ from P to a vertex of U^*. The number of internal vertices in such a shortest path is at most $2s+1$; else, the path contains an induced $sP_1 + P_4$. As R is a subset of P and these internal vertices, we find that $|R| \leq 4 + (2s+1)(s-1) = 2s^2 - s + 3$.

So in all cases R contains at most $2s^2 - s + 3$ vertices and our algorithm is just to consider every such set R and check, in each case, whether $G[U \cup R]$ is connected. Our solution is the smallest set found that satisfies the connectivity constraint. As there are $O(n^{2s^2-s+3})$ sets to consider, and checking connectivity takes $O(n^2)$ time, the algorithm requires $O(n^{2s^2-s+5})$ time. □

We are now ready to prove our second dichotomy.

Theorem 2 (restated). *Let H be a graph. For every $s \geq 0$,* Vertex Steiner Tree *is polynomial-time solvable for H-free graphs if H is an induced subgraph of $sP_1 + P_4$; otherwise even unweighted* Vertex Steiner Tree *is* NP*-complete.*

Proof. If H has a cycle, then we apply Theorem 6 or Theorem 7. Hence, we may assume that H has no cycle, so H is a forest. If H contains a vertex of degree at least 3, then the class of H-free graphs contains the class of claw-free graphs, which in turn contains the class of line graphs. Hence, we can apply Lemma 8. Thus we may assume that H is a linear forest. If H contains a connected component with at least five vertices or two connected components with at least two vertices each, then the class of H-free graphs contains the class of $2P_2$-free graphs. Hence, we can apply Theorem 7. It remains to consider the case where H is an induced subgraph of $sP_1 + P_4$ for some $s \geq 0$, for which we can apply Lemma 10. □

4 Conclusions

We presented complexity dichotomies both for Edge Steiner Tree restricted to (H_1, H_2)-free graphs and for Vertex Steiner Tree for H-free graphs. The latter dichotomy also holds for the unweighted variant, in which case the problems Edge Steiner Tree and Vertex Steiner Tree are polynomially equivalent. In particular, we observed that Edge Steiner Tree can be solved in polynomial time for (H_1, H_2)-free graphs if and only if the class of (H_1, H_2)-free graphs has bounded treewidth. This correspondence is not true in general.

Theorem 8. *There exists a hereditary graph class $\mathcal{G}$ of unbounded treewidth for which* Edge Steiner Tree *can be solved in polynomial time.*

Proof. Let $\mathcal{G}$ consist of graphs G of maximum degree at most 3 such that every path between any two degree-3 vertices in G has at least 2^r vertices, where r is the number of degree-3 vertices in G. As deleting a vertex neither increases the maximum degree of a graph nor decreases the number of vertices on paths between degree-3 vertices, $\mathcal{G}$ is hereditary. As $\mathcal{G}$ contains subdivided walls of arbitrarily large height, the treewidth of $\mathcal{G}$ is unbounded due to Theorem 4.

We solve Edge Steiner Tree on an instance (G, w, U, k) with $G \in \mathcal{G}$ as follows. If G has at most one vertex of degree 3, then G has treewidth at most 2, so we can apply Lemma 5. Otherwise, we apply the following rules, while possible.

Rule 1. There is a non-terminal x of degree 2. Let xy and xz be its two incident edges. We contract xy and give the new edge weight $w(xy)+w(xz)$. If there was already an edge between y and z, then we remove one with largest weight.

Rule 2. There is a terminal x of degree 2 and its neighbours y and z are also terminals. Assume $w(xy) \leq w(xz)$. We observe that there is an optimal solution that includes the edge xy. Hence, we may contract xy and decrease k by $w(xy)$.

Rule 3. There is a vertex x of degree 1. Let y be its neighbour. If x is not a terminal, then remove x. Otherwise, contract xy and decrease k by $w(xy)$.

Let (G', w', U', k') be the resulting instance, which is readily seen to be equivalent to (G, w, U, k). Then G' has r vertices of degree 3 and each vertex of degree at most 2 has a neighbour of degree 3; otherwise, one of Rules 1–3 applies. So, G' has at most $4r$ vertices and thus $O(r)$ edges. It remains to solve Edge Steiner Tree on (G', w', U', k). We do this in $r \cdot 2^{O(r)}$ time by guessing for each edge in G' if it is in the solution and then verifying the resulting candidate solution. As $r \geq 2$, we have $|V(G)| \geq 2^r$. So, the running time is polynomial in $|V(G)|$. □

As the hereditary graph class $\mathcal{G}$ in Theorem 8 has an infinite family $\mathcal{H}_{\mathcal{G}}$ of forbidden induced subgraphs, we pose the following open problem.

Open Problem 1. *Is* Edge Steiner Tree *polynomial-time solvable for any* finitely defined *hereditary graph class $\mathcal{G}$ if and only if $\mathcal{G}$ has bounded treewidth*?

So far, we have not found any counterexample to Open Problem 4, and to increase our understanding we first aim to consider classes of (H_1, H_2, H_3)-free graphs. The graph $S_{h,i,j}$, for $1 \leq h \leq i \leq j$, is the *subdivided claw*, which is the tree with one vertex x of degree 3 and exactly three leaves, which are of distance h, i and j from x, respectively. Note that $S_{1,1,1} = K_{1,3}$ and that both walls and net-walls may contain arbitrarily large subdivided claws. Note also that complete graphs are C_3-free and complete bipartite graphs are C_4-free. As such we pose the following open problem.

Open Problem 2. *For every subdivided claw S, does the class of (C_3, C_4, S)-free graphs have bounded treewidth*?

We also propose to consider Vertex Steiner Tree and unweighted Vertex Steiner Tree for (H_1, H_2)-free graphs as future research. To obtain a dichotomy, we need to answer several open problems, including the next ones.

Open Problem 3. *Does there exist a pair* (H_1, H_2) *such that* VERTEX STEINER TREE *and unweighted* VERTEX STEINER TREE *have different complexities for* (H_1, H_2)*-free graphs?*

Open Problem 4. *For every integer* t*, determine the complexity of* VERTEX STEINER TREE *for* $(K_{1,3}, P_t)$*-free graphs.*

To obtain an answer to Open Problem 4, we need new insights into the structure of $(K_{1,3}, P_t)$-free graphs. These insights may also be useful to obtain new results for other problems, such as the GRAPH COLOURING problem restricted to $(K_{1,3}, P_t)$-free graphs (see [11,13]).

References

1. Bergougnoux, B., Kanté, M.M.: More applications of the d-neighbor equivalence: connectivity and acyclicity constraints. In: Proceeding of the 27th Annual European Symposium on Algorithms, ESA, LIPIcs, vol. 144, pp. 17:1–17:14 (2019)
2. Bern, M.W., Plassmann, P.E.: The Steiner problem with edge lengths 1 and 2. Inf. Process. Lett. **32**, 171–176 (1989)
3. Bodlaender, H.L., Cygan, M., Kratsch, S., Nederlof, J.: Deterministic single exponential time algorithms for connectivity problems parameterized by treewidth. Inf. Comput. **243**, 86–111 (2015)
4. Brandstädt, A., Müller, H.: The NP-completeness of Steiner Tree and Dominating Set for chordal bipartite graphs. Theor. Comput. Sci. **53**, 257–265 (1987)
5. Chimani, M., Mutzel, P., Zey, B.: Improved Steiner tree algorithms for bounded treewidth. J. Discrete Algorithms **16**, 67–78 (2012)
6. Chuzhoy, J.: Improved bounds for the flat wall theorem. In: Proceeding of the 26th Annual ACM-SIAM Symposium on Discrete Algorithms, SODA, pp. 256–275 (2015)
7. Du, D., Hu, X.: Steiner Tree Problems in Computer Communication Networks. World Scientific, Singapore (2008)
8. Farber, M., Pulleyblank, W.R., White, K.: Steiner trees, connected domination and strongly chordal graphs. Networks **15**, 109–124 (1985)
9. Földes, S., Hammer, P.L.: Split graphs. Congressus Numerantium **19**, 311–315 (1977)
10. Garey, M.R., Johnson, D.S.: The rectilinear Steiner tree problem is NP-complete. SIAM J. Appl. Math. **32**, 826–834 (1977)
11. Golovach, P.A., Johnson, M., Paulusma, D., Song, J.: A survey on the computational complexity of colouring graphs with forbidden subgraphs. J. Graph Theory **84**, 331–363 (2017)
12. Karp, R.M.: Reducibility among combinatorial problems. In: Miller, R., Thatcher, J. (eds.) Complexity of Computer Computations, pp. 85–103. Springer, Berrlin (1972)
13. Martin, B., Paulusma, D., Smith, S.: Colouring H-free graphs of bounded diameter. In: Proceeding of the 44th International Symposium on Mathematical Foundations of Computer Science, MFCS, LIPIcs, vol. 138, pp. 14:1–14:14 (2019)
14. Prömel, H.J., Steger, A.: The Steiner Tree Problem: A Tour through Graphs, Algorithms, and Complexity. Springer Science & Business Media, Berlin (2012)
15. Renjith, P., Sadagopan, N.: The Steiner tree in $K_{1,r}$-free split graphs - a dichotomy. Discrete Appl. Math. **280**, 246–255 (2020)
16. Robertson, N., Seymour, P.D.: Graph minors. V. Excluding a planar graph. J. Comb. Theor. Ser. B **41**, 92–114 (1986)

On Some Subclasses of Split B_1-EPG Graphs

Zakir Deniz[1], Simon Nivelle[2], Bernard Ries[3], and David Schindl[3](✉)

[1] Duzce University, Duzce, Turkey
zakirdeniz@duzce.edu.tr
[2] ENS Paris Saclay, Paris, France
simon.nivelle@free.fr
[3] University of Fribourg, Fribourg, Switzerland
{bernard.ries,david.schindl}@unifr.ch

Abstract. In this paper, we are interested in edge intersection graphs of paths in a grid, such that each path has at most one bend. These graphs were introduced in [14] and they are called B_1-EPG graphs. We focus on split B_1-EPG graphs, and study subclasses defined by restricting the paths to subsets of the four possible shapes (⌞, ⌜, ⌝ and ⌟). We first state that the set of minimal forbidden induced subgraphs for the class of split ⌞-path graphs is infinite. Then, we further focus on two subclasses, and provide finite forbidden induced subgraphs characterizations for all possible subclasses defined by restricting to any subset of shapes.

1 Introduction

Golumbic et al. introduced in [14] the notion of *edge intersection graphs of paths in a grid* (referred to as *EPG graphs*). An undirected graph $G = (V, E)$ is called an *EPG graph*, if one can associate a path in a rectangular grid with each vertex such that two vertices are adjacent if and only if the corresponding paths intersect on at least one grid-edge. The authors showed in [14] that every graph is in fact an EPG graph. Therefore, they introduced additional restrictions on the paths by limiting the number of *bends* (a bend is a 90 degrees turn of a path at a grid-point) that a path can have. An undirected graph $G = (V, E)$ is then called a *B_k-EPG graph*, for some integer $k \geq 0$, if one can associate with each vertex a path with at most k bends in a rectangular grid such that two vertices are adjacent if and only if the corresponding paths intersect on at least one grid-edge.

Since the introduction of the notion of B_k-EPG graphs, there has been a lot of research done on these graphs from several points of view (see for instance [1–9,11–13,15–17]). Since B_0-EPG graphs coincide with the class of interval graphs, particular attention has been paid to the class of B_1-EPG graphs. The authors in [15] showed that recognizing B_1-EPG graphs is an NP-complete problem; the same holds for B_2-EPG graphs as recently shown in [17]. In any representation of a B_1-EPG graph, each path can only have one of the following four shapes:

Y. Kohayakawa and F. K. Miyazawa (Eds.): LATIN 2020, LNCS 12118, pp. 625–636, 2020.
https://doi.org/10.1007/978-3-030-61792-9_49

⌞, ⌝, ⌜, ⌟ (a path with only a horizontal part or only a vertical part can be considered as a degenerate path of one of the four shapes mentioned before). In [7], the authors analysed B_1-EPG graphs for which the number of different shapes is restricted to a subset of the set above. They showed that testing membership to each of these restricted classes is also NP-complete. Furthermore, they focused on chordal graphs that are B_1-EPG with the additional restriction that only one particular shape (namely ⌞) is allowed for all paths. In particular, they proposed a conjecture concerning the characterization of split graphs that are B_1-EPG and where only paths with an ⌞ shape are allowed, by a family of forbidden induced subgraphs. Indeed, they presented a list of nine forbidden induced subgraphs and conjecture that these are the only minimal ones. In a more recent paper [9], we disproved this conjecture by providing an additional forbidden induced subgraph, and managed to characterize split graphs that are B_1-EPG and where only paths with an ⌞ shape are allowed. This characterization was not in terms of forbidden induced subgraphs, and such a characterization is still missing.

In this paper, we first show that this characterization has an infinite list of minimal forbidden induced subgraphs. Then, for any subset P of the four possible shapes mentioned above, we investigate inclusion relationships among subclasses of split graphs that are B_1-EPG and where only shapes from P are allowed.

2 Preliminaries

We only consider finite, undirected graphs that have no self-loops and no multiple edges. We refer to [10] or [18] for undefined terminology. Let $G = (V, E)$ be a graph. For a subset $V' \subseteq V$, we let $G[V']$ denote the subgraph of G *induced* by V', which has vertex set V' $\{uv \in E \mid u, v \in V'\}$. For a vertex $v \in V$, we write $G-v = G[V \setminus \{v\}]$ and for a subset $V' \subseteq V$, we write $G-V' = G[V \setminus V']$. The set of vertices adjacent to some vertex u is called the *neighborhood of* u and will be denoted by $N(u)$. The *closed neighborhood of* u is defined as $N[u] = N(u) \cup \{u\}$. A vertex u *dominates* some adjacent (resp. non-adjacent) vertex v if $N[v] \subseteq N[u]$ (resp. if $N(v) \subseteq N(u)$) denoted by $v < u$. Two vertices u, v in G are said to be *comparable* if u dominates v or v dominates u. Two vertices that are not comparable are said to be *incomparable.* A *split graph* is a graph $G = (V, E)$ whose vertex set V can be partitioned into a clique K (i.e., a set of pairwise adjacent vertices) and a stable set S (i.e., a set of pairwise non-adjacent vertices). We say that (K, S) is a *split partition of* G. The vertices in S will be called the *S-vertices* (or just stable vertices).

Let $\mathcal{G}$ be a rectangular grid of size $m \times m'$. The horizontal grid lines will be referred to as *rows* and denoted by $x_0, x_1, \cdots, x_{m-1}$ and the vertical grid lines will be referred to as *columns* and denoted by $y_0, y_1, \cdots, y_{m'-1}$. We call *segment* a collection of consecutive grid edges on a column (or a row) of $\mathcal{G}$. As already mentioned above, in any representation of a B_1-EPG graph, each path can only have one of the following four possible shapes: ⌞, ⌝, ⌜, ⌟. A path with an ⌞-shape will be called an *⌞-path.* In a similar way we define a ⌝-path, a ⌜-path and a ⌟-path. For any subset R of the four possible shapes, we denote by $[R]$ the class of B_1-EPG graphs which admit a representation in which each path has one of the shapes in R. In particular, we denote by $[R]_s$ the class of B_1-EPG split graphs which admit a representation in which each path has one of the shapes in R. For simplicity, if R contains all four shapes, we write B_1-EPG$_s$. A representation of a B_1-EPG graph containing only paths with a shape in R is called a *$[R]$-representation* (or just $[R]$-shape).

Let $G = (V, E)$ be a B_1-EPG graph and let $v \in V$. We denote by P_v the path representing v in a B_1-EPG representation of G. Consider a clique K (resp. a stable set S) in G. Any path representing a vertex in K (resp. in S) will simply be referred to as a *path of K* (resp. *path of S*). Concerning cliques, the following useful lemma has been shown in [14].

Lemma 1. *Let $G = (V, E)$ be a B_1-EPG graph. In any B_1-EPG representation of G, a clique K of G is represented either as an edge-clique or as a claw-clique (see Fig. 1).*

Fig. 1. An edge-clique (a) and a claw-clique (b).

Notice that in an edge-clique, all paths share a common grid-edge, while it is not the case in a claw-clique. For a B_1-EPG representation of a split graph with a split partition (K, S), we will call *base* a minimal (with respect to inclusion) set of consecutive grid edges on a same row or column, such that every path of K uses at least one of these edges. It is easy to see that this set consists of one (horizontal or vertical) grid edge in the case of an edge-clique and two grid edges (both horizontal or both vertical) in the case of a claw-clique. Notice also that in a claw-clique, all paths have a unique common grid-point. We will call this point the *center of the clique.*

A *gem* is a graph with vertex set $\{c_1, c_2, c_3, s_1, s_2\}$ and edge set $\{s_1c_1, s_1c_2, c_1c_2, c_2c_3, c_1c_3, s_2c_2, s_2c_3\}$ (see Fig. 2(a)). It is easy to see that a gem, as an induced subgraph of a split graph $G = (V, E)$ with split partition (K, S), must satisfy $c_1, c_2, c_3 \in K$ and $s_1, s_2 \in S$. A *bull* is a graph with vertex set $\{c_1, c_2, s_1, s_2, s_3\}$ and edge set $\{c_1c_2, c_1s_2, c_2s_2, c_1s_1, c_2s_3\}$ (see Fig. 2(b)). Again, it is easy to see that a bull, as an induced subgraph of a split graph $G = (V, E)$ with split partition (K, S), must satisfy $c_1, c_2 \in K$ and $s_1, s_3 \in S$. In the case where $s_2 \in S$ as well, the bull is called an *S-bull.* Gems and S-bulls have played an central role in [7] and [9]. As we will see, they are also important in our results. In Fig. 2, we also define two additional split graphs we will need for our results: the *double S-bull* and the *3-S-bull.*

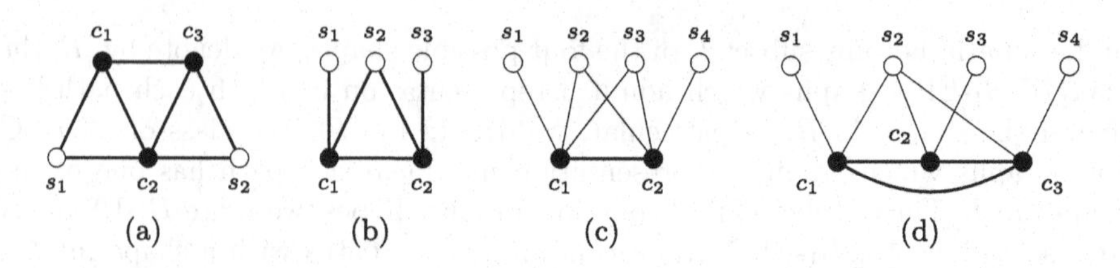

Fig. 2. (a) A gem. (b) An S-bull. (c) A double S-bull. (d) A 3-S-bull.

Let G be a split graph with split partition (K, S). A subgraph H induced by $N[R]$ for a set $R \subseteq S$ is called *stable-component* of G if $N(N(R)) \cap S = R$ and for every $r, t \in R$, there exists a path $r, c_1, s_1, c_2, s_2, \ldots, c_p, t$ for $c_i \in K$, $s_j \in S$ with $i > j$, $i = 1, 2, \ldots, p$. In particular, if all stable vertices of H are pairwise comparable on G, then we will say that H is a *trivial-component* of G. Notice that if a stable-component R does neither contain a gem or an S-bull, then it is trivial.

The following definitions have been introduced in [7] for the study of graphs in $[\llcorner]_s$. Let $G = (V, E)$ be in $[\llcorner]_s$ with split partition (K, S). Consider a $[\llcorner]_s$-representation of G. Clearly, the clique K must be represented as an edge-clique that corresponds to the base, which we already defined above. Without loss of generality, we may assume that the base is vertical. The horizontal parts of the paths representing vertices in K are called *branches*. Let F be the vertical segment which is the union of the vertical parts of all paths representing vertices in K. The part of F below the base is called the *trunk*. The part of F above the trunk is called the *crown*.

The following theorem provides a characterization of graphs in $[\llcorner]_s$.

Theorem 1 ([9]). *Let G be a split graph with split partition (K, S). Then $G \in [\llcorner]_s$ if and only if there exist $S_1, S_2 \subseteq S$ such that:*

(a) each S_i for $i \in \{1, 2\}$ is a set of pairwise comparable vertices;
(b) for every gem in G with vertex set $\{c_1, s_1, c_2, s_2, c_3\}$ (see Fig. 2a), either $s_1 \in S_1$ or $s_2 \in S_1$;
(c) for every S-bull in G with vertex set $\{s_1, c_1, s_2, c_2, s_3\}$ (see Fig. 2b), at least one of s_1, s_2, s_3 belongs to S_1 or $s_2 \in S_2$.

In the proof of Theorem 1, the set S_1 represents the vertices corresponding to the paths lying on the crown in the constructed representation, S_2 represents the vertices corresponding to the paths lying on the trunk, while the paths corresponding to the vertices in $S - (S_1 \cup S_2)$ are actually on the branches since they can be partitioned into trivial-components.

The following simple but general observation will also be useful when we will need to rotate some representations in the sequel. The proof is not difficult and we omit it here.

Observation 1. *For any split graph G in B_1-EPG with split partition (K, S), there always exists a B_1-EPG representation of G such that all paths corresponding to vertices of S contain no bend.*

By Observation 1, we may from now on assume that all paths of S are unbended in any B_1-EPG representation of a split graph $G = (K, S)$.

3 Forbidden Induced Subgraph for $[\llcorner]_s$

In [7], the authors conjectured that the graphs in $[\llcorner]_s$ can be characterized by a list of nine forbidden induced subgraphs. Recently, the authors in [9] disproved this conjecture by presenting an additional forbidden induced subgraph. In this section, we show that a finite forbidden induced subgraph characterization of the graph class $[\llcorner]_s$ does not exist. To do this, we exhibit an infinite set of minimal forbidden induced subgraphs for this class of graphs.

Proposition 1. *There exists an infinite family of minimal split graphs that are not in* $[\llcorner]_s$.

Proof. We construct a minimal split graph that does not belong to the $[\llcorner]_s$ class and whose size can be arbitrarily large. Consider the split graph $G(k) = (C, S)$ as represented in Fig. 3 for $k \geq 2$. The set of stable vertices S of $G(k)$ is decomposed into three classes as $S = X \cup Y \cup Z$, where $X = \{x_1, x_2, \ldots, x_k\}$, $Y = \{y_1, y_2, \ldots, y_k\}$ and $Z = \{z_1, z_2, \ldots, z_6\}$ for $k \geq 2$. For simplicity, the vertices of the clique $C = \{c_1, c_2 \ldots, c_{2k+5}\}$ are depicted as $1, 2, \ldots, 2k + 5$ in Fig. 3.

We construct the graph $G(k)$ as follows:

(i) $N(x_k) = \{c_{2k+4}, c_{2k+5}\}$, and $N(x_i) = \{c_{2i+1}, c_{2i+2}, \ldots, c_{2k+5}\}$ for $i \in [k-1]$.
(ii) $N(y_k) = \{c_{2k+2}, c_{2k+3}, c_{2k+5}\}$, $N(y_{k-1}) = \{c_{2k-2}, c_{2k}, c_{2k+1}, c_{2k+2}, c_{2k+3}, c_{2k+5}\}$, and $N(y_i) = \{c_{2i}, c_{2i+2}\} \cup \{c_{2i+4}, c_{2i+5}, \ldots, c_{2k+5}\}$ for $i \in [k-2]$.
(iii) $N(z_1) = \{c_1, c_3\}$, $N(z_2) = \{2k, 2k+1\}$, and $N(z_i) = \{c_{2k-3+i}\}$ for $i \in \{3, 4, 5, 6\}$.

We can easily see that $G(k)$ has the following properties:

- $x_1 > x_2 > \ldots > x_k$ and $y_1 > y_2 > \ldots > y_k$.
- $x_k, y_k, c_{2k+3}, c_{2k+4}, c_{2k+5}$ induce a gem.
- $x_k, y_{k-1}, c_{2k+3}, c_{2k+4}, c_{2k+5}$ induce a gem.
- For each $i \in [k-1]$, $x_i, y_i, c_{2i}, c_{2i+1}, c_{2k+5}$ induce a gem.
- For each $i \in [k-2]$, $x_{i+1}, y_i, c_{2i}, c_{2i+3}, c_{2k+5}$ induce a gem.
- For each $i \in [k-2]$, y_i dominates x_{i+2}.
- For each $i \in [k-1]$, x_i dominates y_{i+1}.

We first show that $G(k)$ does not admit a $[\llcorner]$-representation and then we prove its minimality. Assume to the contrary that $G(k)$ admits a $[\llcorner]$-representation $\mathcal{G}$ and hence there exist S_1, S_2 as defined in Theorem 1. Recall that for each $i \in [k]$, x_i, y_i are the two S-vertices of an induced gem in $G(k)$. Then by Theorem 1,

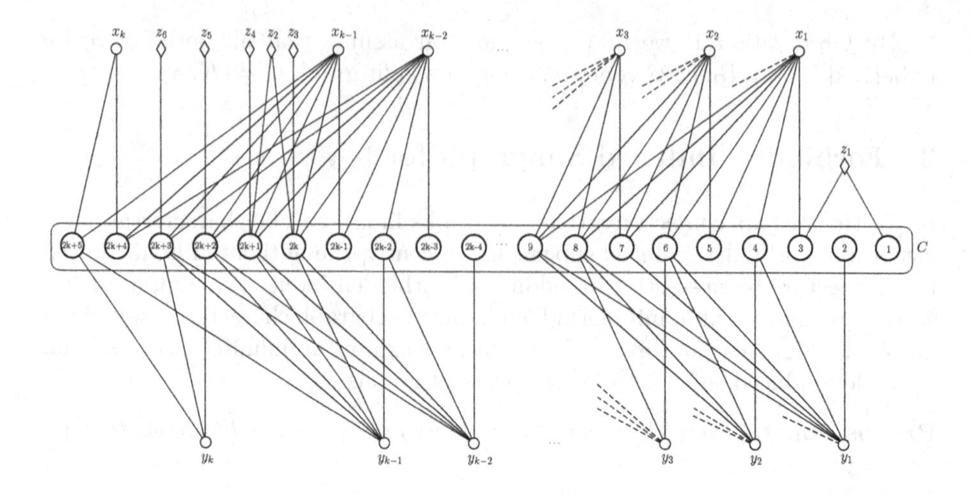

Fig. 3. Forbidden subgraph configuration

one of x_i, y_i should be in S_1. Furthermore x_1, z_1 are the two S-vertices of an induced gem in $G(k)$. Thus, again by Theorem 1, one of x_1, y_1 should be in S_1. Then, we have to put x_1 in S_1 since x_1, y_1, z_1 are pairwise incomparable vertices, and S_1 consists of pairwise comparable vertices. Consequently, since $x_1 \in S_1$, $y_1 \notin S_1$ and y_1 is a gem with x_2, we have $x_2 \in S_1$, $y_2 \notin S_1$. By applying the same argument, we then have $S_1 = \{x_1, x_2, \ldots, x_k\} = X$. Notice that $z_2, z_3, z_4, c_{2k}, c_{2k+1}$ induce an S-bull as well as $z_5, z_6, y_k, c_{2k+2}, c_{2k+3}$ with no stable vertices in S_1. Since S_2 must be a set of pairwise comparable vertices, and since the vertices z_2 and y_k are incomparable, S_2 cannot contain both of them. This is a contradiction with Theorem 1. Therefore, G does not admit a $[\llcorner]$-representation.

For the minimality of $G(k)$, and let $G = G(k)$, we need to check that for every $v \in V(G)$, $G - v$ admits a $[\llcorner]$-representation. It can be done by exhibiting, for each $v \in V(G)$, two subsets of vertices S_1, S_2 satisfying the conditions of Theorem 1 for $G - v$. This is a tedious but not difficult task, and we leave it to the reader. □

We immediately obtain the following:

Theorem 2. *The set of minimal forbidden induced subgraph for the graph class* $[\llcorner]_s$ *is infinite.*

As a consequence, the search for a polynomial-time recognition algorithm consisting in simply checking the existence of a finite list of graphs as induced subgraphs can be discarded. Of course, this does not imply that no polynomial-time algorithm exists and this question remains open.

4 Restricted Shape B_1-EPG Subclasses for (S-bull)-free Split Graphs

In this section, we present forbidden induced subgraph characterizations for all possible subclasses of (S-bull)-free graphs in B_1-EPG$_s$ with respect to shapes. We first mention a simple result which was proven in [9].

Proposition 2 *[9]. Consider a B_1-EPG representation of a gem (see Fig. 2(a)). Let $K = \{c_1, c_2, c_3\}$ and $S = \{s_1, s_2\}$. If K is represented as an edge-clique with base going from (x_i, y_j) to (x_{i+1}, y_j) or if K is represented as a claw-clique with center (x_i, y_j) and no path of K uses the grid-edge going from (x_i, y_{j-1}) to (x_i, y_j), then at least one of P_{s_1}, P_{s_2} intersects paths of K on column y_j.*

The following proposition will be useful in deriving the subsequent results.

Proposition 3. *Let $G = (K, S)$ be a split graph in B_1-EPG. Then three paths corresponding to pairwise incomparable vertices of S can not lie on a same row or column in any B_1-EPG representation of G.*

Proof. Assume by contradiction that there are three such paths corresponding to pairwise incomparable vertices of S lying on some column y_j. If the base is vertical, then at least two of these three paths must lie on the same side (above or below) of it. But by Observation 4 in [9], they must be comparable, a contradiction. Assume then the base is horizontal, say it is on the row x_i. Then all paths of K using a grid edge on column y_j are bended at a grid point $p = (x_i, y_j)$. Then any set of paths of S using grid edges on column y_j above (resp. below) p must be comparable by Observation 4 in [9]. Notice that if there is a path of S using grid edges from (x_{i-1}, y_j) to (x_{i+1}, y_j), it must be comparable with all other paths of S lying on column y_j. We then conclude that there are at most two incomparable vertices of S lying on column y_j; one lies below p, the other lies above p. This completes the proof. □

Using Proposition 2, it is easy to show that the graph V_1 in Fig. 4 is not in B_1-EPG.

Lemma 2. *The graph V_1 is not in B_1-EPG.*

We know from Theorem 1, that if a split graph is $\{S$-bull,gem$\}$-free, by choosing $S_1 = S_2 = \emptyset$, we obtain a [⌞]-representation of G such that every vertex of S lies on a branch of the representation.

Corollary 1. *Any $\{S$-bull,gem$\}$-free split graph admits a [⌞]-representation such that all vertices of S lie on branches of the representation.*

In [7], the authors prove the following.

Theorem 3 *([7, **Theorem 24**]). Let G be a graph with a split partition (K, S) and containing no S-bull. Then G admits a [⌞]-representation if and only if G is $\{U_1, G_4\}$-free (see Fig. 4).*

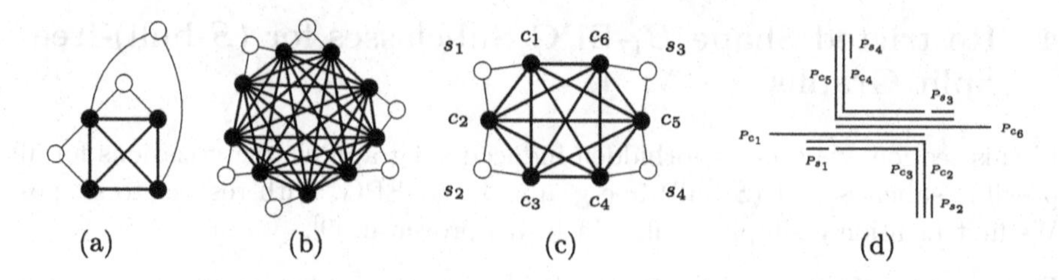

Fig. 4. (a) The graph U_1. (b) The graph V_1. (c) The graph G_4. (d) $[\llcorner, \urcorner]$ representation of G_4.

Consider a $\{S\text{-bull}, U_1, G_4\}$-free split graph G. We know from Theorem 3 that it admits a $[\llcorner]$-representation. But now we can choose $S_2 = \emptyset$ in Theorem 1 because G has no S-bull. Moreover, the representation constructed in Theorem 1 associates S_2 with the paths lying on the trunk. Therefore, we obtain the following.

Corollary 2. *Let G be a $\{S\text{-bull}, U_1, G_4\}$-free split graph. Then G admits a $[\llcorner]$-representation with all vertices in S lying on the crown and the branches of the representation.*

An asteroidal triple (AT) is a stable set consisting of 3 vertices such that for any two of them, there is a path between them which avoids the neighborhood of the third vertex.

Lemma 3 ***(AT Lemma [2, Theorem 9]).*** *In a B_1-EPG graph, no vertex can have an AT in its neighborhood.*

The next result gives a characterization of $\{S\text{-bull}\}$-free split graphs admitting a $[\llcorner, \urcorner]$-representation.

Theorem 4. *Let G be a $\{S\text{-bull}\}$-free split graph. Then G admits a $[\llcorner, \urcorner]$- representation if and only if G is $\{U_1, V_1\}$-free.*

Proof. From [7, Theorem 24], we know that if G is $\{U_1, G_4\}$-free, then G admits a $[\llcorner, \urcorner]$-representation. Therefore, we only need to prove the following:

(a) U_1 is not in $[\llcorner, \urcorner]_s$,
(b) V_1 is not in $[\llcorner, \urcorner]_s$,
(c) If a $\{S\text{-bull}, U_1, V_1\}$-free split graph contains G_4, then it is in $[\llcorner, \urcorner]_s$.

Assertion (a) can be easily deduced from Lemma 3 and (b) is a direct consequence of Lemma 2. Thus, we are left with the proof of (c). First we know that G_4 is in $[\llcorner, \urcorner]_s$ (see Fig. 4). Let G be a $\{S\text{-bull}, U_1, V_1\}$-free split graph containing G_4 whose stable vertices are s_1, s_2, s_3, s_4 as depicted in Fig. 4 and let $C_1 = N(s_1) \cup N(s_2)$, $C_2 = N(s_3) \cup N(s_4)$. For a set S' of vertices belonging to a stable-component, we define its *closure* $\overline{S'}$ as the set of stable vertices of the whole stable-component. Let $S_1 = \{s_1, s_2\}$, $S_2 = \{s_3, s_4\}$.

Claim 1: $\overline{S_1} \neq \overline{S_2}$.

Proof of the Claim. We first claim that $C_1 \cap C_2 = \emptyset$. By contradiction, suppose that $r \in (N(s_1) \cap N(s_3))$. Clearly, $r \neq c_i$, for each $i \in [6]$. If one of s_2, s_4 is in $N(r)$, say $s_2 \in N(r)$, then the vertices $r, s_1, c_1, s_2, c_3, s_3, c_6$ induce U_1, a contradiction. If none of s_2, s_4 is in $N(r)$, then s_2, c_2, s_1, r, s_3 induces an S-bull, again a contradiction. Hence $C_1 \cap C_2 = \emptyset$.

Assume the closure $\overline{S_1}$ contains s_3 (and hence, also s_4). By definition, there is a path

$$s_1, a_1, b_1, a_2, b_2, \ldots, b_{k-1}, a_k, b_k = s_3$$

such that $a_i \in K$, $b_i \in S$, for each $i \in [k]$. Among all those paths between s_1 and s_3, choose one with minimum k. Since $C_1 \cap C_2 = \emptyset$, we have $k \geq 2$. Notice that we could have $b_1 = s_2$ and/or $b_{k-1} = s_4$. By minimality of k, s_1, a_2 are nonadjacent, and a_1, b_2 are nonadjacent. But now s_1, a_1, b_1, a_2, b_2 induce an S-bull, a contradiction. Hence, $\overline{S_1}$ does not contain s_3, s_4, which of course also means that $\overline{S_2}$ does not contain s_1, s_2. This implies that $\overline{S_1} \neq \overline{S_2}$.

Notice that since $\overline{S_1}$ (resp. $\overline{S_2}$) does not contain s_3, s_4 (resp. s_1, s_2), the graph H_1 (resp. H_2) induced by $N[\overline{S_1}]$ (resp. $N[\overline{S_2}]$) is actually G_4-free (otherwise we obtain V_1, a contradiction). Furthermore, H_1 and H_2 are disjoint. By Corollary 2, the graph H_i admits a [⌞]-representation with all vertices in $\overline{S_i}$ lying on the crown and the branches of the representation, for $i = 1, 2$. So we can combine both representations by rotating one by 180 degrees and merging their bases, to obtain a [⌞, ⌝]-representation of $H_1 \cup H_2$, see Fig. 5. Finally, consider the graph H induced by $N[R]$, with $R = S \backslash (\overline{S_1} \cup \overline{S_2})$. It is disjoint from H_1, H_2 since otherwise some vertices of R would belong to $\overline{S_i}$ for $i = 1, 2$. Moreover, the graph H cannot contain a gem (otherwise G contains V_1 as above) and of course it does not contain an S-bull. Hence, H admits a [⌞]-representation such that all vertices of S lie only on branches of the representation, according to Corollary 1. Now it is easy to see that we can combine this representation with the previous one by merging again their bases. This completes the proof of the theorem. □

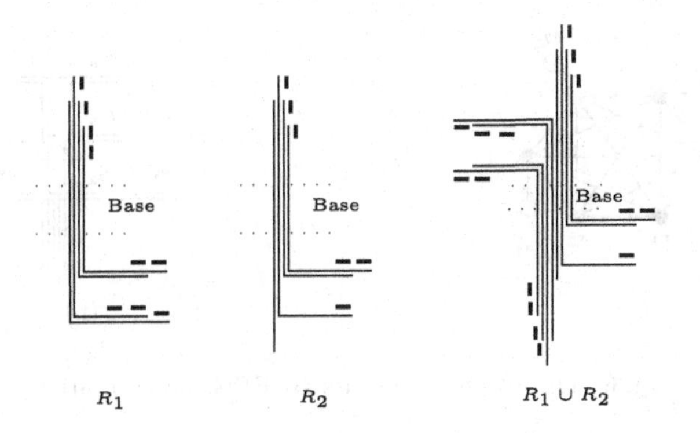

Fig. 5. The combination of two [⌞]-representations R_1, R_2.

In [9], the authors showed that $[\llcorner, \urcorner]_s \subsetneq [\llcorner, \ulcorner]_s \subsetneq [\llcorner, \ulcorner, \urcorner]_s \subsetneq B_1\text{-}EPG_s$. By Theorem 4, U_1 and V_1 are the only minimal forbidden induced subgraphs for $\{S\text{-bull}\}$-free split graphs in $[\llcorner, \urcorner]_s$ and they are also forbidden for B_1-EPG by Lemma 2 and Lemma 3. We therefore conclude that the equalities $[\llcorner, \urcorner]_s = [\llcorner, \ulcorner]_s = [\llcorner, \ulcorner, \urcorner]_s = B_1\text{-}EPG_s$ hold. Furthermore, since G_4 is in $[\llcorner, \urcorner]_s$, but not in $[\llcorner]_s$ by [7, Lemma 20], we have the following result.

Theorem 5. *For $\{S\text{-bull}\}$-free split graphs, we have $[\llcorner]_s \subsetneq [\llcorner, \urcorner]_s = [\llcorner, \ulcorner]_s = [\llcorner, \ulcorner, \urcorner]_s = B_1\text{-}EPG_s$.*

5 Restricted Shape B_1-EPG Subclasses for Gem-Free Split Graphs

In this section, we present a result similar to Theorem 5, but for gem-free split graphs. Due to space restrictions, we decided to omit the intermediate results and proofs. We therefore formulate directly our theorem and refer the interested reader to an extended version of this paper.

Theorem 6. *For all gem-free split graphs, we have*

$$[\llcorner]_s = [\llcorner, \urcorner]_s \subsetneq [\llcorner, \ulcorner]_s = [\llcorner, \ulcorner, \urcorner]_s = B_1\text{-}EPG_s$$

More precisely, we have the following forbidden induced subgraphs characterizations. Apart from the three minimal forbidden induced subgraphs for the class of split graphs (C_4, C_5 and $2K_2$) and the gem:

- *the only minimal forbidden induced subgraph of $[\llcorner]_s = [\llcorner, \urcorner]_s$ is G_5 (see Fig. 6a);*
- *the minimal forbidden induced subgraphs of $[\llcorner, \ulcorner]_s = [\llcorner, \ulcorner, \urcorner]_s = B_1\text{-}EPG_s$ are $G_{10}, G_{11}, G_{12}, G_{13}$ (see Fig. 7).*

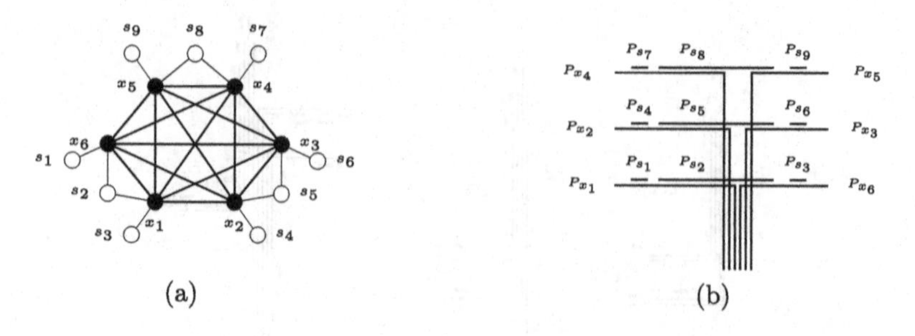

Fig. 6. The graph G_5 and its B_1-EPG representation.

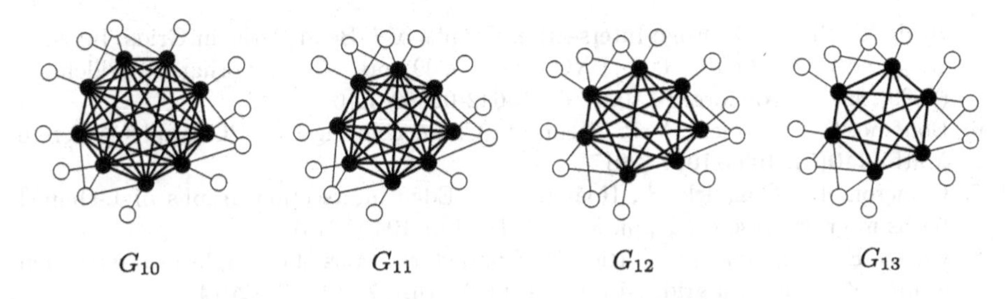

Fig. 7. The graphs G_{10}, G_{11}, G_{12} and G_{13}.

6 Conclusion

In this paper, we were interested in split graphs as edge intersection graphs of single bend paths on a grid. We first exhibited an infinite set of minimal forbidden induced subgraphs for the class $[\llcorner]_s$, showing the non-existence of a straightforward forbidden induced subgraph search as a polynomial detection algorithm for this class. It is still open whether the graphs in $[\llcorner]_s$ can be recognized in polynomial-time.

We were also interested in some subclasses of split graphs that are in B_1-EPG such as gem-free, (S-bull)-free. In [7], the authors have characterized gem-free (resp. S-bull-free) split graph in $[\llcorner]_s$ by a list of minimal forbidden induced subgraphs. We here managed to generalise these results to gem-free (resp. S-bull-free) graphs that are in $[P]_s$, for any subset P of $\{\llcorner, \urcorner, \ulcorner, \lrcorner\}$. Our characterization implies that all these graph classes can all be recognised in polynomial time.

In [9], a relationship was given among subsets of split graphs that are in $[P]_s$, for any subset P of $\{\llcorner, \urcorner, \ulcorner, \lrcorner\}$. We here showed that $[\llcorner]_s \subsetneq [\llcorner, \urcorner]_s = [\llcorner, \ulcorner]_s = [\llcorner, \ulcorner, \urcorner]_s = B_1\text{-}EPG_s$ when we restrict to the set of S-bull free split graphs. In addition, we have obtained a similar result for gem-free split graphs: $[\llcorner]_s = [\llcorner, \urcorner]_s \subsetneq [\llcorner, \ulcorner]_s = [\llcorner, \ulcorner, \urcorner]_s = B_1\text{-}EPG_s$.

Acknowledgements. This work was done while the first author visited the University of Fribourg, Switzerland. The support of the institution is gratefully acknowledged.

References

1. Alcón, L., Bonomo, F., Durán, G., Gutierrez, M., Mazzoleni, M.P., Ries, B., Valencia-Pabon, M.: On the bend number of circular-arc graphs as edge intersection graphs of paths on a grid. Discrete Appl. Math. **234**, 12–21 (2018)
2. Asinowski, A., Ries, B.: Some properties of edge intersection graphs of single-bend paths on a grid. Discrete Math. **312**(2), 427–440 (2012)
3. Asinowski, A., Suk, A.: Edge intersection graphs of systems of paths on a grid with a bounded number of bends. Discrete Appl. Math. **157**(14), 3174–3180 (2009)
4. Bessy, S., Bougeret, M., Chaplick, S., Goncalves, D., Paul, C.: On independent set in B1-EPG graphs. Discrete Appl. Math. **278**, 62–72 (2020)

5. Biedl, T., Stern, M.: Edge-Intersection Graphs of k-Bend Paths in Grids. In: Ngo, H.Q. (ed.) COCOON 2009. LNCS, vol. 5609, pp. 86–95. Springer, Heidelberg (2009). https://doi.org/10.1007/978-3-642-02882-3_10
6. Bonomo, F., Mazzoleni, M.P., Stein, M.: Clique coloring B1-EPG graphs. Discrete Math. **340**(5), 1008–1011 (2017)
7. Cameron, K., Chaplick, S., Hoáng, C.T.: Edge intersection graphs of L-shaped paths in grids. Discrete Appl. Math. **210**, 185–194 (2016)
8. Cohen, E., Golumbic, M.C., Ries, B.: Characterizations of cographs as intersection graphs of paths on a grid. Discrete Appl. Math. **178**, 46–57 (2014)
9. Deniz, Z., Nivelle, S., Ries, B., Schindl, D.: On split B1-EPG graphs. In: Bender, M.A., Farach-Colton, M., Mosteiro, M.A. (eds.) Proceeding of the 13th Latin American Symposium on Theoretical Informatics, LATIN 2018. LNCS, vol. 10807, pp. 361–375. Springer (2018)
10. Graph Theory. PBM. Springer, Cham (2018). https://doi.org/10.1007/978-3-319-97686-0_15
11. Epstein, D., Golumbic, M.C., Lahiri, A., Morgenstern, G.: Hardness and approximation for L-EPG and B1-EPG graphs. Discrete Appl. Math. October 2019
12. Epstein, D., Golumbic, M.C., Morgenstern, G.: Approximation algorithms for B1-EPG graphs. In: Dehne, F., Solis-Oba, R., Sack, J.-R. (eds.) WADS 2013. LNCS, vol. 8037, pp. 328–340. Springer, Heidelberg (2013). https://doi.org/10.1007/978-3-642-40104-6_29
13. Francis, M.C., Lahiri, A.: VPG and EPG bend-numbers of Halin graphs. Discrete Appl. Math. **215**, 95–105 (2016)
14. Golumbic, M.C., Lipshteyn, M., Stern, M.: Edge intersection graphs of single bend paths on a grid. Networks **54**(3), 130–138 (2009)
15. Heldt, D., Knauer, K., Ueckerdt, T.: Edge-intersection graphs of grid paths: the bend-number. Discrete Appl. Math. **167**, 144–162 (2014)
16. Heldt, D., Knauer, K., Ueckerdt, T.: On the bend-number of planar and outerplanar graphs. Discrete Appl. Math. **179**, 109–119 (2014)
17. Pergel, M., Rzazewski, P.: On edge intersection graphs of paths with 2 bends. Discrete Appl. Math. **226**, 106–116 (2017)
18. West, D.B.: Introduction to Graph Theory. Prentice-Hall, United States (1996)

On the Helly Subclasses of Interval Bigraphs and Circular Arc Bigraphs

M. Groshaus[2], A. L. P. Guedes[1(✉)], and F. S. Kolberg[1]

[1] Universidade Federal do Paraná, Curitiba, Brazil
{andre,fskolberg}@inf.ufpr.br
[2] Universidade Tecnológica Federal do Paraná, Curitiba, Brazil
marinagroshaus@utfpr.edu.br

Abstract. A bipartite graph $G = (U, V, E)$ is an interval bigraph if and only if there is a one to one correspondence between $U \cup V$ and a family of intervals on the number line such that two vertices of opposing partite sets are neighbors precisely if their corresponding intervals intersect. Interval bigraphs, as well as many subclasses, have been extensively studied by multiple researchers along the years, and many results on their structural and computational properties have been discovered. A bipartite graph $G = (U, V, E)$ is a circular arc bigraph if and only if there is a one to one correspondence between $U \cup V$ and a family of arcs on a circle such that two vertices of opposing partite sets are neighbors precisely if their corresponding arcs intersect. While it is a generalization of interval bigraphs, it remains a relatively unexplored topic. Few studies about the class and its proper, unit and Helly subclasses have been presented. In this work, we study some subclasses of these classes. We provide forbidden structure characterizations for the Helly subclass of interval bigraphs, as well as the class of non-bichordal Helly circular arc bigraphs. We also prove that Helly interval bigraphs are a subclass of proper interval bigraphs, and that non-bichordal Helly circular arc bigraphs are a subclass of proper circular arc bigraphs.

Keywords: Interval bigraphs · Proper interval bigraphs · Helly · Circular arc bigraphs

1 Introduction

The bipartite graph class of *interval bigraphs* arises as a variation on the class of *interval graphs* [2]. A bipartite graph $G = (U, V, E)$ is said to be an interval bigraph if it admits a one to one correspondence between its vertices and a family of intervals on the number line such that, for any pair of vertices $u \in U, v \in V$, $uv \in E$ precisely if their corresponding intervals intersect. Having been an object of study for at least three decades, characterizations and recognition algorithms

*This work was partially supported by ANPCyT (PICT-2013-2205), CONICET, CAPES and CNPq (428941/2016-8).

Y. Kohayakawa and F. K. Miyazawa (Eds.): LATIN 2020, LNCS 12118, pp. 637–648, 2020.
https://doi.org/10.1007/978-3-030-61792-9_50

for the class [10,11] and its *proper* [3,5] subclass have been presented, as well as studies on the relationship between the class and multiple other classes [3,8].

The class of *circular arc bigraphs* (shortened to CA bigraphs) [1] is a bipartite variation on *circular arc graphs* [9] and a generalization of interval bigraphs. A bipartite graph $G = (U, V, E)$ is said to be a circular arc bigraph if it admits a one to one correspondence between its vertices and a family of arcs on a circle such that, for any pair of vertices $u \in U, v \in V$, $uv \in E$ precisely if their corresponding arcs intersect. While a relatively new topic, multiple characterizations for the class and its *proper* and *unit* subclasses [1,5] exist, as well as polynomial time recognition algorithms for its *Helly* subclass [6].

In this paper, we present forbidden structure characterizations for Helly interval bigraphs and *non-bichordal* Helly CA bigraphs. We also prove that Helly interval bigraphs are a proper subclass of proper interval bigraphs, and that non-bichordal Helly CA bigraphs are a subclass of proper CA bigraphs.

2 Definitions

Denote bipartite graphs as triples (U, V, E), where U, V are the graph's partite sets, and E is its set of edges. We call V the *opposite* partite set to U, and vice versa. For any integer $n > 2$, denote by C_n an induced cycle on n vertices. A bipartite graph is said to be *bichordal* if it admits no induced $C_n, n > 4$. A graph that admits an induced $C_n, n > 4$ is then called *non-bichordal.*

Given a graph G, a *clique* (*biclique*) $K \subset V(G)$ is a maximal subset such that $G[K]$ is a complete (bipartite-complete) graph. For any graph $G = (V, E)$, denote by $G^2 = (V, E^2)$ its *square graph*, where $E^2 = \{vw | d(v, w) \leq 2\}$.

If two vertices $v, w \in V(G)$ are such that their open neighborhoods are equal, they are called *twins.* Given a graph G, the *twin-free version* of G is the graph resulting from removing, for every set of twins in G, every vertex but one. Given a graph G, G^* is the graph G with an isolated vertex added to it.

Given a permutation $(s_1, ..., s_n)$ of a set S, a subset $S' \subset S$ is said to be *circularly consecutive* in the permutation if S' or $S - S'$ are consecutive in it. We then call the *interval* that contains S' the sequence of indices of the elements of S' in the permutation, in numerical order if S' is consecutive. If $S - S'$ is consecutive, then S will be broken in two consecutive subsets in the permutation, and its interval consists of the indices of the second subset in numerical order, followed by the indices of the first subset.

In this paper, we adopt the notation of *bi-interval* and *bi-circular-arc* models from [6]. A *bi-interval model* is a pair of families $(\mathbb{E}, \mathbb{F})$ of intervals on the real line. The *corresponding graph* of a bi-interval model $(\mathbb{E}, \mathbb{F})$ is constructed by creating a vertex for each element of $\mathbb{E} \cup \mathbb{F}$ and, for every pair of intervals $E \in \mathbb{E}, F \in \mathbb{F}$, an edge between the vertices of E and F is added precisely if $E \cap F \neq \emptyset$.

Similarly, a *bi-circular-arc model* is a triple $(C, \mathbb{I}, \mathbb{E})$ such that C is a circle, and $\mathbb{I}, \mathbb{E}$ are arcs over C. The corresponding graph of a bi-circular-arc model is built by creating a vertex v_A for each arc $A \in \mathbb{I} \cup \mathbb{E}$ and, for every pair of arcs $I \in \mathbb{I}, E \in \mathbb{E}$, an edge $v_I v_E$ is added if and only if $I \cap E \neq \emptyset$. It is easy to verify that a bipartite graph is an interval bigraph (CA bigraph) if and only if it is the corresponding graph of a bi-interval (bi-circular-arc) model.

A graph is a *proper interval bigraph* if it admits a bi-interval model $(\mathbb{A}, \mathbb{B})$ such that $\mathbb{A}$ and $\mathbb{B}$ are proper families (i.e. no two distinct elements of the family are comparable) [5]. Analogously, a graph is a *proper CA bigraph* if it admits a bi-circular-arc model $(C, \mathbb{I}, \mathbb{E})$ such that $\mathbb{I}$ and $\mathbb{E}$ are proper families [1].

In this paper, we consider all arcs and intervals to be open unless otherwise stated.

A family $\mathbb{F}$ is said to be *intersecting* if, for every pair $E, F \in \mathbb{F}$, we have $E \cap F \neq \emptyset$. Analogously, a pair of families $\mathbb{E}, \mathbb{F}$ is *bipartite-intersecting* if, for every $E \in \mathbb{E}, F \in \mathbb{F}, E \cap F \neq \emptyset$. A pair of families $\mathbb{E}, \mathbb{F}$ is then said to be *bipartite-Helly* if, for every bipartite-intersecting pair of subfamilies $\mathbb{E}' \subset \mathbb{E}, \mathbb{F}' \subset \mathbb{F}$, there exists an element x such that, for every $F \in \mathbb{E}' \cup \mathbb{F}'$, $x \in F$ [7].

3 Helly Interval Bigraphs

A bipartite graph G is a *Helly interval bigraph* if it admits a bi-interval model $(\mathbb{A}, \mathbb{B})$ that verifies the bipartite-Helly property.

Equivalently, a bipartite graph G is a Helly interval bigraph if and only if it admits a bi-interval model $(\mathbb{A}, \mathbb{B})$ in which, for every biclique $B \subset V(G)$, there is a point p_B on the number line such that, for any interval $A \in \mathbb{A} \cup \mathbb{B}$ that corresponds to a vertex of B, $p_B \in A$. The class can be recognized in quadratic time due to the fact that a bipartite graph is a Helly interval bigraph if and only if its square is an interval graph [6] and interval graphs can be recognized in linear time [2].

In this section, we provide a forbidden subgraph characterization of Helly interval bigraphs, and prove their inclusion in the class of proper interval bigraphs. For the characterization, we use the graphs from Fig. 1 and every even cycle of length greater than 4 as forbidden subgraphs. Note that we may work with twin-free, connected graphs without loss of generality.

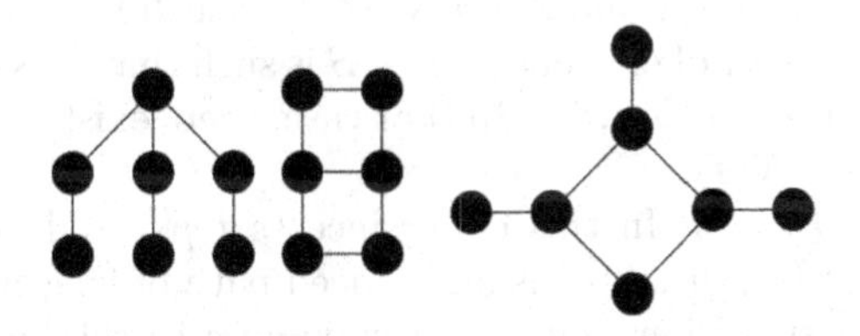

Fig. 1. Forbidden graphs for the class of Helly interval bigraphs. From left to right, the T_2, the *domino*, and the X_2.

Let G be a Helly interval bigraph and $P = (p_1, ..., p_n)$ be an induced path in it. Call a *nuisance* on p_1, p_3 (resp. nuisance on p_{n-2}, p_n) a vertex $v \in V(G) - P$ such that $p_1, p_3 \in N(v)$ (resp. $p_{n-2}, p_n \in N(v)$).

Lemma 1. *If G is a connected twin-free Helly interval bigraph, then there exists in G an induced path of maximum length for which no nuisances exist.*

Proof. By contradiction, we suppose that there isn't such a path. Let $P = (p_1, ..., p_n)$ be a maximum length induced path such that the number of nuisances on p_{n-2}, p_n is minimum. For $|P| < 4$, the proof is trivial. Suppose, then, that the length of P is at least 4. Let $S \subset V(G)$ the set of vertices for which there exists $t \in V(G)$ such that $(s, t, p_3, ..., p_n)$ is an induced path, and let $T \subset V(G)$ the set of vertices for which there exists $s \in S$ such that $(s, t, p_3, ..., p_n)$ is an induced path. We claim that every element of S has at least two neighbors in T. For that, consider these two cases: (1) Every induced path of the form $(q_1, ..., q_n)$ ($\{q_1, ..., q_n\} \subset V(G)$) has nuisances on both q_1, q_3 and q_{n-2}, q_n, or (2) Path $(p_1, ..., p_n)$ has no nuisance on p_{n-2}, p_n.

Start with case 1. Suppose there exists $s \in S$ such that $t \in N(s) \cap T$. The induced path $(s, t, p_3, ..., p_n)$ must contain a nuisance on s, p_3. Let $x \in V(G)$ be said nuisance. Note that x is not neighbor to any element of $\{p_4, ..., p_n\}$, since that would induce either a domino or a cycle of length greater than 4. Therefore, $(s, x, p_3, ..., p_n)$ is also an induced path, implying $x \in T \cap N(s)$.

Now for case 2. The fact that every element of S has at least two neighbors in T follows analogously to the previous case for $|P| > 4$. The case where $|P| = 4$ is special, however, since changing the first two elements of the path changes the vertex in position $n-2$ of the path. Let p_2' be a nuisance on p_1, p_3. Suppose there exists $s \in S$ such that $N(s) \cap T = \{t\}$. The path (s, t, p_3, p_4) must have a nuisance. If there was a nuisance on s, p_3, there'd be more than one vertex in $N(s) \cap T$. Therefore, there is a nuisance on t, p_4. Note that $t \neq p_2$, otherwise, there'd be a nuisance in P on p_2, p_4. Let y be a nuisance on t, p_4. If $p_1 t \notin E(G)$, then (s, t, p_3, p_2, p_1) is an induced path, contradicting the premise that P is a maximum length induced path. Therefore, p_1 is neighbor to t. This being the case, consider the graph induced by p_1, p_2, p_3, p_4, t, y. Note that y cannot be neighbor to p_2, otherwise P would have a nuisance in p_2, p_4, and $p_1 p_4 \notin E(G)$, otherwise P would not be an induced path. With that, however, p_1, p_2, p_3, p_4, t, y induce a domino.

Therefore, in both case 1 and 2, every $s \in S$ is such that $|N(s) \cap T| \geq 2$. We now claim that every pair of elements $s, s' \in S$ is such that $N(s) \cap T$ and $N(s') \cap T$ are comparable. Suppose otherwise. In that case, there exist $t \in T \cap N(s) - N(s')$ and $t' \in T \cap N(s') - N(s)$.

First, suppose $|P| = 4$. In this case, since (s, t, p_3) and (s', t', p_3) are both induced paths, then (s, t, p_3, t', s') is an induced path of length greater than $|P|$, contradicting the assumption that P is a maximum length induced path.

Suppose, therefore, that $|P| > 4$. In this case, the fact that (s, t, p_3) and (s', t', p_3) are induced paths leads to a T_2 in $s, t, s', t', p_3, p_4, p_5$.

Therefore, in all cases, we have it that every $s \in S$ has at least two neighbors in T, and that any pair of elements from S has comparable neighborhoods inside T. Let $S = \{s_1, ..., s_k\}$ such that $N(s_i) \cap T \subseteq N(s_j) \cap T$ when $i < j$. Let $t_1, t_2 \in N(s_1) \cap T$. Note that $N(t_1)$ and $N(t_2)$ contain p_3, plus all of S.

They must not be twins, however, implying there exists a vertex $x \in V(G)$ that is neighbor to t_1 and not t_2, or vice-versa. Suppose the former w.l.o.g.

If $(x, t_1, p_3, ..., p_n)$ is an induced path, then $x \in S$, and therefore x must also be neighbor to t_2. Therefore, $(x, t_1, p_3, ..., p_n)$ is not an induced path, implying x is neighbor to some vertex from $\{p_3, ..., p_n\}$. If x is neighbor to p_4, then $s_1, t_1, t_2, x, p_3, p_4$ induces a domino, and if x is neighbor to $p_i, i > 4$, then there exists an induced cycle of length greater than 4.

Therefore, if every maximum length induced path in G has nuisances, then G is either not twin-free, or contains a forbidden induced subgraph. ■

Lemma 2. *Let G be a connected twin-free Helly interval bigraph, and $P = (p_1, ..., p_n)$ a maximum length induced path in G for which no nuisances exist. Then every vertex in $V(G) - P$ has exactly one neighbor in P.*

Proof. Suppose there exists $v \in V(G) - P$ such that $|N(v) \cap P| \geq 3$. If there exist $p_a, p_b, p_c \in P \cap N(v)$ such that $a = b + 2, b = c + 2$, then $p_a, p_{a-1}, p_b, p_{b-1}, p_c, v$ induce a domino. Otherwise, if there are two elements $p_a, p_b \in P \cap N(v)$ such that $a > b + 2$ and for every $a < i < b$, $p_i \notin N(v)$, then $p_a, p_{a-1}, ..., p_{b+1}, p_b, v$ induce an even cycle of length greater than 4.

Suppose, then, that there exists $v \in V(G) - P$ such that $N(v) \cap P = \emptyset$. Since G is connected, there exists a path from v to P. Let v_1 be the last element of that path to have no neighbors in P, and v_2 the first element that has. Then v_2 has either one or two neighbors in P. If v_2 has two neighbors in P and is not a nuisance, either it is neighbor to two vertices more than two indices apart in the path (inducing a cycle of length greater than 4) or it is neighbor to two vertices p_i, p_{i+2}, $1 < i < n - 2$. In that case, an X_2 is induced with $p_{i-1}, p_i, p_{i+1}, p_{i+2}, v_1, v_2$.

Now suppose v_2 is neighbor to exactly one vertex of P. If v_2 is neighbor to any vertex in $\{p_1, p_2, p_{n-1}, p_n\}$, that implies P is not a maximum length induced path. If v_2 is neighbor to any other vertex, a T_2 is induced.

Therefore, every element of $V(G) - P$ must have at least one neighbor in P.

Now suppose there is a vertex v that is neighbor to exactly two vertices of P. Note that the two neighbors of v in P must be exactly two indices apart, otherwise, a cycle of length greater that 4 is induced. Also, v must not be neighbor to p_1, p_3 or p_{n-1}, p_n, since P has no nuisances. So let v be neighbor to p_i, p_{i+2}, with $1 < i < n - 2$. Since v and p_{i+1} are not twins, either v has a neighbor it does not share with p_{i+1} or vice-versa. Suppose the former w.l.o.g. and let $w \in N(v) - N(p_{i+1})$.

If w is neighbor to no element of P, then an X_2 is induced, otherwise, either a domino or a cycle of length greater than 4 is induced. Therefore, for every $v \in V(G) - P$, $|N(v) \cap P| = 1$. ■

Lemma 3. *Let G be a connected twin-free Helly interval bigraph, and $P = (p_1, ..., p_n)$ a maximum length induced path in G that has no nuisances. Let $v, w \in V(G) - P$, with $\{p_i\} = N(v) \cap P$, and $\{p_j\} = N(w) \cap P$, with i even, j odd, and $i \neq j + 1, j - 1$. Then $vw \notin E(G)$.*

Proof. If $vw \in E(G)$, then $(v, w, p_j, ..., p_i)$ is an induced cycle of length greater than 4. ■

Lemma 4. *Let G be a connected twin-free Helly interval bigraph, and $P = (p_1, ..., p_n)$ a maximum length induced path in G without nuisances. Let $v, w, x \in V(G) - P$, with $\{p_i\} = N(v) \cap P$, $\{p_{i+1}\} = N(w) \cap P$, and $\{p_{i+2}\} = N(x) \cap P$. Then $vw \notin E(G)$ or $wx \notin E(G)$.*

Proof. If $vw, wx \in E(G)$, then $v, w, x, p_i, p_{i+1}, p_{i+2}$ induce a domino. ■

Lemma 5. *Let G be a connected twin-free Helly interval bigraph, and $P = (p_1, ..., p_n)$ a maximum length induced path in G without nuisances. Let $v, w \in V(G) - P$ such that $N(v) \cap P = N(w) \cap P = p_i$ and $N(v) \cap N(p_{i+1}) \neq \emptyset \neq N(w) \cap N(p_{i+1})$ (resp. $N(v) \cap N(p_{i-1}) \neq \emptyset \neq N(w) \cap N(p_{i-1})$). Then $N(v) \subset N(w)$ or $N(w) \subset N(v)$.*

Proof. By the previous lemmas, we know that, for both v and w, all of their neighbors outside P are neighbors of p_{i+1} (resp. p_{i-1}), as otherwise, there would occur an induced domino. If $N(v)$ and $N(w)$ are not comparable, that implies there exist $v' \in N(v) - N(w)$ and $w' \in N(w) - N(v)$ such that $v', w' \in N(p_{i+1})$ (resp. $v', w' \in N(p_{i-1})$. The graph induced by $v, w, v', w', p_i, p_{i+1}$ (resp. $v, w, v', w', p_i, p_{i-1}$) is a domino. ■

Lemma 6. *A bipartite graph is a Helly interval bigraph if and only if it is possible to arrange its bicliques in a linear order such that, for every vertex, the bicliques it belongs to are an interval of the order.*

Theorem 1. *Let G be a bipartite graph such that $V(G)$ is composed of the union of the following subsets, for $k \geq 1, n_2, ..., n_{k-2} \geq 0$:*

- $P = \{p_1, p_2, ..., p_k\}$.
- $V = \{v_2, ..., v_{k-1}\}$.
- $W_i = \{w_{i,1}, ..., w_{i,n_i}\}$ *for all* $1 < i < k - 1$.
- $U_i = \{u_{i,1}, ..., u_{i,n_i}\}$ *for all* $1 < i < k - 1$.

And let the neighborhoods of the vertices in $V(G)$ be the following:

- $N(p_1) = \{p_2\}$, $N(p_k) = \{p_{k-1}\}$, $N(p_i) = \{p_{i-1}, p_{i+1}, v_i\} \cup W_i \cup U_{i-1}$ *for* $1 < i < k$.
- $N(v_i) = \{p_i\}$.
- $N(w_{i,j}) = \{p_i\} \cup \{u_{i,l} \in U_i | l \leq j\}$, *for all* $2 \leq i \leq k - 2, 1 \leq j \leq n_i$.
- $N(u_{i,j}) = \{p_{i+1}\} \cup \{w_{i,l} \in W_i | l \geq j\}$, *for all* $2 \leq i \leq k - 2, 1 \leq j \leq n_i$.

Then G is a Helly interval bigraph. Call graphs that fit this definition interval theorem graphs *(ITG for short).*

Proof. The bicliques of G are the following: $A_2 = \{p_1, p_2, p_3\} \cup W_2$; $A_{k-1} = \{p_{k-2}, p_{k-1}, p_k\} \cup U_{k-2}$; $A_i = \{p_{i-1}, p_i, p_{i+1}, v_i\} \cup W_i \cup U_{i-1}$ for all $1 < i < k$;

$B_{i,j} = \{p_i, p_{i+1}\} \cup \{w_{i,m} | m \geq j\} \cup \{u_{i,l} | l \leq j\}$ for all $1 < i < k-1$, and $1 \leq j \leq n_i$.

We must now prove that it is possible to organize the bicliques in a linear order such that, for every vertex, the bicliques it belongs to are consecutive. Consider the order described as follows:

$$(A_2, B_{2,1}, ..., B_{2,n_2}, A_3, B_{3,1}, ..., B_{3,n_3}, A_4, B_{4,1}, ..., B_{4,n_4}, A_5, B_{5,1}, ..., B_{5,n_5}, A_6, B_{6,1}, ..., B_{k-2,n_{k-2}}, A_{k-1}).$$

Note that, for every vertex, the order presented is such that the family of bicliques it belongs to are an interval. ∎

Corollary 1. *A bipartite graph is a Helly interval bigraph if and only if it does not contain a T_2, a domino, an X_2, or $C_k, k > 4$ as induced subgraphs.*

Proof. Let G be a twin-free, connected bipartite graph without the aforementioned forbidden graphs.

According to the proof of Lemma 1, G admits a maximum length induced path $P = (p_1, ..., p_n)$ without nuisances. According to the proof of Lemma 2, every vertex in $V(G) - P$ is neighbor to exactly one vertex of P. Note that there are no neighbors in $V(G) - P$ to p_1 or p_n, as that would contradict the maximum length of P. Let $V_i = N(p_i) \cap V(G) - P$ for all $1 < i < n$.

According to the proofs of Lemmas 3 and 4, every vertex $v \in V_i$ is such that $N(v) - P \subset V_{i-1}$ or $N(v) - P \subset V_{i+1}$. Also, according to the proof of Lemma 5, if two vertices $v, w \in V_i$ have neighbors in V_{i+1} (or V_{i-1}), then their neighborhoods must be comparable.

Under these restrictions, G is an induced subgraph of an ITG for large enough $k, n_2, ..., n_{k-2}$: the path P would correspond to set P from the definition in Theorem 1, any vertex in $V(G) - P$ that has exactly one neighbor would belong to V, the vertices of $V_i, 1 < i \leq k-2$ that have neighbors in V_{i+1} would belong to set W_i, and the vertices of $V_i, 2 < i \leq k-1$ that have neighbors in V_{i-1} would belong to set U_i. Since ITGs are Helly interval bigraphs, then G also is. ∎

The proof of Theorem 2 depends on the following lemma.

Lemma 7. *A bipartite graph $G = (V, W, E)$ is a proper interval bigraph if and only if there exists a linear order of V such that*

1. *for every $w \in W$, $N(w)$ is an interval in that order, and*
2. *if $w, w' \in W$ are such that $N(w) \subset N(w')$, then the intervals of the order containing $N(w)$ and $N(w')$ either begin or end on the same vertex.*

Proof. (⇒) If G is a proper interval bigraph, then, according to [4], it admits a biadjacency matrix with a *monotone consecutive arrangement.* If we assume that the columns of the matrix represent V, then the order of columns is a linear order with the properties presented in the lemma. (⇐) If there exists a linear order $<$ with the properties presented, then, given a biadjacency matrix of G in which the rows represent W, and the columns represent V ordered according to $<$, there exists a permutation of rows under which the matrix has a monotone consecutive arrangement. ∎

Theorem 2. *The class of Helly interval bigraphs is a subclass of the class of proper interval bigraphs.*

Proof. According to the proof of Corollary 1, every Helly interval bigraph is an induced subgraph of an ITG. Furthermore, every ITG for which k is odd is an induced subgraph of an ITG for which k is even. It suffices to prove, then, that ITGs with even k are proper interval bigraphs, for any values of $n_2, ..., n_{k-2}$.

Let G be an ITG with its vertex set partitioned according to the definition contained in Theorem 1, with k being even. The two partite sets of G are:

- $A = \{p_i \in P | i \text{ odd}\} \cup \{v_i \in V | i \text{ even}\} \cup \bigcup_{i=2,i+2}^{i<k-1} W_i \cup \bigcup_{i=3,i+2}^{i<k-1} U_i$.
- $B = \{p_i \in P | i \text{ even}\} \cup \{v_i \in V | i \text{ odd}\} \cup \bigcup_{i=3,i+2}^{i<k-1} W_i \cup \bigcup_{i=2,i+2}^{i<k-1} U_i$.

It suffices to show that there exists an ordering of A for which the properties in Lemma 7 are observed. Consider an order $<$ of A in which:

- $p_1 < v_2 < w_{2,1}$, $w_{2,n_2} < p_3$.
- For even $i < k-1$ and $j < n_i$, $w_{i,j} < w_{i,j+1}$.
- For odd $i < k-1$ and $j < n_i$, $u_{i,j} < u_{i,j+1}$.
- For odd $i < k-1$, $p_i < u_{i,1} < u_{i,n_i} < v_{i+1} < w_{i+1,1} < w_{i+1,n_{i+1}} < p_{i+2}$.

Note that order $<$ is such that, for every $b \in B$, $N(b)$ is consecutive in $<$, and for any two elements $b_1, b_2 \in B$ such that $N(b_1) \subset N(b_2)$, either the lowest or the highest vertex in $N(b_1)$ and $N(b_2)$ is the same. ∎

4 Non-Bichordal Helly Circular Arc Bigraphs

Similarly to the definition of Helly interval bigraphs, a bipartite graph is a Helly CA bigraph if it admits a bi-circular-arc model $(C, \mathbb{I}, \mathbb{E})$ such that $\mathbb{I}, \mathbb{E}$ verify the bipartite-Helly property. Equivalently, a bipartite graph G is a Helly CA bigraph precisely if it admits a bi-circular-arc model $(C, \mathbb{I}, \mathbb{E})$ such that, for every biclique $K \subset V(G)$ in the graph, there exists a point $p \in C$ such that, for every $v \in K$, the arc corresponding to v contains p.

In [6], a polynomial time recognition algorithm for the class, as well as a forbidden structure characterization for Helly CA bigraphs that admit an induced C_6, were presented. In this section, we present a generalization of that characterization for graphs that admit an induced C_n, for any even $n > 4$. The approach we use to prove this characterization is analogous to the one used for Corollary 1 in the previous section. We also show that every non-bichordal Helly CA bigraph is a proper CA bigraph.

The forbidden graphs we use for the class are the ones in Fig. 2, alongside every ${C_n}^*$ for $n > 4$. Graphs BW_3 and O_4 only turn up in the proofs of the case for $n = 6$ in [6]. Note that, once more, we may assume the graphs are twin-free without loss of generality.

Since the case for which a C_6 is present is treated in [6], in here, we focus on graphs that have a cycle of length greater than 6.

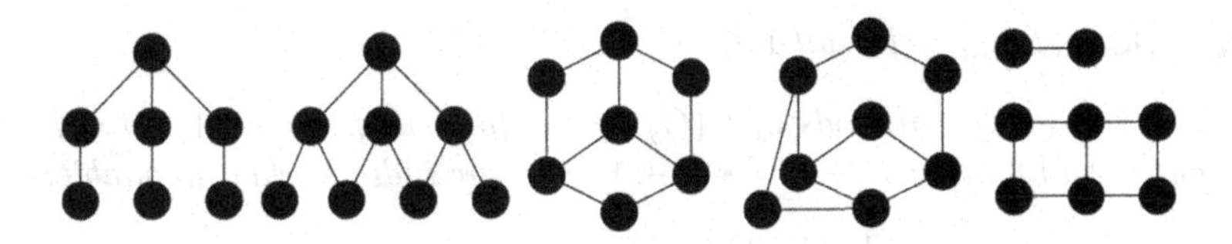

Fig. 2. Forbidden graphs for the class of Helly CA bigraphs. From left to right, T_2, O_2, BW_3, O_4, $L_3 \cup P_2$. Graphs BW_3 and $L_3 \cup P_2$ are proper CA bigraphs.

Lemma 8. *If G is a twin-free Helly CA bigraph with an induced $C_n, n > 6$, then every vertex outside the C_n is neighbor to exactly one vertex of the C_n.*

Proof. Suppose, first, that there exists a vertex v that is neighbor to three or more vertices of the induced cycle $C = (c_1, ..., c_n)$. If there are three neighbors of v of the form c_i, c_{i+2}, c_{i+4}, then G contains an induced O_2. Now, if there are no three neighbors of that form, consider three vertices $c_x, c_y, c_z \in N(v) \cap C$ such that c_x is at a distance of at least 4 from c_y and c_z in $G[C]$. Let $c'_y \in N(c_y) \cap C - N(c_z)$ and $c'_z \in N(c_z) \cap C - N(c_y)$. Note that T_2 is induced by $v, c_x, c_y, c_z, c_{x+1}, c'_y, c'_z$.

Consider, now, that v is neighbor to exactly two vertices of C. Suppose, first, that $N(v) \cap C = \{c_x, c_y\}$ with $x \neq y+2, y-2$. In this case, we have an T_2 with $c_x, c_{x+1}, v, c_{x-1}, c_{x+2}, c_y, c_{x-2}$.

Now, suppose the neighbors of v in C are at a distance 2 in $G[C]$, say, $N(v) \cap C = \{c_1, c_3\}$. Since G is twin-free, there exists a vertex w that is neighbor to either v or c_2, but not both. Suppose that w is neighbor to v w.l.o.g.

If w is neighbor only to v, G contains an induced $C_n{}^*$. If w is neighbor to two or more vertices of C, then $C \cup \{v\} - \{c_2\}$ is an induced C_n such that w is neighbor to three of its vertices, implying $G[C \cup \{v, w\} - \{c_2\}]$ contains an induced O_2 or T_2 as seen in previous paragraphs. If w is neighbor to one vertex of C, that vertex needs to be either c_4 or c_n, otherwise, there'd be an T_2 as shown in previous paragraphs. However, if w is neighbor to either of the two, G contains an $L_3 \cup P_2$. ■

Theorem 3 is analogous to Theorem 1, evidencing a structural similarity between Helly interval bigraphs and non-bichordal Helly CA bigraphs. The proof of Theorem 3 depends on Lemma 9.

Lemma 9 *[6]. A bipartite graph $G = (V, W, E)$ is a Helly CA bigraph if and only if there exists a permutation S of its biclique set such that, for every $v \in V \cup W$, the bicliques to which v belongs are circularly consecutive in S.*

Theorem 3. *Let G be a bipartite graph such that $V(G)$ is composed of the union of the following subsets, for $k \geq 6, n_1, ..., n_k \geq 0$:*

- $C = \{c_1, c_2, ..., c_k\}$.
- $V = \{v_1, v_2, ..., v_k\}$.
- $W_i = \{w_{i,1}, ..., w_{i,n_i}\}$ *for all* $1 \leq i \leq k$.

- $U_i = \{u_{i,1}, ..., u_{i,n_i}\}$ *for all* $1 \leq i \leq k$.

And let the neighborhoods of $V(G)$ *be the following, for all* $1 \leq i \leq k$. *Consider cyclic summation* $(1 - 1 = k, k + 1 = 1)$ *for indices when appliable:*

- $N(c_i) = \{c_{i-1}, c_{i+1}, v_i\} \cup W_i \cup U_{i-1}$.
- $N(v_i) = \{c_i\}$.
- $N(w_{i,j}) = \{c_i\} \cup \{u_{i,l} \in U_i | l \leq j\}$, *for all* $1 \leq j \leq n_i$.
- $N(u_{i,j}) = \{c_{i+1}\} \cup \{w_{i,l} \in W_i | l \geq j\}$, *for all* $1 \leq j \leq n_i$.

Then G *is a Helly CA bigraph. Call graphs that fit this definition* circular theorem graphs *(CTG for short).*

Proof. The graph's bicliques are the following:

- $A_i = \{c_{i-1}, c_i, c_{i+1}, v_i\} \cup W_i \cup U_{i-1}$ for all $1 \leq i \leq k$.
- $B_{i,j} = \{c_i, c_{i+1}\} \cup \{w_{i,m} | m \geq j\} \cup \{u_{i,l} | l \leq j\}$ for $1 \leq i \leq k$, $1 \leq j \leq n_i$.

To prove that G is a Helly CA bigraph, we apply Lemma 9. Consider the following permutation of G's bicliques:

$$(A_1, B_{1,1}, ..., B_{1,n_1}, A_2, B_{2,1}, ..., B_{2,n_2}, A_3, B_{3,1}, ..., B_{3,n_3}, A_4, B_{4,1}, ..., B_{4,n_4}, A_5, B_{5,1}, ..., B_{k-1,n_{k-1}}, A_k, B_{k,1}, ..., B_{k,n_k}).$$

Note that the permutation is such that, for any $v \in V(G)$, the bicliques to which v belongs are circularly consecutive. ■

Theorem 4. *A non-bichordal bipartite graph is a Helly CA bigraph if and only if it does not contain* $T_2, O_2, BW_3, O_4, C_n{}^*$ $(n \geq 6)$ *or* $L_3 \cup P_2$ *as an induced subgraph.*

Proof. Let G be a twin-free non-bichordal bipartite graph that does not contain any of the mentioned induced subgraphs.

Let $C = \{c_1, ..., c_n\}$ be a $C_n (n \geq 6)$ that G contains. By the proof of Lemma 8, every vertex in $V(G) - C$ contains exactly one neighbor in C.

If $n = 6$, it is proven in [6] that G is an induced subgraph of a CTG, so suppose $n > 6$. Let $V_i = N(c_i) - C$ for every $1 \leq i \leq n$. Let $v_i \in V_i, v_j \in V_j$. If $j \neq i + 1, i - 1$, then $v_i v_j \notin E(G)$, otherwise, T_2 or an odd cycle is induced.

For any $1 \leq i \leq n$, let $v_i \in V_i, v_{i+1} \in V_{i+1}, v_{i-1} \in V_{i-1}$, then either $v_i v_{i-1} \notin E(G)$ or $v_i v_{i+1} \notin E(G)$, otherwise, an O_2 is induced.

Therefore, every element $v \in V_i$ is such that $N(v) - C \subset V_{i+1}$ or $N(v) - C \subset V_{i-1}$. Let $V_{i,j}$ be the subset of V_i that contains vertices who are neighbors to elements of V_j ($j = i + 1$ or $j = i - 1$). Suppose two elements $v_1, v_2 \in V_{i,j}$ are such that $N(v_1), N(v_2)$ are not comparable. Without loss of generality, assume $j = i - 1$. Let $w_1 \in N(v_1) - N(v_2)$ and $w_2 \in N(v_2) - N(v_1)$. Note that $c_i, c_{i-1}, c_{i-2}, v_1, v_2, w_1, w_2$ induce a T_2.

Therefore, every pair of elements in $V_{i,j}$ have comparable neighborhoods. Since G is twin-free, that implies it is an induced subgraph of a CTG. Therefore, G is a Helly CA bigraph. ■

The proof of Theorem 5 depends on Lemma 10. In a bipartite graph $G = (V, W, E)$, call $v \in V$ a bi-universal vertex if $N(v) = W$.

Lemma 10. *A bipartite graph $G = (V, W, E)$ without bi-universal vertices is a proper CA bigraph if and only if there exists a permutation of V such that, for every $w \in W$, $N(w)$ is circularly consecutive in the permutation, and for any two vertices $w_1, w_2 \in W$ such that $N(w_1) \subset N(w_2)$, the intervals of the permutation that contain $N(w_1), N(w_2)$ either begin or end in the same vertex.*

Proof. In [1], it is proven that a bipartite graph is a proper CA bigraph if and only if it admits a biadjacency matrix with a *monotone circular arrangement.* Analogously to the proof of Lemma 7, it is easy to show that the existance of such a matrix is equivalent to the existance of a permutation as defined. ■

Theorem 5. *Every non-bichordal Helly CA bigraph is a proper CA bigraph.*

Proof. By the proof of Theorem 4, it suffices to prove that every CTG is a proper CA bigraph. Let G be a CTG with its vertex set partitioned as in the definition contained in Theorem 3 for some values of $k, n_1, ..., n_k$. We apply Lemma 10.

Let $C_1, C_2 \subset C$ be the subset of odd-indexed and even-indexed elements of C, respectively. Also, let $V_1, V_2 \in V$ be the subsets of odd an even index in V.

The partite sets are:

- $X = C_1 \cup V_2 \cup \bigcup_{i=2,i+2}^{k} W_i \cup \bigcup_{i=1,i+2}^{k-1} U_i$.
- $Y = C_2 \cup V_1 \cup \bigcup_{i=2,i+2}^{k} U_i \cup \bigcup_{i=1,i+2}^{k-1} W_i$.

Consider the following permutation of X.

$(c_1, u_{1,1}, ..., u_{1,n_1}, v_2, w_{2,1}, ..., w_{2,n_2}, c_3, ...$
$..., c_{k-1}, u_{k-1,1}, ..., u_{k-1,n_{k-1}}, v_k, w_{k,1}, ..., w_{k,n_k})$.

Note that this permutation satisfies the properties of Lemma 10. ■

5 Relationships Between the Presented Classes

It is easy to verify that Helly interval bigraphs are a subclass of Helly CA bigraphs. Non-bichordal Helly circular arc bigraphs, however, do not include any interval bigraphs, but the clear structural similarities between the classes is notable, as the graphs we named CTG and ITG have a lot in common.

The following result is relevant for the study of the computational properties of Helly interval bigraphs and Helly circular arc bigraphs.

Lemma 11. *If G is a Helly CA bigraph with n vertices, then it has $O(n)$ bicliques.*

Proof. Follows from the proof of Theorem 3 for the non-bichordal case and, for the bichordal case, from the fact that a bichordal bipartite graph is a Helly CA bigraph if and only if its square is a Helly circular arc graph [6]. ■

6 Conclusion

We proved that Helly interval bigraphs are a subclass of proper interval bigraphs, and also that non-bichordal Helly circular arc bigraphs are a subclass of proper interval bigraphs. We provided forbidden structure characterizations for both Helly interval bigraphs and non-bichordal Helly circular arc bigraphs. We also showed that both Helly classes presented are such that their graphs have a linear number of bicliques, thus opening many possibilities for efficient algorithms for biclique-related problems over the classes.

References

1. Basu, A., Das, S., Ghosh, S., Sen, M.: Circular-arc bigraphs and its subclasses. J. Graph Theory **73**(4), 361–376 (2013)
2. Booth, K.S., Lueker, G.S.: Testing for the consecutive ones property, interval graphs, and graph planarity using pq-tree algorithms. J. Comput. Syst. Sci. **13**(3), 335–379 (1976)
3. Brown, D., Lundgren, J.R.: Characterizations for unit interval bigraphs. Congressus Numerantium **206**, 5–17 (2010)
4. Das, A.K., Chakraborty, R.: New characterizations of proper interval bigraphs. AKCE Int. J. Graphs Comb. **12**(1), 47–53 (2015)
5. Das, A.K., Chakraborty, R.: New characterizations of proper interval bigraphs and proper circular arc bigraphs. In: Ganguly, S., Krishnamurti, R. (eds.) CALDAM 2015. LNCS, vol. 8959, pp. 117–125. Springer, Cham (2015). https://doi.org/10.1007/978-3-319-14974-5_12
6. Groshaus, M., Guedes, A.L., Kolberg, F.S.: Subclasses of circular-arc bigraphs: helly, normal and proper. In: Proceeding of the 10th Latin and American Algorithms, Graphs and Optimization Symposium, LAGOS. ENTCS, vol. 346, pp. 497–509 (2019)
7. Groshaus, M., Szwarcfiter, J.L.: Biclique graphs and biclique matrices. J. Graph Theory **63**(1), 1–16 (2010)
8. Hell, P., Huang, J.: Interval bigraphs and circular arc graphs. J. Graph Theory **46**, 313–327 (2004)
9. Lin, M.C., Szwarcfiter, J.L.: Characterizations and recognition of circular-arc graphs and subclasses: a survey. Discrete Math. **309**(18), 5618–5635 (2009), combinatorics 2006. A Meeting in Celebration of Pavol Hell's 60th Birthday, 1–5 May 2006
10. Müller, H.: Recognizing interval digraphs and interval bigraphs in polynomial time. Discrete Appl. Math. **78**(1–3), 189–205 (1997)
11. Rafiey, A.: Recognizing interval bigraphs by forbidden patterns. arXiv e-prints, 1211.2662 (2012)

Author Index